BRAIN DAMAGE AND REPAIR

Brain Damage and Repair

From Molecular Research to Clinical Therapy

Edited by

T. Herdegen
*Institute of Pharmacology,
University of Kiel, Kiel, Germany*

and

J. Delgado-García
*Department of Neurosciences,
University Pablo de Olavide, Sevilla, Spain*

KLUWER ACADEMIC PUBLISHERS
DORDRECHT / BOSTON / LONDON

A C.I.P. Catalogue record for this book is available from the Library of Congress.

ISBN 1-4020-1892-4

Published by Kluwer Academic Publishers,
P.O. Box 17, 3300 AA Dordrecht, The Netherlands.

Sold and distributed in North, Central and South America
by Kluwer Academic Publishers,
101 Philip Drive, Norwell, MA 02061, U.S.A.

In all other countries, sold and distributed
by Kluwer Academic Publishers,
P.O. Box 322, 3300 AH Dordrecht, The Netherlands.

Printed on acid-free paper

All Rights Reserved
© 2004 Kluwer Academic Publishers
No part of this work may be reproduced, stored in a retrieval system, or transmitted
in any form or by any means, electronic, mechanical, photocopying, microfilming, recording
or otherwise, without written permission from the Publisher, with the exception
of any material supplied specifically for the purpose of being entered
and executed on a computer system, for exclusive use by the purchaser of the work.

Printed in the Netherlands.

TABLE OF CONTENTS

Accompanying Words — ix
Preface — xi
Members of the Management Committee — xiii

PART A — THE VULNERABLE NEURON OR WHY NEURONS DIE

Chapter 1 — Introduction to part A:
The aging brain – the burdgen of life
S. Hoyer and K. Plaschke — 1

Chapter 2 — Gene expression and its regulation in neurons
L. Kaczmarek — 23

Chapter 3 — Morphological peculiarities of the neuron
I.P. Johnson — 33

Chapter 4 — Synaptic Plasticity: hyperexcitability and synaptic silencing
L. Martínez-Millán, G. García del Caño and I. Gerrikagoitia — 47

Chapter 5 — The axonal burden
H. Aldskogius — 61

Chapter 6 — The self-destruction of neurons – physiological and pathophysiological decisions for the functional integrity
J. Prehn and D. Kögel — 79

Chapter 7 — The organelles I: Mitochondrial failure
A.D. Ortega and J.M. Cuezva — 95

Chapter 8 — The organelles II: Endoplasmic reticulum and its overload
W. Paschen — 111

Chapter 9 — Transcriptional dysfunctions as pathogenic mechanism of neurodegenerative diseases
B. Kaminska — 123

Chapter 10 — Formation of radicals
M. Götz and M. Gerlach — 135

PART B — THE EXTRANEURONAL MATRIX - SITES OF MULTIPLE DEGENERATIVE PROCESSES

Chapter 11 — Introduction to Part B: — 165

	Axon guidance and repulsion. The molecular code of social life in the brain	
	J.A. del Río, F. de Castro and E. Soriano	
Chapter 12	Microglia and the cerebral defence system	181
	D. van Rossum and U.K. Hanisch	
Chapter 13	Generation and differentiation of astrocytes during central nervous system development and injury	203
	C. Vicario-Abejón and M.J. Yusta-Boyo	
Chapter 14	Oligodendrocytes in health and disease	215
	M. Domercq and C. Matute	
Chapter 15	The myeline glial cell of the peripheral nervous system: The Schwann Cell	231
	D. Echeyarría and S. Martínez	
Chapter 16	The Müller glia: role in neuroprotection	245
	E. Vecino and M. García	

PART C	**THE BRAVE NEURON – SELF PROTECTION, PLASTICITY AND RESTORATION**	
Chapter 17	Introduction to part C: Neural plasticity and regeneration: myths and expectations	259
	J.M. Delgado-García and A. Gruart	
Chapter 18	Neuronal protection against oxidative damage	275
	O.S. Jørgensen and A.S. Nielsen	
Chapter 19	Functional recovery in the peripheral and central nervous system after injury	285
	D. González-Forero, B. Benítez-Temiño, R. R. de la Cruz and A. M. Pastor	
Chapter 20	Neural plasticity and the cell biology of learning	307
	M. Nieto-Sampedro	
Chapter 21	Nueral plasticity and central nervous system lesion repair	323
	M. Nieto-Sampedro	
Chapter 22	Regulation of the intrinsic growth properties in mammalian neurons	349
	F. Rossi	
Chapter 23	Glutamate, calcium and neurodegenerative disease: impact of cytosolic calcium buffers and their potential role for neuroprotective strategies	365
	F. Bergmann and B.U. Keller	
Chapter 24	Postnatal neurogenesis and neuronal regeneration	381
	C. López-García and J. Nacher	

| Chapter 25 | Migration disorders and epilepsy
C. Haas and M. Frotscher | 391 |

PART D   THERAPEUTIC PRINCIPLES

Chapter 26	Introduction to part D: Invasive strategies as therapeutic approaches for central nervous system diseases J.A. Armegol	403
Chapter 27	General aspects of neuropharmacology in relation to brain repair following trauma A. Ribeiro	423
Chapter 28	Stem cells and nervous tissue engineering I. Liste and A. Martínez-Serrano	439
Chapter 29	Cell transplants and artificial guides for nerve repair X. Navarro and E. Verdú	451
Chapter 30	Modern late neurologic rehabilitation: neuroscience and motivating functional rehabilitation P. Bach-y-Rita	473
Chapter 31	Viral gene delivery S. Isenmann	483

PART E   THE NEURODEDENERATIVE DISORDERS: MOLECULAR PATHOGENESIS AND TREATMENT

Chapter 32	Introduction to part E: General remarks on the bridging between basic research and clinics P. Kermer and M. Bähr	501
Chapter 33	Mechanical lesions of the peripheral nervous system C. Krarup	509
Chapter 34	Stroke and ischemic insults V. Ceña, M. Fernández, C. González-García and J. Jordán	525
Chapter 35	Parkinson's disease I: Degeneration and dysfunction of dopaminergic neurons H. Wilms and G. Deuschl	535
Chapter 36	Parkinson's disease II: Replacement of dopamine and restoration of striatal function C. Winkler and D. Kirik	547
Chapter 37	Alzheimer's disease R. Alberca and E. Montes	561

Chapter 38	Amyotrophic lateral sclerosis and related disorders *R. Dengler and J. Buffler*	573
Chapter 39	Pharmacological strategies for neurodegeneration and overview of clinical trials *L. Ley and T. Herdegen*	587
Chapter 40	Monitoring brain dysfunction through imaging techniques *M. Ptito and R. Kupers*	615
Chapter 41	Functional assessment of human brain with non-invasive electrophysiological methods *J. L. Cantero and M. Atienza*	627
Chapter 42	Legal implications of therapeutic options *M. Herdegen*	641
PART F	GLOSSARY	657
PART G	LIST OF CONTRIBUTORS	685
PART H	INDEX	699

ACCOMPANYING WORDS

As chairman of COST B10, I introduce with great pleasure the book *Brain Damage and Repair. From Molecular Research to Clinical Therapy*, edited by Thomas Herdegen and José M. Delgado-García, and published by Kluwer Academic Publishers.

We at COST B10 (an European Union programme devote precisely to the study of brain damage and its possible repair) are proud of this major editing and publishing accomplishment. The two scientist-editors did an extraordinary job of leadership, planning, and execution, bringing together the work of more than 40 top European and American scientists in neurosciences, neurophysiology, neurology and law to produce this comprehensive volume. The editors paid particular attention to build a proper book, and not a simple collection of ancillary research notes, and greatly helped the contributors to construct a transparent and well-shaped chapter.

Brain Damage and Repair is especially gratifying to COST B10 because it is a milestone in our commitment to neural regeneration. This book is an authoritative reference on brain repair and regeneration for a long time to come. It is thanks to the European network COST B10 (http://braindamagerepair.hiim.hr) that this concept could start and many COST B10 members as well as other neuroscientists participated in its preparation. The result is a tribute to our main goal to advance understanding on how the brain can recover after injury and degenerative processes and what mechanisms can be modulate to overcome irreversible damage and injuries.

Our thanks to the editors and to the motivated contributors. It is their life's work that made such a book possible.

Roland Pochet
COST B10 Chairman

PREFACE

The growing incidence of neurological diseases in an aging population provokes an unprecedented challenge for the modern world. Facing the socio-economic consequences, the last decade of the 20th century was declared as the „decade of the brain", and this pretension was reflected by huge investments of both pharmaceutical industry and national research organisations. In consequence, the recent past provided extensive and fascinating insights into the molecular cosmos and the systemic processes of the nervous system. In striking contradistinction to these worldwide efforts and accumulation of information, the progress in novel curative therapies and diagnostic tools is very limited. One reason for this failure is the marginal interconnection between basic research and clinical neurology. On the basis of the European COST program, *Brain Damage and Repair* intents to fill this gap and to bridge between bench and bedside. The concept of *Brain Damage and Repair* leads the reader from the basic molecular features of brain structures to the patho-physiological response due to neuronal damage and chronic degenerative alterations. The comprehensive delineation of complex structural processes sets a counterpoint to the *detailism* of modern information overflow. The contents proceed from cells and organelles to the clinical diseases and therapeutic principles, and are completed by chapters on imaging techniques, viral gene transfer and review of actual neuropharmacological strategies. The final chapter takes up the essential aspect of legal international handling of stem cell transfer and patents of DNA/proteins.

Brain Damage and Repair wants to attract the interest of both, the basic neuroscientists and the clinical neurologists. The editors hope that the intented advance from *molecular research to clinical therapy* will improve our understanding and the cure of neurological diseases resulting in a better quality of life.

Kiel and Sevilla, Spring 2004
                                                                      Thomas Herdegen
                                                                  José M. Delgado-García

MEMBERS OF THE COST B-10 MANAGEMENT COMMITTEE

Austria
Prof. Gert Pfurtscheller
Dr. Christa Neuper

Belgium
Prof. Jean-Marie Godfraind
Prof. Fred Van Leuven

Croatia
Dr. Branimir Jernej
Dr. Srecko Gajovic

Czech Republic
Prof. Frantisek Vozeh
Prof. Eva Sykova

Denmark
Prof. Jens Zimmer
Prof. Ole Stehen Jorgensen

Finland
Prof. Jari Koistinaho

France
Dr. Jean-Baptiste Thiebaut
Dr. Alain Privat

Germany
Prof. W. F. Neiss
Prof. Dr. Thomas Herdegen

Hungary
Dr. Agoston Szel

Ireland
Prof. K.F. Tipton
Dr. Richard K. Porter

Italy
Prof. Piergiorgio Strata
Dr. Lorenzo Magrassi

Norway
Prof. Ole Petter Ottersen
Dr. Clive Bramham

Poland
Dr. Krzysztof Janeczko
Prof. Leszek Kaczmarek

Portugal
Prof. Alexandre Ribeiro
Prof. Alexandre de Mendonca

Serbia and Monte Negro
Dr. Pavle Andjus

Slovakia
Prof. Jan Lehotsky
Prof. Svorad Stolc

Slovenia
Dr. Maja Bresjanac
Dr. Marko Zivin

Spain
Prof. Jose Maria Delgado-Garcia
Prof. Carlos Lopez Garcia
Dr. Carlos Luis Paino

Sweden
Prof. Giorgio Innocenti
Prof. Hakan Aldskogius

Switzerland
Prof. Giorgio Maria Innocenti
Dr. Anne Zurn

The Netherlands
Dr. Frans Vander Werf
Dr. John Van Opstal
Prof. Bram W. Ongerboer de Visser

The COST Scientific Officers of European Commission
Dr. Jasminka Goldoni
Dr. Anne Mandenoff
Prof. Mihail Pascu

S. HOYER[1] AND K. PLASCHKE[2]

[1]Institut of Pathology, University of Heidelberg, Germany

[2]Clinic for Anaesthesiology, University of Heidelberg, Germany

PART A: INTRODUCTION
1. THE AGING BRAIN – THE BURDEN OF LIFE (?)

Summary. Increasing life expectancy has raised health problems with respect to „normal"aging, and particularly to age-related disorders. This holds true for several brain diseases in middle, and more frequently, in old age such as stroke (brain ischemia), Parkinson's disease and dementia of either vascular origin or sporadic Alzheimer type.
Cellular and molecular mechanisms in mammalian brain during adult life and their variations in old age will be discussed in this chapter to get insight into processes which may be risk factors (burden) for age-associated brain diseases or into processes which may be able to prevent such brain disorders. In this respect, focus is laid on oxidative energy and related metabolism what had been demonstrated to meet the functional and structural requirements of the brain.

1. INTRODUCTION

Longevity with both good physical and mental health is one of the oldest desires of humanity. It was Marcus Tullius Cicero (106-43 B.C.) who picked up this point and discussed this unaccomplished desire intensely and in a realistic way in his famous "Cato Maior. De Senectute" in which he demonstrated the fortune and calamity of both young and old age. Some examples may be presented here.

Specificity of young and old age: Cursus est certus aetatis et una via naturale eaque simplex, suaque cuique parti aetatis tempestivitas est data, ut et infirmitas puerorum et ferocitas iuvenum et gravitas iam constantis aetatis et senectutis maturitas naturale quiddam habeat,

however,

Age as a disaster of decreased physical and mental capacities: ...unam quod avocet a rebus gerendis, alteram quod corpus faciat infirmius, tertiam quod privet omnibus fere voluptatibus, quartam quod haud procul absit a morte

but

Fortune of age-associated increase in mental capacity versus minor physical potency: vero multo maiora et meliora facit, non viribus aut velocitate aut celeritate corporum res magnae geruntur, sed consilio auctoritate sententia; quibus non modo non orbari, sed etiam augeri senectus solet,

it follows

The advice to increase mental power in old age: nec vero corpori solum subveniendum est, sed menti atque animo multo magis; nam haec quoque, nisi tamquam lumini oleum instilles, extinguuntur senectute. et corpora quidem exercitationum defatigatione ingravescunt, animi autem se exercendo levantor,

to result in

The honor of old age: habet senectus honorata praesertim tantam auctoritatem, ut ea pluris sitquam omnes adulescentiae voluptates.

It becomes obvious from these few examples that M.T. Cicero leaves open the question whether or not aging of the brain is a burden of life. His view in this respect is reflected by the view of today-gerontologists that old age does not represent a deficit model only. Biological findings at the cellular and molecular levels support this view. Clearly, increasing life expectancy has raised health problems with respect to "normal" aging, and particularly to age-related disorders. This holds true for several brain diseases in middle, and more frequently, in old age such as stroke (brain ischemia), Parkinson's disease and dementia of either vascular origin or sporadic Alzheimer type.

Therefore, in this chapter, cellular and molecular mechanisms in mammalian brain during adult life and their variations in old age will be discussed to get insight into processes which may be risk factors (burden) for age-associated brain diseases or into processes which may be able to prevent such brain disorders. In this respect, focus is laid on oxidative energy and related metabolism what had been demonstrated to meet the functional and structural requirements of the brain.

2. THE NORMAL ADULT BRAIN

Glucose has been found to be the major nutrient of the brain and the source of functionally important metabolites such as acetyl-CoA and ATP. Various mechanisms contribute to the regulation of brain glucose metabolism (Siesjö 1978, Hoyer 1985). Glucose uptake from arterial blood into the extracellular space of the brain across the blood-brain barrier by means of a carrier-mediated transport mechanism, is obviously insulin-dependent. The mRNA of this glucose transport protein 1 has been found to be regulated by neuropeptides in all probability also by insulin. The glucose transport protein (GLUT) 3 mediates the glucose transport from the extracellular space into neurons, and GLUT 5 into glia cells. Some neuron populations in the basal frontal brain, cerebral cortex, hippocampus and cerebellum were found to express the insulin-sensitive GLUT 4 beside GLUT 3.

The glycolytic breakdown of glucose in the nerve cell is controlled by the allosteric enzymes hexokinases and pyruvate kinase working in a concerted way under the influence of phosphofructokinase. The oxidation of the glycolytically formed pyruvate starts by means of the pre-eminent multi-enzyme complex pyruvate dehydrogenase (PDH) yielding the energy-rich compound acetyl-CoA which is used 1. for further oxidation in the tricarboxylic acid cycle to ATP (more than 95% of acetyl-CoA), 2. for the formation of the neurotransmitter acetylcholine (1% to 2% of acetyl-CoA), and 3. for the formation of cholesterol in the 3-hydroxy-3-methylglutaryl-CoA cycle. Cholesterol is the main sterol in membranes, and it also serves as the basic compound from which neurosteroids derive. Glia-derived cholesterol has been demonstrated to promote synaptogenesis in nervous tissue.

PDH activity is reduced by its product acetyl-CoA, for example when fatty acids or ketone bodies are used for oxidation instead of glucose. A small pool of free fatty acids exists in the brain. In the brain, free fatty acids can be metabolized via acetyl-CoA by beta-oxidation. The activity of the enzyme choline acetyltransferase, which catalyzes acetylcholine formation from acetyl-CoA and choline, is closely linked functionally to the PDH complex. Acetylcholine acts 1. as an excitatory neurotransmitter for learning and memory capacities, and 2. as a regulator of regional cerebral blood flow.

From acetyl-CoA, glucose derived carbon is rapidly transferred into amino acids via the tricarboxylic acid cycle (TCAC) and the γ-aminobutyric acid (GABA) shunt. Glutamate, glutamine, aspartate and GABA are formed most abundantly. Dehydrogenating multi-enzymes complexes working in the TCAC provide redox-equivalents which are oxidized in the respiratory chain to yield ATP. ATP represents the driving force for nearly all cellular and molecular work. Some out of many others shall be listed here:

folding of proteins,
sorting of proteins,
transport of proteins,
neurotransmission; the synapse was found to be the site of the highest energy utilization as compared to other cellular compartments,
function of the endoplasmic reticulum/Golgi apparatus
maintenance of a pH of 6,
transport of membrane proteins,
glycosylation of proteins,
maintenance of Na^+ / K^+ transmembranaceous flux.

2.1. Insulin and insulin receptor signaling in the brain: insulin production, insulin receptor distribution

Substantial evidence has been gathered in support both of the transport of peripheral (pancreatic) insulin to the brain and of its production in the CNS (for review Plata-

Salaman 1991). Insulin gene expression and insulin synthesis have been demonstrated in mammalian neuronal cells. Insulin mRNA was found to be distributed in a highly specific pattern, with the highest density in pyramidal cells of the hippocampus and high densities in medial prefrontal cortex, the entorhinal cortex, perirhinal cortex, thalamus and the granule cell layer of the olfactory bulb. Neither insulin mRNA nor synthesis of the hormone was observed in glia cells. Insulin receptors have been demonstrated to be dispersed throughout the brain following also a highly specific pattern, with the highest density detected in olfactory bulb, hypothalamus, cerebral cortex and hippocampus. Two different types of insulin receptors have been found in adult mammalian brain: a peripheral type detected in lower density on glia cells, and a neuron-specific brain type with high concentrations on neurons. It has been shown that the location of phosphotyrosine-containing proteins corresponds to the distribution of the insulin receptor. It has also been established that the insulin receptor substrate (IRS)-1 co-localizes with these phosphotyrosines. IRS-1 and the insulin receptor are found to be co-expressed in discrete neuron populations in rat hippocampus and olfactory bulb whereas their proteins showed the highest densities in the synaptic neuropile.

2.2. Insulin receptor regulation

The major molecular structure and most of the biochemical properties of the neuronal insulin receptor have been demonstrated to be indistinguishable from those ones in non-nervous tissues. However, some differences became obvious in that both the α- and β-subunits of the neuronal insulin receptor are slightly lower in molecular weight as comparted to those ones in non-nervous tissues. Unlike the insulin receptor in non-nervous tissues, the neuronal insulin receptor did not undergo down-regulation after exposure to high concentrations of insulin.

Binding of insulin to the α-subunit of its receptors induces autophosphorylation of the intracellular β-subunit by phosphorylation of the receptor's intrinsic tyrosine residues for activation. The receptor's activity is known to be regulated by the action of both phosphotyrosine phosphatases and serine kinases. Interestingly, glucocorticoids and catecholamines have also been reported to cause insulin receptor densitization either by inhibition of phosphorylation of the tyrosine residues or by phosphorylation of serine residues. The most prominent substrate of the insulin receptor tyrosine kinase is the insulin receptor substrate-1 which transfers the signal to a wide spectrum of cellular and molecular compounds (for review White and Kahn 1994).

Both, insulin binding to its receptor and receptor autophosphorylation are found to be reduced by the derivatives of the amyloid precursor protein, Aβ1-40 and Aβ1-42 via a decrease in the affinity of insulin binding to the insulin receptor (Xie et al 2002). Interestingly, insulin and Aβ are common substrates for the insulin-degrading enzyme. However, the affinity of this enzyme to the substrates are observed to be different: insulin Km ~ 0.1 µM, Aβ analogues Km > 2 µM. There is recent clear evidence that the transfer of insulin across the brain occurs rapidly from the olfactory nerve to CSF.

Recently, first evidence has been provided that insulin and the insulin receptor are functionally linked to cognition in general and spatial memory in particular by up-regulation of the insulin receptor mRNA in the hippocampus and increased accumulation of insulin receptor protein in hippocampal synaptic membranes (for review Park 2001).

2.3. Insulin and glucose / energy metabolism

Acute stimulation of the cerebral insulin receptor was achieved through a single intracerebroventricular injection of insulin. This procedure led to a dose-dependent stimulation of the glycolytic key enzymes hexokinase and phosphofructokinase in the cerebral cortex. Also, acute stimulatory effects of the hormone in the brain have been demonstrated for pyruvate dehydrogenase and choline acetyltransferase. These data may indicate that both glycolytic flux and pyruvate oxidation in the brain are stimulated by insulin paralleling the hormone's effect in non-nervous tissue. Short-term (1 day) or long-term (7 and 21 days) intracerebroventricular infusion of insulin have been found to exhibit a discrete anabolic effect on energy metabolism in the hippocampus as can be concluded from an 11% increase in the concentration of creatine phosphate, the storage form of ATP (Henneberg and Hoyer 1994).

2.4. Insulin and amyloid precursor protein

Normally, the holoprotein of the amyloid precursor protein (APPh), and its derivatives the secreted form of APP ~ APPs and βA4 have been demonstrated to exert beneficial effects. APPh has been found to be an integral plasma membrane protein transducing signals from the extracellular matrix into the cell and stimulating neurite promoting activity. It enhanced neuron viability and modulated neuronal polarity, was shown to interact with intracellular signal transduction, mediated and potentiated neurotrophic activity related to tyrosine phosphorylation of the insulin receptor substrate-1 assuming interactions between APP and the insulin signal transduction pathway (see also above: Xie et al 2002).

APPs has been shown to mediate neuroprotection, modulation of synaptic plasticity along with memory-enhancing capacities, and enhancement of long-term potentiation. Mental activity in an enriched environment increased APPs in brain cortical and hippocampal synapses.

Normally, βA4 is found to be present in picomolar/nanomolar concentrations, and has been detected in plasma and cerebrospinal fluid. In these low concentrations, it exerted cell proliferation, tyrosine phosphorylation and stimulation of protein kinase C. When compared with APPh and APPs, the neuroprotective effect of both these proteins was found to be around 100-times higher than that of βA4.

Both, the intracellular and extracellular concentrations of βA4 are regulated by several β-amyloid metabolizing enzymes such as insulin degrading enzyme (see above), angiotensin-converting enzyme, and neprilysin what may be an indication for the pathological potency of βA4 when its concentration increases to micromolar and millimolar levels.

Recent studies provided clear evidence that βA4 is generated intracellularly in the endoplasmic reticulum/ intermediate compartment to the Golgi apparatus. Intracellular accumulation of both βA1-40 and βA1-42 was found to be reduced by accelerating βAPP/βA transport from the trans-Golgi network to the plasma membrane. This effect, the promotion of its secretion to the extracellular space, and the inhibition of its degradation by insulin-degrading enzyme was mediated by insulin and the tyrosine kinase activity of the insulin receptor. The same holds true for APPs release into the extracellular space. This process was found to be dependent on the activation of phosphatidyl-inositol 3 kinase what has been shown to be activated by physiological levels of βA4. These findings provide clear evidence that the insulin/insulin receptor signal transduction cascade is of pre-eminent significance for the metabolism of APP.

2.5. Insulin and tau-protein

The tau-protein belongs to a family of microtubule-associated proteins which had been found to stimulate the generation and stabilization of microtubules. Normally, the tau-protein is stably phosphorylated at 5 epitopes. Phosphorylation of tau-protein at threonine and serine residues has been demonstrated to be controlled by several protein kinases, dephosphorylation by protein phosphatases. Among the tau-phosphorylating protein kinases are PK^{erk36} and PK^{erk40} and the protein kinase FA/glycogen synthase kinase-3α. In the context discussed here, both PK^{erk36} and PK^{erk40} work in an ATP-dependent manner whereas the phosphorylating power of glycogen synthase kinase-3α has been shown to be insulin-dependent. Two different types of insulin's effect have been found. A short period of insulin treatment (1min) resulted in an increase in tau-protein phosphorylation and an increase in glycogen synthase kinase-3β activity (Lesort et al 1999). Prolonged activation of the insulin receptor induced a down-regulation of the glycogen synthase kinase-3β activity.

2.6. Insulin and cell cycle activity

In non-nervous tissue such as myoblasts, insulin has been found to induce mitogenesis throughout two independent pathways acting in synergy, PI-3 kinase / p 70 S 6 kinase, and p43/p44-MAP kinase along with an increase in the proportion of proliferating cells in both S and G2/M phases of the cell cycle (Conejo and Lorenzo 2001). In mammalian cells, cyclin-dependent kinase (cdk1) kinase regulates the progression from G2 to M phase, and cdk2 kinase from G1 to S phase. First recent data provide some evidence that in all likelihood these mechanisms take place in neurons, too. The mRNAs of both cdk1 kinase and cdk2 kinase have been detected to be expressed in low levels in adult rat brain. However, a neuronal cdc2-like kinase was found to be expressed at high levels in terminally differentiated neurons no longer in the cell cycle suggesting that the cdc2 (also named cdk1) family of kinases may play an important role in cell function other than cell cycle regulation.

With respect to APP, its phosphorylation is maximal at G2/M phase of the cell cycle indicating that cell cycle activity is involved in the regulation of APP metabolism (Suzuki et al 1994). Likewise, the phosphorylation of tau protein is suggested to be mediated by mitogenic stimuli.

The assumed normal function of the neuronal insulin/insulin receptor (I/IR) signal transduction cascade is demonstrated in Figure 1.

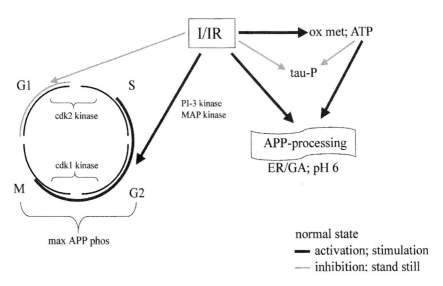

Figure 1. Schematic survey on the normal function of the neuronal insulin/insulin receptor (I/IR) signal transduction cascade. Activation/stimulation is directed to: oxidative energy metabolism leading to the formation of acetyl-CoA and ATP. Acetyl-CoA participates in the generation of acetylcholine and cholesterol. ATP ensures the working pH of 6 of the endoplasmic reticulum (ER) / Golgi apparatus (GA) necessary for protein processing (e.g. APP) in these intracellular compartments, extrusion of APPs and βA4 from the intracellular to the extracellular space, S/G2/M phases of the cell cycle in which APP is phosphorylated. Both I/IR signal transduction and ATP inhibit tau-protein hyperphosphorylation. In the cell cycle, I/IR signal transduction leads to inhibition/stand still in the G1 of the cell cycle.

2.7. Glucocorticoids in the brain

In the brain, the effect of glucocorticoids has been found to be mediated by two types of receptors. Receptor type I is nearly exclusively located on neurons of limbic structures, whereas receptor type II is widely distributed throughout the brain and is mainly activated under stress conditions. Glucocorticoids have been demonstrated to act on neuronal glucose utilization adversely to insulin. In cultured neurons and glia cells, glucose transport was inhibited by glucocorticoids, and glucose uptake into the brain in vivo was diminished. Otherwise, cerebral glucose utilization was found to be enhanced after adrenalectomy. This latter experimental condition caused an increase in both glycolytic flux and glycogen breakdown and a decrease in gluconeogenesis in cerebral cortex but not in hippocampus. A subsequent substitution with corticosterone reversed the effect of adrenalectomy. A long-term application of corticosterone reduced the activities of glycolytic key enzymes equally in both parietotemporal cerebral cortex and hippocampus. However, the diminution of energy-rich phosphates was more severe in hippocampus than in cerebral cortex. In agreement with these findings is the effect of long-term cortisol treatment which damaged nerve cells preferentially in the hippocampus. Hippocampal neurons have been demonstrated to inhibit the HPA-axis-mediated release of cortisol from the adrenal gland. Neuronal damage in the hippocampus then leads to a long-lasting hypercortisolism. It, thus, becomes obvious that insulin and cortisol act as functional antagonists in the brain (Sapolsky 1994).

3. THE AGING BRAIN
AGE AS A RISK FACTOR FOR NEURODEGENERATION: AN OR THE BURDEN OF LIFE?

3.1. Epidemiology

Neurodegeneration has been found to represent the morphobiological basis for leading age-associated brain disorders, Parkinson's disease and sporadic Alzheimer's disease (SAD). Numerically, the latter condition is the predominating brain disorder in old age. The prevalence of SAD increases from 0.5% at the age of 60 years to nearly 50% at 85 years and older. The finding that obviously not more than 50% of people aged 85 years old are affected by SAD, whereas 50% are not, and that in centenarians, the rate of individuals with moderate to severe cognitive deficits was found to be about 60% (i.e. an increase by about 10% within 15 years) whereas 40% of them revealed slight to none cognitive deficits only, point to additional factors beside aging to generate SAD. In this context, susceptibility genes have been demonstrated to participate in the causation of disorders becoming evident late in life then inducing a chronic and progressive course of pathologic conditions. Aging together with such a genetic predisposition may result as an age-associated disorder. This general principle may also hold true for SAD. As most important susceptibility gene, APOE-gen was identified at chromosome 19. More susceptibility genes for SAD are supposed at chromosomes 4, 16, 12, and 20.

Recent findings support this concept at least in part. Infants of diabetic mothers were at increased risk for hippocampal damage and corresponding memory impairment at six months of age. In experimental rats, prenatal stress towards the end of pregnancy corresponding to the third trimenon of human beings induced increased glucocorticoid secretion from young adulthood to senescence accompanied by behavioral abnormalities in aged animals. Related to stress in general, and to psychosocial stress in particular, there is evidence that the regulation of the HPA-axis (for review Blanchard et al 2001), affection of hippocampal neurons and their dendritic morphology and neurogenesis may represent the results of life style events leading to reduced capacities of learning and memory. Interestingly, the level of education was found to be related to the rate of incidence of SAD. In this context, some relationship had been detected between the size of mental capacities and the risk of SAD. Low linguistic capacities in young adulthood were found to be a strong predictor of SAD what became obvious from the nun study. Participants from this study showing intact cognitive function had very low dementia conversion rates while subjects with mild or global memory impairment converted at higher rates.

In a series of studies on experimental animals, clear evidence has been provided that mental capacities decrease with aging. Using the holeboard test system which discriminates long-term memory (reference memory) and short-term memory (working memory), a significant reduction (>50%) in both memory capacities was found (Plaschke et al 1999; see also references herein).

In so far, stress factors with long-lasting effects on cellular and molecular neuronal functions may be assumed to become a burden of life in the aging process contributing to the development of neurodegeneration.

3.2. Morphobiology

In contrast to former studies which reported on a massive cell loss throughout the human brain and severe reduction of dendritic arborization in both human brain and the limbic system with aging, and also partially in contrast to results from age-related animal studies (for review Hoyer 1995), more recent investigations based on carefully conducted quantitative studies yielded less grim data. A small reduction of the number of cortical neurons was found in different but not in all brain areas with aging. In hippocampus, a larger reduction in total cell number was found. Obviously, the subiculum was particularly involved. From a functional point of view, area- and laminar-specific variation of dendrites and synapses would seem to be more important. In layer V pyramidal neurons, a marked regression of the size of the dendritic field was observed accompanied by no changes in layer III pyramidal neurons (Uylings and deBrabander 2002). In contrast, in layer II pyramidal neurons, the dendritic tree showed a lengthening of terminal branches. With respect to the density of dendritic spines and synapses, the effects of aging were found to be specific to brain region and lamina, too. In general, degeneration of spine density of neocortical neurons was observed, but in dentate granule cells no decrease was found. In dentate gyrus, the surface density showed a significant reduction but the area of the remaining synapses increased.

The phenomenon of plasticity in the brain of aged rats is well-documented. Loss of synaptic connection as a consequence of age-related changes in the brain, such as loss of afferent supply and alterations in transmitter concentration, receptors and trophic factors, are followed by a compensatory regrowth of neuronal connections by surviving neurons to replace defective or lost ones. Thus that the aged brain retains the capacity, albeit at a slower rate, to maintain and repair its own circuitry. Even in the absence of overt changes such as those ones occurring after stroke, there is continuous remodelling of the cerebral circuitry, as reflected in the persistent growth and regression of neuronal processes. There is evidence that the capacity to maintain neuronal plasticity in morphological terms may work until the beginning of the senescence, but not in very old age (Popa-Wagner et al 2001).

These morphologic changes were found to be associated with increased membrane rigidity and, thus, decreased membrane fluidity due to a reduction in membrane concentration of polyunsaturated fatty acids. This morphobiology is assumed to cause changes in synaptic function, i.e. synaptic plasticity in functional terms which ensures long-term potentiation (LPT). LPT may represent an essential molecular factor for learning and memory capacities. The age-related changes of synaptic plasticity preferentially occurring in the cholinergic system and involving the cholinergic bouton population apposed to pyramidal neurons may cause the behavioral changes with aging.

Astrocytes have gained serious consideration as active modulators of diverse brain functions: regulation of the ionic environment of the extracellular space, clearance of glutamate from the extacellular space and its transformation into glutamine, active responses to neuronal activity by the release of a variety of modulatory substances, and formation of the blood-brain barrier to name only some out of more functions.

In comparison with neurons, astrocytes do not degenerate but increase in number and become hypertrophic with aging (termed type II astrocytes by Aloys Alzheimer; reactive gliosis or astrogliosis). Reactive astrocytes were found to exhibit both an elevated content of glial fibrillary acidic protein (GFAP) and an increase in oxidoreductive enzyme activities.

Since intense studies on the function of aged astrocytes started just recently, the knowledge of this feature is still very limited. Studies are lacking to show the interrelationship between aged astrocytes and its contribution to neurodegeneration.

3.3. Insulin / insulin receptor signal transduction

As discussed above, the neuronal insulin/insulin receptor signal transduction cascade exerts pre-eminent and broad effects on a number of cellular and molecular processes. As has been demonstrated recently, the neuronal insulin signal transduction system has been found to undergo reduction. The concentration of insulin, density of insulin receptors and its activity of the tyrosine kinase have been shown to fall beyond the age of 60 years. Upon glucose stimulation, the insulin concentration in dialysate of the hypothalamus of aged rats was half the level of young animals.

Both, the age-associated increase in the concentration of circulating cortisol (for more details, see below) and the maintained elevation of cortisol after stress may participate in the diminution of insulin receptor function (Lupien et al 1994). The elevation of cortisol also in the CNS is mirrored by its drastic increase in cerebrospinal fluid during aging as compared to adulthood. Another contributing factor to the reduced function of the neuronal insulin receptor may be the concentration of noradrenaline in the CNS which was found to increase with aging in particular after stimulation. It, thus, becomes obvious that the neuronal insulin signal transduction cascade undergoes changes as a consequence of different metabolic variations all resulting in a diminished receptor function with brain aging.

Reduced insulin/insulin receptor function along with reduced ATP may inhibit the processing of APP in the endoplasmic reticulum/Golgi apparatus resulting in the retainment and accumulation of APPs and βA4 in neurons what is considered to be characteristic for pathologic conditions such as sporadic Alzheimer disease. Furthermore, both lacking insulin and ATP forward the hyperphosphorylation of tau-protein (Hong and Lee 1997). It may be concluded that the pathology of APP and tau-protein may start during normal aging to aggravate and accelerate in sporadic Alzheimer disease.

3.4. Brain blood flow regulation

The maintenance of blood flow to the brain ensures an undisturbed supply of substrates to this organ. The constancy of cerebral blood flow bases on the complex mechanism of its autoregulation. Beside myogenic and metabolic factors, noradrenergic sympathetic and acetylcholinergic parasympathetic innervation of the microvasculature participate in the regulation of cerebral blood flow. The former causes vasoconstriction, the latter vasodilation. The normally existing balance between noradrenergic and acetylcholinergic innervation is dysregulated with aging, during which an increase in the sympathetic tone is present. As pointed out above, noradrenaline concentration increased in cerebral cortex with aging, and the stress-induced release of noradrenaline was found to be prolonged. In contrast, both synthesis and release of acetylcholine were demonstrated to be reduced. In human beings, cerebral blood flow started to fall slowly in the 7th decade of life indicating a threshold phenomenon but deteriorated more rapidly in later life. The imbalance between noradrenergic and acetylcholinergic innervation of small vessels may contribute to their morphologic abnormalities with aging, and in the pathological condition of SAD, too (for review Farkas and Luiten 2001).

3.5. Brain diffusion compartation

As discussed in the previous paragraph, normal aging as well as age-related neurodegenerative disorders are accompanied by a critical reduction of the regional cerebral blood flow (rCBF) in several brain regions. Such changes in perfusion may lead to changes in cellular integrity of the brain tissue. Cerebral atrophy, ventricular enlargement, and polio- and leuko-araiosis may become evident after age of 60 years as perfusion declines. After the age of 70 years, there is an excessive cortical perfusional decrease as well as grey and white matter hypodensities, and cerebral atrophy were found accompanied with cognitive decline.

Beside the possible reasons for the decline of cognitive functions with aging (see above), cortical disconnections may participate in the generation of age-related behavioral changes. This may be supported by post-mortem studies and in vivo investigations during the last decade using functional magnetic resonance imaging (fMRI) and Positron Emission Tomography (PET).

Diffusion-weighted magnetic resonance imaging (MRI) has been proven to be a sensitive diagnostic tool to examine age-associated acute and chronic changes in brain tissue. Diffusion-weighted MR imaging gives insight to microstructural changes in brain tissue caused by changes i) in extracellular space volume and geometry, ii) of cell morphology and, iii) of myelination. Such changes take place with normal brain aging (Sykova et al 2002) as well as in age-related neurodegenerative disorders.

In several studies, mostly performed in volunteers, the age-dependency of the mean apparent diffusion coefficient (ADC) was assessed. ADC characterize the diffusion of water molecules between the extra- and intracellular space as well as changes in brain structure such as nerve cell lost and gliosis. The results of these studies showed either an increase of the ADC with the age or an age-independent ADC or a decrease of ADC with age.

Current animal studies using diffusion-weighted and T2-weighted MR imaging sequences investigated differences in brain diffusion and transverse relaxation time (T2) between young-adult (age: 3 months) and adult (age: 12 months) rats (Heiland et al 2002). The studies showed no differences in T2 relaxation time between both animal groups, either local or global. However, the mean ADC within the whole brain was lower in the adult animals than in the young-adult ones. ADC decrease was mainly found in rat cerebral cortex but not in other brain areas. These results can be attributed to an activity-related or CNS damage-related internal water shift from extracellular to intracellular space without a net increase of water content in brain tissue. These results disagree with the results of different studies in humans, where an age-related ADC increase was found. This ADC increase, however, was only significant for volunteers after their 6th decade of life (O'Sullivan et al 2001). There is no consensus whether there are ADC changes below that age range. These results seem to prove the theory, that loss of neuronal connectivity and demyelination processes should lead to an increase of ADC at higher ages and in senescence.

A gross estimation leads to the assumption, that rat age of 12 months corresponds to middle-aged human life. Therefore, a correlation between neuronal loss and demyelination and diffusion data may not be expected. This is confirmed by the fact, that no T2 changes throughout the brain were found. Normally T2 changes follow soon after changes in ADC. T2 increase can be explained by a variety of histologic changes, including enlargement of the perivascular space, gliosis, and demyelination due to perivascular damage. The lack of T2 changes shows that such degeneration processes have not taken place in rats up to an age of 12 months.

In another study Sykova and co-workers have found a significantly lower diffusion in aged rat brain using the real-time iontophoretic tetramethylammonium method (Sykova et al 2002). The lower diffusion found particularly in cortex, corpus callosum and hippocampus was attributed to an activity- or CNS damage-related cellular swelling which leads to changes in volume and geometry of the extracellular space (Sykova et al 2002). The above shown results (Heiland et al 2002) supply this view because the change in ADC not accompanied by T2 variation. Thus, only an internal water shift from extracellular to intracellular space without a net increase of water content in brain tissue was detected.

3.6. Oxidative energy metabolism

A threshold phenomenon of 70 years of age was also observed with respect to some enzyme activities in the human brain. Reductions of the activities of phosphofructokinase, choline acetyltransferase acetylcholine esterase and also for acetylcholine synthesis have been described. In contrast, no age-related variations have been observed in Na^+-K^+-ATP-ase and Mg^{++}-ATP-ase which utilize ATP. Thus, reduction in the above brain parameters before this threshold of 70 years in human life may, therefore, be disease-related rather than age-related.

From animal studies, it may be deduced that in brain cortex and in some subcortical nuclei glucose consumption diminishes from development to adulthood but not further with advancing age. At the cellular level, a moderate but steady decline in the levels of glycolytic compounds was found from development to senescence. However, the diminution was not evenly distributed with respect to the different periods of life. Glucose and fructose-1,6-diphosphate, as well as ATP, decreased most from development to adulthood, whereas pyruvate and creatine phosphate fell most from adulthood to senescence (Hoyer 1985). The changes in glycolytic compounds may be associated with age-related reductions in the activities of the enzymes hexokinase and phosphofructokinase. The age-related increase in lactate formation and the reduced pyruvate production may indicate changes in cytoplasmic redox potential, and in intracellular pH in terms of an acidic shift.

On the other hand, only slight variations have been found in the oxidative processes of the tricarboxylic acid cycle and the respiratory chain with aging. The decrease in malate levels found in senescent rats may be consistent with the reduced activity of malate dehydrogenase indicating diminished activity of the tricarboxylic acid cycle. Otherwise, no age-related changes were reported in the activities of the multi-enzyme complexes such as pyruvate dehydrogenase (active and total), citrate synthase and NAD^+-isocitrate dehydrogenase whereby the latter finding was inconsistent. Studies on cytochrome a, a3, as the final member of substrate oxidation that reacts directly with molecular oxygen revealed no age-related changes. However, in cerebral cortex of rats, oxygen consumption was found to decrease gradually with age, and $^{14}CO_2$ production diminished by around 30% with senescence.

Acetylcholine as the main neurotransmitter deriving from glucose metabolism also showed reduced levels with aging. Its synthesis declined to 65% in senescence compared with young adulthood, and acetylcholine release to around 25%--50%.

When glucose consumption and energy formation of the aging brain are considered, it has to be noted that in physically and mentally healthy senescent people, a 23% reduction in glucose consumption was found without any change in cerebral oxygen utilization compared with healthy younger people. When ATP formation is calculated on the basis of these data, a slight decrease by around 5% becomes obvious. This also holds true for experimental animals: the availability of energy was found to be reduced by 5% in parietotemporal cerebral cortex of 104-week-old rats compared to 52-week-old animals under resting conditions, but decreased by 15% in very old (130 weeks of age) rats (Dutschke et al 1994).

3.7. Mixed function oxidation and advanced glycation end products

Two processes with great damaging potency have been found to occur during the aging process, and both are related to glucose /energy metabolism: mixed function oxidation (MFO) and the formation of advanced glycation end products (AGEs).
The age-related reduction in energy availability, which becomes particularly obvious under stress conditions (for details see below), may be assumed to compromise energy-dependent processes. Among such processes, one was found to be of central importance during the aging process: mixed function oxidation (MFO), also termed metal-catalized oxidation (MCO). Continuous intracellular protein turnover includes oxidative inactivation by various enzymatic and nonenzymatic MFO processes that precede proteolysis.

The oxidation of amino acids to carbonyl derivatives, mediated by the formation of oxygen free radicals, is synergistically affected by nucleoside phosphates and triphosphates and by bicarbonate. Furthermore, the degradation of proteins has been found to be ATP-dependent. Consistent with these observations, the amount of oxidized proteins has been demonstrated to increase in the brain cortex during normal aging (Smith et al 1991). Thus, the aged brain may be characterized by an imbalance between enhanced formation of oxidized, damaged proteins and the cellular capacity to degrade them because a) the extent of protease activities is reduced and b) the energy deficit is permanently increasing with age (see above).

In long-lasting metabolic processes, glucose and other sugars are able to glycolyse proteins irreversibly via non-enzymatic reactions which finally generate advanced glycation end products (AGEs). These posttranslational modifications were found to be prominent during aging. AGEs show a high affinity to its receptor (RAGE). Receptor binding is followed by an expression of both growth factors and transcription factors as well as by changes of membrane permeability (Vlassara et al 1994). RAGE have been demonstrated to be present in CNS. Interestingly, βA4 has been shown to bind to RAGE. With aging, AGEs have been found to accumulate in the perikarion of large neurons in the hippocampus and the dentate gyrus, and in pyramidal cells of the cortical layer III, V and VI. These cell layers have been demonstrated to be particularly prone to degeneration in sporadic Alzheimer disease.

3.8. Stress conditions during aging

During aging, the brain may be compromised more frequently than in adulthood by external abnormal conditions such as arterial hypoxemia, arterial hypoglycemia and arterial hypotension. These additional stressful events may be assumed to aggravate the changes normally occurring in cerebral oxidative / energy metabolism and related metabolism. In profound arterial hypoxemia of short duration only, changes in cerebral glucose metabolism were more severe in aged than in adult animals pointing to insufficient compensation mechanisms during aging (Degrell et al 1983) and to increased vulnerability to anoxia. Recovery of the energy pool in the cerebral cortex after arterial hypoglycemia was found to be more markedly compromised in aged than in adult animals, as was the restoration of the cerebral amino acid pool. As a marker of the severity of the damage, the concentration of ammonia more than doubled in aged brains when compared to adult ones.

3.8.1. Cerebral oligemia

During aging cerebral oligemia may represent an important stress condition. Thus, in different rat models of global cerebral oligemia [permanent bilateral occlusion of both carotid arteries (BCCAO) and stepwise cerebral four-vessel occlusion model (stepwise 4-vo)] chronic changes in brain function, metabolism and structure and compensatory mechanisms were demonstrated (Plaschke et al 1999, 2000).

Compared to the age of 12 month, rats aged 20-month-old showed a marked decrease in learning and memory abilities such as working and reference memory after permanent BCCAO (Plaschke et al 1999). This is in agreement with other findings in the literature. In addition, after long-lasting BCCAO over 8 month concentration of energy rich phosphate creatine phosphate was significantly decreased in rat cerebral cortex. However, the concentration of ATP was not changed markedly in permanent rat BCCAO model. It is discussed, however, whether or not at least a part of the functional deficit after permanent brain vessel occlusion is more closely related to the 'transmission failure' rather than to the 'energy failure'. According to the 'transmission failure theory' cerebral neurotransmitter deficit may play a role in the process of memory dysfunction. As stated above the concentration of cerebral neurotransmitters such as acetylcholine is reduced during aging. Taken together, the results of the BCCAO studies may conclude that both energy and neurotransmitter deficits may be responsible for stress-related behavioral changes in BCCAO rat brain.

To produce marked changes in rat brain histopathology and to induce substantial nerve cell damage, ligation of both carotid arteries only (BCCAO) would seem to be ineffective in many animals owing to an ample collateral blood flow in chronic studies. Therefore, a chronic rat model in which both carotid and vertebral arteries were permanently occluded in a stepwise manner for at least 8 month would seem to be more appropriate. Chronic cerebral oligemia due to a stepwise 4-vo is characterized by a 30% reduction in cerebral blood flow. At light microscopy no significant changes in rat brain histopathology as well as in receptor densities were obtained. However, marked changes in rat brain capillaries were seen after 8 month of oligemia as compared with controls in electron microscopy. Thus, most capillaries in cerebral cortex and in hippocampus in vessel-occluded animals shrank, the endothelium cells were prominent, and the basal membrane thickened. In contrast to the degenerative changes in brain capillaries seen in 20-month-old rats, pronounced arterial collateralization was disclosed by digital subtraction angiography after chronic brain vessel occlusion (Plaschke et al 2000).

Thus, chronic four cerebral vessels occlusion was characterized by capillary degeneration compensated at least in part through pronounced arterial collateralization.

Cerebral ischemia caused changes of adaptation in glycolytic glucose breakdown, an anaplerotic reaction in the tricarboxylic acid cycle, and a more pronounced fall and loss in the adenosine nucleotide level in aged as compared to adult animals. In the postischemic recirculation period, the delayed decrease in energy-rich phosphates was more severe in aged than in young adult animals, more pronounced in hippocampus as compared to cerebral cortex.

Chronic exposure to lead (Pb) has been proposed as a risk factor for the development of neurodegenerative disorders such as Alzheimer disease and Parkinson disease. Pb has long been known to have effects on a wide range of cellular and molecular mechanisms in mammalian brain including energy metabolism obviously due to cerebral hypometabolism caused by direct inhibition of specific glucose-utilizing enzymes. When the middle-aged brain is under energy depletion, its vulnerability to Pb has been found to be increased. Chronic exposure to low-level lead caused reductions in learning and memory capacities in rats during aging from 1 year to 2 years of age. These abnormalities were accompanied by a slight fall of ATP in parietotemporal cerebral cortex (Yun et al 2000).

Differences in the cerebral energy pool between adult and aged animals became obvious under intense mental activation what increased the level of cerebral cortex high energy phosphates by 16% (ATP) and 32% (creatine phosphate) in adult animals and by 9% (ATP) and 28% (creatine phosphate) in aged animals. These findings may indicate that the age-dependent decline in the size of available energy could not be prevented by short-term mental activation. The diminution of the energy pool with aging was even more marked after mental activation than after mental rest, and the increase of ATP turnover in aged animals was indicative for an elevated energy demand (Dutschke et al 1994). However, in very old senescent animals, ATP turnover decreased drastically although ATP formation was found to be unchanged. These changes are thought to be due to an age-related reduction of the activity of adenine nucleotide translocase.

3.9. Glucocorticoids and HPA-axis

There is clear evidence that a higher basal cortisol concentration becomes effective during aging (Lupien et al 1994). However, circadian rhythm will have to be considered. The increased cortisol concentration in arterial blood has been found to be mirrored in the cerebrospinal fluid of healthy elderly people who showed enhanced CSF cortisol levels as compared to young subjects. In parallel, the number of corticosteroid receptors mainly those ones of type I has been demonstrated to be selectively reduced with aging in the hippocampus, whereas no change or even an increase were found in type II receptors. Since glucocorticoids have been thought to be toxic to hippocampal neurons functional/structural changes may result in the hippocampus as a consequence and may lead to a reduction of its major inhibitory control on cortisol secretion (Sapolsky et al 1994).

The decrease in the hippocampal inhibition function may be assumed to disinhibit the HPA-axis resulting in increased and prolonged response to stress. In parallel, the basal concentration of ACTH has been demonstrated to be increased. Obviously, the age-associated alterations in HPA function are gender-specific in that they are more prominent in women. Social conditions have been found to effect both cortisol and ACTH plasma levels in that the response to CRF was demonstrated to be higher in younger compared to old age monkeys. From these variations, it may be deduced that the basal tone of the HPA-axis has increased with aging leading to hypercortisolemia in a cascade-like manner which finally may compromise the function of the neuronal insulin receptor by a dysregulation of its phosphorylation of the tyrosine residues. Thus, a marked imbalance between insulin and cortisol effects exists in the aging brain (Figure 2).

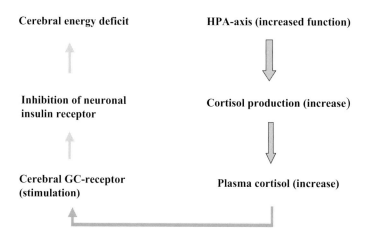

Figure 2. With aging, the HPA-axis exerts an increased function resulting in an enhanced cortisol production and in hypercortisolemia. Either the neurotoxic effect of cortisol itself on neurons, or the stimulation of glucocorticoid (GC, type II) receptors, or both, may inhibit the function of the neuronal insulin receptor resulting in a cerebral energy deficit.

3.10. Cell cycle activity

As has been discussed above (Insulin and cell cycle activity), insulin has been found to induce mitogenesis throughout two independent pathways acting in synergy, PI-3 kinase/p 70 S 6 kinase, and p 43/ p 44- MAP kinase (Virkamäki et al 1999) accompanied by an increase in the proportion of proliferating cells in S and G2/M phases of the cell cycle (Conejo and Lorenzo 2001).

Lacking insulin as well as the inhibition of both the PI-3 kinase and the MAP kinase activities have been demonstrated to result in a reduced incorporation of (^{3}H) thymidine, initiation of apoptosis and in growth arrest in the G0/G1 phases of the cell cycle with a high proportion (90%) of quiescent cells (Conejo and Lorenzo 2001). An abnormality in the insulin signal transduction cascade by a deficit of the insulin receptor substrate-1 has been found to result in the inability of cells to increase DNA synthesis and to enter into the S/G2/M phases of the cell cycle. The clear age-associated fall of both insulin concentration and tyrosine kinase activity of the insulin receptor may be assumed to reduce the action of insulin receptor substrate-1 thus inhibiting the S/G2/M phases and stimulating the G0/G1 phases of the cell cycle (Figure 3). However further studies are needed to substantiate these probable pathways with aging.

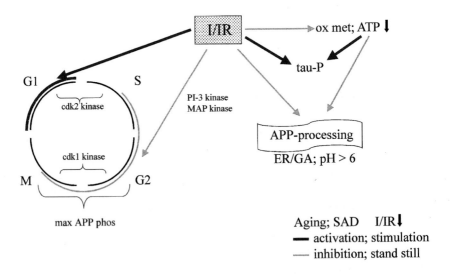

Figure 3. Schematic survey on the function of the neuronal insulin/insulin receptor (I/IR) signal transduction cascade with aging. Inhibition abnormalities are directed to: oxidative energy metabolism leading to decreased concentrations of both acetylcholine and ATP. The fall in ATP may induce an increase of pH in endoplasmic reticulum (ER)/ Golgi apparatus (GA) thus influencing protein processing (e.g. APP) in these intracellular compartments. A reduced availability of ATP may stimulate tau-protein phosphorylation, as does a lack of insulin, both APPs and βA4 may accumulate intracellularly, S/G2/M phases of the cell cycle whereas the G0/G1 phases are activated.

3.11. Gene expression profile

With respect to gene expression with aging, there is now first evidence that the aging process of the brain is accompanied by variations in the gene expression profile. Some genes were switched off whereas other ones are switched on or upregulated eventually in a region-specific manner. The balance between genes expressing proteins working protectively and those ones accelerating atrophy may be assumed to be disturbed. More specific studies showed that the gene expression of some ATP-ases and of proteins working in synaptic transmission was demonstrated to be reduced 2-to 10-fold in the cerebral cortex and the hypothalamus (Jiang et al 2001) what may have impacts on transmembrane ion fluxes as e.g. in the endoplasmic reticulum and Golgi apparatus. In the hippocampus, age-associated changes in gene expression were described to be related to proteins working in energy metabolism and cell cycle regulation. However, exposure to an enriched environment was found to oppositely regulate gene expression as compared to aging at rest (Jiang et al 2001). Another most interesting novel aspect has been reported recently. A stressful event originated to offsprings by maternal separation at postnatal day 9 was found to reduce the expression of brain derived growth factor and NMDA receptor subunits in rat hippocampus at postnatal day 72. The data may indicate that perinatal stress to the fetus/offspring may influence or even determine the quality of the gene expression profile with aging (see also Holness et al 2000).

4. SYNOPSIS AND CONCLUSIONS

In this chapter, the central role of brain glucose metabolism and its control by the neuronal insulin/insulin receptor (I/IR) signal transduction cascade is demonstrated. The normal function of I/IR signal transduction cascade is multifold. The action of acetylcholine the formation of which is controlled by insulin is directed to the microvasculature to maintain cerebral blood flow and substrate supply to the brain. The oxidative metabolism and its product ATP in particular are maintained at constant levels e.g. to ensure a pH of 6 in the endoplasmic reticulum/Golgi apparatus necessary for undisturbed protein trafficking in these intracellular compartments. In a concerted action with ATP, the I/IR signal transduction cascade supports also the APP trafficking in endoplasmic reticulum/Golgi apparatus. Likewise, it may be assumed that both ATP and insulin keep tau-protein in a state of normal phosphorylation. The effect of the I/IR signal transduction cascade in cell cycle function may be assumed to be different. I/IR stimulates processes at the S/G2/M phases of the cycle with inclusion of PI-3 kinase and MAP-kinases, and inhibits processes at the G1 phase. It is suggested that important cell functions such as protein phosphorylation are controlled/stimulated by I/IR rather than cell cycle regulation.

Functionally, the neuronal I/IR signal transduction cascade inclusive both acetylcholine and ATP contribute highly to molecular processes ensuring learning and memory capacities.

With aging, it becomes evident that multiple inherent changes in fundamental metabolic principles at the cellular, and molecular and genetic levels in cerebral glucose/energy metabolism, its control and related pathways are set into motion. Numerous changes are found to be accentuated under stress. Beside changes of single parameters, functional imbalances of regulative systems may develop such as

- energy production (reduced) and energy turnover (increased),
- insulin action (reduced) and cortisol action (increased),
- acetylcholine action (reduced) and noradrenaline action (increased), indicating a sympathetic tone,
- formation of oxidized proteins (increased) and capacity of their degradation (reduced),
- shift in the gene expression profile from anabolic site (reduced) to catabolic site (increased).

These changes/shifts may indicate an uncoupling of synchronization which has been demonstrated to exist in biological systems. This model may correspond to the increase in entropy which is an elemental, inherent principle of chemical and biological processes. In the physical sciences, the term criticality is used respectively to describe a self-organized metalabile steady state (metabile equilibrium in entropy). Smaller additional internal or external events, even one that is ineffective in itself, may change biological and/or biophysical properties of the aging brain. Such events may shift a system from supercriticality to criticality to subcriticality/catastrophic reaction, i.e. a disease in medical terms.

Is the aging brain a burden of life? Even after more than 2000 years, Cicero's view is still actual. Aging in general, and aging of the brain in particular may become a burden of life for many human beings. However, although our present knowledge as to how to maintain mental capacity with aging is still limited, new findings at the cellular, molecular and genetic levels open the promising chance to meet one of the most important problems of human society: longevity with good mental health.

ABBREVIATIONS

ACTH	adrenocorticotropic hormone
ADC	apparant diffusion coefficient
AGE	advanced glycation endproducts
ATP	adenosine triphosphate
APO	apolipoprotein
APP	amyloid precursor protein
BCCAO	bilateral common carotid artery occlusion
GFAP	glial fibrillary acidic protein
GLUT	glucose transport protein
HPA	hypothalamic-pituitary-adrenal axis
I	insulin
IR	insulin receptor
IRS	insulin receptor substrate

LTP	long-term potentiation
MCO	metal-catalized oxidation
MFO	mixed function oxidation
MRI	magnetic resonance imaging
Pb	lead
PDH	pyruvate dehydrogenase
PET	positron emission tomography
PK	protein kinase
RAGE	receptor for advanced glycation endproducts
SAD	sporadic Alzheimer's disease
TCAC	tricarboxylic acid cycle
T2	transverse relaxation time
Vo	vessel occlusion

REFERENCES

Blanchard RJ, McKittrick CR, Blanchard DC (2001) Animal models of social stress: effects on behavior and brain neurochemical systems. Physiol Behav 73: 261-271

Conejo R, Lorenzo M (2001) Insulin signalling leading to proliferation, survival and membrane ruffling in C2 C12 myoblasts. J Cell Physiol 187: 96-108

Degrell I, Krier C, Hoyer S (1983) Carbohydrate and energy metabolism of the aging rat brain in severe arterial hypoxemia. In: Cervos-Navarro J, Sarkander HI (eds) Neuropathology and Neuropharmacology, Aging Series Vol 21, Raven, New York, 289-300

Dutschke K, Nitsch RM, Hoyer S (1994) Short-term mental activation accelerates the age-related decline in brain tissue levels of high-energy phosphates. Arch Gerontol Geriatr 19: 43-51

Farkas E, Luiten PGM (2001) Cerebral microvascular pathology in aging and Alzheimer's disease. Prog Neurobiol 64: 575-611

Heiland S, Sartor K, Martin E et al (2002) In vivo monitoring of age-related changes in rat brain using quantitative diffusion magnetic resonance imaging and magnetic resonance relaxometry. Neurosci Lett 334: 157-160

Henneberg N, Hoyer S (1994) Short-term or long-term intracerebroventricular (icv) infusion of insulin exhibits a discrete anabolic effect on cerebral energy metabolism in the rat. Neurosci Lett 174: 153-156

Holness MJ, Langdown ML, Sugden MC (2000) Early-life programming of susceptibility to dysregulation of glucose metabolism and the development of type 2 diabetes mellitus. Biochem J 349: 657-665

Hong MF, Lee VMY (1997) Insulin and insulin-like factor-1 regulate tau phosphorylation in cultured human neurons. J Biol Chem 272: 19547-19553

Hoyer S (1985) The effect of age on glucose and energy metabolism in brain cortex of rats. Arch Gerontol Geriatr 4: 193-203

Hoyer S (1995) Brain metabolism during aging. In: Macieira-Coelho A (ed) Molecular basis of aging. CRC Press, Boca Raton, pp 493-510

Jiang CH, Tsien JZ, Schultz PG et al (2001) The effects of aging on gene expression in the hypothalamus and cortex of mice. Proc Natl Acad Sci USA 98: 1930-1934

Lesort M, Jope RS, Johnson GVW (1999) Insulin transiently increases tau phosphorylation: involvement of glycogen synthase kinase. F-3β and Fyn tyrosine kinase. J Neurochem 72: 576-584

Lupien S, Lecours AR, Lussier I et al (1994) Basal cortisol levels and cognitive deficits in human aging. J Neurosci 14: 2893-2903

O'Sullivan M, Jones DK, Summers PE et al (2001) Evidence for cortical "disconnection" as a mechanism of age-related cognitive decline. Neurology 57: 632-638

Park CR (2001) Cognitive effects of insulin in the central nervous system. Neurosci Biobehav Rev 25: 311-323

Plaschke K, Yun SW, Martin E et al (1999) Interrelation between cerebral energy metabolism and behaviour in a rat model of permanent brain vessel occlusion. Brain Res 830: 320-329

Plaschke K, Martin E, Bardenheuer HJ et al (2000) Transfemoral digital subtraction angiography for assessment of vertebral artery occlusion in rats. Stroke 31: 1789-1790

Plata-Salaman CR (1991) Insulin in the cerebrospinal fluid. Neurosci Biobehav Rev 15: 243-258

Popa-Wagner A, Schröder E, Schmoll H et al (2001) The role of growth-promoting and growth-inhibiting factors in neurorepair after stroke. Effect of age. In: Current Concepts in Experimental Gerontology. C Bertoni-Freddari/H Niedermüller (ed) pp: 177-201

Sapolsky RM (1994) Glucocorticoids, stress and exacerbation of excitotoxic neuron death. Sem Neurosci 6: 323-331

Siesjö BK (1978) Brain energy metabolism. Wiley, Chichester, pp 1-28, pp 151-209

Smith CD, Carney JM, Starke-Reed PE et al (1991) Excess brain protein oxidation and enzyme dysfunction in normal aging and in Alzheimer's disease. Proc Natl Acad Sci USA 88: 10540-10543

Suzuki T, Oishi M, Marshak DR et al (1994) Cell cycle-dependent regulation of the phosphorylation and metabolism of the Alzheimer amyloid precursor protein. EMBO J 13: 1114-1122

Sykova E, Mazel T, Hasenöhrl RU, Harvey AR, Simonova Z, Mulders WH, Huston JP (2002) Learning deficits in aged rats related to decrease in extracellular volume and loss of diffusion anisotropy in hippocampus. Hippocampus 12(2):269-79

Uylings HBM, deBrabander JM (2002) Neuronal changes in normal human aging and Alzheimer's disease. Brain Cognition 49: 268-276

Vlassara H, Bucala R, Striker L (1994) Biology of disease. Pathogenetic effects of advanced glycosylation: biochemical, biologic, and clinical implications for diabetes and aging. Lab Invest 70: 138-151

White MF, Kahn CR (1994) The insulin signalling system. J Biol Chem 269: 1-4

Xie L, Helmerhorst E, Taddel K et al (2002) Alzheimer's β-amyloid peptides compete for insulin binding to the insulin receptor. J Neurosci 22: RC 221(1-5)

Yun SW, Gärtner U, Arendt T, Hoyer S (2000) Increase in vulnerability of middle-aged rat brain to lead by cerebral energy depletion. Brain Res Bull 52: 371-378

L. KACZMAREK

Nencki Institute, Warsaw, Poland

2. GENE EXPRESSION AND ITS REGULATION IN NEURONS

Summary. This chapter presents an overview of nuclear gene expression and its regulation in neurons. The major steps involved in these processes are (1) relaxation of the chromatin structure to make DNA accessible to transcription regulatory machinery; (2) binding of the transcription factors, proteins capable to interact with specific DNA sequences as well as with other proteins to regulate the transcription; (3) activation of RNA polymerase, an enzyme catalyzing the transcription itself; (4) transcription, resulting in formation (elongation) of a primary transcript; (5) capping (addition of ^{7}methyloGuanine) to the 5' end of the nascent transcript; (6) splicing, i.e., removal of the introns from the primary transcript; (7) polyadenylation (addition of more than a hundred of adenine residues to the 3' end of the transcript); (8) export of the mRNA from the nucleus to the cytoplasm; (9) specific localization of the mRNA to defined cell compartments, such as dendrites; (10) survival of the mRNAs in the cytoplams; and finally (11) the translation. Recent studies show that steps 1 to 8 are dependent on each other, showing thus the enormous complexity of the nuclear events of gene expression. Furthermore, all of the steps 1 to 11 are tightly regulated to meet the needs of complex neuronal physiology. It is suggested that control over gene expression may play the major role in neuronal plasticity, including long-term memory formation.

1. INTRODUCTION

Only about 15 years ago the role of external stimuli-driven gene expression in the brain has become recognized as a major way to regulate neuronal function in response to environmental challenges. Discovery that neuronal activation results in a rapid and transient wave of gene expression has prompted a number of studies on a biological significance of this phenomenon, and parallel enormous increase in our understanding of molecular biology of the neurons. Major lesson coming from these studies appears to be that neurons are essentially not so much different in this regard from other cells in the body. This notion has been extended to a vast array of signal transduction pathways and an overwhelming variety of cellular responses that underlie such phenomena as neuronal plasticity and neurodegeneration, comprising the major foci of present studies in the field of neurobiology. However, it is worthy to bear in mind that brain provides the most complex example of gene expression - out of ca. 30 000 - 40 000 genes, the proportion of those expressed in the brain is estimated to exceed markedly the situation in any other organ, and up to 5 000 genes are believed to be predominantly or exclusively expressed in the brain. The brain complexity with its vast network of neuronal connections could be responsible for this phenomenon, that has been well appreciated by neuroanatomists for a long time, as neurons have been notorious for their high content of (active/potentially active) euchromatin.

2. THE GENES

In the brain cells the genetic material is located in cell nuclei (predominantly) and in the mitochondrion. In the latter there is only a very small number of genes, furthermore their expression appears to be co-ordinately regulated with nuclear genes, in neurons being also responsive to neuronal activity. An interesting example is provided by a multisubunit protein cytochrome oxidase, whose some subunits are encoded by mtDNA and other ones by the nuclear genes. All of them are down-regulated by decrease and upregulated by increase of neuronal activity. Unfortunately, considering vast energetic expense of neuronal function and well recognized role of mitochondrion in apoptosis and neurodegeneration, very little is known about mechanisms of gene regulation in this cell compartment.

The nuclear gene regulation has been a focus of intense studies. There are three classes of genes. Class I codes for major rRNAs, class II for mRNAs and thus proteins and class III for small RNAs (such as tRNAs and RNA molecules involved in splicing). The class II is obviously the most complex, with the highest number of individual species (ca. 15 000 different species in the brain), each coding for a different polypeptide. The class II gene transcripts make up for 58% of the total transcription, and mRNAs represent only about 3% of the total RNA present in each cell.

The class II genes are composite in structure. In the vast majority of them, the mRNA-coding regions (exons) are separated by often much longer stretches of intervening DNA sequences (introns). Each gene contains a number of regulatory areas, involved in control of its expression. The one located immediately upstream the transcription initiation site is called promoter, however, there are regulatory regions found as far as hundreds thousands of DNA base pairs (bp) from the coding sequences, both upstream and downstream. The double helix of DNA in the nucleus is tightly coupled to proteins and nascent RNA in a form of chromatin. Its main unit is nucleosome, composed of 146 bp of DNA surrounding histone octamer. Histones are positively charged (basic) proteins, and by this virtue they display high affinity to a negatively charged DNA. The histone octamer is made of four, so called core histone species (two molecules of each: H2A, H2b, H3, and H4). Between nucleosomes, there are linker regions of a few dozens of bp, that could be bound by histone H1, that is involved in creating higher order chromatin structures. Other proteins are then involved in even tighter packing of the chromatin.

2.1. An overview of gene expression

In the process of activation of gene expression, there is a number of events that can be distinguished: (i) relaxation of the chromatin structure to make it accessible to transcription regulatory machinery; (ii) binding of the transcription factors, proteins capable to interact with specific DNA sequences as well as with other proteins to regulate the transcription; (iii) activation of RNA polymerase (in the case of class II genes it is RNA pol II), an enzyme catalyzing the transcription itself; (iv) transcription, resulting in formation (elongation) of a primary transcript; (v) capping (addition of ^{7}methyloGuanine) to the 5' end of the nascent transcript; (vi) splicing, i.e., removal of the introns from the primary transcript; (vii) polyadenylation (addition of more than a hundred of adenine residues to the 3' end of the transcript); (vii) export of the mRNA from the nucleus to the cytoplasm.

Despite didactic usefulness of dissecting transcription and post-transcriptional RNA processing into aforementioned steps, one has to realize an intricate complexity of those events. Recent studies clearly show that all these steps are dependent one on another. For instance, transcriptional apparatus plays an active role in recruiting the machinery that caps and processes the nascent transcript and the splicing promotes the transcription elongation and is required for efficient mRNA export from the nucleus. Furthermore, all these event apparently overlap also in time. This complexity implies a no less intricacy in regulatory mechanisms involved in transcriptional control, and thus suggest that the multi-step formation of mRNA is the major regulatory process in the gene expression.

2.2. Histone modifications

Tight binding of core histones to DNA in nucleosomes is an obvious obstacle for accessibility of the DNA to transcriptional regulatory and enzymatic machinery. Thus, no wonder that specific covalent modifications of histones play a major role in making chromatin either susceptible or refractory to transcription. Following major modifications are distinguished according to their chemical nature: acetylation, methylation, phosphorylation, and ubiquitination. Specific enzymes, such as histone transacetylases (HAT), histone deacetylasaes (HDAC), histone methyltransferases (HMT), etc. are involved in those modifications. Not of lesser importance are such phenomena as degree of modification (addition of a single or more modifying groups) as well as the exact location of each modification, i.e., which aminoacid residues in which histone molecule are modified. In particular, the transcriptionally active chromatin contains acetylated or even hyperacetylated histones. For example, H4 can be acetylated/hyperacetylated at lysins 5, 8, 12, and 16, whereas H3 at lysins 9, 14, 18, and 23. As far as the methylation is concerned it may occur at Lys 20 in the case of H4, and lysins: 4, 9, 27, 36 in H3. The euchromatin is enriched in histone H3 methylated at Lys4 (this aminoacid residue, when tri-methylated, appears to be the best marker of active chromatin), whereas in heterochromatin H3 is methylated at Lys 9. Phosphorylation may be observed at Ser 1 in H4, and serines 10, and 28 in H3. There is an apparent link between various modifications. For instance, H3 phosphorylation at Ser 10 greatly enhances subsequent H3 acetylation at Lys 14. Notably, all these modifications are concentrated on so called histone tails that stretch out of the nucleosome core and are therefore easy targets for modifications. It has been suggested that they are altogether responsible for the gene regulatory histone code.

Interestingly, the same condition, such as, e.g., status epilepticus may result in differential effect on histone acetylation level at different gene promoters, inducing deacetylation of those that are silenced and simultaneous hyperacetylation of the genes that are upregulated.

2.3. Transcription factors and gene regulation

Relaxation of the chromatin structure that relies mainly on histone modifications as well as activity of the ATP-dependent remodeling proteins (such as components of yeast SWI/SNF chromatin remodeling complex, and their insect and mammalian homologs, like Swi2/Brm1 protein) allows binding of the transcription factors. Those are estimated to be numbered in thousands, and aforementioned unusual complexity of the gene expression in the brain neurons implies that an abundance of transcription factors in this organ should be particularly plentiful. To simplify this picture there are ways to group the transcription factors according to various criteria. One is structural. There are characteristic domains in these proteins, named as "helix-loop-helix", "POU domain", "leucine zipper", "zinc finger", "homeodomain", etc. Unfortunately, those often evoked names do not easily translate to any specific biological significance.

Another, more functionally relevant classification is based on subcellular localization of transcription factors and their mode of activation. There are transcription factors that are constantly present in the cell nuclei, even bound to their target, regulatory sequences (often called "responsive elements") or alternatively to other regulatory proteins. Activation of those TFs often results from their phosphorylation. CREB (cAMP responsive element binding protein) and Elk-1 (interacting with serum responsive factor, SRF that is bound to serum responsive element, SRE) may serve as respective examples in this regard. Members of another class of transcription factors are located in the cytoplasm in a dormant form, usually bound by other proteins. Phosphorylation (or even proteolytic degradation) of the latter allows the TFs to travel to the nucleus to their target genes. NF-κB and NFAT are good examples in this case. Thus, the two classes of TFs described just above share a common feature of being activated (directly or indirectly) by protein kinases (nuclear or cytoplasmic). Function of such kinases is often regulated in response to stimulation of membrane receptors. In the cytoplasm and /or nuclei there are such TFs as steroid receptors. Activation of these receptors occurs after binding to their specific ligands that penetrate into the cells through the cell membrane. Notably, in this case, TF activation may not require the kinases. Members of all three categories of transcription factors are sometimes defined as constitutive ones, i.e., present already in non-stimulated cells. This distinction is often made because of the well studied in the brain inducible transcription factors that are activated secondarily, i.e., as a consequence of enhanced activity of the preexisting, constitutive ones (see below). Notably, these so called constitutive transcription factors are, as a matter of fact, inducible by postranslational modifications.

A special case for transcription factors in the brain is given by inducible transcription factors (ITFs) encoded by immediate early genes (IEGs). The latter term refers to the genes, whose expression occurs early after cell stimulation and is not dependent on de novo protein synthesis. They were first described as activated during G_0-S phase transition in the cell cycle. Thus, by no means they are neuron- or nervous system-specific. However, they have been especially extensively studied in the brain as responsive to a number of physiological and pathological stimuli. Two main transcription transcription factors in this group are AP-1 and ZIF268 (the latter is also known as Egr-1, NGFI-A, Krox 24, TIS 8, and ZENK). AP-1 is a protein dimer composed most often of Jun and Fos proteins. There are three Juns: c-Jun, JunB, and JunD, and five Fos proteins: c-Fos, FosB, ΔFosB (the latter two encoded by the same gene, see below), Fra-1, and Fra-2. Increased expression of c-Fos at either mRNA or protein level has been serving as a prototypical gene response in the brain and often taken as a marker of neuronal activation. Careful analyses of c-Fos expression patterns suggest more caution in this regard, as there are many instances of neuronal activation with no c-Fos expression. It rather appears that c-Fos expression should be regarded as a marker of activated neurons that may undergo a plastic change. Widespread neuronal c-Fos expression in a variety of brain function and dysfunction phenomena would thus imply a great susceptibility of the brain neurons to plastic reorganization.

Transcription factors act in concert with other TFs as well as many other accessory proteins. These interactions help to explain how very distant gene regulatory sequences may colaborate (Fig. 1). In this context it is noteworthy to acknowledge the DNA bending proteins (Fig. 1). Of special importance are proteins displaying enzymatic activity towards histones (see above). Such histone acetyltransferases as CBP, and p300 are examples of such enzymes. Formation of the whole "transcriptome" complex, involving also the RNA polymerase as well as its accessory proteins, such as TBP, TFIIA, TFIIB, TFIID, TFIIE, TFIIF, TFIIH, and in consequence initiation of transcription is thus a very complex event, susceptible to multiple levels of regulation.

2.4. Differential splicing

Especially tightly controlled is the process of intron removal as it may produce a number of differentially spliced mRNAs that could be translated into partially different proteins, all encoded by the same gene. There are plethora of examples that this phenomenon has great functional significance. For instance, the alternative splicing of the NR1 (NMDA receptor subunit) primary transcript may generate as many as eight different mRNAs encoding functionally different NR1 proteins, adding an important complexity to the final form of the NMDA receptor. Another example may be provided by existence of flip and flop variants of AMPA glutamate receptor subunits, so called GluRs 1-4 (or A-B). In each of them there is a protein segment that can occur alternatively either in a flip or flop form, and the decision which one is present is taken at the level of differential splicing of the primary transcript. In the rat, the flop splice forms are expressed at low levels early in postnatal brain, then their expression increases.

The delayed incorporation of flop modules into glutamate operated channels coincides with the peak period of synapse formation throughout the brain. Notably, GluRs incorporating the flip sequence allow more current entry into cells than receptors with flop modules only.

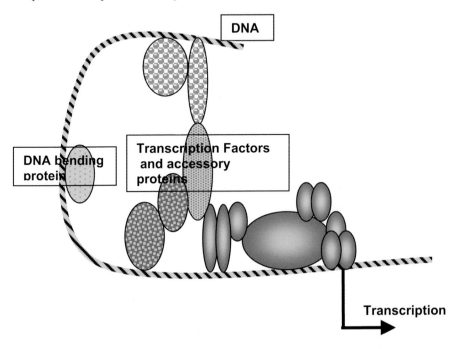

Figure 1. Cooperation of transcription factors and accessory proteins in regulating the gene expression

Therefore, mature forms of these receptors, including the flop form, might depress neuronal excitability. One more from the long list of examples of the prominent functional significance of differential splicing can be provided by two forms of FosB protein that may associate with Jun proteins to form AP-1 transcription factor (see above). The isoform known as ΔFosB, with truncated C-terminus, is much more stable than the full length FosB, and it has been suggested to play an important regulatory role in a process of addiction, because it was found to be accumulated in specific brain structures in response to treatment with drugs of abuse.

2.5. Post-transcriptional gene regulation

After being exported to the cytoplasm, the mRNA ability to be translated into the protein still remains tightly controlled at various levels. The first is the mRNA stability. A half-life of various mRNA species is markedly different, and may last from minutes to many hours. Among the fastest to be turned-over are products of IEGs. Defined mRNA sequences are responsible for this rapid degradation. Recent discovery of micro RNA and small interfering (si) RNA points also to an involvement of these molecules in downregulating specific mRNA species.

Another interesting way of mRNA regulation is provided by means of editing. mRNA editing is a process occurring in the cytoplasm and allowing for a specific change in the mRNA sequence by replacing one nucleotide with another. The most prominent example of RNA editing in the brain concerns AMPA glutamate receptors. In the case of GluR2, the receptor that contains subunit derived from unedited mRNA allows for high Ca^{2+} permeability, whereas the edited mRNA translates into the protein that is Ca^{2+} impermeable.

One more very interesting way to regulate mRNA susceptibility to translation is its specific cytoplasmic localization. Again it is not a neuron-specific phenomenon, as it has been well recognized in developmental biology as a way to distribute specific mRNA to defined compartments of oocytes. However, of special significance for the brain is translocation of the mRNAs towards dendrites. Only recently this phenomenon has been appreciated and considered as potentially important to locally regulate synaptic function, e.g., in learning. A number of mRNAs were found in dendrites, some of them being translocated in response to neuronal activity, suggesting that they may accumulate near activated synapses to subserve the plastic changes. Specific mRNA sequences, such as those located within its 3' untranslated region (3' UTR) play a role in the dendritic localization of mRNAs.

A number of proteins have been implicated in local, dendritic translation, including FMR1. Interestingly some of the dendritic transcripts are poorly polyadenylated and their translation can be regulated by locally operating polyadenylation machinery, acting on so called cytoplasmic polyadenylation elements located at the 3' UTR.

2.6. Functional significance of neuronal gene expression

After describing the main mechanisms of gene expression in the brain, and appreciating complexity of this process, it might be worthwhile to consider this issue in a general manner. What roles could be ascribed to neuronal gene expression, when one considers the expression patterns?

It is obvious that every living cell requires gene expression to make up for the proteins that are lost during physiological metabolic turnover, and thus there is a continuous need for proteins that are responsible for the homeostatic maintenance of nerve cells. The genes that are involved in such processes should be ubiquitously expressed, however their levels may be modulated according to neuronal activity.

Despite the fact that the maintenance explains a lot of examples of neuronal gene expression, it does not elucidate such findings as aforementioned inducible gene expression evoked by external stimuli. However, the term replenishment may capture the events surrounding stimulus-evoked gene expression. It reflects the sudden needs of a neuron following a period of intense activity. By default, the stimulatory conditions involve abrupt bursts of spiking activity - most often following relative quiescence periods. This may result in (1) rapid depletion of neuronal components (e.g., synaptic release machinery, metabolic enzymes, etc.) and (2) the need to replace these exhausted elements. Obviously, one would expect that a replenishment-related wave of gene expression should follow stimulation conditions.

The question remains whether the stimulus-evoked gene expression serves to reinstate the efficacy of synaptic connectivity as it was before the stimulus or it may alternatively provide proteins that are responsible for subserve the plastic changes of the network. In the latter case the gene expression can be viewed as involved in the maintenance of plastic changes. Such plastic change may last at least as long as the proteins are available, or longer if additional stimuli further reinforce the modified connections. In this context it is worthwhile to mention that there are proteins, such as aforemtnioned ΔFosB that can have a lifespan measured in weeks.

This notion about maintenance of plastic changes can be extended further to suggest that stimulation-evoked gene expression may be central to the process of information integration that appears to be the foundation for neuronal plasticity. In this case, the aforementioned multiple gene regulatory events could be activated by different signaling pathways conveying the information to the neurons. Thus, the gene regulatory mechanisms may act as coincidence detectors allowing for a convergence of information. The molecular meaning of this transcriptional coincidence detection may be to trigger and elaborate long-term plastic changes. If this is to be the case, regulation of gene expression could play the pivotal role in learning, allowing for long lasting memory formation.

REFERENCES

Herdegen, T. and Leah, J. D. (1998) Inducible and constitutive transcription factors in the mammalian nervous system: control of gene expression by Jun, Fos and Krox, and CREB/ATF proteins. Brain Research Reviews, 28, 370-490.

Hevner, R.F., and M.T.T. Wong-Riley (1990) Regulation of cytochrome oxidase protein levels by functional activity in the macaque monkey visual system. Journal of Neuroscience 10, 1331–1340.

Huang Y., Doherty J.J., Dingledine R. (2002) Altered histone acetylation at glutamate receptor 2 and Brain-Derived Neurotrophic Factors genes is an early event triggered by status epilepticus. Journal of Neuroscience, 22, 8422-8428.

Job C, Eberwine J.(2001) Localization and translation of mRNA in dendrites and axons. Nataure Reviews Neuroscience, 2, 889-898.

Kaczmarek L., Kossut M. , Skangiel-Kramska J. (1997) Glutamate receptors in cortical plasticity: molecular and cellular biology. Physiological Reviews, 77, 217-255.

Kaczmarek L., Robertson H.A. (Eds.) Handbook of Chemical Neuroanatomy, vol. 19: Immediate early genes and inducible transcription factors in mapping of the central nervous system function and dysfunction. Elsevier, Amsterdam, Boston, London, New York, Oxford, Paris, San Diego, San Francisco, Singapore, Sydney, Tokyo, 2002.

Kelz M.B., Nestler E.J. (2000) deltaFosB: a molecular switch underlying long-term neural plasticity. Current Opinion in Neurolology 13, 715-720.

McManus M.T., Sharp P.A. (2002) Gene silencing in mammals by small interfering RNAs. Nature Reviews Genetics, 3, 737-742.

O'Donnel W.T., Warren S.T. (2002) A decade of molecular studies of fragile X syndrome. Annual Reviews of Neuroscience, 25, 315-338.

Orphanides G., Reinberg D. (2002) Review: A unified theory of gene expression. Cell, 108, 439-445.
Steward O, Schuman EM. (2001) Protein synthesis at synaptic sites on dendrites. Annual Reviews of Neuroscience 24, 299-325.

I. P. JOHNSON

University College, London, UK

3. MORPHOLOGICAL PECULIARITIES OF THE NEURON

Summary. This chapter gives an overview of the morphological features of neurones that distinguish them from other cells, allow them to carry out their unique function in the body. After a brief historical review, the peculiar nature of neuronal size and shape are considered, together with the consequences of this for neuronal classification, connectivity, excitability, intracellular transport and vulnerability to injury. This is followed by a description of the ways organelles are distributed and organised within neurones and the functional correlates of this. Finally, synaptic terminal structure is considered in the context of electrical and trophic interactions

1. INTRODUCTION

While many variants exist, the neuron remains a cell that is specialised for the reception and transmission of electrochemical information. To enable this function, a typical neuron has a highly specialised morphology, with structures associated with general cellular activities localised its cell body and structures associated with information receipt and transfer localised to its cellular processes. This chapter will provide an overview of the morphological features of neurones that equip it to carry out its specialised role. Other chapters in this book will show how such specialisation renders the neurone uniquely vulnerable to insults that would be tolerated by other cell types.

2. HISTORICAL PERSPECTIVE

Heinrich Waldeyer coined the term 'neuron' in 1891. This followed earlier micro dissection studies of large cells taken from nerve tissue by Karl Dieters who described the large nucleolated cell body (soma) from which several branched protoplasmic prolongations and a single unbranched 'axis cylinder' emerged. These cellular processes were termed 'dendrites' and 'axons', respectively. Their various morphological appearances in the light microscope were described extensively by Santiago Ramon y Cajal with the aid of a silver impregnation staining technique developed by his contemporary Camillo Golgi (Mazzarello, 1999 review). Shortly after Cajal's descriptions appeared, there came electrophysiological evidence of unidirectional transmission between neurones, which led Charles Sherrington (Foster & Sherrington, 1897) to coin the term 'synapse' for such intercellular junctions. With the development of electron microscopy, the fine structure of synapses and neuronal organelles was described, and with the development of methods for direct observation of neurones *in vitro* and *in vivo*, information on the morphological correlates of neuronal metabolic processes have been obtained.

3. GENERAL NEURONAL STRUCTURE

Classically, three types of neuron are recognised: unipolar, bipolar and multipolar according to the number of processes that emanate from the cell body. In vertebrates, multipolar neurones are the most common form, followed by a morphological variant of the bipolar neurone (the psuedounipolar neurone) and then by bipolar neurones themselves (Fig. 1). While unipolar neurones are rare in the vertebrate nervous system (e.g. some autonomic ganglia), they are commonly found in the invertebrate nervous system. Bipolar neurons (e.g. retinal ganglion cells) and psueounipolar neurones (e.g. dorsal root ganglion neurones) often have a sensory function. Multipolar neurones have various functions (e.g. interneurones, pyramidal neurones, motoneurones). Golgi classified neurones according to whether their axons projected to distant sites (type I neurones) or arborised within the immediate vicinity of the cell body (type II neurones).

The ability of the axons of neurones to take up and transport trophic as well as toxic materials from their terminal fields to the cell body makes this a useful classification for neuropathological studies. Ramon-Moliner further classified neurons on the basis of their dendritic patterns, which is a significant determinant of the receptive and integrative function of neurones. Taken together, these two methods of classifying neurones (extent of axonal projection and dendritic arbour) reflect the total volume of cytoplasm that needs to be maintained by the cell body. These are important considerations when assessing the effects of cell body stress (e.g. retrograde response to axonal or dendritic injury or direct cell body toxicity) on neuronal maintenance and survival. The unique morphology of neurones is due in large part to the organisation of the neuronal cytoskeleton (*vide infra*) and the existence of an intracellular transport system capable of servicing cellular processes up to 20000 times more distant than the peripheral cytoplasmic regions of a 50 μm diameter cell.

4. SPECIFIC MORPHOLOGICAL FEATURES OF NEURONES

4.1 Size range & implications for excitability

The size of neuronal cell bodies and their processes varies enormously. Some neurones, for example alpha motoneurones, Betz cells and lateral vestibular neurones have cell body diameters typically in the range 40-60μm, whereas cerebellar granule cells and neurones of the substantia gelatinosa and have diameters of the order of 6-15 μm. The nucleus of large neurones is surrounded by abundant cytoplasm, whereas a thin rim of cytoplasm containing very few organelles usually surrounds the nucleus of small neurones.

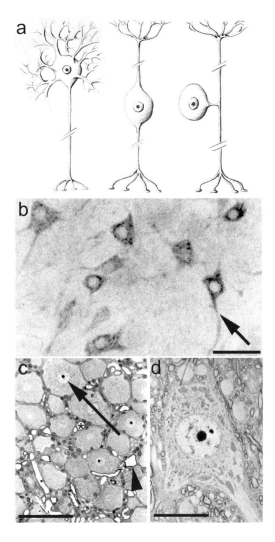

Figure 1. General structure of neurones. a) schematic representation of the three basic neuronal types. From left to right: multipolar; bipolar; psuedounipolar. b) multipolar motoneurones immunostained for calcitonin gene-related peptide in the facial nucleus. Up to four primary dendrites (arrow) can be seen extending from the cell body. Bar = 200μm. c) pseudounipolar neurones in a dorsal root ganglion. Most neuronal cell bodies have a euchromatic nucleus (long arrow) containing a densely stained nucleolus and cytoplasm containing granules of Nissl substance myelinated axons (arrow head), the compact nuclei of satellite cells in the neuropil between the neuronal cell bodies and blood vessels (*) are also visible. Bar = 50μm. d) multipolar motoneurone in the spinal cord, showing prominent cytoplasmic Nissl bodies. Small dark spots in the cytoplasm correspond to lysosomes and similarly sized faint grey spots in dendrites correspond to mitochondria. Bar = 20μm

This feature may have metabolic implications under conditions of neuronal stress, such as ischaemia. In some systems where cell body diameters show marked variation (e.g. alpha and gamma motoneurones, Fig. 2) cell body diameter is a major determinant of excitability (Henneman & Mendell, 1981). The cell body also provides an important surface for the receipt of synaptic contacts because of the proximity of such contacts to the axon hillock and the reduced electrotonic decay of inhibitory postsynaptic potentials. Despite this, the cell body makes up only about 5% of the surface area of a typical multipolar neurone able to accept synaptic input. The remainder of the surface area of such a neurone able to accept synaptic input is provided by its dendrites. The extent and pattern of dendritic branching varies enormously. This has an important effect on neuronal function. Dendritic patterns will largely determine the nature of afferent inputs received, and the effect of dendritic postsynaptic potentials on the soma are affected both by synaptic distance from the cell body and dendrite calibre. Local interneurones in the CNS will typically have processes that extend no more than 10μm, whereas the central and peripheral processes of a lumbar dorsal root ganglion neuron in an adult human can each reach 1000 mm. Small lesions of the nervous system can therefore damage a large proportion of the cytoplasmic volume of interneurones, whereas disruption of intracellular transport renders the long processes of dorsal root ganglia particularly vulnerable.

4.2 Relations with other cell types

Neurones are always closely apposed to- or invested by- the processes of other cells. Most contact with neurones is provided by neuroglial cells (discussed elsewhere in this book); remaining contacts are between neurones themselves or target tissues such as muscle, skin or glands. These contacts may facilitate neurotransmission (e.g. synaptic contacts), provide electrical insulation (e.g. astroglial contacts, pockets within Schwann cells that contain unmyelinated axons, teloglial cells and myelin sheaths), or mediate trophic support of the neurone. Contacts between other neurones are characterised by membrane specialisations forming the synaptic complex.

4.3 Nucleus and nucleolus

Neurones generally have large euchromatic nuclei in which the chromatin is dispersed. Such features are common in cells that are metabolically active. Large neurones tend to have round nuclei, which are located centrally within the cell body, whereas nuclear irregularity (crenation) and a peripheral location (eccentricity) are common in smaller neurones. A membranous envelope that is continuous on the cytoplasmic side with smooth and rough endoplasmic reticulum surrounds the nucleus. This envelope shows periodic attenuations termed nuclear pores. It is possible that connections between the nuclear envelope, the endoplasmic reticulum and the cytoskeleton are responsible for maintaining nuclear shape and holding it in place within the cell body cytoplasm (Peters et al., 1991). In larger neurones, axonal injury is often associated with crenation and eccentricity of the nucleus; this is in turn associated with fragmentation of endoplasmic reticulum changes in the composition of the cytoskeleton (Waxman et al., 1995). Reduced nuclear activity, for example in a neurone about to die, is associated with an increase in nuclear density (pyknosis). The relationship between numbers of nuclear pores and nuclear-cytoplasmic transfer of RNA is unclear. Within the nucleus, one or more dense nucleoli are seen (Fig. 1c, Fig. 4a) which are the primary sites of ribosomal RNA synthesis.

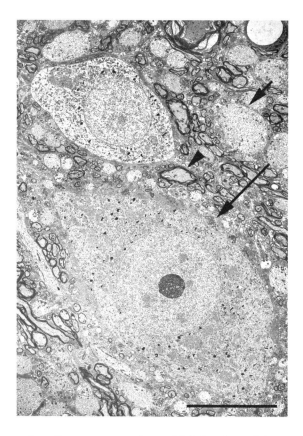

Figure 2. Large and small motoneurones. Large (alpha) and small (gamma) motoneuronal cell bodies in the spinal cord. The larger motoneurone has denser cytoplasm due to a greater density of microtubules and neurofilaments. Synaptic terminals (long arrow) apposing the plasma membrane are much more numerous on the large motoneurone. Dendrites (short arrow) and myelinated axons (arrow head) are seen in the neuropil. Bar = 10μm

4.4 Rough endoplasmic reticulum (RER) and polyribosomes

Over 100 years ago, Franz Nissl described prominent patches of basophilia (Nissl bodies) in neuronal cytoplasm. Nissl bodies have subsequently been equated with aggregations of RER. In some neurones (Fig. 3), Nissl bodies are composed primarily of short disorganised lengths of RER interspersed among larger aggregates of polyribosomes. In other neurones, RER is organised into stacks of parallel lamellae. A similar organisation is seen in secretory cells such as pancreatic acinar cells. In contrast to typical secretory cells, however, neuronal RER has large stretches devoid of abound ribosomes. Also, most of the ribosomes of a neuronal Nissl body are organised as regular rosettes of polyribosomes located between the RER stacks. Membrane bound ribosomes are usually deemed responsible for the synthesis of proteins destined for secretion while free polyribosomes are concerned with the synthesis of proteins destined for intracellular use. The ultrastructure of the neuronal Nissl body, therefore, suggests it is primarily concerned with the synthesis of proteins for intracellular use (Johnson, 1996). Nissl bodies are also a feature of the post-synaptic element of large and rare synaptic terminals found on some neuronal types (Conradi, 1969) and small accumulations of RER and polyribosomes are frequently found at the base of dendritic spines (Stuart et al., 1999). The extent to which changes in the organisation or RER and the relative proportions of polyribosomes reflects changes in the types of proteins synthesised by neurones and the relationship of the spatial distribution of these organelles within the cell body to local protein synthesis remains to be determined.

4.5 Golgi apparatus

In thick sections stained by silver or osmium impregnation, or stained histochemically for nucleoside diphosphatase (NDP) or thiamine pyrophosphatase (TPP), the Golgi apparatus is typically seen as an anastomosing network of cisterns and vesicles encircling the nucleus in the mid-cytoplasmic region. Ultrastructurally (Fig. 4a), it is generally composed of around five crescentic cisternae that are dilated at their ends (Peters et al., 1991). The outer cisterns are usually fenestrated and give way to shorter cisterns and vesicles at the concave face. On the basis of acid phosphatase, NDP and TTPase histochemistry and ultrastructural characteristics, a transitional region between the elements of the inner concave face of the Golgi apparatus and the smooth Endoplasmic Reticulum and Lysosomes has been identified and given the acronym GERL (Novikoff, 1967). In addition to its well-documented role in glycosylation, the Golgi apparatus of neurones (Fig. 4a) plays an important role in the routing of materials to the axon and dendrites (Kandel et al., 2000).

4.6 Lysosomes, multivesicular bodies and lipofuschin

Lysosomes are found in the cell body and seen as 0.25-1μm diameter acid phosphatase-containing vesicles that are strongly osmiophilic in electron micrographs (Fig. 1d, Fig. 4a). They contain many other hydrolytic enzymes and play an important role in the recycling of material from effete organelles and removal of unwanted material in the cell body. Enzymatic deficiencies in lysosomes can result in the lethal accumulation of partially metabolised material in neurones that characterise the lysosomal storage diseases. The capacity of neurones to take up and transport some foreign material from the periphery makes the degradative function of lysosomes important. Retrogradely transported tracers, such as horseradish peroxidase, are destroyed by lysosomes after about 4 days. In contrast, the retrograde tracer Fluorgold remains for many months in neurones although it does induce lysosomal proliferation. In some cases (e.g. retrogradely transported ricin or diphtheria toxin) the lysosomal environment is a necessary stage for toxin activation. Multivesicular bodies, comprising small vesicles contained within a larger (0.5-1μm diameter) vesicle are seen in the neuronal cell body and occasionally in its processes (Fig. 4a). Multivesicular bodies frequently contain retrogradely transported materials and the close association between these organelles and lysosomes, smooth endoplasmic reticulum and the Golgi apparatus (Peters et al., 1991) suggests that they like elements of GERL are involved in the process of intracellular packaging, routing and transport. The post-mitotic nature of neurones means that they are more vulnerable than other cell types to accumulated toxins and metabolites such as oxygen free radicals. This is reflected by the age-related accumulation of lipofuschin granules in the cell body cytoplasm that are believed to be derived from lysosomes (Fig. 4a). Lipofuschin autofluorescence can also be a major problem when using fluorescence methods to study ageing neurones.

4.7 Mitochondria

Mitochondria are found throughout the neuronal cell body cytoplasm and all its processes. They vary in size from 0.1μm to up to 20 μm, although most are in the region of 05-1μm and visible as faint basophilic areas in the light microscope (Fig. 1d). Mitochondria have an outer smooth contoured membrane separated by approximately 6nm from an inner membrane that is thrown up into several cristae that are usually transverse to the long axis of the mitochondrion (Fig. 4a). Mitochondria play a major role in the oxidative metabolism of neurones. Smaller neurones have a greater density of mitochondria compared to larger neurones. This may have metabolic implications, especially in ischaemia. Mitochondria have been observed in vitro to undergo jerky, discontinuous movement within the neuronal cytoplasm and are transported both retrogradely and anterogradely within axons (Waxman et al., 1995). Mitochondria have their own DNA which has been found to code for mitochondrial transfer RNA and mitochondrial ribosomal RNA, as well as a small number of mitochondrial proteins (ATPase, a subunit of ATP synthetase, cytochrome oxidase, cytochrome b, and subunits of NADPH-co-enzyme Q reductase). Most mitochondrial proteins, however, are coded for by nuclear DNA (Kandell et al., 2000).

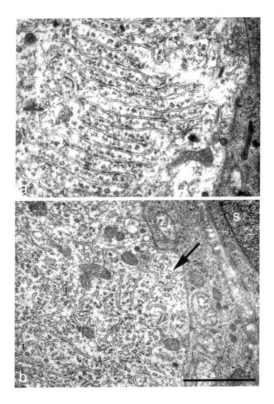

Figure 3. Nissl body ultrastructure in large and small sensory neurones. a) Peripheral cytoplasm of a large 'light' dorsal root ganglion neurone showing organised lamellae of RER with polyribosomes interspersed between the lamellae (Nissl bodies). b) Peripheral cytoplasm of a small 'dark' dorsal root ganglion neurone showing short disorganised fragments of RER and smooth endoplasmic reticulum (arrow) amongst many randomly distributed polyribosomes. S; satellite cell. Bar = 1 μm

4.8 Cytoskeleton

The cytoskeleton composed of microtubules, neurofilaments and microfilaments organised as a lattice and associated with transmembrane proteins. Microtubules are 25 nm diameter tubes assembled from tubulin monomers and can be up to 100nm long (Fig. 4a & b). Neurofilaments are 10 nm diameter fibres of variable length composed of three neurofilament monomers: NF-L (68 kD), NF-M (150 kD) and NF-H (200 kD). Neurofilament density, stability and degree of phosphorylation usually increase with distance from the cell body (Waxman, 1995). Microfilaments are short 4-6 nm diameter structures that contain actin and are wound in a two-strand helix. The neuronal cytoskeleton plays an important role in the elaborate intracellular transport system of neurones. Changes in cell shape and distribution of organelles during development or after injury are associated with changes in composition of the cytoskeleton.

4.9 Dendrites

Dendrites extend from the cell body as 1-8μm diameter proximal dendrites (Fig. 1b). Small cells (e.g. cerebellar granule cells) typically have only one dendrite per cell body, most neurones have four dendrites and large motoneurones can have more than ten (Stuart et al., 1999). Proximal dendrites branch in a manner characteristic of the neurone type until a large dendritic arbour is formed. The intracellular features of proximal dendrites are similar to the cell body cytoplasm. Distal dendrites do not contain lysosomes, lipofuschin, Nissl bodies or Golgi apparatus. Smooth endoplasmic reticulum, mitochondria and cytoskeletal tubules and filaments extend along the length of dendrites. The cytoskeletal elements are orientated along the long axis of dendrites. Microtubules are very prominent in large diameter dendrites whereas neurofilaments are usually sparse (Peters et al., 1991). The smooth endoplasmic reticulum of dendrites forms an anastomosing reticulum. From this reticulum, occasional flattened cisternae arise which are closely related to the inner dendritic membrane. Many synaptic terminals contact the plasma membrane of dendrites (Fig. 4b). Slender prolongations of dendrites (dendritic spines) are often found, especially on 'spiny' neurones such as pyramidal cells. Near the base of these spines are short lamellae of smooth endoplasmic reticulum, termed the spine apparatus, and small clusters of polyribosomes. Presynaptic terminals contact dendritic spines and local changes in spine calibre, possibly involving local protein synthesis, may provide a means of altering synaptic efficacy (Herring & Sheng, 2001).

4.10 Axon

Most neurones have a single axon of 0.1-20μm diameter that emerges from the axon hillock region of the cell body, and then branches in the region of the cell body and near the terminal field. Multiple axons have been reported emerging from the cell bodies of sympathetic ganglia and small CNS neurons, and occasionally from dendrites (Palay & Chan-Palay, 1977). The peripheral process of dorsal root ganglion neurones functions as a dendrite yet retains the morphology of an axon, while the opposite is true for its central process. Characteristic of the cytoplasm of axons are many 10nm neurofilaments (Peters at al., 1991; Waxman et al., 1995). Microtubules and neurofilaments are both orientated parallel to the long axis of the axon. They are usually found in conventional electron micrographs clustered in either microtubule domains or neurofilament domains according to the predominant cytoskeletal element present. It is possible that this appearance is an artefact of fixation, since freeze fracture studies of unfixed axons reveal a lattice of neurofilaments and microtubules which are connected by 3-5nm microfilamentous trabeculae. Mitochondria, which are often much more elongated than those in the cell body and dendrites, are also prominent in the axoplasm. Cytoskeletal elements, mitochondria, variously shaped intra-axonal vesicles and multivesicular bodies form the structural basis upon which most current theories of axonal transport are based (Shea & Flanagan, 2001). High voltage electron microscopy of thick sections by Droz and his school have also revealed an anastomosing endomembranous reticulum extending from the cell body along the length of axons that appears to be involved in slow anterograde axonal transport of materials to the axolemma and to synaptic terminals (Waxman et al., 1995). The presence of ribosomes in invertebrate axons is associated with the presence of mRNA and local protein synthesis (Giuditta et al., 2002).

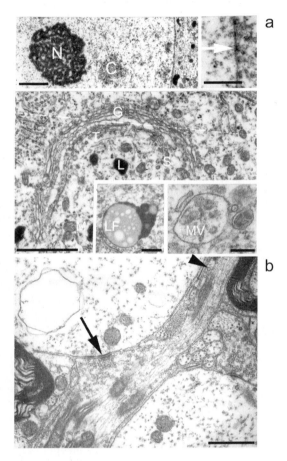

Figure 4. Neuronal organelles and axodendritic synapse. a) Nucleolus (N) and chromatin (.C) (Bar = 1µm); Nuclear envelope and pores (arrow) (Bar = 0.5µm); Golgi apparatus (G), lysosome (L), neurotubules (T) and smooth endoplasmic reticulum (S) (Bar = 1µm); Lipofuschin granule (LF) (Bar = 0.5µm); and multivesicular body (MV). (Bar = 0.1µm).b) Axodendritic synaptic terminal showing pre- and postsynaptic thickening, presynaptic synaptic vesicle clustering and synaptic cleft density characteristic of a synaptic complex (long arrow). Three synaptic vesicles are seen closely associated with a neurotubule in the preterminal axon (arrow head). Bar = 0.5µm.

Ribosomes, however, are very rarely seen in vertebrate axons and this is usually taken to indicate the absence of axonal protein synthesis. This view currently sits uncomfortably against the half-life of some axonally transported proteins that is considerably less than their estimated time in transport. Most axons are located in the CNS, some are partly in the CNS and the PNS and a smaller number (mainly postanglionic autonomic neurones) are in the PNS. Astroglial processes cover the axolemma of unmyelinated CNS axons, while myelination of CNS axons is provided by the plasma membrane extensions of oligodendrocytes. Schwann cell processes, irrespective of whether these processes form myelin sheaths, cover axons in the PNS. Most axons branch to form collaterals and, if myelinated, loose their myelin sheath before forming a presynaptic terminal (Fig. 5). In most cases, several 'en passant' presynaptic terminals are formed from axons rather than single boutons terminaux.

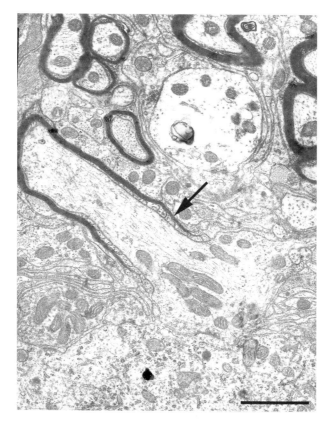

Figure 5. Axosomatic synaptic terminal showing the organisation of the preterminal myelin sheath (arrow) and typical accumulation of mitochondria in the presynaptic bouton. Bar = 1μm.

4.11 Synaptic terminals

The synaptic terminal is the primary way that information is passed rapidly from neurone to neurone and from neurone to effector organs such as muscle or glands. There are three basic elements to the synapse: a presynaptic terminal containing accumulations of synaptic vesicles and mitochondria, a 20-30 nm wide synaptic cleft (30-50 nm at the myoneural junction), and a postsynaptic membrane that may or may not exhibit specialisations. Most presynaptic terminals are 1-2 µm in diameter and occasionally branched or perforated. The presynaptic accumulation of mitochondria, which can be stained, allows the general visualisation of synaptic terminals in the light microscope. Quantitative ultrastructural studies generally reveal a greater numerical density and cover by synapses on dendrites compared to the cell body. Most synapses are of the axo-dendritic (Fig. 4b) or axo-somatic (Fig. 5) form, but there are also many examples of dendro-dendritic and axo-axonic synapses (Craig & Banker, 1994), as well as examples in invertebrates of electrotonic synaptic contacts between neurones which rely on ion passage through gap junctions rather than neurotransmitter release across a synaptic cleft. Synaptic terminals have been classified on the basis of the shape and electron density of their synaptic vesicles, on the basis of the symmetry of pre- and post-synaptic thickening and the presence of post-synaptic specialisations (Conradi, 1969). In some cases, synaptic terminals with known excitatory functions (e.g. myoneural junction of striated muscle) have 40-50nm-diameter round synaptic vesicles and a post-synaptic density that it thicker than the presynaptic density. Other synaptic terminals, that may include inhibitory synapses, have both round and oval synaptic vesicles (pleiomorphic vesicles). Some synaptic terminals have variable numbers of 40-60 nm-diameter dense core vesicles that are generally assumed to contain catecholamines. Neuropeptides and neurotrophic factors often co-localise with neurotransmitters. This, together with uncertainties about the nature of the postsynaptic receptors, makes it difficult to make functional predictions from synaptic morphology alone.

REFERENCES

Kandel, E., Schwartz, J.H. & Jessel, T.M. (2000). Principles of neural science. London: McGraw-Hill.

Conradi, S. (1969). Ultrastructure and distribution neuronal and glial elements on the motoneuron surface in the lumbosacral spinal cord of the adult cat. Acta Physiologica. Scandinavica. (suppl.) 332: 5-48.

Craig, A.M. & Banker, G. (1994). Neuronal polarity. Annual Review of Neuroscience. 17: 267-310.

Foster, M. & Sherrington, C.S. (1897). A textbook of Physiology. Part III: The central nervous system. 7th ed., London: Macmillan.

Henneman, E. & Mendell, L.M. (1981). Functional organisation of motoneuron pools and its inputs. In Handbook of Physiology. Section.1, Volume II, part 1 (Ed. J.M. Brookhart & V.B. Mountcastle), pp.423-507. Bethesda: Maryland. American Physiological Society.

Hering, H. & Sheng, M. (2001). Dendritic spines: structure, dynamics and regulation. Nature Reviews Neuroscience 2: 880-888.

Giuditta, A., Kaplan, B.B., Minnen, J.V., Alvarez, J. & Koenig, E. (2002). Axonal and presynaptic protein synthesis: new insights into the biology of the neuron. Trends in Neurosciences 25: 400-404.

Johnson, I.P. (1996). Target dependence of motoneurones. In: The neurobiology of disease: contributions from neuroscience to clinical neurology (Ed. H. Bostock, P. Kirkwood & A. Pullen), pp 379-394. Cambridge: Cambridge University Press.

Kandel, E.R., Schwartz, J.H. & Jessell, T.M. (2000). Principles of Neural Science, 4th ed. New York: McGraw Hill Company.

Mazzarello, P. (1999). A unifying concept: the history of cell theory. Nature Cell Biology 1: 13-15.

Novikoff, A.B. (1967). Enzyme localisation and ultrastructure of neurons. In: The Neuron (Ed. H. Hyden). pp. 255-318. Amsterdam: Elsevier.

Palay, S. & Chan-Palay, V. (1977). General morphology of neurons and neuroglia. In: Handbook of Physiology, Section 1, Volume I. Part 1. (Ed. E.R. Kandel), pp.5-37. Bethesda. MD: American Physiological Society.

Peters, A., Palay, S.L. & Webster, H. deF. (1991). The fine structure of the nervous system. Oxford: Oxford University Press.

Shea, T.B. & Flanagan, L.A. (2001). Kinesin, dynein and neurofilament transport. Trends in Neurosciences 24: 644-648.

Stuart, G., Spruston, N. & Hausser, M. (1999). Dendrites. Oxford: University Press.

Waxman, S.G., Kocsis, J.D. & Stys, P.K. (1995). The axon. Structure, function and pathophysiology. Oxford: Oxford University Press.

L. MARTÍNEZ-MILLÁN, G. GARCÍA DEL CAÑO AND
I. GERRIKAGOITIA

*Department of Neuroscience, Faculty of Medicine, University of the Basque
Country, Leioa, Spain*

4. SYNAPTIC PLASTICITY: HYPEREXCITABILITY AND SYNAPTIC SILENCING

Summary. Synaptic plasticity is defined as any sort of change in intensity of the synaptic response or in the number of synapses that occurs in nervous structures. Long-lasting increase or decrease of synaptic efficacy — termed long-term potentiation (LTP) and long-term depression (LTD), respectively — has been related to the establishment of memory traces. Experimental induction of both types of long-lasting changes has contributed effectively to the study of the mechanisms that underlie the learning processes. Although the wealth of mechanisms related to synaptic plasticity still remains partially unveiled, at the core of this phenomenon lies the glutamate receptor system together with neurotrophins and their receptors, second-messenger cascades, and changes of gene expression. An increase in the numbers of synapses is termed structural plasticity, and is governed mainly by molecules belonging to the neurotrophin and cytokine families. This form of plasticity participates in the establishment of the intricate arrays of nervous connections during normal development, and in rearrangements of the neuronal circuitries that occur after nervous lesions.

1. INTRODUCTION

A system reacts plastically when an external action applied to it induces any measurable change that persists during a certain period. By contrast, the reaction is termed elastic if the system returns to the original state just after the action ceases. One of Cajal's prophetic proposals was that neural connections or synapses can vary their strength and that this change is the anatomical basis of memory. Konorsky considered the term synaptic plasticity as encompassing a broad spectrum of phenomena such as synaptic strengthening or weakening, activation of silent synapses or silencing of previously active synapses, and changes in the number of synapses that a certain nervous connection possesses. Such increase in the number of synapses occurs during memory-stabilization or after nervous lesions.

The mechanisms that underlie synaptic plasticity are only partially understood and involve a profusion of molecules such as receptors, second messengers, trophic factors, and adhesion molecules, as well as changes of gene expression that participate in the complex orchestration of the synaptic machinery.

The increasing amount of experimental work devoted to studying the diverse aspects of synaptic plasticity is justified by its relation with significant nervous manifestations such as memory or post-lesion reactions.

2. LONG-TERM POTENTIATION

The first example of activity-dependent changes in synaptic efficiency in the mammalian brain was the excitatory connections made by perforant path fibres onto granule cells of the hippocampus. This finding has subsequently been extended to other connections within the hippocampus, the cerebral cortex and other central nervous structures. Although activity-dependent synaptic potentiation occurs within milliseconds, it can persist for many hours in in vitro hippocampal slice preparations and in anaesthetized animals, and for several days when induced in freely moving animals. Therefore, with regard to time span of induction, short-term potentiation lasts from several minutes to a few hours, whereas LTP can be detected during days. LTP consists of a persistent increase in the size of the synaptic component of the evoked response recorded from individual cells or from a population of neurons. It can be induced either by delivering a tetanic stimulus (50-100 stimuli at 100 Hz or more) to the experimental pathway or by applying a modest stimuli paired with a stronger one within certain critical ranges. LTP is characterized by co-operativeness, which indicates the existence of an intensity threshold for induction. Thus, weak stimuli do not induce LTP, intermediate stimuli trigger short-term potentiation, and only strong trains can induce LTP. A weak input signal can be enhanced if it is associated with a convergent input. This effect is a cellular analogue of classical conditioning and it is also a property of Hebb's principle, according to which synaptic potentiation takes place when both pre- and post-synaptic components are coincidentally active. This feature of LTP requires a molecular detector of excitatory input coincidence. The glutamate receptor of the N-methyl-D-aspartate type (NMDAR) is the most plausible candidate. This ubiquously distributed voltage-dependent channel is blocked by Mg2+ at basal membrane potentials. Therefore, Ca2+-influx is allowed only when, in coincidence with L-glutamate agonist binding, the membrane is sufficiently depolarized (excited) to remove Mg2+ from the channel pore. Co-operativeness is required for LTP formation because stimuli that activate only a few fibres — and the level of depolarization provided by the weak inputs — do not produce an adequate reduction of the Mg2+-block. If many fibres are synchronically activated by a strong stimulus, depolarization spreads to neighbouring synapses to enhance the unblocking of their NMDA.

3. CA^{2+} SIGNALLING AND LONG-TERM POTENTIATION

The fact that induction of LTP can be blocked by intracellularly injecting the Ca^{2+}-chelator EGTA emphasizes the role of Ca^{2+}-signalling in LTP induction. Besides Ca^{2+}-influx through NMDA channels, intracellular stores constitute an additional potential source for cytoplasmic increase of Ca^{2+}-concentration. The Ca^{2+}-increase associated with the synaptic activation of NMDARs is considerably reduced by application of drugs such as ryanodine or thapsigargin, which inhibit Ca^{2+}-induced Ca^{2+} release and deplete intracellular Ca^{2+} stores respectively. Under normal conditions, Ca^{2+} enters through NMDA channels, providing a transient signal necessary for the induction of LTP. It is probable that the increase in Ca^{2+}-

concentration is primarily restricted to postsynaptic spines and subsequently amplified by release from intracellular stores.

Other types of glutamate receptor also participate in LTP induction. Specific increase of the synaptic response component mediated by the alpha-amino-3-hydroxy-5-methyl-4-isoxazolepropionate receptor (AMPAR) is associated with LTP, and is an argument for the postsynaptic location of this change or at least that a pre-synaptic changes would result in an increase of AMPAR- as well as NMDAR-mediated components.

4. TRANSITION FROM EARLY TO LONG-TERM POTENTIATION

LTP induction displays two temporal phases that can be distinguished on the basis of the amount of stimuli needed to induce it, and protein synthesis requirements. An early phase lasting 1-3 hours and termed short-term phase, can be induced by a single tetanus train and does not require protein synthesis; and a later one, lasting one to several days and termed long-lasting LTP, can be induced by application of several successive tetanic trains and necessarily involves gene expression at the RNA and protein levels. The Ca^{2+}-influx that results from successive activation of NMDARs activates second-messenger cascades involving Ca^{2+}/calmodulin dependent kinase II (CaM-KII) and adenyl cyclase, which in turn activates the adenosine-3',5'-cyclic monophosphate-dependent (cAMP-dependent) protein kinase. The catalytic subunit of this enzyme then translocates to the neuronal nucleus and phosphorylates the cAMP-response element-binding protein (CREB) that triggers the expression of CREB-dependent genes involved in trophic and structural long-lasting changes. The same second-messenger systems can modify other glutamate receptors during the induction of LTP, providing further postsynaptic molecular mechanisms for synaptic plasticity. Ca^{2+}-influx through the NMDARs, typically localized in dendritic spines, leads to a fast Ca^{2+}-dependent auto-phosphorylation of the NMDAR. In addition, Ca^{2+}-influx activates CaM-KII, which catalyses slow Ca^{2+}-independent auto-phosphorylation and phosphorylation of AMPA receptors on a site that enhances their responsiveness.

5. RETROGRADE MESSENGER

In addition to changes occurring at the postsynaptic level, the quantal analyses of central synapses revealed that the fluctuations in the amplitude of synaptic responses have variable results, indicating that the locus of expression of LTP can be purely pre-synaptic, purely postsynaptic, or a mixture of both pre- and postsynaptic. Glutamate measurements have concluded that LTP induction is followed by an increase in pre-synaptic glutamate efflux. Blockade of the postsynaptic NMDA receptor during tetanic stimulation prevents both LTP formation and the increased glutamate efflux. Diffusible molecules have been proposed as retrograde messengers responsible for the maintenance of a pre-synaptic modification in LTP. These messengers would diffuse across the synaptic cleft, acting upon activated pre-synaptic terminals, and triggering cascades of reactions that would result in an increase release of neurotransmitter. Experimental evidence points to arachidonic acid and nitric oxide (NO) as possible candidates. Exogenous arachidonic acid enhances synaptic transmission in an activity-dependent way and increases presynaptic neurotransmitter release, whereas the inhibition of arachidonic acid production by nordihydroguaiateric acid blocks both LTP formation and the enhancement of neurotransmitter release. The sustained rise in extracellular arachidonic acid concentration that accompanies LTP induction reinforces the hypothesis that this molecule participates as retrograde messenger in synaptic potentiation. In addition, arachidonic acid has been shown to induce glutamate uptake by glial cells, suggesting a link between post-synaptically released arachidonic acid, glial cells, and LTP.

Several studies strongly suggest that NO also participates as retrograde messenger in LTP. For example, LTP formation in the CA1 field of hippocampal slices is blocked by the nitric oxide synthase (NOS) inhibitor L-Nω-arginine and can be reversed by exogenous administration of the NOS substrate L-arginine. Administration of sodium nitroprusside, a spontaneous NO-donor, increases the magnitude of the excitatory postsynaptic potentials (EPSPs). Although some uncertainty remains about the site of NO release and the NO dose-response relationship, the effect of NO could be considered bimodal, inducing depression of synaptic transmission at low concentrations and enhancement at high concentrations.

In summary, stimulation activates postsynaptic NMDARs. The resulting increase of cytoplasmic Ca^{2+} concentration gives rise to a cascade of intracellular events possibly involving the activation of phospholipase A2 and NO synthase. NO released from the postsynapse could act at the presynaptic level, leading to an increase in transmitter liberation in the synaptic cleft. Arachidonic acid released from postsynaptic sites could sustain the increased release of the transmitter induced by postsynaptic NO.

6. METABOTROPIC GLUTAMATE RECEPTORS

The translation of short-term potentiation (or early phase of LTP) to LTP parallels the translation of transient activity involved in learning events into long-lasting memory. Metabotropic glutamate receptors are coupled to second-messengers cascades, and are therefore in an ideal position to be candidates for such translation. Group I mGluRs are positively linked via G-proteins to phospholipase C, and activation of this enzyme results in breakdown of membrane phospholipids with production of the chemical messengers diacylglycerol and inositol 1,4,5-triphosphate, which releases Ca^{2+} from internal stores. However, class II and III mGluRs are negatively coupled to G-protein-mediated activation of adenylate cyclase, which causes a depression of the chemical messenger cAMP. Pre-training intracerebroventricular injection of the competitive antagonist of groups I and II mGluRs (+)-α-methyl-4-carboxyphenylglycine (MCPG) causes amnesia in spatial-alternation, and post-training intraperitoneal injection of groups I and II mGluRs agonist (1S,3R)-1-aminocyclopentane-1,3-dicarboxylic acid (ACPD) enhances retention of hippocampus-dependent learning task. The principal role of mGluRs during behaviour is probably to modulate the signal-to-noise ratio in the nervous system. Activation of mGluRs might produce a signal that increases the signal-to-noise ratio that consequently leads to amnesia.

7. TROPHINS AND SYNAPTIC PLASTICITY

Participation of neurotrophins as molecular mediators of synaptic plasticity is based on the facts that neurotrophins and their receptors are expressed in central areas considered plastic and that nervous activity can regulate both their levels and secretion. BDNF and NT-3, and their corresponding receptors trkB and trkC are highly expressed in central nervous system sites where developmental and adult synaptic plasticity has been repeatedly demonstrated, such as cerebellum, hippocampus and cerebral cortex.

Several lines of evidence indicate that neurotrophin expression is regulated by nervous activity, thus suggesting their role in synaptic potentiation and depression. For example, neuronal growth factor (NGF) and brain-derived neurotrophic factor (BDNF) mRNA levels are rapidly and potently upregulated by epileptiform activity in the hippocampus and cerebral cortex. In the hippocampus, changes in BDNF mRNA levels are particularly intense, increasing more than six-fold in dentate gyrus within 30 minutes of seizure induction. However, the interpretation of mRNA up-regulation has evident limitations due to the lack of either a sensitive method for neurotrophin localization or any predictable relationship between mRNA and protein levels. The role of BDNF in post-synaptic modulation of synaptic transmission and plasticity in hippocampus is clearer. The proposed model considers that the activation of postsynaptic trkB receptors by BDNF causes depolarization of the postsynaptic neuron, probably through opening of Na^+-channels and concomitant activation of Ca^{2+}-channels, resulting in an increase of the cytoplasmic Ca^{2+} concentration. Postsynaptic cell depolarization caused by the activation of trkB receptors enhances NMDA receptor opening probability by removing Mg^{2+} from the receptor channel, leading to facilitation of LTP induction in the postsynaptic dentate granule cells. In order to produce a persistent enhancement of synaptic transmission, similar to that seen with tetanic stimulation, BDNF application must be combined with a weak burst of presynaptic activity that, by itself, would have little effect on synaptic transmission. In this model, the persistent synaptic enhancement is blocked by the Ca^{2+}-channel inhibitor D890 or by an NMDA-receptor antagonist, demonstrating that BDNF-mediated LTP is induced post-synaptically.

The neurotrophin factors BDNF and NT-3 are able to produce a long-lasting enhancement of synaptic transmission in the hippocampus that requires an immediate upregulation of protein synthesis. Plasticity in rat hippocampal slices, in which the synaptic neuropil is isolated from the principal cell bodies, also requires early protein synthesis. In the hippocampus, neurotrophins stimulate trk phosphorylation and increase intracellular Ca^{2+} concentrations. These signalling events may be coupled to protein kinase activity to stimulate protein synthesis. The newly synthesized proteins may act locally to enhance postsynaptic responsiveness, or may communicate with the presynaptic terminal to increase neurotransmitter release. During developmental and adult plasticity, regulated release of neurotrophins and subsequent stimulation of local protein synthesis may permit the site-specific modification of both synaptic transmission and synaptic structure.

8. RECEPTOR TRAFFICKING

During LTP, receptor trafficking and subsequent synaptic deposition have been monitored by optical detection of green-fluorescent-protein-tagged (GFP-tagged) glutamate receptors. Recombinant receptors (GluR-GFP) are functional and their cellular distribution can be monitored with two-photon microscopy. High-frequency synaptic activation generates LTP-induced movement of GFP-tagged receptors to the surface of dendritic shafts and dendritic spines. It has been shown that LTP is rescued by expression of only 10% of the normal amount of GluR1, supporting the view that, in normal conditions, there is an over-abundance of GluR1 available for LTP. Trafficking of receptors to the postsynaptic sites might underlie plastic changes that activate and potentiate synapses. Silent synapses lack AMPAR, and the receptors are rapidly delivered during LTP. Moreover, non-synaptic AMPARs appear to outnumber synaptic AMPARs heavily. A considerable proportion of synapses in CA1 hippocampus lack or have very few. AMPARs, while most synapses have NMDARs. The percentage of synapses lacking AMPARs is greater earlier in development, an observation consistent with the prevalence of silent synapses at these ages. In cultured dissociated neurons, the expression of recombinant tagged AMPAR subunits showed that receptors containing GluR2/3 are delivered continuously to the spine surface with a time constant of about 15 min and unaffected by agents that perturb plasticity. On the other hand, the rate of surface delivery of GluR1-containing receptors is greatly enhanced by stimulation. They appear firs in dendritic extra-synaptic regions and subsequently move into spine regions. Both LTP and increased activity of CAM-KII induce delivery of tagged AMPARs into synapses. The mechanism of anchoring these receptors to the postsynaptic sites was shown in the GluR1 subunit, and depends on the association between GluR1 and a specific domain of the postsynaptic protein complex that integrates the PDZ.

Glial cells can also participate in the synaptic plasticity mechanisms that increase surface expression of AMPARs and, therefore, synaptic efficacy. Tumour necrosis factor-α (TNF-α), a cytokine released by astrocytes causes an increase of the delivery of new AMPARs to the plasma membrane. Preventing the action of endogenous TNF-α has the opposite effect. Thus, the permanent presence of TNF-α is required for preservation or increase of synaptic strength at excitatory synapses.

Similarly to AMPARs, NMDARs undergo dynamically regulated trafficking and targeting. The transport of NMDARs in and out of the synaptic membrane contributes to several forms of long-lasting synaptic plasticity. Given that NMDARs are widely expressed throughout the nervous system, regulation of NMDAR trafficking is a potentially important way to modulate the efficacy of synaptic transmission. Protein kinase C (PKC) regulates the increase of NMDAR deposition in the postsynaptic site by exocytosis of assembled NMDAR from the endoplasmic reticulum. Regulation of NMDAR trafficking contributes to long-lasting synaptic plasticity in the hippocampus. Tetanic stimulation increases the surface density of NMDARs and the splice variant C2-containing NR1 subunits at CA1 synapses of mature animals in a PKC-dependent manner.

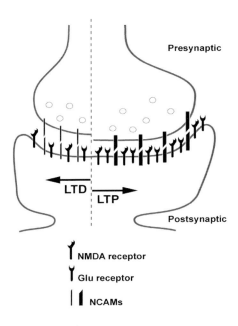

Figure 1. Long-term potentiation is accompanied by an increase of NMDA and AMPA receptors and of neuronal cell adhesion molecules in the synaptic membranes (arrow to the right), triggering second-messenger cascades and retrograde synaptic signals. Removal of the above-mentioned molecules marks the transition towards long-term depression (arrow to the left). LTP, long-term potentiation; LTD, long-term depression.

9. EPH FAMILY AND EPHRINS

Tyrosine kinase receptor proteins of the Eph family and their receptor ephrins are expressed in gradients in the origin and target structures and participate critically in the development of early map formation. In the adult nervous system, they are expressed in structures in which plasticity has been demonstrated. Several experimental results indicate that interactions between EphB receptors and NMDARs play a role in synaptic plasticity. EphB receptors enhance intracellular Ca^{2+} permeability of NMDARs. The NMDAR subunit NR2B is tyrosine phosphorylated in response to EphB activation when co-localized with EphB. The EphB-potentiated glutamate stimulation of NMDAR leads to phosphorylation of

CREB, and increases expression of early genes (c-fos) and genes encoding BDNF and other proteins significant in development and plasticity. A cascade of phenomena could be proposed linking EphB-mediated potentiation of NMDAR function with programs of gene expression involved in synapse maturation and plasticity.

10. NEURONAL CELL ADHESION MOLECULES IN SYNAPTIC PLASTICITY

Neuronal cell adhesion molecules (NCAMs) hold pre- and post-synaptic components in position by homo- and heterophilic interactions. NCAMs are connected with cytoskeleton, and are related with extracellular matrix components and intracellular signalling cascades. When LTP is established, an increase of NCAM deposition in the synaptic membranes (Fig. 1). Regulation of cytoskeletal dynamics by NCAMs could participate in activity-dependent synaptic reorganization. The cytoplasmic domain of N-cadherin interacts with catenins, which bind to the actin cytoskeleton. Upon homophilic NCAM binding, several intracellular signalling cascades are activated, suggesting that NCAMs could regulate molecular events controlling synaptic plasticity by interacting with intracellular signalling reactions. Purified fragments of NCAMs or antibodies against these fragments can modulate G-protein-dependent phosphoinositide turnover and alter intracellular Ca^{2+} levels and pH in cultured neuronal cells. An interesting possibility is that activity-dependent changes in intracellular signalling molecules could transmit signals back across the postsynaptic membrane via trans-membrane NCAMs to modulate cell binding and, thereby, rapidly alter synaptic structure and efficacy.

11. LONG-TERM DEPRESSION

The opposite phenomenon to synaptic potentiation is synaptic depression, which was first described and widely studied in cerebellar Purkinje cells. Simultaneous activation of climbing fibre and parallel fibre inputs to Purkinje cells produces a persistent and input-specific depression upon synaptic contacts between afferent parallel fibre afferents and Purkinje cell dendrites. This phenomenon, termed long-term synaptic depression (LTD), lasts from minutes to hours and has been implicated in memory for several forms of motor learning, including associative eyelid conditioning and adaptation of the vestibulo-ocular reflex. Essentially, three postsynaptic events are required for LTD induction: mGluR1 activation, Ca^{2+} influx via voltage-gated Ca^{2+}-channels and AMPAR activation. These signals converge upon PKC, which becomes activated. The greatest amount of cGMP-dependent protein phosphorylation occur in Purkinje cells, and it has been reported that a cascade involving NO, cGMP and cGMP-dependent protein kinases participates in the induction of cerebellar LTD. Moreover, application of an exogenous PKC activator to produce a LTP-like effect results in internalization of AMPARs in the dendrites of cultured Purkinje cells. This removal of AMPARs from the synaptic membrane explains the decreases in synaptic strength observed in LTD.

LTD has also been detected in hippocampal interneurons. CA3 hippocampal interneurons expressing Ca^{2+}-permeable AMPARs exhibit LTD after tetanic stimulation of CA3 excitatory inputs. This LTD induction requires both Ca^{2+}-influx through post-synaptic AMPARs and activation of pre-synaptic mGluR7. Ca^{2+}-entry through AMPARs might either trigger a synaptic shape change that enables access of released glutamate to presynaptic mGluR7 or cause release of a retrograde messenger that contributes to pre-synaptic mGluR7 activation and subsequent suppression of transmitter release. Given the relevance of the participation of interneurons in shaping the processing of neuronal circuits, changes in synaptic strength of interneurons can result in considerable variation in the outputs of projecting neurons.

12. METAPLASTICITY

The synaptic activity that precedes LTP induction can markedly affect the threshold for such induction. As has been shown in the area CA1 of the hippocampus, the recent history of NMDAR activation can adjust the induction by raising the threshold of LTP, thereby providing a physiological mechanism of transient LTP inhibition. Related to the recent history of synaptic activity is metaplasticity or persistent synaptic plasticity, that is, a change in the ability to induce subsequent synaptic plasticity. Metabotropic GluRs as well as NMDARs, participate in metaplasticity. A brief application of the selective mGluR agonist ACPD can cause a lasting enhancement of pharmacologically isolated NMDAR-mediated excitatory postsynaptic currents and induce a long-lasting depression of GABA-mediated inhibitory postsynaptic potentials (IPSP) that indirectly enhances NMDAR responses. Therefore, the prior synaptic activation can leave a long-lasting trace that affects the subsequent induction of synaptic plasticity.

13. SYNAPTIC PLASTICITY AND MEMORY

If artificially induced LTP were an experimental model of memory, it would be of great relevance to understanding whether learning of information processing in natural circumstances is also related to this form of synaptic plasticity. Tracking the location of an animal demonstrates that the firing patterns of pyramidal cells in the hippocampus are a function of the animal's position in relation to its surrounding. Mutation of LTP components can disrupt the properties of hippocampal pyramidal cells involved in the special placement of the animal. Selective knock-out of the NMDAR subunit NR1 in the pyramidal cells of the CA1 hippocampal area disrupts LTP in the Shaffer-collateral pathway, while the expression of a persistently active form of CaM-KII in the hippocampus interferes with LTP produced at low frequencies of stimulation. These lower frequencies are in the physiological range of a relevant spontaneous rhythm in the hippocampus termed theta rhythm — that occurs while the animal is moving in its environment. In both mutants, positional information is formed, but the absence of LTP interferes with the fine adjustment of positional cells and their long-lasting stability.

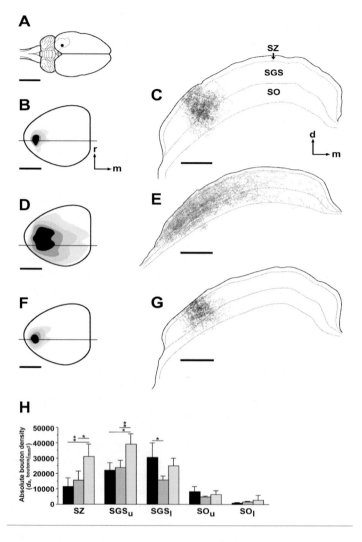

Figure 2. A: Dorsal view of the rat brain; the location of BDA injection site, from which terminal fields represented in B-G result, is marked with a dot. The dashed line on the dorsal surface of the rat brain represents the limits of the primary visual cortex. B, D, F: Location of the adult visual cortico-collicular terminal field on the dorsal surface of a control animal (B), a neonatally enucleated animal (D) and an adult enucleated animal (F). The area in black corresponds to the zone of greatest axonal density through the thickness of the SZ and SGS. Grey areas of decreasing intensity correspond to areas of decreasing axonal density. C, E, G: Camera lucida drawings of the terminal fields in coronal collicular sections at the level of the horizontal lines shown in B, D, F. In normal animals, terminal fields are restricted to a limited portion of superficial strata (B, C). In neonatally enucleated animals, cortico-collicular terminal fields display a considerable tangential expansion occupying almost the totality of the collicular surface (D, E). By contrast, in animals enucleated as adults, no

expansion can be observed. H: Bar graph showing the absolute density of boutons (d_a) corrected for tissue shrinkage in the different strata of the superior colliculus in normal, neonatally enucleated, and adult enucleate rats. Data are means ± SD; n = 4 in each group. Asterisks indicate the significance of the difference between groups for a given stratum (two-way ANOVA followed by post hoc Scheffe's test). *P < 0.05, **P < 0.01. Note the significant bouton density increase in the SZ and SGS_u of adult enucleated animals. r, rostral; m, medial; d, dorsal; SZ, stratum zonale; SGS, stratum griseum superficiale; SO, stratum opticum; SGS_u, upper half of the SGS; SGS_l, lower half of the SGS; SO_u, upper half of the SO; SO_l, lower half of the SO. Scale bars = 4 mm in A; 1 mm in B, D and F; 200 μm in C, E, and G.

Both *in vitro* and in free-moving animals, cortical synaptic plasticity with properties similar to hippocampal synaptic plasticity has been observed. The enhancement of LTP by daily tetanization is reminiscent of cortical memory, in which a slow consolidation process enables new memories to be interwoven with others. In rats, learning of a motor skill has been associated with the strengthening of the horizontal connections in layers II-III of primary motor areas. For example, motor training of one forelimb enhances field excitatory postsynaptic potentials (EPSPs) 20-45 hour later in supragranular layers of motor cortex that controls the active forelimb. Such enhancement of field EPSPs is associated with the occlusion of tetanus-induced LTP, suggesting a participation of an LTP-like mechanism in synaptic plasticity related to learning and memory.

14. STRUCTURAL PLASTICITY

The increase in number of synapses during development or after nervous lesions can be defined as structural synaptic plasticity. Experiments carried out *in vitro* and *in vivo* demonstrate that neurotrophins influence the complexity of axonal arbours. BDNF, NT-3, and NT-4/5 enhance axonal arborisation of retinal axons co-cultured with chicken optic tract, and infusion of BDNF into the optic tectum of *Xenopus* tadpoles triggers a considerable increase in the branch complexity of retinal ganglion cell axons. These effects are rapid, and can be blocked by BDNF antibodies. Neurotrophins also regulate overall synapse numbers. Indirect evidence of these effects comes from sympathetic neurons of transgenic mice over-expressing BDNF and from BDNF knockout mice, which display increased and decreased afferent synaptic innervation, respectively.

Several manifestations of structural plasticity participate in the reorganization of neuronal circuits following nervous lesions. For example, after axotomy, regenerative sprouting is the outgrowth of new neurites from the proximal axonal stump, leading to multiple local axon collaterals and the restoration of normal connectivity. Collateral sprouting, in contrast, is achieved by the growth of axonal branches from uninjured fibres. Reactive synaptogenesis is the formation of new active zones on pre-existing synapses, the innervation of adjacent dendrites, and the formation of new terminals, which in turn form new synapses. The functional status of the extracellular signal transduction systems and synthesizing machinery in the injured neurons, and the environment that surrounds reactive axons at the time of lesion are determinant for the distinct features of the post-injury reaction. Of the environmental factors, neuron-glia and microglial-astroglial interactions as well as the composition of the extracellular matrix, are crucial in this regard. Dennervation results in microglial reaction that precedes astroglial activation. Reactive microglial cells secrete interleukin-1 and transforming growth factor β1, which modulate the proliferation and neurotrophin synthesis by astrocytes. It has also been reported that subsets of microglial cells are able to express neurotrophins *in vivo*.

When a nervous structure is innervated by two afferent groups of fibres coming from different origins, and one of these afferents is axotomized, the other group reacts by either collateral sprouting or reactive synaptogenesis, which may compensate for the loss of synaptic terminals that disappear after axotomy. This compensatory growth has been shown in the septal nuclei that receive connections from the hippocampus via fimbria and from the hypothalamus and aminergic nuclei via the median forebrain bundle. The former input connects almost exclusively on the dendrites of septal cells whereas the latter ends on both dendrites and somata. After sectioning one of these pathways in adult rats, the synapses corresponding to the terminals of the sectioned pathway disappear in a few days, but the full synaptic population is restored after one month. This is due to the occupation of vacated postsynaptic sites by neighbouring terminals of the intact pathway. Neurons of the red nucleus receive afferents from the cerebellar nucleus interpositus and from the motor cortex via the cerebral peduncle. The former afferents end on the soma of the red nucleus neurons, and their stimulation produces short intracellular EPSPs in target neurons, while the latter terminate in dendrites and give rise to a much more slowly rising and declining EPSPs. Destruction of the nucleus interpositus results in degeneration of the synaptic endings on the soma, and a subsequent sprouting of intact peduncular afferents, which occupy a number of postsynaptic sites left vacant by eliminated terminals from the nucleus interpositus. Interestingly, EPSPs registered after stimulation of the sprouted peduncular afferents are a combination of short synaptic and slow dendritic responses, suggesting a physiological effectiveness of this type of structural plasticity.

The hippocampus, and particularly the fascia dentata, is also a suitable model to demonstrate different types of structural plasticity after deafferentiation. Following elimination of the ipsilateral entorhinal cortex, surviving crossed entorhino-dentate, commissural and septo-hippocampal fibres reinnervate the fascia dentata. The crossed entorhino-dentate fibres react, forming axonal collaterals, axonal extensions, and tangles. Sprouted commissural fibres do not trespass beyond their normal limits but their density is increased. A subset of reacting commissural fibres is GABAergic and the eliminated system is glutamatergic. This type of sprouting, termed heterologous, may interfere the functional recovery after trauma, and it appears to be responsible for the reduced capability of dentate granule cells to generate population spike responses upon orthodromic stimulation of the afferents after ipsilateral lesion of the entorhinal cortex. Sprouting of the cholinergic septo-hippocampal fibres in rodents is neurotrophin dependent and results in an increase in the density of afferent fibres within their normal termination territory.

The above-mentioned reactions of structural plasticity have been shown in adult animals, but the plastic capabilities differ from young to adult nervous systems.

Plastic reorganization of the connection from the primary visual cortex to the collicular superficial strata following removal of retinal afferents is an appropriate model to demonstrate these differences. The main afferents to the superficial collicular layers are fibres coming from the contralateral retina and from the ipsilateral visual cortex. Injection of the anterograde tracer BDA into the primary visual cortex has shown that neonatal enucleation causes a considerable expansion of the intact visual cortico-collicular terminal field, to occupy almost the entire collicular surface, suggesting a reactive response that involves axonal sprouting (Fig. 2 A-E). By contrast, following enucleation in adult animals, the size of the terminal fields remains unaltered (Fig 2 F-G), although bouton density increases within the most dorsal parts of superficial collicular strata (Fig. 2H), reflecting a process of reactive synaptogenesis at these levels. Therefore, both neonatal and adult visual cortico-collicular fibres display capacity for structural plasticity, though with distinct characteristics. Molecular factors such as growth-associated cytoskeletal proteins operating in the cortical origin, and extracellular matrix components and myelin-associated axonal growth inhibitors acting on the collicular target, very likely account for these differences.

REFERENCES

Abraham, W.C. and Bear, M.F. (1996). Metaplasticity: the plasticity of synaptic plasticity. Trends Neurosci. 19:126-130.

Carroll, R.C. and Zukin, R.S. (2002). NMDA-receptor trafficking and targeting: implications for synaptic transmission and plasticity. Trends Neurosci. 25:571-578.

Deller, T. and Frotscher M. (1997). Lesion-induced plasticity of central neurons: sprouting of single fibres in the rat hippocampus after unilateral entorhinal cortex lesion. Prog. Neurobiol. 53:687-727.

Fazeli, M.S. (1992). Synaptic plasticity: on the trail of the retrograde messenger. Trends Neurosci. 15:115-117.

Fields, R.D. and Itoh, K. (1996). Neural cell adhesion molecules in activity-dependent development and synaptic plasticity. Trends Neurosci. 19:473-480.

García del Caño, G.,Gerrikagoitia, I. and Martínez-Millán, L. (2002). Plastic reaction of the rat visual corticocollicular connection after contralateral retinal deafferentation at the neonatal or adult stage: axonal growth versus reactive synaptogenesis. J. Comp. Neurol. 446:166-178.

Kandel, E.R. (2000). Cellular mechanisms of learning and the biological basis of individuality. In: Principles of neural science. Chapter 63. Mc-Graw-Hill Inc., New York

Malinow, R. and Malenka, R.C. (2002). AMPA receptor trafficking and synaptic plasticity. Annu. Rev. Neurosci. 25: 103-126.

Martin, S.J.; Grimwood, P.D. and Morris, R.G.M. (2000). Synaptic plasticity and memory: an evaluation of the hypothesis. Annu. Rev. Neurosci. 23:649-711.

Riedel, G. (1996). Function of metabotropic glutamate receptors in learning and memory. Trends Neurosci. 19:219-224.

Schinder, A.F. and Poo, M. (2000). The neurotrophin hypothesis for synaptic plasticity. Trends Neurosci. 23:639-645.

Song, I. and Huganir, R.L. (2002). Regulation of AMPA receptors during synaptic plasticity. Trends Neurosci. 25:578-588.

Steward, O. and Schuman, E.M. (2001). Protein synthesis at synaptic sites on dendrites. Annu. Rev. Neurosci. 24:299-325.

Tsukahara, N. (1981). Sprouting and the neuronal basis of learning. Trends Neurosci. 4:234-237.

H. ALDSKOGIUS

Uppsala University Biomedical Center, Department of Neuroscience, Uppsala, Sweden

5. THE AXONAL BURDEN

Summary. Axonal disintegration is a major cause of neurological disease. Normal axonal integrity is maintained by the combination of intrinsic metabolic activities, delivery of molecules via intra-axonal transport and support from external cellular and extracellular sources. Trauma, ischemia, toxic agents, metabolic disturbances, inflammatory conditions and inadequate axon-glial interactions result in axonal disintegration. This process is characterized by breakdown of the axonal cytoskeleton, mitochondrial failure, and the accumulation of abnormal proteins and protein complexes. Intra-axonal rise in calcium above physiological levels appears to be a key factor in initiating these events. Axonal dysfunction can be attenuated by administration of calcium blockers, a variety of growth factors, immune suppressants, anti-inflammatory agents, as well as free radical scavengers. The recent identification of a gene, which "protects" axons from undergoing degeneration, provides a basis for a genetic approach to the treatment of neurological disorders caused by axonal disease or trauma.

1. INTRODUCTION

Block of impulse conduction, synaptic transmission or abnormal impulse discharges due to loss of axonal integrity are predominating causes of neurological disease. Understanding the mechanisms behind axonal degeneration, and the development of strategies for counteracting this process is therefore a major challenge for basic and applied neurobiological research. At the same time, rapid axonal degeneration is a prerequisite for efficient regeneration and restoration of function, e.g. in the peripheral nervous system. Thus, in certain situations, there may be reasons to promote axon degeneration in a way that does not increase the original damage, but allows regenerative or other repair processes to be instituted as rapidly as possible. This chapter, however, is concerned with characteristics, mechanisms and approaches to attenuate axon degeneration.

2. MAINTENANCE OF AXONAL INTEGRITY

The structural and functional properties of axons are maintained via four sources: i) the parent cell body, ii) the target tissue cells, iii) the surrounding non-neuronal cells and the extracellular environment, and iv) the intrinsic metabolic machinery.

2.1. Somatofugal transport

From the proximal end, the parent cell body conveys molecules and membrane packages as supplies for the continuous turnover of axonal and axon terminal components, as well as for modifications which are induced by alterations in their functional activity. Several systems serve this transport. Rapidly transported vesicles travel at a rate of about 100-400 mm/24h. A substantial proportion of these vesicles contain molecules and components associated with synaptic transmission, and are destined for the axon terminals. Many components destined for the axon itself move at around a 1-8 mm/24h. The slow components of axonal transport includes neurofilament proteins, tubulins and microtubule-associated proteins, but also axoplasmic proteins. Microtubules are the principal conveyors of rapid axonal transport, and therefore of critical importance for the maintenance of the axon and its terminals. Neurofilaments constitute the main structure determining axonal stability and size. Changes in neurofilament protein synthesis, transport and/or assembly will lead to alterations in axonal size and function.

2.2. Target-related trophic support

From the distal end, target-produced molecules are taken up by receptor-mediated endocytosis and retrogradely transported. Many of these molecules belong to various growth factor families, with their main destination being the nerve cell body, where they regulate gene expression. However, an influence on the local axonal metabolism by some of these molecules is possible. At the same time, the extent of the ongoing synaptic activity is likely to influence in particular the state of the axon terminal as a result of its intimate interactions with the postsynaptic target.

2.3. Interaction with non-neuronal cells

The axon is permanently interacting with surrounding cells, the most important ones being Schwann cells in the peripheral nervous system, and astrocytes and oligodendrocytes in the central nervous system. Schwann cells are responsible for myelination in the peripheral nervous system and establish functionally crucial membrane-to-membrane contact with the axolemma. At the nodes of Ranvier, Schwann cell processes and their extracellular matrix form specialized structures and regulate the functional properties of the nodal and paranodal axolemma. Bundles of non-myelinated axons in the peripheral nervous system are enveloped by extracellular matrix, which overlies the plasma membrane of Schwann cells. A corresponding role in the central nervous system is mainly fulfilled by astrocytes.

Oligodendrocytes are responsible for myelination of axons in the central nervous system. In doing so these cells establish close interactions with the internodal and paranodal portions of the myelinated axons. However, at the nodes of Ranvier, the astroglial processes come into contact with the axolemma, influence the development of this site and participate in regulating its homeostasis.

2.4. Intra-axonal protein metabolism

Many proteins and peptides, which are carried down the axon from the nerve cell body, are post-translationally modified. In addition, contrary to a longstanding dogma, compelling evidence demonstrate that synthesis of novel proteins can also take place in the axon and axon terminal (Giuditta et al., 2002). The latter appears to be display a particularly high degree of independence by its content of mRNAs for proteins necessary for mitochondrial maintenance, enzymes for the synthesis of a multitude of peptides and soluble proteins, as well as for cytoskeletal and heat shock proteins. This is obviously not sufficient to prevent the axon from degenerating (see section 3.1). Yet, the potential of this local synthetic machinery may correlate with the length of the period preceding the actual fragmentation and collapse of the axon.

3. FACTORS CAUSING AXONAL DEGENERATION

3.1. Trauma and vascular insults

Transection, crush, ligation of the axon, or immediate destruction of the nerve cell body, promptly initiates degeneration of the entire distal portion of the axon, the so-called Wallerian degeneration. A protracted Wallerian-type degeneration follows when axon injury results in retrograde degeneration and death of the nerve cell body. This process has been termed indirect Wallerian degeneration.

Haematomas interrupt passing axons or destroy their parent cell bodies. Ischemic insults produce a delayed axonal interruption through a multitude of pathways, including energy failure to support rapid axonal transport and ionic pumps.

The stretching, localized pressure changes, fluid movements or tearing, which accompany head or spinal trauma, initially cause diffuse axonal damage with maintained continuity with the parent cell body (Smith and Meaney, 2000). At a later stage, many of these abnormal axons will be disconnected from the cell body.

3.2. Toxicity

Certain chemical agents used in industry and agriculture (acrylamide, pesticides), and heavy metals (e.g. lead, mercury) interfere with axonal integrity. Many of these agents typically cause "dying-back" neuropathies, i.e. degenerative changes starting in the preterminal portion of the axon, and subsequently affecting more proximal regions (Cavanagh, 1979). Impaired microtubule function and disturbed axonal transport appear to underlie e.g. acrylamide neuropathy (Sickles et al., 1996). Anti-tumor drugs such as Vincristine, Cisplatin© and Taxol© also impair microtubule function (Theiss and Meller, 2000), leading to disturbance in the intra-axonal transport necessary for normal axonal structure and function.

3.3. Metabolic disturbances

Several metabolic conditions are accompanied by loss of axonal integrity with ensuing sensory, motor and/or autonomic symptoms, typically affecting distal body sites. Among these disorders are diabetes, uremia, and vitamin deficiencies (thiamine B1, B6, B12 and E). The mechanisms underlying these disturbances are incompletely known.

In diabetic neuropathy, changes in the endoneurial blood vessels may be the primary pathology, which subsequently compromises axonal metabolism and transport, leading to e.g. an increased impact of reactive oxygen species (Rosen et al., 2001). Hyperglycaemia in diabetes is also associated with the accumulation of glycates, i.e. products of reactions between sugars and free amino acids of proteins, and an excessive generation of sugar alcohols via the aldolase reductase pathway (Zochodne, 1999). In addition, the low levels of insulin and changes in insulin-related growth factors are likely to contribute to the process of axonal disintegration in diabetic neuropathy.

3.4. Immune reactions and inflammation

The impact of immune reactions and inflammatory mediators on axons is now well documented. Multiple sclerosis (Bjartmar and Trapp, 2001), the axonal form of the Guillain-Barré syndrome (Chowdbury and Arora, 2001), and axonopathy as a result of HIVinfection (Pardo et al., 2001), are all examples where axon injury is caused by the actions of pro-inflammatory cells and molecules. In this category should probably also be included degeneration of uninjured axons passing in the area of ongoing axons degeneration. This degeneration may be a 'bystander' effect caused by the inflammatory "environment" created by the injury induced nerve fiber breakdown (Yoles and Schwartz, 1998).

3.5. Deficient axo-glial interactions

Axonal damage occurs not only by the active impact of non-neuronal cells and their products, but may also be a consequence of e.g. failure glial regulation of ionic homeostasis, or absence of normal "protection" from surrounding myelin or myelin-forming cells. E.g., hereditary absence of the myelin component, proteolipid protein 1, is associated with segmental axonal degeneration, without associated demyelination (Garbern et al., 2002).

4. CHARACTERISTICS OF AXONAL DEGENERATION

4.1. Light microscopy

In the light microscope the most distinctive feature of axonal degeneration is the appearance in longitudinal sections of axonal fragmentation (Fig. 1). These fragments loose the smooth contour of the intact axon, and are characterized by an irregular outline reflecting regions of swelling and shrinkage, respectively. In myelinated axons these fragments eventually form the core of the myelin ovoids or ellipsoids. The distinct light microscopic features of nerve fiber degeneration can be clearly visualized with specific staining methods (see section 4.5). The axonal components of these profiles are rapidly degraded and eliminated. In contrast, the myelin-derived components of these ovoids are considerably more resistant to degradation, particularly in the CNS. Typically, there is also a variable extent of retrograde degeneration or retraction of the proximal stump. The reactions in the proximal stump may include changes in the pattern of sodium channel deposition, and the appearance of ectopic impulse discharges, which may underlie sensory neuropathies (Waxman, 2001), and contribute to excitotoxic degeneration of the affected neuron.

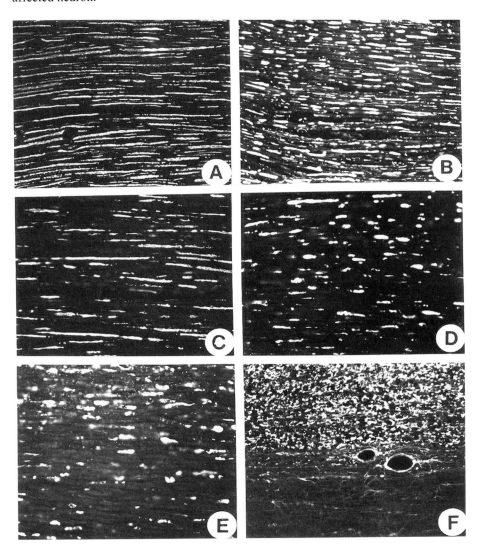

Figure 1. Light microscopy of axon degeneration.

4.2. Ultrastructure

Electron microscopic examination shows a diverse set of changes (Fig. 2). Initially, neurofilaments and microtubules are lost, granular and floccular materials appear, mitochondria become abnormal, and general swelling is common. In myelinated axons, accumulation of intra-axonal organelles and swelling are early and distinct changes particularly at the nodes of Ranvier, i.e. the sites at which the actual breaking up of the injured axon will subsequently take place (Hasegawa et al., 1988; Griffin et al., 1996).

At later stages, the axoplasm shows increased density of "ground substance", usually combined with shrinkage; disruption of mitochondria, and accumulation of vesicular and membranous structures. Often, different sets of changes are present in neighbouring axons, even within a relatively homogeneous population of simultaneously injured axons. In general, the larger, myelinated axons display more prominent and long-standing pathology, while thin, myelinated and non-myelinated axons appear to undergo a more discrete and rapid degeneration process. An interesting observation in this context is that demyelination appears to provide white matter axons with increased resistance to anoxic injury (Imaizumi et al., 1998).

4.3. Spatio-temporal pattern

There has been a longstanding controversy concerning the spatial progression of Wallerian degeneration after axon injury, i.e. whether it i) advances from the site of injury and distally, ii) develops simultaneously along the axon, or iii) proceeds from the distal end towards the injury site. The existing evidence is clearly most compatible with a proximo-distal progression of the degeneration process in the axon itself (George and Griffin, 1994). Exceptions are the preterminal axon and the axon terminals, which, although they may be a considerable distance from the site of injury, display changes at an early stage, even before the distal portions of the stem axon begin to disintegrate (see below).

4.4. The influence of temperature

There is a correlation between temperature and the rate of axonal degeneration (Tsao et al., 1999) as well as for diffuse axonal injury following head trauma (Buki et al., 1999). Changing the temperature surrounding injured axons appears to prolong the period of initial changes, but not the duration of the stage of irreversible disintegration of the axon (Tsao et al., 1999).

4.5. The vulnerability of the distal axon

Degeneration of the axon terminal has unique structural features, which are distinct from those of the axon itself. Generally, axon terminals appear to undergo more rapid deterioration than the main part of the axon, following axonal damage.

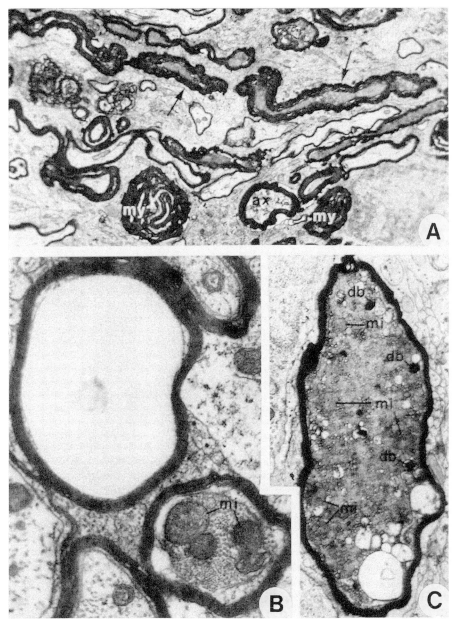

Figure 2. Ultrastructure of axon degeneration.

This may in part reflect the possible existence of specific mechanisms for elimination of synaptic terminals (Gillingwater and Ribchester, 2001). However, axon terminals may just be exceptionally sensitive to disturbances in e.g.

somatofugal fast axonal transport, or in the homeostasis of the local environment. This sensitivity of the axon terminal may initiate a dying-back process of the axon itself (cf. section 3.2), e.g. in neurodegenerative diseases when postsynaptic neurons degenerate.

Also, the preterminal axon is highly sensitive to disturbances of neuronal homeostasis. After dorsal root injury at the lumbar spinal level, changes in preterminal axons in the gracile nucleus in the brainstem are observed at about the same time as nerve fibers are showing signs of degeneration in the lumbar dorsal funiculus (George and Griffin, 1994).

The vulnerability of the axon terminal and preterminal axon is also reflected after injury to the peripheral processes of pseudounipolar sensory ganglion cells. This injury induces striking alterations in the terminal portions of these ganglion cells in the corresponding central termination sites, so-called transganglionic changes (Aldskogius et al., 1992). These changes presumably evolve as a result of modifications in axonal transport in the central processes, which in turn are triggered by the cell body response to the peripheral injury.

The susceptibility of the distal portions of long axons manifests itself also in several other situations, where the nerve cell body is subject of stress. In these situations, survival and the potential for subsequent recovery may require restrictions in the amount or composition of axonally transported components, which conceivably have its greatest impact on the most distally located parts of the neuron.

This mechanism presumably operates in dying back of axons, i.e. conditions where axon degeneration starts distally and progresses towards the cell body. This pathology is caused e.g. by neurotoxic agents, metabolic dysfunction, certain injury situations and aging (Cavanagh, 1979).

4.6. Molecular features

Concomitantly with loss of cytoskeletal elements and intra-axonal organelles the molecules making up these entities will gradually disappear. The resulting breakdown products can be detected for some time with antibodies to neurofilament proteins (Fig. 1a), and with specific silver staining methods, based on the differential sensitivity to oxidation between intact and degenerating axons (De Olmos et al., 1981). This method can be applied to postmortem tissue to determine the extent of nerve fiber degeneration even long periods of time after the original insult.

The basis for this differential sensitivity is presumably the formation of denatured proteins, in which different amino acid groups bind traces of silver and act as nucleation points for an enlargement of the reaction product, which makes it visible in the light microscope (Switzer, 2000). A similar mechanism may underlie the recently developed Fluoro-Jade method, which allows detection of axonal degeneration with a simple fluorescent marker (Fig. 1).

Other cytoskeletal alterations, which may lead to axon degeneration, are abnormal modifications of the microtubule-associated protein tau (taupathies). These modifications cause destabilization of microtubules, and polymerization of tau itself (Avila, 2000). This process underlies the formation of paired helical filaments, the aggregation of which results in the appearance of neurofibrillary tangles, a characteristic feature of Alzheimer's disease and several other neurodegenerative disorders.

Recently, degenerating fibers have been visualized with an antibody toY1-receptor protein of the messenger molecule neuropeptide Y ((Pesini et al., 1999). Although the morphological correlate to this immunostaining has not been clearly established, its appearance suggests that it is connected to axonal disintegration.

As a result of the impaired axonal transport there will also be an accumulation of beta-amyloid peptide; antibody staining for this product can be used to detect axonal damage in experimental (Stone et al., 2000), as well as post-mortem material (Lewén et al., 1995). A clinically useful indicator of axonal degeneration is the loss of the amino acid N-acetyl aspartate, which can be measured by proton magnetic resonance spectroscopy (Bjartmar et al., 2002).

4.7. Physiology

Compound action potentials fail to be propagated in the distal stump of a disintegrating axon within one to several days, and with a marked correlation to surrounding temperature (Levenson and Rosenbluth, 1990; Tsao et al., 1999). The slower rates are usually typical for larger animals. Transmission from nerve to muscle fails prior to the propagation of the compound action potential in the main axon (Levenson and Rosenbluth, 1990), in line with the documented vulnerability of the axon terminal (see section 4.4). Prior to complete cessation of the compound action potential, there may be a moderate decrease in conduction velocity (Levenson and Rosenbluth, 1990).

5. MECHANISMS OF AXONAL DEGENERATION

5.1. Principle mechanisms

From section 2 follows that axon degeneration can be initiated by four principle mechanisms: deficient support from i) the nerve cell body, ii) the target tissue, iii) the surrounding non-neuronal cells, and/or iv) by failure of the intrinsic axonal metabolic machinery.

Traditionally, axon degeneration has been considered as a process of rapid atrophy, caused by maintenance failure. However, the recent information on the role of genetic factors in determining the rate of Wallerian degeneration (section 5.5) suggest that axons may have an intrinsic program for its own removal, which may operate relatively independently from the nerve cell body (Raff et al., 2002).

5.2. Calcium – the final common path

Regardless of etiology, the events leading to axon degeneration appear to converge onto a common pathway, in which calcium and calcium-mediated processes have a key role (George et al., 1995; Lopachin and Lehning, 1997; Lopachin et al., 2000; Smith and Meaney, 2000; Glass et al., 2002). Traumatic, neurotoxic, ischemic, metabolic, immunological and inflammatory insults all disturb the mechanisms regulating axolemmal function, and cause a rise in intra-axonal calcium levels (Fig. 3). This influx appears to largely occur through specific calcium channels (George et al., 1995), which may be tightly associated with the sodium-calcium exchange system (Stys and Lopachin, 1998; Lopachin, 2000).

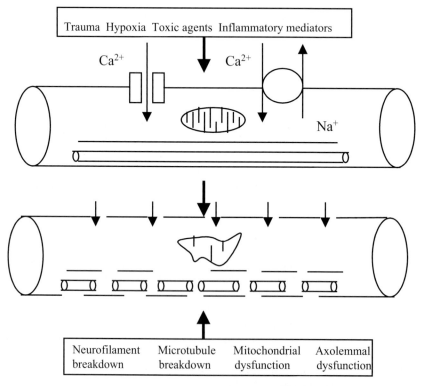

Axon collapse and fragmentation

Figure 3. Calcium – the final common path. Simplified scheme of the central role of calcium influx and, presumably, also calcium mobilization from intracellular stores leading to axonal disintegration.

The intra-axonal accumulation of calcium triggers proteolytic breakdown of neurofilaments and microtubules, with subsequent collapse of the axon and its internal transport system, as well as an increased load on the calcium buffering capacity of mitochondria, with ensuing failure to maintain intra-axonal ionic homeostasis. After traumatic interruption of the axon, this membrane dysfunction spreads proximo-distally at a rate corresponding to that of slow axonal transport (George et al., 1995)

5.3. Mechanisms upstream of calcium dysregulation

5.3.1. Free radicals and energy failure

Axonal integrity is obviously critically dependent on sufficient energy supply for maintaining e.g. axonal transport, and proper ion channel functions. The accumulation of reactive oxygen species during ischemia and trauma may cause elevated calcium influx, e.g. by interfering with mitochondrial function, causing energy failure and subsequent inadequate ion exchange operations. Genetic or acquired mitochondrial deficiencies may trigger axonal degeneration (Carelli et al., 2002).

Nitric oxide and its metabolites may play important roles in these events (Garthwaite et al., 2002). Thus, inhibition of nitric oxide synthase (Koeberle and Ball, 1999), or absence of its gene (Keilhoff et al., 2002) delays axon disintegration following injury. Moreover, electrically active axons, which are exposed to nitric oxide, have an increased risk of degeneration compared to "silent" axons (Smith et al., 2001).

5.3.2. Immune reactions and inflammation

The axon can be the direct target of an immune attack. Thus, sensitization of rabbits with ganglioside GM1, a major component of neuronal membranes, results in the development of an axonal Guillain-Barré syndrome (Yuki et al., 2001). More commonly, axons will suffer as a result from the attack of cytotoxic mediators released during an adjacent immune or inflammatory process. Ischemia, trauma and infections result in activation of endogenous glial cells, and recruitment of hematogenous cells. The mediators released from these cells may exert a direct axonotoxic effect, or enhance the release of other axon damaging molecules (see e.g. Liu et al., 2000; Gebicke-Haerter, 2001).

5.3.3. Apoptosis-like processes

Degeneration of axons sometimes involves processes, which are otherwise associated with apoptotic cell death. A key enzyme in the apoptotic cascade, caspase-3, triggers local axonal pathology following head injury (Stone et al., 2002). Isolated axon terminals are protected from degeneration by caspase inhibition (Gylys et al., 2002). However, the implications of caspase activation in damaged axons is still unclear, and axon degeneration is basically a non-apoptotic process (Raff et al., 2002).

5.3.4. Genetic factors

The spontaneously mutant mouse strain, C57BL/Wld(s) (Ola mouse), displays an extraordinarily slow Wallerian degeneration in the peripheral and central nervous system, a property which resides in the axon itself (Coleman and Perry, 2002). The gene underlying this property has been shown to be a chimeric gene (Ube4b/Nmnat) encoding for an N-terminal fragment of ubiquitination factor E4B (Ube4b) fused to nicotinamide mononucleotide adenylyltransferase (Nmnat) (Mack et al., 2001). A human homologue to this gene has recently been identified (Fernando et al., 2002).

In the mouse, this genetic aberration appears to exert its protective action in the axon indirectly, since the gene product is expressed in the nerve cell body. The known activity of this enzyme suggests that the delayed axon degeneration is a result of changes e.g. in the ubiquitination pathways. Even if axon degeneration is substantially delayed in the Wld(s) mouse, the mechanisms underlying the delayed disintegration is calcium dependent like in wild type animals. Interestingly, the gene defect in Wld(s) mice may also protect against axonal insults caused by toxic agents or disease processes. These observations obviously open up new avenues for the possibility to counteract axon degeneration in trauma and disease (Coleman and Perry, 2002).

Another spontaneously mutant mouse, the progressive motor neuronopathy (pmn) mouse, is characterized by selective dying-back or motor axons (Schmalbruch et al., 1991). The mechanism underlying this process is unknown, but may be associated with insufficient trophic support for the axon and/or axon terminal (Lesbordes et al., 2002).

Apart from the striking deviation from the normal time course of axonal degeneration exhibited by the Wld(s) and pmn mice, more subtle genetic factors are likely to underlie a differential susceptibility to neural insults among different individuals.

6. APPROACHES TO COUNTERACT OR DELAY AXONAL DEGENERATION

As mentioned in the Introduction, there are instances in which rapid nerve fiber degeneration may be advantageous for optimal subsequent repair processes. This aspect will be considered in elsewhere in this volume. The following account will be concerned with approaches to counteract or delay axonal degeneration. Given the profound implications of axon degeneration for neurological disorders, strategies for reversing this process is obviously of great importance. Likewise, a significant attenuation of the degenerative process may be highly beneficial for the patient by allowing time for spontaneous or induced repair of the insult, be it traumatic, ischemic, toxic or metabolic. Clinical aspects on the treatment of these conditions will not be considered here.

6.1. Fusion of separated stumps

A highly attractive approach, from a theoretical point of view, to counteract axonal degeneration in traumatic conditions, is to fuse the stumps so rapidly that anterograde axonal transport and impulse propagation can be restored prior to irreversible axonal damage. Successful attempts to achieve this in mammals have been described (Borgens, 2001). As expected, the requirements are strict, and there is a long way to go until this approach has a chance to become therapeutically useful. The principle is important, however, and its potential should be viewed in the context of the possibility to markedly slow down the progress of axonal degeneration to allow time for the possibility of repair at the injury site, or at sites of focal damage.

6.2. Calcium blockers and calpain inhibitors

Given the central role of calcium and calcium-activated proteases in executing axon degeneration, the anticipation that calcium chelating agents and inhibitors of calpain significantly slow down the rate of Wallerian degeneration has been confirmed (George et al., 1995; Glass et al., 2002).

6.3. Hypothermia

Cooling the affected nerve or tissue retards the rate of axonal degeneration, and may prolong the period from the initiation of axonal insult to the point of irreversible damage. This situation has been used with some success to counteract diffuse axonal injury following head trauma (Maxwell et al., 1999), and to enhance the possibility to fuse separated axon stumps (Marzullo et al., 2002).

6.4. Growth factors/hormones

Growth factors are likely to exert their main axon protecting effects either by influencing nerve cell body metabolism and/or the metabolism of surrounding non-neuronal cells. This rationale is behind the use of growth factors, such as nerve growth factor (NGF), brain-derived neurotrophic factor (BDNF) neurotrophin (NT)-3, insulin-like growth factor (IGF)-1, and glial cell line-derived neurotrophic factor (GDNF) to treat axonal pathology in a variety of neuropathic conditions (Apfel, 2001; Schmidt et al., 2001). A similar basis underlies the use of e.g. melanocortins and melanocortin analogues in toxic or drug-induced experimental neuropathies (Sporel-Ozakat et al., 1990). NGF, BDNF and NT-3 have also been used with some success to counteract axon degeneration in the injured spinal cord (Sayer et al., 2002).

6.5. Immune suppressants, anti-inflammatory agents and free radical scavengers

Several agents, with effects on immune and inflammatory mediators attenuate axon degeneration. These agents include the immunosuppressants Cyclosporin A and FK506 (Gold, 2000; Suchiro et al., 2001), tissue plasminogen activator (Akassaoglou et al., 2000), the tumor necrosis factor-alpha inhibitor, Pentoxifylline (Petrovich et al., 1997), and inhibitor of nitric oxide synthase (Koeberle and Ball, 1999). Also, a "global" treatment such as elimination of macrophages attenuates axon degeneration (Liu et al., 2000).

7. CONCLUDING COMMENTS

Our understanding of the basic molecular mechanisms underlying axon degeneration has increased considerably during the last ten years. The recent identification of a gene and its product, which provide substantial protection to axonal degeneration, is likely to greatly stimulate further progress in this respect. A future prospect of promoting functional recovery in a range of neurological disorders by maintaining axonal integrity with therapeutic interventions may not be too far away.

REFERENCES

Akassaoglou, K., Kombrinck, K.W., Degen, J.L. & Strickland, S. (2000) Tissue plasminogen activator-mediated fibrinolysis protects against axonal degeneration and demyelination after sciatic nerve injury. J. Cell Biol., 149, 1157-1166.

Aldskogius, H. (1974) Indirect and direct Wallerian degeneration in the intramedullary root fibres of the hypoglossal nerve. An electron microscopical study in the kitten. Adv. Anat. Cell Biol., 50, 1-78.

Aldskogius, H., Arvidsson, J. & Grant, G. (1992) Axotomy-induced changes in sensory neurons. In S.E. Scott (Ed.), Sensory Neurons: Diversity Development and Plasticity (pp. 363-383). Oxford Univ. Press.

Apfel, S.C. (2001) Neurotrophic factor therapy – prospects and problems. Clin..Chem. Lab. Med., 39, 351-355.

Avila, J. (2000) Tau aggregation into fibrillar polymers: taupathies. FEBS Lett., 476, 89-92.

Bjartmar, C. & Trapp, B.D. (2001) Axonal and neuronal degeneration in multiple sclerosis: mechanisms and functional consequences. Curr. Opin. Neurol., 14, 271-278.

Bjartmar, C., Battistuta, J., Terada, N., Dupree, E. & Trapp, B.D. (2002) N-acetylaspartate is an axon-specific marker of mature white matter in vivo: a biochemical and immunohistochemical study on the rat optic nerve. Ann. Neurol., 51, 51-58.

Borgens, R.B. (2001) Cellular engineering: molecular repair of membranes to rescue cells of the damaged nervous system. J. Neurosurg., 49, 370-378, Discussion 378-379.

Buki, A., Koizumi, H. & Povlishock, J.T. (1999) Moderate posttraumatic hypothermia decreases early calpain-mediated proteolysis and concomitant cytoskeletal compromise in traumatic axonal injury. Exp. Neurol., 59, 319-328.

Carelli, V., Ross-Cisneros, F.N. & Sadun, A.A. (2002) Optic nerve degeneration and mitochondrial dysfunction: genetic and acquired optic neuropathies. Neurochem. Int., 40, 573-584.

Cavanagh, J.B. (1979) The 'dying back' process. A common denominator in many naturally occurring and toxic neuropathies. Arch. Pathol. Lab. Med., 103, 659-664.

Chowdbury, D. & Arora, A. (2001) Axonal Guillain-Barre syndrome: a critical review. Acta Neurol. Scand., 103, 267-277.

Coleman, M.P. & Perry, V.H. (2002) Axon pathology in neurological disease: a neglected therapeutic target. Trends Neurosci., 25, 532-537.

De Olmos, J.D., Ebbesson, S.O.E. & Heimer, L. (1981) Silver methods for the impregnation of degenerating axoplasm. In L. Heimer & M.J. (Ed.), Neuroanatomical tract-tracing methods. (pp. 117-170). Plenum, New York.

Fernando, F.S., Conforti, L., Tosi, S., Smith, A.D. & Coleman, M.P. (2002) Human homologue of a gene mutated in the slow Wallerian degeneration (C57BL/Wld(s) mouse. Gene, 284, 23-29.

Garbern, J.Y., Yool, D.A., Moore, GJ., Wilds, I.B., Faulk, M.W., Klugmann, M., Nave, K.A., Sistermans, E.A., van der Knaap, M.S., Bird, T.D., Shy, M.E., Kamholtz, J.A. & Griffiths, I.R. (2002) Patients lacking the major CNS myelin protein, proteolipid protein 1, develop lenght-dependent axonal degeneration in the ab sence of demyelination and inflammation. Brain, 125 (Pt3), 551-561.

Garthwaite, G., Goodwin, D.A., Batchelor, A.M., Leening, K. & Garthwaite, J. (2002) Nitric oxide toxicity in the CNS white matter: an invitro study using rat optic nerve. Neuroscience, 109, 145-155.

Gebicke-Haerter, P.J. (2001) Microglia in neurodegeneation: molecular aspects. Mi crosc. Res. Tech., 54, 47-58.

George, E.B., Glass, J.D. & Griffin, J.W. (1995) Axotomy-induced axonal degeneration is mediated by calcium influx through ion-specific channgels. J. Neurosci., 15, 6445-6452.

George, R. & Griffin, J.W. (1994) The proximo-distal spread of axonal degeneration in the dorsal columns of the rat. J. Neurocytol., 23, 657-667.

Gillingwater, T.H. & Ribchester, R.R. (2001) Compartmental neurodegeneration and synaptic plasticity in the Wlds mutant mouse. J. Physiol., 534, 627-639.

Giuditta, A., Kaplan, B.B., van Minnen, J., Alvarez, J. & Koenig, E. (2002) Axonal and presynaptic protein synthesis: new insights into the biology of the neuron. Trends Neurosci., 25, 400-404.

Glass, J.D., Culver, D.G., Levey, A.I. & Nash, N.R. (2002) Very early activation of m-calpain in peripheral nerve during Wallerian degeneration. J. Neurol. Sci., 196, 9-20.

Gold, B.G. (2000) Neuroimmunophilin ligands: evaluation of their therapeutic potential for the treatment of neurological disorders. Expert Opin. Invets. Drugs, 9, 2331-2342.

Griffin, J.W., Li, C.Y., Macko, C., Ho, T.W., Hsieh, S.T., Xue, P., Wang, F.A., Cornblath, D.R., McKhann, G.M. & Asbury, A.K. (1996) Early nodal changes in the acute motor axonal neuropathy pattern of the Guillain-Barre syndrome. J. Neurocytol., 25, 33-51.

Gylys, K.H., Fein, J.A., Cole, G.M. (2002) Caspase inhibition protects nerve terminals from in vitro degradation. Neurochem. Res., 27, 465-472.

Hasegawa, M., Rosenbluth, J. & Ishise, J. (1988) Nodal and paranodal structural changes in mouse and rat optic nerve during Wallerian degeneration. Brain Res., 452, 345-357.

Imaizumi, T., Kocsis, J.D. & Waxman, S.G. (1998) Resistance to anoxic injury in the dorsal columns of adult rat spinal cord following demyelination. Brain Res., 779, 292-296.

Keilhoff, G., Fansa, H. & Wolf, G. (2002) Differences in peripheral nerve degeneration/regeneration between wild-type and neuronal nitric oxide synthase knockout mice. J. Neurosci., 68, 432-441.

Koeberle, P.D. & Ball, A.K. (1999) Nitric oxide synthase inhibition delays axonal degeneration and promotes the survival of axotomized retinal ganglion cells. Exp. Neurol., 158, 366-381.

Lesbordes, J.C., Bordet, T., Haase, G., Castelnau-Ptakhine, L., Rouhani, S., Gilgenkrantz, H. & Kahn, A. (2002) In vivo electrontransfer of the cardiotrophin-1 gene into skeletal muscle slows down progression of motor neuron degeneration in pmn mice. Hum. Mol. Genet., 11, 1615-1625.

Levenson, D. & Rosenbluth, J. (1990) Electrophysiologic changes accompanying Wallerian degeneration in frog sciatic nerve. Brain Res., 523, 230-236.

Lewén, A., Li, G.L., Nilsson, P., Olsson, Y. & Hillered, L. (1995) Traumatic brain injury produces changes of beta-amyloid precursor protein immunoreactivity. Neuroreport, 6, 357-360.

Liu, T., van Rooijen, N. & Tracey, D.J. (2000) Depleteion of macrophages reduces axonal degeneration and hyperalgesia following nerve injury. Pain, 96, 25-32.

LoPachin, R.M. (2000) Redefining toxic distal axonopathies. Toxicol. Lett., 112-113, 23-33.

LoPachin, R.M. & Lehning, E.J. (1997) Mechanism of calcium entry during axon injury and degeneration. Toxicol. Appl. Pharmacol., 143, 233-244.

Mack, T.G., Reiner, M., Beirowski, B., Mi, W., Emanuelli, M., Wagner, D., Thomson, D., Gillingwater, T., Court, F., Conforti, L., Fernandon, F.S., Tarlton, A., Andressen, C., Addicks, K., Magni, G., Ribchester, R.R., Perry, V.H. & Coleman, M.P. (2001) Wallerian degeneration of injured axons and synapses is delayed by a Ube4b/Nmnat chimeric gene. Nat. Neurosci., 4, 1199-1206.

Marzullo, T.C., Britt, J.M., Stavisky, R.C. & Bittner, G.D. (2002) Cooling enhances in vitro survival and fusion-repair of severed axons taken from the peripheral and central nervous systems of rats. Neurosci. Lett., 327, 9-12.

Maxwell, W.L., Donnelly, S., sun, X., Fenton, T., Puri, N. & Graham, D.L. (1999) Axonal cytoskeletal responses to nondisruptive axonal injury and the short-term effects of posttraumatic hypothermia. J. Neurotrauma, 16, 1225-1234.

Pardo, C.A., McArthur, J.C. & Griffin, J.W. (2001) HIV neuropathy: insights in the pathology of HIV peripheral nerve disease. J. Peripher. Nerv. Syst., 6, 21-27.

Pesini, P., Kopp, J., Wong, H., Walsh, J.H., Grant, G. & Hökfelt, T. (1999) An immunohistochemical marker for Wallerian degeneration of fibers in the central and peripheral nervous system. Brain Res., 828, 41-59.

Petrovich, M.S., Hsu, H.Y., Gu, X., Dugal, P., Heller, K.B. & Sadun, A.A. (1997) Pentoxifylline suppression of TNF-alpha mediated axonal degeneration in the rabbit optic nerve. Neurol. Res., 19, 551-554.

Raff, M.C., Whitmore, A.V. & Finn, J.T. (2002) Axonal self-destruction and neurodegeneration. Science, 296, 868-871.

Rosen, P., Nawroth, P.P., King, G., Moller, W., Tritschler, H.J. & Paccker, LL. (2001) The role of oxidative stress in the onset and progression of diabetes and its complications: a sumamry of a Congress Series sponsored by UNESCO-MCBN, the American Diabetes Association and the German Diabetes Society. Diabetes Metab. Res. Rev., 17, 189-212.

Sayer, F.T., Oudega, M. & Hagg, T. (2002) Neurotrophins reduce degeneration of injured ascending sensory and corticospinal motor axons in adult rat spinal cord. Exp. Neurol., 175, 282-296.

Schmalbruch, H., Jensen, H.J., Bjaerg, M., Kamieniecka, Z. & Kurland, L. (1991) A new mouse mutant with progressive motor neuronopathy. J. Neuropathol. Exp. Neurol., 50, 192-204.

Schmidts, R.E., Dorsey, D.A., Beaudet, L.N., Parvin, C.A. & Escandon, E. (2001) Effect of NGF and neurotrophin-3 treatment on experimental diabetic autonomic neuropathy. J. Neuropathol. Exp. Neurol., 60, 263-273.

Schmued, LC. & Hopkins, K.J. (2000a) Fluoro-Jade: novel fluorochromes for detecting toxicant-induced neuronal degeneration. Toxicol. Pathol., 28, 91-99.

Sickles, D.W., Brady, S.T., Testino, A., Friedman, M.A. & Wrenn, R.W. (1996) Direct effect of the neurotoxicant acrylamide on kinesin-based microtubule motility. J. Neurosci. Res., 46, 7-17.

Smith, K.J., Kapoor, R., Hall, S.M. & Davies, M. (2001) Electrically active axons degenerate when exposed to nitric oxide. Ann. Neurol., 49, 470-476.

Smith, D.H. & Meaney, D.F. (2000) Axonal damage in traumatic brain injury. Neuroscientist, 6, 483-495

Sporel-Ozakat, R.E., Edwards, P.M., Van der Hoop, R.G. & Gispen, W.H. (1990) An ACTH-(4-9) analogue, Org 2766, improves recovery from acrylamide neuropathy in rats. Eur. J. Pharmacol., 186, 181-187.

Stone, J.R., Singleton, R.H. & Povlishock, J.T. (2000) Antibodies to the C-terminus of the beta-amyloid precursor protein (APP): a site specific marker for the detection of traumatic axonal injury. Brain Res., 871, 288-302.

Stone, J.R., Okonkow, D.O, Singleton, R.H., Mutin, L.K., Helm, G.A. & Povlishock, J.T. (2002) Caspase-3-mediated cleavage of amyloid precursor protein and formation of amyloid beta peptide in traumatic axonal injury. J. Neurotrauma, 19, 601-614.

Stys, P.K. & LoPachin, R.M. (1998) Mechanisms of calcium and sodium fluxes in anoxic myelinated central nervous system axons. Neuroscience, 82, 21-32.

Suchiro, E., Singleton, R.H., Stone, I.R. & Povlishock, J.T. (2001) The immunophilin ligand FK506 attenuates the axonal damage associated with rapid rewarming following posttraumatic hypothermia. Exp. Neurol., 172, 199-210.

Switzer, R.C (2000) Application of silver degeneration stains for neurotoxicity testing. Toxicol. Pathol., 28, 70-83.

Tsao, J.W., George, E.B. & Griffin, J.W. (1999) Temperature modulation reveals three distinct stages of Wallerian degeneration. J. Neurosci., 19, 4718-4726.

Theiss, C. & Meller, K. (2000) Taxol impairs anterograde axonal transport of microinjected horseradish peroxidase in dorsal root ganglia neurons in vitro. Cell Tiss. Res., 299, 213-224.

Waxman, S.G. (2001) Aquired channelopathies in nerve injury and MS. Neurology, 56, 1621-1627.

Yoles, E. & Schwartz, M. (1998) Degeneration of spared axons following partial white matter lesion: implications for optic nerve neuropathies. Exp. Neurol., 153, 1-7.

Yuki, N., Yamada, M., Koga, M., Odaka, M., Suzuki, K., Tagawa, Y., Ueda, S., Kasama, T., Ohnishi, A., Hayashi, S., Takahashi, H., Kamijo, M. & Hirata, K. (2001) Animal model of axonal Guillain-Barré syndrome induced by sensitization with GM1 ganglioside Ann. Neurol., 49, 712-720.

Zochodne, D.W. (1999) Diabetic neuropathies: features and mechanisms. Brain Pathol., 9, 369-391.

J.H.M. PREHN AND D. KÖGEL

Experimental Neurosurgery, Center for Neurology and Neurosurgery, Johann Wolfgang Goethe-University, Frankfurt, Germany

6. THE SELF-DESTRUCTION OF NEURONS PHYSIOLOGICAL AND PATHOPHYSIOLOGICAL DECISIONS FOR THE FUNCTIONAL INTEGRITY

Summary. A major challenge in current neuroscience research is to understand the molecular mechanisms leading to neuronal apoptosis in neurodegenerative diseases and after acute brain insults. Neuronal apoptosis which has a crucial function for elimination of unwanted neurons in the immature nervous system, for a long time was thought to be restricted to this type of developmental cell death. In recent years however, evidence has accumulated that apoptosis is also activated in the adult nervous system, and may contribute to pathophysiological cell death in both acute and chronic neurological and psychiatric disorders. Based on morphological and biochemical criteria, two major forms of apoptosis can be discriminated. Type I apoptosis is performed via the "classical" execution machinery whose central death mediators are members of the caspase family of cysteine proteases. In contrast, type II apoptosis does not require caspases, but is executed via alternative mechanisms. In this chapter, we will discuss the contribution of type I and type II apoptosis in acute and chronic brain disorders and evaluate the possibilities and drawbacks for the therapeutic rescue of neurons in which the apoptotic death program is already initiated.

1. INTRODUCTION

Half a century ago, Rita Levi-Montalcini and Victor Hamburger discovered nerve growth factor (NGF). Subsequently the concept was developed that target-derived growth factors promotes the survival of neurons, and that neurons not provided with trophic factors die during development (Oppenheim RW 1991). In 1972, Kerr, Wyllie, and Currie proposed the existence of an intrinsic cell death program and introduced the term apoptosis for the execution of this program (Kerr et al., 1972). Apoptosis was believed to be an *active* form of cell death enabling individual cells to commit sucide. In contrast, necrosis is a *passive* form of cell death raised by accidental damage of tissue and does not encompass activation of a specific death program. It is estimated that 50 % of all neurons generated during development undergo apoptotic cell death, and that apoptosis is essential for establishing and maintaining functional neuronal networks. The ground breaking achievements in the model system *C. elegans* have provided fundamental insights into the genes regulating apoptosis in humans (Horvitz, 1999). Most of these genes are highly expressed during brain development. However, this program can also be readily activated in the adult nervous system, and may contribute to pathophysiological cell death in acute and chronic brain disorders.

2. CONTENT

2.1. Apoptosis and programmed cell death

Initial classification of cell death into the apoptosis/necrosis dichotomy was mainly based on morphological criteria: hallmarks of apoptosis include membrane blebbing, cell shrinkage and chromatin condensation and fragmentation. The cell disintegrates into apoptotic bodies which are subsequently phagocytized by macrophages, microglia, or neighbouring cells without eliciting an inflammatory response. Necrosis, on the contrary, is typically associated with early loss of plasma membrane integrity and swelling of the cell body and organelles. The rupture of the plasma membrane and subsequent release of cellular constituents evokes an inflammatory response. In recent years, significant progress has been made identifying key components of the apoptotic cell death machinery and deciphering the signalling pathways in which they are embedded. The caspase family of proteases are central executioners of most forms of apoptotic cell death. Caspase-dependent cell death is also termed "classical" apoptosis or type I apoptosis.

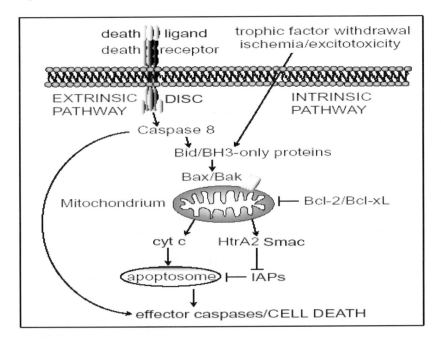

Figure 1: Proposed model for major cell death pathways implicated in type I apoptosis. Classical, type I apoptosis is activated by two major pathways: in the death receptor (extrinsic) pathway, executioner caspases are activated via the death receptor signaling

complex (DISC) whereas the mitochondrial apoptosis pathway is triggered by activation of Bax and/or Bak. This is achieved by activation of BH3-only proteins like Bid, Bim or PUMA. Both pathways are interconnected via caspase 8 (Casp8)-mediated cleavage of Bid. Signalling downstream of mitochondria comprises cytochrome c (cyt c)-mediated activation of the apoptosome, thus triggering effector caspase activation and execution of cell death. In addition, release of Smac and HtrA2 leads to sequestration of inhibitors of apoptosis (IAPs).

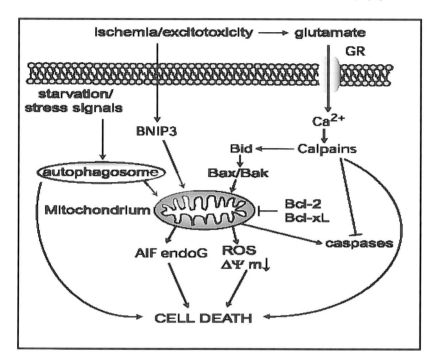

Figure 2: Proposed model for major cell death pathways implicated in type II apoptosis. Similar to type I apoptosis, BH3-only proteins such as Bid and BNIP3 might be activated during type II apoptosis. However, this either does not lead to cytochrome c release and downstream caspase activation (BNIP3) or downstream caspase activation is blocked, e.g. by inhibition through calpains. In this case, caspase-independent mitochondrial death effectors apoptosis inducing factor (AIF) and endonuclease G might trigger cell death by alternative mechanisms. Additionally, other protein degrading programs, such as autophagy might be initiated and kill cells through the activation of lysosomal proteases of the cathepsin family. Type II apoptosis can also be associated with mitochondrial permeability transition which leads to mitochondrial depolarization and increased mitochondrial ROS generation. GR: glutamate receptor.

However, an increasing number of studies substantiates the existence of caspase-independent forms of apoptosis (termed type II apoptosis) that evolved during evolution and may even play a more prominent role in the adult organism, especially in the nervous system (Leist and Jäättelä, 2001). Type I and type II apoptosis can be discriminated by both morphological and biochemical criteria. Type I apoptosis is associated with membrane blebbing, potent chromatin condensation and nuclear fragmentation, phosphatidylserine exposure, disruption of the cell into apoptotic bodies and internucleosomal DNA cleavage. Some of this criteria might also be fulfilled in type II apoptosis, which however usually is devoid of nuclear fragmentation, internucleosomal DNA cleavage and generation of apoptotic bodies. An important hallmark of caspase-independent apoptosis is represented by autophagy, a process normally involved in regulated turnover of long-lived proteins, lipids or even whole organelles. During autophagic cell death, which is characterized by the appearance of abundant autophagic vacuoles in the cytoplasm, cells are destroyed by complete degradation of cellular components via an autophagosomic-lysosomal pathway.

2.2. Caspase activation during apoptosis

Caspases are a family of cytosolic cysteine proteases that cleave proteins after aspartate residues. Caspases reside in the cytosol as dormant proforms that can be activated by proteolytic cleavage at specific aspartate residues. Caspases involved in apoptosis can be subdivided into initiator and effector caspases (Nicholson, 1999). Effector caspases such as caspase-3, -6 and -7 cleave multiple cellular substrates during the death process. These cleavage events result in degradation and reorganization of cellular structures, inactivation or activation of signal transduction pathways, alterations in gene transcription, and inhibition of DNA repair. Initiator caspases predominantly function to activate effector caspases. Activation of initiator caspases requires the binding of adaptor proteins to specific interaction motifs within their prodomain, leading to their dimerization and autoactivation. Two major caspase-activating pathways predominate. Activation of caspases can occur either after ligation of death ligands to their receptors (extrinsic pathway) or after the release of pro-apoptotic factors from mitochondria (intrinsic pathway). In the mitochondrial apoptosis pathway, activation of the initator caspase-9 occurs via binding of adaptor protein apoptotic protease activating factor-1 (Apaf-1) to the caspase recruitment domain (CARD). The association of caspase-9 and Apaf-1 and subsequent apoptosome formation is triggered by the pro-apoptotic factor cytochrome c. Release of cytochrome c from mitochondria is therefore the key regulatory step in the mitochondrial apoptosis pathway (Martinou and Green, 2001). Mice deficient in either APAF-1 or caspase-9 show a pronounced enlargement of brain structures and are non-viable, demonstrating the importance of the mitochondrial apoptosis pathway for brain development.

Death receptor-induced apoptosis is triggered by ligand binding to receptors of the tumor necrosis factor (TNF) receptor superfamily, such as TNF-receptor I, Fas/APO-1/CD95 and death receptor 5. The corresponding ligands (TNF-α, Fas-ligand, TRAIL) are either soluble, secreted proteins or are exposed at the plasma membrane of immunocompetent cells. Upon binding to death receptors, initiator caspases 8 or 10 are recruited via adaptor proteins such as TRADD and FADD. Compared to the immune system, death receptor-triggered apoptosis is believed to be play a less prominent role for the development or maintenance of brain structures, but could become an important process during CNS inflammation. Recent studies suggest that activation of death receptors may also trigger trophic effects in neurons.

Among the executioner caspases, caspase-3 is believed to be the predominant caspase and is responsible for internucleosomal DNA fragmentation, cleavage of specific transcription factors and cytoskletal proteins, and for providing a positive feed-back loop for the activation of caspase-9 and -8 and other caspases (Fischer et al., 2003). However, execution of cell death does not necessarily occur in caspase-independent fashion in cells devoid of caspase-3, since other executioner caspases such as caspase-6 and 7 might in part substitute for caspase-3 activity.

Inhibitor-of-apoptosis proteins (IAPs) are believed to be naturally occuring inhibitors of caspase activation that may have evolved to prevent unwanted caspase activation (Holcik and Korneluk, 2001). Members of this family include XIAP, cIAP1, cIAP2 and NAIP which are also expressed in the nervous system. IAPs have a dual function during apoptosis. They are able to directly inhibit initator and executioner caspases by either sequestering their proforms or by binding directly to the catalytic site of active caspase subunits. Moreover, due to an E3 ligase activity, IAPs trigger the ubiquitination and proteasomal degradation of active subunits and other pro-apoptotic factors. During apoptosis, two other proteins are therefore released from mitochondria that neutralize the anti-apoptotic activity of IAPs: the second mitochondria-derived activator of caspase/direct IAP binding protein with low pI (Smac/DIABLO) and Omi/HtrA2, the mammalian homologue of the *Escherichia coli* heat shock-inducible protein HtrA. Interestingly, the release of both cytochrome c and Smac/DIABLO has been shown to be a prerequisite for apoptotic cell death in several model systems, such as nerve growth factor deprivation-induced cell death of sympathetic neurons.

2.3. Bcl-2 family members

Tight binding of core histones to DNA in nucleosomes is an obvious obstacle for accessibility of the DNA to transcriptional regulatory and enzymatic machinery. Thus, no wonder that specific covalent modifications of histones play a major role in making chromatin either susceptible or refractory to transcription. Following major modifications are distinguished according to their chemical nature: acetylation, methylation, phosphorylation, and ubiquitination. Specific enzymes, such as histone transacetylases (HAT), histone deacetylasaes (HDAC), histone methyltransferases (HMT), etc. are involved in those modifications. Not of lesser importance are such phenomena as degree of modification (addition of a single or more modifying groups) as well as the exact location of each modification, i.e., which aminoacid residues in which histone molecule are modified. In particular, the transcriptionally active chromatin contains acetylated or even hyperacetylated histones. For example, H4 can be acetylated/hyperacetylated at lysins 5, 8, 12, and 16, whereas H3 at lysins 9, 14, 18, and 23. As far as the methylation is concerned it may occur at Lys 20 in the case of H4, and lysins: 4, 9, 27, 36 in H3. The euchromatin is enriched in histone H3 methylated at Lys4 (this aminoacid residue, when tri-methylated, appears to be the best marker of active chromatin), whereas in heterochromatin H3 is methylated at Lys 9. Phosphorylation may be observed at Ser 1 in H4, and serines 10, and 28 in H3. There is an apparent link between various modifications. For instance, H3 phosphorylation at Ser 10 greatly enhances subsequent H3 acetylation at Lys 14. Notably, all these modifications are concentrated on so called histone tails that stretch out of the nucleosome core and are therefore easy targets for modifications. It has been suggested that they are altogether responsible for the gene regulatory histone code.

Interestingly, the same condition, such as, e.g., status epilepticus may result in differential effect on histone acetylation level at different gene promoters, inducing deacetylation of those that are silenced and simultaneous hyperacetylation of the genes that are upregulated.

2.3. BH3 Only proteins

In apoptotic cells, the transcriptional induction or posttranslational activation of Bcl-2-homolgy domain-3 (BH3)-only proteins triggers the activation of Bax and Bak (Huang and Strasser). All members of this subgroup of the Bcl-2 family share a nine amino acid BH3-domain, but otherwise possess very little structural homology. Members of this subgroup include Bim, Bid, Bad, PUMA, Noxa, BNIP3, Hrk, Bcl-G, Bmf, Blk, Bik and Spike. The role of the individual BH3-only family members is to couple specific upstream stress signals to the evolutionary conserved intrinsic pathway of apoptosis. The structural diversity of BH3-only proteins is also reflected by the different mode of their activation: BH3-only proteins have been shown to be activated by either transcriptional induction (PUMA, Noxa, Hrk, Bim) phosphorylation (Bad, Bik, Bim, Bmf), lipid modification (Bid), subcellular translocation (Bim, Bmf), or by proteolytic cleavage by several cell death-related proteases (Bid). There is evidence that the BH3-only proteins can be functionally discriminated into two different subgroups: the so called inducers and enablers. The inducers bind and directly activate pro-apoptotic Bcl-2 family members Bax and Bak at the mitochondria, whereas the enablers bind and sequester anti-apoptotic Bcl-2 family members Bcl-2 and Bcl-xL, thus sensitizing cells to activation of the intrinsic pathway by inducers.

One of the inducers that has been shown to be critical for neuronal apoptosis is Bim (Putcha et al. 2001). The c-Jun NH2-terminal kinase (JNK) stress signaling pathway might have a dual function in activation of Bim. JNK-mediated phosphorylation of Bim which is normally sequestered by dynein motor complexes, leads to release of Bim from the cytoskeleton, thus enabling it to translocate to the mitochondria where it can bind to Bax and Bak. There are three splice variants of Bim, BimEL (extra long), BimL (long) and BimS (short). It has been proposed that transcriptional upregulation of BimEL represents a specific hallmark of neuronal apoptosis. Enhanced BimEL expression after trophic factor withdrawal has been suggested to require the JNK/c-JUN signaling axis. In addition, in neurons deprived from trophic support, Bim might be transcriptionally induced by Forkhead family member FKHRL1 which can no longer be inactivated by the trophic factor-dependent PI3K/Akt survival pathway.

Two other transcriptionally regulated BH3-only family members are Noxa and PUMA both of which are downstream targets of the tumor suppressor protein p53. The human homolog of mouse Noxa is called APR (adult T cell leukemia-derived PMA-responsive), but its role in regulation of cell death has not been established yet. Similar to Bim, transcriptional activation of PUMA also seems to be repressed by the PI3K/Akt survival pathway and can be activated by trophic factor withdrawal in a p53-independent fashion. For both PUMA and Noxa no knockout data regarding their role in neuronal apoptosis are available yet.

Table 1: BH3 only proteins and modes of activation

BH3-only protein	death stimulus	mode of activation
Bim	cytokine withdrawal	transcriptional upregulation, phosphorylation, subcellular translocation
PUMA	DNA damage, cytokine withdrawal	transcriptional upregulation
Noxa	DNA damage	transcriptional upregulation
Bad	cytokine withdrawal	phosphorylation
BNIP3	hypoxia/ischemia	transcriptional upregulation
Bid	death receptor activation, hypoxia/ischemia	proteolytic cleavage, dephosphorylation, myristoylation
Bmf	anoikis	subcellular translocation
Bik	DNA damage	transcriptional upregulation, phosphorylation
Hrk	cytokine withdrawal	transcriptional upregulation
Blk	???	???
Bcl-G	???	???
Spike	death receptor activation	???

The prime example for proteolytic activation of the intrinsic pathway of apoptosis is the inducer Bid which is cleaved by several proteases into its pro-apoptotic form, tBid. tBid is required in most cell types for death receptor-induced apoptosis. Bid can be cleaved by caspase-8, calpains, cathepsins and granzyme B. The expression of Bid is high in the adult nervous system. Interestingly, Bid-deficient mice show an increased resistance to ischemic brain insults, implying an important contribution of tBid to certain forms of neuronal cell death.

In addition to negatively regulating the expression of BH3-only family members, the serine/threonine kinase Akt does also directly phosphorylate the typical enabler

Bad, thus targeting it to the scaffolding protein 14-3-3 under normal conditions. In absence of an activated PI3K/Akt pathway Bad is in its dephosphorylated state allowing it to sequester Bcl-xL (Zha et al. 1996). Of note, Bad has been shown to be activated during seizure-induced neuronal cell death.

BNIP3 is an unusual member of the BH3-only subfamily in that it induces necrosis-like cell death through mitochondrial permeability transition. Cell death triggered by BNIP3 is associated with translocation of BNIP3 to the outer mitochondrial membrane, loss of mitochondrial membrane potential and increased production of reactive oxygen species. However, BNIP3-mediated cell death has been reported to occur independent of Apaf-1, caspase activation and cytochrome c release. It has been suggested that Bcl-2 exerts an anti-necrotic effect by complex formation with BNIP3.

3. CASPASE-INDEPENDENT PATHWAYS

In an individual cell receiving a given apoptotic death signal both caspase-dependent and caspase-independent cell death pathways are activated in parallel. In cells with inhibited caspase activity caspase-independent cell death mechanisms suffice to eventually cause cell death, albeit in a slower, less efficient manner. Although active caspases are not a prerequisite for execution of cell death, the time frame and caspase-dependance of individual events during apoptosis might be stimulus- and cell type-specific. Caspase-independent cell death can be triggered a) by activation of a mitochondrial organelle dysfunction program directly due to the loss of cytochrome c, b) the release of alternative pro-apoptotic proteins from mitochondria, and c) alternative pathways which bypass mitochondria and control of cell death by Bcl-2 family members.

a). While many studies have focused on the role of cytochrome c release to activate the caspase cascade, loss of cytochrome c may also directly affect mitochondrial ATP production and free radical generation. Electron microscopy studies and single-cell analyses of cells expressing cytochrome c-GFP fusion proteins suggested that the release of cytochrome c during apoptosis is rapid and complete (Luetjens et al., 2001). Cytochrome c normally transports electrons between mitochondrial complexes III and IV. A disruption of the mitochondrial electron flow caused by a significant loss of cytochrome c will favor conditions that shift the normal 4-electron reduction of molecular oxygen to a 1-electron reduction, hence generating superoxide. The protective effects of antioxidants, SOD mimetics, and superoxide dismutase overexpression in several apoptosis models suggest that the production of superoxide may play an important role in the execution of cell death in neurons.

Mitochondria that have released their cytochrome c are likewise less capable of producing ATP. Mitochondria are able to maintain a mitochondrial membrane potential after the release of cytochrome c. Evidence has been provided that this is caused by ATP supply via glycolysis and subsequent F_0F_1-ATPase operating in the reverse mode, hence even consuming ATP. Re-addition of cytochrome c to isolated mitochondria that underwent an outer membrane permeability increase likewise restores membrane potential and ATP production. It remains to be shown whether addition of glucose or pyruvate improves the survival of neurons that have lost cytochrome c. Evidence has been presented that the release of cytochrome c is a reversible event in neuronal apoptosis. However, mitochondria may also eventually depolarize after the release of cytochrome c, a process that may be caspase-dependent and caspase-independent. In cultured rat sympathetic neurons deprived of NGF in the presence of caspase inhibitors, cells can be rescued from cell death until the time point of mitochondrial depolarization. Recent studies have shown that the opening of the permeability transition pore is involved in this final depolarization in rat neurons, a finding that has to be confirmed in other species. It has also been shown that the caspase-independent death of neurons is associated with the autophagic destruction of (non-functional) mitochondria that have released their cytochrome c (Xue et al. 2001).

b) There are alternative signalling pathways leading to apoptosis-associated apoptotic events, such as degradation of chromosomal DNA. Apoptosis-inducing factor (AIF) is a mitochondrial pro-apoptotic factor which is released in a Bcl-2-sensitive manner from the mitochondria into the cytosol during execution of apoptosis (Susin et al. 1999). AIF which is most likely the best studied gene product involved in caspase-independent cell death to date belongs to the gene family of oxidoreductases. However, the enzymatic activity of AIF is not required for its cell death-inducing activity. Apparently, upon activation of the intrinsic cell death pathway, both caspase-dependent (cytochrome c) and caspase-independent (AIF) executional pathways can be triggered simultaneously, both leading to distinct nuclear events during cell death. In contrast to caspase-activated DNAses, AIF induces large-scale DNA fragmentation, thus leading to chromatin condensation. This DNA-degrading activity of AIF has been shown to be caspase-independent as partial chromatinolysis and cell death caused by nuclear AIF is not inhibited by the presence of broad spectrum caspase inhibitors. However, recent findings in mammalian cells and in the model system *C. elegans* suggest that AIF release may also be caspase-dependent. In addition, Bcl-2 overexpression has been shown to inhibit AIF translocation from the mitochondria to the cytosol, thus abrogating AIF-triggered cell death. AIF release has been shown to be critically involved in neuronal apoptosis and excitotoxicity, and overexpression of AIF is sufficient to trigger cell death in neurons. Knockout studies revealed that AIF might control early morphogenesis during embryonal development. Not only large-scale fragmentation, but internucleosomal DNA cleavage might also occur in caspase-independent fashion under certain circumstances. Recently, endonuclease G, a novel apoptotic DNAse capable of catalyzing internucleosomal DNA cleavage was characterized. Just like AIF, endonuclease G is released from the mitochondria and translocates to the nucleus during apoptosis. Its contribution to neuronal apoptosis has not yet been elucidated.

c) In addition to caspases, other proteases such as serine proteases, cathepsins and calpains might be involved in alternative apoptotic cell death pathways. Similiar to caspases, calpains and cathepsins have been implicated in cleavage and activation of Bid. Cathepsin B is capable of taking over the role of the dominant execution protease in death receptor-induced apoptosis. Cathepsin D seems to play a central role in neuronal autophagy during development, since the CNS tissues of cathepsin D knockout mice are filled with ceroid lipofuscin-containing autophagosomes, implying an incomplete autophagic proteolysis in these neurons. Interestingly, cathepsin D-deficient animals have been shown to manifest seizures and become blind near the terminal stage.

Elevated cellular Ca^{2+} concentrations during apoptosis, e.g. following mitochondrial dysfunction, may lead to the activation of apoptosis-associated enzymes requiring Ca^{2+} for their activation, such as calpains or death associated protein (DAP) kinase. Like caspases, calpains are a family of cytosolic cysteine proteases, but require binding of Ca^{2+} for their activation. Activation of calpains can be amplified by caspase cleavage of the endogenous calpain inhibitor calpastatin. Calpains have been suggested to be involved in the regulation of caspase activity during apoptosis (Lankiewicz et al. 2000). Cleavage by calpains has been observed in the case of the upstream caspases-9 and -8, as well as in the case of the executioner caspases-3 and -7. Calpain I-mediated proteolysis of procaspase-7 can lead to an activation of this caspase. Similarly, ER stress-induced neuronal apoptosis mediated via murine caspase-12 has been shown to require calpain activation. On the other hand, there is strong evidence supporting the concept that calpain activation negatively regulates caspase activity. Calpain-generated fragments of caspases-7, -8 and -9 were inactive and/or unable to activate downstream executioner caspases, and calpain potently inhibited the ability of cytochrome c to activate executioner caspases in neurons. A recent study also demonstrated calpain cleavage of the cytochrome c-binding protein Apaf-1. It is therefore conceivable that the upstream or concomitant activation of calpains exerts a negative feed-back signal on caspase activation. Interestingly, calpains promote apoptosis-like events during platelet activation and excitotoxic neuron death, including chromatin condensation, phosphatidylserine exposure, caspase substrate cleavage and cell shrinkage, thus mimicking aspects of caspase-mediated apoptosis. Hence, calpains are strong candidates for the execution of type II apoptosis.

3. APOPTOSIS INVOLVEMENT IN NEUROLOGICAL DISORDERS

There is now a large body of evidence supporting the concept that apoptotic pathways contribute to the pathogenesis of acute and chronic neurological and psychiatric disorders (Mattson 2000). Chromatin condensation, DNA fragmentation, cell shrinkage, phosphatidylserine exposure and membrane blebbing have been detected in neurons after focal and global cerebral ischemia, traumatic injury, epileptic seizures, and hypoglycemia. However, during acute brain injury, it may depend on the severity of injury, temporal aspects as well as the intrinsic properties of a given neuronal population whether type I, type II or both types are activated in response to injury. In ischemic stroke, neurons in the core of the infarction, where oxygen supply is severely limited, are are believed to die by necrotic, caspase-independent cell death triggered by energy deprivation. In contrast, apoptotic, caspase-dependent cell death has been reported to occur in the ischemic penumbra where the cellular energy metabolism does not completely collapse. Activation of caspases 1, 3, 8, 9 and 11 as well as mitochondrial cytochrome c release has been observed during cerebral ischemia. In addition, inhibition of neuronal death in the ischemic penumbra by Bcl-2 has been reported. In chronic neurodegenerative disorders, such as Parkinson's Disease, Alzheimer's Disease, amytrophic lateral sclerosis, Huntington's Disease and Creutzfeldt-Jacob Disease, detection of apoptosis has been proven to be more challenging. *In vitro*, overexpression of mutant proteins associated with inherited, familial forms of these disorders has been shown to induce apoptotic cell death. For example, intranuclear huntingtin has been shown to trigger release of cytochrome c from the mitochondria and activate caspase-3. In *in vivo* animal models and in human *post mortem* tissue, detection may be more difficult because only a very limited number of neurons may apoptose at a given time point. However, the full execution of type I or type II apoptosis may be severely inhibited in neurons that have disturbed mitochondrial energetics, protein synthesis or gene expression. From this point of view, chronic neurodegenerative disorders such as Alzheimer's Disease could also be considered disorders of apoptosis *deficiency*. The inability to properly execute apoptosis in these disorders could lead to an inability to remove damaged, unwanted cells, increased necrotic cell death, tissue inflammation, and disease progression. On the other hand, evidence has also been provided that caspases may actively accelerate chronic neurological disorders such as Alzheimer's and Huntington's Disease by generating clevage products that accelerate the disease. Caspase 3, for example is capable of cleaving the amyloid precursor protein (APP) into a fragment which triggers the formation of amyloidogenic ß-amyloid peptides. Conversely, 7 different caspases (caspase 1, 2, 3, 6, 8, 9 and 12) have been implicated in neuronal cell death triggerd by exposure to ß-amyloid peptides. Of note, ß-amyloid peptides, the APP intracellular domain (AICD) and presenilins all have potent effects on cellular Ca^{2+} homeostasis. It has been reported that mutant presenilins and AICD may have a role in pathophysiological ER Ca^{2+} overloading (LaFerla 2002). There is increasing evidence that disturbance of ER function in neurodegenerative diseases, such as Alzheimer's Disease and Parkinson's Disease renders neurons more susceptible to cell death. Endoplasmic reticulum stress triggers the accumulation of malfolded proteins within the secretory pathway and is known to activate a conserved cellular stress responses, the unfolded protein response (UPR) and ER-associated degradation (ERAD), which enable cells to get rid of accumulated proteins and restore ER function. Persisting ER dysfunction, however, leads to irreversible cell injury and apoptotic cell death. Although a number of studies have shown that ER stress can trigger caspase-dependent, type I apoptosis, it is currently not known whether ER stress-mediated cell death can also occur in caspase-independent fashion. It has been suggested that presenilins are required for the unfolded protein response (UPR), since expression of dominant negative presenilin mutants was shown to downmodulate UPR target gene expression and inhibition of translation after induction of ER stress.

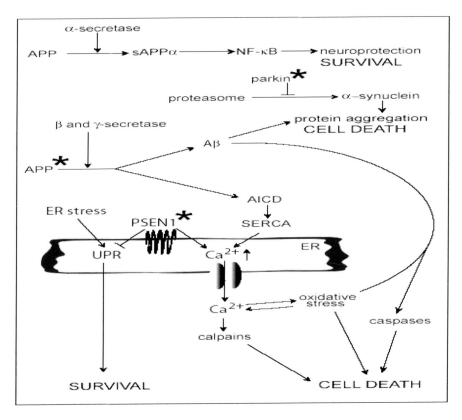

*Figure 3: Molecular mechanisms underlying cell death in neurodegenerative diseases. Processing of the amyloid precursor protein (APP) by α-secretase (non-amyloidogenic pathway) leads to formation of sAPPα which exerts a neuroprotective effect by activation of the transcription factor NF-κB. The Alzheimer Disease-associated amyloidogenic pathway by sequential cleavage of β- and γ-secretases leads to generation of Aβ and AICD. Aβ formation leads to extracellular protein accumulation, triggers oxidative stress by membrane lipid peroxidation and induces activation of several caspases. Both AICD and mutant presenilins are implicated in ER Ca^{2+} overloading and sensitization of neurons to cell death triggered by ER Ca^{2+} store depletion. Enhanced cytoplasmic Ca^{2+} concentrations after store depletion might be associated with enhanced radical production, mitochondrial depolarization and activation of calpains. Mutant presenilins also compromise the unfolded protein response (UPR), thus rendering neurons more vulnerable to ER stress-induced cell death. The E3 ubiquitin ligase parkin is required for proteasomal degradation of substrates associated with neurodegenerative diseases, such as α-synuclein and polyglutamine repeat-containing proteins. Loss of function of parkin leads to aggregation of these target proteins, thus triggering cell death by a yet unknown mechanism. * denotes disease-associated mutant protein.*

In addition, UPR/ERAD attenuation has been implicated in Parkinson's Disease, Huntington's Disease, but also in ischemia. Similar to ischemia, neuronal cell death in neurodegenerative diseases may occur both in caspase-dependent and -independent fashion. Of note, caspase-independent neuronal autophagy has been implicated in several neurodegenerative diseases, such as Alzheimer's Disease, Huntington Disease and Creutzfeldt-Jakob Disease. In most studies, detection of neuronal autophagy was limited to morphological analysis by electron microscopy, staining of lysosomal/autophagosomal vacuoles and *in vitro* application of autophagy inhibitors, such as 3-methyladenine. However, in Alzheimer's Disease, enhanced cathepsin D expression and elevated enzyme activity in the cerebrospinal fluid of Alzheimer patients was observed, suggesting a direct role for cathepsin D in autophagy associated with Alzheimer's Disease.

4. STUDIES WITH APOPTOSIS INHIBITORS

Several studies have employed caspase inhibitors of either biological (viral or cellular caspase inhibitors) or chemical origin (peptide inhibitors). Application of broad-spectrum caspase inhibitors such as zVAD-fmk and BAF has been shown to prevent or delay neuronal cell death in response to ischemic, traumatic or excitotoxic neuron death, as well as in animal models of Parkinson's Disease. These compound binds irreversibly to the catalytic site of caspases, forming a covalent inhibitor/enzyme complex. In addition, other cysteine proteases such as calpains and cathepsins may also be inhibited by these compounds, thus aggravating interpretation of the involvement of type I versus type II apoptosis in these experiments. Moreover, many of these inhibitors also inhibit the formation and secretion of interleukin-1β, and may therefore exhibit potent anti-inflammatory effects. Protective effects have also been observed with more specific inhibitor of executioner caspases such as zDEVD.fmk. Drawbacks of these chemical caspase inhibitors are their limited stability and the gross differences in binding affinity to the individual caspase family members, problems that have meanwhile been largely overcome by efforts from the pharmaceutical industry. Protective effects have also been observed with overexpression of the viral caspase inhibitors CrmA and p35. Some of the most convincing data come from animal studies employing the effects of overexpression of anti-apoptotic Bcl-2 family members Bcl-2 and Bcl-xL, or gene deletion of the pro-apoptotic Bcl-2 family member Bax or the BH3-only proteins Bid or Bim. Neurons from these animals show an increased resistance to trophic factor withdrawal induced-, and ischemic cell death. Likewise, gene deletion of the transcription factor and guardian of the genome, p53, which induces the expression of pro-apoptotic target genes including Puma, Noxa, Bid, Bax, APAF-1, caspases 3 and 9, and Fas, signifcantly reduced brain injury in response to cerebral ischemia. Animals deficient in cytochrome c, APAF-1, or procaspase-9 do not survive the postnatal period, and have thus not yet been tested in animal models of acute brain injury. However, caspase-3-deficient animals which do survive into adulthood display a significantly increased resistance to ischemic brain injury.

5. FUNCTIONAL RECOVERY AFTER CASPASE AND APOPTOSIS INHIBITION

In experimental models of stroke, trauma, and Parkinson's Disease caspase inhibition has been shown to improve functional recovery, such as increased motor or memory performance. Nevertheless, it remains to be shown whether and in which brain disorders it makes sense to therapeutically interfere into a process that eliminates damaged, unwanted cells, and at which level in the death cascade inhibition proves to be successful. Only a few experimental studies have addressed the latter question. It was demonstrated that caspase inhibitors not only inhibit apoptosis of photoreceptor neurons in a model of retinal degeneration, but that these rescued retina cells also significantly regained their visual function (Davidson and Steller, 1998). However, for many chronic neurodegenerative disorders it is not known how neurons perform in which a cell death program was activated but not fully executed. Another aspect that has so far been little investigated is whether the improvement of function by caspase/apoptosis inhibition is caused by enhancing the number of neural stem cells rather than by a direct effect on damaged neurons. Neurons and other cell types of the nervous system are generated throughout life in the subventricular zone and in other brain structures. Neurogenesis increases significantly if brain cells are exposed to insults such as ischemia or epileptic seizures. However, the availability of neural stem cells is determined both by the rate of neurogenesis as well as by the rate of their apoptosis. Protective effect of caspase inhibitors may therefore also be caused by increasing the number of neural stem cells. Apoptosis inhibition might therefore also influence the rate of functional recovery days or weeks after a toxic, metabolic or ischemic insult.

6. SYNAPTIC APOPTOSIS

Evidence has been provided that in non-neuronal cells, once initiated, activation of caspases is a very rapid and efficient process. Maximal caspase activation can be seen 5 to 10 min after the release of pro-apoptotic factors, and maximal cell shrinkage can occur in less than 30 min. However, little information has been made available regarding the kinetics of caspase activation in individual neurons, or the propagation of apoptotic signals along dendrites or axons. Indeed, mitochondria are also localized in neurites, and could in theory also propagate death signals to the soma. Similar to the electrophysiological properties of nerve cells, caspase activation may have to reach a certain threshold before it can reach the soma. Conversely, this suggests that caspases may also only be activated locally, and may play an important role in local degeneration of neurons (Mattson et al. 1998). Evidence has been presented that neurites can undergo a localized injury in response to glutamate, and that calpains function as executioner of localized injury and remodelling in neurites. It remains to be shown whether caspases also play an important role in the degeneration/remodelling of neurites. Interestingly, it has been demonstrated that the delayed death of CA1 rat hippocampal neurons after cerebral ischemia *in vivo* is preceded by caspase activation along the perforant path. However, previous studies are mainly based on immunocytochemical detection of processed caspases or caspase substrates, and may not directly reflect enzyme activity. It will be interesting in the future to determine caspase or calpain activity in the neurites and axons of living cells subjected to a localized or submaximal injury.

REFERENCES

Davidson FF, Steller H. Blocking apoptosis prevents blindness in Drosophila retinal degeneration mutants. Nature. 1998; 391: 587-591.
Fischer U, Janicke RU, Schulze-Osthoff K. Many cuts to ruin: a comprehensive update of caspase substrates. Cell Death Differ. 2003; 10: 76-100.
Holcik M, Korneluk RG. XIAP, the guardian angel. Nat Rev Mol Cell Biol. 2001; 2: 550-556.
Horvitz HR. Genetic control of programmed cell death in the nematode Caenorhabditis elegans. Cancer Res. 1999; 59: 1701s-1706s.
Huang DC, Strasser A. BH3-Only proteins-essential initiators of apoptotic cell death. Cell. 2000; 103: 839-842.
Kerr JF, Wyllie AH, Currie AR. Apoptosis: a basic biological phenomenon with wide-ranging implications in tissue kinetics. Br J Cancer. 1972; 26: 239-257.
LaFerla FM. Calcium dyshomeostasis and intracellular signalling in Alzheimer's disease. Nat Rev Neurosci. 2002 3: 862-872.
Lankiewicz S, Marc Luetjens C, Truc Bui N, Krohn AJ, Poppe M, Cole GM, Saido TC, Prehn JHM. Activation of calpain I converts excitotoxic neuron death into a caspase-independent cell death. J Biol Chem. 2000; 275: 17064-17071.
Leist M, Jäättelä M. Four deaths and a funeral: from caspases to alternative mechanisms. Nat Rev Mol Cell Biol. 2001; 2: 589-598.
Luetjens CM, Kogel D, Reimertz C, Dussmann H, Renz A, Schulze-Osthoff K, Nieminen AL, Poppe M, Prehn JHM. Multiple kinetics of mitochondrial cytochrome c release in drug-induced apoptosis. Mol Pharmacol. 2001; 60: 1008-10019.
Martinou JC, Green DR. Breaking the mitochondrial barrier. Nat Rev Mol Cell Biol. 2001; 2: 63-67.
Mattson MP. Apoptosis in neurodegenerative disorders. Nat Rev Mol Cell Biol. 2000 ; 1: 120-129.
Mattson MP, Keller JN, Begley JG. Evidence for synaptic apoptosis. Exp Neurol. 1998 ; 153: 35-48.
Nicholson DW. Caspase structure, proteolytic substrates, and function during apoptotic cell death. Cell Death Differ. 1999; 6: 1028-1042.
Oppenheim RW. Cell death during development of the nervous system. Ann Rev Neurosci. 1991; 14: 453-501.
Putcha GV, Moulder KL, Golden JP, Bouillet P, Adams JA, Strasser A, Johnson EM. Induction of BIM, a proapoptotic BH3-only BCL-2 family member, is critical for neuronal apoptosis. Neuron. 2001; 29: 615-628.
Ranger AM, Malynn BA, Korsmeyer SJ. Mouse models of cell death. Nat Genet. 2001 ; 28: 113-118.
Susin SA, Lorenzo HK, Zamzami N, Marzo I, Snow BE, Brothers GM, Mangion J, Jacotot E, Costantini P, Loeffler M, Larochette N, Goodlett DR, Aebersold R, Siderovski DP, Penninger JM, Kroemer G. Molecular characterization of mitochondrial apoptosis-inducing factor. Nature. 1999; 397: 441-446.
Xue L, Fletcher GC, Tolkovsky AM. Mitochondria are selectively eliminated from eukaryotic cells after blockade of caspases during apoptosis. Curr Biol. 2001; 11: 361-365.
Zha J, Harada H, Yang E, Jockel J, Korsmeyer SJ. Serine phosphorylation of death agonist BAD in response to survival factor results in binding to 14-3-3 not BCL-X(L) Cell. 1996 87: 619-628.

A. D. ORTEGA AND J. M. CUEZVA

Departamento de Biología Molecular, Centro de Biología Molecular Severo Ochoa, Universidad Autónoma de Madrid, Madrid, Spain

7. THE ORGANELLES, I: MITOCHONDRIAL FAILURE AND NEURODEGENERATION

Summary. Mitochondria are essential cellular organelles required for the generation of biological energy. Mitochondria also play an essential role in the execution of apoptosis, a program of cell death. The molecular components required for the functional activities of mitochondria are very vast and are encoded in both the nuclear and the mitochondrial genomes. Therefore, mutations and/or epigenetic factors that affect the activity of any of the gene products required for mitochondrial function result in pathological phenotypes that have as biological signature the malfunction of mitochondria. Here we review the complex array of neurodegenerative disorders that are associated with defects and alterations of the bioenergetic and apoptotic functions of the mitochondria of the cells of the nervous system.

1. INTRODUCTION

Mitochondria are double-membrane organelles of virtually all eukaryotic cells; they are responsible for harnessing the energy requirements that sustain cell activity. Energy transduction is carried out by protein complexes of the inner mitochondrial membrane. Mitochondria are also involved in apoptosis or programmed cell death. This is a genetically encoded program in which mitochondria act as sensors and executioners of cell death by releasing molecules that are normally confined within the organelle. A feature of the mitochondria is that they contain their own genetic material (mtDNA), which is maternally inherited. Although the mitochondrial genome encodes only a minor fraction of the molecular constituents of the organelle, mtDNA is essential for cell viability. Because of the important function of mitochondria in energy transduction and in apoptosis, it is no surprise that genetic defects in either the nuclear or mitochondrial genomes, or epigenetic factors that impact on mitochondrial function, result in a functional failure of the organelle, with further implications for cell function and/or viability. In this chapter, we describe the basic mitochondrial functions and components involved in energy transduction and apoptosis, and review the neurodegenerative disorders that are associated with defects of the mitochondrial function of nerve cells. The broad spectrum of pathological phenotypes arising as a result of mitochondrial malfunction pointed out the importance of the organelle as a therapeutic target in neurological disorders.

2. THE MOLECULAR COMPONENTS OF MITOCHONDRIA

2.1. Nuclear gene products

Most proteins needed for mitochondrial function are encoded in the nuclear genome (Figure 1). The transcription of these genes is exerted by the same molecular machinery that expresses cell constituents unrelated to the mitochondria. However, the expression of mitochondrial genes occurs in response to cell signals and developmental cues that are orchestrated by specific transcription factors and co-activators involved in defining the function and complement of mitochondria in the cell (Figure 1). The generated mRNAs are further matured, exported from the nucleus, and translated in the cytoplasm by two main pathways that involve mitochondrial-bound and unbound polysomes (Figure1). The characteristics and molecular components defining each pathway remain largely unknown although it appears that the assembly of large ribonucleoprotein complexes in the nucleus are involved in the process (Figure 1). Specific features in some of the transcripts that encode proteins of oxidative phosphorylation, and of the apoptotic machinery, deserve special attention. In fact, some of these mRNAs have a short 3'non-translated region (3'UTR) that defines the metabolic fate of the mRNA. The 3'UTR provides the site for the binding of regulatory proteins (RNABPs in Figure 1) that participate in the localization, translation and stability of the mRNAs, thereby contributing to delineating the functionality of mitochondria within the cell.

After translation, or simultaneously with translation, mitochondrial proteins are imported and sorted to specific sites within mitochondrial compartments (outer membrane, intermembrane space, inner membrane and matrix) in a process that is energy consuming and that requires a complex protein translocation machinery that is assisted by specific molecular chaperones (Figure 1). A feature of most — but not all — mitochondrial proteins that are imported into the organelle is the presence of an N-terminal extension in the primary structure — the pre-sequence or import sequence. The pre-sequence is cleaved off after import by matrix-processing peptidases to generate the mature polypeptides (Figure 1). After maturation the proteins are assembled with other mitochondrial proteins to generate the molecular complexes that participate in the various mitochondrial functions (Figure 1). Mutations affecting the expression, assembly, maturation, or activity of nuclear encoded components required for mitochondrial function are therefore expected to impact at many levels and to follow a Mendelian trait of inheritance.

2.2. Products of the mitochondrial genome

Human mitochondrial DNA (mtDNA) is a circular double-stranded DNA molecule of ~ 16.5 kbp that encodes a small number of essential mitochondrial constituents. Components of the translation machinery of the organelle are encoded by twenty-four genes (22 tRNAs and 2 rRNAs), while hydrophobic protein components of the molecular complexes of the inner membrane that are involved in the respiratory chain and in oxidative phosphorylation are encoded by thirteen genes: seven genes

(ND1-6 and ND4L) of subunits of Complex I (NADH: ubiquinone oxidoreductase), one gene (cyt b) of a subunit of Complex III (ubiquinone: cytochrome c oxidoreductase), three genes (COI-III) of subunits of Complex IV (Cytochrome c oxidase), and two genes (ATPase6 and ATPase8) of subunits of Complex V (H^+-ATP synthase) (Figure 2).

Expression of the mitochondrial genome has several special features. The mtDNA is present in the cell at a very high copy number (estimated 10,000 molecules/cell), far exceeding the representation of the nuclear genes encoding components of the mitochondria. Therefore, mutations in mtDNA result in two populations of mtDNA — mutated and wild-type — a condition known as heteroplasmy. Homoplasmy is the condition where all mtDNA molecules of the individual are identical. In general, non-deleterious pathogenic mutations of mtDNA are heteroplasmic. Therefore, the phenotypic manifestation of a mutation in mtDNA occurs when the proportion of mutated molecules relative to the total mtDNA present in the cell exceeds a certain value, a condition known as the "threshold effect". In addition, and since mtDNA is inherited from the mother, pedigrees harbouring defects in mtDNA exhibit a maternal linage of transmission. Other characteristics of mtDNA are that it is very compact, lacks introns and has overlapping genes, with a genetic code that differs from the universal one.

The replication and transcription of mtDNA (Figure 1) is controlled by organelle specific DNA- and RNA-polymerases and regulatory proteins (mtSSB, mtTFA, TFB1M, TFB2M, mTERF, TAS-associated) and a *cis*-acting regulatory element on mtDNA, located within a non-coding DNA fragment known as the D-loop. This region of mtDNA contains the regulatory elements for the initiation of DNA replication and the three promoters for RNA transcription. Recent findings have suggested that DREF — a transcription factor involved in nuclear DNA replication and cell-cycle control — provides a link for the co-ordination of the replication of nuclear DNA and mtDNA during cell proliferation. In fact, the expression of some of the proteins involved in the replication of mtDNA is also regulated by DREF.

The transcription of mtDNA generates the polycistronic transcripts that are further processed to mature tRNAs, rRNAs, and mRNAs (Figure 1). Two novel transcription factors of the mitochondria (TFB1M and TFB2M) have recently been characterized. These proteins are homologous to bacterial rRNA dimethyltransferases and cooperate with mitochondrial RNA polymerase and mitochondrial transcription factor A (mtTFA) to carry out basal transcription of mtDNA in mammals.

The mitochondrial translation machinery (Figure 1) is at present poorly characterized, with initiation and elongation factors resembling the prokaryotic ones. The synthesized polypeptides are further assembled with other mitochondrial proteins, of both nuclear and mitochondrial origin, to generate the protein complexes of the mitochondria (Figure 1). It appears that regulation of the expression of the nuclear-encoded mitoribosomal proteins during the cell cycle is exerted at the level of translation, and that the overall activity of mitochondrial translation depends on signals and/or factors originated in the cell cytoplasm.

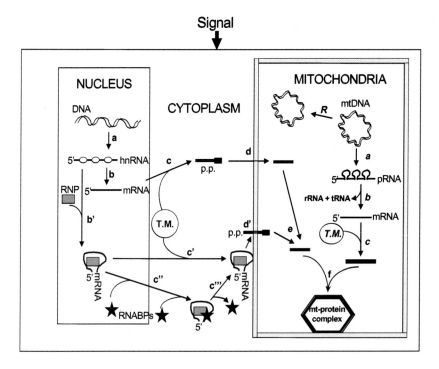

*Figure 1. Cell processes controlling the biogenesis of mitochondria. In the nucleus, transcription (**a**), splicing and maturation (**b**) of the mRNAs. Some mRNAs are assembled in the nucleus into large ribonucleoprotein particles (RNPs) (**b'**). Export of the mRNAs from the nucleus and cytoplasmic translation by mitochondrial-bound (**c'**) or unbound (**c**) translational machinery (T.M.). Some of the mRNAs interact with regulatory proteins (RNABPs) that define the metabolic fate of the transcript (**c''** and **c'''**). Import of the synthesized precursor proteins (p.p.) (**d** and **d'**) into the organelle, where the precursor proteins are processed and matured (**e**). In mitochondria the replication (**R**) and transcription (**a**) of mtDNA occurs. The mitochondrial translational machinery (T.M.) synthesizes mt-encoded polypeptides (**c**), which are further assembled (**f**) with nuclear-encoded proteins into mt-protein complexes. Mutations affecting the activity of molecular components involved in these processes promote the development of some neurodegenerative disorders.*

3. THE ENERGY TRANSDUCTION APPARATUS OF MITOCHONDRIA ALSO GENERATES CELL-DEATH SIGNALS

One of the main functions of mitochondria is the generation of cellular energy by oxidative phosphorylation (Figure 2). The enzymatic machinery required for the oxidation of pyruvate, ketone bodies and fatty acids, as well as for the tricarboxylic acid cycle, are all present in the matrix of the organelle (Figure 2). At this site the electrons of the oxidation reactions are collected on the coenzymes NAD^+ and FAD that serve as shuttles for the transport of the reducing equivalents to the respiratory chain complexes (Figure 2). The respiratory chain complexes are located in the inner mitochondrial membrane and have as overall function the oxidation of NADH and $FADH_2$ whilst the reducing equivalents are vectorially transported along the carriers of the respiratory chain (Complexes I to IV) to oxygen, the final electron acceptor (Figure 2). While electrons are being funnelled in the respiratory chain, the carriers (C-I, C-III, and C-IV) pump out protons from the matrix interior generating the proton gradient that is the electrochemical intermediate ($\Delta\mu_H^+$) in the synthesis of ATP (Figure 2). The establishment of such intermediate therefore requires an inner mitochondrial membrane with low proton conductance — in other words, with a high resistance to proton movement. It is the re-entry of protons to the matrix interior through the proton channel of the H^+-ATP synthase (Complex V) what promotes the conformational changes in this rotatory-engine of the inner membrane, eventually resulting in the net synthesis of ATP from ADP and inorganic phosphate (Figure 2).

However, electron transfer to molecular oxygen is not one hundred percent efficient (Figure 2). It has been estimated that at least 1% of the total rate of electron transport from NADH to oxygen is diverted to the production of superoxide ion, a reactive oxygen species (ROS) that is over-produced in apoptotic cells. The superoxide radical, a strong oxidant of short half-life, is the primary oxygen radical generated by mitochondria in the course of the electron flow down the respiratory chain. Two main respiratory chain complexes participate in the generation of the toxic superoxide radical: Complex I and Complex III (Figure 2). The superoxide radical can combine with nitric oxide to generate reactive-nitrogen species (RNS), such as peroxinitrite — another potent radical. An excess of ROS and RNS alters the REDOX state of the cell and could eventually lead to mitochondrial impairment and apoptosis. The cellular toxicity of the reactive species is due to the covalent modification and inactivation of key mitochondrial enzymes involved in energy transduction. Respiratory complexes I, II, and III, aconitase, and the adenine nucleotide translocator (ANT) are among the enzymes affected by ROS/RNS. The primary targets of the radicals are the thiol groups and iron-sulphur centres of the proteins. In addition, the free radicals produce lipid peroxidation and protein nitrosylation. The covalent modification of the proteins and membrane lipids is thought to be involved in the loss of the membrane potential established across the inner membrane of the organelle, resulting in a diminished production of ATP by oxidative phosphorylation (Figure 2).

To prevent the deleterious effects of the superoxide radical, the cells contain a cytosolic CuZn superoxide dismutase (SOD1) and a mitochondrial Mn superoxide dismutase (SOD2) that transform the superoxide radical into hydrogen peroxide (Figure 2), another toxic molecule that is eliminated from the cell in the form of water by the action of glutathione peroxidase (GPX in Figure 2). The ratio GSH/GSSG (reduced glutathione/oxidized glutathione) is an indicator of the REDOX state of the cell. It has been shown that cell antioxidant mechanisms are overwhelmed in acute oxidative stress insults, such as reperfusion after brain ischaemia. GSH can protect cells from ROS-induced apoptosis. Survival signals transduced by Akt promote the nuclear translocation of NFκB and the transcription of the gene encoding mitochondrial SOD2, contributing to the prevention of oxidative damage and subsequent cell death.

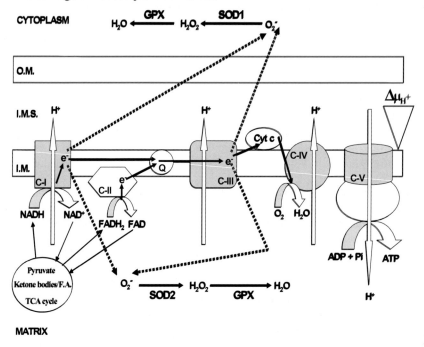

Figure 2. Mitochondrial oxidative phosphorylation and the production of ROS. The oxidation of metabolic substrates in the matrix produces NADH and $FADH_2$. The coenzymes feed the electrons (e^-) to the respiratory chain complexes (black arrows from C-I to C-IV) that are located in the inner mitochondrial membrane (I.M.). Molecular oxygen is reduced to water. In this electron-transfer, C-I, C-III and C-IV pump out protons from the matrix interior to the intermembrane space (I.M.S.) generating the proton gradient ($\Delta\mu_H+$) across the inner membrane. The proton gradient is the driving force used by H^+-ATP synthase (C-V) for the synthesis of ATP. However, at Complex I and III some of the electrons can combine with molecular oxygen to generate the superoxide radical (dashed lines). The toxic superoxide is transformed into hydrogen peroxide by the action of cytosolic (SOD1) and mitochondrial (SOD2) forms of superoxide dismutase. Glutathione peroxidase (GPX) reduces hydrogen peroxide to water. Subunits of grey shaded inner membrane complexes are encoded in both the nuclear and mitochondrial genomes. Q, ubiquinone; Cyt c, cytochrome c; O.M., outer membrane.

4. THE APOPTOTIC MACHINERY OF MITOCHONDRIA

Mitochondria are central cellular regulators of apoptosis, a physiological process conserved from worms to mammals, and which is necessary for the normal development of the organism. The molecular pathways that regulate apoptosis are altered in various human pathologies. In the specific case of neurodegenerative disorders, the apoptotic pathway has become another target of therapies to limit tissue damage and cell death due to the limited ability of the central nervous system to regenerate after various types of insult.

The activation of apoptosis is triggered by various types of apoptotic stimuli, such as DNA damage, ischaemia, deprivation of trophic factors, and chemical, metabolic and oxidative stresses. The program of apoptosis is executed by the activation of members of a family of cell proteases, the caspases, which in non-apoptotic cells exist as non-active zymogens or procaspases (Figure 3). Functionally, caspases can be subdivided into upstream (or initiator) caspases — which are cleaved and activated in response to the apoptotic stimuli — and downstream (or effector) caspases. Activation of the downstream caspases promotes the degradation of key cell components, and eventually results in the characteristic morphological features — nuclear membrane breakdown, DNA fragmentation, and shrinkage of the cell body — of apoptotic cell death. Although it is generally accepted that caspases are the major executioners of apoptosis, cytochrome c and other mitochondrial proteins, such as apoptosis-induced factor (AIF), endonuclease G, Smac/DIABLO and Omi/HtrA2, can induce apoptotic cell death (Figure 3). In the case of AIF and endonuclease G, they directly induce chromatin condensation and DNA fragmentation by caspase-independent mechanisms.

The mitochondrial-initiated, or intrinsic, pathway of caspase activation starts with the release of cytochrome c from the intermembrane space of mitochondria (Figure 3). Once in the cytosol, cytochome c binds the apoptotic protease-activating factor (Apaf-1), which in turn increases its affinity for dATP/ATP. The binding of nucleotide to the Apaf-1/cytochrome c complex promotes its oligomerization, resulting in the macromolecular complex known as the apoptosome (Figure 3). In the apoptosome, Apaf-1 is able to recruit procaspase-9 molecules (Figure 3) through its caspase recruitment domain (CARD). Procaspase-9, and other upstream caspases (1, 2, 4, 5, 11, and 12), contain a large N-terminal prodomain (containing a CARD domain) that is cleaved off when the caspases are activated. The recruitment of procaspase-9 molecules to the apoptosome promotes the auto-activation and subsequent release of active caspase-9 from the complex (Figure 3). Active caspase-9 processes and activates the effector caspases-3, -6, and -7 (Figure 3), which directly degrade key cell components. For instance, caspase-3 cleaves the inhibitor of caspase-activated DNAase (ICAD), releasing CAD, which fragments nuclear DNA.

The precise mechanism by which cytochrome c is released from the mitochondria remains unclear. However, several factors regulating this process have been identified. The main players belong to the protooncogen Bcl-2 family of proteins (Figure 3), which has been classified in three main subgroups: (i) the anti-apoptotic Bcl-2/Bcl-X_L subfamily, (ii) the pro-apoptotic Bax/Bak subfamily, and

(iii) the so-called BH3-only subfamily, also with pro-apoptotic functions. The latter group includes the cytosolic Bad, Bid, Bim and Bik proteins which upon activation translocate to the mitochondria where they interact with Bcl-2 proteins and further promote the release of apoptogenic factors such as cytochrome c, AIF and Endo G.

Pro-apoptotic Bcl-2 members are activated by different and independent pathways. For example, in response to the stimulation of the cells with TNF-α (extrinsic pathway) caspase-8 activation leads to the cleavage of Bid (Figure 3), generating a truncated version of Bid (tBid), the active form that targets mitochondria where it induces the release of apoptogenic factors (Figure 3). The phosphorylation of Bad by Akt in the presence of survival signals maintains Bad sequestered in the cytoplasm, thereby preventing apoptosis. Deprivation of trophic factors, or the alteration in cellular calcium homeostasis, results in Bad dephosphorylation and translocation to the mitochondria, promoting apoptosis (Figure 3).

Mitochondrial Bax/Bak proteins have been shown to form pores large enough to allow the release of apoptogenic factors as a consequence of tBid-induced Bax/Bak oligomerization (Figure 3). Moreover, it has been proposed that Bax/Bak recruits the most-abundant mitochondrial channels (ANT and VDAC) to form a pore big enough to enable the exit of cytochrome c from the mitochondria (Figure 3). In this situation, Bcl-2 anti-apoptotic proteins would block the oligomerization of Bak/Bax by directly binding to them, preventing the opening of the VDAC-Bax pore (Figure 3). Others have suggested the occurrence of a mitochondrial megapore, the permeability transition pore or PTP, connecting the outer and inner mitochondrial membrane, allowing the massive entry of water and solutes into the matrix (Figure 3). This influx would promote mitochondrial swelling and the rupture of the outer mitochondrial membrane resulting in the non-specific release of the content of the organelle's inter-membrane space into the cytoplasm.

In response to an apoptotic stimulus, the cell has additional mechanisms to regulate the activity of active caspases, and therefore to execute or not the cell-death program. This regulation is exerted by the cytosolic inhibitors of apoptosis (IAPs). IAPs bind and inhibit active caspases-9, -3, and -7 (Figure 3) and, in some cases, they may target the substrate caspase to degradation by the proteasome. This regulation is a safeguard mechanism aimed at preventing cell death after transient cytochrome c leakage events that could lead to a point of no return in cell fate. However, mitochondria also contribute to the fine regulation of IAP, since, in response to a sustained apoptotic signal, they release into the cytosol other apoptogenic factors, such as Smac/DIABLO and Omi/HtrA2, which bind IAPs, relieving the inhibition exerted on caspases by these proteins (Figure 3).

The central role of mitochondria in apoptosis is further illuminated by the connection established between the intrinsic and extrinsic pathways of cell death. The extrinsic cell-death pathway is regulated by initiator caspases upon the activation of cell surface death receptors (Figure 3). Stimulation of cells with TNF-α or FAS-ligand promotes the formation of a signalling complex named DISC (death-inducing signalling complex) that recruits adaptor proteins and eventually promotes the activation of upstream procaspase-8 and -10 (Figure 3). Activated caspase-8 and

-10 can cleave downstream caspase-3, -6 and -7 and, more importantly, process Bid generating its active truncated form (tBid), which targets the mitochondria (Figure 3). The existence of a cross-talk between the extrinsic and intrinsic pathways of apoptosis not only reinforces the central role of mitochondria in programmed cell death but also provides a mechanism for the amplification of the apoptotic response when death signalling is not sufficiently intense to trigger the mitochondrial caspase-independent pathway of cell death.

The p53 transcription factor is a key mediator in inducing apoptosis in response to DNA damage, hypoxia, ATP depletion, and loss of survival signals. The activity of p53 is regulated by interaction with cell proteins and by phosphorylation. Active p53 causes the induction of genes involved in DNA repair, cell-cycle arrest, senescence, and differentiation, and of some genes involved in the intrinsic (Apaf-1, Bax, BH3-only proteins) and extrinsic (death receptors FAS and DR5/killer and survival-signalling inhibitors such as PTEN) pathways of apoptosis. Interestingly, and in addition to the activity of p53 as a transcription factor, recent findings revealed the translocation of p53 to mitochondria in apoptotic cells. It has been suggested that p53 interacts with both pro- and anti-apoptotic Bcl-2 proteins in the mitochondria triggering cell death.

Finally, and in addition to the tightly regulated mechanisms in which mitochondrial proteins participate to control the activation of caspases, other proteins, such as AIF and endonuclease G, which are also released from the mitochondria, can induce apoptosis by a caspase-independent pathway (Figure 3). It appears that both caspase-dependent and independent pathways of cell death act in parallel and co-operatively.

Neurodegenerative disorders and brain insults are characterized by the selective loss of specific cell types and/or neurons. This suggests that a particular phenotype of the affected cells is responsible for their vulnerability and eventual fate. It is possible that the mitochondrial phenotype of the cells, which largely depends on the cell type and perhaps on its anatomical situation, contributes to defining the final outcome of the cell in response to an apoptotic stimulus. This would explain the selective degeneration of some neurons and not of others, having the same genetic background and/or being exposed to the same death-stressing signal.

5. WHEN MITOCHONDRIAL FUNCTION GOES WRONG: NEURODEGENERATIVE DISORDERS

Given the complexity of the cellular programs that control the biogenesis and function of mitochondria, and the dual genomic localization of their constituents (Figure 1), it is not surprising that the list of neurodegenerative disorders (ND) involving an impairment of the bio-energetic and/or apoptotic functions of mitochondria is very long. ND could be classified depending on the actual site of action of the defect: as due to (i) mtDNA mutations and (ii) nuclear DNA mutations in mitochondrial proteins. Within the latter group of disorders, mutations can affect proteins of the respiratory chain, proteins involved in the assembly of respiratory complexes, or proteins not-directly involved in oxidative phosphorylation. Finally, there is a heterogeneous group of disorders showing evidence of mitochondrial involvement in the progression of the disease — these disorders present an increase in apoptosis. However, the pathogenic mechanisms that underlie these diseases are not yet well established. In the following, we provide a brief description of the neurodegenerative disorders that fall into this classification.

5.1. ND due to mtDNA mutations

More 100 mtDNA mutations (Figure 1) have been associated with human diseases. The mutations described include deletions, duplications, and point mutations. Mutations on mtDNA can affect the overall translation capacity of the organelle or affect the sequences of protein-coding genes. Within the first category, the mutations include point mutations on tRNA genes (Figure 1) and generally have a multisystemic clinical presentation. Examples of this set of disorders include the mitochondrial encephalomyopathy with lactic acidosis and stroke-like episodes (MELAS) syndrome, the myoclonus epilepsy with ragged-red fibres (MERRF), and the neuropathy ataxia and retinitis pigmentosa (NARP)/maternally-inherited Leigh syndrome (MILS). Examples of disorders that involve mutations in protein-coding genes (Figure 1) are provided by the Leber hereditary optic neuropathy (LHON), essentially affecting subunits (ND1, ND4 and ND6) of Complex I (Figure 2), and the NARP/MILS syndrome, which is associated with mutations in the ATPase-6 gene (Figure 2). As indicated previously, these disorders are maternally inherited and although they all impinge on the same metabolic pathway — the provision of cellular energy — they have very different clinical presentations. An explanation for this situation could be the heteroplasmic nature of the disease and the "threshold effect" that is expressed in some tissues or cells but not in others. The expression of the pathological phenotype most likely depends on the specific energy demand of that particular tissue and/or the possibility that pathogenic mutations accumulate to different extents in different areas of the brain.

5.2. ND due to mutations in nuclear genes encoding mitochondrial proteins

The neurological disorders associated with mutations in nuclear-encoded genes that affect mitochondrial function cover a vast array of biological functions, and have a Mendelian inheritance.

5.2.1. Mutations in proteins of the respiratory chain

There are reported cases of patients with early-onset progressive neurological disorder with lactic acidosis — Leigh syndrome — in which mutations were discovered in some of the thirty-six nuclear-encoded subunits of Complex I (Figure 2). A late-onset neurodegenerative disease has been associated with mutations in the flavoprotein subunit of Complex II (succinate-cytochrome c reductase) (Figure 2) of the respiratory chain. In addition, a variant of adult-onset Leigh syndrome with clinical features has been described, due to a deficiency of ubiquinone, a lipophilic

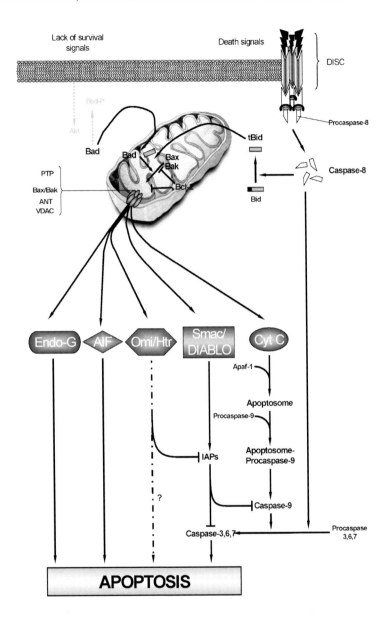

Figure 3. Mitochondria are the central transducers of death signals. Many types of apoptotic stimuli may lead to the translocation of BH3-only proteins (Bad, tBid) to the mitochondria. Once in the organelle they induce Bak/Bax oligomerization and pore formation with the participation of other recruited abundant membrane proteins of mitochondria. This process promotes the release of apoptogenic factors into the cytosol. Pore formation and the release of apoptogenic molecules can be inhibited by anti-apoptotic Bcl-2 proteins. Therefore, mitochondria are responsible for the triggering (or not) of apoptotic cell death based on the existing balance between anti- and pro-apoptotic proteins existing in the organelle. Released factors can induce apoptosis both directly and in a caspase-dependent fashion. For details, see the text.

molecule of the respiratory chain that transfers electrons from Complex I and II to Complex III (Figure 2). The genetic alteration associated with ubiquinone deficiency is currently unknown.

5.2.2. Mutations in proteins involved in the assembly of respiratory complexes

Autosomal recessive Complex IV deficiencies have an early age of onset, the clinical presentation of Leigh syndrome and a fatal outcome. No mutations in any of the 10 nuclear genes that encode the subunits of Complex IV (three additional subunits are encoded in mtDNA) (Figure 2) have been found in these patients. However, mutations responsible for the development of the various phenotypes associated with Complex IV deficiency are found in genes that participate in the assembly of the complex (see **f** in Figure 1).

In this regard, frame-shift, nonsense, and splice-site mutations in the SURF1 gene, encoding a hydrophobic protein of the inner mitochondrial membrane, account for most cases of Leigh syndrome cases. SURF1 is required for the early stages of the assembly of Complex IV. BCSL1 is a mitochondrial inner-membrane protein that functions as a chaperone in the assembly of Complex III (see **f** in Figure 1). Pathogenic mutations in BCSL1 have been shown in infantile cases of Complex III deficiency associated with encephalopathy.

5.2.3. Mutations in other mitochondrial proteins

There is a set of neurodegenerative disorders in which the mutated genes are not obviously involved in the energy transduction pathway of the organelle. For instance, mutations in the intron 1 of the frataxin gene, which encodes a mitochondrial iron-storage protein involved in the assembly of iron-sulphur clusters of mitochondrial proteins (see **e** in Figure 1), are responsible for Freidrich's ataxia, a progressive spinocerebellar ataxia. The mutation, an abnormal GAA-repeat expansion, causes a reduction of frataxin mRNA levels and the accumulation of iron in the tissues of the patient. A likely consequence of the accumulation of iron is that the production of the toxic hydroxyl radical (another ROS) could be increased, resulting in cellular oxidative damage.

Mutations in the paraplegin gene, encoding a mitochondrial ATP-dependent metalloprotease belonging to the AAA-ATPases (see **e** in Figure 1) — a protein family involved in a variety of cell activities — are associated with spastic paraplegia. Likewise, truncation or deletion of a mitochondrial protein of the protein-import machinery of the organelle (see **d,d'** in Figure 1) is responsible for the X-linked deafness-dystonia syndrome (Mohr-Tranebjerg syndrome). Mutations in the OPA-1 gene, which encodes a mitochondrial dynamin, presumably involved in the cellular distribution of mitochondria, are associated with a dominant optic atrophy.

Additionally, there is a set of mitochondrial disorders in which multiple deletions of mtDNA resulting from pathogenic mutations in nuclear genes that exert their function within the organelle, are observed in the patient. Autosomal dominant progressive external ophthalmoplegia is an example. The mutations described to date affect any of the three following genes: the muscle isoform of the adenine nucleotide translocase, a putative mtDNA helicase, and the catalytic subunit of mtDNA polymerase γ.

5.3. ND with evidence of the involvement of apoptosis

Neurodegenerative disorders may also arise when excessive death of a certain neuron population occurs. In such cases, cell death could be the result of (i) mutations in cell proteins that in one way or another participate in defining mitochondrial cell activity, (ii) mechanisms that are at present poorly characterized, or (iii) traumatic brain injury. The involvement of apoptosis in the development of neurodegenerative disorders is today widely accepted. Various signals can activate mitochondrial-geared cell death, such as the lack of neurotrophic growth factors, glutamate-triggered excitotoxicity, oxidative stress by oxygen free-radicals, and lack of energy substrates. In the following, we summarize a set of these disorders.

5.3.1. Huntington's disease

In Huntington's disease, striatal medium spiny neurons selectively degenerate. The clinical presentation of the disease is characterized by ataxia, chorea, and dementia, with onset in early adulthood. Evidence that mitochondria are involved in the development of the disease comes from the observation of defects in Complex II, III, and IV of the respiratory chain (Figure 2) in putamen and caudate and an impaired energy metabolism in the brain of the patient. The mutations associated with the disease affect the huntingtin gene, a protein unrelated to mitochondria. Expression of mutant huntingtin in the brain of rats, using viral vectors, results in neuronal cell death (Figure 3), whereas expression of the wild-type protein protects from mitochondrial-triggered apoptosis. Therefore, it has been suggested that the involvement of mitochondria in the progression of the disease is likely to be a consequence of excessive cell death triggered by oxidative stress and excitotoxicity.

5.3.2. Parkinson's disease

In Parkinson's disease, dopaminergic neurons of the substantia nigra selectively degenerate. Clinical features of the disease include bradykinesia, rigidity, and tremor, with symptom onset between the fifth and seventh decade of age. A selective Complex I (Figure 2) deficiency has been reported in the substantia nigra in sporadic Parkinson's disease. The role of mitochondria in the pathogenesis of Parkinsonism became clear when it was observed that a Parkinsonian syndrome that develops in drug abusers was caused by the inhibition of Complex I activity by a metabolic product of the drug. This compound produces a selective loss of dopaminergic neurons as in idiopathic Parkinsonism. Both environmental toxins and genetic factors may sensitize dopaminergic neurons of the substantia nigra to induce selective apoptosis by oxygen radicals and/or energy deficits. Although the genetic background of Parkinson's disease is unknown, it has been observed that mutations in the α-synuclein gene, a protein found in intracytoplasmic inclusion bodies of the surviving neurons, are responsible for a small percentage of Parkinsonian cases. Interestingly, ectopic expression of the mutant protein induces apoptosis (Figure 3) in cells in culture.

5.3.3. Amyotrophic Lateral Sclerosis

Amyotrophic lateral sclerosis (ALS) patients suffer progressive paralysis as a result of the degeneration of motor neurons in the spinal cord. The majority of ALS cases are sporadic, and about ten percent are autosomal dominantly inherited (family ALS). Mutations in the cytosolic isoform of superoxide dismutase (SOD1) (Figure 2) have been found in familiar ALS. Expression of the mutated SOD1 gene in mice results in a spinal cord pathology similar to that of ALS patients. The mutations described do not affect SOD1 activity, but appear to provoke a gain of function in pro-apoptotic activity of the motor neuron. Evidence that mitochondrial function is involved in ALS comes from the observation of morphological abnormalities in mitochondria, as well as from deficits in Complex IV (Figure 2) activity in the motor neurons of ALS patients. Likewise, evidence of the activation of the mitochondrial pathway of apoptosis comes from findings that in ALS patients and SOD1 mutant mice, the pro-apoptotic Bax is increased in the mitochondria, while the levels of the antiapoptotic Bcl-2 are decreased in motoneurons. Overall, the findings suggest that the pathogenesis of the disease is the result of malfunction of the apoptotic pathway, induced by oxidative stress. It appears that oxidative damage renders the neurons more vulnerable to the over-activation of glutamate receptors and the subsequent cellular calcium overload. In fact, over-expression of the anti-apoptotic Bcl-2 in SOD1 mutant mice delays the degeneration of motor neurons.

5.3.4. Alzheimer's disease

Alzheimer's disease is the most common neurodegenerative disorder. It is characterized by progressive impairment of higher cortical functions (memory, cognition, judgement, orientation, and emotion) due to the degeneration and death of neurons in limbic structures of the cerebral cortex. Histopathologically, the affected neurons show aggregates of the hyperphosphorylated microtubule-associated Tau protein (neurofibrillary tangles) and extracellular β-amyloid plaques, formed by aggregates of the amyloid β-peptide. Mutations in the genes of the β-amyloid precursor protein (β-APP) and presenilin are associated with the family (early-onset) version of the disease. Mutations in the β-APP gene alter the proteolytic processing of the protein product, whereas mutations in the presenilin gene increase the production of the amyloid β-peptide. Certain polymorphisms in the ApoE gene have been associated with the sporadic form of Alzheimer's disease. Alterations in the energy metabolism of the brain, as well as a decreased activity of Complex IV in cortical areas of the brain of Alzheimer's patients suggest that mitochondrial function is involved in the disease. Furthermore, increased activity of caspases and of the expression levels of the pro-apoptotic genes have been found in neurons of Alzheimer's patients, suggesting that excessive apoptosis (Figure 3) underlies the progression of the disease. In fact, both the amyloid β-peptide and a carboxy-terminal peptide derived from β-APP as a result of the activity of caspases induced apoptosis of neurons. The β-amyloid increases the vulnerability of neurons to apoptotic cell death by oxidative and metabolic stresses by a mechanism that involves the perturbation of the correct functioning of the neuronal plasma membrane. Antioxidants, Ca^{2+} buffers and neurotrophic factors protect neurons from β-amyloid-induced cell death.

5.3.5. Stroke

Stroke is an acute neurodegenerative disorder in which mitochondria are also implicated. It results from a deficit in oxygen supply to specific brain regions after the occlusion of a blood vessel, or from trauma. In such regions, there is a combination of necrotic and apoptotic cell death. The damaged tissue presents two regions, depending on the severity of the ischaemic insult to cells: the ischaemic core — in which the cells die by necrosis — and the penumbra, where the degree of hypoxia is less because of some collateral blood flow. Apoptosis in the penumbra affects neurons, glia, and vascular cells, and is explained on the basis of three main mechanisms: excitotoxicity, oxidative stress, and apoptotic-like cell-death pathways. At the subcellular level, the main traits observed after ischaemia are ATP depletion, organelle dysfunction, and Ca^{2+}-induced cytotoxicity. In addition, Ca^{2+} and ADP induce ROS production in mitochondria (Figure 2), which largely overcomes the cell antioxidant systems. Therapies aimed at reducing tissue damage and neurological dysfunctions are effective when targeted at the cells of the penumbra.

REFERENCES

Bates G. (2003) Huntingtin aggregation and toxicity in Huntington's disease. *Lancet* 361, 1642-1644.
Blomgren K., Zhu C., Hallin U., Hagberg H. (2003) Mitochondria and ischemic reperfusion damage in the adult and in the developing brain. *Biochem. Biophys. Res. Commun.* 304, 551-559.
Cuezva J.M., Ostronoff L.K., Ricart J., López de Heredia M., Di Liegro C.M., Izquierdo J.M. (1997) Mitochondrial biogenesis in the liver during development and oncogenesis. *J. Bioenerg. Biomembr.* 29, 365-377.
DiMauro S., Schon E.A. (2001) Mitochondrial DNA mutations in human disease. *Am. J. Med. Genet.* 106, 18-26.
Fernández-Checa J.C. (2003) Redox regulation and signaling lipids in mitochondrial apoptosis. *Biochem. Biophys. Res. Commun.* 304, 471-479.
Friedlander R.M. (2003) Apoptosis and caspases in neurodegenerative diseases. *N. Engl. J. Med.* 348, 1365-1375.
Garesse R., Vallejo C.G. (2001) Animal mitochondrial biogenesis and function: a regulatory cross-talk between two genomes Gene. 263, 1-16.
Kroemer G., Reed J.C. (2000) Mitochondrial control of cell death. *Nat. Med.* 6, 513-519.
Lo E.H., Dalkara T., Moskowitz, M.A. (2003) Mechanisms, challenges and opportunities in stroke. *Nat. Rev. Neurosci.* 4, 399-415.
Marx J. (2001) Neuroscience. New leads on the 'how' of Alzheimer's. *Science* 293, 2192-2194.
Mattson M.P. (2000) Apoptosis in neurodegenerative disorders. *Nat. Rev. Mol. Cell. Biol.* 1, 120-129.
Orth M., Schapira A.H. (2001) Mitochondria and degenerative disorders. *Am. J. Med. Genet.* 106, 27-36.

Raha S., Robinson B.H. (2001) Mitochondria, oxygen free radicals, and apoptosis. *Am. J. Med. Genet.* 106, 62-70.

Suomalainen A., Kaukonen J. (2001) Diseases caused by nuclear genes affecting mtDNA stability. *Am. J. Med. Genet.* 106, 53-61.

Tatton W.G., Chalmers-Redman R., Brown D., Tatton N. (2003) Apoptosis in Parkinson's disease: signals for neuronal degradation. *Ann. Neurol.* 53, S61-70.

Van Gurp M., Festjens N., Van Loo G., Saelens X., Vandenabeele P. (2003) Mitochondrial intermembrane proteins in cell death. *Biochem. Biophys. Res. Commun.* 304, 487-497.

Vila M., Przedborski S. (2003) Targeting programmed cell death in neurodegenerative diseases. *Nat. Rev. Neurosci.* 4, 365-375.

Wang X. (2001) The expanding role of mitochondria in apoptosis *Genes Dev.* 15, 2922-2933.

Yuan J., Yankner B.A. (2000) Apoptosis in the nervous system. *Nature* 407, 802-809.

Zeviani M, Spinazzola A., Carelli V. (2003) Nuclear genes in mitochondrial disorders. *Curr. Opin. Genet. Dev.* 13, 262-270.

W. PASCHEN

Laboratory of Molecular Neurobiology, Max-Planck-Institute for Neurological Research, Cologne, Germany

8. THE ORGANELLES II: ENDOPLASMIC RETICULUM AND ITS OVERLOAD

Endoplasmic reticulum dysfunction in various brain disorders

Summary. The endoplasmic reticulum (ER) is a subcellular compartment playing a central role in calcium storage and calcium signalling. Furthermore, all newly synthesized membrane and secretory proteins are folded and processed in this subcellular compartment. These are strictly calcium-dependent processes that need for correct functioning a calcium activity in the range close to that of the extracellular space. Under conditions associated with ER dysfunction, unfolded proteins accumulate in the lumen of the ER. This is the warning signal for activation of highly conserved stress responses, including the unfolded protein response (UPR), necessary to restore normal functioning of the ER, and the ER-associated degradation (ERAD) to degrade unfolded proteins at the proteasome. In acute pathological states of the brain, such as stroke, neurotrauma, epileptic seizures, and in degenerative diseases ER functioning is impaired in multiple ways. These include disturbances of ER calcium homeostasis, impairment of ERAD and UPR, and insufficient proteasome functioning which triggers secondary ER dysfunction. Therapeutic interventions designed to suppress the pathological process culminating in neuronal cell death in acute and degenerative diseases of the brain could therefore focus on strategies aimed at improving the capability of neurons to withstand conditions associated with ER dysfunction.

1. INTRODUCTION

The possible mechanisms underlying the various pathological processes culminating in neuronal cell death in acute disorders of the brain such as ischemia, hypoxia, hypoglycemia or epileptic seizures have been extensively studied [1]. Neurons located in certain defined areas of the brain have been shown to be particularly sensitive to metabolic disturbances associated with these disorders, indicating that some common pathomechanisms may be at play here. At first glance, it would appear that energy failure elicits a common response. However, the pattern of disturbances of energy metabolism varies considerably among the acute disorders mentioned above. While severe depletion of high energy phosphates, a marked increase in lactate levels and low tissue pH are associated with ischemia, changes in energy charge are minimal in status epilepticus. In severe hypoglycemia, on the other hand, energy depletion, low lactate levels and a rise in tissue pH are found [1].

More than 20 years ago, a calcium hypothesis of neuronal cell death was put forward [2]. It was proposed that a marked rise in cytoplasmic calcium activity triggers events leading to neuronal cell death. This hypothesis has since been modified several times, and it is now believed that mitochondrial, as opposed to cytoplasmic, calcium overload is the main forerunner of neuronal cell injury. Furthermore, neuronal vulnerability to calcium is thought to be determined by influx pathways rather than by the overall cytoplasmic calcium load [3]. Finally, the endoplasmic reticulum calcium hypothesis, put forward most recently, proposes that depletion of ER calcium stores is one of the triggering events leading to neuronal cell death [4-6]. This hypothesis has been supported by observations indicating that ER functioning may be impaired in multiple ways: by a depletion of ER calcium stores and disruption of ER-resident processes driven by high calcium activity, by direct disturbance of ER resident processes or by a blocking of the ubiquitin/proteasomal pathway.

2. ER FUNCTIONING

The ER compartment plays a pivotal role in vital neuronal functions [7]. Besides calcium storage and signalling, the processing and folding of all newly synthesized membrane and secretory proteins takes place in the ER lumen [8]. The importance of these processes for normal cell functioning is indicated by the observation that their blocking is potentially lethal for cells.

2.1. ER calcium homeostasis

ER calcium activity is several orders of magnitude above cytoplasmic calcium activity, i.e. close to that of the extracellular space [9]. ER calcium homeostasis is controlled by two receptors, the ryanodine and the IP_3 receptors, which upon stimulation induce calcium release from ER stores, and by sarcoplasmic/endoplasmic reticulum Ca^{2+}-ATPase (SERCA), which pumps calcium ions against a steep concentration gradient back into the ER lumen. This latter process is energy-requiring and is necessarily blocked under conditions associated with impaired energy metabolism. After being released from ER stores into the cytoplasm, given a sufficient energy supply, calcium ions are taken up by mitochondria and transported by membrane calcium pumps out of the cytoplasm into the extracellular space. At this stage, cytoplasmic calcium levels are insufficient to replenish ER calcium stores. An inward calcium current is therefore activated (capacitative calcium current, calcium release-activated calcium current (I_{crac})) that provides enough calcium ions to refill ER calcium stores to physiological levels. IP_3 and ryanodine receptor, SERCA and I_{crac} are the main components controlling ER calcium homeostasis and calcium signaling.

Various different factors are involved in the fine-tuning of ER calcium homeostasis and calcium signalling [10]. Calcium-binding proteins such as calreticulin and calmodulin, located in the ER lumen and in the cytoplasm respectively, are involved in ER calcium storage and termination of calcium signals triggered by ER calcium release. The pattern of ER calcium release triggered by cell activation is also modulated by a number of different factors. The IP_3 and ryanodine receptors are both influenced by cytoplasmic calcium activity in such a way that the opening of the respective calcium channel is dependent on cytoplasmic calcium activity. ER calcium release is also induced by arachidonic acid (AA) and cyclic ADP-ribose (cADPR). Levels of both compounds are increased after cell activation, suggesting that AA and cADPR are involved in modulating temporal profiles of activity-dependent calcium signalling. Furthermore, homer proteins, a family of factors linking metabotropic glutamate receptors at the cell membrane with IP_3 and ryanodine receptors located at the ER membrane, modulate receptor-activated calcium signalling. Two different homer forms have been identified, a long form harbouring a coiled-coil (CC) domain necessary for protein/protein interaction, and a short form (homer 1a) lacking the CC domain. Homer 1a exerts dominant-negative activity with respect to long form homer proteins.

2.2. ER protein processing

The ER plays a central role in protein folding and processing. All newly synthesized membrane and secretory proteins are folded and processed within the lumen of the ER. Membrane proteins are glycoproteins, their carbohydrate side-chain is synthesized and attached to the polypeptide backbone in the ER lumen (processing reaction). Protein folding and processing are both calcium-dependent and need a high calcium activity to be performed correctly. A high ER intraluminal calcium activity is therefore required to ensure appropriate folding and processing capacity. The folding and processing of newly synthesised membrane and secretory proteins is of fundamental importance for cells. The ER compartment is therefore particularly well developed in cells with a high rate of synthesis of secretory proteins, such as pancreatic beta cells secreting insulin. Blocking of these processes, either by depleting ER calcium stores or by interfering directly with folding and processing reactions constitutes a severe form of stress [8]. This results in cell death when affected cells are not able to withstand the pathological conditions associated with ER dysfunction by activating a stress response that is specifically triggered under conditions associated with ER dysfunction. When ER functions are impaired unfolded proteins accumulate in the ER lumen. This is the signal activating the stress response, hence termed the unfolded protein response (UPR). Another stress response activated under conditions associated with ER dysfunction is the ER-associated degradation of unfolded proteins (ERAD). This is required for degradation of these proteins through the ubiquitin/proteasomal pathway.

2.3. Induction of UPR

UPR is characterized by the activation of two ER-resident kinases, the PKR-like ER kinase (PERK) and IRE1 [11]. In the physiological state activation of PERK and IRE1 is blocked by the 78 kDa glucose-regulated protein (GRP78) bound to both kinases (Figure 1 + 2).

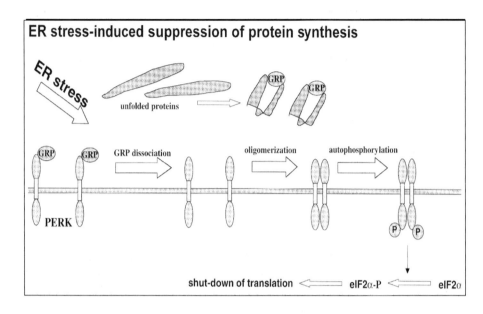

Figure 1: Scheme of ER stress-induced activation of PERK and shut-down of translation

GRP78 is an ER luminal chaperon required for the folding of unfolded or misfolded proteins. Under conditions of ER dysfunction, when unfolded proteins accumulate in the ER lumen, GRP78 dissociates from PERK and IRE1 to bind and assist refolding of the unfolded proteins. This results in oligomerization and activation of both kinases. The main function of UPR is to increase the folding capacity under conditions where unfolded proteins accumulate in the ER lumen. This is done by parallel induction of two signal transduction pathways initiated by activation of PERK and IRE1, both of which are activated by autophosphorylation. PERK-P phosphorylates the alpha subunit of the eukaryotic initiation factor 2 (eIF2α) resulting in a block of the initiation step of protein synthesis (Figure 1). The main function of this shut-down of translation is to suppress the new synthesis of proteins that cannot be correctly folded in the ER lumen. The activated form of IRE1 (IRE1-P) is turned into an active endonuclease that cuts out a sequence of 26 bases from the coding region of xbp1 mRNA. This leads to a shift in the open-reading frame of xbp1 mRNA.

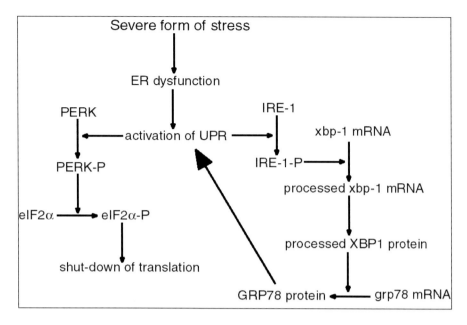

Figure 2: Scheme of UPR. Under conditions of ER dysfunction, IRE1 and PERK are both activated, resulting in shut-down of translation and induction of the expression of genes coding for ER stress proteins. Once GRP78 synthesis is upregulated, UPR is suppressed

While xbp1 mRNA is translated into an unstable XBP1 protein of 33 kDa, processed xbp1 mRNA is translated into a much more stable 54 kDa processed XBP1 protein (XBP1^{proc}). XBP1^{proc} functions as a transcription factor that specifically activates the expression of ER stress genes including grp78, grp94 and calreticulin. Grp78 mRNA is then translated into the respective protein. Once enough GRP78 protein has been synthesized to bind and refold the accumulated unfolded proteins and to bind and inactivate PERK and IRE1, UPR is shut-down and ER function is restored. While protein synthesis is blocked after activation of PERK, translation is required to turn processed xbp1 and grp78 mRNA into the respective proteins. Cells can therefore withstand conditions associated with ER dysfunction only when, in the course of UPR activation, the rate of protein synthesis is not held below the level required to fully activate the IRE1-mediated signal transduction pathway resulting in new synthesis of GRP78 protein.

2.4. Induction of ERAD

ER-associated degradation of proteins (ERAD) is activated under conditions associated with ER dysfunction on accumulation of unfolded proteins in the ER lumen [12]. Like UPR, ERAD is an ER stress response required to remove accumulations of unfolded proteins from the lumen of the ER under pathological conditions where ER functioning is impaired. In the course of ERAD, unfolded proteins are poly-ubiquitinated and degradated at the proteasome. The importance of this stress response for normal cellular functioning is indicated by the fact that blocking proteasomal functioning induces ER stress and activates UPR. This suggests that even in the physiological state some proteins processed in the ER lumen need to be degraded at the proteasome to avoid accumulation and disruption of ER functioning.

3. ER DYSFUNCTION IN PATHOLOGICAL STATES OF THE BRAIN

3.1. ER functioning is impaired in acute pathological states of the brain including ischemia, trauma and epileptic seizures

Energy is needed to maintain ER calcium homeostasis, because calcium released from ER stores during activation of the IP_3 and ryanodine receptors or by leakage currents must be pumped out of the cytoplasm back into the ER lumen in an energy-requiring process. ER calcium stores cannot therefore be refilled when energy producing metabolism is impaired. Other mechanisms besides energy failure through depletion of ER calcium stores are thought to be involved in ER dysfunction induced in various acute pathological states of the brain such as ischemia, trauma or epileptic seizures [5]. The ER calcium pump has been found to be sensitive to free radicals, toxic metabolic waste products produced in the course of these disorders. A major role may be played here by nitric oxide, excessive synthesis of which has been shown to contribute to pathological processes culminating in neuronal cell injury. Exposure of neuronal cell cultures to nitric oxide (NO) leads to inactivation of SERCA, depletion of ER calcium stores and suppression of protein synthesis, implying that NO triggers activation of UPR [13]. The depletion of ER calcium stores observed after transient cerebral ischemia in vulnerable neurons may also be caused, at least in part, by activation of nitric oxide synthase (NOS) and excessive synthesis of NO. After transient global cerebral ischemia ER calcium stores in vulnerable neurons of the hippocampal CA1 subfield were found to be depleted, and refilling of neuronal ER calcium stores was observed only in animals pretreated with neuroprotective doses of an NOS inhibitor [14]. This implies that ischemia-induced depletion of ER calcium stores is indeed triggered by NO. Inactivation of SERCA has been observed in cerebral ischemia and in an experimental model of epileptic seizures, indicating that it has a direct, energy-independent effect on SERCA activity [15, 16].

One way to identify whether ER functioning is disturbed in certain pathological states is to investigate whether UPR is activated. One acute pathological state of the brain that has been most intensively studied for signs of induction of UPR is transient cerebral ischemia. The stress response of cells to transient cerebral ischemia is in many respects similar to the response of neurons to isolated events disrupting ER functioning. In both cases, the initiation step of protein synthesis is blocked, as indicated by an observed phosphorylation of eIF2α and disaggregation of polyribosomes. The hypothesis has therefore been put forward that transient cerebral ischemia triggers ER dysfunction [4]. This view is supported by the observation that PERK is phosphorylated and thus activated after transient cerebral ischemia, implying ischemia leads to an activation of UPR [17]. Furthermore, ischemia-induced activation of eIF2α phosphorylation was not changed in transgenic animals where the eIF2α kinases PKR, HRI or GCN2 had been knocked-out. It can therefore be concluded that ischemia-induced suppression of protein synthesis is brought about through an activation of PERK, triggered by ER dysfunction.

The IRE1-mediated signal transduction pathway, leading to expression of genes coding for ER stress proteins, has also been found to be activated after transient cerebral ischemia. The first step in the signal transduction cascade triggered by activated IRE1 is the processing of xbp1 mRNA [18]. This is markedly stimulated after both transient global and focal cerebral ischemia [19]. After 1h of reperfusion processed xbp1 mRNA levels were found to be maximally increased indicating that this is a rapid response of cells to a transient interruption of blood supply. Since energy is required to maintain ER calcium homeostasis, one would expect ER calcium stores to be depleted and ER functions to be disrupted already during ischemia. To trigger UPR, ATP is needed for phosphorylation of PERK and IRE1 to their activated forms. During ischemia, however, cells are depleted of high energy phosphates so that phosphorylation of PERK and IRE1 may be hindered by low ATP levels. A marked increase in levels of phosphorylated eIF2α is indeed found during early reperfusion and not during ischemia, implying that ischemia-induced energy depletion hinders cells from activation of UPR [20]. Whether processing of xbp1 mRNA is induced already during ischemia remains to be established.

Protein synthesis is required to translate processed xbp1 into processed XBP1 protein. This step is necessarily blocked in cells in which shut-down of translation by activated PERK is so severe that it prevents translation of processed xbp1 mRNA. As expected, processed XBP1 protein levels are not raised in the early stages of reperfusion after transient cerebral ischemia, even though processing of xbp1 mRNA is maximally activated at that time [19]. This indicates that the shutdown of translation does indeed limit activation of the signal transduction pathway triggered by IRE1-P. Further evidence that the suppression of protein synthesis mediated by activated PERK leads to an uncoupling of ER dysfunction and IRE1-P activation of the signal transduction pathway comes from the observation that mRNA levels of the ER stress genes grp78, grp94 and gadd153 are only slightly increased, if at all, after ischemia, despite a massive rise in processed xbp1 mRNA levels [19]. Insufficient synthesis of new GRP78 protein may indeed be seen as the main obstacle to the recovery of vulnerable neurons after transient ischemia, as it hinders the recovery of cells from ischemia-induced ER dysfunction. A marked rise in processed xbp1 mRNA levels in the presence of almost unchanged grp78/grp94 mRNA levels is also found after traumatic brain injury, suggesting common underlying mechanisms involved in ischemia- and trauma-induced cell death (unpublished observations).

3.2. ER dysfunction in degenerative diseases of the brain

A common feature of neurodegenerative disorders is the accumulation of unfolded proteins that are poly-ubiquitinylated and form potentially toxic protein aggregates. Another striking feature is the fact that many genes, mutations of which result in the development of neurodegenerative disorders, code for ER-resident proteins or proteins that are invloved in the ubiquitin/proteasomal pathway. This suggests that dysfunctioning of the ER and/or the ubiquiting/proteasomal pathway is a central features of neurodegenerative diseases [5, 21].

Many observations pointing to a key role of ER dysfunction in degenerative disorders of the brain are based on experimental studies performed on various cell culture systems. In these studies cells are transfected with genes carrying those mutations identified in familial forms of these diseases. For example, cells transfected with mutant presenilin genes were taken as an in-vitro model of Alzheimer's disease. Mutant presenilins were found to exaggerate stimulation-induced ER calcium release, to impair capacitative calcium influx required to restore ER calcium levels after activation-induced calcium release, and to increase levels of the pro-apoptotic transcription factor gadd153, expression of which is specifically activated under conditions associated with activation of UPR. Furthermore, mutant presenilin expressing cells have been shown to exhibit markedly increased levels of ryanodine receptors. Collectively these findings suggest that mutant presenilins impair ER calcium homeostasis [22]. In mutant presenilin-transfected cells the extent of cell death has been found to be blocked or increased by agents that suppress or activate ER calcium release, respectively. GRP78 protein plays a role in ER calcium homeostasis, as indicated by the observation that cells in which GRP78 levels have been greatly dimished by using the antisense technique exhibit markedly increased ER calcium release on exposure to glutamate, while this response is suppressed in GRP78 over-expressing cells. All these observations imply that Alzheimer's disease is triggered, among other things, by impairment of ER calcium homeostasis, possibly caused by decreased levels of GRP78 protein. Reduced GRP78 protein levels have indeed been found in the brains of Alzheimer's disease patients [23]. The in-vivo situation was successfully mimicked in cell cultures transfected with the respective mutant presenilins: transfected cells exhibited reduced GRP78 levels and were particularly sensitive to conditions associated with ER dysfunction. Furthermore this pathological process was completely blocked in cells overexpressing grp78. As in cells expressing mutant presenilins, exaggerated glutamate-induced calcium release has been found in spinal cord motor neurons expressing amyotrophic-lateral-sclerosis (ALS)-linked Cu/Zn superoxide dismutase mutation. This suggests that impairment of ER calcium homeostasis again plays a role in motor neuron loss in ALS [24].

Signs of ER dysfunction have also been found in cultured cells transfected with toxic expansions of CAG trinucleotide repeats coding for polyglutamine [poly(Q)], a characteristic feature of many autosomal dominant neurodegenerative disorders including Huntington's disease. In these cells intracellular aggregates were formed and grp78 expression and caspase 12 were activated, both hallmarks of ER stress [25]. ER stress triggered by neurotoxic expansions of CAG repeats may be induced by proteasomal dysfunction. Protein aggregates have been shown to obstruct the ubiquitin/proteasomal pathway [26], and impairment of this pathway has been found to cause ER stress [27]. Extrapolating from these observations, impairment of the ubiquitin/proteasomal pathway and ER dysfunction might perhaps contribute to neurodegeneration or neuronal cell death in all those disorders of the brain characterized by the formation of protein aggregates.

Mutations in the parkin gene is the cause of autosomal recessive juvenile Parkinson's disease, and depletion of parkin protein has been observed in an experimental model of transient cerebral ischemia [27]. Parkin protein is an E3 ubiquitin ligase involved in the ubiquitin/proteasomal pathway. In a neuroblastoma cell line exposed to ER stress conditions, parkin expression has been found to be markedly upregulated at both the transcriptional and translational level, and upregulation of parkin protein has been shown to protect cells from injury induced by ER dysfunction [28]. In primary neuronal cell cultures exposed to conditions associated with ER dysfunction, parkin protein levels were however not upregulated [27], while in mixed neuronal/glial cultures an upregulation was induced in glial cells but not in neurons [29]. The inability of neurons to respond to ER dysfunction with an upregulation of the transcription and translation of parkin may therefore be one of the reasons why neurons are particularly sensitive to pathological states of the brain associated with ER dysfunction.

4. CONCLUDING REMARKS

Signs of ER dysfunction have been found in various acute pathological states of the brain, such as transient global or focal ischemia, trauma and epileptic seizures, and in degenerative diseases. These include depletion of ER calcium stores, inactivation of ER Ca^{2+}-ATPase, activation of PERK, processing of xbp1 mRNA, a decrease in GRP78 protein levels, and an accumulation of protein aggregates in the ER lumen. Each of the pathological states of the brain listed above is characterized by appearance of protein aggregates, suggesting dysfunction of the ER/ubiquitin/proteasomal pathway. Under such conditions vulnerable neurons may be trapped in a self-aggravating process (Figure 3): ER dysfunctioning and impairment of the ubiquitin-proteasomal pathway result in an accumulation of unfolded proteins that form potentially toxic protein aggregates. These protein aggregates impair proteasomal function, which in turn causes further ER stress. This pathogenic scenario is well documented from experimental studies performed on cell cultures. Whether the same processes are activated in the intact brain in-vivo has still to be fully elucidated.

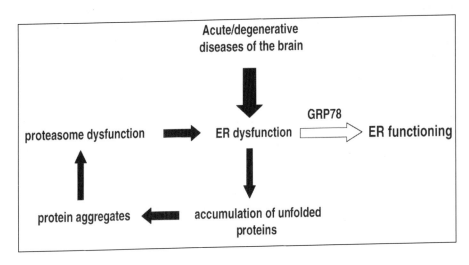

Figure 3: Scheme of the putative self-aggrevating process in which neurons are trapped in acute and degenerative diseases of the brain

Furthermore, the pathological processes underlying these disorders and culminating in neuronal cell death are complex ones. Various signal transduction pathways may be involved, including activation of the apoptotic pathway triggered by mitochondrial cytochrome c realease. Since, however, ER dysfunction has been found to induce secondary release of mitochondrial cytochrome c, impairment of ER functioning may also be the primary process in those pathological states of the brain where mitochondrial dysfunction is believed to play a pivotal role. If the ER compartment is indeed a primary target in acute disorders and in neurodegenerative diseases of the brain, therapeutic strategies designed to make the ER more resistant to stressful conditions or to restore ER functioning may serve to protect vulnerable neurons from irreversible damage.

REFERENCES

[1] Siesjö BK (1978) Brain Energy Metabolism. New York, John Wiley
[2] Siesjö BK (1981) Cell damage in the brain: a speculative synthesis. J Cereb Blood Flow Metabol 1:155-185
[3] Sattler R, Charlton MP, Hafner M, Tymianski M (1998) Distinct influx pathways, not calcium load, determine neuronal vulnerability to calcium neurotoxicity. J Neurochem 71:2349-2364
[4] Paschen W (1996) Disturbances in calcium homeostasis within the endoplasmic reticulum may contribute to the development of ischemic cell damage. Med Hypoth 47:283-288
[5] Paschen W, Doutheil J (1999) Disturbances of the functioning of endoplasmic reticulum: a key mechanism signaling from the lumen of the endoplasmic reticulum: coordination of gene transcriptional underlying neuronal cell injury? J Cereb Blood Flow Metabol 19:1-18
[6] Paschen W, Frandsen A (2001) Endoplasmic reticulum dysfunction - a common denominator for cell injury in acute and degenerative diseases of the brain? J Neurochem 79:719-725
[7] Berridge MJ (1998) Neuronal calcium signaling. Neuron 21:13-26
[8] Kaufman RJ (1999) Stress and translational controls. Genes Dev 13:1211-1233

[9] Hofer AM, Machen TE (1993) Technique for in situ measurement of calcium in intracellular inositol 1,4,5-triphosphate-sesitive stores using the fluorescent indicator mag-fura-2. Proc Natl Acad Sci USA 90:2598-2602
[10] Galione A, Churchill GC (2002) Interactions between calcium release pathways: multiple messengers and multiple stores. Cell Calcium 35:343-354
[11] Kaufman RJ (2002) Orchestrating the unfolded protein response in health and disease. J Clin Invest 110:1389-1398
[12] Hampton RY (2002) ER-asociated degradation in protein quality control and cellular regulation. Curr Opinion Cell Biol 14:476-482
[13] Doutheil J, Althausen S, Treiman M, Paschen W (2000) Effect of nitric oxide on endoplasmic reticulum calcium homeostasis, protein synthesis and energy metabolism. Cell Calcium 27:107-115
[14] Kohno K, Higuchi T, Ohta S, Kohno K, Kumon Y, Sakaki S (1997) Neuroprotective nitric oxide synthase inhibitor reduces intracellular calcium accumulation following transient global ischemia in the gerbil. Neurosci Lett 224:17-20
[15] Parsons JT, Churn SB, Delorenzo RJ (1997) Ischemia-induced inhibition of calcium uptake into rat brain microsomes mediated by Mg^{2+}/Ca^{2+}-ATPase. J Neurochem 68:1124-1134
[16] Parsons JT, Churn SB, Kochan LD, DeLorenzo RJ (2000) Pilocarbin-induced status epilepticus causes N-methyl-D-aspartate receptor-dependent inhibition of microsomal Mg^{2+}/Ca^{2+}-ATPase-mediated increases in Ca^{2+} uptake. J Neurochem 75:1209-1218
[17] Kumar R, Azam S, Sullivan JM, Owen C, Cavener DRC, Zhang P, Ron D, Harding HP, Chen J-J, Han A, White BC, Krause GS, DeGracia DJ (2001) Brain ischemia and reperfusion activates the eukaryotic initiation factor 2α kinase, PERK. J Neurochem 77:1418-1421
[18] Calfon M, Zeng H, Urano F, Till JH, Hubbart SR, Harding HP, Clark SG, Ron D (2002) IRE1 couples endoplasmic reticulum load to secretory capacity by processing XBP-1 mRNA. Nature 415:92-96
[19] Paschen W, Aufenberg C, Hotop S, Mengesdorf T (2003) Transient cerebral ischemia activates processing of xbp1 mRNA indicative of endoplasmic reticulum stress. J Cereb Blood Flow Metabol, in press
[20] Althausen S, Mengesdorf T, Mies G, Oláh L, Nairn AC, Proud CG, Paschen W (2001) Changes in phosphorylation of initiation factor eIF2α, elongation factor eEF-2 and p70 S6 kinase after transient focal cerebral ischemia in mice. J Neurochem 78:779-787
[21] Chung KKK, Dawson VL, Dawson TM (2001) The role of the ubiquitin-proteasomal pathway in Parkinson's disease and other neurodegenerative disorders. Trends Neurosci 24 (11, Suppl. 1):S7-S14
[22] Mattson MP (2002) Oxidative stress, perturbed calcium homeostasis and immune dysfunction in Alzheimer's disease. J Neurovirol 8:539-550
[23] Katayama T, Imaizumi K, Sato N, Miyoshi K, Kudo T, Hitomi J, Morihara T, Yoneda T, Gomi F, Mori Y, Nakano J, Takeda J, Tsuda T, Itoyama Y, Murayama O, Takashima A, St George-Hyslop P, Takeda M, Tohyama M (1999) Presenilin-1 mutations downregulate the signalling pathway of the unfolded protein response. Nature Cell Biol 1:479-485
[24] Roy J, Minotti S, Dong LC, Figlewicz DA, Durham HD (1998) Glutamate potentiates the tocixity of mutant Cu/Zn-superoxide dismutase in motor neurons by postsynaptic calcium-dependent mechanisms. J Neurosci 18:9673-9684
[25] Kourroku Y, Fujita E, Jimbo A, Kikuchi T, Yamagata T, Momoi, MY, Kominami E, Kuida K, Sakamaki K, Yonehara S, Momoi T (2002) Polyglutamine aggregates stimulate ER stress signals and caspase-12 activation. Hum Mol Gen 11:1505-1515
[26] Bence NF, Sampat RM, Kopito RR (2001) Impairment of the ubiquitin-proteasomal system by protein aggregation. Science 292:1552-1555
[27] Mengesdorf T, Jensen PH, Mies G, Aufenberg C, Paschen W (2002) Down-regulation of parkin protein in transient focal cerebral ischemia: A link between stroke and degenerative disease? Proc Natl Acad Sci USA 99:15042-15047
[28] Imai Y, Soda M, Takahashi R (2000) Parkin suppress unfolded protein stress-induced cell death through its E3 ubiquitin-ligase activity. J Biol Chem 275:35661-35664
[29] Ledesma MD, Galvan C, Hellias B, Dotti C, Jensen PH (2002) Astrocytic but not neuronal increased expression and redistribution of parkin during unfolded protein stress. J Neurochem 83:1431-1440

B. KAMINSKA

Laboratory of Transcription Regulation, Dept. Cellular Biochemistry, Nencki Institute of Experimental Biology, Warsaw, Poland

9. TRANSCRIPTIONAL DYSFUNCTIONS AS PATHOGENIC MECHANISM OF NEURODEGENERATIVE DISEASES

Summary. Critical steps regulating transcription of DNA into mature messenger RNA include transcription initiation, elongation, termination and pre-mRNA processing. Multiple proteins involved in RNA transcription and pre-mRNA processing can interact with general transcription factors and transcriptional activators, which associate with polymerase at gene promoters. Huntington's disease, dentatorubralpallidoluysian atrophy and five spinocerebellar ataxias (SCAs 1, 2, 3, 6, 7) are inherited neurodegenerative diseases caused by expansion of trinucleotide (CAG) repeats encoding polyglutamine (polyQ). Interactions between polyQ tracts and short polyglutamine tracts in ubiquitous transcription factors such as Sp-1, CREB, CBP can sequester basic transcription factors in the cytoplasm affecting transcription initiation. Alternative splicing of mRNA precursors can contribute to the generation of increased protein diversity and determine the subcellular location, molecular interactions, or function of proteins. Neuron-specific proteins involved in RNA splicing and metabolism are affected in several neurological disorders. Defects in RNA processing proteins, such as SMN protein, may lead to the Spinal Muscular Atrophy and aberrant splicing of glutamate receptors is linked to the pathology of Amyotrophic Lateral Sclerosis. Fragile X Mental Retardation Protein (FMRP) is mRNA binding protein and may regulate translation of dendrically localised mRNA influencing local protein synthesis and affecting synaptic structure and plasticity.

1. INTRODUCTION

Transcription of DNA into messenger RNA is one of the most highly regulated processes in the cell and depends on the molecular machinery consisting of more than 100 proteins. Genes are switched on and off through the tightly directed interactions of large numbers of proteins. These proteins interact with each other and with regulatory DNA elements that specify the activity of each gene in the genome. The RNA polymerase II (RNA pol II) must be instructed by regulatory proteins to bind to a specific region of DNA. The biosynthesis of mature mRNAs is coordinated within large multifunctional complexes called transcriptosomes. Critical steps regulating transcription of DNA into messenger RNA include transcription initiation, elongation, termination and pre-mRNA processing. Multiple proteins involved in pre-mRNA processing can interact with general transcription factors and transcriptional activators, which associate with polymerase at gene promoters. Integration of transcriptional and post-transcriptional activities, many of which once were considered to be functionally isolated within the cell, occurs within the nucleus.

2. TRINUCLEOTIDE REPEAT EXPANSION DISEASES AND TRANSCRIPTION FACTOR DEFICIENCY

To date, eight CAG-repeat diseases have been identified: Huntington's disease (HD), dentatorubralpallidoluysian atrophy (DRPLA) and five spinocerebellar ataxias (SCAs 1, 2, 3, 6, 7). These inherited neurodegenerative diseases are caused by expansion of trinucleotide (CAG) repeats encoding polyglutamine (polyQ). Several mechanisms have been postulated as a pathogenic process for neurodegeneration caused by the expanded polyglutamine tract. Although cytotoxicity of expanded polyQ stretches is implicated, the molecular mechanisms of neurodegeneration remain unclear. In addition to cellular toxicity, truncated and expanded polyglutamine tracts have been shown to form intranuclear inclusions (NI). Neuronal intranuclear inclusions that contain mutant huntingtin as well as other regulatory proteins have been observed in the nuclear compartment of HD brain cells. Given that this is the cellular compartment where transcription occurs, these aggregates may nonspecifically alter gene expression. Neuronal intranuclear inclusions may also contribute to the pleiotropic deregulation of gene expression.

2.1. Sp-1 transcription factor and Huntington disease

Huntington's disease is an inherited neurodegenerative disorder characterized by progressive motor and cognitive deficits, leading to death. A mutant form of the huntingtin protein has been identified as the cause of HD. Presence of multiple CAG trinucleotide repeats in the HD gene results in an expanded stretch of glutamine amino acids (>37 units) in mutant protein huntingtin. Full-length huntingtin is predominantly distributed in the cytoplasm, whereas mutant huntingtin with expanded polyglutamine tracts accumulates in the nucleus. Huntingtin fragments containing expanded polyglutamine tracts are toxic to cells.

Recent studies suggest that glutamine expansion may enable mutant huntingtin to deregulate normal transcription in neurons in the human brain. In fact, intranuclear huntingtin alters the expression of a number of genes, both in HD cells and in transgenic animals serving as HD model. Studies using DNA arrays to profile ~6000 striatal mRNAs in transgenic mice (R6/2), revealed diminished levels of mRNAs encoding components of the neurotransmitter, calcium and retinoid signaling pathways at both early and late symptomatic periods (6 and 12 weeks of age). Similar changes in gene expression were observed in another HD mouse model (N171-82Q). It proves that mutant huntingtin directly or indirectly reduces the expression of a set of genes involved in signaling pathways known to be critical to striatal neuron function.

The known regulatory sequences of these genes contain binding sites for the transcription factor Sp1, suggesting that huntingtin may interfere with Sp1-mediated transcription. Specificity protein 1 (Sp1) is the first of many sequence-specific transcriptional activators isolated from human cells. Extensive biochemical and molecular characterization of Sp1 revealed that it binds to GC-box DNA elements present in promoters targeting specific genes. Sp1 contains distinctive glutamine-rich activation domains, that selectively bind and target core components of the transcriptional machinery such as TFIID, a multiprotein complex composed of the TATA-box binding protein (TBP) and multiple TBP-associated factors (TAFIIs). Sp1-dependent transcription requires various TAF subunits of TFIID. Association of glutamine-rich proteins represents a major class of protein-protein interactions that enable transcription factors to regulate the expression of specific genes.

Growing evidence suggests that an early step in the development of HD may involve deregulation of specific transcriptional programs in the brain. A specific interaction between huntingtin and Sp1 in the brains of genetically engineered HD mice was observed. Mutant huntingtin in human HD brain cells has ability to associate with Sp1 and to disrupt a specific activator-co-activator interaction. By blocking the specific interaction of Sp1 with TAFII 130 in brain cells, Dunah and colleagues (2002) found that mutant huntingtin interferes with the normal patterns of Sp1-mediated gene expression.

Many pieces of evidence support links between the glutamine expansion in mutant huntingtin and a negative effect on Sp1-dependent transcription in the brain. First, the enhanced association of mutant huntingtin with Sp1 in extracts from the brains of asymptomatic HD individuals was observed. Second, the association of Sp1 with TAFII130 is reduced in HD brains compared with brains of healthy individuals. The enhanced association of mutant huntingtin with Sp1 also blocked the binding of Sp1 to promoter DNA. Such deleterious effects of mutant huntingtin disrupt Sp1-mediated transcription in HD brain cells. Mutant huntingtin decreases the expression of several Sp1-dependent neuronal genes, including the dopamine D2 receptor gene, whose expression is compromised in HD brains. Third, overexpression of both Sp1 and its normal target co-activator TAFII 130 can overcome inhibition of dopamine D2 receptor gene expression by mutant huntingtin. Neither alone was sufficient to restore normal transcription. Concomitant overproduction of Sp1 and TAFII130 reversed the cellular toxicity associated with the mutant huntingtin protein in brain cells.

2.2. CREB and CREB binding proteins in Huntington disease

CREB binding protein (CBP) is a co-activator for CREB-mediated transcription and contains an 18 (human) or 15 (mouse) glutamine residues. CREB-mediated gene transcription promotes cell survival, and CBP is a major mediator of survival signals in mature neurons. Control postmortem cortex stained with CBP does not contain inclusions and show diffusely nuclear staining. CBP was present in the HD nuclear inclusions in human cortex and sequestered from its normal nuclear location. Huntingtin with a normal glutamine repeat had no effect on CBP-mediated transcription, whereas huntingtin with an expanded repeat significantly inhibited transcription. CBP overexpression rescues cells from toxicity mediated by huntingtin or atrophin-1 with expanded polyglutamine repeats. Gene expression screens have shown that genes potentially regulated by CBP, such as enkephalin and Jun, are down-regulated in HD transgenic mice and HD postmortem brain tissue. Also BDNF (brain-derived neurotrophic factor), a protein implicated in neuronal survival whose expression is regulated by CREB, is down-regulated in HD patient tissue.

These findings suggest that huntingtin becomes a hyperactive co-repressor that disturbs the normal interactions between the transcriptional activators: Sp-1, TAFIID, CREB, or CBP. Such disruption of activators-mediated transcription may be one of the earliest deleterious consequences of accumulating mutant huntingtin and may lead to cell death. Alteration of CBP-regulated gene expression may mediate neuronal dysfunction and neuronal toxicity.

2.3. Disruption of interactions between basic transcription factors and activators as a common mechanism in diseases caused by glutamine expansions

Several genetic diseases are caused by glutamine expansions in various proteins. Dentatorubral-pallidoluysian atrophy is an autosomal dominant neurodegenerative disorder characterized by various combinations of cerebellar ataxia, choreoathetosis, myoclonus, epilepsy, dementia and psychiatric symptoms. In 1994 the gene for DRPLA has been described. Expression of truncated DRPLA proteins in cultured cell has been shown to result in aggregate body formation and apoptosis. Studies of transgenic mice for DRPLA (Q129 mice) showed the development of a severe neurological phenotype characterized by ataxia, myoclonus and seizures. Mutant DRPLA protein (also called atrophin-1) contains an expanded polyglutamine tract that targets TAFII130 and disrupts transcription by CREB. CREB, like Sp1, is a transcriptional activator protein known to engage TAFII130 as a co-activator partner.

Spinal and bulbar muscular atrophy (SBMA) is an X-linked neurodegenerative disease caused by the expansion of a trinucleotide CAG repeat in the first exon of the androgen receptor (AR) gene. Cell culture and transgenic mouse studies have implicated the nucleus as a site for pathogenesis, suggesting that a critical nuclear factor or processes are disrupted by the pQ expansion. McCampbell et al. (2000) demonstrated that CREB-binding protein (CBP), a transcriptional co-activator is incorporated into nuclear inclusions formed by polyglutamine-containing proteins in cultured cells, transgenic mice and tissue from patients with SBMA. The soluble levels of CBP were reduced in cells expressing expanded polyglutamine despite increased levels of CBP mRNA. Overexpression of CBP rescues cells from polyglutamine-mediated toxicity in neuronal cell culture. Moreover, CBP incorporation into nuclear inclusions was demonstrated in a cell culture model of spinocerebellar ataxia type 3 (SCA type 3) thus providing further evidence for the role of basic TF deficiencies in polyglutamine-mediated diseases. Interactions between expanded polyQ tracts and short polyglutamine tracts in ubiquitous transcription factors such as CBP can provide an explanation for the polyglutamine disease, because the effects on CBP celullar localization, sequestration, and transcription are specific only to the disease protein.

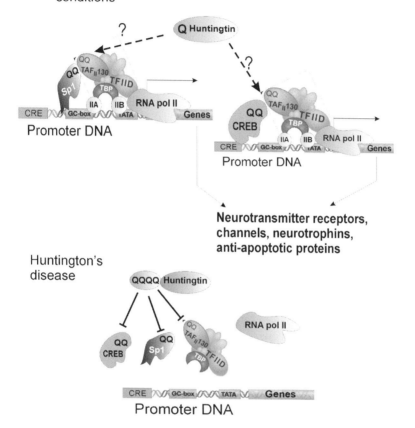

Fig. 1. Disruption of interactions between basic transcription factors and activators as a common mechanism in diseases caused by glutamine expansions. Under normal conditions, specific interactions between the glutamine-rich (QQ) regions of transcription factors such as Sp-1 or CREB and the TAFII130 subunit of TFIID is necessary for recruitment of the general transcriptional machinery. These interactions are required for recruitment of RNA polymerase II and transcription initiation. In Huntington's disease, Huntingtin (mutant protein with an expanded polyglutamine tracts) sequesters transcription factors blocking their interactions with the general transcriptional machinery. Without proper targeting by the general transcription factors, RNA polymerase II cannot properly locate promoter regions and genes cannot be transcribed.

3. MISREGULATED ALTERNATIVE SPLICING AND NEUROLOGIC DISORDERS

Alternative splicing of mRNA precursors can contribute to the generation of increased protein diversity and determine the subcellular location, molecular interactions, or function of proteins. Neuron-specific proteins involved in RNA splicing and metabolism are affected in several neurological disorders.

Table 1. Misregulation of alternative splicing and neurologic diseases

Disease	Protein target	Misregulation of alternative splicing
Frontotemporal dementia (FTDP-17)	Microtubule binding protein, Tau	Missense mutation affects alternative splicing in exon 10 tau
Amyotrophic lateral sclerosis	Glutamate transporter EAAT2	Intron retention and exon skipping of EAAT2 mRNA
Quaking viable mutant mouse; defects in myelination	Myelin associated glycoprotein	Reduction of alternatively spliced form of MAG in oligodendrocytes;
Neurodegeneration Paraneoplastic opsoclonus myoclonus ataxia (POMA)	Neurotransmitter receptors GlyRα2 and GABA γ2	Splicing defects in GlyRα2 and GABA γ2 mRNAs due to NOVA-1 deficiency
Cerebral cavernous malformation	KRIT1, a protein product of cerebral cavernous malformation 1 (CCM1) gene	Point mutations activate cryptic splice-donor sites, causing aberrant splicing
Familial neonatal convulsions and autosomal dominant nocturnal frontal lobe epilepsy	Neuronal nicotinic acetylcholine receptors alpha4 subunit	Exon skipping in mRNA alpha4 subunit neuronal nicotinic acetylcholine receptors

3.1. Deficiencies of neuronal splicing factors and neurodegeneration

Paraneoplastic opsoclonus myoclonus ataxia (POMA) is a neurologic disorder, in which patients harbor systemic tumors and develop immune responses against onconeural antigens that are expressed both by their tumors and by neurons. One POMA disease antigen, termed Nova-1, has been identified as a neuron-specific RNA-binding protein. Nova-1 expression is not only specific to neurons, but to a subset of neurons in the hypothalamus, brainstem, and spinal cord. Clinical reports showed progressive neurological deficits in some POMA patients. Recent studies indicate that Nova-1, a neuron-specific RNA binding protein regulates alternative splicing in neuronal cells and deficiency of Nova-1 leads to neuronal death in spinal and brainstem neurons. Nova-1 null mice show specific splicing defects in mRNAs for two inhibitory receptors: Glycine α2 and the $GABA_A$. Such defects in inhibitory receptor splicing could lead to alterations of receptor function. Altered glycine and/or GABA receptors can become toxic by disturbing a balance between the neuronal inhibition and excitation that leads to excitotoxic cell death.

Recent reports have implicated splicing defects as a cause of motor neuron death in the spinal cord. The Survival of Motor Neurons (SMN) protein, the product of the spinal muscular atrophy-determining gene, is part of a large macromolecular complex that functions in the ordered assembly of spliceosomal small nuclear ribonucleoproteins (snRNPs). The SMN protein is associated with snRNPs to form the machinery carrying out transcription and pre-mRNA splicing. The SMN protein ensures that the entire complex assembles only on correct RNA targets and prevents their promiscuous association with other RNAs. Thus, the SMN complex functions as a specificity factor essential for the efficient assembly of RNA processing proteins and likely protects cells from potentially deleterious, nonspecific binding of snRNPs proteins to RNAs. Reduced levels of SMN protein result in SMA and it have been suggested that the fine regulation of splicing is necessary for postnatal survival of motor neurons.

3.2. Splicing defects in the glutamate transporter EAAT2 are associated with ALS

Amyotrophic lateral sclerosis (ALS) is an adult-onset chronic neuromuscular disorder clinically characterized by muscle wasting, weakness, and spasticity, reflecting degeneration of cortical motor neurons and spinal/bulbar motor neurons. 95% of all cases are sporadic, the remaining 5% of cases show an autosomal dominant inheritance. About 15%–25% of familial ALS is due to mutations of $Cu2+/Zn+$ superoxide dismutase; these mutations are not present in the sporadic ALS population. Multiple mechanisms have been proposed as a cause of the disease, including excitotoxicity and oxidative injury. Many observations suggest that excitotoxicity can contribute to motor neuron degeneration in ALS: abnormalities in glutamate metabolism and high affinity glutamate transport; susceptibility of motor neurons to glutamate toxicity; and clinical efficacy of anti-glutamate agents.

The level of glutamate is regulated by sodium-dependent glutamate transport. EAAT1 and EAAT2 are glutamate transporters specific for astrocytes, while neuronal glutamate transport is mediated by EAAT3 and EAAT4. Impaired glutamate transport in ALS is associated with a loss of the EAAT2 protein. Up to 60%–70% of sporadic ALS patients have a 30%–95% loss of EAAT2 protein. This is not due to decrease of overall transcription of the EAAT2 mRNA or genomic mutations of EAAT2 in familial or sporadic ALS. It has been reported that splicing defects in the glutamate transporter EAAT2 are associated with ALS. Aberrant EAAT2 mRNA is present in sporadic ALS patients. About 65% of all postmortem specimens have the partial intron 7–retention or/and exon 9–skipping transcripts. The relative quantity of these abnormal EAAT2 mRNA correlates with the degree of

EAAT2 protein loss in ALS tissue. Study of EAAT2 exon–intron splice sites in sporadic and familial ALS did not reveal the presence of mutations, ruling out intragenic DNA mutations as a cause of the aberrant mRNAs. *In vitro* studies suggest that proteins translated from aberrant EAAT2 mRNAs may undergo rapid degradation resulting in loss of protein and activity.

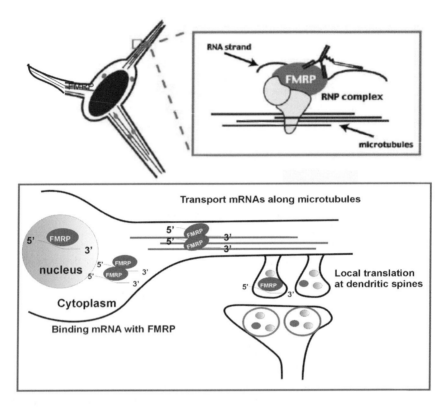

Fig. 2. The fragile X syndrome protein FMRP associates with RNA, regulates its transport and the translation of specific mRNAs at synapses. FMRP may bind specific mRNAs in the nucleus, transport them to the cytoplasm, where it may form a complex with other RNA binding proteins and mRNAs. Such complex attached to an anterograde motor is shuttled along microtubules to target dendritic spines or dendritic filopodia. These are sites where FMRP may play a role in synaptogenesis or act as a repressor of protein synthesis (at the spine). FMRP may regulate synthesis of proteins for spine structure and maintenance in an activity-dependent manner.

The presence of the aberrant EAAT2 mRNAs was restricted to motor cortex and spinal cord, the primary regions of neurodegeneration. Defects in mRNA processing may arise from oxidative stress, which damages DNA coding for RNA processing proteins or directly the proteins responsible for RNA processing. Damaged RNA processing proteins could lead to loss of RNA processing fidelity. It has been proposed that acquired damage to RNA processing proteins, such as SMN protein, may lead to aberrant splicing of EAAT2 in affected ALS brains

4. POSTTRANSCRIPTIONAL EDITING OF MRNA AND NEUROLOGIC DISEASES

Posttranscriptional editing of mRNA generates heterogeneity of isoforms and functional diversity. In the mammalian brain, RNA editing results in significant changes in the functional properties of receptors for the important neurotransmitters glutamate and serotonin. These changes result from site-specific deamination of single adenosines in the pre-messenger RNA encoding these receptors. The arginine residue in the GluR-B subunit (at 586 position) makes AMPA glutamate receptor channels impermeable to calcium. The codon for this arginine is introduced at the precursor messenger RNA (pre-mRNA) stage by site-selective adenosine editing of a glutamine codon. Trangenic mice bearing an editing-incompetent GluR-B allele prodeced unedited GluR-B subunits and expressed AMPA receptors with increased calcium permeability in neurons and interneurons. Such transgenic mice developed seizures and died postnatally.

Rare, hereditary epilepsies are linked to the mutations in genes encoding ion channels or neurotransmitter receptors. Besides mutations in particular genes, alterations of RNA editing seems to play a role in pathological processes, which contribute to epileptogenesis and seizures. It has been demonstrated that the editing efficiency (a ratio of the unedited to edited form) for the kainate receptor subunits GluR5 and GluR6 is significantly higher in temporal cortex of patients with pharmacoresistant temporal lobe epilepsies than in normal controls.

5. LOSS OF RNA BINDING PROTEINS AND DISTURBANCES IN DENDRITIC MRNA TRANSLATION

Dendritic local protein synthesis provides a mechanism for targeting proteins to activated synapses, which may control a synaptic structure and function. Polyribosomes localize within dendrites and are often concentrated beneath synapses and at dendritic spines.

The most common form of inherited mental retardation, Fragile X Mental Retardation syndrome is caused by an amplification of the trinucleotide CGG repeats in the 5' untranslated region of the FMR1 gene. In consequence, a FMR1 gene is not expressed which results in the loss of FMRP, a mRNA binding protein with multiple binding domains. FMRP binds specific mRNAs in the nucleus, forming a ribonucleoprotein complex, which is translocated to the cytoplasm, transported to dendrites, and locally translated in response to stimuli. Many mRNAs that bind FMRP, encode proteins important for neuronal development and synaptic function. FMRP was shown to be in a complex with MAP1B mRNAs, and dendritically localised mRNAs: CaMKII [alpha] and Arc mRNAs. In brain fractions of Fmr1 knockout mice, the levels of these FMRP-associated mRNAs were increased and increased protein levels were also observed in whole-brain and/or synaptosomal fractions, providing support for a role of FMRP in translational repression at synapses. The translation of specific mRNAs, which is upregulated in the absence of FMRP, may result in excessive local protein synthesis at the synapse. Recent studies using cortical synaptosomes demonstrated that FMRP itself is translated at the synapse. FMRP level can be regulated by synaptic stimulation in vivo and mGLUR stimulation in vitro which suggests a role of activity-driven local protein synthesis. Another function for FMRP stems from the fact that FMRP binds to cytoplasmic FMRP interacting protein (CYFIP) which binds to Rac-1, a member of the Ras family of GTPases. Involvement of Rac-1 in the maturation and maintenance of dendritic spines has been postulated. Altered morphology of the dendritic spines (spines are longer, thinner and more abundant) of FMR patients have been described. Similar immature spine morphology was observed in cortical organotypic cultures of *FMR1* knockout mice during the critical period of plasticity. Additionally, observations of axonal defects in Drosophila *fmr1* null mutants suggest a role for FMRP in the regulation of axonal outgrowth.

In Fragile X syndrome, trinucleotide expansion in the 5' untranslated region of the FMR1 gene leads to hypermethylation of the repeats and the adjacent CpG-rich promoter. Hypermethylation decreases expression of the gene in which the expanded repeat is located by the formation of transcriptionally silent chromatin. This modification may prevent the binding of the transcription factor alpha-Pal/NRF-1 and possibly other factors. In Friedreich's ataxia, the most common inherited ataxia, nucleotide (GAA TTC) expansion in the intron of the FRDA gene reduces its expression by interfering with transcription elongation.

REFERENCES

Antar LN, and Bassell GJ. (2003) Sunrise at the synapse: the FMRP mRNP shaping the synaptic interface. Neuron. 37, 555-558.

Bentley D. (1999) Coupling RNA polymerase II transcription with pre-mRNA processing. Curr Opin Cell Biol. 11, 347-351.

Dredge BK, Polydorides AD, Darnell RB. (2001) The splice of life: alternative splicing and neurological disease. Nat Rev Neurosci 2, 43-50.

Dunah AW, Jeong H, Griffin A, Kim YM, Standaert DG, Hersch SM, Mouradian MM, Young AB, Tanese N, Krainc D. (2002) Sp1 and TAFII130 transcriptional activity disrupted in early Huntington's disease. Science 296, 2238-2243.

Finkbeiner S, Tavazoie SF, Maloratsky A, Jacobs KM, Harris KM, Greenberg ME. (1997) CREB: a major mediator of neuronal neurotrophin responses. Neuron. 19,1031-1047.

Fischer,U., Liu,Q. and Dreyfuss,G. (1997) The SMN–SIP1 complex has an essential role in spliceosomal snRNP biogenesis. Cell 90, 1023–1029.

Jensen KB, Dredge BK, Stefani G, Zhong R, Buckanovich RJ, Okano HJ, Yang YY, Darnell RB. (2000) Nova-1 regulates neuron-specific alternative splicing and is essential for neuronal viability. Neuron 25, 359-371.

Kaytor MD, Orr HT. (2001) RNA targets of the fragile X protein. Cell 107, 555-557.

Kortenbruck G, Berger E, Speckmann EJ, Musshoff U. (2001) RNA editing at the Q/R site for the glutamate receptor subunits GLUR2, GLUR5, and GLUR6 in hippocampus and temporal cortex from epileptic patients. Neurobiol Dis. 8, 459-468.

Li SH, Cheng AL, Zhou H, Lam S, Rao M, Li H, Li XJ. (2002) Interaction of Huntington disease protein with transcriptional activator Sp1. Mol Cell Biol. 22, 1277-1287.

Lin, C.L.G., Bristol, L.A., Jin, L., Dykes-Hoberg, M., Crawford, T., Clawson, L., and Rothstein, J.D. (1998). Aberrant RNA processing in a neurodegenerative disease: the cause for absent EAAT2, a glutamate transporter, in amyotrophic lateral sclerosis. Neuron 20, 589–602.

Luthi-Carter, R., A. Strand, N. L. Peters, S. M. Solano, Z. R. Hollingsworth, A. S. Menon, A. S. Frey, B. S. Spektor, E. B. Penney, G. Schilling, C. A. Ross, D. R. Borchelt, S. J. Tapscott, A. B. Young, J. H. Cha, and J. M. Olson. (2000) Decreased expression of striatal signaling genes in a mouse model of Huntington's disease. Hum. Mol. Genet. 9, 1259-1271.

Martindale, D., A. Hackam, A. Wieczorek, L. Ellerby, C. Wellington, K. McCutcheon, R. Singaraja, P. Kazemi-Esfarjani, R. Devon, S. U. Kim, D. E. Bredesen, F. Tufaro, and M. R. Hayden. (1998) Length of huntingtin and its polyglutamine tract influences localization and frequency of intracellular aggregates. Nat. Genet. 18,150-154.

McCampbell A, Taylor JP, Taye AA, Robitschek J, Li M, Walcott J, Merry D, Chai Y, Paulson H, Sobue G, Fischbeck KH. (2000) CREB-binding protein sequestration by expanded polyglutamine. Hum Mol Genet. 9, 197-202.

Nucifora, F. C., Jr., M. Sasaki, M. F. Peters, H. Huang, J. K. Cooper, M. Yamada, H. Takahashi, S. Tsuji, J. Troncoso, V. L. Dawson, T. M. Dawson, and C. A. Ross. (2001) Interference by huntingtin and atrophin-1 with cbp-mediated transcription leading to cellular toxicity. Science 291, 2423-2428.

O'Donnell WT, Warren ST (2002) A decade of molecular studies of fragile X syndrome. Annu Rev Neurosci. 25, 315-338.

Shimohata T, Nakajima T, Yamada M, Uchida C, Onodera O, Naruse S, Kimura T, Koide R, Nozaki K, Sano Y, Ishiguro H, Sakoe K, Ooshima T, Sato A, Ikeuchi T, Oyake M, Sato T, Aoyagi Y, Hozumi I, Nagatsu T, Takiyama Y, Nishizawa M, Goto J, Kanazawa I, Davidson I, Tanese N, Takahashi H, Tsuji S. (2000) Expanded polyglutamine stretches interact with TAFII130, interfering with CREB-dependent transcription. Nat Genet. 26, 29-36.

Shyu,A.B. and Wilkinson,M.F. (2000) The double lives of shuttling mRNA binding proteins. Cell 102, 135–138.

Sprengel R, Higuchi M, Monyer H, Seeburg PH. (1999) Glutamate receptor channels: a possible link between RNA editing in the brain and epilepsy. Adv Neurol 79, 525-534.

Zalfa F, Giorgi M, Primerano B, Moro A, Di Penta A, Reis S, Oostra B, Bagni C. (2003) The fragile X syndrome protein FMRP associates with BC1 RNA and regulates the translation of specific mRNAs at synapses. Cell 112: 317-327.

M.E. GÖTZ[1] AND M. GERLACH[2]

[1]*Department of Pharmacology and Toxicology, University of Würzburg, Germany*

[2]*Clinical Neurochemistry, Department of Child and Youth Psychiatry and Psychotherapy, University of Würzburg, Germany*

10. FORMATION OF RADICALS

Summary. Cerebral formation of oxygen and nitrogen centered radicals including superoxide, hydroxyl radical, nitric oxide, and peroxynitrite is a physiological process originating from enzyme catalyzed redox reactions triggered by the turnover of endogenous and exogenous substrates. Due to the high reactivity of radicals, covalent modifications of lipids, proteins and DNA are likely to occur if radicals are not trapped by scavengers such as tocopherol, ascorbate of glutathione. Lipid peroxidation gives rise to cytotoxic aldehydes that have to be detoxified by glutathionylation. Oxidized proteins are metabolized by proteases and radical-mediated DNA-base modifications may be repaired by specific glycosylases. Incomplete repair of DNA and proteins may however, result in altered transcriptional response and protein aggregation. In chronic neurodegenerative diseases including Alzheimer's and Parkinson's disease increased levels of biomarkers of oxidative and nitrosative stress have been identified in the brain. Some of which such as hydroxyalkenals and isoprostanes may even gain diagnostic value for the determination of the degree of neurodegeneration. However, scientific efforts have to be continued to further elucidate the roles of radicals for disease onset and disease progression in neurodegeneration. The search for novel and selective biomarkers of radical-mediated brain damage might provide new diagnostic tools and perspectives for drug development to combat the progression of neurodegeneration.
Summary on relevant findings of free radical research in Alzheimer's disease. In AD neurodegeneration is pathologically first detected in distinct cortical areas including the entorhinal cortex and the hippocampus, leading to the destruction of long axon bearing cholinergic neurons. The majority of degenerating neurons is moderately myelinized, indicating a disturbance of lipid metabolism and selective vulnerability of these neurons. The accumulation of risk factors i.e. alterations in lipid transport by apolipoproteins, as well as amyloid precursor protein (APP) and presenilin mutations overadditively increase the risk for AD. In AD cortical and hippocampal brain regions as well as cerebrospinal fluid show increased susceptibility to damage by reactive species (reactive oxygen and reactive nitrogen species). Enzymatic defence and repair of damage induced by reactive species is upregulated already at early stages of the disease especially in patients with an apolipoprotein-Eε4 genotype. The deposition of amyloid plaques might lead to local formation of reactive species by enhancing transition metal catalyzed redox reactions and to the activation of astrocytes and microglial cells releasing chemokines, cytokines and superoxide. Vice versa plaque formation may be facilitated by reactive species and protein coss links with reactive aldehydes resulting from lipid peroxidation. The products of lipid peroxidation, hydroxyalkenals and isoprostanes, might further react with DNA, alter cellular signaling (hyperphosphorylated Tau protein) interfere with de novo protein synthesis, transform formally non pathogenic molecules to pathogenic ones (beta amyloid aggregation) e.g. by triggering the formation of advanced glycation end products, and by changing membrane lipid constituents, membrane functions, and fluidity. As a result of fatty acid peroxidation and deacylation of fatty acid peroxides from lipids by phospholipases, increases in levels of prostaglandins and isoprostanes occur in AD that are already considered as peripheral markers in cerebrospinal fluid. Thus, it appears very likely that reactive species play a crucial role for neural degeneration in AD, and therapeutic approaches involving the concept of antagonizing oxidative stress in AD are in progress.

Summary on relevant findings of free radical research in Parkinson's disease: Parkinson's disease (PD) is a progressive neurodegenerative disorder characterized by the inability to initiate, execute and control movement. Neuropathologically there is a striking loss of dopamine-producing neurons in the substantia nigra pars compacta, accompanied by depletion of dopamine in the striatum. As the disease progresses other neurotransmitters are lost to a minor degree including norepinephrine and serotonin. In PD decreased activities of glutathione peroxidases and catalase, as well as decreased glutathione levels, increased activities of superoxide dismutases, and elevated levels of non-ferritin-bound iron concomitant with a high turnover of catecholamines, may participate in the production of reactive species. All these factors may render substantia nigra cells in PD more susceptible to hereto undefined toxic noxae and may provoke lipid peroxidation as shown by increased levels of hydroxynonenal-modified proteins, consequently lead to increased levels of intracellular Ca^{2+}, activation of proteases, lipases, and endonucleases that ultimately execute cell death. Predominantly neuromelanin containing neurons are lost in PD. This suggests the involvement of neuromelanin in the pathogenetic process. It is hypothesized that the antioxidant capacity of neuromelanin is overcome by neuromelanin-bound iron. Interestingly, carbonyl content in the normal human substantia nigra is twofold increased as compared to other brain areas. This points toward an increased carbonyl stress that might be related to catecholamine oxidation and the formation of fatty acid peroxidation products including hydroxynonenal in the substantia nigra. Carbonyl compounds in the form of advanced glycation end products have been identified in Lewy bodies, a pathological hallmark in PD. Another pathological characteristic of Lewy bodies is the ubiquitously expressed presynaptic protein α-synuclein. It has been reasoned that mutations in the α-synuclein gene in autosomal dominant forms of PD might impair storage of presynaptic dopamine making the neuromelanin containing neurons more vulnerable to catechol toxicity. The ocurrence of continuous oxidative stress leads to the inhibition of the proteasome by oxidatively modified proteins. This may further trigger the formation of neurotoxic protein aggregates and extent neuronal damage beyond catecholaminergic brain areas.

1. INTRODUCTION

This is to focus on the oxidative stress hypothesis of neurodegeneration, as a very likely pathogenetic factor of cell death in chronic neurodegenerative diseases. Concerning progressive chronic diseases, such as Parkinson's and Alzheimer's disease, a gradual impairment of cellular defence mechanisms might lead to cell damage because of oxidative stress, a situation mediated by excess formation of toxic compounds, so called "reactive oxygen species (ROS)". This point of view brings into consideration the possibility that the pathogenetic process of neurodegeneration might be fueled by a cellular dysbalance of reactive intermediates and their detoxification either triggered by exogenous compounds, or initiated by genetic defects. It is not yet established whether oxidative stress is a major cause of cell death or simply a consequence of an unknown pathogenetic factor. However, it is generally accepted that ROS are mediators of cell death in a wide variety of cytotoxic situations, in aging and diseases including chronic progressive neurodegenerative disorders, since markers for ROS mediated damage on biomolecules have been unequivocally identified.

2. REACTIVE OXYGEN SPECIES (ROS)

The brain is metabolically one of the most active organs in the body. This is reflected by cerebral O_2 consumption in normal, concious, young men which amounts to 3,5 ml O_2 per 100 g brain per minute. Thus, two percent of total body weight accounts for 20 percent of the resting total body O_2 consumption. Nearly all O_2 is utilized for the oxidation of carbohydrates and results in an estimated steady state turnover of approximately 4×10^{21} molecules of adenosine triphosphate (ATP) per min in the entire human brain. O_2 maintains brain function and is crucial for life. However, oxygen supplied at concentrations greater than those in normal air are highly toxic. High pressure O_2 can lead to convulsions which is attributed to an inhibition of the enzyme glutamate decarboxylase by O_2 or reactive oxygen species (ROS). Even normal O_2 consumption could lead to toxic cellular reactions mediated by oxidative stress.

"Oxidative stress" is an expression used for a process which implicates reactions of O_2 or derived substances with biomolecules. If a reaction is thermodynamically feasible, its reaction rate depends primarily on the concentrations of the reacting partners. Thus, to evaluate effects of ROS on biomolecules, their concentrations and sites of production have to be considered.

Groundstate O_2 is in the triplet or diradical electronic configuration, having two unpaired electrons each located in a different pi* antibonding orbital. These two electrons have the same spin quantum number ("parallel spins") in contrast to singlet O_2, which has antiparallel spins. A prerequisite for exergonic reactions is the rule that reacting electrons in an energetic groundstate have to have antiparallel spins. Thus, in order to achieve spin conversion, groundstate O_2 must react in a two step, energy-dependent process. This is the reason for the slow reactivity of groundstate O_2 when this energy is not provided by enzymes or light. Among all oxygen species superoxide $(O_2)^{*-}$, hydroxyl radical $(HO)^*$, nitric oxide (NO), peroxynitrite (ONOO) and H_2O_2 are supposed to be the most abundant ROS in biological systems (Fig. 1).

2.1. Superoxide $(O_2)^{*-}$

$(O_2)^{*-}$ is mainly produced in biological systems through one-electron reduction of triplet O_2 mediated by enzymes. In brain xanthine oxidase and aldehyde oxidase, located in nuclear membranes and cytoplasm can increase $(O_2)^{*-}$ production if activated.

As well, these two enzymes generate H_2O_2 and are dependent on pO_2 and pH. Xanthine oxidase can oxidize a variety of substrates including aldehydes, pteridines, purines and hypoxanthine. Xanthine oxidase can be converted from a dehydrogenase (non-$(O_2)^{*-}$-producing) to the oxidase form during tissue hypoxia by proteolysis catalyzed by calpains. The relative rates of $(O_2)^{*-}$ production by these enzymes may vary with the concentrations of the enzymes in various cell types and the availability of substrates and cofactors. $(O_2)^{*-}$ can behave as a free radical, a weak nucleophile, a one-electron oxidant or as a one-electron reductant. Free radicals usually are very likely to abstract hydrogen or to add to double bonds, however, as for $(O_2)^{*-}$, these reactions are slow. The most important reaction in terms of biological effects is the dismutation of $(O_2)^{*-}$, which proceeds rather slowly in physiological conditions but can be considerably favoured by superoxide dismutases (SOD). $(O_2)^{*-}$ can act as a reductant of peroxides in the presence of transition metals or of quinones. On the other hand $(O_2)^{*-}$ can act as a one-electron oxidant, oxidizing e.g. hydroquinones to semiquinone radicals, or oxidizing ascorbate or epinephrine with concomitant production of H_2O_2.

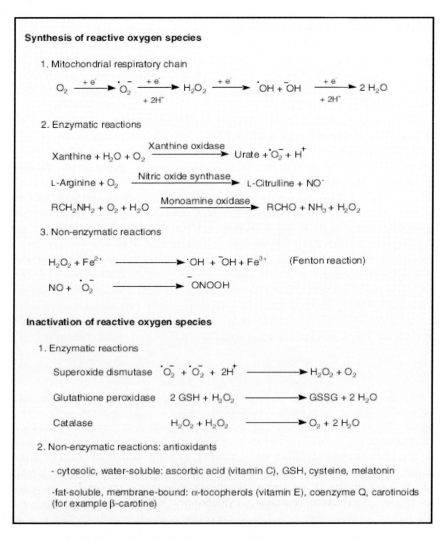

Figure 1: The cellular production and inactivation of reactive oxygen and reactive nitrogen species. Modified according to Gerlach et al., 2003b with permission.

2.1.1. Microsomal production of superoxide

The body is continually threatened by toxic compounds which are inhaled, dermally absorbed or ingested. Thus, enzymatic detoxification systems have been developed for elimination of toxic compunds. For these purposes, and for the catalysis of oxidations of fatty acids and steroids, microsomal membranes (including membranes from lysosomes, peroxisomes and endoplasmic reticulum) possess a large number of enzymes. Microsomes contain at least two electron transport systems, one being dependent on NADH and consisting of NADH cytochrome b_5 reductase and cytochrome b_5 (needed for fatty acid acyl coenzyme A desaturase system), and the other involving NADPH dependent cytochrome P-450 reductase (also referred to as NADPH cytochrome c reductase) and many isoenzymes. Cytochrome P-450 is involved in the oxidation of a wide range of substrates at the expense of O_2 (known as monooxygenation or mixed function oxidation) and requiring a reducing agent (normally NADPH). However, normally more NADPH is oxidized and more oxygen is consumed than needed. The excess of O_2 can lead to production of $(O_2)^{*-}$. It can be generated both from dissociation of the oxygenated complex of reduced cytochrome P-450 and from the autooxidation of cytochrome P-450 reductase containing FAD and FMN. Like other flavoproteins reaction of flavinsemiquinone with O_2 generates $(O_2)^{*-}$. Although liver is the major organ involved in the P-450 mediated metabolism, monooxygenases have also been detected in the brain. Immunocytochemical study of the rat brain P-450 reductase using an antibody to the rat liver enzyme revealed the presence of the enzyme in catecholaminergic neurons in the substantia nigra (SN), the locus coeruleus (LC) and the ventrolateral medullary region. Spectral quantification revealed that the level of cytochrome P-450 in brain microsomes is approximately 0.25 % of that found in the liver microsomes of control rats.

2.1.2. Superoxide production in mitochondria

In brain, the oxidation of NADH and $FADH_2$ produced in the Krebs-cycle from various substrates in mitochondria, is mediated by an electron transport chain consisting of flavoproteins [E-FMN], non-heme iron-sulfur proteins [Fe-S_x], iron- and copper-containing cytochromes (b_{562}, b_{566}, c_1, c, a, a_3) and coenzyme Q (ubiquinone, oxidized form of Q; ubiquinol, reduced form of Q) located in the inner mitochondrial membrane of mitochondria. The electron transfer chain can be resolved into four catalytically active complexes by fractionation with detergents and salt. These are complex I, the NADH ubiquinone reductase containing [E-FMN] and [Fe-S_x]; complex II, succinate-ubiquinone reductase; complex III, ubiquinone cytochrome c reductase containing cytochromes b_{562}, b_{566}, c_1, and iron-sulfur proteins; and complex IV, cytochrome c oxidase containing cytochromes a, a_3 and copper. Of the redox centers that have been implicated in electron transport, only Q and cytochrome c are not firmly associated with one of these complexes. Free radicals are formed during activity of the mitochondrial electron transfer chain with the rate of $(O_2)^{*-}$ formation being proportional to mitochondrial O_2 utilization and amounting to 2 % of total oxygen consumption. Considerable amounts of $(O_2)^{*-}$ are produced when the electron flow is inhibited. There are two separate sites of $(O_2)^{*-}$ production; the flavoprotein NADH dehydrogenase (located in complex I) and the ubiquinone cytochrome b segment. Whether $(O_2)^{*-}$ formation is coupled with autooxidation of ubisemiquinone or with autooxidation of cytochrome b_{566} is still unclear. In contrast, ubiquinone has been shown to act as a potent protectant against free radical damage to subcellular membranes *in vitro*. It is assumed that under "normal" conditions little of the $(O_2)^{*-}$ formed escapes the mitochondria due to the high levels of manganese dependent superoxide dismutase (MnSOD) within the matrix. On the other hand, NO may diffuse into mitochondria and react with $(O_2)^{*-}$ to form $ONOO^-$ as emphasized by Stewart and Heales 2003 (see sections 1.5. and 1.6.).

2.2. Hydrogen peroxide

Enzymes known in the liver to generate H_2O_2 are assumed to be present in human brain, too. These oxidases bear flavins (FMN, FAD) or pyridoxale phosphate (P_i) and metal ions as prosthetic groups. Pyridoxamine phosphate oxidase triggers H_2O_2 production.

The most prominent oxidase in brain tissue, however, is the flavin containing monoamine oxidase (MAO) which preferentially deaminates primary, secondary and tertiary monoamines and is located at the outer mitochondrial membrane. The A form is mainly responsible for the deamination of serotonin and norepinephrine and is pharmacologically defined by its sensitivity to inhibition by clorgyline. MAO-B, on the other hand, is known to deaminate predominantly non-polar amines, phenethylamine and methylhistamine. The B-form is characterized by its high sensitivity to inhibition by selegiline. Both enzymes metabolize DA and tyramine. In the human brain, MAO-B increases with increasing age due to an increase in MAO-B concentration. Moreover, this increase in MAO-B activity in brain is further accelerated in neurodegenerative disorders such as AD. Astrocytes have been shown to be rich in MAO-B activity. Astrocytosis has been demonstrated in senile brains and, interestingly, MAO-B activity was detected in astrocytes of senile plaques. The less consistant changes in MAO-A activities may reflect neuronal loss, but this is still uncertain.

Moreover, a great source of H_2O_2 production in the intact cell appears to be generated by the autooxidation of chemically reactive compounds during reductive processes associated with the mitochondrial and microsomal electron transport systems and during the action of SOD. H_2O_2 or $(O_2)^{*-}$ may diffuse some distance from their sites of production. Consequently, radical generation by subcellular compartments may be a threat for the whole cell.

2.3. Hydroxyl radical and Fenton chemistry

If H_2O_2 is not detoxified by catalase (CAT) or peroxidases (Px), one electron reduction results in the formation of the hydroxyl radical $(HO)^*$. This species is assumed to be the most toxic reactive oxygen radical, with an approximate intracellular half-life of 10^{-9} seconds so that reactions with biomolecules become diffusion controlled. In a variety of biological phenomena including aging, cancer, diabetes, phagocytosis and cataractogenesis, ischemia-reperfusion injury, quinone toxicity, and radiation injury the (OH)* is assumed to contribute to or to cause toxic events. In the brain strong water-soluble electron donors such as NADPH, catechin, hydroquinone, ascorbic acid or glutathione (L-γ-glutamyl-L-cysteinyl-glycine; GSH) can promote formation of (HO)* from H_2O_2 in the presence of Cu^+ or some iron complexes (e.g. Fe^{2+}-adenosine diphosphate complexes). *In vivo* formation of (HO)* is determined by measurement of hydroxylated salicylic acid following microdialysis. Involvement of oxygen radicals will often be detectable by means of electron spin resonance spectroscopy using spin traps like 5,5-dimethyl-1-pyrroline-N-oxide if care is taken of possible pitfalls. This method has been successfully used to determine that mainly low molecular weight complexes of iron catalyze formation of substantial amounts of (HO)*. Iron can promote peroxidation of biological macromolecules due to its reactions with ROS and thus is of high toxic potential for cells, if it is not kept in a toxicologically inactivated form bound to specific proteins. Only when iron is tightly bound to a chelator is its capacity for promoting the peroxidation of biomolecules minimal. Amongst synthetic chelators of iron, DTPA (Bis-(2-aminoethyl)-amine-N,N,N',N'-pentaacetic acid), desferrioxamine, o-phenanthroline and batho-phenanthroline are able to complex Fe^{3+} and thus slow down reduction of Fe^{3+} to Fe^{2+} by reductants including ascorbic acid or $(O_2)^{*-}$ *in vitro*, but EDTA (ethylenediaminetetraacetic acid) is ineffective. Desferrioxamine was originally developed for the treatment of iron overload disease because it binds Fe^{3+} rather selectively, but there are current efforts to create more specific iron chelators that pass the blood-brain barrier. As long as iron is correctly bound to ferritin, it seems that it does not initiate peroxidation of biomolecules. Ferritin, an iron storage protein holding 4500 moles of Fe^{3+} per mole of protein, is able to stimulate lipid peroxidation (LPO) by releasing Fe^{2+}, and ascorbate enhances the rate of ferritin stimulated LPO. Especially catechols very effectively release iron from ferritin.

Iron is an essential participant in many metabolic processes including DNA, RNA and protein synthesis, the formation of myelin and the development of the neuronal dendritic tree, and as a cofactor of many heme and non-heme enzymes. A deficiency in iron metabolism would therefore be expected to alter some or all of these processes, but excesssive accumulation of tissue iron may lead to oxidative stress via the formation of ROS. Cytotoxicity of iron was confirmed by studies *in vitro* to cultured neurons or by intranigral injection of iron into rats. So, in order to be toxic to cells, iron has to be present in brain in a more or less loosely bound form. This could mean that minimizing the amount of non-heme iron in biological systems is an important part of antioxidant defence. If iron or copper were causally involved in neurodegenerative diseases, transition metal distribution in brain should ideally reflect neuropathological changes and perhaps explain the region specific cell loss. Thus, many investigators were and are still concerned with the question of metal distribution in normal and pathological brain. Using histochemical techniques, the presence of iron in the brain was first detected at the end of the last century and subsequently was a subject of intense investigation. More recent detailed studies of the anatomical distribution of bulk iron in non-pathologic human *post mortem* brain confirmed and extended the findings of earlier studies, showing highest levels of stainable iron in the extrapyramidal system (globus pallidus, SN, red nucleus and myelinated fibres of the putamen). Moderate staining with Perl's technique was found in thalamus, cerebellar cortex and SN. Microscopically the non-heme iron appears to be predominantly located in glial cells as fine cytoplasmic granules. Neurons, in general, show low reactivity for iron, and this is difficult to discern due, often, to the higher reactivity of the surrounding neuropil. In the globus pallidus and SN pars reticulata however, neurons with highly stainable iron content are found with granular cytoplasmic iron reactivity similar to that seen in the local glial cells. Although there seem to be no apparent correlations of iron staining with known transmitter systems, the extrapyramidal system is favoured in iron uptake and storage. This could point towards involvement of iron in pathogenesis of disorders involving striatonigral degenerations, such as Hallervorden-Spatz disease, multi system atrophy, progressive supranuclear palsy, amyotrophic lateral sclerosis, Huntington's disease, and PD. Increased iron levels are likely to be involved in neurodegenerative diseases affecting basal ganglia. However, it is still questionable whether iron acts as primary initiator of nerve cell death in PD or represents a secondary response to another yet unknown pathological cause. Nevertheless, iron in the SN may exist in a form capable of contributing to the toxic processes occurring in PD by stimulating formation of ROS.

2.4. Calcium and oxidative stress

Cells overstimulated by excitotoxic inputs or suffering from decreased levels of ATP react by taking up sodium and water, resulting in swelling. Subsequently the cells are exposed to an increase of cytoplasmic free Ca^{2+} via channel-mediated influx, mobilization of Ca^{2+} from internal stores resulting from activation of second messengers and alterations in Ca^{2+} clearance due to depletion of energy reserves or of ATP resynthesis

In ischemia reperfusion injury blocking Ca^{2+} entry to cells may be of superior relevance as compared to other neuroprotective strategies since downstream many enzymes are activated including proteases, lipases, endonucleases, kinases and phosphatases. They contribute to cellular damage after excitotoxic receptor activation. Calpain inhibition attenuates neuronal death triggered by exogenous excitotoxins *in vitro* and following transient global ischemia in rodents. Potent inhibitors of calpains decrease infarct volume after focal ischemia even when administered six hours postocclusion. Cytosolic Ca^{2+} following glutamate receptor activation triggers the formation of multiple free radical species, which are deleterious to lipids, proteins and DNA. Antioxidants reduce neuronal death induced by exogenous excitotoxins in culture or by intrastriatal injections of excitotoxins *in vivo*. Oxidative stress follows through activation of constitutively expressed neuronal nitric oxide synthase (nNOS, see section 1.5.). In the presence of $(O_2)^{*-}$, NO forms $ONOO^-$ (see section 1.6.) that subsequently can degrade to nitrogen dioxide and (OH)*. Further enhancement of oxidative stress may originate from uncoupled mitochondria following Ca^{2+} overload. $(O_2)^{*-}$ can be produced as well during arachidonic acid metabolism catalyzed by cyclooxygenase, COX. As a matter of fact, COX-2 inhibitors decrease excitotoxicity in cortical neuronal cultures. Whenever ROS are involved in membrane damage, Ca^{2+} must be suspected as a participant.

2.5. Nitric oxide

NO is synthesized from the amino acid L-arginine catalyzed by NOS using molecular oxygen, NADPH and the cofactor tetrahydrobiopterin (H_4B). The first step is a two-electron oxidation of arginine to hydroxy-arginine. This reaction is accelerated by H_4B, requires calcium and calmodulin as activators and can be blocked by carbon monoxide. The second oxidation yields citrullin and NO using the same cofactors and activators. Three isoforms of NOS have been classified, the neuronal (nNOS), endothelial (eNOS) and inducible (iNOS) that are localized predominantly in neurons, endothelium and activated macrophages and microglial cells, respectively. Whilst nNOS and eNOS are constitutively expressed, iNOS requires *de novo* synthesis. iNOS does not need calcium for activation. Induction of iNOS is mediated by cytokines and lipopolysaccharides. Although NO originating from endothelial cells does not activate specific membrane receptors, it does have specific intracellular targets such as the important interaction with the heme moiety of the soluble guanylyl cyclase that leads to the activation of cyclic GMP production in smooth muscle cells of the vascular wall. Thus, besides many other actions NO plays an important role in the maintenance of the normal vascular tone. However, NO is acting as well as an atypical neurotransmitter. Atypic, because NO is not stored in vesicles and synthesized on demand. However, NO may be relevant in pathophysiological events such as ischemia-reperfusion injury, inflammation and neurodegenerative diseases. It is known that NO inhibits several enzymes, including the complexes I and II of the mitochondrial electron transport chain, the citric acid cycle enzyme aconitase, and the rate-limiting enzyme in DNA replication ribonucleotide reductase (Stewart and Heales, 2003). Due to the high affinity of NO to iron, ferritin-iron may be released and become available to promote LPO. The reaction of NO with thiols forms S-nitroso compounds that may have deleterious or protective effects on enzyme activities or on receptors. For example, inhibition of caspases and calpains, enzymes that bear critical cysteine residues for proteolytic activity, may be in some instances favorable for neurons, if inhibition of apoptosis is essential to maintain or regain normal physiological function. Nitrosylation of N-methyl-D-aspartate (NMDA) - receptors leads to decreased probability of channel opening. This may suppress excitotoxicity in ischemia reperfusion injury, but may not be advantageous if long term potentiation is needed to increase cognitive functions. Thus, NO has many different but important regulatory functions ranging from neuroprotective to neurodestructive ones. The neurodestructive features become dominant if a significant amount of $ONOO^-$ is formed.

2.6. Peroxynitrite

The diffusion controlled reaction of NO with $(O_2)^{*-}$ generates peroxynitrite (ONOO⁻), a short-lived compound that is cytotoxic. Because of the longer half-life of NO as compared with $(O_2)^{*-}$ it is conceivable that ONOO⁻ is formed and will further react with biomolecules within that cellular compartment where $(O_2)^{*-}$ is present e.g. mitochondria. L-arginine-depleted NOS may also contribute to the cytosolic formation of ONOO⁻ because under this condition nNOS is generating NO as well as $(O_2)^{*-}$ which consequently may react in the vicinity of NOS to form ONOO⁻. ONOO⁻ may diffuse only small distances because of its short half-life. It may pass membranes through anion channels, though. Reaction with carbon dioxide generates the nitroso-peroxocarboxalate that rapidly decompose to carbonate radical anion and nitrogen dioxide. These reaction products can promote secondary oxidations. ONOO⁻ may become protonated and may consequently isomerize to nitrate or decomposes to nitrogen dioxide and $(OH)^*$. Targets for reaction with ONOO⁻ are sulfhydryls, transition metals and carbon dioxide. Nitration of aromatic residues such as tyrosine generates 3-nitrotyrosine, a compound that is under focus for use as a peripheral marker for nitrosative stress. Neurotoxicants such as methamphetamine and MPTP may exert neurotoxicity through ONOO⁻ and pathological inclusions in neurodegenerative synucleinopathies are strongly nitrated. Thus, ONOO⁻ is proposed to contribute to the pathology in ischemia reperfusion and in neurodegenerative disorders. NADPH diaphorase - positive glial cells and NO radicals have been identified in the SN. Nitrotyrosine residues and elevated levels of nitrosylated proteins including α-synuclein have been detected and levels of nitrite are increased in the cerebrospinal fluid of parkinsonian patients. As well in AD, NOS activity and the number of NADPH diaphorase-positive neurons are increased. Nitrotyrosine residues have been detected in brain of AD patients but not in age-matched controls indicating the presence of ONOO⁻. Since NO can critically interfere with key mitochondrial enzymes and thiols, its role in neurodegeneration appears very likely (Stewart and Heales, 2003).

3. CONSEQUENCES OF ROS

3.1. Lipid peroxidation (LPO)

One hypothesis to explain mechanisms of cellular aging and chronic progressive cell degeneration suggests the impairment of enzymatic and/or nonenzymatic antioxidant defence, resulting in uncontrolled damage of biomolecules by ROS. Primary targets of ROS depend on sites of their formation. Since compartmentalization is crucial for cell viability, severe damage to membrane structure could be an irreversible step towards cell death. Impairment of membrane function can be triggered either directly, by oxidation of polyunsaturated fatty acids of lipids called lipid peroxidation (LPO) or indirectly, by mechanisms leading to decreased lipid synthesis, decreased fatty acid desaturation, impaired redox equilibrium or increased activities of lipases.

LPO involves the direct or metal catalyzed reaction of oxygen and unsaturated fatty acids associated with polar lipids, generating free radical intermediates and semistable peroxides. Since subcellular membranes in brain cells contain high amounts of polyunsaturated fatty acids, formation of a single carbon-centered radical within a membrane can lead to peroxidation of many fatty acids. This can occur when O_2 is present. The complex process of LPO is commonly described by three stages. A, Initiation: The generation of a radical with sufficient reactivity to extract hydrogen atoms from methylene groups of fatty acids [(HO)*; (HO_2)*], B, Propagation: Reaction of these radicals to yield another radical which likewise is capable of generating more radicals (radical chain reaction), and C, Termination: Recombination of two radicals or reactions yielding stabilized radicals no longer capable of propagating chain reactions. (HO_2)* and (HO)*, but not (O_2)*⁻ at pH 7,4, are able to extract hydrogen from allylic or bis-allylic positions of polyunsaturated fatty acids. The carbon radicals tend to be stabilized by molecular rearrangements to form conjugated dienes. In the presence of sufficient amounts of O_2, peroxyl radicals are formed. In media of low hydrogen-donating capacity, the peroxyl radical is free to react further by competitive pathways, resulting in cyclic peroxides, double bond isomerization or formation of dimers and oligomers. Thus, random peroxidation of e.g. arachidonic acid could give a complex mixture of isomers of cyclic peroxides and hydroperoxides. If peroxidation of free fatty acids is driven enzymatically by cyclooxygenases or lipoxygenases, stereospecific hydroperoxides and endoperoxides are produced which are precursors of eicosanoids (prostaglandins, thromboxanes, leukotrienes). If the peroxyl radical extracts a hydrogen atom from an adjacent fatty acid to yield another lipid radical (L*), which subsequently reacts with O_2, a hydroperoxide chain reaction is propagated. Other peroxyl radical reactions are the ß-scission, intermolecular addition and self-combination. These reactions and those of phenols such as α-tocopherol (vitamin E), aromatic amines and conjugated polyenes such as β-carotene, with various radicals (carbon- and oxygen- centered) can terminate radical chain reactions. If lipid hydroperoxides (LOOH) are not removed by GSH-dependent peroxidases, transition metal ions, especially iron and copper, can catalyze the decomposition of peroxides to form either alkoxyl- (LO*), alkyl- (L*), or (OH)*-radicals. These radicals could initiate a secondary propagation of radical chain reactions known as LOOH-dependent LPO. Consequently iron, or complexes of iron with low molecular iron chelators stimulate LPO by lipid decomposition reactions. In contrast to popular belief, alkoxyl radicals of polyunsaturated fatty acids do *not* significantly abstract hydrogens, but rather are channeled into epoxide formation through intramolecular rearrangement. Moreover, besides homolytic reactions of polyunsaturated fatty acids, one has to keep in mind the susceptibility of hydroperoxides to heterolytic transformations such as nucleophilic displacement and acid catalyzed rearrangement. In 1990, Babbs and Steiner published a computational model of kinetics of LPO in a two-compartment model system (membrane and cytosol) assuming an iron-catalyzed, (O_2)*⁻-driven Fenton reaction as the initiator of LPO. Kinetic interactions of up to 109 simultaneous enzymatic and free radical reactions, thought to be involved in the initiation, propagation and termination of LPO, were calculated using rate constants from the literature. From these model studies it was concluded that the segregation and concentration of lipids within membrane compartments promote chain propagation, that in the absence of antioxidants, computed concentrations of lipid hydroperoxides increase linearly at a rate of 40 µM/min during oxidative stress, that LPO is critically dependent on O_2 concentration, that LPO is rapidly quenched by the presence of tocopherol-like antioxidants, SOD and CAT, that only small amounts of "free" iron (1 to 50 µM) are required for initiation of LPO, and that substantial LPO occurs only when cellular defence mechanisms have been weakened or overcome by prolonged oxidative stress. Hence understanding of the balance between free radical generation and antioxidant defence systems is critical to the understanding and control of free radical reactions in biology and medicine.

Dependent on the fatty acid hydroperoxide (primary product of oxidation of unsaturated fatty acids with O_2), and on catalytic degradation by either iron complexes or by NADPH cytochrome P-450 reductase, a huge range of secondary products of LPO is formed (Fig. 2). These include conjugated dienes, hydrocarbon gases (e.g. ethane, ethene from linoleic acid) and carbonyl compounds (e.g. malondialdehyde, MDA; alkenals; alkadienals; and alpha-ß-unsaturated aldehydes). Carbonyl compounds are formed by ß-scission of alkoxyl radicals or thermic or metal catalyzed degradation of cyclic endoperoxides. The latter process produces MDA. In addition, it is suggested that MDA can also be formed *in vivo* as a byproduct of eicosanoid biosynthesis. Various techniques exist to evaluate products of LPO in tissues but all are limited either with respect to sensitivity, specificity or practicability, since the most accurate assays for measuring lipid peroxides are the most chemically sophisticated, requiring sample preparation under inert gas to ensure no further peroxidation during handling of lipid material (e.g. gas- or liquid-chromatography coupled to mass spectrometry). Whenever possible, a combination of methods measuring primary and secondary, as well as tertiary products of LPO (amino acid adducts, nucleotide adducts and glutathionyl conjugates) is advisable.

3.2. Oxidation of proteins

ROS can directly oxidize free or protein bound amino acids leading to deactivation of enzymes (Stadtman, 2001). Cysteine, methionine, histidine and tryptophan are preferentially oxidized resulting in sulfenic, sulfinic or sulfonic acids from thio-containing amino acids and in histidine- and tryptophan - endoperoxides, which subsequently degrade. Oxidation of thiols in proteins is often involved in regulation of enzyme activity such as glucose-6-phosphate dehydrogenase; pyruvate kinase; brain adenylate cyclase; gamma-glutamyl-synthetase; and others. Carbonyl compounds can be attacked by amino groups. Increase of MDA *in vivo* could result in both intra- and intermolecular cross links of proteins, giving fluorescent products such as conjugated imines. Proteins which have been oxidatively modified may loose function. Such post-translational modifications mark proteins for degradation by proteases.

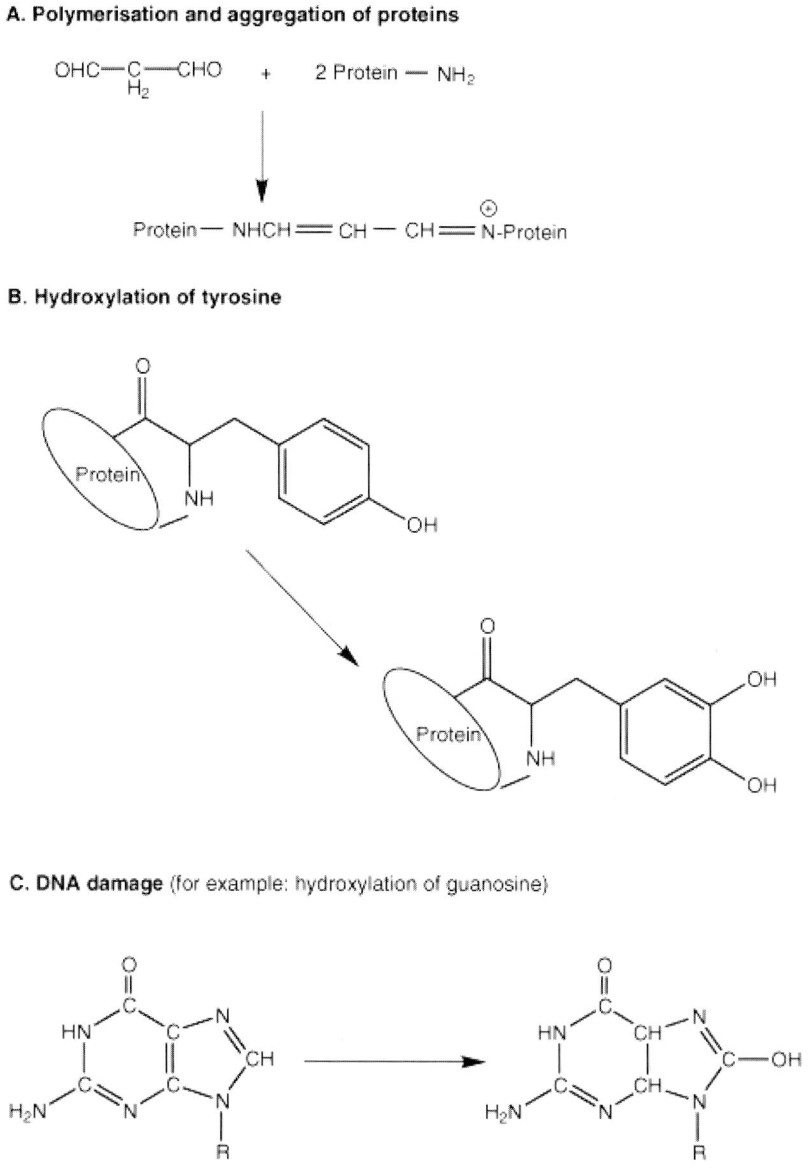

10. FORMATION OF RADICALS

A. Polymerisation and aggregation of proteins

$$OHC-\underset{H_2}{C}-CHO \quad + \quad 2\ Protein-NH_2$$

$$\downarrow$$

$$Protein-NHCH=CH-CH\overset{\oplus}{=}N\text{-}Protein$$

B. Hydroxylation of tyrosine

C. DNA damage (for example: hydroxylation of guanosine)

Figure 2: Some possible consequences of oxidative stress and lipid peroxidation to proteins and DNA. A, Cross-link of malondialdehyde and ε-Lys-amino groups of proteins. B, Hydroxyl radical attack on protein tyrosine resulting in protein-bound dihydroxyphenylalanine. C, 8-hydroxy-2'deoxyguanosine formation following hydroxyl radical attack on 2'-deoxyguanosine in DNA. Modified according to Gerlach et al., 2003b with permission.

High molecular weight proteolytic complexes called ingensin, macropain, macrosin, proteasome, multicatalytic protease or macroxyproteinase are responsible for the degradation of oxidatively modified proteins providing amino acids for *de novo* synthesis. However, oxidations of proteins may also lead to conformational changes that lead to denaturation and subsequent increase in protein hydrophobicity, giving rise to protein aggregation or precipitation. This process may lead to protein structures that are not any more easily digested by proteases and may lead to the activation of microglia and phagocytosis.

Trans-4-hydroxy-2-nonenal (HNE) is the most toxic α-β-unsaturated hydroxyalkenal. Hydroxyalkenals are mainly detoxified by alcohol,- and aldehyde dehydrogenases or by forming adducts with cysteine or GSH. The latter process is catalyzed by GSH transferases. In addition, etheno- and propano- adducts of trans-4-hydroxy-2-nonenal with nucleosides have been identified (Marnett, 2000). Thus, oxidative damage to lipids can affect both classes of biomolecules, proteins and DNA (Fig. 3) by the formation of adducts with secondary products of LPO.

3.3. Nucleic acid damage caused by ROS

There is increasing interest in the potential role of ROS as mediators of metal-catalyzed carcinogenesis and in genetic changes occurring as a consequence of ionizing radiation, chemical carcinogens and tumor promotors (e.g. phorbolesters). Besides ribonucleic acids, DNA is an important factor damaged by ROS *in vivo*. This may result in the disruption of transcription, translation and DNA replication. The amount of oxidative damage, even under normal physiological conditions, may be quite extensive, with estimates as high as 1 base modification per 130,000 bases in nuclear DNA. Damage to mitochondrial DNA is estimated to be as much as 1 per 8000 bases. DNA-DNA and DNA-protein cross-links, sister chromatid exchange, single or double strand breaks and base modifications are reported to occur due to reactions of ROS with DNA. In principle, all four DNA bases can be oxidatively modified, thymidine being most susceptible to ROS. As for the reaction mechanisms, it is thought that H_2O_2 interacts with metal ions (Fe, Cu) on DNA bases and the sugar backbone, causing site-specific (OH)*-mediated DNA damage (Marnett, 2000).

The nucleosides thymidine glycol and 8-hydroxy-2'-deoxyguanosine are considered to be biomarkers of DNA damage by ROS. These are specific since, in contrast to the free bases, they are not easily metabolised and can be measured by HPLC in urine using electrochemical detection. In eucaryotes several glycosylases which act on DNA oxidation products have been characterized. Not only do exogenous factors like ionizing radiation or chemicals (bleomycin, adriamycin, benzo(a)pyrene) increase urinary levels of hydroxylated nucleosides, but high dietary caloric intake and high metabolic rates correlate with urinary thymidine glycol and 8-hydroxy-2'-deoxyguanosine excretion. It has been further documented that DNA repair is less efficient in older organisms. In contrast to the known inherited metabolic disorders, to date there is little evidence of oxidative damage of nuclear DNA in relation to the pathophysiology of PD or AD.

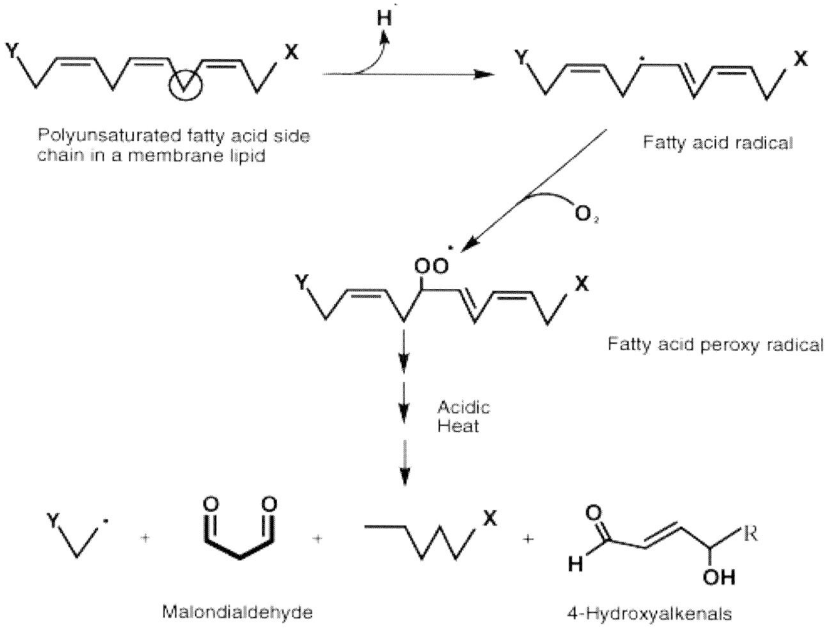

Figure 3: Simplified process of lipid peroxidation resulting in the formation of the reactive aldehydes malondialdehyde and 4-hydroxyalkenals. Modified according to Gerlach et al., 2003b with permission.

Mitochondria are one of the main generators of ROS. Consequently, at the sites of cellular free radical generation, the enzymes of the respiratory chain and the mitochondrial DNA (mtDNA) are particularly susceptible to damage by ROS. The rate of mitochondrial $(O_2)^{*-}$ and H_2O_2 generation increases with age in houseflies and in the brain, heart and liver of the rat. In PD, several groups of investigators have reported mitochondrial respiratory dysfunctions in brain, muscle and platelets. The 16,596-bp human mtDNA codes for two ribosomal ribonucleic acids, 22 transfer ribonucleic acids, and 13 peptides which are part of enzyme-complexes of the respiratory chain in the inner mitochondrial membrane. They proliferate independently of the cell cycle. In mammals, mtDNA mutates much faster than nuclear DNA possibly because mtDNA is not covered by histones, and is at least transiently attached to the inner mitochondrial membrane, where comparatively large amounts of ROS are produced. Therefore, mtDNA is particularly susceptible to oxidative damage. The steady-state level of oxidized bases in mtDNA is about 16 times higher than in nuclear DNA. ROS generate strand breaks in mtDNA, and DNA repair in mitochondria is much less efficient than in the nucleus. These mammalian organelles do not have significant recombinational repair, but may excise damaged bases. The role of genetic factors in the etiology of sporadic cases of PD remains to be determined. Changes in mitochondrial 8-oxodG levels or other biomarkers of mtDNA damage such as adducts of mitochondrial nucleotides with products of LPO (e.g. HNE) are to be investigated in PD, and in the brain of Alzheimer patients. However, there is still the possibility that endogenous reactive intermediates could act directly on mitochondrial proteins producing the different mitochondrial dysfunctions reported in PD and AD (Stewart and Heales, 2003) without the need for an interaction of ROS with nuclear or mitochondrial DNA.

3.4. Effects of radicals on signal transduction and gene expression: Implications for cell death

Within the last decade the effects of ROS and reactive nitrogen species (RNS) on the regulation of signal transduction and gene expression has become an exponentially growing area of research (Dalton et al., 1999). The regulation of gene expression by oxidants, antioxidants, and the overall cellular redox equilibrium attracted greater interest when scientists showed that even in mammalian cells protooncogene expression is induced by oxidative stress. Since protooncogenes play central roles in growth and differentiation, it was concluded that radicals must be of importance not only for gene expression in bacteria but in mammalian cells as well.

It has been extensively shown that changes in signaling and gene expression affect cell proliferation, differentiation and survival. As well, the mode of cell death is differentially regulated depending on the extent and duration of oxidative stress and may express itself morphologically as apoptotic or necrotic cell death. Not only oxidants but the overall redox equilibrium modulates mechanisms of neuronal survival and neuronal death. The intracellular signaling phase in neurons is exquisitely sensitive to redox changes, in particular through reactions with protein kinases, protein posphatases, and redox-sensitive transcription factors. Apparently, too many radicals are deleterious for cells and lead to apoptosis or necrosis but complete depletion of radicals by scavengers might constitute a reductive stress that may become harmful to cells as well.

Apoptosis is an active process in response to an intra- or extracellular death signal. It involves cross talk between pro- and anti-apoptotic factors released from mitochondria, proteases ,and DNases (Hengartner, 2000). This cross talk eventually amplifies the death signal and triggers a cascade of events leading to nuclear chromatin condensation and proteolysis without disruption of the cytoplasmic membrane. The resulting apoptotic bodies will be removed by microglial phagocytosis in the brain. In contrast, necrosis results from a disruption of the cell membrane with a rapid cessation of cellular functions and death. High doses of ROS and RNS often lead to necrosis because of breakdown of mitochondrial energy production, wheras low levels of ROS and RNS may also result in apoptotic cell death escpecially if the duration of exposure is short. A threshold dose for ROS and RNS that is triggering the switch from apoptotic to necrotic cell death is cell type specific and a critical function of the experimental setting. Primary neurons in culture appear to be the most sensitive cells towards H_2O_2. Important executors of the apoptotic mode of cell death are caspases and calpains. Their activation can be effectively delayed by ROS. This has been attributed to the oxidation of cysteine thiols at the proteolytic site of these enzymes. The rate of mitochondrial respiration, and hence the rate of superoxide formation, is predominantly determined by the coupling state of the mitochondria, which in turn is regulated by intracellular calcium and the oxidation state of thiols and pyridine nucleotides. An increase in cytosolic calcium causes mitochondria to take up excess calcium by an electrogenic mechanism linked to the proton gradient that stimulates electron transfer to oxygen. Thus, energy uncoupling of mitochondria by calcium stimulates the rate of mitochondrial respiration, oxygen utilization and superoxide formation (Dalton et al., 1999). Mitochondrial ROS production, enhanced by uncoupling agents or excess mitochondrial calcium, may lead to the release of cytochrome c from the inner mitochondrial membrane, an initiating signal for apoptosis. Together with dATP, apoptosis activating factor 1 (APAF1) and procaspase 9, cytochrome c is forming a complex called apoptosome that enables cleavage of procaspase 9 to active caspase 9 which then may activate caspase 3. The latter caspase is responsible for the cleavage of a huge variety of proteins including the DNA repair enzyme poly-ADP-ribose-polymerase (PARP), the nuclear membrane protein lamin and the cytoskeletal protein fodrin, to name only few of them (Hengartner, 2000). Cytochrome c release can be stimulated by oxidants and subsequently activates caspases. On the other hand, mitochondrial $(O_2)^{*-}$ production is increasing eight times above normal in heart muscle cells after cytochrome c release. In cerebellar granule neurons, and astrocytoma cells (D384) caspase activation by H_2O_2 however, is not the reason for nuclear fragmentation and cell death in contrast to the situation in T-cells because in some types of neurons, caspase inhibitors do not block fragmentation of DNA and cell death. In addition to cytochrome c release, the release of apoptosis inducing factor (AIF) and the release of DNases from lysosomes but not the activation of caspase acitvated DNases (CAD) have been postulated to account for nuclear DNA fragmentation in cerebellar granule cells and D384 cells.

A recent microarray analysis of the short term (4h) transcriptional response in a human retinal pigmented epithelium cell line following single dosing of 300 µM H_2O_2 revealed significant downregulation of nearly 200 genes, whereas only 13 were upregulated at a cell survival of 85 % (Weigel et al., 2002). Interestingly, short term response did not involve genes commonly expected to be changed in oxidative stress but genes involved in apoptosis, cell cycle regulation, cell-cell communication, signal transduction and transcriptional regulation.

ROS dependent redox cycling of protein thiols is critical for the establishment of protein-protein and protein-DNA interactions, for intracellular signaling via kinases and phosphatases, and for the initiation of transcription. Redox cycling of cysteinyl residues is an important, however, only one of several possible oxidant-dependent mechanisms that regulate the activity of transcription factors such as nuclear factor kappa B (NFκB), activator proteins-1 (AP-1) or protein phosphatases.

NFκB is an important transcription factor for genes involved in immunological response and cellular defence including cytokines and cell adhesion molecules. The transcriptionally active factor is a dimer of p50 and p65. In the cytoplasm NFκB is bound to inhibitor proteins of the IκB family. If the latter proteins are phosphorylated the NFκB-IκB complex is falling apart and the released p65/p50 dimer can diffuse into the nucleus and bind to DNA at specific promoters of NFκB-responsive genes. Phosphorylation of IκB via ubiquitin dependent ROS-sensitive IκB-kinases is inhibited by dithiocarbamates, suggestig a role for redox active cysteines in phosphorylation activity. NFκB is selectively activated as a result of H_2O_2 since in cells which overexpress catalase NFκB is hardly activated. Pharmacological and genetic inhibition of transcription factor NFκB protect T-cells from H_2O_2-elicited death that is presumably mediated by death effector genes such as p53.

As is the case with NFκB, AP-1 binding to DNA depends on the presence of critical cysteine residues in the FOS/JUN proteins. AP-1 is a term for a family of basic domain / leucine zipper (bZIP) transcription factors selectively binding to the 12-O-tetradecanoyl phorbol-13-acetate (TPA) response element (TRE). AP-1 is a heterodimer of the protein products of individual members of the FOS and JUN immediate-early response gene families, or a homodimer of JUN proteins. Expression of c-JUN and c-FOS is quickly induced by mitogens and by phorbol esters such as TPA, and by H_2O_2, UV-A, ionizing radiation, asbestos and dioxins. The key signaling events responsible for AP-1 activation are reversible oxidation and reduction of FOS/JUN proteins, oxidant-induced changes in calcium mobilization, production of arachidonate metabolites and shifts in the activities of protein kinases and protein phosphatases. The level of GSH is a key regulator of the induction of stress activated signal transduction pathways including JUN-kinases (Dalton et al., 1999).

On the other hand, increased levels of intracellular H_2O_2 may lead to increased formation of GSSG by the action of GSH-peroxidases. Subsequent interaction of the free cysteine thiol of an active protein phosphatase with GSSG leads to a mixed disulfide and to the inactivation of the enzyme. In turn, intracellular GSH and thioredoxin activity cause the reduction of the mixed disulfide back to the active form of protein phosphatase. The inactivation of these protein phosphatases, which may serve as specific negative regulators of their cognate kinases, can result in a continued and heightened state of phosphorylation and activation of specific signal transduction pathways. In case of the p38-mitogen activated protein kinases, an abolition of the negative regulation by redox inactivation can result in a protracted expression of inflammatory genes and also in the fulfillment of apoptotic programs.

These few examples however, can only give a rough impression of the complex interactions of various sites of signal transduction with ROS and RNS via modification of redox sensitive protein cysteine residues in kinases, phosphatases and transcription factors which may lead to alterations in signal transduction to the nucleus and consequently in altered gene expression.

Thus, neuronal differentiation and survival in the brain depends on a well balanced redox equilibrium which has to be maintained by a variety of factors including oxidants, antioxidants, and enzymes. Some of which shall be discussed below with respect to neurodegeneration in Parkinson's and Alzheimer's disease. It has to be emphasized however, that the importance of NFκB and jun signaling, caspase activation and apoptosis as causative factors for neurodegeneration in AD and PD remains to be elucidated.

4. FACTORS SCAVENGING ROS

4.1. Enzymes

Several enzymes exist to remove H_2O_2 within cells, the heme-protein CAT, present in most aerobic cells, which catalyzes the degradation of H_2O_2 to triplet O_2 and water, and the heme- or selenocysteine-bearing peroxidases utilizing electron donors to reduce H_2O_2 to water. H_2O_2 detoxification prevents formation of reactive oxidants including (HO)*. However, in contrast to liver and erythrocytes, which contain high levels of CAT (about 1300 U/mg protein), the brain appears to contain less than 20 U/mg protein. CAT is predominantly located in small peroxisomes (microperoxisomes). Studies of the regional distribution of CAT revealed highest activity in hypothalamus and SN; lowest activity in striatum and frontal cortex. The distribution was reported to correspond to the localization of CAT to catecholaminergic nerve cell bodies.

GSH is an essential tripeptide present in virtually all animal cells. It is synthesized by the consecutive actions of gamma-glutamyl-cysteine synthetase and GSH synthase. The rate of the former enzyme is regulated through feedback inhibition by GSH. During action of GSH-transdehydrogenases and GSH-peroxidases (GSH-Px) glutathione disulfide (GSSG) is formed. GSH is regenerated via glutathione disulfide reductase (GSSG-Rd) utilizing NADPH resulting from nicotinamide adenine dinucleotide (NADH) by transdehydrogenation and, mainly, from glucose-6-phosphate dehydrogenase, which is very specific for $NADP^+$ and is regulated by intracellular contents of ATP, NADPH and ribose-5-phosphate. Due to its high nucleophilicity, GSH forms conjugates with endogenous compounds such as estrogens and leukotriene A, or with xenobiotics and products of LPO. These reactions are often catalyzed by GSH transferases. GSH-Px catalyzes the reductive destruction of H_2O_2 and organic hydroperoxides (ROOH), using GSH as an electron donor. GSH-Px activity of human brain amounts to approximately 70 U/mg protein. GSH-Px is a selenium dependent enzyme and accounts for about one fifth of total brain selenium. In perfused rat brain, activities of GSH-Px and GSSG-Rd are highest in the striatum. GSH-Px but not GSSG-Rd activity is high in SN. This points towards a high need for peroxide detoxifying enzymes in dopaminergic neurons. However, GSH-Px activity is more pronounced in glial cells than in neurons. Comparing primary cultures of murine astrocytes and neurons with respect to their content of total glutathione (GSH + GSSG), differentiated astrocytes contain about 16-fold higher levels (16 nmol/mg protein) than neurons. The overall concentration of GSH in rat brain is about 2 millimolar. The ratio GSH / GSSG is roughly 10 to 1 or higher in favour of the reduced form. Since MAO is also predominantly localized in glial cells deamination of catecholamines appear to be linked to GSH-Px content. Moreover, a specific phospholipid hydroperoxide GSH-Px has been described which reduces directly the hydroperoxide moiety of the still esterified fatty acid to phosphatidylglycerol without the necessity of phospholipase A_2 activity. This prevents successive formation of prostanoids and lysolipids which otherwise would affect cellular metabolism and destabilise membranes. GSH is located in a key position of cellular defence against free radical mediated injury. Protective potency against membrane protein oxidation, lipid oxidation, and chelation of free heme iron has been ascribed to GSH. In addition, maintenance of the proper GSH/GSSG ratio may be of significance in the metabolic regulation of the cell. Since regeneration of GSH is dependent on NADPH, activity of glucose-6-phosphate dehydrogenase could be rate limiting for the activity of GSSG-Rd, and since GSH synthase is dependent on ATP, the overall pool of GSH is hence linked to oxidative phosphorylation implying that impairment of mitochondrial respiration could lead to decreased synthesis of GSH. Thus, a proper balance of antioxidant enzyme activities and reducing equivalents (NADH, NADPH, GSH, ascorbate) is crucial for optimal cell function and resistance to oxidative stress.

SODs are metalloenzymes that are widely distributed among oxygen-consuming organisms (yeasts, plants, animals). McCord and Fridovich discovered $(O_2)^{*-}$ to be a substrate for a copper- and zinc-containing protein in which copper is associated with enzymatic activity, while zinc serves as a stabilizer of protein structure. Interestingly, a manganese- (MnSOD) and an iron-dependent SOD were first characterised in *Escherichia coli* (FeSOD). Localization of these enzymes is very different, indicating functional changes during the evolutionary history of SOD. MnSOD in eucaryotic cells is strictly a mitochondrial enzyme in the inner membrane and synthesized by nuclear genes. It resembles the FeSOD found in procaryotes while cytosolic and peroxisomal CuZnSOD are different from MnSOD with respect to amino acid sequence and secondary structure supporting the idea for an endosymbiotic origin for mitochondria. In rat brain, SOD is homogeneously distributed with respect to brain region. In human grey matter, CuZnSOD amounts are almost equaling the amount of CuZnSOD in liver. The subcellular localization of brain SOD is highest in cytoplasm. Mitochondria and microsomes show only 13-15% of cytoplasmic levels. Glial cells from rat cortex contain higher specific activity of SOD than neurons. Several studies have demonstrated a direct toxicity of $(O_2)^{*-}$ without invoking $(HO)^{*}$ produced by the metal catalyzed reactions. Failure of SOD could result in increased production of $(O_2)^{*-}$, for example during respiratory burst of phagocytic leukocytes, and could aggravate inflammatory processes and reperfusion injury. Thus, biosynthesis of CAT, peroxidases and SODs have to be rigorously controlled to ensure protection. Since spontaneous or catalytic dismutation of $(O_2)^{*-}$ by SOD provides cells with H_2O_2, cellular response to oxidative stress must not only elevate SOD activity to counteract $(O_2)^{*-}$ toxicity, but those of CAT and of GSH-Px as well. In contrast to GSH-Px, levels of CuZnSOD mRNA and protein as well as the susceptibility to LPO increase with age in mouse suggesting involvement of ROS in ageing, trisomy 21 (Down syndrome) and possibly neurodegenerative diseases.

In PD patients the non-GSH-dependent peroxidase is decreased in homogenates from SN, caudate and putamen (reduction 50% of controls) but not in other brain regions. Furthermore a lower but significant decrease of GSH-Px in frontal cortex, putamen, external globus pallidus and SN (reduction 20% of controls) is reported. Since GSH-transport seems not to be affected, decreases in peroxidase and GSH-Px activity would imply a possible increase in susceptibility to oxidative stress of some brain regions in PD. There are significant reductions in the activity of CAT, the other important antioxidative enzyme, in PD in SN and putamen. It is evident that the extent of damage would be increased if H_2O_2 degrading enzyme activities were decreased or if production of H_2O_2 was enhanced. SOD generates H_2O_2. Interestingly, within the SN, CuZnSOD gene is preferentially expressed in the neuromelanin-pigmented neurons.

In AD patients using immunohistochemical methods high levels of CuZnSOD protein exist in large pyramidal neurons of the hippocampus, which are known to be susceptible to degenerative processes in AD. Biochemical pathways leading to $(O_2)^{*-}$ generation might be especially active in these neurons requiring an active transcription of CuZnSOD gene. Alternatively, a high cellular CuZnSOD activity might also, by promoting H_2O_2 production, contribute to the vulnerability of these neurons, in particular within compartments low in GSH-Px or CAT activity. In the AD brain normal or slightly increased levels of GSH were shown, indicating that stress in AD might not result from decreased detoxification but from increased formation of reactive biological intermediates.

4.2. Glutathione counteracts quinone toxicity in Parkinson's disease

GSH-Px needs GSH as a substrate, and many attempts have been made to evaluate the levels of GSH and GSSG in PD. In SN, there is a 40 % to 50 % reduction of GSH. By contrast, no differences in GSH levels were observed in other brain regions. There were no changes in the levels of GSSG. Activities of glutathione transferases and γ-glutamyl cysteine synthetase activity are not altered in PD. A decrease in levels of GSH and of GSH-Px activity could provide a source for H_2O_2 accumulating in cells. Thus, GSH concentration could be a key factor determining the fate of a cell at the threshold between life and death. However, it has been argued that, in order to be really a cause of cell death, GSH levels would have to be decreased to about 10% of that value existing in healthy cells which is 1μmol/g GSH and about 10 nmol/g GSSG. GSH depletion to a lesser extent, renders cells more susceptible to impairments of cellular metabolism. Interestingly, the levels of GSH are decreased to the same extent in incidental Lewy body (LB) disease, considered to reflect early presymptomatic stages of PD as in advanced PD, despite a far less severe neuronal loss in the SN. This could be an indication that impairment of GSH/GSSG equilibrium is an early event in neuronal degeneration. The importance of GSH for mental function is underlined by the observation that patients with GSH synthetase deficiency show gradual neurological deterioration of motor functions, retardation of movement, tremor and rigidity, and psychomotor retardation beginning in childhood. Since GSH is predominantly localized in non-neuronal cells the severe loss of GSH in SN in PD has to be, at least in part, attributed to impairment of glial functions or to extensive neuronal loss. The latter seems unlikely because, in LB disease cell loss is very moderate but GSH depletion is nearly 40%. In PD, there is no alteration in levels of α-tocopherol in serum, nor in various brain regions including SN when compared to control subjects. In addition, brain levels of ascorbate are not altered in PD.

An increased dopamine (DA) turnover in PD as a consequence of a 80-90% loss of nerve cells in the SN and amplified DA-liberation and -reuptake in remaining axons of the striatum are all factors that may contribute to the progressive loss of DA neurons in PD. An elevation in DA turnover may be a compensating mechanism in PD to overcome the effects of the loss of dopaminergic neurons. This could indicate that in aging or pathological states surviving neurons contain higher concentrations of catecholamines. It is known that DA is unstable in solution at neutral pH and easily undergoes autooxidation. In the presence of transition metal complexes catechols enhance the formation of (HO)* from H_2O_2. Thus, the presence of catechols in cells could provide a threat to cell viability, especially if low molecular weight iron complexes are present. The more heavily pigmented neurons of the SN appear to be preferentially lost in PD and during the course of aging, whilst both iron and melanin are known to increase with age. When comparing the SN of control and Parkinsonian brains, the population of DA neurons containing neuromelanin show higher vulnerability to the neurodegenerative process of PD, and there is a direct relationship between the distribution of pigmented neurons normally present and the distribution of cell loss in the SN of individuals dying with the disease. In addition, an inverse relationship was observed between the percentage of surviving neurons in PD compared to controls and the amount of neuromelanin they contain. Moreover, the largest pigmented neurons in SN are lost preferentially in PD suggesting that the vulnerability of the dopaminergic neurons is related to their neuromelanin. It accumulates during life in pigmented brain-stem nuclei, appears first in cells of the LC around the time of birth and in SN around the age of 18 months. In normal subjects the intracellular content of neuromelanin was shown to increase with age, up to 60 years. Then it begins to decrease, presumably due to destruction of melanin-containing cells (Zecca et al., 2001). Many investigators have suggested that DA is a precursor of neuromelanin consisting of cysteinylcatechol-polymers and proteins, appearing as a dark brown pigment, mainly located in the cell bodies of SN and LC. In the skin, the formation of melanin is catalyzed by tyrosinase, which is a bi-functional enzyme, oxygenating tyrosine to 3,4-dihydroxyphenyl-alanine (DOPA) and oxidizing DOPA to DOPA-quinone. However, this enzyme has not yet been detected in the SN. Thus, autooxidative mechanisms may play a certain role in neuromelanin formation. The exact chemical composition of this pigment is still unknown and debated. Depending on conditions, melanin can significantly increase or decrease the yield of reactive products of iron-catalyzed decomposition of H_2O_2 *in vitro*, as determined by spin trapping of the products. Up to 20 % of the total iron content of SN from normal subjects is bound to neuromelanin. Increased tissue iron found in parkinsonian SN may saturate iron-chelating sites on neuromelanin resulting in increased probability for iron catalyzed ROS formation (Double et al., 2002; Gerlach et al., 2003a). Thus, it is possibly not the presence of melanin *per se* but the interaction between catechols, iron and H_2O_2 that determines the vulnerability of melaninized DA neurons to neurodegeneration in PD.

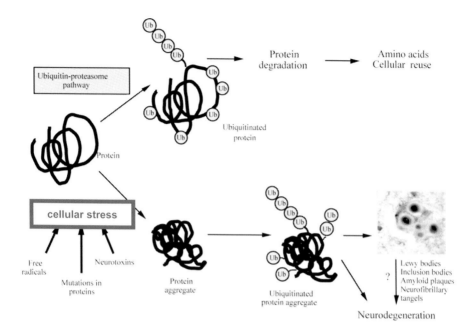

Figure 4: Alterations of endogenous protein digestion due to post-translational protein modifications following oxidative stress suggesting involvement in protein aggregate formation including Lewy body inclusions, neurofibrillary tangles and amyloid plaques. Modified according to Gerlach et al., 2003b with permission.

The commonly accepted cellular marker of parkinsonian pathology, the LB (consisting of pathologically phosphorylated proteins, ubiquitin, phospholipids and sphingomyelin; Fig. 4), is not confined to neurons containing neuromelanin and is not correlated with a selected neurotransmitter system. Neuropathologically, PD is now classified to the group of α-synucleinopathies. Quinones stabilize synuclein fibrils, suggesting a role for oxidation of catechols in the process of neurodegeneration in PD and a role for GSH as the most important detoxification system by forming adducts with reactive quinones that may result from DA or norepinephrine.

In conclusion, in PD decreased activities of GSH-Px and CAT, as well as decreased GSH levels, increased activity of SOD, and elevated levels of non-ferritin-bound iron concomitant with a high turnover of catecholamines, may participate in production of ROS. All these factors may render SN cells in PD more susceptible to hereto undefined toxic noxe and may provoke LPO as shown by increased levels of hydroxynonenal-modified proteins and/or, as a consequence, lead to increased levels of intracellular Ca^{2+}, with activation of proteases, lipases and endonucleases (Koutsilieri et al., 2002).

4.3. Lipid peroxidation in Alzheimer's disease

Although dementias are considered of multigenetic origin with many, not yet identified exogenous factors modulating disease onset and severity, recently, some of those factors could be specified that significantly increase the risk for AD. Among those, changes in lipid metabolism, amyloid and tau pathology as well as microglia activation and astrocyte stimulation all contribute to AD progression that can be pathologically first detected in distinct cortical areas including the entorhinal cortex and the hippocampus, leading to the destruction of long axon bearing cholinergic neurons. The majority of degenerating neurons is moderately myelinized, indicating a disturbance of lipid metabolism or selective vulnerability of these neurons. The accumulation of risk factors i.e. alterations in lipid transport (APOE lipoproteins) as well as amyloid precursor protein (APP) - and presenilin mutations overadditively increase the risk for sporadic AD. At present there is fairly convincing evidence that oxidative stress, a dysbalance of the production and detoxification of ROS, and alterations in lipid transport and metabolism are risk factors for the development and enhanced progression of both, familial and sporadic forms of AD (Montine et al., 2002; Butterfield et al., 2002). In sporadic AD cortical and hippocampal brain regions as well as cerebrospinal fluid (CSF) show increased susceptibility to damage by ROS, and enzymatic defence against ROS or ROS mediated damage is upregulated already at early stages of the disease especially in patients with an apolipoprotein-Eε4 genotype (Ramassary et al., 2000). In the studies of Montine et al., (2002) lipid peroxidation in AD correlated to the degree of neurodegeneration however there was no association with APO-E genotype. The deposition of amyloid plaques might lead to local ROS formation by enhancing transition metal catalyzed redox reactions and to the activation of astrocytes and microglial cells releasing chemokines, cytokines and superoxide. Changes of brain redox state in APP overexpressing animals are attenuated by concomitant overexpression of superoxide dismutase 1 gene, pointing to a role of superoxide for enhancing the progression of amyloid induced brain damage. Dependent on the site of generation, ROS react with multiple targets resulting mainly in protein and fatty acid oxidation. The products of which, hydroxyalkenals and isoprostanes, might further react with DNA, alter cellular signaling, protein phosphorylation (hyper-phosphorylated Tau protein) and de novo protein synthesis, transform formally non pathogenic molecules to pathogenic ones (beta amyloid aggregation) by triggering the formation of advanced glycation end products (AGE), and change membrane lipid constituents, membrane functions, and fluidity. Fatty acid peroxides become rapidly deacylated by phospholipase A_2. The resulting lysophospholipid lead to labilisation of the lipid membrane if it is not removed or reacylated. Levels of glycerophospholipids, plasmalogens and polyphosphoinositides are markedly decreased in patients with AD compared with age matched control subjects. This decrease in glycerophospholipids is correlated with elevations of phospholipid metabolites such as glycerophosphocholine, phosphocholine and phosphoethanolamine in autopsy samples of AD patients. These changes are associated with elevated activities of lipolytic enzymes in AD. As a result of deacylation of fatty acids from lipids, increases in levels of prostaglandins and isoprostanes occur in AD that are already considered as peripheral markers in CSF (Teunissen et al., 2002). Thus, it appears very likely that ROS play a crucial role for neural degeneration in AD, and therapeutic approaches involving the concept of antagonizing oxidative stress in AD are in progress (Behl and Moosmann, 2002).

4. ANTIOXIDATIVE THERAPEUTIC STRATEGIES

Oxidative stress, a cellular imbalance between production and elimination of ROS, such as NO, $(O_2)^{*-}$, H_2O_2, $(OH)^*$, and $ONOO^-$, is considered to be of major pathophysiological relevance for a variety of pathological processes, including ischemia-reperfusion injury and chronic progressive neurodegenerative diseases. This hypothesis has prompted research efforts to identify compounds which might act as antioxidants, i.e. compounds that antagonize the deleterious actions of ROS on biomolecules. The modes of action of these compounds could be either to directly scavenge ROS or to trigger protective mechanisms inside the cell, resulting in an improved defence against ROS. Based on the phenolic A-ring, estrogens are potent free radical scavengers. Their lipophilic character is a prerequisite for membrane interaction and penetration. In order to exert protective properties at least a phenolic moiety is necessary. It is hypothesized that these estrogen derivatives interfere with the redox status of membrane proteins and lipids thereby protecting the cell from more severe damage leading to death. In line with this hypothesis are data reporting that 17β-estradiol protects rat cortical synaptosomes from amyloid peptide $Aβ_{25-35}$ and $FeSO_4$ - induced membrane LPO and prevents the oxidative stress-related impairment of Na^+/K^+-ATPase activity, glutamate transport and glucose transport.

Estra-1,3,5(10),8-tetraene-3,17α-diol, J 811, a $Δ^{8,9}$-dehydro derivative of 17β-estradiol, is known to exhibit antioxidative activity by altering iron redox status and inhibiting the formation of $(O_2)^{*-}$ *in vitro*. The advantage of this compound is based on a more potent radical scavenging activity than that of the naturally occurring 17β-estradiol. The stronger antioxidant properties of J811 as compared to 17β-estradiol might be linked to the more extended delocalized π-electron system capable of stabilizing radical adducts. The future will show whether estradiol derivatives devoid of hormonal function can replace vitamin E, a naturally occurring lipophilic antioxidant as potentially neuroprotective agents. At least *in vitro*, vitamin E is in some experimental conditions more potent as antioxidant than 17β-estradiol. Vitamin E is the term used for eight naturally occurring fat-soluble nutrients. Four compounds bear a saturated phytyl side chain and differ only with respect to number and position of methyl groups at the chromanol ring (α,β,γ,δ-tocopherols). However, α - tocopherol predominates in many species. The phytyl side chain in the 2-position facilitates incorporation and retention of tocopherol in biomembranes, while the active site of radical scavenging is the 6-hydroxyl group of the chromanol ring. The most widely accepted physiological function of tocopherol is its role as a scavenger of free radicals. Thus it prevents oxidant injury to polyunsaturated fatty acids and thiol rich proteins in cellular membranes and cytoskeleton. It is thought to preserve the structure and functional integrity of subcellular organelles. Tocopherol is transferred from circulating lipoproteins to the brain, spinal cord and peripheral nerves and muscle by unknown mechanisms. In contrast to other brain regions, the cerebellum is particularly active in the utilization of tocopherol. During experimental tocopherol deficiency, nerve tissue retains a greater percentage of tocopherol than do serum, liver and adipose tissue. Morphological and functional studies performed on experimental tocopherol deficient rats have revealed axonal dystrophy and degeneration of peripheral nerve. This can be aggravated by increasing dietary polyunsaturated fatty acids providing increased quantities of peroxidisable substrate and attenuated by feeding synthetic antioxidants. These experiments provide evidence in favour of an antioxidant role for tocopherol in the nervous system. In brain, tocopherol is predominantly localized in the mitochondrial, microsomal and synaptosomal fractions suggesting that protection by tocopherol from peroxidative damage to subcellular membranes may be important for mitochondrial energy production or microsomal enzyme activity. Recent *in vitro* studies on neuronal cultures support the important role of vitamin E for neuroprotection especially following amyloid Aβ-peptide mediated neuronal damage (Butterfield et al., 2002). Although vitamin E does not inhibit fibril formation from Aβ peptides, it protects from oxidative stress, induced by mechanisms following Aβ peptide exposure. Moreover, vitamin E may even provide neuroprotection *in vivo* through suppression of signaling events necessary for microglial activation. The impressive antioxidant potentials of phenolic structures including vitamin E and estradiol derivatives in various *in vitro* models were summarized by Behl and Moosmann (2002). A placebo-controlled, clinical trial of vitamin E in patients with moderately advanced Alzheimer's disease was conducted by the Alzheimer's Disease Cooperative Study Group. Vitamin E 2000 IU (1342 alpha-tocopherol equivalents / d) slows the functional deterioration leading to nursing home placement in Alzheimer subjects according to that study. These data are encouraging to design clinical trials using antioxidants aiming at investigating the delay of cognitive function in patients with mild cognitive disorders (Grundman, 2000). In contrast, it was the Parkinson Study Group (1993) that diminished hope that lipophilic antioxidants such as vitamin E (2000 IU) would be neuroprotective in PD. In the DATATOP clinical trial vitamin E did not reduce the requirement for L-DOPA. The potential of vitamin E to counteract oxidative stress may be restricted to certain lipophilic compartments and thus be not sufficient for radical defence in PD. Vitamin E may need support for antioxidant defence by hydrophilic compounds. And in fact, the water soluble antioxidant Trolox, a vitamin E analogue, is capable to decrease neuronal death *in vivo*.

Vitamin E antioxidant action may be as well facilitated by ascorbic acid, also called vitamin C. Ascorbic acid is an extremely water-soluble antioxidant essential for humans, primates and guinea pigs but not for rodents, which can synthesize it from glucose. Ascorbic acid serves as a cofactor in several iron-dependent hydroxylases, important for collagen synthesis, (prolyl- and lysyl- hydroxylases), for carnitine biosynthesis (6-N-trimethyl-L-lysine-hydroxylase) and for catabolism of tyrosine (4-hydroxyphenyl-pyruvate-hydroxylase). Two major further functions of ascorbate are support of the synthesis of norepinephrine, and α-amidation of neurohormones, explaining in part its higher concentrations in brain and endocrine tissues (adrenal gland). Following one-electron oxidation of ascorbate the semidehydroascorbate radical is formed. It decays by disproportionation to ascorbate and dehydroascorbate (the latter subsequently degrades to oxalic acid and L-threonic acid), rather than acting as a reactive free radical. However, ascorbic acid is readily oxidized by $(O_2)^{*-}$. Furthermore, an important protective action of ascorbic acid is its ability to act synergistically with tocopherol in the inhibition of various oxidation reactions *in vitro* and *in vivo*. Although the antioxidant ascorbic acid does not penetrate the blood brain barrier, its oxidized form, dehydroascorbic acid, enters the brain by means of facilitated transport. Unlike exogenous ascorbic acid, dehydroascorbic acid confers *in vivo*, dose-dependent neuroprotection in reperfused and non-reperfused cerebral ischemia at clinically relevant times. As a naturally occurring interconvertible form of ascorbic acid with blood brain barrier permeability, dehydroscorbic acid represents a promising adjunct pharmacological therapy for stroke. Recently a dehydroascorbate reductase has been found in the rat brain that regenerates ascorbate after it is oxidized during normal aerobic metabolism. This enzyme can be found in endothelial cells, perivascular astrocytes and in neuronal cytosol and nuclei. These data may speak in favour of a neuroprotective effect of dehydroascorbate following intracerebral reduction to ascorbate as long as dehydroascorbate reductase is functional.

Coenzyme Q_{10} is an important electron transducer in mitochondria, microsomal membranes, Golgi complex, nucleus, and cytoplasmic membrane. Furthermore coenzyme Q is a potent lipophilic antioxidant capable of scavenging lipophilic radicals within these membranes, as well as in cytosol and plasma when bound to lipoproteins. The semiquinones and quinones produced upon oxidation of quinols are substrates for reductases so that the quinols are regained. This redox cycle can be interrupted if oxygen is oxidizing the hydroquinone or semiquinone yielding $(O_2)^{*-}$. Thus coenzyme Q may under certain conditions become a prooxidant, as it is the case with ascorbate. However, as long as sufficient reductive equivalents such as NADH are formed, reduction of semiquinones is assured. In the brain roughly 80-90 % of coenzyme Q are in the quinol form. Recently, a highly significant correlation between the level of CoQ_{10} and the activities of both complexes I and II/III of the mitochondrial electron transfer chain has been reported. By using mitochondrially transformed cells (cybrids) from PD patients a 26 % deficiency of complex I activity could be detected although cytosolic calcium and energy levels of those transformed cells were normal. These subtle alterations may reflect an increased susceptibility of cells by impaired mitochondrial electron transport under circumstances not ordinarily toxic to those cells containing mitochodria from parkinsonian patients.

The redox ratio of cerebral coenzyme Q can be augmented by intraperitoneal injections of lipoic acid, and in several experimental conditions producing striatal lesions in the rat by malonate, 3-nitropropionic acid, MPTP, and methamphetamine coenzyme Q confers protection. In contrast to these encouraging results, similar to vitamin E, coenzyme Q monotherapy (200 mg/d) turned out not to be protective in PD (Strijks et al., 1997) in an open label three months trial. However, to date new galenic forms of administration of coenzyme Q are considered that should increase bioavailability of coenzyme Q in the brain. Again pharmacokinetic aspects may have to be more thoroughly addressed and reevaluated in advance of further clinical trials and before final conclusions concerning the neuroprotective activity of coenzyme Q in man can be drawn.

The oxidized form of alpha-lipoate (thioctic acid, 1,2-dithiolane-3-pentanoic acid, 1,2-dithiolane-3 valeric acid, 6,8-dithiooctanoic acid) is a low molecular weight substance that is absorbed from the diet and crosses the blood-brain barrier. Alpha-lipoate is taken up and reduced in cells and tissues to dihydrolipoate, which is also exported to the extracellular medium. Thus antioxidant protection is possible intra- and extracellularily. Lipoate and dihydrolipoate exert antioxidant activity by reducing dehydroascorbate or glutathione disulfide to raise intracellular glutathione levels. The most important thiol antioxidant glutathione has to be synthesized in the brain since it will not cross the blood brain barrier following systemic administration. Dihydrolipoate is therefore an interesting alternative to increase cerebral antioxidant potential in the cytosol and the extracellular space. In addition to the antioxidant features of dihydrolipoate, it serves as a covalently bound coenzyme in alpha-ketoacid dehydrogenases, such as the important mitochondrial pyruvate dehydrogenase and alpha-ketoglutarate dehydrogenase. Because of these characteristics it is reasonable to assume a considerable neuroprotective potential.

And indeed, lipoate and dihydrolipoate exert neuroprotection in experimental cerebral ischemia-reperfusion injury, excitotoxic amino acid brain injury, mitochondrial dysfunction and diabetic neuropathy. Despite the broad antioxidant activity of lipoate and dihydrolipoate, it has been argued, that other modes of action might contribute to the remarkable neuroprotective effects observed in rodents and diabetic patients. There are indications that lipoic acid triggers both heat-shock and phase II responses following activation of certain signaling proteins capable of detecting oxidants and electrophiles. Recently, it has been reported that lipoate induces time and concentration dependently the activity of NAD(P)H:quinone oxidoreductase (NQO1) and of glutathione transferases in C6 astroglial cells. Thus, upregulation of phase II detoxification enzymes may highly contribute to lipoates neuroprotective potential. In a first randomized study including nine patients with AD lipoic acid was successfully administered for up to one year demonstrating constant scores for cognitive performance in two neuropsychological tests. Thus, lipoate might evolve as an adjunct therapeutic option in age related disorders and dementia (Hager et al., 2001).

Recently, subtype-selective inhibitors for iNOS and nNOS became available to investigate the putative neuroprotective effects of NOS inhibition in stroke. This therapeutic approach might be conceptualized as well for the treatment of chronic neurodegenerative diseases if adverse reactions of NOS inhibitors can be minimized. Another promising approach may be the development of inhibitors of stress activated kinases of the c-Jun Amino Kinase pathway.

ABBREVIATIONS

AD	Alzheimer's disease
APAF 1	apoptosis activating factor 1
AP-1	activator protein 1
APOE	apolipoprotein E
ATP	adenosine triphosphate
Ca^{2+}	calcium ion
$[Ca^{2+}]_i$	intracellular calcium ion concentration
CAT	catalase
CNS	cyclooxygenase
DA	deoxyribonucleic acid
DTPA	bis-(2-aminoethyl)-amine-N,N,N',N'-pentaacetic acid
EAA	excitatory amino acids
EDTA	ethylenediamine tetraacetic acid
$FADH_2$	flavin adenine dinucleotide
FMN	flavin adenine mononucleotide
GMP	guanosine monophosphate
GSH	glutathione
GSH-Px	glutathione peroxidase
GSSG	glutathione disulfide
GSSG-Rd	glutathione disulfide reductase
H_2O_2	hydrogen peroxide
H_4B	high performance liquid chromatography
IκB	inhibitor of κB
LB	Lewy body
LC	locus coeruleus
LPO	lipid peroxidation
MAO	monoamine oxidase
MPTP	1-methyl-4-phenyl-1,2,3,6-tetrahydropyridine
mtDNA	mitochondrial deoxyribonucleic acid
NADH	nicotinamide adenine dinucleotide
NADPH	nicotinamide adenine dinucleotide phosphate
NFκB	nuclear factor κB
NMDA	N-methyl-D-aspartate
NO	nitric oxide
NOS	nitric oxide synthase
$ONOO^-$	peroxynitrite
O_2	groundstate triplet dioxygen
$(O_2)^{*-}$	superoxide
$(OH)^*$	hydroxyl radical
PD	Parkinson's disease
ROS	reactive oxygen species
SN	substantia nigra
SOD	superoxide dismutase
TBARS	thiobarbituric acid reactive substances

TPA   12-O-tetradecanoyl phorbol-13-acetate
TRE   TPA response element

REFERENCES

Babbs, C.F. and Steiner M.G. (1990) Simulation of free radical reactions in biology and medicine: A new two-compartment kinetic model of intracellular lipid peroxidation. *Free Radic. Biol. Med.* 8, 471-485.

Behl, C. and Moosmann, B. (2002) Oxidative nerve cell death in Alzheimer's disease and stroke: antioxidants as neuroprotective compounds. *Biol. Chem.* 383, 521-536.

Butterfield, D.A., Castegna, A., Lauderback, C.M. and Drake, J. (2002) Evidence that amyloid beta-peptide-induced lipid peroxidation and its sequelae in Alzheimer's disease brain contribute to neuronal death. *Neurobiol. Aging* 23, 655-664.

Dalton, T. P., Shertzer, H.G. and Puga, A. (1999) Regulation of gene expression by reactive oxygen. *Annu, Rev. Pharmacol. Toxicol.* 39, 67-101.

Double, K.L., Ben-Shachar, D., Youdim, M.B.H., Zecca, L., Riederer, P. and Gerlach, M. (2002) Influence of neuromelanin on oxidative pathways within the human substantia nigra. *Neurotoxicol. Teratol.* 24, 621-628.

Gerlach, M., Double, K.L., Ben-Shachar, D., Zecca, L., Youdim, M.B.H. and Riederer, P. (2003a) Neuromelanin and its interaction with iron as a potential risk factor for dopaminergic neurodegeneration underlying Parkinson's disease. *Neurotox. Res.* 5, 35-44.

Gerlach, M., Reichmann, H. and Riederer, P. (2003b) Die Parkinson-Krankheit: Grundlagen, Klinik, Therapie, 3rd ed., Springer, Wien NewYork.

Grundman, M. (2000) Vitamin E and Alzheimer disease: the basis for additional clinical trials, *Am. J. Clin. Nutr.* 71, 630S-636S.

Hager, K., Marahrens, A., Kenklies, M., Riederer, P. and Münch, G. (2001) Alpha-lipoic acid as a new treatment option for Alzheimer type dementia. *Arch. Gerontol. Geriatr.* 32, 275-282.

Hengartner, M.O. (2000) The biochemistry of apoptosis. *Nature* 407, 770-776.

Koutsilieri, E., Scheller, C., Grünblatt E., Nara, K., Li, J. and Riederer, P. (2002) Free radicals in Parkinson's disease. *J. Neurol.* 249 Suppl.2; II1-5.

Marnett, L.J. (2000) Oxyradicals and DNA damage. *Carcinogenesis* 21, 361-370.

Montine, T.J., Neely, M.D., Quinn, J.F., Beal, M.F., Markesbery, W.R., Roberts, L.J. and Morrow, J.D. (2002) Lipid peroxidation in aging brain and Alzheimer's disease. *Free Radic. Biol. Med.* 33, 620-626.

Ramassamy, C., Averill, D., Beffert, U., Theroux, L., Lussier-Cacan, S., Cohn, J.S., Christen, Y., Schoofs, A., Davignon, J. and Poirier, J. (2000) Oxidative insults are associated with apolipoprotein E genotype in Alzheimer's disease brain. *Neurobiol Dis* 7, 23-37.

Stadtman, E. R. (2001) Protein oxidation in aging and age-related diseases. *Ann. N.Y. Acad. Sci.* 928, 22-38.

Stewart, V.C. and Heales, S.J.R. (2003) Nitric oxide-induced mitochondrial dysfunction: implications for neurodegeneration. *Free Radic. Biol. Med.* 34, 287-303.

Strijks, E., Kremer, H.P. and Horstink, M.W. (1997) Q_{10} therapy in patients with idiopathic Parkinson's disease. *Mol. Aspects Med.* 18(Suppl), S 237-240.

Teunissen, C.E., de Vente, J., Steinbusch, H.W.M. and De Bruijn, C. (2002) Biochemical markers related to Alzheimer's dementia in serum and cerebrospinal fluid. *Neurobiol. Aging* 23, 485-508.

Weigel, A. L., Handa, J.T. and Hjelmeland, L.M. (2002) Microarray analysis of H_2O_2-, HNE-, or tBH-treated ARPE-19 cells.*Free Radic. Biol. Med.* 33, 1419-1432.

Zecca, L., Gallorini, M., Schünemann, V., Trautwein, A.X., Gerlach, M., Riederer, P., Vezzoni, P., and Tampellini, D. (2001) Iron, neuromelanin and ferritin content in the substantia nigra of normal subjects at different ages: consequences for iron storage and neurodegenerative processes. *J. Neurochem.* 76, 1766-1773.

J.A. DEL RÍO[1], F. DE CASTRO[2] AND E. SORIANO[1]

[1]*Department of Cell Biology, Parque Científico, University of Barcelona, Barcelona, Spain*

[2]*Institut of Neuroscience of Castilla y León, University of Salamanca, Salamanca, Spain*

PART B: INTRODUCTION
11. AXON GUIDANCE AND REPULSION. THE MOLECULAR CODE OF SOCIAL LIFE IN THE BRAIN

Summary. The generation of a functional nervous system is dependent on precise pathfinding of axons during their development. This pathfinding is directed by the distribution of local and long-range guidance cues. Gradients of long-range guidance cues have been associated with growth cone function for over a hundred years, but their developmental roles are still poorly understood. Here we review the developmental roles and the complex intracellular signaling pathways enabling chemoattractive or repulsive axon-guidance molecules like netrins, semaphorins and ephrins to regulate axon growth. Recent research findings suggest that these molecules signal through specific receptors leading to local cytoskeletal rearrangements in the growth cone or cell leading edge; and through intracellular kinases, which have the potential to alter gene expression changes in the developing neuron. In addition, we also summarize the increasing body of knowledge on their roles in the inhibition of axon regeneration after adult lesions in the central and peripheral nervous system.

1. AXON GUIDANCE MOLECULES: AN OVERVIEW

One of the major challenges facing contemporary neurobiology is to explain the molecular mechanisms responsible for the formation of neural connections in the central and peripheral nervous systems. The mammalian brain contains approximately 10^{12} neurons which, during development, extend axons for long distances along specific pathways to their correct target tissues. The establishment of neuronal connections involves periods of axonal elongation and retraction, both of which are crucial for the directional navigation of the axon through the neural parenchyma. During such navigation, the tip of the growing axon (the growth cone) explores the surrounding extracellular matrix (ECM) and the surfaces of other cells, sensing and responding to a large number of guidance cues (see also Giger and Kolodkin, 2001).

Since the development of the chemotropic hypothesis, considerable information has been accumulated regarding the identification and characterization of the cues that influence axonal growth and guidance, and an increasing number of ligand and receptor families have been described. These molecules can be broadly categorized as either long-range (diffusible) or short-range (non-diffusible or membrane bound), and further sub-categorized as inhibitory/repulsive or non-inhibitory/attractive (Tessier-Lavigne and Goodman, 1996). However, there are exceptions, and a molecule which inhibits the growth cone of one particular neuron may promote the elongation of others. Thus, a given molecule may be regarded as both attractive and repulsive, as which it is depends not only on its identity but also on the molecular environment of the receptor/s located in the growth cone. Thus, there is a combinatorial regulation of axon guidance receptor signalling as recently reviewed by Yu and Bargmann 2001. In addition, recent studies have revealed an extensive transcriptional and post-transcriptional regulation of guidance cues and their receptors, which generate an astonishingly varied set of neuronal and axonal responses.

All guidance molecules have been implicated in numerous axon guidance or targeting events. Moreover, the functions of these molecules are not restricted to axonal guidance, but also concern the regulation of other biological processes such as cell migration, organ morphogenesis, and angiogenesis.

The expression of most guidance molecules is developmentally regulated; some of them are present in the adult nervous system and they may, or be re-expressed by non-neuronal cells following injury. Although their function in the adult is still unclear, they are likely to be involved in the plasticity of the adult nervous system and may also play a role in modulating neuronal regeneration after injury (Pasterkamp and Verhaagen, 2001).

This chapter concerns the mechanisms and molecules that direct axons to their targets. In particular, it explores recent advances in our understanding of the molecular mechanisms controlling the biological responses of the migrating axon during development and following injury in adults. We will focus on the functions of four conserved families of axon guidance molecules: netrins, slits, semaphorins and ephrins. For a comprehensive review of the effects of cell adhesion molecules regulating axon guidance and fasciculation during CNS development (see Walsh and Doherty, 1997)

2. NETRINS, DCC/UNC RECEPTORS AND AXON GUIDANCE

Netrins are a family of secreted proteins (~600 aa) that were first identified as a floorplate-derived chemoattractant for commissural axons navigating towards the spinal cord midline (Serafini et al., 1994, 1996, Kennedy et al., 1994)(Figure 1). They are related to laminins and contain a short basic C-terminal domain which confers diffusible properties. Four members of this family (netrin-1/–4) have so far been identified in vertebrates (Tessier-Lavigne and Goodman, 1996), although the existence of a mammalian netrin-2 has yet to be demonstrated. Netrin-1 and –2 bind similarly to DCC (deleted in colorectal cancer) and neogenin, and to UNC5H-1, -2 and –3 (the known receptors for netrins in mammals), whereas netrin-3 shows a preferential binding to the latter (reviewed by Livesey et al., 1999). The binding properties of netrin-4 remain poorly understood. Netrins have been shown to play a dual role in axon guidance, depending on the expression of DCC or UNC receptors in the target neuron. The interaction of netrin-1 and DCC/neogenin produces a chemoattractive response, while the binding of netrin-1 with members of the UNC5 family of receptors induces chemorepulsion. In the event of co-expression of both types of receptors, DCC/neogenin and UNC5H, a chemorepellent effect will predominate (see Cooper, 2002 for a comprehensive review). Moreover, research has shown that both types of receptors are used by netrin-1 to exert a survival-promoting (trophic) effect on neurons in the ventricular zone of the brainstem (Llambi et al., 2001). Netrin-1 attracts axons of cortical and hippocampal projections, habenular, thalamocortical, and precerebellar neurons, retinal ganglion cells, and commisural interneurons of the spinal cord. In contrast, it repels troclear axons, cranial motorneurons and the parallel fibres of cerebellar granule cells (see Cooper, 2002, for review).

Mutant mice with netrin-1 or DCC loss of function showed a similar phenotype with severe deficits in the formation of the main forebrain commissures (corpus callosum, hippocampal commissures and optic chiasm) and commissural axon extension at the floorplate (Serafini et al., 1996; Braisted et al., 2000; Barallobre el al., 2000). Moreover, mice lacking UNC5 showed a rostral migratory malformation with ectopic positioning of cerebellar neurons (Ackerman et al., 1997). In addition to these long-range effects, secreted netrins have short-range functions in driving neuritic outgrowth and branching, and are also involved in neuro-oligodendroglial interactions (Spassky et al., 2002). Recent studies have demonstrated that the extracellular distribution of netrin-1 depends on its binding activity with respect to extracellular matrix proteins and other guidance molecules (such as Slit) or receptors (such as Frazzled, the DCC Drosophila orthologue). Frazzled is required by netrin-mediated axonal chemoattraction, but not by axonal repulsion in Drosophila. Thus, it is thought that Frazzled sequestering netrin-1 cooperates in presenting the chemoattractive molecule to other target cells (reviewed by Cooper, 2002).

Despite these compelling data, the intracellular signalling mechanism responsible for netrin-1 activity is still not fully understood. However, recent studies have demonstrated that netrin-1/DCC-mediated axon guidance requires the activation of mitogen-activated protein kinase (MAPK) by extracellular signal-regulated kinase (ErK1-2)(Forcet et al., 2002). Moreover, the chemoattractive and repulsive effects of netrin-1 are also regulated by cAMP and cytosolic Ca2+ levels. The small GTPases Rac1 and Cdc42 also play key roles in intracellular signalling (Li et al., 2002a). Furthermore, an adaptor protein, NcK-1, has recently been described as linking netrin-1 receptors and cytoskeleton with the cytoplasmatic tail of the DCC receptor via SH3 domains, thus regulating the interaction of the receptor with the actin-based cytoskeleton (Li et al., 2002b). These data show that the cell transduction machinery responsible for netrin-1 is adaptive to netrin-1 gradients, as was clearly demonstrated in a recent study by Poo et al., (Ming et al., 2002). In this in vitro study, the growth cones of spinal cord neurons underwent consecutive phases of desensitization and resensitization during their migration towards a local source of netrin-1. These adaptive phases correlate well with modifications of the intracellular transduction machinery (Ca2+ levels and MAPK activation) and may also be regulated by receptor phosphorylation, as has been suggested for UNC5 and RCM-receptors.

3. SEMAPHORINS AND THEIR RECEPTORS AS AXON GUIDANCE CUES

Semaphorins have been identified as chemorepellent axon guidance molecules in the developing nervous system. They constitute a large conserved family of secreted and membrane-associated molecules. Secreted semaphorins comprise classes II (invertebrates), III (vertebrates) and V (viral), while the other classes (I, IV-VII) are membrane-associated. All semaphorins share a "Sema" domain of ~500 aa containing 12-16 cysteine residues, and this confers their binding specificity (see Nakamura et al. 2000 and Raper 2000 for details) (Figure 1). This Sema domain is necessary and sufficient to trigger their biological activity. Initially, secreted semaphorins were characterized as chemorepulsive cues for hippocampal (Figure 2A-D), olfactory, and pontocerebellar axons in the forebrain, as well as for sympathetic, sensory and motor axons in the peripheral nervous system. However, recent studies have demonstrated that some semaphorins can also act as attractants for cortical dendrites to the pial surface (for example, Sema 3A), or for cortical axons (Sema 3C) or mitral cell axons from the olfactory bulb (e.g., Sema 3B) (Bagnard et al., 1998, De Castro et al., 1999).

Of the above classes, mammalian secreted semaphorins (class III) are the most widely studied and characterized. They bind selectively to two known members of the neuropilin family, neuropilin-1 and –2 (Np1–2) (e.g., Neufeld et al., 2002). These receptors – simultaneously identified in 1997 by the laboratories of Tessier-Lavigne and Kolodkin - are transmembrane proteins with a short cytoplasmatic domain (~40 aa) lacking signalling-competent motifs. There is evidence that both Np1 and Np2 form dimers and heterodimers. It is interesting to note that Np-1 can bind with Sema 3A, 3B and 3C, although only Sema 3A can induce chemorepulsion in Np-1-expressing neurons. This suggests competitive receptor binding between several semaphorins and it may be that Np-1expressing axons can navigate in Sema 3A-expressing areas if other semaphorins act as antagonists. Np2 binds exclusively to Sema 3F (see Tamagnone and Comoglio, 2000, for review).

Plexins, transmembrane molecules which bind transmembrane semaphorins, can also combine with neuropilins to form a receptor complex in which neuropilins (Raper 2000). All four classes of plexins have been found to interact with semaphorins and neuropilins. Thus, in the Np-1-PlexinA1 receptor complex, Np-1 is the binding site for Sema 3A, while PlexinA1 serves as a signal transducer. Receptors for vascular endothelial growth factor (VEGF), cell adhesion molecules such as L1, or neurotrophin receptors (for example, trkA has been reported to combine with neuropilin receptors (and most probably with Plexins) to form a multimeric receptor complex (see Neufeld et al., 2002).

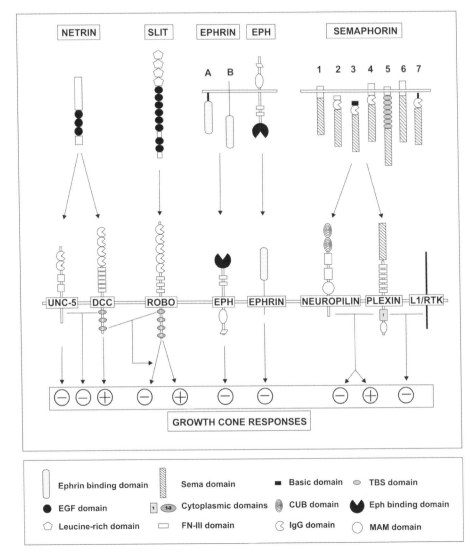

Figure 1. Summary of the axon guidance molecules reviewed in the present chapter. A domain key table is included (bottom). The effect of the guidance molecules and their receptor/s is indicated as (+) in case of chemoattraction and/or (-) or chemorepulsion

The effects described for secreted semaphorins in axonal pathfinding are almost exclusively repulsive. In some cases, these molecules collapse growth cones, an effect directly related to repulsion, and may even promote the apoptosis of some neuronal populations (e.g., Shirvan et al., 2000). An opposite trophic effect has recently been observed in oligodendrocyte progenitors (Spassky et al., 2002)(Figure 2E-F). Sema 3A exclusively binds neuropilin-1, and the main effects observed in knockout mutant mice appear in the peripheral nervous system (where many groups of sensory neurons are found to be lacking) and, to a lesser extent, in the laminar pattern of hippocampal connections (Behar et al., 1996; Pozas et al., 2001). These minor effects suggest there is an overlap in the action of different semaphorins belonging to the same class.

The cytoplasmic mediators of semaphorin signalling are still poorly understood. The downstream cytoplasmic mediators CRMPs (collapsin response mediator proteins) are a family of cytosolic phosphoproteins expressed exclusively in the nervous system. CRMPs are believed to play a key role in semaphorin signalling in conjunction with CRAM (a novel CRMP-associated protein). Moreover, several kinases have been described as participating in semaphorin-mediated growth cone signalling, including among others glycogen synthase kinase (GSK-3)(Eickholt et al., 2002), Fes/Fps (Mitsui et al., 2002), the tyrosine kinase receptor family member Offtrack (OTK)(Winberg et al., 2001), P21-activated kinase (PAK) and LIM kinase, (Aizawa et al., 2001) as well as members of the Rho family of GTPases (for example, Rac1)(Negishi and Katoh, 2002). A recent study showed that MICAL, a large cytosolic protein expressed in axons which contain flavoprotein mono-oxygenase domains, interacts with plexins (PlexinA1) and that mono-oxygenase inhibitors abrogate Sema3A-mediated axonal repulsion. These are the first experiments to link oxidoreductase processes with axon guidance (Terman et al., 2002). Taken together, these results illustrate the complexity of semaphorin/neuropilin/plexin signalling.

4. THE SLIT/ROUNDABOUT GUIDANCE MOLECULES

The third large family of chemotropic molecules is Slit, which were identified at the spinal cord midline (see Kaprelian et al., 2000 for a review). In mammals, known members of this family are Slit-1, -2 and –3. These secreted molecules are 200 amino acids in length and contain an N-terminal leucine-rich region which is vital for their biological functionality. Although they are yet to be demonstrated in vertebrates, Slits may be less diffusible than other secreted molecules due to the low diffusibility of their N-terminal fragment. Unlike semaphorins, the biological receptors of Slits, known as Robos (derived from 'roundabout', the Drosophila mutation from which the cue was originally isolated) have a large cytoplasmatic domain which makes them available for intracellular transduction. In vertebrates, Robo-1, Robo-2 and Rig-1 are the three known receptors of the Robo family. As in the case of Drosophila, a promiscuity in the binding of Slits to Robos (each Slit binds each Robo with a similar affinity), has also been reported in mammals, the exception being Rig-1 whose relationship to Slits is not completely clear. Interestingly, in invertebrates the slit-2/netrin-1 (two secreted ligands) interaction is as strong as Slit-2/Robo-2 (one ligand and its receptor), which makes the overall picture even more complicated (e.g., Guthrie, 2001). Slits also bind glypican-1 (a heparan sulfate proteoglycan) and Robos interact with other receptors, such as DCC (a netrin receptor) and CXCR4 (a chemokine receptor) (see Cooper, 2002 for review).

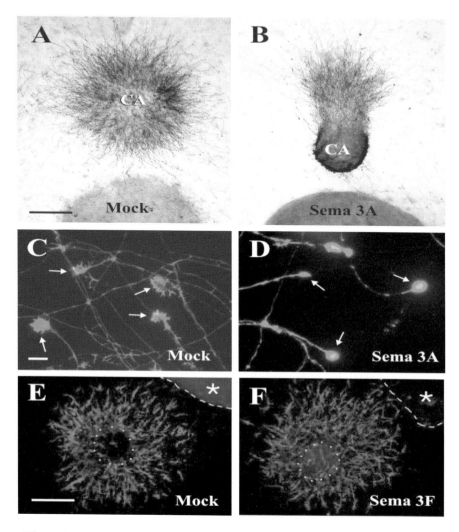

Figure 2. Examples of chemorepulsive and chemoattractive effects of secreted semaphorins. A-B) Sema3A producing cells induce strong chemorepulsion of embryonic hippocampal (CA) axons (stained with antibodies against TUJ-1; B) compared to controls (Mock transfected cells in A) in collagen matrices. C-D) Microphotograph showing growth cone collapse of hippocampal axons induced by Sema 3A. Explants from embryonic hippocampus (E16) were cultured on laminin-coated coverslips, treated with Sema3A during 1 hour, fixed and stained with Faloidin-TRITC. The addition of Sema3A into the culture media induces both filopodial and lamelipodial retraction in axonal ending tips (arrows in D). E-F) Examples of chemoattractive properties of semaphorins. Oligodendrocyte progenitors migrate towards the Sema3F source (asterisk in F) whereas they migrate radial when

confronted to Mock transfected cell aggregates (asterisk in E). Scale bars A and E = 200 μm pertains to B and F respectively; C = 10 μm pertains to D.

It has recently been demonstrated that Slit2/Robo-1 interactions need cell-surface heparan sulfate to exert maximal repulsive activities. Furthermore, the expression patterns of Slits and Robos, which are complementary and/or overlapping in many CNS structures, have led some researchers to propose that they may also act in an autocrine fashion: one cell secretes the cue to which it is responding to. This autocrine function has also been suggested for secreted semaphorins, although in neither case is there direct evidence of such a function in invertebrates or vertebrates.

Results published so far indicate that Slits exclusively repel growth cones (Nguyen-Ba-Charvet and Chedotal, 2002), this being the case for motor neurons and commissural neurons of the spinal cord (only after crossing the midline; see below), retinal ganglion cells, mitral cells from the olfactory bulb, thalamic neurons projecting into the cortex, cortical axons projecting into the corpus callosum and neurons projecting from the dentate gyrus. In some of these cases, Slits also induce collapse of the growth cones, as was described above for semaphorins. It has also been suggested that Slit-2 can induce axonal elongation and branching in neurons of the dorsal root ganglia of mammals (reviewed by McAllister, 2002).

The cytoplasmic mediators of Slit/Robo signalling are largely understood. However, recent studies have reported that GTPase activity regulate neuronal migration mediated by Robo receptors (e.g., Grose and van Vactor, 2002).

5. THE EPHRIN / EPH RECEPTOR GUIDANCE SYSTEM

Ephrins are a family of membrane-anchored molecules which bind to members of the Eph tyrosine kinase receptors (RTK). There are two classes of ephrins, depending on their anchoring to the cell membrane: Type A ephrins are GPI-linked proteins, while type B comprise transmembrane proteins with a short citoplasmic domain (Gale et al., 1996). To date, 5 ephrin-A ligands (ephrin-A1 to A5), 3 ephrin-B molecules (ephrin-B1 to B3) and 14 Eph receptors have been identified in vertebrates. The Eph receptors are categorized into two groups: A-subclass of Eph receptors (EphA1 to EphA8) and B-subclass of Eph receptors (EphB1 to EphB6). Although binding between receptors and ligands is promiscuous, ephrin-A ligands mainly bind only EphA receptors, and ephrin-B´s bind EphB´s, with the major exception of EphA4, which binds with similar affinities ligands from type-A or B subclasses. Ephrins and Eph receptor signalling can be uni or bidirectional. Thus, under special circumstances, Eph receptors can play roles as ligands, as well as ephrin family can function as receptors (Cowan and Henkemeyer, 2002).

Ephrins and their receptors have been involved in different events during development of nervous system (Wilkinson, 2001): determination of regional identities during regionalization of the neural tube, guidance of migrating cells, formation of axonal tracts and topographic projection maps, and axonal branching. Ephrins and Eph receptors were identified as repulsive molecules for growing axons and classified as major contact-dependent repellents. However, It is well known that different responses can be elicited by some ephrins if truncated Eph receptors are expressed playing roles of endogenous dominant-negative suppressors promoting cell adhesion instead axonal repulsion (Yue et al., 2002). For example, ephrins and Eph receptors prevent corticospinal tracts to cross the midline at the spinal cord (Imondi et al., 2000), while they canalize commissural axons to force them to cross midline at forebrain and midbrain (Orioli et al., 1996; Park et al., 1997).

Ephrins/Eph receptors play a capital role in the formation of topographic projection maps. The classical example is the development-of the retinotectal projection: axons from retinal ganglion cells of the temporal retina, project to the anterior part of the tectum, while axons coming from the nasal retina project to the posterior part of the optic tectum (Cheng et al., 1995; Drescher et al., 1995). The expression patterns of some ephrins and Eph receptors match with the topographic projection. Ephrin-A2 and –A5 are expressed in a high posterior-low anterior gradient, while EphA3 (in chick) or EphA5 (in mouse) are expressed in higher levels in the neurons of the temporal retina and in low levels in the nasal retina. Axons sensitive to ephrins (temporal axons) stop in the anterior part of the tectum, while those unsensitive to ephrins (nasal axons) colonise the posterior part of this structure. Studies in homozygous null mutants for either ephrin-A2 and/or ephrin-A5, as well as the ectopic expression of EphA3 in subsets of retinal ganglion cells corroborate the important role of ephrin/Eph in the establishment of this topographic map (Frisén et al., 1998; Brown et al., 2000; Feldheim et al., 2000). Similarly, ephrins/Eph receptors have been involved in the formation of many other topographic maps: from the retina to the lateral geniculate nucleus from the vomeronasal organ to the accessory olfactory bulb, nuclei, thalamocortical, from the hippocampus to the forebrain septum and the topographic peripheral projection from the motor neurons in the spinal cord to muscular groups (see also Knöll and Drescher, 2002). Ephrins has been also involved in axonal branching. They are suppresors of axonal branching from retinal ganglion cells (Roskies et al., 1994), while ephrin-A5 and ephrin-A2 could underlie the proper branching of hippocampal axons (Gao et al., 1999; Man et al., 2002). Ephrin-A5 shows a dual effect: selectively inhibits the branching of EphA5-expressing cortico-cortical axons projecting to ephrin-A5-non-expressing cortical layers, and promotes axonal branching of those neurons which project to ephrin-A5-expressing cortical layers (Castellani et al., 1998).

As happens with other axon guidance molecules, intracellular signalling for ephrins/Eph receptors remains poorly understood. EphAs activation results in inactivation of integrins via FAK (focal adhesion kinase involved in the integrin signalling pathway), which, in turn, results in reduction in cell adhesion and repulsion (Miao et al., 2000). Conversely, Ephrin-As activation acts via β1-integrin and src-kinases increasing cell adhesion and attraction of axons (Davy and Robbins, 2000; Huai and Drescher, 2001). Opposite effects of EphA receptors and ephrin-As on MAP kinase signalling have been also shown: EphAs inactivate ERK1 and ERK2 which results in the inactivation of myosin light chains and ephrin-A ligands activate-it promoting of cell adhesion and axonal grow (Davy and Robbins, 2000). Type B ephrins and Eph receptors contained a carboxy-terminal extreme which binds to proteins containing PDZ domains, which drives the activation of Ras-kinase (see Cowan and Henkemeyer, 2002). Moreover, some metalloproteases (e.g., Kuzbanian) or truncated Eph receptors could modulate the intracellular signalling by binding to extracellular domains of the EpH receptors (Hottori et al., 2000; Yue et al., 2002).

6. ROLE OF AXON GUIDANCE MOLECULES IN NEURAL REGENERATION

Unlike peripheral axons, those of the adult central nervous system are unable to regenerate spontaneously after lesion. CNS injuries are followed by persistent histological alterations that include cell death, degeneration of transected nerve fibres, infiltration and proliferation of non-neuronal cells (meningeal and fibroblast cells), and deposition of extracellular matrix components at the lesion site (Fawcett and Asher, 1999). Thus, despite local sprouting responses of lesioned fibres, CNS axons are unable to regrow and therefore fail to reconnect to their former target cells. Meningeal and reactive astroglial cells interact at the lesion site forming a barrier or glia limitant-like structure which greatly inhibits axon regrowth. This non-permissive "milieu", also termed "glial scar", contains inhibitory molecules, such as Nogo, and myelin-associated glycoproteins (MAG or MOG), as well as a huge number of extracellular matrix elements, such as sulphated proteoglycans and collagen (reviewed by Horner and Gage, 2000).

Recent studies have demonstrated that stab lesions in the forebrain and spinal cord induce the re-expression of several semaphorins by infiltrating meningeal cells at the lesion site (Pasterkamp and Verhaagen 2001). In contrast, in contusion lesions semaphorin expression is restricted to the meningeal sheet. A recent study by Winter et al.(Winter et al., 2002), showed that descending spinal fibres (cortico- and rubrospinal tract axons) whose neuropilin and plexin expression remains unchanged after spinal cord lesion are not able to re-enter the semaphorin-positive area of the meningo-glial scar. Sema 3A is not present at the lesion site after peripheral lesion, but motor neurons downregulate their Sema 3A expression after peripheral nerve injury, which coincides with the period of axonal re-growth (Pasterkamp et al., 1998).

Miranda et al., 1999 have demonstrated the over-expression of other axon guidance molecules in the "glial scar" and reactive glia. For example, Eph B3 is re-expressed by reactive astrocytes after contusive spinal cord lesion. In contrast, EphB receptors are unchanged in other lesion models (such as after optic nerve transaction), while EphA receptors levels are reduced after deafferentation (Knoll et al., 2001). Taken together, these data demonstrate that each CNS lesion is unique in terms of cellular and molecular responses and these growth inhibitory molecules have specific roles in different lesions.

Netrins and their receptors have also been shown to be regulated after lesion, there being a 40-fold increase in the expression of netrin-1 in Schwann cells after peripheral nerve lesion (Madison et al., 2000). Furthermore, several studies have demonstrated a down-regulation of netrin receptors (DCC and UNC5) in retinal ganglion cells after optic nerve lesion transections following peripheral nerve grafting in fish (Petrausch et al., 2000) and rats (Ellezam et al., 2001), as well as in regenerating olfactory axons following unilateral bulbactomy (Astic et al., 2002). In addition, UNC5 receptors are up-regulated exclusively in neurons whose axons regenerate poorly after spinal cord lesion in lamprey (Shifman and Selzer, 2000). Further studies are needed to elucidate the physiological role of netrins and their receptors in neural regeneration. These data strengthen the rationale for studying the role of developmental guidance molecules during CNS regeneration.

7. THE SYNERGIC EFFORT OF AXON GUIDANCE CUES: THE DEVELOPMENT OF HIPPOCAMPAL CONNECTIONS AS A MODEL

The hippocampus and fascia dentada are discrete cortical areas with a unique laminar organization of cell layers and afferent connections. The cell bodies of the principal neurons (pyramidal and granule cells) are grouped into two cell layers, and the major neural connections (entorhinal, commissural and septal projections) terminate in different lamina along the dendrites of these principal neurons (Amaral and Witter, 1995). The segregated termination of these afferents raises the question of how these accurate connections are established and specified during development. In addition, commissural and septal afferents strongly react after the lesion of the entorhino-hippocampal connections, and sprout into the deafferented hippocampal layers establishing new ectopic synaptic contacts (Amaral and Witter, 1995).

In vitro developmental studies have shown that entorhinal and commissural axons are repelled by secreted semaphorins (3A and 3F) (see above) (Chedotal et al., 1998, Steup et al., 1999; Pozas et al., 2001). In addition, netrin-1 attracts hippocampal commissural axons in collagen gel cultures (Barallobre et al., 2000, Steup et al., 2000, see also Skutella and Nitsch 2001, for review). In situ hybridization studies have shown that semaphorins and their receptors are expressed in the entorhinal cortex and hippocampus. Moreover, the netrin-1 receptor (DCC) is expressed by hippocampal neurons, and the presence of netrin-1 mRNA close to the fimbria, in the ventricle. These data suggest that entorhinal axons repelled by cortical semaphorins are directed towards the hippocampus. In addition, semaphorin expression in the entorhinal cortex prevents the invasion of commissural axons which are attracted towards the fimbria by netrin-1.

Another member of the semaphorin family, (Sema 3C), repels septal neurites but not those of entorhinal or hippocampal origin, thus regulating the entry of septohippocampal afferents (Steup et al., 2000). In addition, ephrin A3 and its ligand EphA5 are involved in the lamina-specific ingrowth of entorhinal afferents (Stein et al., 1999). Lastly, a recent study has reinforced the role of the Eph family of guidance factors in regulating the hippocampo-septal connections (Yue et al., 2002). Eph receptors and their ligands are expressed in opposing patterns in the hippocampus and the lateral septum and the ectopic expression of truncated Eph receptors alters the hippocampo-septal topographic map (Gao et al., 1999, Yue et al., 2002).

Slits and Robos are also present during the development of hippocampal connections. Thus, Slit1 and Slit2 are expressed in the entorhinal cortex, while their receptors (Robos) are expressed in the dentate gyrus of the hippocampus (Nguyen Ba-Charvet et al., 1999). The putative role of the Slits/Robos may be in preserving the laminar organization of the intrinsic hippocampal connections.

Taken together, these data demonstrate that several axon guidance molecules and their receptors play active roles in promoting outgrowth and/or guidance of hippocampal connections. In addition, synergic and coordinated actions with other molecules, such as neurotrophins or cell adhesion molecules, modulate their attractive/chemorepulsive properties.

REFERENCES

Ackerman SL, Kozak LP, Przyborski SA, Rund LA, Boyer BB, Knowles BB. (1997) The mouse rostral cerebellar malformation gene encodes an UNC-5-like protein. Nature. 386:838-842.

Aizawa H, Wakatsuki S, Ishii A, Moriyama K, Sasaki Y, Ohashi K, Sekine-Aizawa Y, Sehara-Fujisawa A, Mizuno K, Goshima Y, Yahara I (2001) .Phosphorylation of cofilin by LIM-kinase is necessary for semaphorin 3A-induced growth cone collapse. Nat Neurosci. 4:367-373

Astic L, Pellier-Monnin V, Saucier D, Charrier C, Mehlen P (2001) .Expression of netrin-1 and netrin-1 receptor, DCC, in the rat olfactory nerve pathway during development and axonal regeneration. Neuroscience. 109:643-656.

Bagnard D, Lohrum M, Uziel D, Puschel AW, Bolz J (1998) Semaphorins act as attractive and repulsive guidance signals during the development of cortical projections. Development. 125:5043-5053

Barallobre MJ, Del Rio JA, Alcantara S, Borrell V, Aguado F, Ruiz M, Carmona MA, Martin M, Fabre M, Yuste R, Tessier-Lavigne M, Soriano E. (2000) Aberrant development of hippocampal circuits and altered neural activity in netrin 1-deficient mice. Development.127:4797-4810.

Behar O, Golden JA, Mashimo H, Schoen FJ, Fishman MC.(1996) Semaphorin III is needed for normal patterning and growth of nerves, bones and heart. Nature. 383:525-528

Braisted JE, Catalano SM, Stimac R, Kennedy TE, Tessier-Lavigne M, Shatz CJ, O'Leary DD. (2000) Netrin-1 promotes thalamic axon growth and is required for proper development of the thalamocortical projection. J Neurosci.20:5792-5801.

Brown, A. *et al.* (2000) Topographic mapping from the retina to the midbrain is controlled by relative but not absolute levels of EphA receptor signaling. Cell 102:77-88.

Castellani V, Yue Y, Gao P-P, Zhou R, Bolz J (1998) Dual action of a ligand for Eph receptor tyrosine kinases on specific populations of axons during the development of cortical circuits. J. Neurosci. 18:4663-4672.

Cheng H-J, Nakamoto M, Bergemann AD, Flanagan JG (1995) Complementary gradients in expression and binding of ELF-1 and Mek4 in development of the topographic retinotectal projection map. Cell 82:371-381

Cooper, H.M. (2002) Axon guidance receptors direct growth cone pathfinding: rivalry at the leading edge. Int. J. Dev. Biol., 46:621-631.

Cowan CA, Henkemeyer M (2002) Ephrins in reverse, park and drive. Trends Cell Biol. 12: 339-346.

Davy A, Robbins SM (2000) Ephrin-A5 modulates cell adhesion and morphology in an integrin-dependent manner. EMBO J. 19:5396-5405.

De Castro F, Hu L, Drabkin H, Sotelo C, Chedotal A.Chemoattraction and chemorepulsion of olfactory bulb axons by different secreted semaphorins.J Neurosci. 1999 Jun 1;19(11):4428-4436.

De Winter F, Oudega M, Lankhorst AJ, Hamers FP, Blits B, Ruitenberg MJ, Pasterkamp RJ, Gispen WH, Verhaagen J.(2002) Injury-induced class 3 semaphorin expression in the rat spinal cord. Exp Neurol. 175:61-75.

Drescher U, Kremoser C, Handwerker C, Loschinger J, Noda M, Bonhoeffer F. (1995) *In vitro* guidance of retinal ganglion cell axons by RAGS, a 25 kDa tectal protein related to ligands for Eph receptor tyrosine kinases. Cell 82:359-370.

Eickholt BJ, Walsh FS, Doherty P.(2002) An inactive pool of GSK-3 at the leading edge of growth cones is implicated in Semaphorin 3A signaling. J Cell Biol. 157:211-217

Ellezam B, Selles-Navarro I, Manitt C, Kennedy TE, McKerracher L.(2001) Expression of netrin-1 and its receptors DCC and UNC-5H2 after axotomy and during regeneration of adult rat retinal ganglion cells. Exp Neurol. 168:105-115

Fawcett JW, Asher RA.(1999) The glial scar and central nervous system repair. Brain Res Bull. 49:377-391.

Feldheim DA, Kim YI, Bergemann AD, Frisén J, Barbacid M, Flanagan JG (2000) Genetic analysis of ephrin-A2 and ephrin-A5 shows their requirement in multiple aspects of retinocollicular mapping. Neuron 25:563-574.

Forcet C, Stein E, Pays L, Corset V, Llambi F, Tessier-Lavigne M, Mehlen P. (2002). Netrin-1-mediated axon outgrowth requires deleted in colorectal cancer-dependent MAPK activation. Nature. 417:443-447.

Frisen J, Yates PA, McLaughlin T, Friedman GC, O'Leary DD, Barbacid M (1998) Ephrin-A5 (AL-1/RAGS) is essential for proper retinal axon guidance and topographic mapping in the mammalian visual system. Neuron 20:235-243.

Gale N, Holland S, Valenzuela D, Flenniken A, Pan L, Ryan T, Henkemeyer M, Strebhardt K, Hirai H, Wilkinson D, Pawson T, Davis S, Yancopoulus G (1996) Eph receptors and ligands comprise two major specificity subclasses and are reciprocally compartmentalized during embryogenesis. Neuron 17:9-19.

Gao PP, Yue Y, Cerretti DP, Dreyfus C, Zhou R.(1999) Ephrin-dependent growth and pruning of hippocampal axons. Proc Natl Acad Sci U S A. 96:4073-4077.

Ghose A, Van Vactor D.(2002) GAPs in Slit-Robo signaling. Bioessays. 24:401-404.

Giger RJ, Kolodkin AL (2001) Silencing the siren: guidance cue hierarchies at the CNS midline. Cell 105:1-4.

Guthrie S (2001).Axon guidance: Robos make the rules. Curr Biol. 11:300-303.

Hattori M, Osterfield M, Flanagan JG (2000) Regulated cleavage of a contact-mediated axon repellent. Science 289:1360-1365.

Horner PJ, Gage FH.(2000) Regenerating the damaged central nervous system. Nature. 407:963-970.

Huai J, Drescher U (2001) An ephrin-A-dependent signalling pathway controls integrin function and is linked to the tyrosine phosphorylation of a 120 kDa protein. J. Biol. Chem. 276:6689-6694.

Imondi R, Wideman C, Kaprielian Z (2000) Complementary expression of transmembrane ephrins and their receptors in the mouse spinal cord: a possible role in constraining the orientation of longitudinally projecting axons. Development 127:1397-1410.

Kaprielian Z, Imondi R, Runko E (2000). Axon guidance at the midline of the developing CNS. Anat Rec. 261:176-197.

Kennedy, T.E., Serafini, T., de la Torre, J.R Tessier-Lavigne, M. (1994) Netrins are difusable chemotrophic factors for commissural axons in the embryonic spinal cord. Cell 78:425-435.

Knöll B, Drescher U (2002) Ephrin-As as receptors in topographic projections. Trends Neurosci. 25:145-149.

Knöll B, Isenmann S, Kilic E, Walkenhorst J, Engel S, Wehinger J, Bahr M, Drescher U.Graded expression patterns of ephrin-As in the superior colliculus after lesion of the adult mouse optic nerve. Mech Dev. 2001 Aug;106(1-2):119-127

Li X, Meriane M, Triki I, Shekarabi M, Kennedy TE, Larose L, Lamarche-Vane N (2002). The Adaptor Protein Nck-1 Couples the Netrin-1 Receptor DCC (Deleted in Colorectal Cancer) to the Activation of the Small GTPase Rac1 through an Atypical Mechanism. J Biol Chem. 277:37788-37797.

Li X, Saint-Cyr-Proulx E, Aktories K, Lamarche-Vane N (2002) .Rac1 and Cdc42 but not RhoA or Rho kinase activities are required for neurite outgrowth induced by the Netrin-1 receptor DCC (deleted in colorectal cancer) in N1E-115 neuroblastoma cells. J Biol Chem. 277:15207-15214.

Livesey, F.J. (1999) Netrins and netrin receptors, Cell Mol. Life Sci. 56:62-68.

Llambi, F., Causeret F., Bloch-Gallego E and Mehlen P (2001). Netrin-1 acts as a survival factor via its receptors UNC5H and DCC. EMBO J. 2001 20:2715-2722.

Madison RD, Zomorodi A, Robinson GA.(2000) Netrin-1 and peripheral nerve regeneration in the adult rat. Exp Neurol. 161:563-570.

Mann F, Peuckert C, Dehner F, Zhou R, Bolz J (2002) Ephrins regulate the formation of terminal axonal arbors during the development of thalamocortical projections. Development 129:3945-3955.

McAllister AK.(2002) Conserved cues for axon and dendrite growth in the developing cortex. Neuron.33:2-4.

Miao H, et al. (2001) Activation of Eph receptor tyrosine kinase inhibits the Ras/MAPK pathway. Nat. Cell. Biol. 3: 527-530.

Ming GL, Wong ST, Henley J, Yuan XB, Song HJ, Spitzer NC, Poo M (2002) Adaptation in the chemotactic guidance of nerve growth cones. Nature.417:411-418.

Miranda JD, White LA, Marcillo AE, Willson CA, Jagid J, Whittemore SR.(1999) Induction of Eph B3 after spinal cord injury. Exp Neurol.156:218-222.

Mitsui N, Inatome R, Takahashi S, Goshima Y, Yamamura H, Yanagi S (2002).Involvement of Fes/Fps tyrosine kinase in semaphorin3A signaling. EMBO J. 21:3274-3285.

Nakamura F, Kalb RG, Strittmatter SM (2000).Molecular basis of semaphorin-mediated axon guidance. J Neurobiol. 44:219-229.

Neufeld G, Cohen T, Shraga N, Lange T, Kessler O, Herzog Y.(2002) The neuropilins: multifunctional semaphorin and VEGF receptors that modulate axon guidance and angiogenesis.Trends Cardiovasc Med. 12:13-19.

Nguyen Ba-Charvet KT, Brose K, Marillat V, Kidd T, Goodman CS, Tessier-Lavigne M, Sotelo C, Chedotal A.Slit2-Mediated chemorepulsion and collapse of developing forebrain axons. Neuron. 22:463-473.

Nguyen-Ba-Charvet KT, Chedotal A (2002).Role of Slit proteins in the vertebrate brain. J Physiol (Paris). 96:91-98.

Orioli D, Henkemeyer M, Lemke G, Klein R, Pawson T. (1996) Sek4 and Nuk receptors cooperate in guidance of commissural axons and in palate formation. EMBO J. 15:6035-6049.

Park S, Frisen J, Barbacid M (1997) Aberrant axonal projections in mice lacking EphA8 (Eek) tyrosine protein kinase receptors. EMBO J. 16:3106-3114.

Pasterkamp RJ, Giger RJ, Verhaagen J.(1998) Regulation of semaphorin III/collapsin-1 gene expression during peripheral nerve regeneration. Exp Neurol. 153:313-327.

Pasterkamp, R.J. and Verhaagen J. (2001). Emerging roles for semaphorins in neural regeneration. Brain Res. Rev. 35:36-54.

Petrausch B, Jung M, Leppert CA, Stuermer CA.(2000) Lesion-induced regulation of netrin receptors and modification of netrin-1 expression in the retina of fish and grafted rats. Mol Cell Neurosci. 16:350-364.

Pozas E, Pascual M, Nguyen Ba-Charvet KT, Guijarro P, Sotelo C, Chedotal A, Del Rio JA, Soriano E. (2001) Age-dependent effects of secreted Semaphorins 3A, 3F, and 3E on developing hippocampal axons: in vitro effects and phenotype of Semaphorin 3A (-/-) mice. Mol Cell Neurosci. 18:26-43.

Raper JA. (2000) Semaphorins and their receptors in vertebrates and invertebrates. Curr Opin Neurobiol. 10:88-94.

Roskies AL O'Leary DDM (1994) Control of topographic retinal axon branching by inhibitory membrane-bound molecules. Science 265:799-803.

Serafini T, Colamarino SA, Leonardo ED, Wang H, Beddington R, Skarnes WC, Tessier-Lavigne M (1996). Netrin-1 is required for commissural axon guidance in the developing vertebrate nervous system. Cell 87:1001-1014.

Serafini T., Kennedy T.E., Galko M.J., Mirzayan, C., Jessell, T.M. and Tessier-lavigne M. (1994) The netrins define a family of axon outgrowth-promoting proteins homologous to C. elegans UNC-6. Cell, 78:409-424.

Serafini, T., Colamarino, S.A., Leonardo, E.D., Wang, H., Beddington R y cols., (1996). Netrin-1 is required for commissural axon guidance in the developing vertebrate nervous system. Cell, 87:1001-1014.

Shifman MI, Selzer ME.(2000) Expression of the netrin receptor UNC-5 in lamprey brain: modulation by spinal cord transection. Neurorehabil Neural Repair. 14:49-58.

Shirvan A, Shina R, Ziv I, Melamed E, Barzilai A.(2000) Induction of neuronal apoptosis by Semaphorin3A-derived peptide. Brain Res Mol Brain Res. 83:81-93.

Skutella T, Nitsch R.(2001) New molecules for hippocampal development. Trends Neurosci. 24:107-113.

Stein E, Savaskan NE, Ninnemann O, Nitsch R, Zhou R, Skutella T (1999).A role for the Eph ligand ephrin-A3 in entorhino-hippocampal axon targetin J Neurosci. 19:8885-8893.

Spassky N, de Castro F, Le Bras B, Heydon K, Queraud-LeSaux F, Bloch-Gallego E, Chedotal A, Zalc B, Thomas JL (2002). Directional guidance of oligodendroglial migration by class 3 semaphorins and netrin-1. J Neurosci 22:5992-6004.

Steup A, Lohrum M, Hamscho N, Savaskan NE, Ninnemann O, Nitsch R, Fujisawa H, Puschel AW, Skutella T.(2000) Sema3C and netrin-1 differentially affect axon growth in the hippocampal formation. Mol Cell Neurosci. 15:141-155.

Steup A, Ninnemann O, Savaskan NE, Nitsch R, Puschel AW, Skutella T.Semaphorin D acts as a repulsive factor for entorhinal and hippocampal neurons. Eur J Neurosci. 1999 11:729-734.

Tamagnone L, Comoglio PM.(2000) Signalling by semaphorin receptors: cell guidance and beyond. Trends Cell Biol.10:377-383.

Terman JR, Mao T, Pasterkamp RJ, Yu HH, Kolodkin AL (2002).MICALs, a family of conserved flavoprotein oxidoreductases, function in plexin-mediated axonal repulsion.Cell. 109:887-900.

Tessier-Lavigne, M. and C.S. Goodman (1996). The molecular biology of axon guidance. Science, 274:1123-1133.

Walsh, F.S. and P. Doherty (1997) Neural cell adhesion molecules of the immunoglobulin superfamily: Role in axon growth and guidance. Annu. Rev. Cell Biol. 13:425-456.

Wilkinson DG (2001) Multiple roles of Eph receptors and ephrins in neural development. Nature Rev. Neurosci. 2:155-164.

Winberg ML, Tamagnone L, Bai J, Comoglio PM, Montell D, Goodman CS (2001).The transmembrane protein Off-track associates with Plexins and functions downstream of Semaphorin signaling during axon guidance. Neuron. 32:53-62.

Yu TW, Bargmann CI.(2001) Dynamic regulation of axon guidance. Nat Neurosci. 2001 4:1169-1176.

Yue Y, Chen ZY, Gale NW, Blair-Flynn J, Hu TJ, Yue X, Cooper M, Crockett DP, Yancopoulos GD, Tessarollo L, Zhou R (2002) Mistargeting hippocampal axons by expression of a truncated Eph receptor. Proc Natl Acad Sci U S A. 99:10777-10782.

Zisch AH, Stallcup WB, Chong LD, Dahlin-Huppe K, Voshol J, Schachner M, Pasquale EB (1997) Tyrosine phosphorylation of L1 family adhesion molecules: implication of the Eph kinase Cek5. J Neurosci Res. 47:655-665.

D. VAN ROSSUM[1] AND U.K. HANISCH[2]

[1]*Institute of Neuropathology, University of Göttingen, Germany*

[2]*Department of Cellular Neuroscience Max Delbrück Center (MDC) for Molecular Medicine Berlin and University of Applied Sciences Lausitz, Senftenberg, Germany*

12. MICROGLIA AND THE CEREBRAL DEFENCE SYSTEM

Summary: Mechanisms of microglial activation do not only concern neuroscientists with specialized interest in glial physiology. Microglial cells are gaining attraction as these cells represent the primary sensors and response elements in virtually any neuropathology. As macrophage-like cells they represent a CNS-intrinsic element of innate defence mechanisms. The inducible synthesis of chemoattractive and immunoregulatory factors and the ability to present antigen help the recruitment of leukocytes and the engagement of adaptive immune responses. Even though the primary functions of microglia serve the defence and protection of the nervous tissue the beneficial contributions of these cells are only now being increasingly acknowledged. Indeed, microglial cells appear to safeguard and support neuronal functions, always being ready to transform into alerted states upon challenges by foreign material or disturbances in the CNS homeostasis. Deregulated or dysfunctional microglia can, on the other hand, be critical or at least instrumental in a harmful way. Excessive and chronic reaction of microglia can then fuel destructive cascades upon trauma or during inflammatory and neurodegenerative processes. As the signals and mechanisms of microglial activation become illuminated hope builds up on strategies for selective interference with its undesired outcomes in favour of the overall beneficial potential.

1. INTRODUCTION

Microglia represents the nervous system equivalent of a tissue macrophage and is considered its major resident immunocompetent cell. For a long time, these small inconsiderable cells had been the most enigmatic cell type of the CNS. A functional significance remained obscure. With the late nineteen eighties, the representation of microglia within the research efforts of basic and clinically oriented neurosciences changed dramatically. Over the recent decade, more than 1700 publications carried 'microglia' in the title. The total number of reports relating to microglia is even larger. Why is there such a sudden increase of interest? First, critical information had accumulated revealing the ontogenetic routes and very macrophage-like nature of these cells. Improved techniques to stain microglia in tissue slices and to prepare cultures allowed for *in vitro* and *in vivo* analyses of morphological, antigenic and physiological properties. Experimental manipulations delivered valuable data about the functional changes upon microglial challenges. Induced *in vivo* by an array of still mostly unidentified signals, the process of 'microglial activation' can also be triggered *in vitro* by using defined agents, such as bacterial cell wall components (Fig. 1). The activation process then consists of a stepwise transformation turning the cells into highly motile, secretory active and potentially toxic phagocytes. Stages of activation from alerted to fully reactive cells could be correlated with alterations in morphology, the expression pattern of cell surface antigens and a repertoire of executive functions, such as production and secretion of potent mediators or potential toxins. Second, an enormous body of histological evidence based on CNS specimen from virtually any neuropathological scenario revealed moderate to massive microglial activation. These findings paired with the *in vitro* demonstrations of neurotoxic potential suggested that microglia involvement would necessarily mean disease promotion. Microglial cells can, indeed, contribute to and aggravate destructive processes, by multiple means, but this dominant prejudice got weakened. Improved techniques and refined models have been meanwhile revealing neurotrophic abilities of microglia. Third, with a more balanced view of potentially dangerous, but primarily beneficial activities of microglia current research is approaching a question of huge practical value and therapeutic relevancy: Why and how can the phylogenetically approved microglial program to defend and protect neighbouring neurons (and glia) occasionally escape into disastrous outcomes? Having raised this question, efforts are taken to identify molecular targets and cellular principles for the development of pharmacological interference with undesired or overshooting microglial activity. This chapter provides an introduction into the neurobiology and clinical relevancy of microglia. For further reading, we may suggest some recent review collections on microglial topics as published in book or special issue form (see the first three entries of the reference list). Historical information on the concept of microglia is excellently summarized Rezaie and Male (2002).

2. THE MICOGLIAL ENVIRONMENT

Microglia is an integral cellular component of the CNS (Giulian, 1995; Streit *et al.* 2000). The CNS is known to differ in many aspects from most other tissues as it regards the mechanisms of innate and adapted immunity, a situation termed 'immune privilege'. Still, local inflammatory processes can develop and specific immune reactions can occur, but the vulnerable neuronal structures require that the defence system activities are tightly controlled (Streit *et al.*, 2000). Mammalian central neurons have — despite of the need for functional plasticity and structural adaptations of their connections — only limited capacity for repair and regeneration. The CNS cannot tolerate all the consequences of a defence battle, including extensive tissue swelling and collateral damage from soluble compounds and cell-mediated attacks. Therefore, the CNS appears to impose some specific regulations on these mechanisms. Reluctance to generate inflammatory and immune reactions — involving the collaboration of activated microglia — is an actively maintained status rather than deficiency to host such events (Lazarov-Spiegler *et al.*, 1998).

Microglia faces this situation and has to interact with a unique ensemble of neighbouring cell types it serves to protect (Raivich *et al.*, 1999; Streit *et al.*, 2000).

Parenchymal microglia is adapted to this community and its intercellular signalling. Molecules employed for neuronal signalling, such as certain (co)transmitters, may provide microglia with information about normal brain activity. Direct physical connections between microglia and neuronal cells may fulfil similar functions.

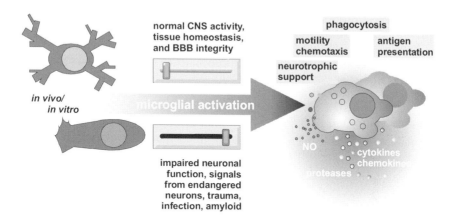

Figure 1. Microglial activation is pictured as the stepwise transition of cells from the resting status of the healthy adult brain tissue (in vivo) or an unstimulated control condition of a culture (in vitro) to an alerted, activated and finally reactive state. Activated cells can enhance the expression of cell surface receptors, increase their release of a plethora of factors, including neurosupportive, immuno/neuroregulatory and toxic compounds, gain high motility, perform phagocytosis and present antigen. Endogenous and exogenous influences control the activity status, triggering transition upon increased or decreased signalling through microglial receptor systems.

An exciting field within neurosciences deals with the communication between neurons and glia, data demonstrating that glial cells, such as astrocytes, do not only provide metabolic and homeostatic support. Glia more and more appears to be an active partner of neuronal information processing (Polazzi and Contestabile, 2002). Neuron-glia signalling may involve the coordinated exchange of multiple molecular signals between the various cell types (Bezzi and Volterra, 2001). It thus takes no wonder that glial cells show, at closer inspection, regional variation in their cellular and molecular features, such as receptor equipment. In addition to the CNS-specific conditions in general, microglial properties can also vary with the subregion, gender and age.

The blood-brain barrier (BBB) and blood-cerebrospinal fluid-barrier warrant an organized exchange of molecules and limit the passage of cells between the CNS and extraneural compartments. Parenchymal microglia being embedded in the CNS milieu is thus normally not confronted with certain circulating compounds, namely constitutive or facultative serum proteins.

Lymphocyte patrolling of the brain and spinal cord tissue is apparently a rare process, when compared to other body regions. Still, microglia of the healthy brain may have occasional contact to cellular components of the immune system as part of, at least, some immune surveillance. More intensive exposure to leukocytic populations and their release products occurs when these cells invade the CNS as part of an inflammatory process and immune response. Microglia can then present antigen and seems to be rather efficient in re-stimulating lymphocytes.

Another difference between microglia and its macrophage counterparts of other tissues concerns the life span (turnover). Parenchymal microglia shows little signs of proliferation. Replenishment is apparently limited. Transient activation episodes with increased local cell numbers might be followed by a return to a resting state. Such transition is observed in organotypic cultures. While proliferation/migration increase microglial numbers, cell death would deplete a regional population. Loss of microglia could result in locally impaired production of supportive factors and a drop in protective capacity as dependent on a critical microglial density.

3. MICROGLIA AS A SENSOR OF CNS HOMEOSTASIS

Microglial cells of the healthy adult tissue typically reveal a ramified morphology. Accounting for 5 to 20 % of the total glial population, cells are found in both grey and white matter, but somewhat unevenly distributed throughout the CNS. Regional differences in morphology may extend to functional aspects. Still, each cell takes its own territory after amoeboid cells populated the brain during early development to participate as phagocytes in neural tissue maturation. Thereafter, having acquired the ramified ('resting') state, microglia remains a rather stable population. While such a static view may not hold true completely, as the plasticity of this system is still not fully understood, it is still important to consider resting microglia as an established CNS cell which resides unobtrusively among its neighbours. Even though the term 'resting' could imply functional quiescence, it should instead be seen as a 'standby mode' at which cells permanently monitor their environment. This is an active process. Such a functional profile requires a sensory equipment, i.e. receptors and ion channels that report on the appearance, unusual concentration or the altered features of soluble and insoluble factors as they indicate danger or damage (Fig. 2). As soon as signs of homeostatic disturbance reach a microglial cell it can rapidly respond with a complex program of supportive and protective activities, safeguard innate defence mechanisms and assist in adaptive (i.e. antigen-specific) immune reactions. Microglia must also harbour an efficient intracellular apparatus for integrating various signals. The cellular 'activation' program includes sensory adjustments. Several receptors and ion channels show an increased or a *de novo* expression — or just the opposite, i.e. suppression of receptor signalling capacities, in order to shut off certain external influences.

Table 1. Signals and response modifiers in microglial activation

Structures of infectious agents	Neurothrophic factors (NF)
bacterial cell wall proteoglycans	brain-derived NF (BDNF)
gp41 (viral coat protein)	glial-derived NF (GDNF)
gp120	nerve growth factor (NGF)
lipopolysaccharide (LPS)	neurothrophin 3 (NT-3)
prion protein (PrP)	neurothrophin 4 (NT-4)
Tat (viral nuclear protein)	
teichoic and lipoteichoic acid (LTA)	**Neurotransmission-related compounds**
	β-adrenergic agonists
AD-related proteins	ATP (and related purines)
β-amyloid (aggregates)	glutamate
Aβ25-35	kainate
Aβ40/Aβ42	NMDA
Cytokines	**Ions**
GM-CSF	K^+
IL-1β	Mn^{2+}
IL-2	
IL-4	**Other proteins and peptides**
IL-6	albumin
IL-10	apolipoprotein E (ApoE)
IL-12	CD40L
IL-15	complement factor C1q
IL-18	complement factor C5a
IFNγ	melanocyte stimulating hormone
M-CSF	endothelin
TGFβ	S100B
TNFα	thrombin
	vasoactive intestinal peptide (VIP)
Ligands for the chemokine receptors*	
CCR3	**Other compounds**
CCR5	cannabinoids
CXCR2	ceramide
CXCR3	gangliosides
CXCR4	lysophosphatidic acid (LPA)
CX3CR1	melatonine
IL-8R	opiods/endomorphines
	platelet-activating factor (PAF)
Immunoglobulins (Ig)	prostaglandine E_2 (PGE_2)
IgG	steroids/steroid hormones
IgM	Vitamin D_3

*due to their promiscuous nature, chemokine receptors may accept several chemokines

Structures of viral or bacterial nature are among the factors with demonstrated potential for triggering microglial activation. Viral envelopes and nuclear proteins, bacterial cell wall components and other infectious agents (prion protein) can cause tremendous activation. Lipopolysaccharide (LPS) from Gram-negative strains has been especially useful in macrophage/microglia research. It is frequently employed as a tool both to mimic infections as well as to trigger 'activation' *per se* in order to study modulatory influences of other factors. Proteoglycans or (lipo)teichoic acid (LTA) as the cell wall determinants of gram-positive bacteria are powerful inducers of macrophage/microglia responses as well. Bacterial DNA has own contributions. Platelet activating factor, lysophosphatidic acid, gangliosides, prostaglandins and other lipids or fatty acid-derived and inflammatory mediators are all stimulating or modifying microglial activity. The range of structural and physiological diversity of these factors is wide (Nakamura, 2002) (Table 1).

Normally, serum proteins are denied CNS entry. However, when the integrity of the BBB is compromised, e.g. at sites of a tissue injury, blood content can inundate the tissue. A series of proteins can then carry microglia-activating information as immediate trauma signal, notably coagulation factors, proteases, their inhibitors or complexes of both, lipid-loaded albumin, complement factors or immunoglobulins. Some proteins fulfil functions as accessory factors for the interaction of bacterial toxins with the microglial cell surface (e.g., LPS-binding protein). CNS injury may also release factors with latent microglia-activating character that are bound to the extracellular matrix in a functionally 'silent pool'.

Endangered neurons emit signals that instruct gradual microglial transformation. Neurons may call for (enhanced) neurotrophic support or temporary removal of excitatory inputs ('synaptic stripping') to recover from a partial injury or impaired vitality. Some of these signals may be liberated or generated from the cytosol or the membranes of stressed cells. Such signals can be subtle, affecting microglia in the vicinity of neuronal somata while the primary insult is set farther away. While search for these signals is still on the way first candidates are nominated. Microglial activation might be initiated or modified by molecules commonly (co-)released in neurotransmission. Low levels of ATP or its degradation products could 'convince' microglia of normal neuronal activity and even counteract features of activation. In contrast, much higher ATP levels or extracellular K^+ concentrations exceeding the normal range would indicate rather abnormal firing activity of neurons, irreversible damage and ongoing disintegration (Fig. 3).

Cytokines are of particular importance in the regulation and pathophysiological contribution of microglia. Several interleukins (IL), namely IL-1 family members, interferon-γ (IFNγ), tumor necrosis factor α (TNFα) or the macrophage as well as granulocyte/macrophage colony stimulating factors (CSF) are established to induce or to modify (enhance or attenuate) the expression of microglial surface receptors, adhesion molecules, ion channels and soluble compounds, such as reactive oxygen species, nitric oxide and cytokines. They control microglial motility, chemotactic movement and phagocytotic performance. Chemokines are pivotal for the migration and attraction of microglia. These powerful chemoattractive cytokines have other functions as well but they are indispensable for the guided recruitment of peripheral cells and microglia to a site of need.

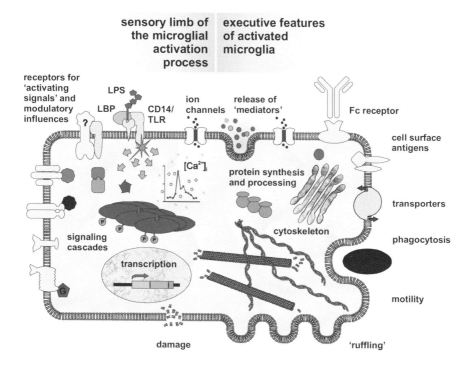

Figure 2. Sensory and functional properties of resting and activated microglia. Cells are equipped with an array of receptors to collect information about the homeostatic conditions of the surrounding tissue. Receptors (and ion channels) allow for the monitoring of neuronal activity and the detection of unusual types or concentrations of molecules. Activating signals may arise from functional impairment or damage of neurons, tissue destruction, leakage of the blood-brain barrier or a confrontation with foreign material, such as bacterial cell wall components. Shown as an example, gram-negative lipopolysaccharide (LPS) in combination with serum-derived LPS binding protein (LBP) can bind to a CD14/Toll-like receptor (TLR) complex to trigger intracellular signalling cascades which organize for microglial reactions. LPS is often employed as a tool to simulate a bacterial challenge and to study the microglial activation process. As a consequence of the exposure to an activating signal, microglial cells undergo drastic morphological changes, migrate along chemotactic gradients, release a plethora of 'mediators', including cyto/chemokines, neurotrophic factors and nitric oxide, become phagocytotic and present antigen. The functional properties require massive protein synthesis to guarantee secretion and an increased or de novo expression of ion channels, transporters, receptors and other 'cell surface antigens'. Microglial activation is also subject to modulatory influences, often involving autocrine loops or a communication with invading immune cells. Besides initiating the executive features the activation process associates with some sensory adjustments as well, such as increased or decreased susceptibility to external influences. (Adapted from Hanisch 2002, Glia 40: 140-155)

Amyloid β (Aβ) accumulating in the plaques of Alzheimer disease (AD) brains is a typical example for a protein with a disease-related production, processing and aggregation (Akiyama et al., 2000). Aβ peptides in soluble and — especially — in fibrillary form were shown to affect microglia, including stimulation of release properties. Other stimuli can act in conjunction with Aβ to trigger enhanced reactions. Several of the synergistic factors associate with the AD plaques, suggesting massive microglial irritation. Clusters of activated microglia will then produce factors (IL-1, TNFα) that drive neurotoxic cascades and directly or indirectly recruit more microglia.

While the above mentioned factors would have to appear or to increase in the vicinity of microglia in order to trigger a transformation to alerted, activated or full-blown reactive states, other factors could induce microglial activation by a kind of 'off-signalling'. They may usually maintain the resting status and probably keep an attenuating influence. Interruption of their calming input would then result in activation, by unchaining an autonomous program for installing increased readiness to provide support or to prepare for an attack. The neuronal/endothelial/lymphoid glycoprotein CD200 is an example of factors constitutively providing an inhibitory influence on macrophages/microglia. Fractalkine forms or NCAM may represent candidates for cyto/chemokines and adhesion molecules. Moreover, suppression of neuronal firing ability causes an upregulation of major histocompatibility complex (MHC) class II molecules on microglia, a structure involved in antigen presentation to lymphocytes. An intimate communication between neurons and microglia is thus likely based on permanent glia-tropic signs of neuronal well-being which counteract spontaneous activation but stimulate microglial responses upon interruption.

4. MICROGLIA AS PART OF THE CNS DEFENCE POTENTIAL

The microglial receptor equipment organizes for the sensory limb but information about a threatening situation would remain without consequence if not followed by a set of cellular reactions. Microglia is in a key position to not only sense and integrate molecular and cellular inputs but also to immediately respond as well as to organize for responses. The fact that microglia has many features in common and even shares the myeloid-lineage origin with (other) macrophages points already to some functional implication in the 'first-line', i.e. the innate defence mechanisms. Moreover, microglia provides a link to the adaptive immunity as it can serve to some extent as a CNS-resident antigen-presenting cell (APC). Finally, microglia guarantees the intrinsic supply of a plethora of soluble mediators, mostly on request or challenge, which activate, guide and control recruited immune cells. In addition, the microglial synthesis potential covers factors that act more or less primarily on neurons, other glial cell types or the endothelium.

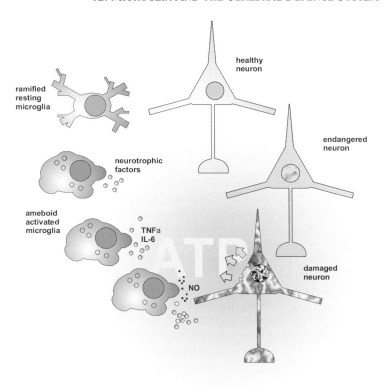

Figure 3. Microglia 'sensing' neuronal firing activity. Release of cotransmitters, such as ATP, may play a critical role in neuron-to-microglia communication. Low ATP levels could indicate normal neuronal activity, whereas high levels would signal excessive firing or cell damage. Microglia will respond to the elevation in ATP concentrations with a successive production of trophic and — eventually — toxic compounds to support endangered neurons or to 'facilitate' cell death, followed by a subsequent removal of debris, when a neuronal recovery is anyway impossible. (Adapted from Inoue 2002, Glia 40: 156-163).

4.1. Microglia as a macrophage-like cell

Despite the share of essential macrophage features the unique ramified morphology of microglia in the healthy adult brain tissue is probably indicative of differences in its physiology. Microglia is a cell of macrophage potential which needs appropriate stimulation to enter the stepwise transformation to a full-blown macrophage.

During development, 'amoeboid' microglia participates in the maturation of the nervous tissue by removing cellular debris resulting from the physiological shaping of the neuronal network, i.e. the elimination of excessive cells and aberrant axons. Phagocytosis is an important, but not the only contribution of microglia. Microglia may induce apoptosis in neurons that are destined to be removed from the transitory circuitry. Microglia also appears to take decisions on neuronal development. These cells seem to influence the neuronal commitment to a certain transmitter type. Data indicate a promotion of cholinergic and GABAergic differentiation.

Phagocytosis is certainly a facet of the antimicrobial activity repertoire as it involves attachment, engulfing, incorporation and destruction of invading germs. Nevertheless, phagocytotic attacks of myelin sheets is one of the microglial contributions to multiple sclerosis (MS). Furthermore, microglia is aggravating the demyelination process through a production of leukocyte-influencing signals. Cyto- and chemokines of microglial origin may attract, guide and support immune cell invasion — and probably even affect the bias of Th1 *versus* Th2 lymphocyte subset engagement.

Microglial cells are themselves readily able to migrate to sites of injury or infection herds. They follow the chemotactic gradients as build up by endogenous factors (e.g., from already activated local microglia) or factors released from the infectious material. This feature is seen by the microglial concentration at affected brain loci as well as in experimental settings using chemoattractive substances. Proliferation and migration can thus together rise local microglial cell numbers.

Microglial activation includes a macrophage-like program for the synthesis of countless soluble factors. Nevertheless, capacities and regulatory mechanisms may differ between macrophage populations and microglia, as they are required and instructed by the tissue environment. Table 2 offers a selection of molecules which are typically induced, upregulated or released in microglial activation.

4.2. Microglia as an antigen-presenting cell (APC)

In CNS diseases that involve the activity of antigen-specific lymphocytes material of pathogenic origin or originating from cell and tissue breakdown has first to be captured, processed and presented as antigen to $CD4^+$ and $CD8^+$ T cells in the context of MHC class II and I molecules. This is usually a function assigned to tissue macrophages and dendritic cells. When T cells subsequently travel through the tissues they will be presented the respective antigen by a tissue-resident APC or an infiltrated macrophage. Parenchymal microglia may have the capacity to act as an APC, but perivascular cells (macrophages) are also predestined for such a function, the latter being situated in the very vicinity of the vasculature (Thomas, 1999). Moreover, CNS-infiltrating macrophages (and dendritic cells) are certainly able of serving as APC. However, it has been quite difficult to distinguish between microglia 'proper' and inflammatory macrophages as many markers (cell surface antigens) are shared by both, and only quantitative differences and expression ratios ($CD45^{low/high}$) could tell one from the other type. New markers are being reported now which allow to identify microglia among macrophages (e.g., glucose transporter 5). Regardless of uncertainties of cell type recognition in the past, microglia has convincingly been shown to offer the MHC class structures for antigen presentation, including the co-stimulatory molecules required for a proper T cell interaction. Situations or defined stimuli causing microglial activation will thus induce (increase) the expression of MHC class II, B7.1/B7.2, CD40 and ICAM-1 molecules for successful antigen presentation (O'Keefe *et al.*, 2002) (Table 2).

While certain factors enhance expression of critical structures (TNFα, IFNγ), others dampen the ability of antigen presentation (IL-4, NO, PG). Microglia also harbours an effective machinery for the proteolytical antigen processing, a requirement for successful presentation. Internalisation and degradation of a potential antigen may, however, be strongly regulated. Moreover, exposure of antigens may not always trigger an effective T cell response when costimulatory signals are insufficient. The turning into an APC is apparently influenced by the genetic background, the history of the cell as well as the situational context.

4.3. Microglia as a source of soluble mediators

Microglia uses a whole repertoire of releasable tools and messengers to instruct and communicate with resident or infiltrating cells, especially in emergency situations. Countless cellular interactions within and between tissues of the body are based on the exchange of cyto- and chemokines. Historically associated with haematopoietic cells and the physiology of the immune system, these small signalling proteins are produced and effective in most cell types. Neurodevelopmental roles are postulated and some cytokines modulate neuronal activities in the mature CNS or participate in neuro-immune-endocrine communication.

CNS expression and/or effects were reported for TNFα, interferons, IL-1, -2, -3, -4, -6, -10, -12, -15 and –18, transforming growth factor β (TGFβ), CSFs, platelet-derived (PDGF), epidermal (EGF), fibroblast (FGF), insulin-like growth factors (IGF) and — not surprisingly — neurotrophic factors, such as nerve growth factor (NGF), neurotrophins (NT-3 and –4) or brain-derived neurotrophic factor (BDNF). An increasing body of evidence relates to the ever growing family of chemokines. While involvement in the day-to-day activity of the brain is still poorly understood, it is accepted that many cyto/chemokines are pivotal to the process and outcome of traumatic, infectious or degenerative challenges. Microglia represents a major, if not the prominent endogenous source for cyto- and chemokines (Table 2).

As an immunostimulatory and pro-inflammatory signal IL-1 has a strategic position in innate defence and immune responses. Lymphoid and myeloid lineage cells (microglia in the CNS) are main sources. IL-1α is cell-associated, IL-1β the main soluble form. Targets include T and B cells, monocytes, macrophages and microglia. IL-1 receptor antagonist (IL-1ra), the third family member, serves in the control of IL-1 activities. IL-1 appears as a common link in processes leading to neuronal death in most variable disorders. IL-18 (IFNγ-inducing factor, IGIF) is IL-1-like by structure and activities in innate and adaptive immunity. In the CNS, astrocytes and microglia appear to be a source. Increased levels may also result in cell death without obvious participation of immune cells.

IL-4, -10, -13 and TGFβ share some anti-inflammatory, immunosuppressive and neuroprotective features. Many of them can be attributed to a downregulation of (micro)glial production of cytokines, e.g. IL-1 and TNFα, or the attenuation of their secondary release effects as well as an altered microglial receptor expression.

IL-6 is produced by T, B, bone marrow and endothelial cells, macrophages and certain CNS cells, including microglia. Like IL-1 and TNFα, IL-6 is considered a proinflammatory cytokine important in limiting the spread of infectious agents. IL-6 particularly helps to initiate and regulate acute phase responses, a complex of adjustments in metabolic and executive organ functions (liver, immune cells) and circulating serum components assisting in host defence. Depending on the presence of other factors, IL-6 has both pro- and anti-inflammatory outcomes. Microglia may provide IL-6 especially in early phases of CNS insults, IL-6 subsequently acting on astrocytes to orchestrate attempts for tissue repair.

While IFNα/β play roles in the resistance of mammalian cells against viral infection, IFNγ ('immune interferon') is mostly triggered by antigenic stimulation and T cell activation. $CD4^+$ and $CD8^+$ T and natural killer (NK) cells serve as major sources. Macrophages can synthesize IFNγ upon costimulation with IL-12 and IL-18. IFNγ has a critical role in the decision over Th1 *versus* Th2 immune responses. In microglia, IFNγ induces or upregulates many cell surface molecules, including MHC class I and II, ICAM-1, immune-accessory B7 molecules, leukocyte function-associated molecule 1, LPS receptor, Fc and complement receptors, changes in the proteasome composition as well as production of NO, cytokines (TNFα, IL-1, -6) and complement (C1q, C2, C3, C4). IFNγ effects on the cytotoxic, phagocytic and APC features of microglia are typical for macrophage support, while it can also induce apoptosis through simultaneous Fas/FasL upregulation.

TNFα is a proinflammatory cytokine of monocytes/macrophages, dendritic cells and lymphocytes and acts on many cells to regulate their growth, differentiation and viability. CNS production is attributed to neurons, astrocytes and microglia. Glia is often a donor as well as recipient, with autocrine loops maintaining activated states, but also aggravating destructive activities under pathological conditions. Increased levels of TNFα in the brain have been observed especially after injury or ischemia, in bacterial and viral infections as well as in multiple sclerosis and AD. Microglial TNFα can boost glutamate release in astrocytes, whereas microglia can also release soluble TNF receptors, which likely affect the bioavailable TNFα pool. Harmful outcomes of TNFα relate to a promotion of inflammation and edema and to direct toxic effects for neurons and myelin. In contrast, TNFα can also promote neuronal survival, probably via NGF. TNFα thus stands for occasionally opposing actions of cytokines in the CNS. TNFα belongs to a whole family of ligands, including also lymphotoxins, Fas ligand (FasL), CD40 ligand (CD40L) and TRAIL that signal through a family of respective receptors (including those for NGF/neurotrophins) to cause most heterogeneous effects, ranging from apoptosis to enhanced proliferation. These ligand-receptor systems are very crucial for cell selection and programmed cell death, also during neural development and synaptogenesis. They participate in microglia-mediated cell communication and toxicity. Extravasation and tissue trafficking of leukocytes is depending on a coordinated expression of cytokines and, in particular, chemokines, which still expand in terms of identified structures and functions. Leukocyte recruitment is a necessary process but may occasionally leave deleterious net consequence, since activities and release products of invading cells can be toxic to glia and neurons. Interference with leukocyte invasion is often beneficial in CNS infections.

Table 2. Cell-associated molecules and soluble factors of activated microglia

Surface structures
  B7.1 (CD80/CD28R)
  B7.2 (CD86/CD28L)
  C3bi receptor (CR3/CD11b/Mac1)
  CD14/TLR (LPS receptor/signalling complex, pattern recognition receptor)
  CD68
  immunoglobulin receptors (FcR I, II, III)
  intercellular cell adhesion molecule-1 (ICAM-1/CD54)
  leucocyte common antigen (LCA/CD45)
  major histocompatibility complex II (MHC II)
  several cyto- and chemokine receptors

Ion channels
  Kv1.3 outward K^+ channel

Intracellular factors
  granule cell death-10 gene (*gcd-10*) product
  Iba-1 Ca^{2+}-binding protein
  microglia response factor (MRF-1)

Cytokines and chemokines
  growth regulated oncogene (GROα)
  IL-1α/-1β and IL-1 receptor antagonist (IL-1ra)
  IL-3
  IL-6
  IL-8
  IL-10
  IL-12 ($p40_2$, p70)
  IL-15
  IL-18, also interferon-γ inducing factor (IGIF)
  gamma interferon inducible protein-10 (IP-10)
  monocyte chemoattractant protein-1 (MCP-1)
  macrophage colony stimulating factor (M-CSF)
  macrophage derived chemokine (MDC)
  macrophage inflammatory protein-1α/-1β/-2/-3β (MIP-1α/1β/-2/-3β)
  transforming growth factor (TGFβ)
  TNFα
  regulated on activation, normal T cell expressed and secreted (RANTES)

Neurotransmitters, neuromodulators, neurotoxins
  glutamate
  nictric oxide (NO)
  reactive oxygen species

Besides meningeal and perivascular macrophages, ependymal cells and astrocytes, activated microglia can contribute to leukocyte recruitment and guidance. Microglia can produce a battery of chemoattractants in response to stimulation by bacterial agents, Aß peptides or cytokines, such as TNFα or IL-1 (Table 2).

Mechanical injury, viral and bacterial infections, tumors, neuroinflammatory processes, demyelinating diseases, stroke or neurodegenerative processes, such as in AD, seem all to contain chemokine actions via their constitutive or upregulated G protein-coupled receptors. Although its respective receptor expression is not strictly complementary to the inducible profile of chemokines, activated microglia could attract further microglia. Via chemokines, microglia can also affect (e.g., support) neurons and astrocytes. Certain ligand-receptor systems may prove to be crucial for the CNS responses to injury, such as secondary lymphoid tissue chemokine (SLC)/CXCR3. The fractalkine-CX3CR pair has also special implications for microglia. Fractalkine seems to support microglial survival and to produce a 'calming' effect on microglia, although reciprocal glial-neuronal expression of the system and various molecular forms of the ligand imply a complex interplay.

Endangered neurons depend on support for survival, and microglia can produce neurotrophic factors (NGF, BDNF, NT-3 and NT-4). Expression reveals regional specificity and heterogeneity. Constitutive or inducible production could be largely responsible for the primarily neurosupportive function of microglial cells, whereas neurotoxic features would dominate in situations when signals from the neuronal environment indicate irreversible cell damage. In turn, some of these factors also cause microglial proliferation and phagocytosis.

The microglial repertoire includes many more candidates with immuno- and neuromodulatory or toxic effects. Proteases and protease inhibitors, glutamate, prostanoids and oxygen radicals develop tremendous impacts on glia and neurons. Most notably, NO is a known immuno- and neuromodulatory agent which, at higher concentrations, causes damage. Produced by microglia in defence and attacks it is certainly a central tool through which a multitude of direct and indirect actions are mediated. Table 2 should be taken as a short list only to inspire further reading.

5. MICROGLIAL DYSREGULATION AND EXCESSIVE ACTIVATION

Activated microglia is observed in many diseases, ranging from neurodegenerative processes in AD, amyotrophic lateral sclerosis (ALS) or Parkinson disease (PD) and immune system-driven demyelination of multiple sclerosis (MS), to infectious CNS complications, such as in HIV, meningitis, cerebral malaria or Creutzfeldt-Jakob disease, to destructive cascades following stroke or trauma (Akiyama et al., 2000; Emsley and Tyrell, 2002; Heppner et al., 2001; McGeer and McGeer, 2002; Nau and Brück, 2002; Nelson et al., 2002; Orr et al., 2002; Perry et al., 2003; Stoll and Jander, 1999). Is the microglial engagement thus accompanying, contributing to, aggravating or even provoking the structural and functional impairments in various pathologies? Or does it reflect attempts to cope with a primary insult in order to limit the damage? As stressed before, microglia was 'created' for protection. Still, the microglial control mechanisms may occasionally fail. Harmful engagement can be attributed to neuropathologies of varying etiology. There is no reason to assume that a single molecule can be identified as a sole trigger of microglial misbehaviour. While the starting point to a damaging action may depend on genetic and epigenetic factors some common themes in the subsequent cascades can be certainly extracted as executive invariants of a microglial response. Release of (neuro)toxic substances and inflammatory mediators or phagocytosis belong to this list. Yet, the common view of an only stereotype reaction of these cells to any kind of challenge may not hold true. As more information becomes available on variable responses, such as a context-dependent reaction to defined experimental stimulation or unique patterns of inducible genes as relating to different diseases, microglial activation will prove a much more adaptive process. In the following section, scenarios are illustrated that relate to undesired activities of microglia.

5.1. Excessive acute activation of microglia

Ideally, microglia participates in defence mechanisms to wall-off (further) invasion by infectious agents, protects neurons from excitatory stress, provides trophic aid to endangered neurons or accelerates the termination of vital functions in neurons which are irreversibly prone to die. Subsequently, microglia may contribute to the CNS' attempts of repair and regeneration or support a scar formation for wound closure and filling. In many, if not most of the acute confrontations, microglial engagement must have a positive net effect on the CNS — when this cell just shows a phylogenetically rewarded behaviour.

Upon excessive challenges, however, the microglial response may turn out to be exaggerated. The production of neurotoxic compounds and inflammatory mediators may overshoot, e.g., as a result of massive bacterial load or a sudden inundation by microglia-activating material following bacterial autolysis or antibiotic treatment. The quantitative side of a challenge, such as tissue concentrations of an activating signal, is probably not the only aspect influencing the course of the response. The particular type of such a factor may dictate the executive program. Variations may exist for the kinetics, duration and the pattern of inducible cell surface structure expression and release of soluble compounds.

Challenging agents of exogenous or intrinsic origin may show a more or less strong impact on microglia with a variation in the context — in other words, the ensemble of coincidental signalling. Experimental observations indicate differences in the efficacy (e.g., the magnitude of a response or the required exposure period to induce a response) or the response profile (e.g., ratios of released mediators) upon challenges with material of different bacterial strain background. Most traditional (*in vitro*) models with single-factor treatment can reflect only isolated aspects. The complex influences under *in vivo* conditions are only inadequately represented. Is it difficult to imagine a variation in the susceptibility for activating signals and the interpretation of threatening situations? Still, the response ability of these cells carries the stigma of a stereotype. Activated cells may look rather alike, but it is now becoming clear that correlations between morphological and other activation-associated properties, such as release activity, are rather weak. In addition, BBB leakage may cause aggravating synergism of the microglial stimulus and serum factors. When the magnitude of microglia reactions escapes the limits of the defensive purpose substantial damage can be the outcome. Activated microglia can thereby be assisted by invading immune cells. Above the tolerable limit, microglial mechanisms leading to cell death and tissue destruction could even further spread the damage and eventually result in organ failure.

5.2. Chronic activation of microglia

Neurodegenerative diseases are characterized by the progressive loss of neurons in certain CNS regions, resulting in an increased functional impairment. Initial events in the pathogeneses of AD, ALS or PD certainly differ but it appears that microglia has at least some exacerbating contributions. Activated cells associate with regions of damage, still leaving the causal relation unresolved. While microglial recruitment to disintegrating neurons is likely, more active roles can appear at closer inspection. There is reason to assume that microglial activation may terminate due to regulatory feedbacks. It may be possible that exhaustion or impaired vitality of microglia leads to a depletion of its function, when the cells themselves suffer from the pathological changes. Alternatively, the activation could mostly fade with the disappearance of the stimulating signals. In section 6.4. we will briefly address the potential value of facilitating microglial contributions to the neutralization of (e.g.) infectious agents based on biochemical degradation and phagocytotic removal. By its phagocytotic activity, microglia (as a population, not necessarily as an individual cell) may thus eliminate its own stimulus, i.e. cause of activation. However, when the irritant is continuously delivered or resists a removal, microglial activation would probably remain — and so the production of potentially harmful mediators. Taking immuno- and neuroregulatory cytokines just as an example, detrimental consequences are particularly strong when elevated tissue levels persist over extended periods of time. Although it is not clearly established how long a microglial cell could tolerate a sustained activation, vicious cycles of cell damage and death provoked by and causing excessive microglial activity could attract more cells to the site of irritation.

Abnormal production and processing of amyloid precursor protein (APP) as followed by deposition of Aβ in plaques are a hallmark in AD. Brains of patients also reveal signs of robust inflammation, indicated by the presence of characteristic molecules, namely complement and various cyto- and chemokines. Although the causal relations between inflammatory factors, Aβ and neuronal death are not yet resolved inflammation-driven cascades seem to exacerbate disease progression and even sensitize for AD development. Clusters of activated microglia associate with plaques, the microglial release products being thought to mediate some significant portion of the neurotoxic potential. Receptors for fibrillary and non-fibrillary forms of Aβ peptides are postulated that trigger in conjunction the activation, including release of cyto- and chemokines (e.g., IL-1β, IL-6, TNFα, MIP-1α, MIP-1β and IL-8). Inflammation-related molecules can synergistically join Aβ to enhance their production. While microglial products likely contribute to the destruction and loss of neuronal structures material released from dying neurons could fuel further microglia activation and the emission of microglia-attracting and -activating signals, which renews the release of potentially neurotoxic agents at high rates.

It is intriguing to speculate that the AD plaque material endures the microglial attempts for removal. Apparently, microglia clears plaque material in vitro, whereas such activity is not sufficiently present in the AD brains. Arrested and incapable of eliminating the source of activation, microglia is condemned to continue with the 'inflammatory' response component. Is the phagocytic machinery paralysed? Is there a lack of pro-phagocytotic signals? Are there factors involved which suppress it? TGFβ1 is a phagocytosis-facilitating and anti-inflammatory factor. Interestingly, Aβ could block its action as it seems to interfere with TGFβ binding, and there are other interesting molecules being discovered with proposed down-regulating effects on the microglial uptake. High-mobility group protein-1 (HMG-1), a chromosomal transcription and replication regulator which is macrophage-secreted and increased in AD, was found to suppress Aβ uptake in microglia and to stabilize Aβ peptide oligomerization. This candidate may illustrate that critical factors could be at key position. One should, however, neither reduce a microglial contribution in AD and other progressive neuropathologies to impaired phagocytosis nor be distracted from a more complex consideration of disturbed microglial function in other diseases. In human immunodeficiency virus (HIV) infection, severe neurological dysfunction develops, also known as AIDS dementia. Infecting peripheral monocytes and then perivascular macrophages, the virus subsequently exploits microglia as a CNS cell compartment, a cell also producing toxins for the propagation of neuronal damage.

Despite their characteristic clinical symptoms and the specificity in the primary functional deficits as relating to afflicted anatomical structures, neurodegenerative diseases, such as AD, ALS or PD share certain molecular and cellular disturbances which could prove rather relevant to the pathogeneses and — most importantly — potential treatments. At biochemical level, an abnormal protein-protein interaction, marked aggregation and finally intra- or extracellular deposition (inclusion bodies, plaques) appears to be such a common theme. The extremely complex morphology of projecting neurons may make these cells especially susceptible for intracellular protein deposition and a related impairment of cytoskelettal structures (and hence axonal transport). Genetic background (mutations), environmental risk factors, viral infection or specific metabolic conditions (namely the ability to cope with oxidative stress) could also render certain neuronal populations in certain individuals more vulnerable. Selective but progressive cell death would then cause local engagement of microglial cells. Indeed, an accumulation of reactive microglia is observed in the degenerating areas of ALS. While microglial activation in the spinal cord and motor cortex may thus appear to be the result of motor neuron decline the presence of activated microglia will certainly add to the oxidative stress burden. Gathering of activated microglia in the degenerating areas together with multiple signs of local inflammation could suggest a rather disease-aggravating role in ALS. In addition, one should bear in mind that microglial release products cover many potentially neurotoxic compounds, glutamate being an interesting candidate for the excitotoxic hypothesis of ALS.

Presence of activated microglia at sites of ongoing neuronal degeneration is seen also in PD. Interestingly, the recruitment and activation of microglia is a consistent feature in various animal models even though different neurotoxins are used for the selective induction of dopaminergic cell loss in the nigrostriatal system. Microglial contributions likely consist again in form of an increased oxidative stress and the release of multiple compounds with direct and indirect (via further engagement of microglia) toxic outcomes. Anti-inflammatory or other means of interference with microglial activation can thereby reduce the neuronal damage, pointing to some pathogenic involvement of these cells. It should be stressed that such a focus on microglial activation in order to broaden the view from a mere consequence to the initiation and progression of neurodegenerative diseases must not ignore their much more complex etiologies. Nevertheless, it comes with the value of practical yield if pharmacological manipulation of a perpetual microglia activation turns out to have beneficial effects.

5.3. Aging of microglia

With aging, the microglial performance and response profile may change as it does for other CNS cells. First, this notion considers a decline in the supportive activity for neurons, even though this activity is not yet fully understood. With the guarding and caring services fading small neuronal insults or stressful situations become less counterbalanced. It is possible that most episodes of microglial alert never surface to clinical signs until deprivation or dysfunction of microglia cannot be tolerated by the neuronal neighbourhood anymore. Abandoned neurons would degenerate, their death triggering activation of the microglia. Second, microglial senescence could reduce the sensitivity for extracellular signals that have a constant calming effect. Soluble and cell-associated factors of such nature have been described. Neurons are likely a source. Aging microglia may thus enter levels of alert due to uncoupling from regulatory influences. Drastic consequences may especially arise whenever opportunistic activations cannot be supervised or terminated properly anymore. Unfortunate combinations of factors could thus orchestrate microglial contributions to the onset and progression of age-related diseases.

6. MICROGLIA AS A TARGET OF PHARMACOLOGICAL INTERFERENCE

The concept of considering microglia for therapeutic intervention is certainly not new. At present, however, combined efforts of experimental neurobiology, clinical as well as basic molecular and cell biological research have given new cues to base potential strategies for pharmaceutical development on actual targets. Three major levels of the microglial engagement may be defined for prevention or modulation of undesired activity, i.e. the extracellular signals causing activation of microglia, the products of the activated microglia and the intracellular processes organizing for the conversion of sensory influences into executive actions.

6.1. Interference with activating signals

Receptors for activating signals could be in focus. Recognition of non-self patterns, e.g. on bacterial surfaces, is well-known to lead to microglial reactions as part of an evolutionary old host defence. Receptor systems, such as the CD14 complex, serve in macrophages/microglia and other mammalian cells for a response initialisation against Gram-negative and –positive strains. So-called binding proteins of host origin are required for optimal signalling through the respective receptors and may offer targets of interference with 'successful' receptor occupation. Promising efforts could also aim at endogenous or modified antagonists and binding proteins for other microglia-relevant factors. The families of cytokines with their spectrum of receptor antagonists (IL-1), soluble receptors (e.g. TNFα, IL-2) or binding proteins (e.g. IL-18) may be cited. Viruses use strategies of mimicry not only for a confusion of the immune system but also for the docking to the surface of mammalian cells (e.g. CD4, chemokine receptors). In such a case, prevention of the receptor occupation may prevent the infection of the cell.

6.2. Interference with microglial products

Focussing on the output side of microglial activation the excessive production of potentially neurotoxic compounds (e.g., oxygen-derived radicals, NO, glutamate, proteases) as well as immuno- and neuroregulatory mediators as affecting resident and infiltrating cell populations is a conceivable target. Even though many of these factors are not necessarily restricted to microglial production their containment could be useful when recruitment of microglia already occurred. Certain cytokines have been suggested to be implicated in neuropathological scenarios of varying etiology. Activated glia, namely microglia, and invading immune cells can serve as major sources upon infection, ischemia, stroke, excitotoxicity and mechanic injury. In neurotrauma, IL-1 was found to be very rapidly released, faster than expected for a *de novo* synthesis of a protein. IL-1(β) and probably other members of its growing family (e.g. IL-18) may prove as pivotal effector molecules and common links through which most heterogeneous CNS insults can finally funnel into functional disturbance and cell death. Suppression of endogenous IL-1, e.g. by IL-1ra, has thus already been reported for neuroprotective effects.

6.3. Interference with the intracellular signalling during microglial activation

Intracellular cascades as they translate receptor stimulation into microglial reactions finally offer another level for pharmacological manipulation. Several signalling pathways with characteristic enzyme activities, adapter proteins and transcription factors have been identified to be most essential for microglial responses, including the multiple induction of genes. While most of these pathways are not restricted to this cell type moderate manipulation may especially affect the activated microglia. Protein phosphorylation pathways through the mitogen-activated protein kinases (MAPK) p44/42 and p38 are apparently critical to the cyto/chemokine induction. They are recruited by many microglia-activating stimuli, ranging from Aβ peptides, prion protein and bacterial components to cytokines, proteases and excitatory amino acids. In addition, the c-Jun N-terminal kinase (JNK) pathway is thought to play important roles as well. The NF-κB system controls many genes during microglial activation. While manipulation of more central elements in these pathways would guarantee a multiple impact on the executive performance refined hits at specialized intracellular factors could offer control over individual features. Conceivable sites of action for new drug development relate to the large ensemble of kinases and phosphatases which are at present only partially assigned to cytosolic and nuclear functions. Inhibition of certain protein tyrosine kinase activities, for example, has been shown not only to calm an overshooting microglial production of potentially toxic cytokines, while mostly sparing the respective endogenous antagonists. The very same kinase inhibition had obviously no toxic side effects and was even found to be directly neuroprotective. Search for similar drug candidates may provide new tools for selective manipulations, especially in adjuvant therapies of CNS infection.

Considering the microglia as the major resident cell type to fuel inflammatory processes in the CNS available pharmacological means should help in containing the microglia-related damage. An anti-inflammatory drug treatment is regarded as beneficial when the CNS mounts an inflammatory reaction that can aggravate an emerging brain damage. Glucocorticoids (GC) are traditionally useful for their anti-inflammatory effects but could also exacerbate hippocampal neurotoxicity and even increase inflammatory cell numbers. On the other hand, treatment of patients with non-steroidal anti-inflammatory drugs (NSAID) has been reported for reducing the prevalence of an AD development. Nevertheless, some similar double-edged sword situation may apply here as well. NSAID apparently do help impeding the onset and slowing the course of AD, but they will hit at extraneural sites, too. If so, an anti-inflammatory strategy should ideally build up on microglia-centered measures. However, it should be added, that — as paradox it may appear at first glance — an inflammatory component may contribute a rather neuroprotective or regenerative influence via lymphocytic and/or macrophage functions.

6.4. Support for desired activities of activated microglia

The latter aspect directly leads to the idea of considering conscious enhancement of beneficial microglial activities. Autolysis and the use of antibiotics during bacterial infections release masses of cell wall fragments and intracellular material which carry biological activity. Those compounds can be as dangerous as living germs in causing functional CNS failure and damage. Microglia is the first resident candidate for phagocytotic cleaning of the extracellular space. It appears attractive to promote this activity. Nevertheless, physical interactions of microglia when approaching the material for incorporation will stimulate production of inflammatory mediators. It is thus worth searching for pharmacological tools to allow for selective attenuation of excessive inflammatory responses while sparing the phagocytotic performance.

Plaque material in AD is another candidate for such hypothesis. Could an effective phagocytotic removal of amyloid deposits help in slowing the progression of the disease? Drug development may concentrate on proteases that are responsible for detrimental APP processing, namely β- and γ-secretases, whereas a support for

Aβ-degrading enzymes or substances interfering with aggregation would facilitate Aβ turnover and hinder plaque formation. Vaccination experiments in transgenic mouse models resulted in both prevention as well as clearance of plaque load and improved CNS function. The findings suggest phagocytosis and protein degradation in the clearance of existing deposits. Apparently, 'tagging' of the amyloid material by antibodies helps in the process. Nevertheless, a vaccination strategy might still bear the risk of challenging attacks against APP-expressing cells, whereas passive immunization could run into an undesired activation of defence cells by antibody-antigen complexes. On the other hand, a reported promotion of microglial amyloid clearance by TGFβ1 suggests that phagocytotic performance can also be enhanced by alternative mechanisms.

These examples illustrate the need for a better understanding of the cellular mechanisms controlling individual features of microglial functions. Can desired and undesired activities be dissected, and can they also be separately manipulated? The picture is still controversial. One opinion may still claim that phagocytosis of AD plaque material is inevitably tight up with an induction of inflammatory mediator production. In contrast, microglia serves in the normal physiological process of ontogenetic CNS tissue maturation by a phagocytotic removal of cells debris — without boosting an inflammatory battle. Microglia of the adult tissue can show such 'peaceful' activity as well, i.e. in certain situations of a neuronal injury. Regardless how precise the 'pharmacological surgery' could once differentiate between the various facets of microglial activation any manipulation should aim at moderation, rather than global diminishing of microglial function. At its extreme, functional depletion of microglia could be devastating, even though compensating mechanisms may exist. Recent experiments point to some microglial repopulation via recruitment of peripheral cells — a phenomenon stimulating theories for Trojan horse delivery of therapeutic substances to the CNS. Calling-in for microglial action may thus also exceed 'classical' administration of drugs and even exploit adoptive therapies with manipulated cells. Lessons from the past, however, should raise the respect for a complex friend-and-foe nature of the CNS' defence machinery before infringing its unpublished rules.

REFERENCES

Microglia in the degenerating and regenerating CNS (Streit WJ, ed), Springer Verlag, New York, (2002).
Special issue Microglia (Hanisch UK, Kohsaka S, Möller T, guest eds), Glia 40 number 2 (2002).
Special issue Neuroinflammation (Lassmann H, Wekerle H, Hickey WF, Antel J, guest eds), Glia 36 number 2 (2001).
Akiyama H, Barger S, Barnum S, Bradt B, Bauer J, Cole GM, Cooper NR, Eikelenboom P, Emmerling M, Fiebich BL, Finch CE, Frautschy S, Griffin WS, Hampel H, Hull M, Landreth G, Lue L, Mrak R, Mackenzie IR, McGeer PL, O'Banion MK, Pachter J, Pasinetti G, Plata-Salaman C, Rogers J, Rydel R, Shen Y, Streit W, Strohmeyer R, Tooyoma I, Van Muiswinkel FL, Veerhuis R, Walker D, Webster S, Wegrzyniak B, Wenk G, Wyss-Coray T (2000) Inflammation and Alzheimer's disease. Neurobiol Aging 21: 383-421.
Bezzi P, Volterra A (2001) A neuron-glia signalling network in the active brain. Curr Opin Neurobiol 11: 387-394.

Emsley HCA, Tyrell PJ (2002) Inflammation and infection in clinical stroke. J Cer Blood Flow Metabol 22: 1399-1419.
Giulian D (1995) Microglia and neuronal dysfunction. In: Neuroglia (Kettenmann H, Ransom BR, eds), pp 671-684. New York: Oxford University Press.
Heppner FL, Prinz M, Aguzzi A (2001) Pathogenesis of prion diseases: possible implications of microglial cells. Prog Brain Res 132: 737-750.
Lazarov-Spiegler O, Rapalino O, Agranov G, Schwartz,M. (1998) Restricted inflammatory reaction in the CNS: a key impediment to axonal regeneration? Mol Med Today 4: 337-342.
McGeer PL, McGeer EG (2002) Inflammatory processes in amyotrophic lateral sclerosis. Muscle Nerve 26: 459-470.
Nakamura Y (2002) Regulating factors for microglial activation. Biol Pharm Bull 25: 945-953.
Nau R, Brück W (2002) Neuronal injury in bacterial meningitis: mechanisms and implications for therapy. Trends Neurosci 25: 38-45.
Nelson PT, Soma LA, Lavi E (2002) Microglia in diseases of the central nervous system. Ann Med 34: 491-500.
O'Keefe GM, Nguyen VT, Benveniste EN (2002) Regulation and function of class II major histocompatibility complex, CD40, and B7 expression in macrophages and microglia: implications in neurological diseases. J NeuroVirol 8: 496-512.
Orr CF, Rowe DB, Halliday GM (2002) An inflammatory review of Parkinson's disease. Progr Neurobiol 68: 325-340.
Perry VH, Newman TA, Cunningham C (2003) The impact of systemic infection on the progression of neurodegenerative didease. Nat Rev 4: 103-112.
Polazzi E, Contestabile A (2002) Reciprocal interactions between microglia and neurons: from survival to neuropathology. Rev Neurosci 13: 221-242.
Raivich G, Bohatschek M, Kloss CU, Werner A, Jones LL, Kreutzberg GW (1999) Neuroglial activation repertoire in the injured brain: graded response, molecular mechanisms and cues to physiological function. Brain Res Rev 30: 77-105.
Rezaie P, Male D (2002) Mesoglia and microglia: a historical review of the concept of mononuclear phagocytes within the central nervous system. J Hist Neurosci 11: 325-374.
Stoll G, Jander S (1999) The role of microglia and macrophages in the pathophysiology of the CNS. Prog Neurobiol 58: 233-247.
Streit WJ, Walter SA, Pennel NA (2000) Reactive microgliosis. Prog Neurobiol 57: 563-581.
Thomas WE (1999) Brain macrophages: on the role of pericytes and perivascular cells. Brain Res Rev 31: 42-57.

C. VICARIO-ABEJÓN AND M.J. YUSTA-BOYO

Group of Growth Factors in Vertebrate Development, Centro de Investigaciones Biológicas, Consejo Superior de Investigaciones Científicas, Madrid, Spain

13. GENERATION AND DIFFERENTIATION OF ASTROCYTES DURING CENTRAL NERVOUS SYSTEM DEVELOPMENT AND INJURY

Summary. Astrocytes are the most abundant cell type of the central nervous system. Their functions comprise the formation of the blood brain barrier, the supply of energetic substrates and neurotransmitters to neurons and the participation as active components of some synapses. In addition, some astrocytes in the adult brain have stem cell features. Vimentin is the major intermediate filament protein present in immature astrocytes whereas in differentiated astrocytes, vimentin is replaced by GFAP. The molecular mechanisms controlling astrocyte generation and differentiation from multipotent cells include the actions of diffusible factors (CNTF and BMP, among others), cell-cell interactions (Delta/Notch) and bHLH transcription factors.

1. INTRODUCTION

Astrocytes are the largest glia among the macroglial cells of the nervous system, and they outnumber neurons. Their perikarya (18-20 μm in diameter) possess large glycogen granules, and in some cases prominent filaments which end in swellings at their extremities called *end feet*. Two distinct types of astrocyte have been recognized: *fibrous* astrocytes and *protoplasmic* astrocytes. The latter occur mainly in grey matter and contain little glial fibrillary acidic protein (GFAP), an intermediate filament protein. In contrast, fibrous astrocytes occur mainly in white matter and express high levels of GFAP.

Astrocytes have diverse functions in the central nervous system, including the formation of the blood brain barrier, control of the extracellular ion composition, and the supply of energy substrates and neurotransmitters to neurons. These cells are active cellular components of some synapses. Astrocytes may also regulate the formation of new neurons, and some astrocytes in the adult brain have stem-cell features.

Vimentin is the major intermediate filament protein present in immature astrocytes, whereas in differentiated astrocytes, vimentin is replaced by GFAP. Anti-GFAP antibodies are used to label mature astrocytes, although some astrocytes are not labelled with GFAP. Additional markers include $S100\beta$ and the glutamate/aspartate transporter (GLAST). Within a single brain region, GFAP-positive astrocytes may comprise a functionally and molecularly diverse population.

Several types of precursor cells can give rise to astrocytes during central nervous system development. These include neural stem cells, radial glia, glial-restricted progenitor cells, and astrocyte-restricted progenitors. The molecular mechanisms controlling astrocyte generation and differentiation from multipotent cells are being elucidated and include the actions of diffusible factors (ciliary neurotrophic factor, CNTF, and bone morphogenetic protein, BMP, among others), cell-cell interactions (Delta/Notch), and basic helix-loop-helix (bHLH) transcription factors. These aspects will be reviewed in depth in the present chapter.

2. ASTROCYTE GENERATION IN THE BRAIN AND SPINAL CORD

The formation of the nervous system in most vertebrate species involves a fundamental cell diversification, i.e., the generation of neurons and glia. In the cerebral cortex, for example, most neurons arise on a precise time schedule from the ventricular zone, a germinal layer of neuroepithelial cells. Ventricular zone cells also generate a second germinal region, the subventricular zone, from which the majority of astrocytes arise. In the mammalian central nervous system, including humans, most astrocytes are generated after neurons are born. In the mouse cerebral cortex, neurogenesis commences around E11, whereas astrocytes are first detected around E15. It has been shown, however, that a proportion of glial cells might be specified as early as neuronal cells. Astrocytes are generated from dorsal as well as from ventral domains of the neuroepithelium and, during development, they can be generated from several types of precursor cell, including neural stem cells, radial glia, glial-restricted progenitors, and astrocyte-restricted progenitor cells.

The development of fibrous and protoplasmic astrocytes may be under the control of different regulatory molecules. In support of this, mice with a targeted deletion in the *Fgfr3* locus strongly upregulate GFAP in grey-matter (protoplasmic) astrocytes, implying that signalling through FGFR3 normally represses GFAP expression *in vivo* in protoplasmic but not in fibrous astrocytes.

3. MECHANISMS PROMOTING ASTROCYTE GENERATION AND DIFFERENTIATION

3.1. Ciliary neurotrophic and leukaemia inhibitory factors

The CNTF and the leukaemia inhibitory factor (LIF) cause embryonic hippocampal and cortical stem cells to differentiate into astrocytes, as measured by GFAP expression. CNTF can induce a cell-fate switch on neural stem cells or promote differentiation of committed astrocyte progenitors. The CNTFR complex contains a unique CNTF-receptor-α (CNTFRα) component, a LIF-receptor-β (LIFRβ) subunit and gp130. Binding of CNTF to CNTFRα results in heterodimerization with LIFRβ and gp130 and subsequent activation of several intracellular signal transduction cascades (Figure 1). The best known are the janus kinase/signal transducer and activator of transcription (JAK-STAT) pathway and the mitogen-activated protein kinase (MAPK) pathway. Upon activation of the receptor complex by CNTF, receptor-associated JAK kinases phosphorylate tyrosine residues on LIFRβ, gp130, and the JAK kinases themselves. This leads to the recruitment of proteins containing Src homology 2 (SH2) domains, including STAT1 and STAT3 proteins. The STATs are in turn tyrosine phosphorylated, causing them to dimerize, translocate to the nucleus, and act as transcription factors which induce several genes, including GFAP (Figure 1). Activated JAK2 can also cause the activation of MAPK, possibly through a Ras-Raf pathway.

The CNTF-induced differentiation along the glial lineage in embryonic and adult central nervous system stem cells is thus mediated by activation of JAK1, STAT1 and STAT3. Activation of the MAPK pathway is required early in the astrocytic differentiation process, while activation of STATs is required for complete differentiation. One of the nuclear targets of CNTF may be the nuclear co-repressor (N-CoR). Stimulation of cortical neural stem cells with CNTF results in phophorylation of N-CoR via a phosphatidyl-inositol-3-OH kinase/AKT1-dependent mechanism. This decreases N-CoR ability to maintain the state of neural stem cells, and in turn promotes astrocytic differentiation.

In support of a role of LIFRβ and the gp130 receptor system in astrocyte generation and differentiation, astrocyte numbers, as evaluated by GFAP expression, were reduced in the hippocampus, spinal cord and brain stem of LIFRβ and gp130 knockout mice. Moreover, differentiation of neuroepithelial precursor cells into astrocytes was severely reduced in the mutant mice. In addition, mice overexpressing CNTF contained more GFAP-positive cells within the olfactory bulbs than did wild-type mice. Analysis of the brain and spinal cord of LIF knockout adult mice did not reveal, however, a marked reduction in the numbers of astrocytes, suggesting the existence of other ligands which — acting throughout the LIFRβ and gp130 receptor complex — would promote the differentiation of astrocytes. Interestingly, BMP-2 induced the generation of astrocytes in precursor cells from the LIFRβ$^{-/-}$, indicating that several signalling cascades promote the differentiation of astrocytes (Figure 1).

In contrast to the above results, a recent study reported that signals acting throughout the CNTFRα/LIFRβ/gp130 receptor complex, promoted the maintenance of neural stem cells rather than their differentiation into astrocytes. An explanation for these distinct results is that CNTF and LIF could promote GFAP expression and astrocyte differentiation into embryonic neural stem cells. In postnatal and adult neural stem cells, however, these factors could act to maintain the population of stem cells. The fact that a proportion of postnatal and adult stem cells are GFAP-expressing astrocytes may reconcile these apparently conflicting observations.

3.2. Bone morphogenetic proteins

BMP-2 and 4 exert multiple actions in central nervous system cells. They cause apoptosis early in development, promote neurogenesis in mid-gestation of central nervous system precursors, and promote glial differentiation in late embryonic or adult central nervous sytem precursors. The action of BMPs is mediated by the activation of a complex of type I and type II serine/threonine kinase receptors (Figure 1).

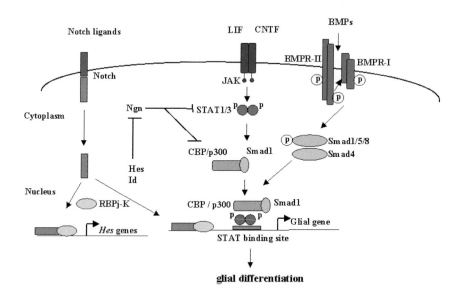

Figure 1 Mechanisms promoting astrocyte generation and differentiation. An interplay between diffusible factors (LIF, CNTF, and BMP), cell-cell signalling mechanisms (Notch), and transcription factors regulates astrocyte generation and differentiation from neural stem cells. Activation of CNTF and LIF receptors triggers phosphorylation of STAT1/3 while activation of BMP receptors phosphorylates Smad1/5/8. In both cases, a common transcriptional complex (STAT-CBP/P300-Smad1) is formed which induces the expression of GFAP by binding at the STAT site. Upon ligand binding, the intracellular fragment of the Notch receptor is cleaved, and then translocates into the nucleus, where it forms a complex with RBPj-k. This complex activates expression of Hes and GFAP genes. Hes proteins repress neurogenin inhibition of the STAT signalling pathway.

The type II receptor is the primary ligand-binding complex. Type I receptors are essentially responsible for transducing the signal into the cell. Activated receptors phosphorylate the DNA-binding proteins Smad1, Smad5, and Smad8 on serines. They then form a complex with a common mediator, Smad4, and the complex is translocated into the nucleus to activate transcription of specific genes.

The co-operation between Smads and STATs inducing GFAP expression appears to be facilitated by a family of coactivator proteins termed CBP/p300 (Figure 1). CBP (CREB binding protein) and p300 are involved in the transcriptional coactivation of many different transcription factors. As known, CREB is the cyclic-AMP-response-element-binding protein. STATs and Smads bind to different domains of CBP/p300; the STAT/p300/Smad complex, acting at the STAT binding element in the astrocyte-specific GFAP promoter, is particularly effective at inducing astrocyte differentiation into neural stem cells.

3.3. Epidermal growth factor

Neural stem cells become progressively more biased towards a glial fate during development, coincident with an increase in expression of the epidermal growth factor (EGF) receptor. Other studies, however, suggest that the increased bias towards glial differentiation during development does not depend on EGFR signalling since the propensity of neural stem cells to generate astrocytes augmented with time in culture even in the absence of EGFR signalling. Although the extent to which EGF promotes astrocyte differentiation needs to be further assessed, recent data shows that developmental changes in EGFRs regulate the competence of neural stem cells to respond to LIF as an astrocyte-inducing signal. In support of a role of EGF in the differentiation of astrocytes, cerebral cortices from EGFR mutants contain lower numbers of GFAP-positive cells.

3.4. Fibroblast growth factor-2 (FGF-2)

The fibroblast growth factor-2 (FGF-2) exists in several isoforms that vary in size as well as in cell and tissue distribution. In particular, the 18kD form and its receptors FGFR1-3 are highly expressed during mitotically active phases of vertebrate nervous system development. A number of *in vitro* and *in vivo* studies have established that FGF-2 is mitogenic for stem/precursor cells, and is a differentiation and survival factor for cortical and hippocampal neurons. Furthermore, FGF-2 may act in collaboration with CNTF, EGF, or Notch to induce astrocytic cell fate and differentiation of multipotent precursors. Interestingly, these distinct effects can be modulated by different FGF-2 concentrations; low levels of extracellular FGF-2 are neurogenic, whereas higher levels promote gliogenesis. In support of a role of FGF-2 in astrocyte differentiation, injection of FGF-2 into the lateral ventricles produces an increase in the number of cells expressing S100β in the cortex, and mice lacking FGF-2 possess lower numbers of S100β-positive cells.

3.5. Insulin-like growth factor-I (IGF-I)

The insulin-like growth factor-I (IGF-I) is a secreted protein related to IGF-II as well as to insulin and its unprocessed precursor, proinsulin. IGF-I and its receptor (IGF-IR), as well as the insulin receptor (InsR), are expressed in many regions from early stages of rodent brain development. It has recently been shown that exogenous IGF-I increases the number of GFAP-positive astrocytes differentiating from olfactory bulb neural stem cells. Supporting a role of IGF-I in the regulation of astrocyte numbers, neural stem cells prepared from the olfactory bulb of *Igf-I* knockout mice generated significantly fewer astrocytes than wild-type cells.

3.6. Pituitary adenylate-cyclase-activating polypeptide (PCAP)

The pituitary adenylate-cyclase-activating polypeptide (PACAP) is a member of the vasoactive intestinal peptide (VIP)/secretin/glucagon family. In the rodent developing cortex, both PACAP and the PAC1 receptor are expressed at significantly high levels during late gestation and postnatally. Under tissue culture conditions, PACAP acts as an extracellular signal to trigger cortical precursor-cell differentiation into astrocytes via stimulation of intracellular cAMP production.

3.7. Notch/Delta signalling pathway

The Notch family of proteins are cell-surface transmembrane receptor proteins that are activated by the cell-bound ligands, Delta and Jagged/Serrata, presented on neighbouring cells. Upon ligand binding, the intracellular portion of the Notch receptor is cleaved and, as a consequence, this domain is released from the membrane and translocates to the nucleus (Figure 1). In the nucleus, it acts to control expression of downstream genes, most prominently the hairy/enhancer of split (Hes) genes, which encode a set of bHLH transcription factors. Hes proteins inhibit the activity of proneural transcription factors such as the neurogenins, Ngn1 and 2, and Mash1. The astrogliogenic role of Notch may also be mediated by direct binding of the Notch intracellular domain to the DNA-binding protein, RBPj-k, forming a transcriptional activation complex on the GFAP gene.

In Drosophila, Notch signaling has been characterized for its actions inhibiting neuronal differentiation and maintaining cells in an undifferentiated state through a mechanism called lateral inhibition or lateral specification. More recently, evidence in mammals has suggested that Notch signalling induces glial differentiation. In the rat retina, introduction of an active form of Notch and of Hes-1 promotes the acquisition of Müller glial identity. Furthermore, expression of dominant-negative forms of Hes1 and 5 inhibits glial cell generation, and the Hes5-deficient retina shows a 30-40% decrease in Müller glia. Similarly, an activated form of Notch introduced into mouse telencephalic precursors *in vivo* at embryonic day E9.5 promoted radial glial morphology. This role for the Notch signalling pathway promoting gliogenesis is further supported by experiments showing that Notch1 and 3 restrict adult hippocampal neural stem cells to an astroglial fate.

However, other studies using neural stem cell cultures from the embryonic cortex and ganglionic eminence of mice lacking Notch or molecules of its signaling pathway such as RBP-jk, Hes1 and Hes5, have concluded that the Notch pathway is important for the maintenance of neural cells in a stem or precursor state. As already mentioned, the maintenance of a precursor state and the promotion of glial identity may not be mutually exclusive as cells having glial features — such as radial glia in the embryonic cortex, Müller glia in the adult retina, and some astrocytes in the adult subventricular zone and in the hippocampus — have properties of neural stem and precursor cells. Thus, depending on the developmental stage and the environmental context, the promotion of neural stem cell maintenance, and radial glial as well as astrocytic cell fate by Notch might be partly a reflection of the same phenomenon at different times.

4. MECHANISMS INHIBITING ASTROCYTE GENERATION AND DIFFERENTIATION

4.1. Basic helix-loop-helix (bHLH) transcription factors

Neurogenins 1 and 2 (Ngn1 and Ngn2) are members of the bHLH family of transcription factors. In the developing mammalian cerebral cortex, these genes are highly expressed in the cortical ventricular zone. Both Ngn1 and 2 dimerize with ubiquitous bHLH proteins, such as E12 or E47. These heterodimers then bind to DNA sequences that contain an E-box consensus motif. This binding has been found to be critical in bHLH-proteins activation of tissue-specific gene expression that promotes neuronal differentiation. It has recently been reported that Ngn1, in addition to inducing neurogenesis by functioning as a transcriptional activator, inhibits the differentiation of neural stem cells into astrocytes via two mechanisms: by sequestering the CBP-Smad1 transcription complex away from astrocyte differentiation genes, and by inhibiting the activation of STAT transcription factors (Figure 1). In support of a role of neurogenins inhibiting astrocytogenesis, double mutant mice for *Ngn2* and *Mash1* (a bHLH gene related to the Drosophila *achaete-scute* complex, *as-c*) show a premature and excessive generation of astrocytes in the cerebral cortex.

Other bHLH transcription factors, *Olig1* and *Olig2*, are also involved in the repression of astrocyte generation. In double-knockout mice that lack expression of the two genes, bipotent precursor cells that normally produce motoneurons and oligodendrocytes in the spinal cord generate astrocytes and interneurons.

4.2. Sonic hedgehog

Sonic hedgehog (Shh) is a member of the hedgehog multigene family that encodes signalling proteins involved in induction and patterning in vertebrate and invertebrate embryos. In the spinal cord, Shh released from the notochord and the floor plate induces the generation of motor neurons and oligodendrocytes. The effects of Shh are in part mediated through the inhibition of BMP signalling. This inhibition could also act during the generation of astrocytes, since Shh blocked the effects of BMP-2, promoting astroglia differentiation of neural progenitor cells in culture.

5. MECHANISMS CONTROLLING NEURON/ASTROCYTE FATE CHOICE BY NEURAL STEM CELLS

As already mentioned, during development most neurons are generated before astrocytes are born. In accord with this fact, early cortical precursors are biased towards neuronal differentiation when plated in cell culture. Moreover, CNTF (or LIF) has no gliogenic effect on E12.5 rat cortical precursor cells, in spite of the presence of downstream signaling molecules, but efficiently promotes astrocyte differentiation from E14.5-derived neural stem cells. Below are summarized some mechanisms that have been proposed to explain the switch from neurogenesis to gliogenesis as development proceeds.

The temporal switch from neurogenesis to gliogenesis may involve a progressive domination of Hes and STAT activity at the expense of neurogenic activity induced by neurogenins and Mash. During early phases of development, when neurogenin levels are high, neurogenesis is actively promoted by neurogenin/CBP/p300 complexes, while gliogenesis is blocked by the sequestration of these transcriptional coactivators and by inhibition of STAT phosphorylation (Figure 1). At later developmental stages, when neurogenin levels are lower, gliogenesis can occur following STAT phosphorylation and the recruitment of p300/CBP by STAT to glial promoters. Similarly, a decrease in the levels of Ngn favours the ability of BMP to induce the association of Smad1 with CBP/p300 and STAT. Thus, the abundance levels of p300/CBP appear to be limiting in cortical stem cells, suggesting that competition between neurogenin and STAT for these transcriptional coactivators is a potentially viable mechanism for alteration of fate choice.

The transition from neurogenesis to astrocytogenesis may also be reinforced by BMP's ability to induce the expression of the inhibitors of DNA binding, Id1 and Id3, and of Hes5 in late cortical progenitor cells. Upregulation of these proteins leads to a reduced activity of neurogenin1 and Mash1 (Figure 1). Concordantly, Id1 and Id3 are expressed in the developing subventricular zone during astrocytogenesis.

DNA methylation of glia-specific genes, such as the one occurring in the GFAP promoter, suppresses the activation and function of the astrocytogenic JAK-STAT signalling pathway. This phenomenon has been observed during active phases of central nervous system neurogenesis.

The sequential generation of neurons and astrocytes may also be regulated by the C/EBP family of transcription factors. This family is composed of basic leucine zipper DNA-binding proteins that activate transcription of neuron-specific genes and inhibit gliogenesis. These factors may in part collaborate with neurogenins and Mash at the level of the transcriptional coactivator complex CBP/p300.

6. ASTROCYTES AS NEURAL STEM CELLS

The neuroepithelium of the central nervous system in mammals such as rats, mice or humans is composed of precursor cells. Depending on the developmental stage, varying proportions of precursors have stem-cell features, i.e., self-renewal and the potential to generate neurons, astrocytes and oligodendrocytes. These processes lead to the formation of a functional central nervous system; in parallel there is a marked reduction in the number of stem/precursor cells. However, in discrete adult brain areas, specific populations of stem cells remain. Although no completely specific molecular markers of neural stem cells have been found, a general feature of embryonic these stem cells is the expression of nestin, an intermediate filament protein, and the absence of markers for neurons, astrocytes, or oligodendrocytes. These facts supported the notion that precursor cells did not express markers of differentiated cells.

It was proposed in the 80s and recently confirmed that some glial cells — the radial glia — can function as neural precursor cells giving rise to astrocytes and neurons. Radial glia is not the only cell with glial characteristics to possess neurogenic potential. As mentioned above, analysis of the adult subventricular zone and the subgranular cell layer of the dentate gyrus of the hippocampus indicates that some astrocytic cells (GFAP-positive cells) have neural stem cells features. In particular, astrocytes located in the adult subventricular zone generate neuroblasts that differentiate into olfactory bulb interneurons. These studies have challenged the existing notion that precursor and differentiated cells express completely different molecular markers. Interestingly, the subventricular zone astrocytes appear to maintain a process that extends to the ventricular wall, a feature also occurring in radial glial cells. What might the origin of these astrocytic-type of stem cells be? One hypothesis is that during development, some neuroepithelial nestin-positive stem cells become transformed into radial glial cells, and these into astrocytes. Some astrocytes would retain the potentiality of precursor cells to generate a diverse differentiated progeny.

7. THE ROLE OF ASTROCYTES IN BRAIN DAMAGE AND REPAIR

Most lesions in the central nervous system are accompanied by a phenomenon termed reactive gliosis. This is characterized by the hypertrophy and hyperplasia of astrocytes, as well as by the proliferation of microglial cells. Reactive astrocytes have also been shown to occur in neurodegenerative disorders such as Alzheimer's disease, and as a normal consequence of aging. The formation of a glial scar by reactive astrocytes is considered one of the major impediments of axonal regeneration in the central nervous system. Fibroblasts from the adjacent conjuctive tissue proliferate on top of the layer of astrocytes (mostly fibrous) and deposit collagen, creating a barrier to axonal regeneration. Although the glial scar isolates the still-intact central nervous system tissue from secondary lesions, reactive astrocytes are nevertheless a non-permissive substrate for axon growth. However, molecules that could promote neuronal survival and axonal regeneration such as neurotrophic factors and extracellular matrix molecules are released in the glial scar. Thus, manipulation of the glial scar to reduce both hyperplasia and hypertrophy is a target for therapeutic intervention. At the molecular level, the hallmark of reactive gliosis is the upregulation of intermediate filament proteins, particularly vimentin, GFAP, and nestin. Of the three, GFAP upregulation maybe critical, obstructing axonal growth as deletion of GFAP expression in astrocytes improves neuronal survival and neurite growth.

Little is known about the cellular and molecular events involved in the enhancement of astrocyte number and size during reactive gliosis. Reactive astrocytes re-express nestin, a marker of neuroepithelial stem cells, and upregulate expression of GFAP, a marker of some adult neural stem cells. These facts suggest that precursor cells, including stem cells residing in the adult brain, could be the source of reactive astrocytes. If this is the case, the formation of a glial scar during central nervous system injury would involve, at least in part, the upregulation of programs signalling cell fate and cell differentiation during development. In support of this hypothesis, mice overexpressing CNTF or interleukin-6 (IL-6) develop spontaneous gliosis. In addition lesion-induced STAT3 phosphorylation was found in reactive astrocytes of the fascia dentata following entorhinal cortex lesion. Other factors, such as FGF-2 and the transforming growth factor β (TGFβ) could also be involved in the formation of the glial scar.

REFERENCES

Alvarez-Buylla A, Garcia-Verdugo JM, and Tramontin AD (2001). A unified hypothesis on the lineage of neural stem cells. Nat. Rev. Neurosci. 2, 287-293.

Campbell K and Götz M (2002). Radial glia: multi-purpose cells for vertebrate brain development. Trends in Neurosci. 25, 235-238.

Gaiano N and Fishell G (2002). The role of Notch in promoting glial and neural stem cell fates. Annu. Rev. Neurosci. 25, 471-490.

Giménez y Ribotta M, Menet V, and Privat A (2001). The role of astrocytes in axonal regeneration in the mammalian CNS. In *Progress in Brain Research*, vol. 132 (Castellano López B and Nieto-Sampedro M, eds.), Elsevier Science, pp. 587-610.

Grandbarbe L, Bouissac J, Rand M, Hrabé de Angelis M, Artavanis-Tsakonas S, and Mohier E (2003). Delta-Notch signaling controls the generation of neurons/glia from neural stem cells in a stepwise process. Development 130, 1391-1402.

Hermanson O, Jepsen K, and Rosenfeld MG (2002). N-CoR controls differentiation of neural stem cells into astrocytes. Nature 491, 934-939.

Lee JC, Mayer-Proschel M, and Rao MS (2000). Gliogenesis in the central nervous system. Glia 30, 105-121.

Lundkvist J and Lendahl U (2001). Notch and the birth of glial cells. Trends in Neurosci. 24, 492-494.

Matthias K, Kirchhoff F, Seifert G, Hüttmann K, Matyash M, Kettenmann H, and Steinhäuser C (2003). Segregated expression of AMPA-type glutamate receptors and glutamate transporters defines distinct astrocyte populations in the mouse hippocampus. J. Neurosci. 23, 1750-1758.

Ménard C, Hein P, Paquin A, Savelson A, Yang XM, Lederfein D, Barnabé-Heider F, Mir AA, Sterneck E, Peterson AC, Johnson PF, Vinson C, and Miller F (2002). An essential role for a MEK-C/EBP pathway during growth factor-regulated cortical neurogenesis. Neuron 36, 597-610.

Miller RH, ffrench-Constant C, Raff M (1989). The macroglial cells of the rat optic nerve. Ann. Rev. Neurosci. 12, 517-534.

Panchision DM and McKay RDG (2002). The control of neural stem cells by morphogenic signals. Current Opinion in Genetics and Development 12, 478-487.

Pringle, N. P., Yu, W-P., Howell, M., Colvin, J., Ornitz, D. M. and Richardson, W. D. (2003). *Fgfr3* expression by astrocytes and their precursors: evidence that astrocytes and oligodendrocytes originate in distinct neuroepithelial domains. Development 130, 93-102.

Ridet JL, Malhotra SK, Privat A, and Gage FH (1997). Reactive astrocytes: cellular and molecular cues to biological function. Trends in Neurosci. 20, 570-577.

Sauvageot C and Stiles CD (2002). Molecular mechanisms controlling cortical gliogenesis. Current Opinion in Neurobiology 12, 244-249.

Schuurmans C and Guillemot F (2002). Molecular mechanisms underlying cell fate specification in the developing telencephalon. Current Opinion Neurobiol. 12, 26-34.

Turnley AM and Bartlett PF (2000). Cytokines that signal through the leukemia inhibitory factor receptor-□ complex in the nervous system. J. Neurochem. 74, 889-899.

Vetter M (2001). A turn of the helix: preventing the glial fate. Neuron 29, 559-562.

Vicario-Abejón C, Yusta-Boyo MJ, Fernández-Moreno C, and de Pablo F (2003). Locally-born olfactory bulb stem cells proliferate in response to insulin-related factors and require endogenous IGF-I for differentiation into neurons and glia. J. Neurosci. 23, 895-906.

Viti J, Feathers A, Phillips J, and Lillien L (2003). Epidermal growth factor receptors control competence to interpret leukemia inhibitory factor as an astrocyte inducer in developing cortex. J Neurosci. 23, 3385-3393.

M. DOMERCQ AND C. MATUTE

Department of Neurosciences, University of País Vasco, Leioa, Spain

14. OLIGODENDROCYTES IN HEALTH AND DISEASE

Summary. Oligodendrocytes are the myelinating cells of the CNS. Here we review recent advances in oligodendrocyte proliferation, differentiation, and maturation, both during development and in the mature CNS. Understanding how oligodendrocyte development proceeds in health is fundamental to discovering new therapeutic targets for demyelinating diseases. We also summarize some relevant advances in the etiology of multiple sclerosis, the most frequent disease involving oligodendrocyte and myelin loss, and new concepts as to how white-matter damage during ischemia contributes to the clinical outcome of stroke in humans.

1. BASIC FEATURES OF OLIGODENDROCYTES

Glia, first described by Virchow as the connective tissue of the brain, constitute the majority of cells in the central nervous system (CNS), and their relative quantitative contribution increases in higher species. Taking advantage of the Golgi silver staining method, Ramón y Cajal first described astrocytes and another cell type that he called the "third element". Years later, his disciple Del Río Hortega identified this third element as being two sub-types of glial cell: interfascicular cells, later termed oligodendrocytes, and microglia (Del Río Hortega, 1928). He observed the morphological and functional complexity of oligodendroglia, which compared to Schwann cells, and classified into 4 main sub-types (see Figure 1). Subsequent ultrastructural analysis has confirmed oligodendrocyte heterogeneity in terms of the cells' morphology and the size and thickness of the myelin sheath they form.

There are at least two types of oligodendrocyte. Myelinating oligodendrocytes are postmitotic cells derived from oligodendrocyte precursor cells that migrate into developing white matter from their germinal zones. The main function of these oligodendrocytes is to form myelin, a fatty insulating material composed of modified plasma membrane which ensheaths axons and allows rapid and efficient conduction of electrical impulses along them. There are also satellite oligodendrocytes, which are located perineuronally, and are not directly connected to the myelin sheath. Although their function is still unclear, they may serve to regulate the microenvironment around the neurons. At the morphological level, oligodendrocytes are small cells with highly condensed chromatin, and a polarized cytoplasm that lacks intermediate filaments and glycogen deposits.

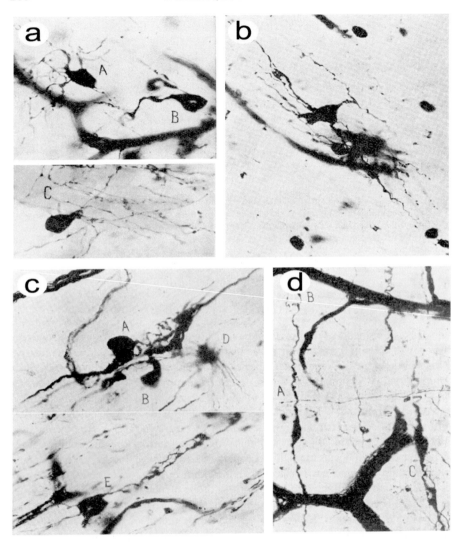

Figure 1. Oligodendrocyte heterogeneity according to Del Río Hortega (1928) **a)** *Type I oligodendrocytes from canine cerebral white matter. Oligodendrocytes A and B are close to capillaries.* **b)** *Type II oligodendrocytes from cat cerebellar white matter.* **c)** *Type III oligodendrocytes from cat cerebral white matter (except D, a fibrous astrocyte).* **d)** *Type IV oligodendrocytes from cat spinal cord white matter.*

Mature oligodendrocytes extend several processes which contact and ensheath stretches of axons, with subsequent condensation of the multispiral membrane-forming myelin.

Myelination is an essential feature of vertebrate evolution due to its role in saltatory nerve conduction and in supporting axonal function and integrity.

Disruption of CNS myelin leads to severe functional deficits, and frequently a reduction in life span. In most mammals, myelination takes place during late embryonic and early postnatal life. Correct myelination involves several steps. First, the neuroepithelial cells must differentiate into oligodendrocyte precursor cells (OPCs) and migrate into the developing white matter. OPCs then proliferate, and subsequently differentiate into mature myelinating oligodendrocytes as they lose their proliferative and migratory potential. From the capacity of OPCs to differentiate *in vitro*, in the absence of axons, together with the possibility of identifying the oligodendroglial lineage at specific developmental stages, initial insights have been gained into the mechanisms that control oligodendrocyte proliferation, differentiation, and maturation. However, the signals that regulate myelination remain largely unknown, and represent a major goal for neurobiology. New approaches, capable of looking at thousands of genes activated sequentially during oligodendrocyte development, promise to shed new light on this issue.

2. OLIGODENDROCYTE DEVELOPMENT

2.1. Oligodendrocyte origin

The earliest oligodendrocyte precursor cells originate at early stages in highly restricted ventricular domains of the neural tube. They then migrate long distances away from these zones, and populate the developing brain to form the white matter. In the optic nerve, OPCs are restricted to small foci of cells in the floor of the third ventricle. In the spinal cord, oligodendrocyte progenitors arise mainly from the ventral germinative zone around the ependymal canal, just above the floor plate. The initially restricted location of OPCs is also observed in mid- and forebrain (Miller, 2002). Later on, the subventricular zone (SVZ) generates most glial precursors, which migrate into white matter and cortex, giving rise to oligodendrocytes and astrocytes. The SVZ enlarges during the period of gliogenesis, between P5 and P20, and then shrinks, but persists in adults (Baumann and Pham-Dinh, 2001).

The initial appearance of oligodendrocyte precursors depends crucially on spatial information present in normally developing CNS. In fact, the ventral origin of OPCs in spinal cord is not the result of any exclusive potential of this region to generate OPCs, but on ventrally-located signals from the notochord, a midline transient embryonic structure that specifies the production of OPCs. Sonic hedgehog (shh), the vertebrate homologue of the *Drosophila hedgehog* gene, is necessary and sufficient to induce oligodendrogenesis along the entire neural axis (Qi et al., 2002; Miller, 2002). *Shh* contributes to the initial commitment of neuroepithelial cells to the oligodendrocyte lineage or to motor neurons through the induction or repression of cell-type-specific transcription factors. Expression of transcription factor Olig2 and the basic helix-loop-helix transcription factors neurogenin-1 and 2 in neuroepithelial cells results in motor neuron specification. Towards the end of motor neuron specification in the spinal cord neurogenins are downregulated, which allows the expression of other transcription factors such as Nkx2.2. The co-expression of Olig2 and Nkx2.2 determines the commitment of neuroepithelial cells into oligodendrocytes. Another transcription factor involved in oligodendrogenesis is Olig1, which is required for the later development of oligodendrocytes, particularly in rostral regions of the CNS (Miller, 2002).

Other development studies, however, suggest that oligodendrocyte precursors originate from a more immature cell, termed glial restricted precursor (GRP), which is capable of giving rise to two distinct types of astrocyte and to oligodendrocytes. GRP cells are more widely distributed in spinal cord during development, and — unlike motor neuron/oligodendrocyte precursors — are not confined to ventral regions. The model suggests that the initial step in neuroepithelial cell differentiation is the separation of neuronal and glial fates. Subsequently, the GRPs in dorsal and intermediate regions of the spinal cord give rise predominantly to astrocytes, while those in ventral regions are influenced by local *shh* to generate oligodendrocytes at the expense of astrocytes (for further reading on oligodendrocyte development, see Miller, 2002; Rogister et al., 1999).

The first oligodendrocyte precursor was identified by Martin Raff and colleagues in the rat optic nerve, and named O-2A progenitor because of its bipotentiality (Barres and Raff, 1999). A further defining characteristic of the O-2A progenitor cell is its capacity for proliferation, self-renewal, and motility. The bipotentiality of O-2A progenitor was demonstrated in culture experiments where O-2A progenitors differentiated into oligodendrocytes in the absence of serum-derived signals. In contrast, in the presence of serum O-2A cells acquire astrocyte markers whilst retaining progenitor-related gangliosides. The latter cells were called type-2 astrocytes to differentiate them from type-1 astrocytes, which do not express gangliosides. However, there is no evidence for the presence of type-2 astrocytes *in vivo*. O-2A progenitors have been identified in different regions of the brain by the expression of specific markers characteristic of this stage of development. Specific markers of O-2A progenitors — later denominated oligodendrocyte precursor cells (OPCs) — are nestin (an intermediated filament protein also expressed in other proliferating cells such as radial glia), platelet-derived growth factor receptor alpha (PDGF-Rα), the GD3 ganglioside and other gangliosides recognized by the monoclonal antibody A2B5, the membrane chondroitin sulfate proteoglycan NG2, and myelin transcription factor 1.

2.2. Oligodendrocyte migration

Oligodendrocyte precursors originated in ventricular or subventricular areas have to migrate long distances to achieve their final destination and correctly myelinate axons. The capacity for migration is a characteristic of oligodendrocyte precursors that is lost as they mature. OPC mobility *in vitro* is promoted by PDGF, which is ubiquitously expressed in CNS and is therefore unlikely to guide the migration of these cells. In the optic nerve, OPC migration is axophilic and utilizes the retinal ganglion cell axons. Other factors that play an instructive role in the control of OPC migration are cell surface components such as adhesion molecules and extracellular matrix receptors. OPCs express a limited repertoire of integrin receptors, which regulate migration through specific interactions with the extracellular matrix components and other cell surface receptors. As OPCs migrate, they extend their processes along the extracellular matrix, in a similar fashion to the extension of neurites from neuronal cell bodies, with the aid of metalloproteinases such as MMP-9 (Miller, 2002; Baumann and Pham-Dinh, 2001).

Recent studies in optic nerve have demonstrated that OPC migration is unidirectional from the chiasm to the retina, and that it is regulated by a combination of chemorepellents and chemoattractants. The chiasm-derived chemorepellent netrin-1 appears to control OPC migration in optic nerve and spinal cord. There are also chemorepellent signals, such as the extracellular matrix molecule tenascin-1, which stop OPC migration at the head of the optic nerve and prevent their entry into the retina. In spinal cord, the chemokine CXCL1, signaling through the CXCR2 chemokine receptor, regulates spatial and temporal migration of OPCs by inhibiting cell motility, as well as by promoting cell proliferation (Miller, 2002).

We are now beginning to understand the mechanisms that control OPC migration. The OPC pool in adult mammalian CNS provides an important potential cell source for remyelination after demyelinating lesions. Identifying the specific molecular cues that guide OPCs will therefore have important therapeutic applications in demyelinating diseases. Therapeutic strategies based upon ectopic placing of cells that express guidance cues and facilitate OPC migration are now within the scope of preclinical studies.

2.3. Control of oligodendrocyte number

The final number of OPCs in postnatal white matter tracts depends on the number of oligodendrocytes that migrate into the white-matter, and the number of times each precursor divides before it finally differentiates. The proliferation of oligodendrocyte precursors is regulated in part by the availability of PDGF and basic FGF (bFGF), which are potent mitogens and inhibit differentiation. Basic FGF increases the expression of PDGFα receptor in immature cells, extending the period that OPCs respond to PDGF (Miller, 2002).

Oligodendrocyte precursor proliferation and differentiation are closely coupled processes, since these cells lose their capacity to proliferate as they differentiate. TGFβ arrests the proliferation of oligodendrocytes and promotes their maturation. As in other cell types, proliferation is regulated by the balance of stimulatory proteins such as cyclins or cdks, which normally drive the cell cycle, and inhibitory proteins such as cdk inhibitors, which normally suppress cell-cycle progression. Accumulation of the cdk inhibitor $p27^{kip1}$ leads to cell-cycle arrest. However, p27 inhibitory regulation is not exclusive to oligodendrocytes, since p27 null mice are larger than normal and have more cells in all the organs analyzed (for further reading on the topic, see Miller, 2002; Rogister et al., 1999).

The final number of oligodendrocytes also depends on their survival. Indeed, although there is a large increase in the number of OPCs in mice overexpressing the mitogen PDGF in neurons, excess OPCs in these animals later die through apoptosis, suggesting that the final oligodendrocyte population is definitively arranged to match the number of axons to be myelinated. Thus, oligodendrocyte survival following differentiation relies in part on the availability of axons, since only 50% of oligodendrocytes normally find an axon and survive (Barres and Raff, 1999). Survival of oligodendrocytes in culture is promoted by PDGF, insulin-like growth factor, ciliary neurotrophic growth factor (CNTF), and neurotrophin-3 (NT-3). Such factors are probably present in limited amounts in the developing nerve, because an exogenous supply improves oligodendrocyte survival. It is therefore possible that normally occurring oligodendrocyte death may also be a reflection of competition for these survival signals, which are secreted mostly by astrocytes (Barres and Raff, 1999). It appears that both astrocytes and axons are critical to oligodendrocyte survival: astrocytes sustain the viability of immature oligodendrocytes, and as these cells progress into differentiation, they require new survival signals that are provided by axons (e.g., neuregulin). Oligodendrocyte number is also negatively regulated by death signals such as nerve growth factor, which induces oligodendroglial cell death by binding to its p75 receptor.

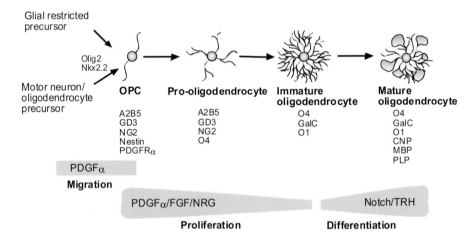

Figure 2. Schematic representation of the major developmental stages of oligodendrocytes. Oligodendrocyte precursor cells (OPC) originate from glial restricted and/or motor neuron/oligodendrocyte precursors, migrate into developing white matter, and proliferate to generate oligodendrocytes that myelinate axons. As OPCs differentiate, they lose their proliferative capacity. Each developmental stage can be identified morphologically and immunochemically using antibodies to specific markers. OPCs are bipolar cells characterized by the expression of gangliosides and PDGFRα, whereas fully differentiated oligodendrocytes are characterized by the expression of myelin proteins such as MBP, CNPase, and PLP. Oligodendrocyte development is controlled by multiple factors including those highlighted. PDGFα, platelet-derived growth factor α; FGF, fibroblast growth factor; NRG, neuregulin; TRH, thyroid receptor hormone.

2.4. Oligodendrocyte differentiation into myelinating oligodendrocytes

OPCs settle randomly along white matter and generate processes passing through different developmental stages, each identified by morphological features and by the expression of specific markers (see Figure 2 and the review by Baumann and Pham-Dinh, 2001). Once they contact axons, the processes extend longitudinally along the axonal surface, forming myelin, a membrane with a defined composition and specific lipid-protein interactions. Myelination therefore requires activation of the intracellular machinery to synthesize myelin glycolipids and proteins and for their export to appropriate sites (a description of myelin composition and the enzymes responsible for its synthesis is reviewed in Baumann and Pham-Dinh, 2001).

Oligodendrocyte maturation seems to depend on both soluble and cell-mediated signals from adjacent axons. A key regulator of oligodendrocyte differentiation is thyroid hormone. In culture, 3,3',5-triiodothyronine (T3), the active form of this hormone acting on its nuclear receptors, stops OPC proliferation and regulates the timing of oligodendrocyte differentiation. TH receptors are differently regulated during oligodendrocyte development, suggesting that they have different roles. Their role in myelination *in vivo* is supported by the hypomyelination observed in hypothyroidism (for the theory of T3 as a "time clock" for oligodendrocytes, see Barres and Raff, 1999). In addition, activation of erbB2 by neuregulins expressed or secreted by axons is essential to final differentiation and to induce myelination in spinal cord oligodendrocytes.

The Jagged-Notch-Hes pathway determines the early stages of axon/oligodendrocyte crosstalk and regulates the timing of oligodendrocyte differentiation and myelination. Premyelinating oligodendrocytes express Notch1 receptors, while neurons/axons express the ligand Jagged1. Contact-mediated activation of Notch1 by Jagged1 induces the expression of the inhibitory basic helix-loop-helix protein (bHLH) Hes5, which blocks the maturation of oligodendrocytes. As development proceeds, expression of Jagged1 in axons is downregulated, allowing oligodendrocyte precursors to differentiate and myelinate. The signals that induce Jagged1 downregulation in axons are not known. However, myelination of axons begins after they reach their target, raising the possibility that target innervation activates signals to downregulate Jagged1 in axons, and subsequently triggers myelination (John et al., 2002). Similarly, PSA-NCAM expression in axons negatively regulates myelination, and its downregulation or removal allows the onset of myelination (Miller, 2002).

Recent studies have also identified transcriptional factors implicated in oligodendrocyte differentiation. A key role in the latter phase of oligodendrocyte differentiation is played by the transcription factor Sox10. In general, oligodendrocyte differentiation is associated with declining levels of the bHLH proteins Id2 (inhibitor of DNA-binding1), Id4, and Hes5, and of the zinc-finger protein myelin transcription factor 1, and with increasing levels of AP-1, the bHLH protein Mash1, the homeodomain factor Gtx, and the peroxisome proliferator-activated receptor-δ (PPAT-δ) (Franklin, 2002).

The interaction between axons and myelinating cells is bidirectional and complex, and leads to systemic changes in the cytoskeleton of the oligodendrocyte in order to direct the growth of processes around the axon. Phosphorylation of Fyn kinase seems to play a key role in oligodendroglial cytoskeleton rearrangements to form myelin. Cell-surface candidates are NCAM 120 and F3 adhesion molecules, which are compartmentalized with Fyn kinase. Ligation of glial F3 via an axonal ligand such as L1 activates Fyn kinase in raft microdomains, leading to increased binding of tau. Lipid rafts are gsl- and cholesterol-rich domains that form in the trans-Golgi network, sort proteins and lipids to their appropriate destination within the cells during myelogenesis, and act as transduction platforms via selective inclusion/exclusion of signaling elements. Another mediator of axon/oligodendrocyte interactions is myelin-associated glycoprotein (MAG), a member of the family of sialoadhesins, which play a role in the formation of spiraling loops, the number of myelin loops, and the maintenance of myelin (Buttery and ffrench-Constant, 2001). Extracellular matrix molecules also regulate myelination by their signaling through integrins. Integrins are formed by two transmembrane glycoprotein subunits (α and β), and act by linking ECM molecules to cytoskeleton elements. There are many different heteromeric receptors whose expression changes during oligodendrocyte development. αvβ1 integrin is involved in OPC migration, while αvβ3, αvβ5, and α6β1 increase during differentiation and may regulate the final phase of differentiation by promoting myelin-sheath formation (Buttery and ffrench-Constant, 2001).

In sum, oligodendrocyte differentiation into myelin-producing cells is a complex process that requires multiple and coordinated interactions that are beginning to be clarified. However, the formation of the compact and spiraled myelin sheath remains largely obscure (for further reading on signals controlling oligodendrocyte differentiation, and myelination see Buttery and ffrench-Constant, 2001; Miller, 2002; and Baumann and Pham-Dinh, 2001).

3. OLIGODENDROCYTE VULNERABILITY

3.1. Oligodendrocytes express neurotransmitter receptors

Besides their support to neurons, glial cells integrate neuronal inputs and modulate synaptic activity. Astrocytes, which express different neurotransmitter receptors, are able to release neurotransmitters and, due to their perisynaptic location, modulate different neuronal functions. In addition, astrocytes are interconnected by gap-junction hemichannels forming a glial syncytium, allowing transfer of information and possibly serving to synchronize the activity of neighboring cells (Gallo and Ghiani, 2000). Gap junctions also occur between oligodendrocytes and between astrocytes and oligodendrocytes. The main gap-junction protein in oligodendrocytes is connexin 32, and gap-junction communication in oligodendrocytes serves to buffer K^+ around myelinated axons (Baumann and Pham-Dinh, 2001)

Oligodendrocytes also express many voltage-sensitive ion channels and neurotransmitter receptors similar to those present in neurons; however, they lack the membrane properties required to fire action potentials. Oligodendrocyte progenitors and mature oligodendrocytes express glutamate receptors and transporters in gray- and white-matter tracts, where they could serve to sense glutamate released by synapses and/or axons and to maintain physiological concentrations of the excitatory neurotransmitter respectively (see Matute et al., 2001; Gallo and Ghiani, 2000). In addition, hippocampal neurons form terminals onto OPCs that activate AMPA receptors in these cells and respond to Schaffer collateral stimulation (reviewed in Gallo and Ghiani, 2000).

One of the functions attributed to glutamate receptors in oligodendrocytes is the regulation of their proliferation and differentiation. Glutamate inhibits OPC proliferation by increasing the levels of the cyclin-dependent kinase inhibitors $p27^{kip1}$ and $p21^{Cip1}$, and cell-cycle arrest in G1 phase (reviewed in Gallo and Ghiani, 2000), an effect that is also caused by membrane depolarization. Intriguingly, glutamate induces OPC cell-cycle arrest without leading to differentiation. In turn, purinergic receptor activation by adenosine inhibits OPC proliferation, stimulates differentiation, and promotes the formation of myelin (Stevens et al., 2002). The functions mediated by other neurotransmitter receptors present in the oligodendrocyte lineage are not known at present.

3.2. Excitotoxicity in oligodendrocytes

Oligodendrocytes are highly vulnerable to glutamate signals. Early studies of this feature showed that toxicity was not mediated by glutamate receptors but rather by a transporter-related mechanism involving the inhibition of cysteine uptake, resulting in glutathione depletion and cell vulnerability to toxic free radicals. More recently however, it was shown *in vitro*, as well as in different models *in situ* and *in vivo*, that prolonged activation of AMPA and kainate receptors is toxic to oligodendrocytes.

Ca^{2+} influx via AMPA and kainate receptors alone is sufficient to initiate cell death in oligodendrocytes, and it does not require the entry of calcium via other routes such as voltage-activated calcium channels or the plasma membrane Na^+-Ca^{2+} exchanger (reviewed in Matute et al., 2002). Various features may render oligodendrocytes vulnerable to overactivation of AMPA and kainate receptors. First, the subunits that constitute the receptors endow them with higher Ca^{2+} permeability. Second, oligodendrocytes do not express several of the Ca^{2+}-binding proteins present in neurons, and this makes the cells more resistant to excitotoxicity (Matute et al., 2001).

As in neurons, a rapid Ca^{2+} influx through AMPA and kainate channels may result in mitochondrial Ca^{2+} overload and in the production of oxygen radicals, leading to neuronal death. Mitochondrial alterations subsequent to excitotoxic insult in oligodendrocytes can trigger apoptotic and necrotic death. Thus, mild insults induce prompt activation of caspases, including caspase-3, and alteration of mitochondrial function which leads to apoptosis, whereas severe insults result in necrotic cell death (Matute et al., 2002).

Immature oligodendrocytes appear to be more sensitive than their mature counterparts to glutamate-mediated excitotoxicity. This could be due to regulation of glutamate receptor expression by growth factors such as bFGF and PDGF which also control OPC proliferation and differentiation. The functional properties of glial glutamate receptors are determined mainly by their molecular composition, by the level of subunit editing and by electrical activity (Liu and Cull-Candy, 2000). Thus, changes in the different properties of glutamate receptor channels that affect their function could take place during oligodendrocyte development and/or after contacting axons, and could explain the greater sensitivity of OPC cells to glutamate-mediated insults. Such changes remain to be determined.

4. MULTIPLE SCLEROSIS: AN OLIGODENDROCYTE DISEASE

4.1. Pathology and etiology

Multiple sclerosis is the most common demyelinating disease, affecting nearly one million people worldwide. The disease usually begins in young adulthood, and is more prevalent among women than men. In most cases, MS starts with immune attacks, lasting from days to weeks, followed by remission lasting months to years. This period is known as the relapsing-remitting (RR) phase, and can last between five and ten years. Paralysis, sensory disturbances, lack of co-ordination, and visual impairment are common features. Over time, the number of relapse episodes decreases, but most patients enter a secondary chronic progressive phase (SP) characterized by the development of progressive neurological deficits. In 10-20% of patients, MS begins with a primary progressive (PP) course without acute relapses. MS lesions are focused on myelinated areas of the brain and spinal cord. The RR phase is characterized by a disturbance of the blood brain barrier, local edema, immune-cell infiltration, and demyelination, features that are compatible with an inflammatory process. By contrast, such inflammation is less evident when progressing to the secondary phase and in patients with primary progressive disease. Oligodendroglial cell death and degeneration of myelin and of axons are more dominant in this phase, and appear to correlate with disability (Keegan and Noseworthy, 2002).

The etiology of MS is still uncertain; however, it is generally assumed that MS is an autoimmune disease against myelin proteins. The contribution of autoimmunity to MS has been studied using experimental autoimmune encephalitis (EAE), an animal model of MS, which is induced by immunization with myelin antigens and Freund's adjuvant, or by passive transfer of autoreactive T-lymphocytes from affected animals. However, studies of the immune responses in patients with MS has revealed that autoantibodies to several myelin components as well as autoreactive T cells, specific for myelin, are present in the CNS and serum in normal conditions as well as in inflammatory neurodegenerative diseases (Hemmer et al., 2002). Although autoimmunity is not unequivocally linked to MS, the EAE model has proved to be an important tool for research on MS therapy.

There is also evidence that MS is associated to viral infections (for a list of infectious agents see Keegan and Noseworthy, 2002). Infectious agents are known to cause demyelinating diseases in animals as well as in humans (e.g., virus causes progressive multifocal leukoencephalopathy). Exposure to an infectious agent in a genetically susceptible individual or within a critical time in development can be the primary cause of MS, while an inadequate immune response probably contributes to disease progression. Acute attacks are possibly related to reactivation of a latent infection. However, a clear association of any pathogen to MS is still lacking (Hemmer et al., 2002).

4.2. Oligodendrocyte excitotoxicity as a primary cause in multiple sclerosis

Although the immune reaction appears to be important in the RR phase of the disease, some other mechanism must mediate the chronic progressive phase. The oligodendroglial and axonal damage in PP disease can also be observed in the absence of acute episodes of inflammation, suggesting that the immune response could be a secondary event in MS. In addition, the benefits of the current therapy — anti-inflammatory and immunosuppressive agents — are limited, and consist mostly of a slowing of the progression of the disease (Hemmer et al., 2002).

A recent finding which may contribute to clarifying MS etiology is the sensitivity of oligodendrocytes to excitotoxicity. Glutamate receptor agonists kill oligodendrocytes *in vitro* and induce lesions *in vivo* that have the major features of MS. In addition, mild prolonged excitotoxic insults result in the long term in massive atrophy and a profound demyelination, which — in contrast to acute insults — can not be repaired by endogenous mechanisms (Matute et al., 2001). In this experimental paradigm, it appears that oligodendrocytes die and myelin degenerates before the appearance of axonal damage on exposure to GluR agonist.

The contribution of excitotoxicity to MS etiology was further confirmed by experiments in EAE with antagonists of AMPA and kainate receptors. The blockade of these receptors ameliorates EAE and prevents clinical relapses, without affecting the immune response (Steinman, 2000). Studies in EAE suggest that the immune response somehow produces an alteration in glutamate homeostasis that ultimately results in an excessive activation of glutamate receptors present in oligodendrocytes. An important element in this picture could be microglia, which as a result of the inflammatory reaction, change to a reactive state characterized by the production of different toxins to oligodendrocytes, such as cytokines and glutamate. In conclusion, these studies suggest that excitotoxicity secondary to the immune response contributes to MS pathology.

However, the early loss of oligodendrocytes and neurons in MS also points to excitotoxicity as a primary cause of the disease (see Matute et al., 2001; Hemmer et al., 2002). Damage to the CNS is commonly associated with microglial activation, cytokine production, gliosis and infiltration of leukocytes. Consistently, white-matter damage by overactivation of glutamate receptors induces inflammation similar to that observed in other neurodegenerative diseases and traumatic lesions. Thus, the generation of cellular debris as a result of oligodendrocyte death and myelin destruction may prime an autoimmune reaction (see Figure 3A). In keeping with the idea of excitotoxicity as a major cause in MS, alterations in the levels of glutamate as well as in other glutamatergic markers have been observed in MS patients and the EAE model (Matute et al., 2001).

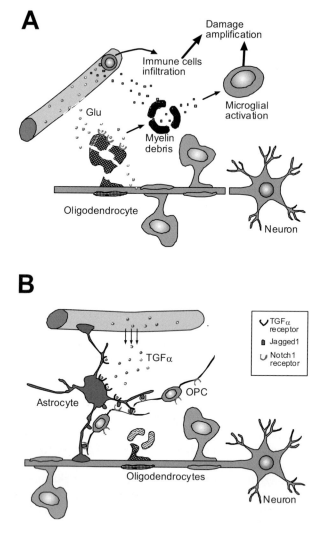

Figure 3. A) Excitotoxic damage to oligodendrocytes induces activation of microglia and the immune system. Blood brain barrier disruption, allowing glutamate entry into the CNS, or an excessive release of glutamate from axons and microglial cells, can overactivate glutamate receptors and kill oligodendrocytes. Myelin debris generated by dying oligodendrocytes triggers microglial activation and primes T- and B-cells to autoimmunity. Infiltration of immune cells into the CNS and neurotoxins secreted by reactive microglia, can in turn amplify oligodendroglial and myelin damage. B) Molecular mechanisms underlying the failure of remyelination in multiple sclerosis. TGFα, released from blood capillaries or microglial cells, induces astrogliosis and the expression of Jagged1, which binds to the Notch1 receptor expressed in oligodendrocyte precursors cells (OPCs). As a result of Notch receptor activation, OPC differentiation into mature oligodendrocytes is blocked and remyelination prevented.

However, none of these studies has definitely associated the disease to a specific alteration of glutamatergic neurotransmission in MS patients.

In conclusion, different ideas about the etiology of MS are under scrutiny, including autoimmunity, viral infection, and excitotoxic degeneration. Although the nature of the primary event leading to MS remains unresolved, activation of immune cells and CNS infiltration of primed T- and B-cells are major hallmarks of the disease (for a detail explanation of the immune response, see Hemmer et al., 2002).

4.3. Failure of remyelination in multiple sclerosis

The remyelination capacity in the CNS is very limited. Moreover, it decreases with MS progression, thereby worsening the clinical symptoms of the disease. Remyelination is carried out by oligodendrocytes generated de novo from OPCs present in the adult CNS. Oligodendrocyte generation in the mature brain and spinal cord recapitulates many of the features described during development.

Remyelination enables axons to recover conduction properties that were lost as a result of demyelination. However, new myelin sheaths are thinner and shorter than might be expected for the diameter of the axon, and yield to pale myelin staining (the so-called "shadow plaque"). Yet its composition, the manner in which it is formed, and the transcriptional regulators mediating remyelination are similar to those in developmental myelination (for an excellent review on remyelination see Franklin, 2002). Whereas in experimental animal models of MS remyelination occurs efficiently, remyelination in human CNS is limited, and the reasons for this failure are unknown. Numerous strategies involving promoting endogenous repair mechanisms or transplanting an exogenous source of myelinating cells have been successfully tested in animal models (Stangel and Hartung, 2002). However, initial trials in humans have been disappointing.

Efficient remyelination involves the recruitment of a sufficient number of OPCs, and their subsequent differentiation into myelinating oligodendrocytes. Failure to remyelinate appears not to be due to depletion of the source of OPCs, because numerous OPCs are present within and around MS demyelinated plaques (Franklin, 2002). Therefore, it appears that OPC differentiation is not correctly accomplished, thus preventing remyelination. An explanation for this feature has been proposed in a recent study showing the re-expression of a developmental pathway that restricts oligodendrocyte maturation (John et al., 2002). Using cDNA microarray screen and immunohistochemical analysis, John and colleagues elegantly demonstrated that Jagged1, which is expressed during development in axons and inhibits oligodendrocyte development (see Section 2.d), is upregulated in reactive astrocytes in MS lesions. As oligodendrocytes express Notch1 receptor and Hes5, such upregulation could restrict oligodendrocyte differentiation and the subsequent remyelination (see Figure 3B). However, in view of the complex mechanism and the multiple signals that regulate OPC migration, proliferation, and differentiation, it is possible that other factors yet to be identified may contribute to the failure of remyelination.

Microglia also play a role in MS. Classical studies considered that microglial activation generally renders these cells harmful to surrounding tissue, because they release cytokines known to induce oligodendrocyte death and axonal damage. This conclusion derives from an immune-mediated model of the disease, in which microglial activation is secondary to the immune response. However, recent studies suggest that activated microglia may in fact be beneficial for remyelination (see Franklin, 2002). Thus, macrophage depletion impairs the ability of oligodendrocytes to remyelinate following toxin-induced demyelination. Remyelination is also impaired in knockout mice lacking the pro-inflammatory cytokines IL-1β and TNFα, which are produced by reactive microglia (Arnett et al., 2001). Such findings could explain why therapies aimed at blocking TNFα in multiple sclerosis have produced little improvement, and in some patients have even led to a worsening of symptoms (see Arnett et al., 2001). In the light of these results, anti-inflammatory therapy must also be questioned.

5. OLIGODENDROCYTE AND WHITE-MATTER DAMAGE AFTER HYPOXIC/ISCHEMIC INSULTS

Excitotoxicity appears to be the predominant mechanism underlying ischemic damage. Like neurons, immature and differentiated oligodendrocytes are vulnerable to oxygen and glucose deprivation. The mechanisms triggering ischemic oligodendrocyte death include Ca^{2+} overload and the activation of AMPA and kainate receptors. Indeed, blockade of these receptors attenuates hypoxic-induced oligodendroglial death in culture, and preserves white-matter function in brain slices (Goldberg and Ransom, 2003). An important observation in one of these studies is that ischemic insults were made on pure cultures of oligodendrocytes, indicating that these cells release glutamate under ischemic conditions possibly by reverse functioning of glutamate transporters (Matute et al., 2001).

Consistent with *in vitro* studies, permanent middle cerebral artery occlusion and brief transient global ischemia induce rapid oligodendroglial death, and in certain brain regions these cells are even more vulnerable than neurons. Recent studies have reported white-matter-damage protection (or reduced damage) by AMPA and kainate receptor antagonists in spinal cord ischemia and transient cerebral ischemia (Goldberg and Ransom, 2003). In traumatic spinal cord lesions, glutamate spill-over from affected cells or axons is deleterious to the oligodendroglial population, and some protection is achieved by blockade of AMPA and kainate receptors (Matute et al., 2001). Increasing evidence indicates that oligodendrocytes are vulnerable to anoxic and ischemic insults in humans after stroke, cardiac arrest, and vascular dementia in the aging brain. Perhaps the most dramatic cases of hypoxia-/ischemia-related diseases with white-matter damage are cerebral palsy and periventricular leukomalacia, the latter characterized by selective white-matter damage with little or no cortical injury. These two conditions are characterized by isolated foci or diffuse areas of gliotic tissue that appear to be formed as a consequence of oligodendrocyte death, possibly caused by glutamate toxicity. The importance of white-matter damage should be considered in new approaches to treating ischemia-related diseases. The fact that NMDA receptors are not expressed in white matter could explain the failure of NMDA antagonists in human clinical trials of acute stroke. Future approaches to reducing both gray- and white-matter damage after ischemia, including the use of AMPA and kainate receptor antagonists, promise to be more successful.

6. CONCLUSIONS

Important advances have been made in understanding the development of oligodendrocytes, the mechanisms regulating their differentiation, and the complex process of myelination. New knowledge about oligodendrocyte biology and function will undoubtedly lead to innovative therapies for pathologies in which oligodendrocytes are severely damaged. The vulnerability of oligodendrocytes to glutamate appears to be central to demyelinating diseases and to white-matter damage in ischemia. These novel ideas are ripe for testing in clinical trials.

REFERENCES

Arnett H.A., Mason J., Marino M., Suzuki K., Matsushima G.K. and Ting J.P.-Y. (2001) TNFα promotes proliferation of oligodendrocyte progenitors and remyelination. Nat. Neurosci. 4; 1116-1122.

Barres B. and Raff M.C. (1999) Axonal control of oligodendrocyte development. J. Cell Biol. 147; 1123-1128.

Baumann N. and Pham-Dinh D. (2001) Biology of oligodendrocyte and myelin in the mammalian central nervous system. Physiol. Rev. 81; 871-927.

Buttery P.C. and ffrench-Constant C. (2001) Process extension and myelin sheet formation in maturating oligodendrocytes. Progress in Brain Res. 132; 115-130.

Del Río Hortega P. (1928) Tercera aportación al conocimiento morfológico e interpretación funcional de la oligodendroglía. Mem. R. Soc. Esp. Hist. Nat. 14; 5-122.

Franklin R.J.M. (2002) Why does remyelination fail in multiple sclerosis? Nat. Rev. Neurosci. 3; 705-714.

Gallo V. and Ghiani C.A. (2000) Glutamate receptors in glia: new cells, new inputs and new functions. TIPS 21; 252-258.

Goldberg M.P. and Ransom B.R. (2003) New light on white matter. Stroke 34; 330-332.

Hemmer B., Archelos J.J. and Hartung H.P. (2002) New concepts in the immunopathogenesis of multiple sclerosis. Nat. Rev. Neurosci. 3; 291-301.

John G.R., Shankar S.L., Shafit-Zagardo B., Massimi A., Lee S.C., Raine C.S. and Brosnan C.F. (2002) Multiple sclerosis: Re-expression of a developmental pathway that restrictis oligodendrocyte maturation. Nat. Med. 8; 1115-1121.

Keegan B.M. and Noseworthy J.H. (2002) Multiple sclerosis. Annu. Rev. Med. 53; 285-302.

Liu S-Q.J. and Cull-Candy S.G. (2000) Synaptic activity at calcium-permeable AMPA receptors induces a switch in receptor subtype. Nature 405; 454-458.

Matute C., Alberdi E., Domercq M., Pérez-Cerdá F., Pérez-Samartín A. and Sánchez-Gómez M.V. (2001) The link between excitotoxic oligodendroglial death and demyelinating diseases. TINS 24; 224-230.

Matute C., Alberdi E., Ibarretxe G. and Sánchez-Gómez M.V. (2002) Excitotoxicity in glial cells. Eur. J. Pharmacol. 447; 239-246.

Miller R.H. (2002) Regulation of oligodendrocyte development in the vertebrate CNS. Progress in Neurobiol. 67; 451-467.

Qi Y., Stapp D. and Qiu M. (2002) Origin and molecular specification of oligodendrocytes in the telencephalon. TINS 25; 223-225.

Rogister, B., Ben-Hur T. and Dubois-Dalq M. (1999) From neural stem cells to myelinating oligodendrocytes. Mol. Cell. Neurosci. 14; 287-300.

Stevens B., Porta S., Haak L.L., Gallo V. and Fields R.D. (2002) Adenosine: A neuron-glial transmitter promoting myelination in the CNS in response to action potentials. Neuron 36; 855-868.

Stangel M. and Hartung H.-P. (2002) Remyelinating strategies for the treatment of multiple sclerosis. Progress in Neurobiol. 68; 361-376.

Steinman L. (2000) Multiple approaches to multiple sclerosis. Nat. Med. 6; 15-16.

D. ECHEVARRÍA AND S. MARTÍNEZ

Institut of Neuroscience, University Miguel Hernandez (UMH-CSIC), Alicante, Spain.

15. THE MYELIN GLIAL CELL OF THE PERIPHERAL NERVOUS SYSTEM: THE SCHWANN CELL

Summary. The work reviewed in this chapter shows that Schwann cells and precursors act as a source of diverse developmental signals. These signals influence both cell differentiation and cell survival in early nerves. Schwann cells regulate the development of the three main cell types in nerves: neurons, connective tissue cells, and, via autocrine circuits, the Schwann cells themselves. Myelination depends on the coordination of a battery of glial structural proteins that are responsible for the architecture of compact myelin. We have also reviewed a number of transcription factors that play important roles in Schwann cell differentiation and progression through the promyelinating stage of differentiation. Schwann-cells are the fundamental elements in promoting peripheral nerve regeneration and have been used in variousmodels of axonal regeneration in the central nervous system. An understanding of the main processes that regulate Schwann-cell biology is a requisite for the use of this interesting cell population — with proven neurotrophic and regenerative properties — in cell therapy for human diseases

1. INTRODUCTION

One of the major and most-evolutionary features of the animal kingdom in the last 450 million years is the production of the myelin sheath. The myelin sheath is a structure unique to the nervous system of vertebrates. In its mature form, it consists of broad regions of compact plasma membranes bordered by narrow channels filled with cytoplasm. Myelination conferred the irrefutable evolutionary advantage of allowing great speed of conduction while minimizing the size of the axon.

The cells that are able to form and produce myelin are two glial cell types: the oligodendrocyte glial cells in the central nervous system (CNS) and the Schwann glial cells in the peripheral nervous system (PNS; see Figure 1). The two synthesize similar sets of proteins, but each has unique proteins as well that may account, at least in part, for the differences in myelin structure and formation characterizing each nervous system subtype. Oligodendrocytes ensheath many axons via multiple processes, providing further conservation of space by enabling a single cell body to control the myelination of several internodal segments. In the PNS, this constraint of space does not exist: each Schwann cell myelinates only a single axonal segment.

This chapter focuses on the embryological process of the origin and development of Schwann cells. We will draw on the many works in this field to discuss the phenotypic characteristics of these cells during development, and the accumulating evidence that developing Schwann cells act as regulators in the development of neurons, connective tissue cells, and the Schwann cells themselves — the three main components of peripheral nerves. The characteristics and properties of Schwann cells in injury and regeneration processes are also described.

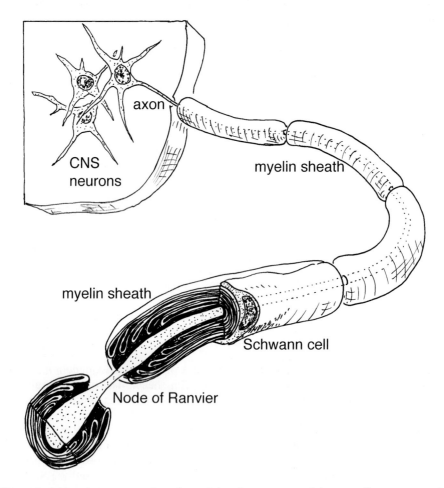

Figure 1: Schematic representation of a peripheral motor nerve. Schwann cells wrap around the axon, forming the myelin sheaths. One Schwann cell generates a myelin segment and the Ranvier nodes are localized in the intersegmental spaces.

2. ORIGINS AND LINEAGE OF SCHWANN CELLS

The origin and cell linEage of Schwann cells ARE now well established. Most Schwann cells come from neural-crest cells (Figure 2; also see Le Douarin et al., 1991). These neural-crest cells — not yet determined and differentiated but yet identifiable — suffer a number of transitions in their lineage towards what we know as the Schwann cell. Two intermediate steps have been identified during embryonic development: the Schwann-cell precursor, present around mouse embryonic day 12 and 13 (E12 and E13), and the immature Schwann cell, present from E14.5 to the time of birth (Mirsky and Jessen, 1996). During this transition stage, immature Schwann cells proliferate, migrate, invade, bundle and sort the axonal fibers of the embryonic nerves (Figure 2). For several days following birth, the immature Schwann cells differentiate and diverge, generating the myelin-forming Schwann cells that wrap the large-diameter axons, and the non-myelin-forming Schwann cells that accommodate small-diameter axons in shallow tubular structures along their surface (Hahn et al., 1978).

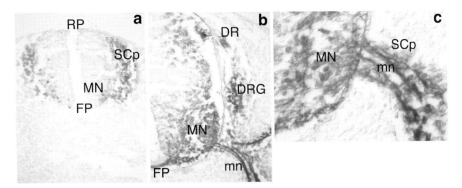

Figure 2: The neural-crest cells migrate from the dorsal midline of the neural tube to produce their cellular derivatives. In this figure, Schwann-cell progenitors (SCp) have been detected in a transgenic mouse that express X-gal under the promoter of the plp/dm20 gene (Spassky et al., 1998). a) Transverse section of the spinal cord of an E12.5 embryo, showing the migrating SCp at the lateral side of the neural tube. b,c) At E13.5 SCp are localized around dorsal root (DR) and motoneuron (MN) axons, as well as around dorsal root ganglia (DRG) sensorial neurons. FP: floor plate; RP: roof plate; mn: motor axons.

Thus, the Schwann-cell lineage involves a number of steps: i) pluripotent neural crest stem cells must acquire a glial-cell fate, ii) this glial-cell population expands and invades the nerve bundles, iii) immature cells adopt a special differentiation and specialization configuration with the axons, and cease proliferation and iv) those becoming myelin-forming cells initiate and complete myelogenesis, while non-myelin-forming cells further (appearing later) engulf multiple smaller-diameter axons in cytoplasmic cuffs (Dong et al., 1999).

What are the signals controlling the first choice for glial fate by these crest cells? ß-neuregulin-1 is clearly important in regulating these events, but does not appear to be necessary for the establishment of PNS glia (see next section). A more fundamental transcription factor that plays a key role in establishing Schwann-cell lineage can be attributed to *Sox10*, since — in mice lacking this gene — Schwann-cell precursors, mature Schwann cells, and satellite cells are missing (Britsch et al., 2001). Another marker that can be used as early glial marker is myelin protein zero (MP0, also named P_0). P_0 (a glycosilated 30kDa membrane protein) is the major structural myelin component of the adult Schwann cells, which account for half the myelin component. During embryogenesis it is expressed, albeit at low levels, in immature Schwann cells, Schwann-cell precursors, and some crest cells. Observations that in early development P_0 is also expressed outside the nervous system — notochord and by cells in non-sensory regions of the ear — clearly show that its mRNA is not restricted to PNS glia. Thus, P_0 might be considered as a marker of further maturation of crest cells. However, due to the complex pattern of P_0 gene expression, several issues need to be considered in using this putative glial marker as cell-lineage candidate for Schwann-cell fate commitment (Lee et al., 2001).

3. SCHWANN-CELL DEVELOPMENT AND MYELIN-SPECIFIC GENES

Numerous studies have indicated that the differentiation of Schwann cells requires continued axonal contact. In particular, it has been shown by gene-ablation studies in mice that the survival of immature Schwann cells, and perhaps their generation from the neural crest, critically depends on axonal derived ß-neuregulin-1 (ß-NG-1) receptor complex expressed on the surface of Schwann cells. The isolation of recombinant ß-NG-1 isoforms (I, II and III) and the generation of mice that lack either ß or its receptors ErbB2 and ErbB3 have revealed crucial roles for these molecules in the generation of Schwann cells. From this experimental analysis, two different mechanisms of action of ß-NG-1 have been proposed. One mechanism concerns the functions as a *survival factor* for cell precursors and early Schwann cells. ß-NG-1 is expressed in vivo by dorsal root ganglial (DRG) neurons and in motoneurons during embryonic nerve development, and released forms of ß-NG-1 proteins accumulate along the axonal tracts. Recent observations in knockout isoform III mice corroborate this role of ß-NG-1 function in these animals, although the number of precursor is not very different from normal in early peripheral nerves (E11.5), at E14.5 the number of these cells is severely depleted, a stage when precursor are converting rapidly to Schwann cells (Dong et al., 1999; Wolpowitz et al., 2000). The other mechanism relates to the ability of ß-NG-1 to stimulate Schwann-cell migration and proliferation. *In vitro*, ß-NG-1 promotes migration of neonatal Schwann cells away from DRG explants (Morris et al., 1999), and in ErbB3 knockout mice, failure of migration of neural-crest cells along the ventromedial pathway appears to be the reason why sympathetic ganglia do not fomr properly (Britsch et al., 1998). It is, therefore, a plausible suggestion that the dramatic absence of precursor and Schwann cells in ß-NG-1 and ErbB3 knockout mice might be due to reduced migration combined with the death of those cells that take up positions in peripheral nerves.

Intracellular signaling pathways which transduce ß-NG-1 signals have revealed that ß-NG-1-induced proliferation in Schwann cells involves mitogen-activated protein (MAP) kinase activation (Meier et al., 1999) by raising intracellular cAMP levels after application; moreover, pharmacological inhibition of the PI3 kinase pathway causes Schwann-cell death (Dong et al., 1999). These results indicate that the survival of Schwann-cell precursor cells depends on input from ß-NG-1 through both intracellular MAP and PI3 kinase pathways.

Two members of the Erg (early growth response factor) gene family (*Krox20* and *Krox24*) are expressed during Schwann-cell development. These Zinc finger transcription factors have been implicated in a diverse array of processes, including commitment to mitogenic, differentiation and apoptotic pathways. Comparison of *Krox20* and *Krox24* expression patterns during embryogenesis shows that initially the two genes are regulated in a successive and largely mutually exclusive manner — *Krox24* being restricted to Schwann-cell precursors and *Krox20* to immature Schwann cells. Both are activated around E10.5 in peripheral nerves. At E15.5, important changes in pattern expression occur: *Krox20* is activated along the peripheral nerves in a subpopulation of immature Schwann cells, while *Krox24* is downregulated. This suggests that the myelination decision coincides with the period of conversion of Schwann-cell precursors into immature Schwann cells, and implicates *Krox20* as the earliest specific myelination marker (Mirsky et al., 2001).

Another transcription factor that plays a role in cell-fate lineage and differentiation is *Oct-6*. *Oct-6* is a member of a family of nuclear proteins characterized by a highly conserved POU domain. In vivo, *Oct-6 is* first expressed in Schwann-cell precursors and immature Schwann cells, but not in neural-crest cells from which they derive. During postnatal development, *Oct-6* expresses strongly in promyelinating and actively myelin-forming cells, and in low levels in non-myelin-forming cells. This expression is gradually extinguished in nearly all Schwann cells during the first postnatal weeks. The downregulation is dependent on nerve maturation and on Oct-6 expression itself, suggesting that *Oct-6* is involved in a negative feedback loop during terminal differentiation of Schwann cells (Ghazvini et al., 2002). Within the nerve, *Oct-6* expression is strictly dependent on continued axonal contact (see section 17.4). In cultured Schwann cells, the differentiation-inducing effect of axonal contact can be mimicked in part by elevating intracellular cAMP levels, suggesting that *Oct-6* expression may function in initiating the myelination program (Monuki et al., 1989). Surprisingly, co-transfection experiments in Schwann-cell cultures reveal *Oct-6* as a repressor of P_0 (the major myelin component) gene (Monuki et al., 1993). Using the gene knockout technology for *Oct-6* has suggested that *Oct-6* serves as a competence factor in promyelinating cells for their timely differentiation into myelin forming cells (Jaegle and Meijer, 1998). *Oct-6* is required in myelin-forming cells for the downregulation of its own gene, either directly, or indirectly since the existence of auto-regulatory loops has been established for other POU genes such as *pit-1* and *cf1a/drifter/ ventral veinless* (Certel et al., 1996). Moreover, the expression of *Krox20* depends on *Oct-6* levels (Zorick et al., 1999).

Taking these data together a model for the function of *Oct-6* maybe proposed. In response to axonal cues, *Oct-6* stimulates a repertoire of genes, including *Krox20*, driving the differentiation of Schwann cells through the promyelinating stage into the myelinating stage. In myelin-forming cells, *Oct-6* functions as a repressor of its own gene, and possibly of other genes. Thus, *Oct-6* acts as activator of promyelinating cells and as repressor of myelin-forming cells (Ghazvini et al., 2002).

Finally, we should mention the role of the *Pax3* gene. It is one of the nine members of a family of vertebrate control genes encoding nuclear proteins characterized by paired box, a conserved amino-acid motif with DNA-binding activity. The expression pattern profile of this gene needs to be confirmed, since the only two studies on the *Pax3* expression pattern are controversial on the time of appearance during embryogenesis (Kioussi et al., 1995; Blanchard et al., 1996). Expression of *Pax3* in the embryonic nerves is thought to be in Schwann-cell precursors. However, it has been shown that rat embryonic sciatic nerves contain significant numbers of cells with pluripotent differentiation potential, resembling neural-crest stem cells (Morrison et al., 1999). During postnatal development and in adult life, *Pax3* expression is maintained in the non-forming-myelin Schwann cells and depends, again, on axonal contact, as nerve crush leads to rapid down-regulation on the gene and rapid re-initiation of its expression after axon/Schwann-cell contact. Therefore, it is hypothesized that *Pax3* is influencing the switch between myelin-forming and non-myelin-forming cells, favoring the generation of non-myelin-forming Schwann cells.

4. DEGENERATION AND REGENERATION OF MYELIN SHEATH AFTER AXOTOMY

Wallerian degeneration is a feature of any insult that causes axonal degeneration, although it has been considered too as mechanical injuries in experimental studies. It encompasses all the events that occur distal to the site of axotomy. A notable feature of the Schwann-cell phenotype is how labile it remains throughout life (Dupin et al., 2003). The signal that triggers Wallerian degeneration is unknown. If a nerve is transected, the myelinating and non-myelinating Schwann cells in the distal stump will promptly undergo numerous and radical alterations in morphology and gene expression (Terenghi, 1999; Scherer, 1997). This process involves the differentiation or developmental regression of individual Schwann cells and myelin breakdown. The outcome of these complex, transient responses is the generation of an apparently single population of cells showing a state of differentiation comparable to that of immature cells prior to generation of myelinating and non-myelinating Schwann cells. If these cells re-contact the proper axon targets or specific axons, they re-express a mature phenotype, indicating that axonal signals are required to maintain the plastic phenotype of Schwann cells (Jessen and Mirsky, 1999). After axotomy, the injured axons fragment and disappear, and the surrounding myelin sheaths separate (denervation) and form what are called myelin ovoids. During the next few weeks, these ovoids are phagocytosed — in part by Schwann cells, but mainly by macrophages that invade the degenerating nerve. Neither the whole Schwann-cell population nor the basal lamina is degenerated within the injury: they remain during these degenerative events. As we will discuss below, both of them play an important role during regeneration. Inmediately after axotomy, the denervated myelinating and non-myelinating Schwann cells undergo a stage of active proliferation, with high levels of ß-NG-1 (see section 15.2) and changes in the levels of their cognate mRNAs. Genes characteristic of myelinating Schwann cells (i.e., protein zero, P_0; myelin basic protein, MBP; peripheral myelin protein, PMP22; conexin32, etc.) are downregulated, whereas genes of non-myelinating Schwann cells (NCAM, NGFR/p75, GAP-43) are upregulated.

The pathway taken by regenerating axons depends largely on the nature of the lesion. When a crush or freezing injury occurs, the basal lamina of the Schwann cells remains intact at the site of injury. Proximal to the injury site, axons give rise to one or more sprouts, each of which is tipped by a growth cone. For axonal regeneration to be successful, growth cones must first reach the distal nerve stump; therefore they must cross the gap between the proximal and distal nerve stump. Upon reaching the distal nerve stump growth cones enter the Schwann tubes — the Schwann cells and their basal lamina — which provide the sole pathway for regenerating axons in the distal nerve stump (Figure 3; also see Ramon y Cajal, 1928). In the process of regeneration the axon/Schwann-cell interaction is fundamental. Schwann cells initially surround bundles of regenerating axons, and eventually segregate the larger fibers into 1:1 relationship, create a new basal lamina, and form the myelin sheath. With time, remyelinated axons may enlarge to nearly normal size, but the thickness of myelin sheaths and length of the myelin internodes are not similar to those of the normal situation (Beuche and Friede, 1985). During this process, lipoproteins and fatty acids are supplied to the myelinating Schwann cell by macrophages and fibroblasts. The specificity of axonal regeneration depends on the integrity of the Schwann tubes and, consequently, axonal misdirection is common after nerve transection. Hence, although PNS axonal degeneration can take place in the presence of extracellular matrix solely, Schwann cells provide important additional trophic factors that significantly enhance axonal growth (Sketelj et al., 1989).

It is well known from the work of Ramon y Cajal (1928) that the degeneration and regeneration of injured nerves depend on specific trophic factors or combinations of them. Schwann cells appear to be the main source of some factors including brain-derived neurotrophic factor (BDNF). Fibroblasts and macrophages probably contribute to trophic factor activity, both directly and through autocrine and paracrine interactions (Be'eri et al., 1998; Tolwani et al., 2002). Trophic factors in the distal nerve stump appear to serve as a temporary, surrogate source of target-derived growth factors. As detailed below, the production of multiple trophic factors by denervated Schwann cells matches the diversity of PNS neurons: different kinds of neuron require different trophic factors.

The neurotrophin family in mammals has four members: NGF, BDNF, NT-3, and NT-4. These neurotrophins have distinct but overlapping tissue distributions, and promote neurite outgrowth and survival of sensory, motor, and autonomic neurons.

NGF is normally produced by the target sympathetic and sensory neurons. Its mRNA levels increases after axotomy. Schwann cells appear to be the main source of NGF. After a nerve crush, the NGF produced by the distal nerve stump is transported retrogradely by regenerating axons. If the transport is blocked, changes occur in the neurons, mainly in the levels of mRNA. For example, p75 (NGF receptor), neurofilament subunits, and substance P and calcitonin gene-related peptide decrease. Exogenous administration of NGF to the lesion site prevents those changes (Rush et al., 1995). Recently it has been shown that NGF given to the axotomized nerve site in wild type animals resulted in a 2.6-fold increase in Schwann-cell apoptosis in the distal nerve stumps compared with axotomy alone (Petratos et al., 2003).

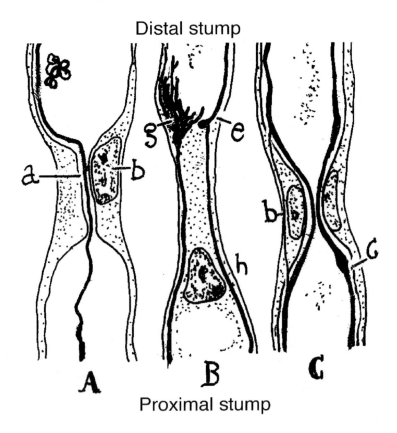

Figure 3: Reproduction of Ramon y Cajal's drawing showing details of the sprouts growth among the protoplasmic masses of the Schwann cells of the peripheral stump. A) Fine sprout detected in the axis of the protoplasmic mass (a) of Schwann cells (b); B) a marginal sprout with membraniform terminal (g) and a bud which arises from a fiber (e); C) other marginal fibers (b) and thick axial fibers (c). (Ramon y Cajal, 1928)

BDNF, NT-3, and NT-4 have been reported as neurotrophins with various effects on axotomized motoneurons and sensory neurons. These effects are probably mediated by the appropriate Trk receptors. Thus, motoneurons express TrkB and TrkC (but not TrkA), and they respond to BDNF, NT-3, and NT-4 (but not to NGF). Exogenous administration of BDNF or NT-4 ameliorates the loss of choline transferase, and NT-3 or NT-4 prevents the decrease of motor-nerve conduction velocity (Mendell et al., 1999). BDNF and NT-4 mRNA levels increase after axotomy, although much more slowly than that of NGF. The level of NT-3 mRNA in the nerve is lower that those of BDNF and NT-4 (Funakoshi et al., 1993). BDNF, NT-3 and NT-4 are transported retrogradely by normal motor neurons and sensory neurons, probably by their Trk receptors as well as by non-selective neurotrophin p75, and this transport increases following axotomy.

Denervated Schwann cells express high levels of p75. The increase of this neurotrophin is a consequence of the loss of axonal contact. When axons regenerate, axon/Schwann-cell interactions reduce the expression of p75. However, the role of p75 in Schwann cells is still unknown. It has been shown recently that myelin formation is inhibited in the absence of functional p75 and enhanced when TrkC activity is blocked. Moreover, the enhancement of myelin formation by endogenous brain-derived neurotrophic factor (BDNF) is mediated by the p75 receptor, whereas TrkC receptors are responsible for NT-3 inhibition. Thus, p75 and TrkC receptors have opposite effects on myelination (Cosgaya et al., 2002).

Transforming growth factor-ß (TGF-ß) is comprised of three forms termed TGF-ß1, -ß2, and -ß3. The TGF-ß superfamily contains, in addition, a large number of homologous proteins belonging to the activin, bone morphogenetic protein (BMP), and glial-derived neurotrophic factor (GDNF) families. They require activation by extracellular proteases, and in many tissues play crucial roles in regulating cell growth and differentiation during development after injury (McLennan and Koishi, 2002). All TGF-ß members are synthesized by Schwann cells. Their expression is regulated by axonal interaction: during Wallerian degeneration, the level of TGF-ß1 mRNA increases, whereas the expression of TGF-ß3 mRNA decreases. Conversely, regenerating axons inhibits the expression of TGF-ß1 mRNA. Hypotheses based on the molecular anatomical evidence can thus be further tested with genetically modified mice. This type of evidence suggests that TGF- ß1 is an autocrine regulator of Schwann cells. Motoneuron survival is also regulated by multiple sources of TGF-ßs, with TGF-ß2 being the most important isoform (see McLennan and Koishi, 2002). After mechanical injury of the adult rat spinal cord, upregulations of NGF and GDNF mRNA occurred in meningeal cells adjacent to the lesion. BDNF and p75 mRNA increased in neurons, GDNF mRNA increased in astrocytes close to the lesion, and GFRalpha-1 and truncated TrkB mRNA increased in astrocytes of degenerating white matter (Hoke et al., 2003).

CNTF, leukemia inhibitory factor (LIF), and interleukin (IL-6) are a family of structurally related cytokines that share a common set of receptors with an even larger family of cytokines. In multiple sclerosis, myelin repair is generally insufficient, despite the relative survival of oligodendrocytes within the plaques, and the recruitment of oligodendrocyte precursors. Promoting remyelination appears to be a crucial therapeutic challenge. Different neurotrophic factors have been screened for their ability to enhance myelination: neurotrophins (NGF, NT-3, NT-4/5, BDNF, GDNF, neurturin) and growth factors (such as PDGF-AA, FGF-2, and insulin) did not increase myelinogenesis. In contrast, certain factors belonging to the CNTF family (CNTF, LIF, cardiotrophin-1, and oncostatin M) induced a strong promyelinating effect. CNTF acts on oligodendrocytes by favoring their final maturation; this effect is mediated through the 130 kDa glycoprotein receptor common to the CNTF family and transduced through the Janus kinase pathway (Stankoff et al., 2002).

We have already mentioned that Schwann cells and their basal lamina provide the main, if not the sole, pathway for regenerating axons. In the absence of Schwann cells, growth cones adhere to the inner surface of the basal lamina, but most growth cones prefer to adhere somehow to remaining Schwann cells (Martini, 1994; Agius and Cochard, 1998). Extracellular-matrix molecules and cell-adhesion molecules are expressed at the distal nerve stump, and seem to promote axonal outgrowth. Of these, inmmunoglobulin-like cell-adhesion molecules (Ig-CAMs), include several of the best-characterized adhesion molecules, such as P_0, NCAM, L1, and MAG. These molecules promote neurite outgrowth and probably neuronal survival by more complex mechanisms that are outside the scope of this chapter (see Doherty et al., 2000). The cadherins, a family of calcium-dependent molecules that mediate adhesion via homophilic mechanisms, are linked intracellularly to the actin cytoskeleton. E-cadherin is present in non-compacted regions of myelin sheaths in the peripheral nervous system. There, it is localized to electron-dense structures — between membranes of the same Schwann cell — referred to as autotypic adherens junctions. It has been proposed recently that E-cadherin is required for the proper establishment and/or maintenance of the outer axonal membrane in myelinated PNS fibers, but for proper nerve function (Young et al., 2002). Integrins bind to extracellular matrix components, and in some cases to cells. Integrins are linked to the cytoskeleton and like Ig-CAMs, have been implicated in the cell-signaling events to transmit signals that co-ordinate cell-cycle progression and initiation of differentiation, including myelin-specific gene expression (Taylor et al., 2003).

In experimental models of spinal cord section, bridges constructed of rat Schwann cells or genetically modified Schwann cells to secrete human BDNF, promote axonal regeneration across the injury (Xu et al., 1997; Menei et al., 1997). In the brain, retinal ganglion cells can regenerate their axons into a peripheral-nerve graft to reinnervate primary brainstem visual nuclei (Villegas-Perez et al., 1988). A combination of both Schwann-cell grafts and neurotrophin infusion into lesions of central and peripheral nervous system could be necessary to improve the efficacy in axonal growth and regeneration of neural connections.

5. CONCLUDING REMARKS AND FUTURE DIRECTIONS

The work reviewed in this chapter shows that Schwann cells and precursors act as a source of diverse developmental signals. These signals influence both cell differentiation and cell survival in early nerves. Schwann cells regulate the development of the three main cell types in nerves: neurons, connective tissue cells and, via autocrine circuits, the Schwann cells themselves. Myelination depends on the co-ordination of a battery of glial structural proteins that are responsible for the architecture of compact myelin. Mutations of myelin genes illustrate how glial differentiation and motor development proceed in the absence of the corresponding myelin protein. An on-going major challenge is to establish the functional interactions of proteins in the myelin sheath. Certain severe diseases are caused by aberrant gain-of-function effects, usually dominantly inherited. Mouse genetics will continue to be an important research tool in elucidating the mechanism(s) of axon-glia interactions and the role of myelinating Schwann cells in long-term axonal function.

We have also reviewed a number of transcription factors that play important roles in Schwann-cell differentiation. *Oct-6* and *Krox20* are required sequentially for the progression of Schwann cells through the promyelinating stage of differentiation. However the exact mechanisms by which these genes work is mostly unclear. Potential candidate target genes for the two factors are now beginning to emerge, and we expect that many more will be identified in the near future. It is expected that the use of cell- and stage-specific gene-knockout strategies and microarray techniques for the genome-wide screening of cDNAs expressed in peripheral nerves — at different developmental stages and in different genetic backgrounds — will soon clarify our understanding of the exact role of the transcription factors and targeted genes discussed here.

The nervous system originates embryologically as an epithelium, so it is interesting that ancient adhesion mechanisms seem to have been conserved and adapted for the use of the myelin sheath. Particularly important in terms of myelination are the immunoglobulin (Ig-CAMs) and cadherin superfamilies, both of which are strongly represented in epithelia.

REFERENCES

Agius, E. and Cochard, P. (1998) Comparison of neurite outgrowth induced by intact and injured sciatic nerves: a confocal and functional analysis. J. Neurosci. *18*, 328-338.

Be'eri, H., Reichert, F., Saada, A., and Rotshenker, S. (1998) The cytokine network of wallerian degeneration: IL-10 and GM-CSF. Eur. J. Neurosci. *10*, 2707-2713.

Beuche, W. and Friede, R.L. (1985) Millipore diffusion chambers allow dissociation of myelin phagocytosis by non-resident cells and of allogenic nerve graft rejection. J. Neurol. Sci. *69*, 231-246.

Blanchard, A.D., Sinanan, A., Parmantier, E., Zwart, R., Broos, L., Meijer, D., Meier, C., Jessen, K.R., and Mirsky, R. (1996) Oct-6 (SCIP/Tst-1) is expressed in Schwann cell precursors, embryonic Schwann cells, and postnatal myelinating Schwann cells: comparison with Oct-1, Krox-20, and Pax-3 J Neurosci. Res. *46*, 630-640.

Britsch, S., Goerich, D.E., Riethmacher, D., Peirano, R.I., Rossner, M., Nave, K.A., Birchmeier, C., and Wegner, M. (2001) The transcription factor Sox10 is a key regulator of peripheral glial development. Genes Dev. *15*, 66-78.

Britsch, S., Li, L., Kirchhoff, S., Theuring, F., Brinkmann, V., Birchmeier, C., and Riethmacher, D. (1998) The ErbB2 and ErbB3 receptors and their ligand, neuregulin-1, are essential for development of the sympathetic nervous system Genes Dev. *12*, 1825-1836.

Certel, K., Anderson, M.G., Shrigley, R.J., and Johnson, W.A (1996) Distinct variant DNA-binding sites determine cell-specific autoregulated expression of the Drosophila POU domain transcription factor drifter in midline glia or trachea. Mol. Cell Biol. *16*, 1813-1823.

Cosgaya, J.M., Chan, J.R., and Shooter, E.M. (2002) The neurotrophin receptor p75NTR as a positive modulator of myelination. Science *298*, 1245-1248.

Doherty, P., Williams, G., and Williams, E.J. (2000) CAMs and axonal growth: a critical evaluation of the role of calcium and the MAPK cascade. Mol. Cell Neurosci. *16*, 283-295.

Dong, Z., Sinanan, A., Parkinson, D., Parmantier, E., Mirsky, R., and Jessen, K.R. (1999) Schwann cell development in embryonic mouse nerves. J. Neurosci. Res. *56*, 334-348.

Dupin, E., Real, C., Glavieux-Pardanaud, C., Vaigot, P. and Le Douarin, N. (2003) Reversal of developmental restrictions in neural crest lineages:Transition from Schwann cells to glial-melanocytic precursors in vitro. PNAS. 100, 5229-5233.

Funakoshi, H., Frisen, J., Barbany, G., Timmusk, T., Zachrisson, O., Verge, V.M., and Persson, H. (1993) Differential expression of mRNAs for neurotrophins and their receptors after axotomy of the sciatic nerve. J. Cell Biol. *123*, 455-465.

Ghazvini, M., Mandemakers, W., Jaegle, M., Piirsoo, M., Driegen, S., Koutsourakis, M., Smit, X., Grosveld, F., and Meijer, D. (2002) A cell type-specific allele of the POU gene Oct-6 reveals Schwann cell autonomous function in nerve development and regeneration. EMBO J. *21*, 4612-4620.

Hahn, A.F., Chang, Y., and Webster, H.D. (1987). Development of myelinated nerve fibers in the sixth cranial nerve of the rat: a quantitative electron microscope study. J. Comp. Neurol. *260*, 491-500.

Hoke, A., Ho, T., Crawford, T.O., LeBel, C., Hilt, D., and Griffin, J.W. (2003) Glial cell line-derived neurotrophic factor alters axon Schwann cell units and promotes myelination in unmyelinated nerve fibers. J. Neurosci. *23*, 561-567.

Jaegle, M. and Meijer, D. (1998) Role of Oct-6 in Schwann cell differentiation 1. Microsc. Res. Tech. *41*, 372-378.

Jessen, K.R. and Mirsky, R. (1999) Developmental regulation in the Schwann cell lineage. Adv. Exp. Med. Biol. *468*, 3-12.

Kioussi, C., Gross, M.K., and Gruss, P. (1995) Pax3: a paired domain gene as a regulator in PNS myelination. Neuron *15*, 553-562.

Le Douarin, N., Dulac, C., Dupin, E., and Cameron-Curry, P. (1991) Glial cell lineages in the neural crest. Glia *4*, 175-184.

Lee, M.J., Calle, E., Brennan, A., Ahmed, S., Sviderskaya, E., Jessen, K.R., and Mirsky, R. (2001) In early development of the rat mRNA for the major myelin protein P(0) is expressed in nonsensory areas of the embryonic inner ear, notochord, enteric nervous system, and olfactory ensheathing cells. Dev. Dyn. *222*, 40-51.

Martini, R. (1994). Myelin-associated glycoprotein is not detectable in perikaryal myelin of spiral ganglion neurons of adult mice. Glia *10*, 311-314.

McLennan, I.S. and Koishi, K. (2002). The transforming growth factor-betas: multifaceted regulators of the development and maintenance of skeletal muscles, motoneurons and Schwann cells. Int. J. Dev. Biol. *46*, 559-567.

Meier, C., Parmantier, E., Brennan, A., Mirsky, R., and Jessen, K.R. (1999) Developing Schwann cells acquire the ability to survive without axons by establishing an autocrine circuit involving insulin-like growth factor, neurotrophin-3, and platelet-derived growth factor-BB. J. Neurosci. *19*, 3847-3859.

Mendell, L.M., Johnson, R.D., and Munson, J.B. (1999) Neurotrophin modulation of the monosynaptic reflex after peripheral nerve transection. J. Neurosci. *19*, 3162-3170.

Menei, P., Montero-Menei, C., Whittemore, S.R., Bunge, R.P. and Bunge, N.B. (1997) Schwann cells genetically modified to secrete human BDNF promote enhanced axonal regrowth across transected adult rat spinal cord. Exp. Neurol. *140*,218-229.

Mirsky, R. and Jessen, K.R. (1996) Schwann cell development, differentiation and myelination. Curr. Opin. Neurobiol. *6*, 89-96.

Monuki, E.S., Weinmaster, G., Kuhn, R., and Lemke, G. (1989) SCIP: a glial POU domain gene regulated by cyclic AMP. Neuron *3*,783-793.

Monuki, E.S., Kuhn, R., and Lemke, G. (1993) Repression of the myelin P0 gene by the POU transcription factor SCIP. Mech. Dev. *42*, 15-32.

Morris, J.K., Lin, W., Hauser, C., Marchuk, Y., Getman, D., and Lee, K.F. (1999) Rescue of the cardiac defect in ErbB2 mutant mice reveals essential roles of ErbB2 in peripheral nervous system development. Neuron *23*, 273-283.

Morrison, S.J., White, P.M., Zock, C., and Anderson, D.J. (1999). Prospective identification, isolation by flow cytometry, and in vivo self-renewal of multipotent mammalian neural crest stem cells. Cell *96*, 737-749.

Petratos, S., Butzkueven, H., Shipham, K., Cooper, H., Bucci, T., Reid, K., Lopes, E., Emery, B., Cheema, S.S., and Kilpatrick, T.J. (2003) Schwann cell apoptosis in the postnatal axotomized sciatic nerve is mediated via NGF through the low-affinity neurotrophin receptor. J. Neuropathol. Exp. Neurol. *62*, 398-411.

Ramon y Cajal, S. (1928) Degeneration and Regeneration of Nervous System, R.M. May, ed. (Oxford: Oxford University Press).

Rush, R.A., Mayo, R., and Zettler, C. (1995) The regulation of nerve growth factor synthesis and delivery to peripheral neurons. Pharmacol. Ther. *65*, 93-123.

Scherer, S.S. (1997) Molecular genetics of demyelination: new wrinkles on an old membrane. Neuron *18*, 13-16.

Sketelj, J., Bresjanac, M., and Popovic, M. (1989) Rapid growth of regenerating axons across the segments of sciatic nerve devoid of Schwann cells. J. Neurosci. Res. *24*, 153-162.

Spassky, N., Goujet-Zalc, C., Parmantier, E., Olivier, C., Martínez, S., Ivanova, A., Ikenaka, K., Macklin, W., Cerruti, I, Zalc, B., and Thomas, J-L. (1998) Multiple restricted origin of oligodendrocytes. J. Neurosci. 18:8331-8343.

Stankoff, B., Aigrot, M.S., Noel, F., Wattilliaux, A., Zalc, B., and Lubetzki, C. (2002) Ciliary neurotrophic factor (CNTF) enhances myelin formation: a novel role for CNTF and CNTF-related molecules. J. Neurosci. *22*, 9221-9227.

Taylor, A.R., Geden, S.E., and Fernandez-Valle, C. (2003) Formation of a beta1 integrin signaling complex in Schwann cells is independent of rho. Glia *41*, 94-104.

Terenghi, G. (1999) Peripheral nerve regeneration and neurotrophic factors. J Anat. *194*,1-14.

Tolwani, R.J., Buckmaster, P.S., Varma, S., Cosgaya, J.M., Wu, Y., Suri, C., and Shooter, E.M. (2002) BDNF overexpression increases dendrite complexity in hippocampal dentate gyrus. Neuroscience *114*, 795-805.

Villegas-Perez, M.P., Vidal-Sanz, M., Bray, G.M., Aguayo, A.J. (1988) Influences of peripheral nerve grafts on the survival and regrowth of axotomized retinal ganglion cells in adult rats. J. Neurosci. *8*,65-80.

Wolpowitz, D., Mason, T.B., Dietrich, P., Mendelsohn, M., Talmage, D.A., and Role, L.W. (2000) Cysteine-rich domain isoforms of the neuregulin-1 gene are required for maintenance of peripheral synapses. Neuron *25*, 79-91.

Xu, X.M., Chen, A., Guenard, V., Kleitman, N. and Bunge, M.B. (1997) Bridging Schwann cell transplants promote axonal regeneration from both the rostral and the caudal stumps of transected adult rat spinal cord. J. Neurosci. 26:1-16.

Young, P., Boussadia, O., Berger, P., Leone, D.P., Charnay, P., Kemler, R., and Suter, U. (2002) E-cadherin is required for the correct formation of autotypic adherens junctions of the outer mesaxon but not for the integrity of myelinated fibers of peripheral nerves. Mol. Cell Neurosci. *21*, 341-351.

Zorick, T.S., Syroid, D.E., Brown, A., Gridley, T., and Lemke, G. (1999) Krox-20 controls SCIP expression, cell cycle exit and susceptibility to apoptosis in developing myelinating Schwann cells. Development *126*, 1397-1406.

E. VECINO AND M. GARCÍA

Department of Cell Biology, Faculty of Medicine, University of País Vasco. Leioa, Spain

16. THE MÜLLER GLIA: ROLE IN NEUROPROTECTION

Summary. The Müller cells are the predominant glial element of the retina and the major supportive glia for neurones, performing many of the functions subserved by oligodendrocytes, astrocytes and ependymal cells in other regions of the central nervous system. Glial cells are thought to protect neurons from various neurologic insults. Thus, when the retina has an excitotoxic damage, Müller cells are responsible for diminishing the extracelular glutamate, protecting the neurons from dead. Moreover, Müller cells play an important role in neuroprotection, inflammatory processes or phagocytosis after damage, developing important processes for tissue repair after injury. These aspects and others developed by the Müller cells will be reviewed in the present chapter.

1. INTRODUCTION

The Müller cell is the predominant element of the retinal glia. In contrast to glia in the brain and spinal cord, radial glia from the retina — the Müller cells — maintain a radial appearance throughout adult life and interdigitate with every class of retinal neurons, glia, or retinal vessels.

Considering their strategic location, Müller cells are in a position to influence and be influenced by neuronal activity. It has been shown that Müller cells are the major supportive glia for neurons in the adult retina, and perform many of the functions performed by oligodendrocytes, astrocytes, and ependymal cells in other regions of the central nervous system.

In pathological conditions, Müller cells can play a protective role in the retina, since these glial cells express growth factors, neurotransmitter transporters, and antioxidant agents that prevent excitotoxic damage to retinal neurons. Moreover, Müller cells release growth factors or make contact with endothelial cells to facilitate the neovascularization process during hypoxic conditions. Finally, recent studies have pointed to a role of Müller cells in retinal regeneration after damage.

2. STRUCTURE AND FUNCTION OF MÜLLER CELLS

The vertebrate retina contains four types of glia. Müller cells are the main glial element, comprising 90% of the retinal glia. After Müller cells, astrocytes and microglia are the two glial types most frequently seen in the retina. These cells have different embryological origins, and are found predominantly in species with vascularized retinas. The fourth type of glia, the oligodendrocyte, is seen in the retina only when myelinated ganglion cell axons are present in the nerve fiber layer.

Müller cells are radially oriented and cross the retina from its inner border to the distal end of the outer nuclear layer. In the adult retina, their somata lie within the inner nuclear layer, where they may form a distinct median sublayer in some species. Although Müller cells of different species vary considerably in shape, some features are fairly universal. At the level of the outer limiting membrane, Müller cells extend apical microvilli into the subretinal space between the inner segments of photoreceptor cells. Apicolaterally, Müller cells are connected to their neighboring Müller and photoreceptor cells by specialized junctions to form the outer limiting membrane. In rat, these junctions are *zonulae adherens*, while in fish they are tight junctions. Müller cells send side branches into the two plexiform layers of the retina, where they form sheaths around neuronal processes and synapses. Smooth (and sometimes quite long) processes are sent from the somata of the Müller cells through the nerve fiber layer to terminate in a basal endfoot, which lies adjacent to the inner limiting membrane. This membrane is a basal lamina, produced at least partly by Müller cells (Sarthy and Ripps, 2001).

Müller cells are found in all retinal regions, except the optic nerve head, of every vertebrate studied. However, the population density and morphology of Müller cells vary in different parts of the retina, reflection of the functional requirements of different retinal regions and the special properties of the microenvironment in which the cells develop.

There is little variation in the number of ganglion cells per Müller cell. Likewise, the number of neurons in the inner nuclear layer (and even the number of cone photoreceptors) per Müller cell seems to be fairly similar among species. By contrast, the number of rods per Müller cell may vary widely. This relationship has led to a hypothesis describing the generation of retinal cells by two distinct types of progenitor cell. According to this hypothesis, an early type of progenitor cell produces — by only one or two mitoses — a small number of neurons. These early-generated cells, such as ganglion cells, cones, amacrine cells, and horizontal cells, belong in general to the phototypic system. The hypothesis predict that after a few mitoses of the second progenitor cell — which has the capacity to produce bipolar cells, rods, and Müller cells — small repetitive groups of retinal cells, the so-called "retinal units" appear. The idea is almost accepted that each Müller cell subserves many of the metabolic, ionic, and extracellular buffering requirements of those neurons with which it shares a common progenitor.

Like other glial cells, Müller cells express a wide variety of voltage-gated ion channels and many types of neurotransmitter receptor, including a gamma-aminobutyric acid (GABA) receptor and several types of glutamate receptor. Studies on Müller cells have provided some of the clearest examples of glial-cell control of the neuronal microenvironment. It is known that Müller cells have the capacity to modulate extracellular fluid bathing retinal cells. The concentrations of neurotransmitters and potassium, for example, are regulated by glial-cell homeostatic mechanisms. Müller cells possess high-affinity uptake carriers for many transmitters, and are believed to regulate extracellular transmitter levels in the retina.

This uptake is essential for terminating synaptic transmission as well as preventing the spread of transmitters away from the synaptic cleft (Newman and Reichenbach, 1996).

Müller cells are also involved in the structural organization of the blood-retina barrier. Blood capillaries are ensheathed by Müller cell processes, acting as communicating system for metabolic exchange between vasculature and neurons in much the same manner as postulated for brain astrocytes. However, the close physical association between Müller cell processes and the retinal vessels leaves questions open as to whether the Müller cell *per se* is an essential component of the blood-retina barrier, and whether it is capable of affecting the latter's permeability. Although some findings clearly point to a potential role for Müller cells in the formation of the barrier properties of vascular endothelium, other studies have demonstrated that these glial cells are not involved in this mechanism (Sarthy and Ripps, 2001).

2. MÜLLER CELLS AND NEUROPROTECTION

Glial cells are thought to protect neurons from various neurologic insults. When retinal conditions become abnormal or when there is injury to retina, Müller cells undergo significant morphological, cellular and molecular changes. Some of these changes reflect Müller cell involvement in protecting the retina from further damage.

2.1. Role of Müller cells in excitotoxicity and neuroprotection

2.1.1. Müller cells and glutamate transporters

The excitatory amino acid neurotransmitters, and particularly glutamate, can have potent neurotoxic activity if their extracellular concentration becomes elevated in the central nervous system, including the retina. Excessive glutamate is thought to be responsible for a variety of acute neurologic insults, including ischemia and anoxia, hypoglycemia, trauma, and several chronic neurodegenerative diseases.

Increased glutamate levels in the extracellular fluid could be due to excessive glutamate release from presynaptic terminals, leakage from cellular compartments, or impairment of glutamate transporters. In retina subjected to trauma or anoxia, there is a rise in extracellular K^+ and glutamate (the latter to toxic levels). It has been demonstrated that the level of K^+ is increased in the anoxic brain and during spreading depression; this increase leads to the depolarization of glutamatergic nerve terminals and calcium-dependent release of glutamate. Moreover, the depolarizing effect of K^+ on Müller cells and neurons inhibits glutamate uptake, and induces calcium-independent release of glutamate by reversed uptake. All these mechanisms lead to a further increase in extracellular glutamate concentration. In addition, glutamate activation of N-methyl-D-aspartate (NMDA) receptors on second-order neurons causes the release of arachidonic acid, another source of uptake inhibition.

Lastly, the rise in glutamate will cause neurons to depolarize further, release more K^+, and stimulate an even greater efflux of glutamate.

From their DNA sequence, pharmacology, and channel properties, five types of glutamate transporter (also called excitatory aminoacid transporter, EATT) have been described: EAAT1 (GLAST), EAAT2 (GLT-1), EAAT3 (EAAC1), EAAT4 and EAAT5. The primary glutamate transporter expressed by retinal astrocytes and Müller cells is GLAST, which has been postulated as contributing to the clearance of glutamate, protecting retinal ganglion cells from glutamate neurotoxicity (Sarthy and Ripps, 2001).

Retinal neurons are susceptible to glutamate-induced damage; subcutaneous injections of sodium L-glutamate in adult and neonatal mice resulted in severe destruction of ganglion cells and partial loss of cells in the inner nuclear layer. The excitotoxic effect of glutamate and its relative selectivity for neurons of the inner retina when this amino acid was injected intravitreally have also been demonstrated.

It has been shown that Müller cells can protect against the excitotoxic effects of glutamate in the whole retina, and increase survival of ganglion cells in culture. This protective effect seems to involve uptake of glutamate from the medium through the GLAST transporter.

Changes in the expression of GLAST have been found during some pathologies but not in others. Thus, an increase in GLAST activity has been reported during glaucoma in monkeys. A small rise in the concentration of glutamate, when maintained for long periods of time, is toxic to retinal ganglion cells. Some works showed a relatively small (micromolar) rise in the vitreous of human and monkey eyes with glaucoma. However, a recent study has not found an increase in vitreal glutamate concentration in glaucomatous eyes from monkeys. This finding does not exclude the role of glutamate excitotoxic damage in glaucoma, since the authors reported an increase in glutamine, glutathione, and GLAST activity. Increases in glutamine, glutathione, and GLAST content in glaucomatous monkey eyes indicate a raised and enhanced glutamate transport and metabolism, but the possibility of excitotoxic damage to ganglion cells is not excluded (Carter-Dawson et al., 2002). Increases in GLAST/GluT-1 mRNA following transient retinal ischemia have also been found.

In contrast, after occlusion of the central retina artery no changes in the expression of GLAST are observed, although a reduction in the ability to transport D-aspartate into Müller cells has been reported.

In diabetes, the glutamate transporter activity is decreased, by a mechanism involving oxidation, in retinal Müller cells (Li and Puro, 2002). It has been shown that oxidative stress causes an increase in glutamate levels in the diabetic retina and that oxidizing agents decrease GLAST function *in vitro*.

Whereas some works have demonstrated that the protective effect of Müller cells against glutamate neurotoxicity is not dependent on glial-cell/retinal-ganglion-cell contact (Kawasaki et al., 2000), others have suggested that this neuroprotective effect needed cell-cell contact (Heidinger et al., 1999). The protective effect of Müller cells also depends on confluence of glial cells, since confluent retinal glial cells are able to take up and metabolize excessive glutamate, whereas medium from co-cultures of retinal ganglion cells and non-confluent retinal Müller cells showed an increase in concentration of glutamate. Therefore, retinal Müller cells under proliferative conditions may express fewer receptors and possess limited enzyme activity possibly exerting neurotoxic effects of retinal glial cells on retinal ganglion cells.

2.1.2. Müller cells and glutamine synthetase

In addition to glutamate transporters, glutamine synthetase levels are also important for regulating glutamate toxicity in the retina. Moreover, it has been shown that glutamate uptake by Müller cells in intact retina is strongly influenced by the activity of glutamine synthetase.

In physiological conditions, after its release from neurons, glutamate is quickly taken up by glial cells and amidated to form the non-neuroactive compound glutamine. This amidation is catalyzed by the enzyme glutamine synthetase, which is present only in glia and is confined exclusively to the Müller glial cells in the retina. Glutamine is then released by the glial cells and taken up by neurons for conversion back to glutamate. Glutamine synthetase is involved directly in neuroprotection, since it has been found that chick retinas treated with cortisol to induce high levels of glutamine synthetase are more resistant to glutamate-induced damage than retinas expressing low levels of the enzyme. Moreover, using mixed cultures of neuronal cells seeded onto glial cells it has been demonstrated that glutamate induces an increase in glutamine synthetase activity in both mature and immature mixed retinal cell cultures, although only adult glia induce protection (Heidinger et al., 1999). Increases in glutamine synthetase activity occur only in mixed neuronal-glial cultures, indicating that neuron/glial-cell contact is required in the regulation of the enzyme. All these works therefore indicate that the excitatory amino acids by themselves do not upregulate glial glutamine synthetase expression, but that neuron/glial-cell contact is required. The existence is possible of a diffusible factor, released by neurons damaged under excitotoxic conditions, capable of activated glutamine synthetase activity. In any case, neuronal-glial interactions seem to be important in astroglial and Müller glial responses to ischemia.

Decreased activity of glutamine synthetase has been observed in pathologies such as diabetes; however, no changes in glutamine synthetase expression in Müller cells is observed after optic nerve crush in rats. In the latter case, the enzyme shows a transient shift in its cellular distribution, and translocates from the cell body to the inner and outer Müller processes and (particularly) to the basal endfeet located in the ganglion cell layer (Chen and Weber, 2002). Trophic factors can downregulate the activity of glutamine synthetase, since it has been observed that bFGF may produce a direct down-regulation of the enzyme, and thereby exacerbate glutamate-mediated neurotoxicity.

2.1.3. Müller cells and glutathione

Reactive oxygen species are generated in the retina under various conditions such as anoxia, ischemia, and reperfusion. One of the crucial substances protecting the retina against reactive oxygen species is glutathione, a tripeptide constituted of glutamate, cysteine, and glycine. Glutathione in the retina of the rabbit has been located in Müller cells, suggesting that glia play a critical role in regulating the content of potentially damaging oxidative species in the retina. Glutathione is synthesized extramitochondrially, and transported into the mitochondrial matrix of Müller cells; glutamate is the rate-limiting substance in glutathione synthesis.

Physiological concentrations of glutathione can protect Müller cells from oxidative injury. Both Na^+-dependent and Na^+-independent transport systems for glutathione exist in Müller cells, and the Na^+-dependent glutathione transporter may be involved in the protective role of glutathione.

Glutathione transfer from Müller cells to neurons has been observed under ischemic conditions in rat retina.

During impaired glutamate uptake, as is observed during total ischemia, a glutamate deficiency occurs in Müller cells. In this condition, the amino acid is preferentially delivered to the glutamate-glutamine pathway, at the expense of glutathione. This mechanism may contribute to the finding that total ischemia causes a depletion of glial glutathione. The ischemia-induced lack of glutathione is particularly dangerous, considering the increased production of reactive oxygen species under this condition.

2.1.4. Müller cells and neurotrophins

Glia may protect neurons against excitatory amino-acid-induced degeneration or anoxia via the release of neurotrophic factors.

Release of neurotrophic factors from glial cells increases the long-term survival of developing retinal ganglion cells in culture. It has been postulated that survival of retinal ganglion cells in the Müller-conditioned medium declines as the differentiation of neurons proceeds, since Müller cells in culture increase the survival of retinal ganglion cells from rats at birth but not of retinal ganglion cells from 6-day-postnatal rats (Raju and Bennett, 1986). However, a recent work has demonstrated in cultures of adult pig retinas that Müller cells increase the survival and enhance the neuritogenesis of retinal ganglion cells. This protective effect seems to be mediated not only by substrate but also by some factor(s) secreted by retinal Müller glia (García et al., 2002). Other authors have found that Müller cells have a neurite-promoting effect on retinal ganglion cells only when there is a direct cell-cell contact: soluble factors released by Müller cells do not induce significant neurite outgrowth in retinal ganglion cell (Raju and Bennett, 1986). These differences could be due to the age of the Müller-cell cultures, since it has been demonstrated by proteomic technique, that Müller cells in culture change their protein expression dramatically. Thus, when the cultures are older than seven days, the molecular characteristics of the cells are no longer the same as in recently dissociated cells.

The low-affinity p75 neurotrophin receptor is expressed in Müller glia, as has been demonstrated both *in vivo* (figure 1A) and *in vitro* (figure 1B). This receptor seems to be involved in neuroprotection in the retina, since an upregulation of the low-affinity p75 nervous growth factor receptor on Müller cells is observed in pathologies such as diabetes, and after ischemia and reperfusion in rats (Vecino et al., 1998).

Neurotrophic factors from glial cells diminish the degeneration of retinal ganglion cells after optic nerve damage in rat. Moreover, Müller cells have been implicated in the determination of photoreceptor survival. It has been demonstrated that blockade of p75 prevents the basic fibroblast growth factor (bFGF) reduction mediated by nerve growth factor, resulting in an increase of both structural and functional photoreceptor survival *in vivo*. Moreover, brain-derived neurotrophic factor (BDNF), ciliary neurotrophic factor (CNTF), and fibroblast growth factor (FGF-2) may exert their effects on photoreceptors by acting indirectly through activation of Müller cells (Wahlin et al., 2001).

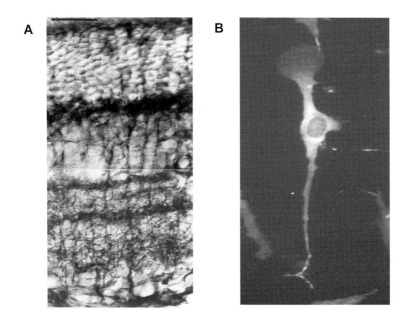

Figure 1: A) Distribution of the p75 receptor within the Müller cells and their processes in rat retina. B) Expression of p75 in Müller cell of pig retina in culture. Scale bar: 20 μm

Müller cells are also involved in the neuroprotection of bipolar cells, since in dissociated and cultured retinas from rats, brain-derived neurotrophic factor may induce Müller cells to produce a secondary factor, perhaps FGF-2, which directly rescues bipolar cells.

2.2. Müller cells and immunity

In the nervous system, astrocytes, oligodendrocytes, and microglia secrete cytokines, have the capacity to respond to them, and present antigens to T-lymphocytes. Antigen-presenting cells in the eye may be classified into bone-marrow-derived cells, such as uveal dendritic cells and retinal microglia, or non-bone-marrow-derived cells, represented by retinal pigment epithelium, Müller cells, and vascular endothelial cells (Sarthy and Ripps, 2001). Histopathological studies have shown that Müller cells are involved in the immune response during pathologies such as subretinal fibrosis and uveitis syndrome.

Experimental autoimmune uveoretinitis induced by retinal antigens is mediated by CD4+ lymphocytes. Generation of autoreactive CD4+ cells requires the processing and presentation of autoantigen by antigen-presenting cells in combination with MHC (major histocompatibility complex) class II antigen. It has been demonstrated that several ocular cells express class II antigens during inflammation, while other cells — such as Müller cells — inhibit antigen presentation *in vitro*.

Inflammatory-cell products secreted in the course of intraocular immune processes induce the expression of surface antigens by Müller cells. Expression of MHC class II antigen in equine recurrent uveitis is observed in proliferating Müller cells. Müller cells can also be induced to express MHC class II determinants in culture, when exposed to supernatant from activated lymphocytes. Interactions between the Müller cells and T-lymphocytes can be diverse, and even opposite in different conditions, suggesting that Müller cells could play a determining role in the course of immune reactions at the level of the neuroretina. This dual effect of Müller cells has been observed *in vitro* on autoimmune T-helper lymphocytes. In a co-culture system, Müller cells have a primary inhibitory effect on the proliferation of T lymphocytes. In conditions where their inhibitory action is suppressed, Müller cells stimulate T-cells.

In vivo, retinal diode laser photocoagulation stimulates a wound-healing response in the outer retina and choroid. The cellular infiltrate includes macrophages and activated CD4T cells. Müller cells express MHC Class II antigen and the intercellular adhesion molecule ICAM-1, an accessory molecule required for an efficient presentation of antigen to T-cells.

In addition to this immunoregulatory role, Müller cells may participate in healing and scar formation following experimental allergic uveitis. These glial cells are involved in the formation of the outer limiting membrane, which can potentially act as an anatomical barrier, preventing entry of circulating lymphocytes into the retina when the blood-retina barrier is compromised. If antigen-specific T cells escape into the retina, Müller cells may become activated and suppress T helper cell proliferation by direct cell-cell contact.

The potential role of Müller cells in the immune response is based mainly on experimental evidence obtained using *in vitro* cell culture systems. The role of Müller cells *in vivo* is not clear; Ishimoto et al. (1999) have demonstrated that bone-marrow-derived antigen-presenting cells are the ones likely to be involved in initiating uveoretinitis. Moreover, analysis of the presence of antigen-presenting cells in the retina in an experimental intraocular inflammation in rats shows that vascular endothelia, retinal pigment epithelia, Müller cells, and astrocytes do not express either class II molecules or macrophage markers. Only a subpopulation of retinal microglia (those derived from bone marrow), can express MHC class II.

2.3. Müller cells and phagocytosis

Phagocytosis of exogenous particles, cell debris, and hemorrhagic products may be an important mechanism in tissue repair after injury.

Friedenwald and Chan (1932) reported the first evidence of phagocytic activity by Müller cells. After the injection of a suspension of melanin granules into the vitreous of albino rabbits, they noted that the pigment was accumulated in monocytic phagocytes and Müller cells. Subsequently, it was shown that materials such as erythrocyte debris and subretinal hemorrhagic products are phagocytosed by Müller cells.

In the developing retina, fragmenting DNA is phagocytosed mainly by microglia and Müller cells. In postnatal retinas, microglia are the predominant phagocytes for cells dying in the ganglion-cell and inner nuclear layers of the retina, whereas Müller cells appear able to phagocytose cells dying in any retinal layer, including dying photoreceptors. In some retinal diseases, and following transplantation of retinal pigment epithelium, melanin granules are liberated into the subretinal space. It has recently been found that implantation of melanin granules in the subretinal space of albino rabbits may induce a phagocytic cell response in macrophages and Müller cells.

The phagocytic activity of Müller cells has been analyzed in cultures. Mano and Puro (1990) found that human Müller-cell cultures from post-mortem eyes were able to phagocytose retinal fragments, as well as latex beads, through a mechanism similar to that observed in macrophages. Rabbit retinal Müller cells *in vitro* also show an intense phagocytosis of latex beads.

However in other species, such as fish, Müller glial cells are phagocytic for latex beads when the cells are isolated in cultures, but not when they are in contact with other cells within the retina.

2.4. Reactive gliosis

Reactive gliosis refers to cytological changes, such as hypertrophy, enlargement of the cell body, and increase in cell number, observed in glial cells after nerve damage. Virtually every disease of the retina is associated with a Müller-cell reactive gliosis, which may either support the survival of retinal neurons or accelerate the progress of neuronal degeneration.

Swelling of cell processes and enlargement of the cell body characterize reactive gliosis in the Müller cell. In addition, the fine processes are retracted or lost, the distal processes become more tubular, the cell body becomes rounded, and the endfeet become enlarged. A displacement of the Müller cell nucleus from the inner nuclear layer to the outer nuclear layer is also observed as a consequence of retinal detachment. Other changes observed during reactive gliosis of Müller cells include an increase in the number of glycogen granules, Golgi cisternae, and rough endoplasmic reticulum (Sarthy and Ripps, 2001).

Glial cell hypertrophy is often followed by cell proliferation. Müller cell proliferation has been reported in proliferative diabetic retinopathy and massive retinal gliosis. Although Müller cells are postmitotic in the mature retina, they retain the capacity to undergo mitotic activity under pathological conditions. Following neuronal damage, Müller cells proliferate and undergo changes in gene expression. Downregulation of tumor suppressor protein $p27^{kip1}$ and re-entry into the cell cycle occurs after retinal injury. Thereafter, Müller glial cells upregulate genes typical of gliosis, and downregulate cyclin D3, in concert with an exit from mitosis (Dyer and Cepko, 2000).

Alterations in the expression of genes other than those involved in cell-cycle regulation has been observed in Müller cells in response to injury. Proteins such as glial fibrillar acidic protein (GFAP), GLAST transporter, and ciliary neurotrophic factor are upregulated under pathological conditions, whereas other proteins such as glutamine synthetase seem to be down-regulated (Sarthy and Ripps, 2001).

During Müller cell gliosis, an increase in intermediate filament content is observed. The major intermediate filament expressed by reactive Müller cells is GFAP; this filament is expressed at a low level or is not detectable in mammalian Müller cells in physiological conditions. Müller cells upregulate their GFAP expression in a wide variety of retinal pathological states, including age-related retinal degeneration, retinal ischemia, inherited retinal dystrophy, retinal hyperoxia, induced retinal degeneration, glaucoma, and diabetes.

Increased GFAP expression is due to transcriptional activation of the GFAP gene in Müller cells. The loss of cell-cell contact between retinal neurons and Müller cells has been suggested as one possible factor inducing GFAP expression, but this hypothesis is not clear. Neuronal loss does not appear to be a prerequisite for GFAP induction, although some reports have found that increased rate of neuronal loss was paralleled by the presence of greater numbers of more-active Müller cells in a model of glaucoma in rats. Moreover, Müller cell changes accompanied photoreceptor degeneration in time and location in the rat retina.

GFAP induction could be mediated by the action of extracellular, diffusible substances, since it has been shown that a focal damage to retina produces GFAP accumulation in Müller cells remote from the site of injury. There is some evidence that growth factors, such as bFGF, and cytokines, such as CNTF, are signaling molecules involved in GFAP induction. Free radicals could also be involved in GFAP induction since it has been found that free-radical scavengers inhibit the expression of GFAP in degeneration of photoreceptors. The cellular source of the extracellular signaling could be neurons undergoing degeneration, or activated macrophages that invade the retina.

Although the specific function of reactive gliosis is not clear, it has been postulated as being involved in phagocytosis to clear neuron-degeneration products. Glial cells undergoing reactive gliosis also upregulate the production of cytokines and neurotrophic factors that could be important for neuronal survival. Another possible function of reactive gliosis is in restoring the blood-brain barrier by scar formation. The increase in GFAP expression may be related to membrane-remodeling events at the outer limiting layer or may be involved in stabilizing Müller-cell contacts with the retinal pigmented epithelium or Bruch's membrane in pathologies such as retinitis pigmentosa or atrophic macular lesions. Finally, by analogy with other intermediate filaments, GFAP could provide added mechanical stability to the retina. When rabbits are dosed with sodium iodate, large expanses of retinal pigment epithelium and photoreceptors are destroyed, and a subretinal scar, consisting mainly of the ascending processes of Müller cells, replaces them. Experimental retinal holes in rabbit retina are filled with tissue consisting of Müller cell processes playing a role in retinal hole closure.

3. MÜLLER CELLS AND REGENERATION

The vertebrate retina is derived from paired evaginations of the neural tube in embryonic development; it is initially produced by progenitor cells similar to those that generate the neurons and glia of other areas of the central nervous system. Studies of retinal progenitor cells have led to the identification of several factors that control their proliferation. It has been found that cell type determination in the rodent retina is independent of lineage, and that during the generation of retinal cell types, the cessation of mitosis and the cell type determination are independent events, controlled by environmental interactions. Therefore, both intrinsic properties and extrinsic cues direct the path of cell fate. Progenitors pass through intrinsically determined competence states, during which they are capable of giving rise to a limited subset of cell types under the influence of extrinsic signals (Livesey and Cepko, 2001). The different retinal progenitor cells are heterogeneous with respect to their expression of cell-cycle regulators.

Each type of retinal cell is generated in a characteristic order, with Müller glia, rod photoreceptors, and bipolar interneurons being born last. Some of the molecular signals that can influence the production of glia by retinal progenitor cells have been described, thus, the basic helix-loop-helix gene, neuroD, negatively regulates gliogenesis. NeuroD promotes the development and/or survival of rods and amacrine cells, while suppressing two of the last-born cell types: Müller glia and bipolar neurons. Recent work has focused on the role of $p27^{kip1}$, a cyclin-dependent kinase inhibitor, in the genesis of Müller glia, and it has been suggested that $p27^{kip1}$ may collaborate with the notch pathway. Rax, Hes1 and notch1 are expressed in retinal progenitor cells, and it has been found that Rax is expressed in differentiating Müller glia in the postnatal rodent retina. Moreover, the two other genes that are downregulated in neurons — notch1 and Hes1 — continue to be expressed in glial cells early in development. Following neuronal damage, Müller cells undergo reactive gliosis, characterized by proliferation and changes in gene expression. Downregulation of $p27^{kip1}$ and re-entry into the cell cycle occurs within 24 hours after retinal injury, whereas accumulation of $p27^{kip1}$ correlates with cell-cycle withdrawal and differentiation. Therefore, Müller glia require $p27^{kip1}$ to maintain their differentiated state.

In lower vertebrates, the retina continues to grow throughout the animal's life. New retinal cells are added at the ciliary margin of the eye, from the mitotic activity of neural/glial stem cells in a region known as the germinal zone. In cold-blooded vertebrates, neural retina partially or totally destroyed is reconstituted by regenerative neurogenesis. Two processes are involved in the retinal regeneration in fish and amphibians: transdifferentiation of retinal pigment epithelial cells into retinal neural progenitors and alteration in the fate of photoreceptor progenitors intrinsic to the retina (Hitchcock and Raymond, 1992). Thus, one source of regenerated retinal cells in teleost fish is a population of scattered proliferating cells located in the outer nuclear layer within the differentiated retina; these proliferating cells are modified neuroepithelial cells termed rod precursors (because in the intact retina they produce only rod photoreceptor cells). When retinal neurons are destroyed, rod precursors cease producing rods and give rise to clusters of primitive neuroepithelial cells which divide and reconstitute the retina, following a pattern that mimics the process of normal development.

Neural stem cells have been identified in the central nervous system and retina of adult birds and mammals. It has been reported in chick retina that stem cells at the retinal margin continue to produce new neurons through postnatal development and into adulthood. However, there are few reports of neural regeneration following acute damage in the central nervous system of warm-blooded vertebrates.

After neurotoxin-induced damage, proliferation is not increased at the chicken retinal margin, but the central retina undergoes a massive proliferative response. Proliferating cells are Müller glia that re-enter the cell cycle, de-differentiate, acquire progenitor-like phenotypes, and produce new neurons and glia (Fischer and Reh, 2001). These findings suggest that some proliferating Müller glial cells are capable of de-differentiating into retinal progenitors and subsequently forming new retinal neurons, whereas others differentiate into Müller cells. The notion that glia may give rise to neurons is not without precedent; it has been proposed that transformation of Müller cells into neural progenitors in the injured retina might occur in the fish retina. After laser ablation of photoreceptors in adult goldfish retina microglia, Muller cells and retinal progenitors proliferate in the inner nuclear layer. The nuclei of Müller glia and associated retinal progenitors migrate from the inner to the outer nuclear layer. The proliferating Müller cells, which express Notch-3 and N-cadherin, do not generate extra glial cells in the region of the lesion so they must either die or transform into another cell type. It is possible that the progeny of the dividing Müller cells regenerate cone photoreceptors and then rod photoreceptors (Wu et al., 2001).

Evidence exists that in other regions of the nervous system, glial cells can give rise to neurons. Thus, it has been found that neural stem cells in the adult mammalian brain are actually a subclass of glial cell, either specialized astrocytes in the subventricular zone or ependymal cells at the ventricular surface. Radial glial cells in the developing cerebral cortex in mammals can also behave as neural progenitors, even as they continue to express glial-specific markers, such as GFAP.

Retina of postnatal chickens has the potential to generate new neurons. In response to damage, Müller glia lose their phenotype and de-differentiate into retinal progenitors. Proliferating Müller cells may express several genes expressed by progenitors, including Pax6 and Chx10. Newly formed cells became distributed throughout the inner and outer nuclear layers of the retina. Some of them differentiate into retinal neurons, a few form Müller glia, and most of them remain undifferentiated. The cells that do not differentiate, co-express Pax6 and Chx10, and are in an arrested state. It is possible that the cues inducing embryonic progenitors to divide and differentiate are absent in mature retina, so that most of these cells remain in an arrested state. Amacrine and bipolar cells can be generated from damage-induced progenitors; however, no other retinal cell types are generated. Thus, progenitors at the retinal margin of postnatal chickens produce primarily amacrine and bipolar cells. This finding is consistent with the hypothesis that the microenvironment required to generate all cell types may be absent in the postnatal retina.

REFERENCES

Carter-Dawson L, Crawford ML, Harwerth RS, Smith EL, Feldman R, Shen FF, Mitchell CK, Whitetree A. Vitreal glutamate concentration in monkeys with experimental glaucoma. *Invest Ophthalmol Vis Sci.* 2002; 43: 2633-2637.

Chen H, Weber AJ. Expression of glial fibrillary acidic protein and glutamine synthetase by Müller cells after optic nerve damage and intravitreal application of brain-derived neurotrophic factor. *Glia.* 2002; 38: 115-125.

Dyer MA, Cepko CL. Control of Müller glial cell proliferation and activation following retinal injury. *Nat Neurosci.* 2000; 3: 873-880.

Fisher AJ, Reh TA. Müller glia are a potential source of neural regeneration in the postnatal chicken retina. *Nat Neurosci.* 2001; 4: 247-252.

Friedenwald JS, Chan E. Pathogenesis of retinitis pigmentosa with a note on the phagocytic activity of Müller fibers. *Arch Ophthalmol.* 1932; 8: 173-181.

García M, Forster V, Hicks D, Vecino E. Effects of Müller glia on cell survival and neuritogenesis in adult porcine retina *in vitro*. *Invest Ophthalmol Vis Sci.* 2002; 43 : 3735-3743

Heidinger V, Hicks D, Sahel J, Dreyfus H. Ability of retinal Müller glial cells to protect neurons against excitotoxicity *in vitro* depends upon maturation and neuron-glial interactions. *Glia.* 1999; 25: 229-239.

Hitchcock PF, Raymond PA. Retinal regeneration. *Trends Neurosci.* 1992; 15: 103-108.

Kawasaki A, Otori Y, Barnstable CJ. Müller cell protection of rat retinal ganglion cells from glutamate and nitric oxide neurotoxicity. *Invest Ophthalmol Vis Sci.* 2000; 41: 3444-3450.

Ishimoto S-I, Zhang J, Gullapalli VK, Pararajasegaram G, Rao NA. Antigen-presenting cells in experimental autoimmune uveoretinitis. *Exp Eye Res.* 1999; 67: 539-548.

Livesey FJ, Cepko CL. Vertebrate neural cell-fate determination: lessons from the retina. *Nat Rev Neurosci.* 2001; 2: 109-118.

Li Q, Puro DG. Diabetes-induced dysfunction of the glutamate transporter in retinal Müller cells. *Invest Ophthalmol Vis Sci.* 2002; 43: 3109-3116.

Mano T, Puro DG. Phagocytosis by human retinal glial cells in culture. *Invest Ophthalmol Vis Sci.* 1990; 31: 1047-1055.

Newman E, Reichenbach A. The Müller cell: A functional element of the retina. *Trends Neurosci.* 1996; 19: 307-312.

Raju TR, Bennett MR. Retinal ganglion cells at a particular developmental stage are maintained by soluble factors from Müller cells. *Brain Res.* 1986; 383: 165-176.

Sarthy V, Ripps H. *The retinal Müller cell. Structure and function.* Blakemore C, ed. Kluwer Academic/Plenum Publishers: New York; 2001: 1-34.

Vecino E, Caminos E, Ugarte M, Martín-Zanca D, Osborne NN. Immunohistochemical distribution of neurotrophins and their receptors in the rat retina and the effects of ischemia and reperfusion. *Gen Pharmacol.* 1998; 30: 305-314.

Wahlin KJ, Adler R, Zack DJ, Campochiaro PA. Neurotrophic signaling in normal and degenerating rodent retinas.Exp Eye Res. 2001; 73: 693-701

Wu DM, Schneiderman T, Burgett J, Gokhale P, Barthel L, Raymond PA. Cones regenerate from retinal stem cells sequestered in the inner nuclear layer of adult goldfish retina. *Invest Ophthalmol Vis Sci.* 2001; 42: 2115-2124.

J. M. DELGADO-GARCÍA AND A. GRUART

Division of Neuroscience, University Pablo de Olavide, Sevilla, Spain

PART C: INTRODUCTION
17. NEURAL PLASTICITY AND REGENERATION: MYTHS AND EXPECTATIONS

Summary. In this chapter, we will make a short revision of the concepts of neural plasticity and regeneration, and their relationships with processes such as motor learning and functional recovery following a lesion of the central or peripheral nervous systems. Particular attention will be paid to the actual morphological and physiological limits between plastic and regenerative processes in the adult mammal's brain, as well as to their potential functionality and adaptability. A precise delimitation will be trace between regenerative phenomena and compensatory mechanisms and other cognitive activities, which are sometimes confused with neural plastic processes. As a practical support to the concepts proposed here, some illustrative examples, collected from animal experimentation carried out in our laboratory, will be described briefly.

1. INTRODUCTION

A basic principle underlying contemporary neuroscience is that the brain changes each time we learn a new motor or cognitive spatial-temporal relationship (see for example the introductory chapter in Kandel et al., 2000). This principle is supported by the so-called *neural plasticity*; that is, the property of the nervous system enabling it to change in both structure and functioning, inducing at the same time specific modifications in behavioral and/or cognitive phenomena. *Sensu stricto*, neural plasticity refers to a persistent change in the functional properties of a group, or circuit, of neural elements (Carlin and Siekevitz, 1983). It is not a totally new concept: since the need for a modifiable anatomical substrate underlying behavioral changes and motor or cognitive learning was proposed by Ramón y Cajal almost a century ago (Cotman and Nieto-Sampedro, 1984; Ramón y Cajal, 1911). It may be opportune, however, to demarcate some conceptual presumptions from actual data collected from animal experimentation and clinical work. This text has been translated and extensively modified from a previous on-line publication (see Delgado-García, 2003).

2. IS NEURAL PLASTICITY AN AXIOM OR A PROPERTY OF THE NEURAL TISSUE?

In physical terms, a plastic substance is one that when modified (for example, by heating) remains in this new state for an indefinite period of time. It is more than probable that this is not the concept that we really want to apply to the nervous system, because if the changes introduced in the system are permanent, then we are denying the main proposal — that the brain can always be changed in structure and/or function, depending on the circumstances. Perhaps a more precise term to be applied to a permanently malleable neural tissue is that of elastic. At the very least, elastic properties allow both change and return to the previous state. Elastic properties allow learning a phone number and forgetting it, the brain changing during the acquisition period and returning to its previous state after the oblivion. More importantly, elastic properties make room for another basic property of the nervous system: regeneration (to a certain degree) of its structural organization, even a return to its previous state, as, for example, after a traumatic or chemical lesion of the peripheral or central neural elements.

Whether mainly plastic or mainly elastic, the brain's putative malleability must have specific limits. If our brain were extraordinarily plastic, we would also be extraordinarily different ourselves, enough to be unidentifiable as members of the same species. Moreover, clinical work would be extremely difficult, since it would be almost impossible to observe the same symptom or disease on different occasions (cf. Llinás, 2001). Even histologists and physiologists would run into problems: how to define the location and characteristics of layer V of the cerebral cortex in a, let's say, super-plastic brain; or how to characterize spinal motoneurons as the neural elements that precisely encode the position and velocity of innervated muscle fibers.

Leaving aside the basic problems inherent in the very concept of neural plasticity, it will be worthwhile to consider the available information on animal experimentation and clinical studies. In fact, the nervous system of adult mammals is not susceptible, under normal conditions, to an enormous reorganization of its neural centers and pathways. Actual neural changes thus far described are preferentialy restricted to the private microenvironment of presynaptic axon terminals and/or the underlying post-synaptic neural elements (Cotman and Nieto-Sampedro, 1984; Delgado-García, 1998; Delgado-García et al., 1998; Haydon and Drapeau, 1995; Kirkwood et al., 1995). Neural remodeling in the central nervous system (CNS) of adult mammals is restricted to the microcosmos of synaptic connectivity, as reported following discrete and well-localized neural damage, hormonal changes, environmental modifications, and motor learning (Bliss and Collingridge, 1993; Cotman and Nieto-Sampedro, 1984; Edwards, 1995; Malenka, 1995; Tsukahara, 1981). Even so, it is convenient to remember that in adult mammal brains, available postsynaptic sites are usually occupied; that is, there is no such thing as a surplus of free (empty) synaptic sites. Accordingly, permissible synaptic reorganization (without becoming pathological changes) must be rather limited.

There are other conceptual considerations to be made. To begin with, a change in behavior (for example, the acquisition of a new motor ability) will perhaps not require a tremendous synaptic reorganization in the parent brain, for the following reason. In an adult behaving cat, the neurons of the cerebellar posterior interpositus related with the corneal reflex generate about 2-3 millions of action potentials per day (Gruart et al., 2000; Morcuende et al., 2002). It could be expected that if such animal is trained to acquire a conditioned eyeblink response (e.g., to close the lids in response to a tone advising of the proximate arrival of a corneal air puff), those posterior interpositus neurons should change their firing activity substantially. However, once the appropriate response in acquired, the total number of action potentials per day is increased by less than 1/10000 (Gruart et al., 2000). It can be expected that only a similar percentahe of synaptic contacts is modified — that is, one terminal bouton out of the available 10000 (mean) terminal synaptic contacts assumed to be received by interpositus neurons: in short, a nightmare for the histologist. The conceptual difficulty presented here in finding out the expected subcellular changes is remarkable in comparison with the extreme facility by which those synaptic changes are found, and described, in the available literature (Gruart et al., 2000; Geinisman et al., 2001).

In summary, the concept of plasticity should be restricted very parsimoniously, if it is to be approached experimentally with sufficient efficacy and reliability. From the Platonic notion of a brain capable of all possible modifications, we should probably move on to examine the subtle morpho-functional neural mechanisms underlying motor and cognitive learning in humans and proximate species (Delgado-García, 1998; Delgado-García and Gruart, 2002; Gruart et al., 2003).

3. RELATIONSHIPS BETWEEN PLASTIC AND REGENERATIVE MECHANISMS

It has been suggested above that if motor and cognitive learning implies a permanent change in the circuitry or, more particularly, in the subcellular organization of specific neural elements, regeneration of the neural tissue means a return to the previous state, mainly when referring to partial lesions of limited neural structures. Therefore, the complete process of neural regeneration of the neural tissue in adult mammals implies two successive states: the loss of a given structure and/or function, and its subsequent recovery. A good example of this, widely used in animal experimentation, is the complete section of a peripheral motor or sensorimotor nerve (de la Cruz et al., 1996; Gruart et al., 2003).

In the light of present knowledge, the neural elements destroyed by a traumatic, chemical, or neurotoxic lesion are never recovered. Although some neurogenetic capability has been described for specific cortical and sub-cortical neural structures (e.g., the hippocampus in the rat), there is not enough information to accept this mechanism as an intentional process of morphological (or functional) recovery after a brain lesion (Schwab, 2002). On the other hand, the proposal that all neural elements of the CNS die following their (partial) lesion — for example, after their axotomy — is also highly questionable. According to available information (de la

Cruz et al., 2000; Pastor et al., 2000), CNS neurons survive their axotomy if, in a given time span (weeks to months, depending upon the neuronal type), they are able to find their initial cell target, or a new one. This survival capability has been widely described for brain-stem and spinal motoneuronal pools. Apart from in new-born mammals, brain-stem and spinal motoneurons survive their axotomy indefinitely (with the condition that the axotomy is not carried out too close to the cell body), and are able to reinnervate the muscle fibers that their axons find during their regrowth. It is also well known how axotomized motoneurons recover their functional capabilities — that is, their ability to generate synaptic and action potentials almost simultaneously with the reinnervation of a new muscle target.

Again, it is necessary to clear up some basic concepts that are sometimes confused, mainly because of finalistic simplification of neural regenerative processes taking place in the CNS of adult mammals. The questions are the following: Is the neural regenerative process aimed at functional recovery or at the survival of lesioned neural elements? Is the regenerative process the same for the peripheral nervous system (PNS) asfor the CNS? We will try to answer these questions in the following sections.

4. NEURAL CELLS NEED A TARGET IN THE ADULT MAMMAL BRAIN: FUNCTIONAL CONSIDERATIONS

Synaptic connectivity is fundamental for adequate communication between neural elements, and between them and peripheral sensory receptors, or motor and gland effectors. Specific neural characteristics, such as soma size, dendritic arborization, presence of specific ionic channels, membrane receptors and neurotransmitters, and firing properties, depend upon the presence of an adequate (neuron, muscle fiber, etc.) target, even in adulthood (Czéh et al., 1978; de la Cruz et al., 1996; Levi-Montalcini, 1982; Purves, 1986; Sofroniew et al., 1990; Takata, 1993).

An additional consequence of brain damage is target loss for a given neural pool. As will be explained in detail below, the survival and putative functional recovery of targetless neurons depend on the reinnervation of the original target or, eventually, of a new target with a minimum of subcellular and molecular requisites (Barron, 1989; Delgado-García et al., 1988; Kuno et al., 1974; Lieberman, 1971; Selzer, 1980; Sunderland, 1991). It would not be logical that neurons survived in the mammalian CNS without any connection with another neuron, receptor, or effector. Even so, we should analyze in detail whether this tendency of neurons to maintain some connectivity with other cell elements of the nervous system is aimed at the mere survival of the disconnected elements, or at the recovery of the lost function. It is also important to consider whether a functional re-specification of the innervating neuron is possible following its reinnervation of a functionally-different new target.

Following complete sectioning, the proximal stump of an axotomized motoneuron grows until reinnervating its parent muscle, orin fact, any nearby muscle. It must be remembered that once development is finished, mammalian motoneurons do not retain the capability of identifying their parent muscle. Therefore, almost any pool of muscle fibers will be useful (if successfully reinnervated) for motoneuron survival. We usually assume that the axons of axotomized motoneurons regrow to recover the lost function. However, it will be more parsimonious to accept for the moment that the axons grow to avoid motoneuron degeneration, because of the lack of putative trophic factors supplied by the muscle target (De la Cruz et al., 1996; Levi-Montalcini, 1982; Purves, 1986). On the other hand, it is well known that functional properties of chromatolytic motoneurons are recovered upon reinnervation; however, are those functional properties adaptable to the functional needs of the reinnervated target? We will consider this question next.

The trophic theory of neuronal connectivity (Levi-Montalcini, 1982; Purves, 1986), developed during the second half of the past century, proposes that neural elements depend on a permanent contact with their cell targets. Targets release specific neurotrophic factors, the list of which is too long to be considered here, but can be consulted elsewhere (Bothwell, 1995; de la Cruz et al., 1996). In mammals, dependence on trophic signals is more determinant during development than in the adult brain, but post-embryonic neurons still need an ill-determined functional and/or molecular interaction with their natural targets, in order to survive and to carry out their functional tasks. An elusive question, already touched on, is: what will it happen if a neuron switches to a new target following the loss of its original one? According to the trophic theory, targetless neurons need to find a new cell target able to supply them with the necessary trophic factor. In the most than plausible event that the new target has a different function, will the neuron adapt to the function of the (new) target? An alternative possibility is that the neuron stops its characteristic firing to avoid any interference with the (different) function of its new target. In order to answer these questions, let's have a look at some experimental approaches.

Recent experiments carried out by our group (Gruart et al., 1996; 2003) have largely confirmed the proposals Sperry and co-workers advanced almost 60 years ago (Sperry, 1945). Thus, adult mammal motoneurons seem unable to readapt their firing properties (i.e., their physiology) to the functional needs of a muscle target different from the one innervated during development. For example, both the hypoglossal and facial nerves can be experimentally sectioned in adult cats. In this situation, the proximal stump of the hypoglossal nerve can be firmly sutured to the distal stump of the facial nerve, and vice versa. We have been able to follow in chronic cats the whole process of muscle reinnervation and the functional processes accompanying them, up to one year following surgery (Gruart et al. 1996; 2003). It was shown that hypoglossal motoneurons are able to reinnervate facial muscles fully and to supply a tonic activity to them. In contrast, they were unable to generate the appropriate firing programs to accomplish the blink reflex (see Figure 1). Moreover, it was very easy to see that when the animal ate, the upper eyelid moved in a perfect synchrony with tongue movements (see Figure 2).

It seems convenient to mention here the steps necessary for a complete recovery of the corneal reflex, as well as for the complete disappearance of lingual motor commands, following a successful hypoglossal-facial anastomosis: i.e., when the complete ipsilateral hypoglossal pool is forced to reinnervate facial motor muscles. First, for a complete recovery of the corneal reflex, it will be necessary that second-order trigeminal neurons (whose cell bodies are located in the trigeminal nucleus) retract their axon terminals from the facial motor complex and redirect them to the ipsilateral hypoglossal nucleus. However, it has already been pointed out that such massive axonal redistributions do not happen in the CNS of adult mammals. Second, in order that hypoglossal motoneurons lose their incoming inputs with masticatory motor commands, the whole premotor network projecting monosynaptically on them would have to retract its presynaptic contacts from the plasmatic membrane of those motoneurons. This does not happen as illustrated in Figure 2. Apparently, reinnervation of (any) available muscle is a necessary and sufficient signal for brainstem motoneurons to recover the synaptic contacts retracted from their plasmatic membrane during the axotomized period (Delgado-García et al., 1988; de la Cruz et al., 1996). Once synaptic boutons are back onto the appropriate sites of the motoneuron, the cell recovers its functional capabilities, with similar characteristics to those available before axonal sectioning.

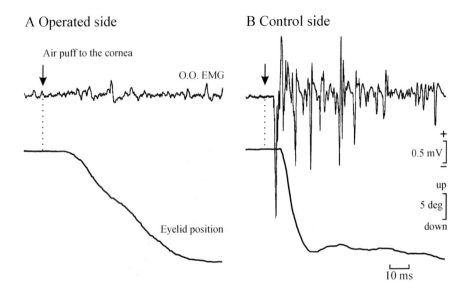

Figure 1. Examples of eyelid responses evoked by an air puff to the ipsilateral cornea in a cat with a left hypoglossal-facial anastomosis. The stimulus consisted of an air puff of 3 kg/cm^2 of pressure, lasting for 100 ms. Recordings were carried out seven months after the facial muscles were successfully reinnervated by the hypoglossal nerve. A. Evoked response in the anastomosed (left) side. The electromyographic (EMG) activity of the orbicularis oculi (OO) muscle and the actual downward displacement of the eyelid are shown. B. Evoked response in the control (right) side. In both cases, the arrow and the dotted line indicate stimulus presentation. The evoked movement on the anastomosed side was clearly not the result of the activity of the OO muscle, as illustrated by the lack of EMG activity. The evoked response in the anastomosed side was in fact a compensatory response produced by the sole action of the retractor bulbi muscle (not shown). Also, note that the compensatory movement presented a

larger latency than and a different profile to, that of the true eyelid reflex response recorded on the control side (reproduced with permission from Gruart et al., 1996).

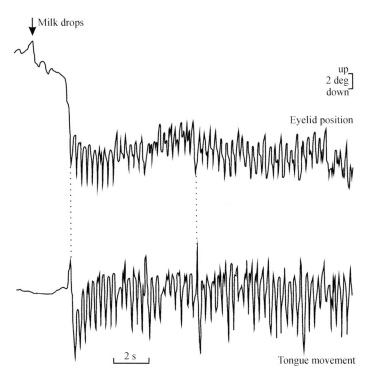

Figure 2. Eyelid movements in a cat with an ipsilateral hypoglossal-facial anastomosis. Recordings were made six months after reinnervation of the orbicularis oculi muscle by hypoglossal axons. The top trace corresponds to eyelid oscillatory movements during the ingestion of a few drops of milk. The bottom trace was recorded simultaneously, and corresponds to tongue displacements during licking. Up and down indicate the direction of eyelid and tongue displacements (reproduced with permission from Gruart et al., 1996).

5. FUNCTION RECOVERY BY COMPENSATORY MECHANISMS

In the case of a paralytic condition involving the orbicularis oculi muscle in humans, there is no other synergistic motor system able to close the eyelids actively. It is still possible to close the eyelids by a complete relaxation of the levator palpebrae muscle, because in the resting position (i.e., when no muscle is acting on them) the eyelids are closed, even if we are in an upside-down position (Gruart et al., 1995). In other species, such as the cat, the result of a facial paralysis on eyelid responses could be quite different. Following the complete section of the zygomatic branch of the facial nerve, i.e., the one innervating the orbicularis oculi muscle, the cat is still able to close the lids using the retractor bulbi motor system. This system is present in all terrestrial species having a nictitating membrane; primates are devoid of such a third lid, and thus of the retractor bulbi system. The action of the retractor bulbi system is to pull the eyeball back into the orbit. In cats, this system is usually active in response to strong, or harmful, corneal stimulations; but its activation threshold is significantly decreased following zygomatic nerve transection (Gruart et al., 1996, 2003). This is a good example of a *compensatory mechanism*; that is, a motor system that increases its gain when a similar (although not equal) motor system is not functioning properly, or at all. The way the retractor bulbi system increases its response to previously sub-threshold stimulation of corneal receptors seems to be by decreasing the activation threshold of corneal mechano- and chemo-nociceptors following facial paralysis (Pozo and Cerveró, 1993). This decrease in receptor threshold evokes a trigeminal hyperreflexia, able to bring retractor bulbi motoneurons to their firing level in response to previously sub-threshold stimulus.

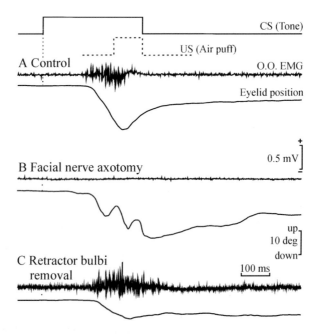

Figure 3. Different eyelid responses obtained by a classical conditioning procedure in three groups of rabbits: A, controls; B, animals with a complete transection of the facial nerve; and C, animals from which the retractor bulbi muscle has been removed. At the top are illustrated the conditioned stimulus (a 600 Hz, 90 dB tone lasting for 350 ms), the unconditioned stimulus (a 3 kg/cm^2, 100 ms air puff applied to the ipsilateral cornea). The recordings illustrated here were carried out in response to the sole presentation of the conditioned stimulus, to allow a better representation of the learned response evoked in each experimental situation (reproduced with permission from Delgado-García and Gruart, 2002).

The presence of true compensatory mechanisms is often misinterpreted as a functional recovery of the damaged motor system. Compensatory motor systems never perform movements equally, for two main reasons. First, all motor systems, even those with a synergistic function, are different in the specific motor commands arriving at their respective motor pools, in the innervation pattern of their parent muscles, and in the resonant properties of their respective muscles. Second, when a motor system is functionally removed, the functional properties of the compensatory system usually change, as indicated above for the retractor bulbi system. The presence of a trigeminal hyperreflexia makes the retractor bulbi system more sensitive than usual to corneal stimulation.

Figure 3 illustrates the eyelid displacement in rabbits during classically evoked responses in a delay paradigm (see Delgado-García and Gruart, 2002, for details). The conditioning consisted of the presentation of a tone (600 Hz, 90 dB) lasting for 350 ms. This conditioned stimulus was followed 250 from its beginning by an air puff (3 kg/cm^2, 100 ms) as an unconditioned stimulus. The unconditioned stimulus (i.e., the air puff) co-terminated with the conditioned stimulus (i.e., the tone). In this situation, the rabbit learned to close the eyelid in prevention of the forthcoming air puff. Animals were divided in three groups: i) controls; ii) animals with the facial nerve completely sectioned; and iii) animals with complete removal of the retractor bulbi muscle. It can be observed in Figure 3 that each motor system (Figure 3B and C) produced a conditioned response different in latency, profile, and kinematics. Even when the two systems acted together (Figure 3A, controls), the resulting conditioned response was not the simple addition of the conditioned responses evoked by each motor system when acting alone.

6. MOTOR LEARNING FOLLOWING A NEURAL LESION

It has already been pointed out that hypoglossal-facial anastomosis in cats allows the recovery of muscle tonicity in the affected facial musculature, but not the recovery of the corneal reflex. At the same time, the eyelid started moving in synchrony with tongue movements, such as during licking and eating. Both findings were the result of hypoglossal motoneuron motor commands arriving at the newly reinnervated facial muscles. This experimental model offered additional information regarding the role of compensatory motor systems. In cats, we never observed a corneal lesion following facial nerve transection, because the retractor bulbi system is able to move the eyelid passively in the downward direction by retracting the eyeball into the orbit (Gruart et al., 1996, 2003). This motor system does not exist in humans, so we cannot take advantage of its synergistic role to protect the cornea following facial nerve damage. Nevertheless, we can say to a patient with a hypoglossal-facial anastomosis "If you need to protect your eye in windy conditions the best way to close youreyelid is to move your tongue". This situation means using abstract motor commands (sometimes reproducible in the experimental animal via instrumental conditioning procedures) to apply a given motor system in an unusual way. It is obvious that this type of practice does not imply any sort of structural and/or functional reorganization of the hypoglossal motor system — just a peculiar use of this motor system, accessible in humans though the appropriate use of linguistic symbols, and in experimental animals though Skinnerian procedures.

7. REGENERATION PROCESSES IN THE CENTRAL NERVOUS SYSTEM: LIMITATIONS TO THEIR STUDY, AND FUTURE STRATEGIES

The precise and detailed study of the mechanisms underlying regenerative processes present in the CNS of mammals is rather complicated from an experimental point of view. Usually, a neuronal group in the CNS does not have an exclusive cell target, but, due to the abundant divergence of neuronal projections, several targets. On the other hand, each neuronal group receives afferents from many brain sites. In this situation, it is not easy to disconnect a neuronal group from its usual outputs or inputs completely. Incomplete disconnection can introduce severe experimental bias, and is frequently a cause of misinterpretation of the results. If we add to this technical limitation the presence of intentional manipulations aimed at recovering lost function, such as cell implants, or the injection of antibodies or neurotrophic factors, etc., it will be very difficult to analyze the experiment properly, or to obtain conclusive results (Gudino-Cabrera et al., 2000; Privat et al., 1997; Schnell and Schwab, 1990).

During the past few years, our group has used the population of internuclear interneurons located in the abducens nucleus and projecting onto the medial rectus subdivision of the contralateral oculomotor complex as an experimental model for the study of regenerating processes taking place in the CNS. These internuclear interneurons project almost exclusively onto medial rectus motoneurons. We have studied in them the effects of target removal. For this purpose, we injected a lectin (from *Ricinus communis* agglutinin II) into the contralateral medial rectus muscle. This lectin is captured by the motoneuron axon terminals, and is transported back to motoneuron somata. Lectins kill medial rectus motoneurons in about 24 hours, but they lack any (neuronal or glial) carrier in CNS cell elements. For this reason, the lectin is a very good tool for specific target removal, without the side-effects of direct chemical injections into specific brain sites, or electrolytic lesions (see de la Cruz et al., 1996 for details). The response of abducens internuclear neurons to target loss is similar to that of motoneurons to axotomy. Indeed, presynaptic neurons projecting onto internuclear neurons retract their terminal boutons from somata and dendrites of internuclear cells. This fact was confirmed physiologically, because the abducens internuclear neurons have noticeable alterations in their synaptic potentials from vestibular origin, and in the generation of action potentials during appropriate eye movements in the alert behaving cat (de la Cruz et al., 2000; Pastor et al., 2000). Nevertheless, abducens internuclear neurons recover their synaptic inputs and their regular firing as soon as they innervate available synaptic sites on nearby neurons located in the oculomotor nucleus. It is important to clarify that available sites probably correspond to vacancies produced by the disappearance of axon collaterals from the killed medial rectus motoneurons. At the same time, for their survival abducens internuclear neurons need to maintain only up to 10% of their previous terminal boutons. This facilitates the task of finding empty sites for synaptic contacts. As already indicated for hypoglossal motoneurons reinnervating *de novo* facial muscles, CNS neurons do not re-specify their physiology when reinnervating a new cell target. Abducens internuclear neurons recover their usual firing properties as soon as they reinnervate a target. However, it is almost impossible for the target to be related functionally with them, because the total population of medial rectus motoneurons was killed and this motor pool in the oculomotor complex is exclusively for horizontal eye movements. Hence, the new targets for abducens internuclear neurons unquestionable had functional activities very different from those normally characterizing them.

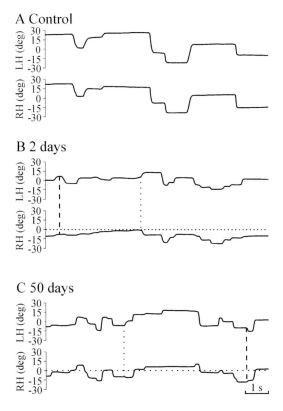

Figure 4. Spontaneous horizontal eye movements recorded before and after the transection of the right medial longitudinal fascicle. A, Eye movementes before transection. B, Oculomotor deficits in the horizontal position of the right eye (RH) two days after transection, as compared with horizontal movements of the left eye (LH). Note that the right eye was unable to move in the adducting direction and to cross the midline (i.e., the resting position of the eye in the orbit). C, Horizontal eye movements were partially recovered 50 days after transection, although an evident deficit was still present for disconjugate eye movements (reproduced with permission from de la Cruz et al., 2000).

Complete sectioning of abducens internuclear neuron axonal projections at the contralateral medial longitudinal fascicle yielded similar results. Thus, internuclear interneurons remained almost silent, and devoid of appropriate synaptic inputs from vestibular semicircular canals, during the period they remained without a target. Nevertheless, they managed to reinnervate unknown cells near the sectioned fascicle. In this situation, they were able to regain normal firing as if they were reinnervating their parent medial rectus motoneurons (de la Cruz et al., 1996). Morphological, synaptic, and functional recovery of a CNS neuron following its axotomy (or the loss of its target) is therefore not related to the finding of a functionally related target, but to the finding of any target able to supply it with sufficient of the trophic factor necessary for survival.

As already described for axotomized motoneurons, it is very important to distinguish what morpho-functional processes are taking place in the axotomized neuron, and what can be observed, from a functional point of view, in the alert behaving animal. For instance, hypoglossal motoneurons followed the expected process of regeneration even when forced to reinnervate a different (i.e., the facial) muscle. As explained, hypoglossal motoneurons did not adapt there firing profiles to the motor needs of their new target. However, if we restrict our experimental approach to recording eyelid responses in alert behaving cats during air-puff presentations to the operated side, we will observe the eyelid moving downward (see Figure 1). Accordingly, in these experimental models addressed to the study of morpho-functional changes involved in the regenerative processes of the neural tissue, it is vital to have a complete picture of the circuits involved, and of the presence of compensatory mechanisms. In this example, the presence of the retractor bulbi system compensated for the lack of trigeminal information arriving at the hypoglossal motor complex.

Going back to the use of abducens internuclear interneurons as an experimental model for studying regenerative processes in CNS neurons, it is also essential to distinguish what processes are taking place in the axotomized internuclear interneurons against what can be observed in the behavior of the whole animal. As is well known fromclinical practice, a mechanical lesion of the medial longitudinal fascicle immediately evokes a strabismus, with a characteristic abducting-directed deviation of the eye ipsilateral to the lesioned fascicle. However, the cat will be able to recover the normal position of both eyes in about five to six weeks (Figure 4). This is another example of a compensatory mechanism evoked by the over-activation of the mono-synaptic projection from Deiters' vestibular nucleus neurons onto medial rectus motoneurons. It has been shown that neurons located in Deiters' nucleus carry eye position and velocity signals, i.e., those signals not arriving at medial rectus motoneurons following the axotomy of abducens internuclear neurons. Interestingly enough, Deiters' nucleus neurons increase their gain (in an evident plastic process) to compensate for the loss of abducens internuclear interneuron afferents to medial rectus motoneurons (de la Cruz et al., 1996; Pastor et al., 2000). Here again, it is very important to have a complete picture of the experimental model before drawing erroneous conclusions from the results. If one is not aware of the presence of this extra source of eye-position and velocity signals represented by Deiters' nucleus neurons, the results observed in the whole animal following a complete section of the medial longitudinal fascicle (i.e., an early recovery from the evoked squint condition) will be badly misinterpreted. The appropriate study of processes taking place in the axotomized population of abducens internuclear interneurons will confirm that these cells are unable to reinnervate medial rectus motoneurons when their axons are sectioned at the medial longitudinal fascicle. Hence, they cannot be the origin of eye-position signals observed in medial rectus motoneurons after the lesion of the medial longitudinal fascicle. The use of complex sensory-motor systems as experimental models of regenerative processes should thus be exercised with caution. A precise knowledge of the involved neural circuits is highly recomended before hasty conclussions are drawn. A parsimonious approach to each experimental model will avoid the very frequent confusion about the origin of functional recoveries observed in the whole animal. Thus, we should clearly differentiate recoveries produced by compensatory mechanisms from those due to a truly morpho-functional regeneration of lesioned pathways.

8. CONCLUSIONS AND PERSPECTIVES

Neurons of the PNS of adult mammals are able to survive their axotomy —in particular, spinal and brain-stem motoneurons have been extensively studied in this context. They are also able to regrow their sectioned axon and find a target muscle. This target does not need to be the original one; even, dermic tissue can be used as a new target (Johnson et al., 1995). However, motoneurons lack the necessary plasticity to adapt their firing properties to the functional needs of the new target muscle (de la Cruz et al., 1996; Gruart et al., 2003). This limitation will produce the complete disappearance of a given function if the latter can be carried out only by a sole motor system. An example of this situation appears if we remove one of the extraocular recti muscles, for example the lateral rectus. In this situation, the experimental animal will never be able to make abducting movements with the operated eye; the only compensatory response possible is to contract the neck muscles appropriately, to direct the gaze toward the affected side. In other cases (hypoglossal-facial anastomosis), the activity of compensatory mechanisms (retractor bulbi system, co-contraction of the four extraocular recti muscles) is not as obvious, and can easily be confused with that of a re-specification of motoneuronal firing properties.

In the experience of our group, brain-stem neurons are also unable to re-specify their function when reinnervating a new neuronal target (de la Cruz et al., 1996; 2000; Pastor et al., 2000). We have concentrated on the study of the internuclear interneuron population, located in the abducens nucleus, and which projects monosynaptically onto medial rectus motoneurons located in the contralateral oculomotor complex. These internuclear interneurons survive the selective removal of their target cells (by killing medial rectus motoneurons with lectin injection in the parent medial rectus muscle) and their axotomy (by complete section of the medial longitudinal fascicle). In both cases, internuclear interneurons look for new neural targets, located either in the oculomotor complex or in quite different brainstem regions, but they seem unable to adapt their firing profiles to the physiology of the (newly) reinnervated targets.

In conclusion, it is advisable in each experimental model (as well as in each clinical case) to determine precisely whether the functional recovery observed following a neuronal lesion (of mechanical, chemical, or neurotoxic origin) is due to a complete regenerative process and/or to the presence of synargistic compensatory mechanisms. The presence of cognitive mechanisms (involving the cerebral cortex) also allows the use of motor systems for a completely different function (e.g., holding a pencil with the teeth, for writing). In the long run, a correct interpretation of the collected results is more important than a bright headline, which usually dulls with time.

REFERENCES

Barron KD (1989) Neuronal responses to axotomy: consequences and possibilities for rescue from permanent atrophy and cell death. In: Neural Regeneration and Transplantation (Ed. FJ Seil), pp. 79-100. Alan R Liss Inc, New York.

Bliss TVP and Collingridge GL (1993) A synaptic model of memory: long-term potentiation in the hippocampus. *Nature*, 361, 31-39.

Bothwell M (1995) Functional interactions of neurotrophins and neurotrophin receptors. *Annual Review of Neuroscience*, 18, 223-253.

Carlin RK and Siekevitz P (1983) Plasticity in the central nervous system: do synapses divide? *Proceedings of the National Academy of Sciences (USA)*, 80, 3517-3521.

Cotman CW and Nieto-Sampedro M (1984) Cell biology of synaptic plasticity. *Science*, 225, 1287-1294.

Czéh G, Gallego R, Kudo N and Kuno M (1978) Evidence for the maintenance of motoneurone properties by muscle activity. *The Journal of Physiology (London)*, 281, 239-252.

De la Cruz RR, Pastor AM and Delgado-García JM (1996) Influence of the postsynaptic target on the functional properties of neurons in the adult mammalian central nervous system. *Reviews in the Neurosciences*, 7, 115-149.

De la Cruz RR, Delgado-García JM and Pastor AM (2000) Discharge characteristics of axotomized abducens internuclear neurons in the adult cat. *Journal of Comparative Neurology*, 427, 391-404.

Delgado-García JM (1998) Output-to-input approach to neural plasticity in vestibular pathways. *Otolaryngology-Head and Neck Surgery*, 119, 221-230.

Delgado-García JM (2003) *Plasticidad y regeneración neuronal: mitos y expectativas*. Medicina Intensiva Online 3, 289-295.

Delgado-García JM and Gruart A (2002) The role of interpositus nucleus in eyelid conditioned responses. *The Cerebellum*, 1, 289-308.

Delgado-García JM, del Pozo F, Spencer R and Baker R (1988) Behavior of neurons in the abducens nucleus of the alert cat—III. Axotomized motoneurons. *Neuroscience*, 24, 143-160.

Edwards FA (1995) LTP—a structural model to explain inconsistencies. *Trends in the Neurosciences*, 18, 250-255.

Geinisman Y, Berry RW, Disterhoft JF, Power JM and Van der Zee EA (2001) Associative learning elicits the formation of multiple-synapse boutons. *Journal of Neuroscience*, 21, 5568-5573.

Gruart A, Blázquez P and Delgado-García JM (1995) Kinematics of spontaneous, reflex, and conditioned eyelid movements in the alert cat. *Journal of Neurophysiology*, 74, 226-248.

Gruart A, Guillazo-Blanch G, Fernández-Mas R, Jiménez-Díaz L and Delgado-García JM (2000) Cerebellar posterior interpositus nucleus as an enhancer of classically conditioned eyelid responses in alert cats. *Journal of Neurophysiology*, 84, 2680-2690.

Gruart A, Gunkel A, Neiss WF, Angelov DN, Stennert E and Delgado-García JM (1996) Changes in eye blink responses following hypoglossal-facial anastomosis in the cat: evidence of adult mammal motoneuron unadaptability to new motor tasks. *Neuroscience*, 73, 233-247.

Gruart A, Streppel M, Guntinas-Lichius O, Angelov DN, Neiss WF and Delgado-García JM (2002) Motoneuron adaptability to new motor tasks following two types of facial-facial anastomosis in cats. *Brain*, 126, 115-133.

Gudino-Cabrera G, Pastor AM, de la Cruz RR, Delgado-Garcia JM and Nieto-Sampedro M (2000) Limits to the capacity of transplants of olfactory glia to promote axonal regrowth in the CNS. *Neuroreport*, 11, 467-471.

Haydon PG and Drapeau P (1995) From contact to connection: early events during synaptogenesis. *Trends in the Neurosciences*, 18, 196-201.

Johnson RD, Taylor JS, Mendell LM and Munson JB (1995) Rescue of motoneuron and muscle afferent function in cats by regeneration into skin. I. Properties of afferents. *Journal of Neurophysiology*, 73, 651-661.

Kandel ER, Schwartz JH and Jessell TM (2000) *Principles of Neural Science*. McGraw-Hill, New York.

Kirkwood A, Lee HK and Bear MF (1995) Co-regulation of long-term potentiation and experience-dependent synaptic plasticity in visual cortex by age and experience. *Nature*, 375, 328-331.

Kuno M, Miyata Y and Muñoz-Martínez EJ (1974) Differential reaction of fast and slow α-motoneurones to axotomy. *The Journal of Physiology (London)*, 240, 725-739.

Levi-Montalcini R (1982) Developmental neurobiology and the natural history of nerve growth factor. *Annual Review of Neuroscience*, 5, 341-362.

Lieberman AR (1971) The axon reaction: a review of the principal features of perikaryal responses to axon injury. *International Review of Neurobiology*, 14, 49-123.

Llinás R (2001) *I of the Vortex. From Neurons to Self*. The MIT Press, Cambridge, MA.

Malenka RC (1995) LTP and LTD: dynamic and interactive processes of synaptic plasticity. *The Neuroscientist*, 1, 35-42.

Morcuende S, Delgado-García JM and Ugolini (2002) Neuronal premotor networks involved in eyelid responses: retrograde transneuronal tracing with rabies virus from the orbicularis oculi muscle in the rat. *Journal of Neuroscience*, 22, 8808-8818.

Pastor AM, Delgado-García JM, Martínez-Guijarro FJ, López-García C and de la Cruz RR (2000) Response of abducens internuclear neurons to axotomy in the adult cat. *Journal of Comparative Neurology*, 427, 370-390.

Pozo MA and Cerveró F (1993) Neurons in the rat spinal trigeminal complex driven by corneal nociceptors: receptive-field properties and effects of noxious stimulation of the cornea. *Journal of Neurophysiology*, 70, 2370-2378.

Privat A, Chauvet N and Gimenez y Ribota M (1997) Repousse axonale et obstacle glial. *Revieu de Neurologie (Paris)*, 153, 515-520.

Purves D (1986) *The trophic theory of neural connections*. Harvard University Press, Cambridge, MA.

Ramón y Cajal S (1911, 1972) Histologie du système nerveux de l'homme et des vertébrés, Vol 2. C.S.I.C., Madrid.

Schnell L and Schwab ME (1990) Axonal regeneration in the rat spinal cord produced by an antibody against myelin-associated neurite growth inhibitors. *Nature*, 343, 269-272.

Selzer ME (1980) Regeneration of peripheral nerve. In: *The Physiology of Peripheral Nerve Disease* (Ed. AJ Summer), pp. 358-431. WB Saunders Company, Philadelphia.

Sofroniew MV, Galletly NP, Isacson O and Svendsen CN (1990) Survival of adult basal forebrain cholinergic neurons after loss of target neurons. *Science*, 247, 338-342.

Sperry RW (1945) The problem of central nervous reorganization after nerve regeneration and muscle transposition. *Quarterly Review of Biology*, 20, 311-369.

Sunderland SS (1991) *Nerve Injuries and their Repair. A Critical Appraisal*. Churchill Livingstone, Edinburgh.

Schwab ME (2002) Repairing the injured spinal cord. *Science*, 295, 1029-1031.

Takata, M (1993) Two types of inhibitory postsynaptic potentials in the hypoglossal motoneurons. *Progress in Neurobiology*, 40, 385-411.

Tsukahara N (1981) Synaptic plasticity in the mammalian central nervous system. *Annual Review of Neuroscience*, 4, 351-379

O. S. JØRGENSEN[1] AND A. S. NIELSEN[2]

[1]*Laboratory of Neuropsychiatry, Department of Pharmacology. University of Copenhagen, Copenhagen, Denmark*

[2]*Department of Neurorehabilitation, Copenhagen University Hospital, Hvidovre Hospital, Hvidovre, Denmark*

18. NEURONAL PROTECTION AGAINST OXIDATIVE DAMAGE

Summary. The continous function of brain neurons depends heavily on elaborate antioxidant defences. Although the oxygen molecule qualify itself as a free radical by having two unpaired eletrons it is not very reactive. However, in the presence of catalytic concentrations of free metal ions or reactive oxygen species, oxidative damage may occur. Such damage may be counteracted by superoxide dismutases, glutathione peroxidase, and metallothioneins. Antioxidant protectors can be generated by ascorbic acid and α-tocopherol. Oxidative mechanisms in neurodegenerative diseases are exemplified by Alzheimer's disesase.

1. INTRODUCTION

The need for oxygen as an essential fuel for efficient energy production by means of the electron transport chain in eukaryotic mitochondria obscures the fact that oxygen is a toxic gas and aerobes only survives because they have antioxidant defences to protect against it. Consequently, like in all cells, cellular components in neurons are subject to continuous oxidative damage both from superoxide ions produced as an inevitable by-product of the mitochondrial respiration and from other reactive oxygen species (ROS), for example the reactive hydroxyl radical formed in the iron-catalysed destruction of hydrogen peroxide, the Fenton reaction (Fig. 1).

$$H_2O_2 + Fe^{2+} \rightarrow Fe^{3+} + {}^{\bullet}OH + OH^-$$

Figure 1: The Fenton reaction

T. Herdegen and J.M. Delgado-García (eds.), Brain Damage and Repair, 275–284.

However, unlike the situation in most other tissues, adult brain neuronal tissue has only very limited possibilities for self renewal from stem cells and the continuous function of brain neurons depends heavily on elaborate antioxidant defences.

Nevertheless, the defences may break down and neurons may then die by two basic processes. During necrosis a catastrophic failure of cellular homeostasis may directly follow an insult, like mechanical damage or anoxia. The cells then splits open and spill their contents into the interstitial space. Alternatively, in apoptosis the cells commit active suicide, using their own cellular mechanisms to initiate a series of molecular events that eventually lead the cells to digest away many of its components from the inside without disrupting the plasma membrane. An extreme point of view on this matter could be, that all non-dividing cells, and therefore all mature neurons, are committed to apoptosis all the time but are only temporarily prevented from initiating the process by the action of various trophic factors and other molecules that suppress the apoptotic molecular machinery.

2. OXYGEN AS AN OXIDANT

The free oxygen concentration in human venous blood is only about 53 µM (5 kPa), interstitial space oxygen concentration may be 10-20 µM and free oxygen concentration is decreasing gradually from the cell plasma membrane to the oxygen-consuming mitochondria (<0.5 µM; 0.05 kPa). However, as oxygen is five to eight times more soluble in organic solvents than in water, the local concentration of oxygen within the hydrophophic environment of the cell plasma membrane might be relatively high, explaining the prevalent oxidative damage to membrane lipids.

Four Fe^{2+}-heme cytochrome groups in cytochrome oxidase, located in the mitochondria inner membrane, tightly bind most of the oxygen within the cells. The bound oxygen is reduced to water, one electron at a time, by the Fe^{2+}-heme cytochromes. With age the efficiency decreases and oxidative damages accumulate (Shigenaga et al., 1994). Most of the 10-15% of oxygen taken up by aerobic eukaryotes that is not consumed by mitochondria is used by various oxidases and oxygenases. Of relevance for brain damage and repair is the superoxide ion producing xanthine oxidase, which uses oxygen to oxidise xanthine and hypoxanthine into uric acid (Atlante et al., 1997).

The oxygen molecule, more correctly called dioxygen, qualify itself as a free radical by having two unpaired electrons and therefore ought to be written $^{\bullet\bullet}O_2$. But it is not very reactive because the two unpaired electrons have parallel spin and can only easily react with a single unpaired electron (from *e.g.* Fe^{2+} or Cu^+) or with another molecule having two unpaired electrons with parallel spin. The latter situation is a rare in nature and therefore aerobes can exist in an atmosphere of oxygen in spite of the negative free energy associated with combustion of organic material. However, in the presence of either catalytic concentrations of free metal ions able to transfer single electrons or ROS able to initiate oxidative chain reactions, the metastable balance is disturbed until the cellular protective programs inactivate the catalysts, abort the catalysed oxidative chain reaction or, ultimately, protect the whole organism by initiating cellular apoptosis.

3. SINGLET OXYGEN

A more reactive form of oxygen, known as singlet oxygen, can be generated by an input of energy, for example by UV irradiation. Having no unpaired electron, singlet O_2 in its ground state is not a free radical *per se* but nevertheless an extremely reactive ROS. Although UV irradiation is irrelevant in relation to brain damage, it is presently not known whether the proportion of singlet oxygen to normal oxygen could be increased by molecular orbital overlap between dioxygen and orbitals from peptide side chain residues like methionine and tryptophan.

4. SUPEROXIDE ION RADICAL

Probably the most important source of the superoxide ion radical $^{\bullet}O_2^-$ in neurons is the electron transport chain. At physiological oxygen levels, it has been suggested that about 3% of the oxygen reduced in mitochondria may form superoxide ions (Halliwell & Gutteridge, 1999). The low rate of leakage is probably both due to the low intra-mitochondrial free oxygen concentration and to the arrangement of the electron carriers into complexes that facilitates electron movement to the next component in the chain rather than electron escape to oxygen. However, when mitochondrial organisation is damaged, leakage increases and formation of superoxide ions increases, resulting in damage to mitochondrial proteins, lipids and DNA.

Although the superoxide ion as such is not strongly reactive with non-radical species, it may generate the very reactive hydroxyl radical $^{\bullet}OH$ in the Haber-Weiss reaction of a Fe^{2+}/Fe^{3+}-catalysed reaction with hydrogen peroxide (Fig. 2).

$$^{\bullet}O_2^- + Fe^{3+} \rightarrow Fe^{2+} + O_2$$
$$Fe^{2+} + H_2O_2 \rightarrow Fe^{3+} + {}^{\bullet}OH + OH^-$$
$$\overline{^{\bullet}O_2^- + H_2O_2 \rightarrow {}^{\bullet}OH + O_2 + OH^-}$$

Figure 2: The Haber-Weiss reaction

It is therefore of major importance for neuronal protection against oxidative damage that copper-zinc superoxide dismutase (CuZnSOD) in the cytosol and in the space between the inner and outer mitochondrial membranes removes the superoxide formed (Fig. 3; Halliwell & Gutteridge, 1999).

$$O_2^- + CuZnSOD\text{-}Cu^{2+} \rightarrow CuZnSOD\text{-}Cu^+ + {}^{\bullet\bullet}O_2$$

$$CuZnSOD\text{-}Cu^+ + {}^{\bullet}O_2^- + 2H^+ \rightarrow CuZnSOD\text{-}Cu^{2+} + H_2O_2$$

$$2{}^{\bullet}O_2^- + 2H^+ \rightarrow H_2O_2 + {}^{\bullet\bullet}O_2$$

Figure 3: The copper zinc dismutase reaction.

Another important superoxide dismutase is manganese superoxide dismutase (MnSOD), which is an essential mitochondrial enzyme catalysing fundamentally the same reaction as CuZnSOD. In neurons the activity of MnSOD has not been measured but mice with the gene for MnSOD knocked out do not survive for more than 10 days and display decreased mitochondrial enzyme activities (Li et al., 1995). However, in the presence of low molecular weight scavengers of superoxides such mice survive longer but display brain degeneration, probably because the scavengers do not cross the blood-brain barrier (Melov et al., 1998).

5. HYDROGEN PEROXIDE

The hydrogen peroxide H_2O_2 formed by superoxide dismutase and by other oxidase may be removed by catalase and by peroxidases. However, in the brain catalase activity is very low, so neurons are better protected from hydrogen peroxide by glutathione peroxidase (Marklund et al., 1982). Glutathione peroxidase consists of four protein subunits, each of which contains one atom of the element selenium (Se) at its active site (Maiorino et al., 1995). It is present in the active site as selenocysteine. The overall reaction is an oxidation mediated by peroxide (either by hydrogen peroxide or by an organic peroxide) of two reduced glutathions (GSH) to one oxidised glutathione (GSSG). To continuously keep the ratio of reduced to oxidised glutathion high as it is in neurons (Halliwell & Gutteridge, 1999), glutathione reductase reduces oxidised glutathione to glutathione on the expense of NADPH from the cytoplasmic pentose phosphate pathway metabolism of glucose.

Selenium deficiency in animals produces a variety of serious diseases consistent with the importance of selenium in the antioxidant defence of the body. Likewise, low levels of human liver glutathione has been associated with various liver diseases and Wilson's disease and low levels in substantia nigra has been associated with Parkinson's disease (Riederer et al, 1989; Uhlig & Wendel, 1992).

6. REACTIVE THIOL SPECIES

The protection against thiyl radicals is much dependent on thioredoxins. Thioredoxins are polypeptides, MW 12,000, especially concentrated in endoplasmic reticulum but also found in the plasma membrane. They have two adjacent -SH groups in the reduced form, which are converted to a disulphide in oxidised thioredoxin. They can undergo redox exchanges with multiple proteins and NADPH can reduce the oxidised thioredoxins, just like oxidised glutathion (Chae et al. 1994).

7. IRON AND COPPER AS OXIDATIVE CATALYSTS

Although iron and copper are essential in the mammalian body as prosthetic factors in the enzymatic synthesis of many metabolites in respiration and redox reactions, these metals are also potentially toxic because they are able to undergo catalytic single electron re-cycling, Fe^{2+}/Fe^{3+} and Cu^{+}/Cu^{2+}, and to interact with oxygen and hydrogen peroxide and catalyse the formation of the extremely reactive hydroxyl radical. Being uncharged, the hydroxyl radical can initiate chain reactions with membrane lipids and other molecules.

Two thirds of the about 4 g body iron is found in haemoglobin. As far as haemoglobin stays packaged within the erythrocytes, it only slowly produces superoxide from the bound oxygen, a process which is normally counteracted by the local antioxidant protection. However, when haem is released from haemoglobin following local haemolysis of erythrocytes in the course of brain haemorrhage, the resulting free iron may act as an efficient catalyst for ROS because in brain extracellular fluid, including CSF, the concentration of antioxidant defence proteins is low. Non-haem iron is mainly transported in the circulation bound to transferrin as Fe^{3+}. Iron-transferrin complexes may be bound to cells by specific transferring receptors, the iron removed by unknown iron-binding agents, and surplus iron stored in ferritin as insoluble oxides (Harrison & Arosio, 1996).

Copper is transported in the circulation bound to ceruloplasmin, which also has a ferroxidase activity and oxidises Fe^{2+} to Fe^{3+}. Ceruloplasmin is not only produced in the liver but is also seen in a plasmamembrane bound form in astrocytes and neurons in neurodegenerative disorders (Loeffler et al, 1996) and may play a protective role towards iron-mediated free radical injury in the brain (Patel et al., 2002). However, free copper ions may catalyse formation of hydroxyl radicals in a copper-catalysed Haber-Weiss reaction (Halliwell & Gutteridge, 1990).

Within cells, iron appears to be in constant transit between ferritin and its destination in functional proteins. Oxidative stress can lead to more iron release and possibly also more copper release. Metallothioneins are cytosolic peptides encoded by genes on chromosome 16 with MW 6,500. Two isoforms are found in all animal tissues and a third isoform, MT-III, only in brain. Metallothioneins are rich in cystein and therefore represent a significant proportion of total cell thiols. Each metallothionein can bind 5-7 ions of copper, zinc or other transition metals ions like silver, cadmium and mercury (Dunn et al., 1987). They do not appear to play a major role in binding iron but have been suggested to have antioxidant properties.

8. REGENERATION OF ANTIOXIDANT PROTECTORS, BY ASCORBIC ACID AND α-TOCOPHEROL

Low-molecular-mass antioxidants may have a major role in the regeneration of the neuronal antioxidant protective systems. Some antioxidants are biosynthesised in the human body but many have to be delivered by the diet. Bilirubin, the degradation product of haem, is a powerful scavenger of ROS but will never directly reach neurons from the liver. Female sex hormones, like estradiol, have in vitro been shown to have antioxidant properties through the phenolic hydroxyl group. It can, however, be questioned whether the in vivo brain concentration of estrogens reach levels where antioxidant properties might be of importance (Liehr, 1996). Lipoic acid (1,2-dithiolane-3-pentanoic acid; thiooctic acid) is an essential prosthetic group in the enzymes catalysing decarboxylations in the Krebs cycle. It has antioxidant properties in vitro and it can bind iron and copper ions but its concentration in body fluids is very low. Coenzyme Q plays an essential role in the mitochondrial electron-transport chain, undergoing simultaneous oxidation and reduction to ubiquinol via a free-radical intermediate, ubisemiquinone. Ubiquinol can scavenge superoxide ions, forming ubisemiquinone, and inhibit lipid peroxidation. Furthermore it can regenerate α-tocopherol from its radical in lipoproteins and membranes. However, the overall contribution of ubiquinol to antioxidant defence in vivo is uncertain (Halliwell & Gutteridge, 1999).

From the diet we have the major antioxidants ascorbic acid (vitamin C) α-tocopherol (vitamin E) and plant phenols. Ascorbic acid is required as a coenzyme for several enzymes, including proline hydroxylase, lysine hydroxylase and dopamine-β-hydroxylase. However, copper- and iron-induced oxidation of ascorbate produces hydrogen peroxide and the reactive hydroxyl radical. It is thus not very obvious whether ascorbate in vivo is an anti-oxidant or a pro-oxidant, although ascorbate is able to regenerate α-tocopherol from its radical and also scavenge nitroxide radicals. Moreover, the concentration of ascorbate in human CSF (150 µM) is higher than in plasma and it is even more concentrated in neuronal and astrocytic cytoplasm (Reiber et al., 1993).

Being fat soluble, α-tocopherol is probably the most important inhibitor of the free-radical chain reaction of lipid peroxidation in mammals, also because it scavenge lipid peroxyl radicals much faster than these radicals can react with other fatty acid side-chains or with membrane proteins (Halliwell &Gutteridge, 1999). In addition, α-tocopherol may react with singlet oxygen and hydroxyl radicals, however only slowly with the superoxide ion. The α-tocopheryl radical formed have to be recycled back to α-tocopherol, often by an ascorbate-dependent recycling mechanism (Fig. 4).

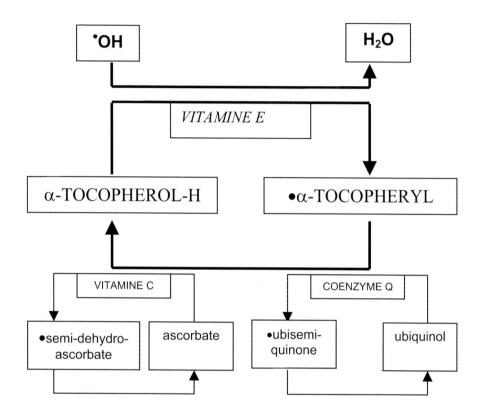

Figure 4: Vitamine E in neuronal protection against oxidative damage by hydroxyl radicals.

Whereas evidence for β-carotene to play an antioxidant role in vivo is scarce, good evidence exists for plant phenols to be inhibitors of lipid peroxidation in vitro and in vivo. These include flavanols as epicatechin in green tea, flavonols and anthocyanidins in red wine and curcumin in curry.

9. REPAIR OF OXIDATIVE DAMAGE TO LIPIDS AND PROTEINS

Peroxides within neuronal membranes can be converted to alcohols by phospholipid hydroperoxide glutathione peroxidase enzymes (Maiorino et al., 1995) or cleaved from the membranes by phospholipases followed by reduction with glutathione peroxidase.

Oxidative damage to proteins by ROS can to some degree be repaired. Glutathion and thioredoxin can reduce disulphide bridges formed by aberrant cross-linking of cystein side-chains.

Oxidative damage to DNA may to a certain degree be repaired by DNA proof-reading enzymes. However interesting, this subject will not be further considered in the present chapter.

Protein methionine is readily oxidized to methionine sulphoxide by the hydroxyl radical and other oxidants, resulting in many instances in blockade of biologic activity. An enzyme, peptide methionine sulphoxide reductase, stereoselectively reduces the sulphoxide back to methionine (Weissbach et al., 2002).

10. OXIDATIVE MECHANISMS IN NEURODEGENERATIVE DISEASES EXEMPLIFIED BY ALZHEIMER'S DISESASE

β-amyloid peptide, which has a fundamental role in Alzheimer's disease pathogenesis, has repeatedly been shown to be cytotoxic in tissue culture with the cytotoxicity associated with β-sheet formation in protofibrils (Rosenblum 2002). β-amyloid peptide shows evidence of formation of free radicals in solution, associated with the presence of methionine in the peptide (Varadarajan et al., 2001). Although not directly considered by these authors, the possible stabilisation of singlet oxygen in the tight β-sheet environment of amyloid might offer an explanation to the experimentally found ROS in the apparently pure solution of β-amyloid peptides and might thus be of potential relevance for understanding Alzheimer's disease pathogenesis. Furthermore, β-amyloid peptide seems to be a copper-binding metalloprotein, which catalytically converts molecular oxygen into hydrogen peroxide (Lynch et al., 2000).

The relevance of oxidative mechanisms in the pathogenesis of Alzheimer's disease has important therapeutic implications. Treatment of patients with Alzheimer's disease of moderate severity with α-tocopherol has been found to slow disease progression (Sano et al., 1997). Currently the copper and iron-binding agent clioquinol is undergoing clinical trials as a promising novel Alzheimer therapeutic (Bush 2003).

Finally apolipoprotein E might have isotype-dependant antioxidant properties. Bound to a lipoprotein receptor in the neuronal plasma membrane, the most common isoform apoE3 has one exposed cystein, whereas apoE4 has none and apoE2 has two. In addition to an isotype-specific effect of apolipoprotein E in cholesterol transport during the astrocyte-mediated neuronal membrane recycling in synaptic remodelling (Poirier, 1994), this could pose a second mechanism underlying the risk conferred by apoeE4 in the development of Alzheimer's disease (Stritmatter & Roses, 1996).

ACKNOWLEDGEMENT

Support for this study came in part from grants from Lions Club Denmark (Røde Fjer) and NeuroScience PharmaBiotec.

REFERENCES

Atlante, A., Gagliardi, S., Minervini, G.M., Ciotti, M.T., Marra, M.T., Marra, E. & Calissano, P. (1997). Glutamate neurotoxicity in rat cerebellar granule cells: A major role for xanthine oxidase in oxygen radical formation. *Journal of Neurochemistry, 68*, 2038-2045.

Bush, A. (2003). The metallobiology of Alzheimer's disease. *TRENDS in Neurosciences, 26*, 207-214.

Chae, H.Z., Robison, K., Poole, L.B., Church, G., Storz, G. & Rhee, S.G. (1994). Cloning and sequencing of thiol-specific antioxidant from mammalian brain: Alkyl hydroperoxide reductase and thiol-specific antioxidant define a large family of antioxidant enzymes. *Proceeding of the National Academy of Science USA, 91,* 7017-7021.

Dunn, M.A., Blalock, T.L. & Cousins, R.J. (1987). Metallothionein. *Proceedings of the Society for Experimental Biology and Medicine*, N.Y., 185, 107-119.

Halliwell, B. & Gutteridge, J.M.C. (1999). *Free radicals in biology and medicine.* (3rd ed.). Oxford. Oxford University Press.

Harrison, P.M. & Arosio, P. (1996). The ferritins: molecular properties, iron storage function and cellular regulation. Biochimica biophysica acta, 1275, 161-203.

Li, Y., Huang, T.T., Carlson, E.J., Melov, S., Ursell, P.C, Olseon, J.L., Noble, L.J., Yoshimura, M.P., Berger, C., Chan, P.H. et al. (1995). Dilated cardiomyopathy and neonatal lethality in mutant mice lacking MnSOD. *Nature Genetics, 11,* 376-381.

Liehr, J.G. (1996) Role of estrogen and iron in neurodegeneration: Open peer commentary to "Astrocytes, brainaging, and neurodegeneration" by H. Schipper. *Neurobiology of Aging, 17,* 481-482.

Loeffler, D.A., LeWitt, P.A., Juneau, P.L., Sima, A.A.F., Nguyen, H.-U., DeMaggio, A.J., Brickman, C.M., Brewer, G.J., Dick, R.D., Troyer, M.D. & Kanaley, L. (1996). Increased regional brain concentrations of ceruloplasmin in neurodegenerative disorders. *Brain Research, 738,* 265-274.

Lynch T., Cherny, R.A. & Bush, A.I. (2000). Oxidative processes in Alzheimer's disease: the role of Aβ-metal interactions. *Experimental Gerontology, 35,* 445-451.

Maiorino, M., Aumann, K.D., Brigelius-Flohe, R., Doria, D., van-den-Heuvel, J., Roveri, A., Ursini, F. & Flohe, L. (1995). Probing the presumed catalytic triad of Se-containing peroxidases by mutational analysis of $PHGP_x$. *Biological Chemistry of the Hoppe-Seyler's, 376,* 651-660.

Marklund, S.L., Westman, N.G., Lundgren, E. & Roos, G. (1982). Copper- and zinc-containing superoxide dismutase, manganese-containing superoxide dismutase, catalase, and glutathione peroxidase in normal and neoplastic human cell lines and normal human tissues. *Cancer Research, 42,* 1955-1961.

Melov, S., Schneider, J.A., Day, B.J., Hinerfeld, D., Coskun., P., Mirra., S.S., Crapo, J.D. & Wallace, C. (1998). A novel neurological phenotype in mice lacking mitochondrial MnSOD. *Nature Genetics, 18,* 159-164.

Patel, B.N., Dunn, R.J., Jeong, S.Y., Zhu, Q., Julien, J.-P. & David, S. (2002). Ceruloplasmin regulates iron levels in the CNS and prevents free radical injury. *The Journal of Neuroscience, 22,* 6578-6586.

Poirier, J. (1994). Apolipoprotein E in animal models of CNS injury and in Alzheimer's disease. *Trends in Neuroscience, 17,* 525-530.

Reiber, H., Ruff, M. & Uhr, M. (1993). Ascorbate concentration in human CSF and serum. Intrathecal accumulation and CSF flow rate. *Clinical Chemical Acta, 217,* 163-173.

Riederer, P., Sofic, E., Rausch, W.D., Schmidt, B., Reynolds, G.P., Jellinger, K. & Youdim, B.M.H. (1989). Transition metals, ferritin, glutathione, and ascorbic acid in parkinsonian brains. *Journal of Neurochemistry, 52,* 515-520.

Rosenblum, W.I. (2002). Structure and location of amyloid beta peptide chains and arrays in Alzheimer's disease: new findings require reevaluation of the amyloid hypothesis and tests of the hypothesis. *Neurobiology of Aging, 23,* 225-230.

Sano, M., Ernesto, C., Thomas, R.G., Klauber, M.R., Schafer, K., Grundman M., Woodbury, P., Growdon, J., Cotman, C.W., Pfeiffer, E., Schneider, L.S. & Thal, L.J. (1997). A controlled trial of selegiline, alpha-tocopherol, or both as treatment for Alzheimer's disease. The Alzheimer's Disease Cooperative Study. *New England Journal of Medicine, 336,* 1216-1222

Shigenaka, M.K., Hagen, T.M & Ames, B.M. (1994). Oxidative damage and mitochondrial decay in ageing. Procedings of the National Academy of Science USA, 91, 10771-10778.

Strittmatter, W.J. & Roses, A.D. (1996). Apolipoprotein E and Alzheimer's disease. *Annual Review of Neuroscience,* 19, 53-77.

Uhlig, S. & Wendel, A. (1992). The physiological consequences of glutathione variations. *Life Sciences, 51,*1083-1094.

Varadarajan S., Kanski, J., Aksenova, M., Lauderback, C. & Butterfield, D.A. (2001). Different mechanisms of oxidative stress and neurotoxicity for Alzheimer's Aβ(1-42) and Aβ(25-35). *Journal of the American Chemical Society, 123,* 5625-5631.

Weissbach, H., Etienne, F., Hoshi, T., Heinemann, S.H., Lowther, W.T., Matthews, B., St John, G., Nathan, G. & Brot, N. (2002). Peptide methionine sulfoxide reductase: structure, mechanism of action, and biological function. *Archives of Biochemistry and Biophysics, 397,* 172-178.

D. GONZÁLEZ-FORERO, B. BENÍTEZ-TEMIÑO,
R.R. DE LA CRUZ AND A.M. PASTOR

Department of Physiology and Zoology, Faculty of Biology, University of Sevilla, Sevilla, Spain

19. FUNCTIONAL RECOVERY IN THE PERIPHERAL AND CENTRAL NERVOUS SYSTEM AFTER INJURY

Summary. Although function is generally irreversibly impaired following brain damage, homeostatic and hebbian plastic mechanisms that occur during development can be reactivated in adulthood leading to variable degrees of recovery. Knowledge and manipulation of such potential intrinsic mechanisms are being used to promote functional reorganization of injured brains. Likewise, recent advances in transplantation, neurogenesis and cell differentiation open new possibilities based on substitutory therapies to reconstruct damaged neural circuits, otherwise lost due to regenerative failure in the adult CNS. In this chapter we will review and discuss the contributions and limitations of intrinsic plastic mechanisms, as well as the perspectives about potential extrinsic interventions to promote functional recovery after brain injury.

1. INTRODUCTION

Differences in the ability of peripheral and central fibers to regenerate following damage have been recognized for more than a century. Thus, the loss of the original function after axonal transection is generally reversible in the peripheral nervous system (PNS) and permanent in the central nervous system (CNS). The intrinsic plasticity of the developing brain acts on the architecture and the wiring pattern in use-dependent manners. In the mature CNS, these mechanisms partially persist. In this respect, although neuronal excitability and synaptic efficacy are continuously modulated through learning and memory processes, properties such as permissiveness for axonal regrowth and genesis of new neurons are lost or spatially restricted at mature stages. This has led in recent years to considerable efforts to reconnect damaged circuits or to replace dead neurons. We are also beginning to understand different mechanisms of plasticity acting on single neurons or on larger extents of the nervous system, which partially compensate neuronal or behavioral dysfunction, respectively, after injury. However, although functional reorganization appears to be useful for improving functional outcome, in some cases its significance remains unclear or even detrimental. In this chapter we will focus on 1) disruption of determinants maintaining the functional state of adult neurons after synaptic loss and regenerative responses; 2) compensatory mechanisms of plasticity in lesioned pathways; 3) functional reorganization of non-lesioned brain regions after permanent destruction of neural pathways; and 4) therapeutics for functional recovery.

2. NEURONAL FUNCTION IS MAINTAINED BY CONNECTIVITY

Traumatic lesions and neurodegenerative diseases affecting the nervous system lead to varying degrees of afferent or efferent disconnection as a result of axonal injury or neuronal death. Likewise, a great variety of natural venoms and neurotoxins impair synaptic function, leading to modified levels of electrical activity. A crucial question in understanding the potential for neuronal recovery after brain damage is to what extent electrophysiological properties are regulated by the target. It is well established that discharge activity of peripheral and central neurons depends on two factors — intrinsic electrophysiological properties and afferent presynaptic influences — which in turn are critically sustained by retrograde signaling from the target. Different experimental approaches have been used to study the physiological alterations following target or afferent disconnection and the factors influencing the functional phenotype of the neuron. For the widely studied neuronal response and mechanisms of recovery of target-disconnected neurons, axotomy is likely the procedure most used. Following interruption of their axons, neurons show several functional alterations affecting intrinsic membrane properties, afferent synaptic efficacy, and discharge patterns. As first shown by Eccles and colleagues and then by Kuno and Llinás in spinal motoneurons, axotomy-induced changes in electrical properties tend to increase the neuronal membrane excitability (see below). Other alterations observed in axotomized motoneurons include reduction in axonal conduction velocity and shortening of the afterhyperpolarization duration. These findings led Kuno to propose that axotomy, or rather muscle disconnection, induces a functional dedifferentiation on motoneurons, towards a postnatal-like stage. He concluded that motoneuron properties during adulthood depend partly upon some target-derived trophic factors related to the activity of the muscle.

In spite of the increased excitability, axotomized neurons usually show a depressed and unmodulated firing activity that has been associated with a massive loss of afferent inputs. Taken together, these findings indicate that the maintenance of functional connectivity with the target supports both the electrical and the synaptic properties of parent neurons. However, axonal section not only means the loss of contact with the target, but is also a direct physical lesion to the cell. In this respect, an important question arises: are these functional changes due to target disconnection, to axonal injury itself, or do they reflect a complex response to both circumstances? To answer this question, several experimental approaches that involve functional disconnection or selective target ablation, without direct axonal damage have been used. Thus, as reported by us, the specific removal of central targets elicits morpho-functional changes similar to axotomy on premotor internuclear neurons of the abducens nucleus. Likewise, axotomy-like changes in motoneuronal electrical properties have been observed after functional blockade of the neuromuscular transmission with botulinum neurotoxin or following tetrodotoxin application to the motor nerve. These observations indicate that the absence of effective neurotransmission to the target is sufficient to give rise to axotomy-like functional alterations.

The identity of target-derived signals with trophic influences on presynaptic neurons is beginning to be known. The first strong evidence emerged from Purves' experiments on sympathetic neurons. He reported axotomy-like changes — including synaptic potential attenuation and increased excitability — in postganglionar neurons treated with colchicine to inhibit retrograde axonal transport, which is the transport pathway for neurotrophin-receptor complexes. It was later shown that the exogenous application of nerve growth factor (NGF) after postganglionic axotomy largely prevented those changes. In contrast, treatment with anti-NGF antibodies induces synaptic depression and retraction of terminals on non-axotomized postganglionar neurons. Thus, NGF was found to combine two biological actions: neurotrophic and synaptotrophic. All these observations led to the formulation of the trophic theory of neuronal connections, which establishes the functional role of target-derived neurotrophic factors in the development and maintenance of neuronal properties, including presynaptic connections. After NGF, other target-derived neurotrophic factors have been demonstrated to exert similar actions on specific neuronal populations: in particular, brain-derived neurotrophic factor (BDNF), neurotrophin 3 (NT-3) and neurotrophin 4/5 (NT-4/5), which form along with NGF the so-called family of neurotrophins. Furthermore, it has been demonstrated for these and other trophic factors, such us CNTF (ciliary neurotrophic factor), GDNF (glial-cell-line-derived neurotrophic factor), bFGF (basic fibroblast growth factor), IGF (insulin-like growth factor), or TGF-β (transforming growth factor-beta), that, when applied exogenously after injury, they stimulate regeneration in adulthood and rescue developing neurons from death. In conclusion, normal synaptic and electrical functioning of mature neurons is strictly dependent on the appropriate supply of target-derived neurotrophic factors.

Target reinnervation following axonal regeneration and, consequently, restoration of the trophic supply, generally re-establishes the original neuronal and behavioral function. To achieve this, the injured neuron has to regrow its axon through the environment and also find its correct target. The first condition is easily met by neurons sending axons to the periphery, but is a limiting factor for the reconstruction of central circuits after damage. Chemical and physical barriers, derived mainly from glial cells at the lesion site, strongly inhibit axonal regrowth in the CNS. Nevertheless, central neurons have the intrinsic ability to regenerate their axons, as first demonstrated by David and Aguayo bridging the CNS milieu with segments of peripheral nerve, and more recently by Silver and colleagues by micro-transplanting adult isolated neurons into adult spinal tracts to minimize glial scarring. In both cases, severed axons grew for long distances in CNS tissue. In the PNS, regeneration after axonal section is largely successful. As noted above, reconnection with original targets and establishment of functional synaptic contacts constitute the second condition for functional recovery. Furthermore, the target has to remain viable and permissive for reinnervation. Munson's group demonstrated that nine months after section of medial gastrocnemius nerve muscle and self-reinnervation, the membrane electrical properties of motoneurons, the contractile characteristics of their motor units, and the normal relationship among them were restored. This experiment confirmed the hypothesis that functional connection with the target muscle is necessary for the expression of normal properties in motoneurons. However, axons usually do not reinnervate exactly the same original target cells and some contacts between regenerated axons and their targets could be inappropriate.

So, when the severed nerve is a mixed nerve containing motor and cutaneous

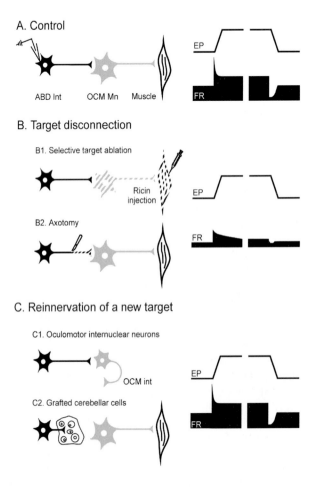

Figure. 1. Schematic diagram showing the discharge profile (right) of abducens internuclear neurons (ABD Int) during spontaneous eye movements in the on- and off-direction for the activation of the cell (EP, eye position; FR, firing rate). Three different situations are illustrated. In control (A), ABD Ints project onto the medial rectus motoneurons of the oculomotor nucleus (OCM Mn). After target disconnection induced by either selective target removal with toxic ricin (B1) or axotomy (B2), ABD Ints show drastic changes in their firing pattern. When ABD Ints reinnervate a new target, such as the oculomotor internuclear neurons after selective target removal (C1) or grafted cerebellar cells after axotomy (C2), they recover normal firing characteristics.

fibers, the chances of erroneous reinnervation are high. Therefore, peripheral reinnervation is not a completely selective process, and usually implies anomalous motor output or sensory perception. Nevertheless, although inappropriate target innervation may lead to a poor functional recovery at the behavioral level, it usually includes re-establishment of the original neuronal properties. Thus, foreign target innervation by providing trophic support rescues physiological properties of neurons from axotomy, and does not re-specify the phenotypic characteristics of neurons in accordance with the new target. For instance, both axotomized muscle afferents and motoneurons recover their proper firing and synaptic characteristics after skin reinnervation. Likewise, cross-reinnervation of tibial anterior and gastrocnemius motoneurons after sciatic nerve section restores motoneuronal properties and, consequently, leads to abnormal electromyographic activity. Non-specific reinnervation associated with functional recovery has also been observed in central neurons. As reported by the authors, selective target ablation or axotomy affecting the internuclear neurons of the abducens nucleus induces similar alterations in the discharge characteristics and synaptic coverage. However, the innervation of either neighboring non-natural targets or grafted cerebellar neurons is accompanied by the re-establishment of afferent synaptic connections and normal firing patterns (Fig. 1). In summary, the functional mature phenotype in central and peripheral neurons can be restored after efferent disconnection when reinnervation, even of foreign targets, is accomplished. The significance of this intrinsic ability for functional recovery through non-specific target reinnervation is not well understood, particularly if it does not always lead to behavioral recovery. It illustrates a loss of specificity in the adult nervous system regarding axonal fate, when compared with the developing brain. At least for the mammalian CNS it could represent an "hereditary vestige" from lower vertebrates, including lamprey, fish, and some amphibians, whose central axons can regenerate through a much more permissive environment. So, the degree of axonal regeneration and recovery of function after CNS lesion depends on both ontogeny and phylogeny, since in immature "higher" vertebrates, regeneration and functional recovery can also occur.

The loss of afferent inputs also affects the normal functioning of postsynaptic neurons. Deafferentation not only implies a reduction in the amount of synaptic bombardment, which will undoubtedly modify the integrative membrane properties and postsynaptic firing modulation, but also unbalances inhibitory and excitatory influences, leading to abnormal levels of activity. Presynaptic afferent and postsynaptic firing activities have been shown to regulate different electrical, synaptic, and discharge characteristics of postsynaptic neurons. Thus, when they are modified, compensatory changes take place affecting intrinsic membrane properties, synaptic efficacy, or synaptic balance (see below). Since lesioned afferents in the CNS do not regenerate, local sprouting from initially non-dominant or residual inputs can occur after incomplete deafferentation, which in some cases contributes to functional restoration. Additionally, although a retrograde mechanism of action has been demonstrated for most of the neurotrophic factors (retrophins), recent evidence suggests anterotrophin actions for these molecules in the CNS. Thus, it is possible that the maintenance of the functional differentiation of adult neurons is supported not only by target-derived trophic supply, but also by bi-directional trophic interactions between connected cells.

3. THE CONSERVATIVE NEURON. COMPENSATION AFTER DYSFUNCTION

Neurons exhibit complex responses to abnormal functioning, involving compensatory changes in intrinsic membrane properties, synaptic strength, and innervation. Occasionally, functional modifications are followed by parallel structural changes. Underlying mechanisms of neuronal plasticity that promote compensatory modifications have been termed *homeostatic plasticity*. Regulation of neural excitability is essential to maintain the stability of neurons and circuits and, consequently, the functioning of the nervous system. Homeostatic processes are thought to regulate synaptogenesis during postembryonic development, but in the adult, nervous tissue subjected to modifications of excitability (axotomy or deafferentation), homeostatic plasticity also operates, although not necessarily implying functional recovery. Davis and Bezprozvanny explain homeostatic neural mechanisms as a nuclear monitoring of the activity levels that would be transduced into regulated changes in synaptic and membrane function.

As described above, peripheral and central axotomy not only induces massive deafferentation and reduced firing activity, it also increases membrane excitability. Thus, rheobase current and, consequently, threshold for discharge decrease, while membrane time constant and input resistance increase. This is related to a higher specific membrane resistivity, a decreased cell size and an altered dendritic geometry. Likewise, in axotomized motoneurons the ratio between discharge frequency and current intensity increases its gain and becomes monotonic, in contrast to the case of control motoneurons, which present two or three different slope regions along their firing frequency range. Additionally, axotomy leads to changes in the density and distribution of sodium channels, which adopt aberrant locations on the dendritic membrane, giving rise to the appearance of anomalous dendritic electrogenesis. In contrast, certain potassium conductances are downregulated after axotomy. Thus, axotomized neurons undergo alterations in intrinsic membrane properties that would tend to make them more responsive to excitatory inputs. Eccles suggested that dendritic spikes, which appear as dome-like partial responses superimposed on excitatory synaptic potentials, could increase the efficacy of synaptic transmission and, in this way, compensate the deficiency in ongoing excitatory drive.

Although these electrophysiological changes constitute a general response extendable to both peripheral and central neurons, the significance of this "attempted firing" is not clear, particularly if target cells are physically disconnected. Furthermore, while the threshold for spike generation is reduced in the somatodendritic membrane, conduction of axonal action potentials is impaired after nerve injury. Thus, it is difficult to envisage that discharge facilitation after axotomy could serve primarily as a signal to cells other than the axotomized neuron. This increased excitability after peripheral lesions accounts for the spontaneous activity that contributes to the generation of neuropathic pain in humans as an additional pathological disorder. Barres and colleagues, however, demonstrated in cultured retinal ganglion cells that axonal elongation is not only promoted by neurotrophic signaling, but also is profoundly potentiated by physiological patterns of electrical activity. They showed that, under trophic stimulation, active neurons elongate axons better than silent or less active neurons do. Gordon and colleagues have obtained *in vivo* demonstrations for the effect of electrical activity on the speed of axonal regrowth in adulthood. They observed that after femoral nerve transection, electrical stimulation had a double benefit: accelerating axonal regeneration and promoting reinnervation of appropriate targets. Furthermore, they demonstrated that the positive effect of electrical stimulation was mediated via the cell body, probably through the activation of growth programs. Consequently, increased excitability in axotomized neurons could constitute a compensatory neuronal response to sustain a certain level of electrical activity, which in turn would facilitate a faster and more-appropriate target reinnervation.

Neuronal deafferentation after physical or functional afferent disconnection induces activation of compensatory mechanisms. This does not generally involve growth of the injured fibers, which will eventually degenerate, but rather the reorganization and replacement by axonal sprouting of intact axons over the vacant space left on the denervated postsynaptic membrane. This phenomenon, the so-called *reactive synaptogenesis*, has attracted interest, since it could mean a possible mechanism of functional recovery after injury in the mature nervous system. Response to deafferentation shows similarities and differences between CNS and PNS neurons, fundamentally in the capacity for regeneration of the lesioned afferents. In both cases, postsynaptic neurons can survive and be reinnervated, recovering synapse number and, partially, their discharge characteristics. However, in the PNS, injured axons regenerate and are able to occupy the denervated space, whereas in the CNS, axons do not regenerate, and the mechanism of reinnervation is local — that is, from non-injured and frequently residual afferences. Reinnervation is not a risky process; on the contrary, it is carried out through competitive phenomena. Proctor and colleagues demonstrated that somatic motor axons establish synaptic contacts on postganglionar parasympathetic neurons. These synapses are later eliminated when regeneration of preganglionar axons is facilitated, demonstrating selectivity in the process of competitive reinnervation. Several studies suggest that deafferentation leads to pre- and postsynaptic changes tending to modify neuronal excitability and the response to afferent activity in the opposite direction to that induced by denervation. Thus, monosynaptic action of primary spindle afferents on spinal motoneurons increases after dorsal rhizotomy or rostral spinal cord transection. These observations would explain the hyperreflexia produced in caudal segments after a spinal cord lesion. Various mechanisms, including new synapse formation from lesioned or intact afferences, denervation supersensitivity, or reduction of the electrotonic distance in the deafferented neuron, have been proposed as underlying such synaptic facilitation. Furthermore, in spite of Hebbian rules for the use-dependent modification of synapses, it has been shown that prolonged synaptic inactivity or disuse increases synaptic efficacy in peripheral and central synapses. This phenomenon is termed "disuse hypersensitivity", and is suggested to underlie diverse pathologies such as tardive dyskinesia, glaucoma, neuropathic pain, or cocaine addiction. As recently demonstrated by Murthy and colleagues in hippocampal cultures, ultrastructural correlates of increased synaptic strength at chronically silenced synapses affect the size of active zones, boutons, and postsynaptic densities, as well as the total and docked number of vesicles. So, synaptic inactivity after functional or physical deafferentation would induce a compensatory increase in the efficacy of previously inactive or silent synapses, which in turn could partially contribute to functional restoration.

Partial deafferentation frequently disrupts the existing balance between excitation and inhibition, leading to an abnormal discharge activity that occasionally exceeds the functional limits. Experiments carried out on cultured neurons and embryonic and neonatal preparations demonstrate that homeostatic mechanisms could govern the electrical behavior of neuronal networks re-establishing normal firing levels after synaptic decompensation. This idea was initially conceived by O'Donovan and colleagues from studies with developing chick spinal cord. They showed that the blockade of excitatory activity for several hours is counteracted by an increment in the efficacy of different systems of excitatory neurotransmitters that would mediate the re-establishment of spontaneous activity. Similar results have been obtained in cultured hypothalamic neurons after disbalancing excitatory and inhibitory synaptic influences by the chronic blockade of ionotropic glutamate receptors. In this case, excitatory depression is compensated through an increase in cholinergic neurotransmission and in the number of acetylcholine receptors. Thus, compensation of synaptic strength would oppose the changes in firing activity induced by the loss of selective inputs.

Not only synaptic strength and balance are homeostatically regulated. The electrical properties of neurons are also subject to these control mechanisms that act by modifying neuronal excitability. Desai and colleagues showed that tetrodotoxin treatment for two days induces an increased sensitivity to current injection by selectively regulating voltage-dependent conductances in cultured cortical neurons. Therefore, neurons regulate their intrinsic excitability and synaptic composition to promote firing stability in response to activity changes, such as those occurring during synaptogenesis or after deafferentation. The question now is whether similar homeostatic mechanisms of plasticity are expressed in the adult brain in response to injury. Various experimental manipulations are furnishing proof suggesting the prevalence of homeostatic control mechanisms in the mature nervous system that are activated after lesion to restore normal functioning. For example, the postural adjustments and long-term recovery that follow ablation of the vestibular apparatus are mediated by compensatory mechanisms. Deafferentation of one labyrinth in humans and other mammals results in a syndrome characterized by vertigo, nausea, vomiting, spontaneous nystagmus, and deficits of the gaze-stabilizing and postural reflexes. Within the post-injury period of two to three days, some of these oculomotor and postural symptoms disappear in the function-recovery process known as vestibular compensation. During this process, intrinsic membrane properties and the relative sensitivity to inhibitory and excitatory inputs are modified, contributing to restore the spontaneous discharge that vestibular neurons exhibit in normal situations (Fig. 2).

A. Control

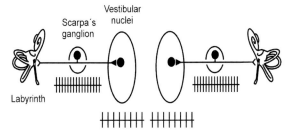

B. Labyrinthectomy

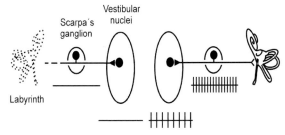

C. Compensation

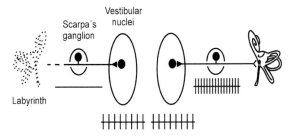

Figure 2. Homeostatic plasticity underlying vestibular compensation after labyrinthectomy. (A) At rest, in absence of head movements, primary vestibular afferents of both sides show a spontaneous discharge activity, which in turn makes vestibular neurons fire. (B) Unilateral labyrinthectomy causes deafferentation of vestibular neurons, which cease firing activity since they become deprived of their major excitatory input. Imbalance between left and right vestibular activities induces deviation of head, eyes, and body to the lesioned side. (C) Behavioral symptoms subside after 48 hours. Homeostatic mechanisms upregulating membrane excitability and modifying the relative sensitivity to inhibitory and excitatory inputs lead to the reestablishment of normal discharge activity in vestibular neurons. Adapted from http://www.umh.ac.be/~neuro/research.htm#COMP web page of the Laboratory of Neurosciences with permission of Prof. Emile Godaux.

Thus, medial vestibular neurons lacking their main excitatory input rapidly up-regulate their intrinsic membrane excitability and down-regulate their responsiveness to GABA agonists, counteracting the disfacilitation and excessive commissural inhibition. Therefore, firing activity destabilization following synaptic loss in the adult brain activates regulatory systems, which change the number, composition, and strength of synapses, as well as the neuronal excitability, which would compensate deafferentation-induced alterations and partially restore normal discharge activity.

There is not sufficient evidence about how the regulatory and activity-dependent systems of homeostatic control operate, and what signaling mechanisms are implicated in this form of plasticity. Recent evidence obtained by Turrigiano's group proposes BDNF as one molecule involved in the homeostatic regulation of synaptic strength and in the maintenance of the balance of excitation and inhibition in cultured cortical neurons. They showed that activity blockade decreased GABA-mediated inhibition of pyramidal neurons and raised pyramidal neuron firing rate, while these effects were prevented by incubation with BDNF. Thus, firing activity regulates synaptic inhibition through postsynaptic expression and release of BDNF. Additionally, in a different experiment they demonstrated that activity-dependent regulation of intrinsic excitability after tetrodotoxin application similarly depends on BDNF. Therefore, BDNF may exemplify a signal controlling a coordinated homeostatic regulation of synaptic and intrinsic properties. However, information is lacking about the role of neurotrophins in homeostatic plasticity mechanisms prevailing at mature stages. Investigations in this field could open up new therapeutic possibilities not only to restore circuit activity after disconnection induced by traumatic lesions or degeneration, but also to develop specific treatments against neuropathologic disorders characterized by anomalous neuronal activities.

4. THE MODIFIABLE ENVIRONMENT. FUNCTIONAL REORGANIZATION AFTER PERMANENT DESTRUCTION OF NEURAL PATHWAYS

Following the permanent destruction of neuronal pathways, variable degrees of functional recovery are frequently observed. This could be difficult to understand from the viewpoint of "localization of function" — how can a function recover after its specific brain region has been permanently damaged? Relevant work in this field supports the concept that significant *functional reorganization* can occur in the non-damaged nervous system, leading occasionally to partial recovery of the original function. Mechanisms underlying such processes would be different from additional substitutory or "learned" compensatory strategies to overcome a functional deficit. Dramatic network rearrangements associated with functional reorganization have been reported in the cortex following central lesions. Thus, by recording the receptive fields of neurons in the somatosensory cortex after focal lesions in area 3b in monkeys, Jenkins and Merzenich showed the emergence of hand skin areas previously encoded by the damaged region on adjacent cortical regions. Later, they demonstrated that these changes of cortical representation were correlated with the reacquisition of sensorimotor skills. Similar results were obtained in the motor cortex of adult primates after inducing focal ischemic infarcts. Reorganization of the undamaged motor cortex and functional recovery require that animals receive rehabilitative training. In humans, there is also evidence for cortical reorganization of maps. Positron emission tomography (PET) studies in patients with selective infarct lesions of the internal capsule demonstrated that recovery of hand movement is associated with activation of the face area. Thus, plasticity mechanisms in the brain prompt functional reorganization of areas in the CNS, taking over the functions previously performed by damaged regions. Lesions can also indirectly affect distal or proximal structures that would be deprived of afferent inflows from the injured area, a phenomenon termed *diaschisis*. However, recovery from these specific alterations usually occurs because of a spontaneous reduction in the metabolic depression in these areas. For example, memory impaired following septal infarction is usually recovered. PET studies have demonstrated that this transient memory loss is related to decreased activity in the hippocampus, which receives septal afferences via the fornix. Nevertheless, most of the recovery after the first two weeks is likely due not to short-term secondary effects of the injury, but rather to plasticity mechanisms involving functional reorganization of non-damaged areas. Proposed mechanisms implicated in the functional reorganization and recovery after CNS lesions are diverse and include: a) redundancy of parallel pathways mediating the same function, providing the CNS with a "safety factor"; b) the unmasking of vicarious functions, since the undamaged and initially inactive or silent pathway may have the latent capacity to replace the lost function in part; and c) the sprouting of fibers from the surviving neurons with formation of new synapses. Changes induced by the growth of new connections are extendable to larger brain regions over time. For example, as reported by Fouad and colleagues, four weeks after transection of the corticospinal tract in the lower thoracic spinal cord, the cortical motor representation of hind-limb muscles changed, since stimulation of this field yielded fore-limb, whisker, and trunk responses. This shift in cortical motor representation does not imply functional recovery, but it could represent a compensatory response improving motor control on regions just above lesion.

Peripheral deafferentation or sensory deprivation also induces functional reorganization of central sensory and motor systems, sometimes involving structural and physiological changes at several cortical and subcortical levels. In such cases, functional re-establishment is often incomplete, since deafferentation is irreversible, as in limb amputations or blindness. Nevertheless, although the original function is lost, adaptive changes could compensate associated sensory and motor deficits, enhancing perceptions of other sensory modalities or improving control of specific muscle groups.

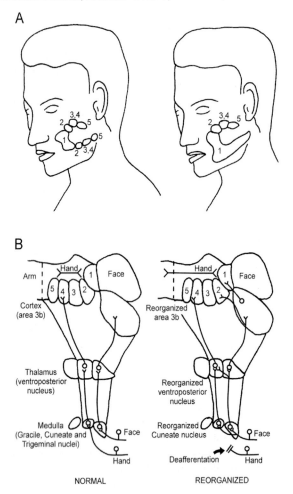

Figure. 3. Functional and structural reorganization of neuronal pathways as substrate for phantom limb sensations. (A) Topographic distribution of the reference fields relative to the thumb (1) and digits (2, 3, 4 and 5) on the face of the same arm-amputated patient mapped on two different occasions with a time interval of 6 months (left and right representations, respectively). Note that receptive fields were not stable, but were modified probably in relation to changes in the afferent activity. (B) Schematic somatosensory representations of the hand and face in normal (left) and reorganized CNS of owl monkeys following arm amputation (right). In the control situation, the extension of the peripheral representation in the CNS is limited by the segregate distribution of hand and face inputs. However, if inputs from the hand are lost, afferents from the face sprout and grow into the hand regions in the cuneate nucleus, thalamus, and cortex. Therefore, under this situation the previously deafferented hand cortex becomes activated by both, face and intact arm inputs. Panel A adapted from Ramachandran and Hirstein (1998) by permission of Oxford University Press. Panel B adapted from Kaas et al. (1999) by permission of Elsevier Science.

Rapid and reversible changes may be induced in the cortical representation after transient peripheral deafferentation by epidural block. Neurons previously responsive to stimulation of the anesthetized area become activated by stimulation of adjacent and non-anesthetized areas. Larger and longer-lasting sensory deprivation induces more-extensive changes in the topography of cortical maps. This was demonstrated by Pons and colleagues studying the effects of long-lasting sensory deafferentation of the arm on the somatotopic cortical organization in the monkey. They observed that sensory deafferentation induced elimination of hand, wrist and arm representations in area 3b, which now was responsive to inputs from the ipsilateral face. Additionally, following amputation, the cortical map of the stump is expanded on originally limb-responsive areas. Therefore, sensory deprivation causes reduction in the cortical area representing the lost input, and enlarging of neighboring spared representations, which indicates that the cortical maps are not fixed, but show use-dependent plasticity.

Although most of the studies have been focused on functional and structural changes at cortical level, reorganizations at other subcortical levels sometimes explain, large-scale cortical remodeling observed after long-standing deafferentation. Since subcortical projections to cortex are divergent, and representations are smaller, limited subcortical changes may lead to extensive cortical reorganization. Therefore, reorganization following peripheral deafferentation can occur at multiple levels, including the cortex, thalamus, brainstem, spinal cord and peripheral nerves. For example, Jain and colleagues showed that, in adult monkeys, the face afferents from the trigeminal nucleus sprout and grow into the cuneate nucleus following lesions of the dorsal columns of the spinal cord or after arm amputations. This sprouting may contribute to the extensive expansion of the face representation into the hand region of the somatosensory cortex that follows such deafferentation.

Similar reorganizations have been observed in humans suffering arm amputation. It is difficult to think that representational cortical reorganizations could partially contribute to restitute or compensate the lost function. Even though increased somatosensory representation of the stump may increase perception and discrimination in the proximal areas, no clear functional benefits could be imagined when hand and face are represented in the same cortical region. Thus, if the face is touched, a patient suffering arm amputation will feel the double sensation of being touched both on the face and on the lost hand (Fig. 3). This phenomenon is commonly referred to as "phantom limb", and in some cases as "phantom pain". Phantom limb and pain have been correlated with plasticity in the somatosensory cortex. Therefore, plasticity changes following injury are not always related to functional adaptations, but rather they may lead, in some situations, to perceptual errors or harmful sensations. At this point, an interesting question arises: can these large-scale functional reorganizations be reverted? Recent experiments by Giraux and colleagues have ratified such possibility. They investigated the dynamics of cortical reorganization in a patient's motor cortex before and after bilateral hand transplantation, observing that amputation-induced cortical reorganization was reverted after transplant. Furthermore, as reported by Florence and colleagues in monkeys, original receptive field organization is restored more precisely if appropriate retraining is supplied.

Dynamic changes of brain sensory representations have been demonstrated for the different sensory systems. However, it has been found that functional reorganization of sensory processing may take place not only within individual sensory systems but also across different sensory modalities. Contrasting with the prevailing view that the brain loses its *cross-modal plasticity* early during development, it has been demonstrated in animal and human studies that cross-modal plastic changes occur even in the adult brain. Thus, following complete sensory loss (e.g., blindness or deafness), the brain area deprived of its normal input may become responsive to stimuli presented via another sensory modality. Brain mapping and electrophysiological techniques have made it possible to study such plastic changes in sensory functions in blind and deaf humans and animals. For example, Newton and colleagues recently observed that individual neurons recorded in the primary visual cortex become responsive to tactile stimulation after monocular enucleation in the adult rabbit. These observations are in agreement with PET studies of early-blinded adult humans demonstrating that the visual cortex is activated along with the somatosensory cortex during Braille reading. Transient interference of the occipital cortex with trans-cranial magnetic stimulation during Braille reading induces errors and distorts tactile perceptions. In addition, Braille reading has been found to activate both striate and extra-striate cortices of late-blinded subjects. Cross-modal processing has also been demonstrated in auditory cortex, which is activated by visual stimulation in early-deafened subjects. These results indicate that removal of one sensory modality leads to neural reorganization in the deprived cortex, which may be recruited into a role in the processing of a different sensory modality. Even if this form of brain plasticity cannot compensate for the loss of a sensory modality, it could serve to enhance perceptions from other sensory modalities, additionally allowing the brain to utilize deafferented sensory cortex that otherwise would remain disabled.

5. PROMOTION OF FUNCTIONAL RECOVERY. FUTURE OUTLOOK

Unlike most tissues, the adult brain lacks regenerative capacity, making it especially vulnerable to injury. Several factors contribute to regenerative restriction in the CNS. Perhaps the most important of these are axonal regrowth limitations and inability to generate new neurons that substitute for the injured or dead cells. They represent important obstacles to functional restoration, and explain the devastating permanence of symptoms in brain-damaged patients. For that reason, repair of the CNS following injury is a long-standing and vitally important area of neuroscience. Most treatments have hitherto been aimed at relieving these symptoms and limiting further damage. However, in the last 20 years, research advances in this field obtained from animal studies have yielded new potential therapies and restorative treatments to promote recovery after brain damage. Some investigations pursue the promotion of axonal regrowth by overcoming the inhibitory environmental influences of the CNS.

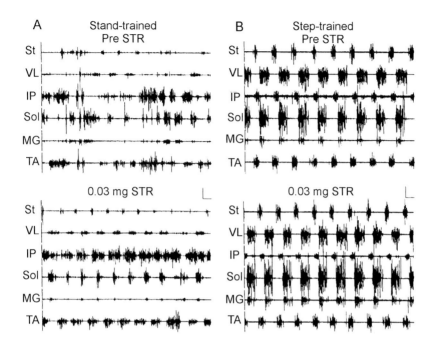

Figure 4. Effects of training on locomotor recovery in spinal cats. EMG activity recorded from the soleus and tibialis anterior muscles during treadmill stepping in spinalized cats retrained for 12 weeks to stand (A) or to perform stepping (B). Disorganized EMG burst patterns initially observed in the stand-trained cat during treadmill test changed to a more consistent and improved stepping following spinal disinhibition with strychnine (STR). Step-trained cats performed organized stepping, which was not significantly modified by strychnine. Upregulated spinal inhibition following spinalization was reduced by training. Taken from de Leon et al. (1999) with permission of the American Physiological Society.

Others have centered on the possibility of substituting the injured or dead neurons by tissue implantation, which will probably not re-establish the original connection patterns, but can contribute to functional recovery. Finally, numerous approaches have focused on the facilitation of functional compensatory strategies through rehabilitative training or pharmacological treatments to compensate neurochemical deficits following brain damage.

A) Pharmacotherapy plus brain retraining: Experimental work over the past decades indicates that pharmacological intervention to enhance functional recovery may be possible in some cases following irreversible CNS injury in humans, especially when used in conjunction with behavioral and physical rehabilitation. In this field, treatments have attempted to replace neurotransmitters or neuromodulators lost in the disease process. An example is L-dopa, which is the biological precursor of dopamine. This drug has been used for many years as the only effective treatment to relieve Parkinsonian symptoms. However, L-dopa becomes ineffective over time, and its prolonged use may promote the development of side-effects, such as cardiovascular and sleep problems or depression. Likewise, some of the symptoms of Alzheimer's disease can be treated with drugs that enhance the cholinergic function, since although most of the major neurotransmitter systems have been implicated in its etiology, the loss of forebrain cholinergic function is the most dramatic and consistent neurobiological abnormality.

From previously described experimental studies in spinal animals, it has been known for several years that patients with spinal cord injury benefit from specific locomotor training. In the absence of supraspinal input, the adult cat lumbar spinal cord can "learn" to generate efferent motor patterns that enable independent stepping. Unlike cats, adult spinal rats do not spontaneously develop consistent stepping, but this can be facilitated by training. Locomotor recovery does not involve the formation of new pathways, rather the functional reorganization in the existing neural networks. The cat spinal cord can more readily reacquire the motor ability for stepping, lost following spinal cord transection with repetitive exposition to step and stand (Fig. 4). Therefore, adaptative modifications in the spinal networks are driven by sensory signals from moving legs. De Leon and colleagues have provided evidence supporting the observation that, following spinal cord transection in the cat, inhibitory influences on the spinal neural networks are upregulated. They demonstrated that locomotor training has a facilitatory effect on the neural pathways controlling stepping, since it reduces spinal inhibition. Furthermore, in congruence with this idea, pharmacological reduction of spinal inhibition with strychnine or bicuculline facilitates stepping in non-trained cats (Fig. 4). Pearson suggested that, in addition to disinhibition, the enhancement of reflex function following supraspinal input loss may contribute in the training of the central pattern-generating networks to modulate coordinated locomotor activity. Pharmacological agents that increase the activity in locomotor-generating pathways and that substitute neurotransmitter-deprived systems may overcome upregulated inhibition, and so induce locomotor step-like activity. For example, noradrenaline-precursors (such as L-dopa) or agonists (such as clonidine) and serotonin agonists, improve locomotor activity in spinal cats, and recovery is even more complete when pharmacological application is combined with training.

Similar automatic step-like movements are present at birth and in anencephalic children. The oscillating neuronal circuit responsible for this task is dependent on peripheral sensory feedback. Recent evidence suggests that at least some of the circuits necessary to generate step-like motor activity have been conserved in primates. Thus, the human spinal cord can modulate by itself motor activities which facilitate locomotion, when appropriate sensory information is given. Rehabilitative treatment in patients with chronic spinal cord injury relies on either intense physical therapy using a treadmill or electrical stimulation of leg muscles. It has been demonstrated that locomotor function in incomplete paraplegic patients improves by specific training on a treadmill with partial body-weight support, and even in patients with complete loss of supraspinal inputs, a locomotor pattern can be induced by training.

B) Faciliting axonal re-growth: If injured pathways are located peripherally, considerably recovery may occur due to axonal re-growth and target re-connection capacities. CNS axons, however, do not regenerate following their interruption. Intrinsic differences between peripheral and central fibers in regenerating can be observed in dorsal root ganglion neurons (DGF), which project one axon to the periphery and the other to the spinal cord and brain. Whereas the peripheral axon regenerates following injury, this does not occur in the axon entering in the spinal cord. However, a conditioning injury firstly affecting the peripheral fiber facilitates subsequent central fiber regeneration. As reported recently by Qiu and colleagues, the facilitatory effect to regenerate central axons following peripheral injury in the same neuron is mediated by elevation of AMPc levels, since intraganglionar injection of AMPc analogs mimic this effect.

Although peripheral fibers can regrow and establish new functional synaptic contacts, reinnervation is frequently inappropriate, and does not contribute to functional restoration. In this respect, intensive research has been aimed at promoting and accelerating axonal re-growth and increasing the specificity for proper target reinnervation. Therefore, to obtain a significant degree of regeneration, a number of molecular and cellular responses must be overcome, including 1) injury-induced retrograde degeneration and nerve cell death; 2) the formation of scar tissue and the expression of chemorepulsive proteins in the lesioned neural tissue blocking axonal growth; and 3) inappropriate target reinnervation. Neurotrophic factor support not only keeps the cells from dying and prevents afferent synaptic depression following axotomy, but also promote sprouting and regeneration of severed axons *in vivo*. Thus, the rate of regeneration of axotomized peripheral and motor axons may be accelerated by targeted expression of a multifunctional chimeric neurotrophin. However, although peripheral nerve axons regenerate faster under trophic supply, sprouting is also promoted, and consequently the degree of axonal branching and number of supernumerary axons. This can lead to simultaneous projection along several different fascicles, and erroneous reinnervation by single axons. Recent demonstrations by Streppel and colleagues support this idea, since focal application of neutralizing antibodies to BDNF and NGF reduces the level of axonal branching. Alone or combined with additional treatments, neurotrophic administration can also promote or amplify axonal growth in the CNS. As reported by Goldberg and colleagues, axonal elongation in cultured retinal ganglion cells is greatly potentiated by combined trophic signaling and electrical activity. Likewise, axotomized adult corticospinal and rubrospinal axons regenerate following graft transplant of genetically modified fibroblasts to express NT-3 or BDNF, respectively, associated to sensorimotor recovery in both situations. These results demonstrate that besides scar and myelin inhibition at the injury site, the lack of trophic support in axotomized central neurons contributes decisively to regenerative failure.

While the potential to increase the intrinsic capacity of CNS neurons to re-activate cellular programs supporting regeneration is crucial, it is also evident that a more successful regeneration requires modification of the environment at the site of injury to overcome inhibitory glial influences. Adult CNS fails to provide a stimulating environment for axonal regrowth. Studies by Schwab and colleagues have shown that oligodendrocytes inhibit the growth of regenerating axons, and that these CNS glial cells express myelin-associated proteins (NI-35 and NI-250), which cause growth cone collapse. Regeneration of corticospinal axons is significantly improved when neutralizing antibodies to these proteins are present. Myelin-associated glycoprotein is also a potent inhibitor of axonal regeneration. Astrocytes present at the lesion site have a critical role in inhibiting axonal regrowth, not only by the physical barrier they constitute, but also because they express chondroitin sulfate-bearing proteoglycans, which are deposited within the extracellular matrix acting as inhibitory molecules. Procedures that abolish inhibitory influences of proteoglycans, such as chondroitinase treatment, render the environment of the damaged CNS more permissive to axon regeneration. All these non-permissive molecules are absent in the adult peripheral nervous system. Likewise, during development these signals cannot be inhibitory to growing axons, since some degree of axonal regrowth is possible; the capacity decreases with age. Two undemonstrated hypothesis could explain this process: either the lack of receptors for these inhibitory molecules in embryonic CNS neurons, or the later myelin development and maturation of most oligodendrocytes relative to the establishment of neuronal connections. However, as shown by Davies and colleagues, both embryonic and adult axons may regenerate in the adult CNS white matter when transplanted with minimal tissue damage. They also demonstrated that in response to injury, proteoglycans produced by astrocytes accumulate in the extracellular matrix. Thus, it is possible that normally inhibitory oligodendrocytic and astrocytic molecules are not externally located on the myelin and external matrix, respectively, but they could be externally exposed after tissue disruption. This could explain the observed plasticity mechanisms and large-scale structural and functional reorganizations implying axonal growth over long distances in the intact adult CNS following peripheral deafferentation. However, when the adult CNS is directly injured, glial inhibition of axonal growth becomes evident. Since peripheral Schwann cells constitute a more permissive environment to regrowth, multiple experimental repair strategies have used segments of peripheral nerve as "bridges" grafted into central lesioned pathways to promote regeneration and act as a guide towards the target cells. In this case axons have to be bridged up to their terminal fields in order to contact with original targets, since otherwise, axonal elongation is arrested at the end of the graft at the PNS/CNS interface. Interesting results have also been obtained transplanting olfactory ensheathing cells at the injury site, as promoters of axonal regeneration. Compared with Schwann cells, these glial cells integrate better within the host CNS, migrate along the white matter tracts of the host, and can also form myelin. Altogether, it seems that regenerative strategies will probably require both removal of glial inhibition and stimulation of intrinsic regrowth ability to achieve greater effectiveness.

C) Replacing cells: Significant neuronal and glial loss frequently takes place in the course of neurodegenerative diseases, as well as in stroke and trauma. Cowdry, in 1952, classified neurons as "postmitotic cells" since they do not retain the capacity to proliferate, and no new elements can be generated during adult life.

Nevertheless, neurogenesis has been observed in several regions of the adult primate and human brain, including the subventricular zone in the wall of the lateral ventricle and the subgranular zone of the dentate gyrus, where there is a continuous turnover of interneurons and granule cells. Various manipulations, as well as environmental enrichment, exercise or selective lesions, increase neurogenesis in these areas, suggesting that newly generated neurons may have a function in plastic processes such as learning, memory, or recovery from injury. It has been reported that in the neocortex of the adult mouse there is also a continuous formation of new cells. These cells express neuronal markers, acquire characteristic morphology, and extend processes to the original target sites following selective apoptosis induction in a subset of pyramidal neurons. Therefore, newly formed neurons could be implicated in the structural and functional remodeling of damaged circuits. Evidence for this hypothesis has come from experiments performed recently by van Praag and colleagues. They demonstrated that newly formed cells in the dentate gyrus of the adult mouse hyppocampus mature into functional neurons displaying morphological, electrophysiological, and synaptic characteristics similar to those found in mature dentate granule cells. Despite these findings, the presence of neurogenesis in only these restricted areas limits the potential of plasticity and functional recovery following larger lesions, or those affecting other CNS areas. However, such endogenous repair mechanism has opened up new prospects towards future therapeutic interventions, since it may represent a latent capacity for functional recovery if precursor cells could be mobilized and functionally differentiated into appropriate neuronal types in response to injury.

Neural transplantation is a relatively new field and has attracted interest in recent years for its potential as replacement therapy. In the 1970s, it was realized from the works of Das and colleagues and Björklund and colleagues that transplantation was more viable if fetal tissue was used. Moreover, the brain is a favorable site for transplantation because the immune response to grafts is relatively mild. Various experiments using different animal models and cell-based therapies for brain damage and neurological disorders have been performed. McDonald and colleagues have demonstrated that neural embryonic mouse stem cells transplanted into lesioned rat spinal cord survive and are able to differentiate into specific neuronal or glial populations. They also integrate, form functional synaptic contacts, reconstitute neural circuitry, and promote functional recovery. A fast development towards human application in many diseases has followed relevant transplant assays in animal models. For example, after observations in substantia-nigra-lesioned rodents and non-human primates that grafting dopamine-containing fetal cells into the striatum reverses the motor disturbances, human trials in Parkinson's patients were initiated, transplanting human fetal dopaminergic neurons. Although not all the patients improved, in the best cases transplantation had positive and lasting effects. Autopsies and PET scanning revealed both survival of grafted cells and increased dopamine function in the striatum. However, neurotransplantation approaches are not restricted to fetal neural tissues, since the implantation of non-neuronal adult or cultured cells, which provide trophic support and/or functionally substitute dead neurons, has been found to promote functional recovery in animal models of neurodegenerative diseases. Thus, autologous striatal implants of dopaminergic-releasing cells, obtained from the carotid bodies or adrenal tissue, produce significant improvements in experimental models of Parkinson's disease and in human patients. Neurotransplantation has been employed in humans as potential therapy for numerous other brain diseases or disorders, such as Alzheimer's and Huntington's diseases or schizophrenia, as well as for stroke or infantile cerebral palsy. In some cases, it has constituted a successful treatment, or at least provided moderate and definite improvements.

Neural stem cells are "omnipotent" cells with the ability to proliferate and develop into any sort of brain cell and are the current focus of intensive research, which promises tangible applications. Encouraging progress is being made in characterizing the signals, such as growth factors and culture conditions, that selectively promote differentiation of stem cells into specific glial or neuronal cell types. In vitro manipulation of stem cell fate, prior to transplantation and modification of the host environment, may help to control the terminal lineage of the transplanted cells, and to obtain a functionally higher significant number of neurons. However, successful functional reversion from brain damage will probably need multiple therapeutic interventions. Prevention of secondary greater damage, promotion of axonal regeneration modifying both intrinsic and extrinsic environmental factors, rehabilitative training to enhance residual functions, and cell loss replacement, could all contribute to more-complete functional recovery.

REFERENCES

Bavelier D. and Neville H. J. (2002) Cross-modal plasticity: where and how? *Nat. Rev. Neurosci.* 3, 443-452.
Björklund A. and Lindvall (2000) Self-repair in the brain. *Nature* 405, 892-895.
Bregman B. S. (1998) Regeneration in the spinal cord. *Curr. Opin. Neurobiol.* 8, 800-807.
Chen R., Cohen L. G. and Hallett M. (2002) Nervous system reorganization following injury. *Neuroscience* 111, 761-773.
Cotman C. W., Nieto-Sampedro M. and Harris E. W. (1981) Synapse replacement in the nervous system of adult vertebrates. *Physiol. Rev.* 61, 684-784.
Davis G. W. and Bezprozvanny Y. (2001) Maintaining the stability of neural function: a homeostatic hypothesis. *Annu. Rev. Physiol.* 63, 847-869.
de la Cruz R. R., Pastor A. M. and Delgado-García J. M. (1996) Influence of the postsynaptic target on the functional properties of neurons in the adult mammalian central nervous system. *Rev. Neurosci.* 7, 115-149.
Edgerton V. R, Leon R. D., Harkema S. J., Hodgson J. A., London N., Reinkensmeyer D. J., Roy R. R., Talmadge R. J., Tillakaratne N. J., Timoszyk W. and Tobin A. (2001) Retraining the injured spinal cord. *J. Physiol.* 533, 15-22.
Fitzsimonds R. M. and Poo M. M. (1998) Retrograde signaling in the development and modification of synapses. *Physiol. Rev.* 78, 143-170.
Geuna S., Borrione P., Fornaro M. and Giacobini-Robecchi M. G. (2001). Adult stem cells and neurogenesis: historical roots and state of the art. *Anat. Rec.* 265, 132-141.
Goldberg J. L. and Barres B. A. (2000) The relationship between neuronal survival and regeneration. *Annu. Rev. Neurosci.* 23, 579-612.
Goldberger M. E. (1980) Motor recovery after lesions. *TINS.* 3, 288-1291.
Hamilton R. H. and Pascual-Leone A. (1998) Cortical plasticity associated with Braille learning. *TICS* 2, 168-174.
Jones L.L., Oudega M., Bunge M. B. and Tuszynski M. H. (2001) Neurotrophic factors, cellular bridges and gene therapy for spinal cord injury. *J. Physiol.* 533, 83-89.

Kaas J. H., Florence S. L. and Jain N. (1999) Subcortical contributions to massive cortical reorganizations. *Neuron* 22, 657-660.

Mendell L. M. (1984) Modifiability of spinal synapses. *Physiol. Rev.* 64, 260-324.

Raisman G. (2001) Olfactory ensheathing cells - another miracle cure for spinal cord injury?. *Nature Rev. Neurosci.* 2, 369-375.

Ramachandran V. S and Hirstein W. (1998). The perception of phantom limbs. The D. O. Hebb lecture. *Brain* 121, 1603-1630.

Titmus M. J. and Faber D. S. (1990) Axotomy-induced alterations in the electrophysiological characteristics of neurons. *Prog. Neurobiol.* 35, 1-51.

Turrigiano G. G. (1999) Homeostatic plasticity in neural networks: The more things change, the more they stay the same. *TINS* 22, 221-227.

M. NIETO-SAMPEDRO

Institute Cajal, CSIC, Madrid, Spain and Unit of Experimental Neurology, SESCAM, Hospital Nacional de Paraplejicos, Toledo, Spain

20. NEURAL PLASTICITY AND CELL BIOLOGY OF LEARNING

Summary. Changes in the number, type, and function of nervous system connections, in the morphology and function of glia, and in neuron-glia interactions, are at the center of vertebrate adjustments to changing environmental and physiological conditions. Collected together under the term "neural plasticity", these changes underlie adaptations as apparently different as learning, the physiological response to dehydration, or injury repair. In this chapter, the concepts of synapse renewal and reactive synaptogenesis are revised, as well as the cellular and molecular steps involved in synapsis renewal. A brief review is also made regarding the present understanding of long-term potentiation, and its relationship with synaptic structural changes.

1. INTRODUCTION

The term *plasticity* was introduced in 1890 by William James to describe the susceptibility of human behavior to modification. It was also used by Marinesco (1907) and Minea (1909), in their work on transplantation of sensory ganglia, to describe the changes in morphology undergone by the transplanted neurons. Cajal was aware of their work, and believed that the behavioral modifiability reported by James must have an anatomical basis, as reported by Marinesco and Minea. However, after Cajal's death, researchers adopted a rigid view of the adult central nervous system (CNS), assuming that once development had finished, CNS anatomy was unchangeable, except for degenerative processes. Against this general view, Liu and Chambers showed in 1958 that axon sprouting occurred in the adult CNS, and overwhelming evidence has accumulated in the last three decades confirming their finding. The nervous system maintains the capacity of functional and anatomical modification throughout life. The neuronal networks that make up the nervous system of mammals remain plastic —modifiable — during their whole life, and plasticity is one of their main adaptations. *Plasticity,* denominated *neural* to remind us that it refers to both neurons and glia, is today considered a fact, and the questions now concern its celular and molecular basis. The stimuli that induce neural plasticity are experiences, environmental pressures, physiological modifications, or lesions of all kinds. I propose that the cellular and molecular processes that underlie neural plasticity are the same, regardless of the precise activity involved. This article presents a general view of plasticity, its origin, function, mechanisms, and possible clinical manipulation. Assuming that neural plasticity mechanisms apply to the whole CNS, then I propose that we use the information on the physiological principles that apply to learning, an activity mammals are quite good at — to enhance the very limited mammalian CNS lesion repair.

2. NEURONS AND GLIA: A UNIT OF FUNCTION

The main cell types in nervous tissue are neurons and glial cells. Neurons are cells highly specialized for the rapid receipt and transmission of messages. They have a relatively small body and multiple ramifications that cover an extensive surface, allowing them to maximize intercommunication. The human cerebrum contains more than ten thousand million neurons, while the cerebellum, between ten and one hundred thousand million. The *synapses*, or 'synaptic contacts', are the sites where a neuron transmits its message to another neuron. A typical CNS neuron frequently receives tens of thousands of synaptic contacts, although cerebellar Purkinje cells may receive up to 200,000. Connections between neurons give rise to neuronal circuits and neural plasticity is, to a large extent, synaptic plasticity — that is, the susceptibility to modification of the type, form, number, and function of the synapses, and hence, of the neuronal circuits. Processes as diverse as learning and memory, the response to physiological situations such as pregnancy or thirst, and the response to lesions, have synaptic plasticity and neural plasticity as a common basis.

Today's consensus is that nervous tissue function may be understood only taking into account the other cells characteristic of this tissue: the glial cells. Their number exceeds that of neurons about 10-fold, and they constitute about half the nervous tissue mass (Pope, 1978). The original description of glia by Virchow in 1859 as nervous glue, accorded glia a static image, maintained mainly by neuroanatomists and neuropathologists during the following 100 years. This view has changed notably during the last 25 years, and the neuronal-dominated viewpoint of nervous function has been widened to take in neural development and nervous activity, maintenance, and pathology, based on the neuron-glia as a unit of function. The idea of the neuron-glia as a dynamic unit of function has been proposed independently by various researchers in the last 20 years, but was formulated explicitly in detail by Arenander and de Vellis (1983) and, later by Nieto-Sampedro (1988). The glial types in the CNS are astroglia, oligodendroglia, and microglia, of which astroglia and macroglia are probably most-directly related to neural plasticity.

Astrocytes are intimately associated to both neurons and the whole organism. They envelope central synapses and form the *glia limitans*, the boundary between the CNS and other tissues, particularly blood vessels, with which their endfeet make intimate contact in high-conductance regions (Newman, 1986). Thereby, astrocytes monitor the blood content of nutrients, oxygen, vitamins, and hormones. They are sensitive to ions, especially potassium, and can bind, transport, and metabolize neurotransmitters. Astrocytes respond to excitatory neurotransmitters by depolarizing, and some may conduct action potentials (refs. in Arenander and de Vellis, 1983, and Nieto-Sampedro, 1988a). In addition, they all intercommunicate directly by "gap-junctions" and similar mixed junctions probably link them with neurons (Sontheimer, 1995). As found by Smith's team (Cornell-Bell et al., 1990), astrocytes also intercommunicate by means of Ca^{2+} 'waves', which affect intracellular Ca^{2+} concentration in the neurons in contact with them (Nedergaard, 1994). In addition, astrocytes synthesize glutamate (the most-abundant CNS excitatory neurotransmitter), store it, and liberate it in a non-vesicular manner (Parpura et al., 1994). They also synthesize NO, a highly diffusible neuromodulador (Garthwaite, J. 1991; Murphy et al., 1993) that strongly affects the physiology of both neurons and astrocytes. All these properties make astrocytes capable of continuously monitoring (detecting, receiving and interpreting) the activity of neurons and modifying it as necessary. Astrocytes function as feedback controllers of the neural environment, with pointers set to the normal neuronal activity. Any modification of the normal tissue composition triggers compensatory glial responses; for example, by eliminating excess neuroexcitatory molecules before they reach excitotoxic levels, or by producing neurotrophic factors that allow effective buffering of intracellular Ca^{2+}, avoiding neuronal apoptosis.

There is less information on the cell biology of microglia, although great progress has been made in the last decade, and they certainly play an essential role in communication of the nervous and immune systems (see Microglia, 1993).

To summarize, nervous tissue is made up of functional units consisting of societies of groups of neurons and glial cells. Glial cells are the controllers of the environment of these dynamic ensembles as regards ionic composition, neurotransmitter levels, and growth factor supply. The response of the nervous system to perturbations can be correctly understood only as the coordinated reaction of these cellular ensembles.

3. SYNAPSE RENEWAL AND NEURAL PLASTICITY

The plasticity of the nervous system (functional and anatomical modifiability) is denominated *neural* to remind us that it refers to both neurons and glia. This capacity is to a large extent the plasticity of the networks that make up the nervous system, which, during the whole lifetime of the organism, remain modifiable in response to all kinds of experiences. Such experiences may be environmental pressures, physiological modifications, or lesions. The essential elements of neural networks are the connections between their neuronal elements, the synapses, and much neural plasticity is synaptic plasticity. The maximum expression of synaptic plasticity is observed during developmental synaptogenesis, when synapses go through cycles of formation and regression. One of the most elegant exemples of synaptic plasticity was observed by Clark C. Speidel in the sensory endings of the living tadpole (Speidel, 1941). Speidel observed in the same sensory arborization resting terminals, growing terminals, and nerve endings in frank regression. Depending on time and environmental conditions, some resting terminals became growth cones and some growth cones became stable or degenerating terminals. In summary, what Speidel observed was that in the developing nervous system, synapses are dynamic structures.

The renewal or turnover of synapses is maintained in adult mammals in a manner that is more limited, yet sufficient for changes in type, form, and/or number to mediate physiological and behavioural adaptations. Renewal of a synapse population implies breaking a set of synaptic contacts and substituting them by new ones. These are population changes and individual synapses may disappear without being substituted, or a new synapse may form where there was none before. In general, the renewal process includes four stages: 1) disconnection of synapses; 2) initiation and growth of new axons; 3) formation of new synaptic contacts; and 4) maturation of the new synapses, i.e., appearance of synaptic vesicles and pre- and post-synaptic densities. In every one of these steps, glia may or must intervene actively.

4. SYNAPSE RENEWAL STEPS: NOMENCLATURE

The presynaptic terminals that take part in synapse turnover arise from pre-existing axons, a process called axon *sprouting*. Axon sprouts are named depending on their point of origin in the axon: sprouts called *terminal or ultraterminal* are extensions of the presynaptic terminal. The so-called *collateral* sprouts arise as a new axonal branch, independent of other pre-existing nerve endings. When a collateral sprout originates in the node of Ranvier of a myelinated axon, it is usually called *nodal*. If the sprout originates as a renewed continuation of a sectioned axon stump, it is denominated a *regenerative* sprout. Axon sprouting is independent of subsequent synapse formation. In fact, in the CNS sprouts frequently degenerate and never form synapses. The term axon sprout simply designates a type of growth response. It may or may not be the first step in the formation of a new synapse.

Synapse renewal is, in any case, at the origin of neural plasticity and may have evolutionary importance. Greater synapse renewal capacity and hence greater neural circuit plasticity, is an advantageous adaptation for a nervous system, and may facilitate its selection. The pituitary regulation of the state of hydration, parturition, or lactation described next, supplies clear examples of both neuron-glia unit function and the adaptive advantages that efficient synapse renewal gives mammals relative to other classes. The secretory axons of the neurohypophysis originate in the magnocellular neurons of the supraoptic nucleus of the hypothalamus and terminate in the spaces surrounding fenestrated capillaries, where they discharge their secretion products, hormonal peptides. The actions of two of these peptides, oxytocin and vasopressin, are well-characterized as controlling water retention and smooth muscle contraction. The neurons of the supraoptic nucleus of well-hydrated female rats that are neither pregnant nor lactating, are separated from one another by astrocytes. Similarly, their axonal endings are isolated from capillaries by pituicytes, a glial type surrounding these terminals. Water deprivation for four or more hours, lactation, advanced pregnancy stages or parturition, initiate the following cascade of events: (i) withdrawal of glial processes and, consequently, appearance of contacts between cell bodies and dendrites of neighboring neurons, leading to electrotonic coupling; (ii) the appearance of synaptic contacts between adjacent magnocellular neurons; (iii) the retraction of pituicytes, allowing axon terminal access to perivascular space; (iv) the substitution of the slow, irregular electrical activity of supraoptic neurons by continuous rapid rhythmic activity, with occasional high-frequency discharges; (v) the synthesis of proteins, particularly the hormones previously mentioned and their precursors. All these changes occur concomitantly with the appropriate physiological response (i.e., water retention in the kidneys or mammary pressure increase), are completely reversible, and allow the organism exhibiting them successful survival and reproduction in environments of variable humidity (Hatton, 1985).

It is difficult to demonstrate experimentally synapse renewal in the CNS of adult mammals which, unlike Speidel's tadpole tail, are not easily accessible to repeated microscopic observation *in vivo*. Turnover was inferred from the simultaneous observation, in the same animal, of structures that imply synapse renewal such as degenerating axons together with growth cones, or vacated postsynaptic densities. Because these are not time sequential observations, the evidence is only indirect. Electrophysiological recordings could help to detect synapse renewal in living animals. Unfortunately, the multiplicity of alternative explanations is such that clear conclusions can rarely be reached. Concomitant anatomical and electrophysiological studies, such as those described by Hatton (1985) or Tsukahara (1985), would be necessary.

5. EXPERIMENTAL SYNAPSE RENEWAL: REACTIVE SYNAPTOGENESIS

Because of the technical difficulties in demonstrating spontaneous CNS synapse renewal in mammals, research on its cellular and molecular mechanisms have used systems where renewal is initiated experimentally and the subsequent responses observed. The experimental stimulus that elicits the largest, most reproducible, and best-studied responses are lesions. Synapse formation evoked by lesions or other stimuli that are not part of the normal developmental program of the organism is called *reactive synaptogenesis.*

The most detailed studies of reactive synaptogenesis have been carried out in the dentate gyrus of the hippocampus (reviewed by Nieto-Sampedro and Cotman, 1985; Collazos-Castro and Nieto-Sampedro, 2002). The hippocampus is a brain cortex bilateral formation with two major types of neuron: pyramidal in the hippocampus proper and granule in the dentate gyrus. The most-abundant afferents extrinsic to the hippocampus originate in the entorhinal cortex; these fibers are excitatory and terminate on the outer approximately 2/3 of the dendrites of the dentate granular cells (the so-called molecular layer). Although the projection is bilateral, only a small number of afferents arrives from the contralateral cortex. The second most important of extrinsic afferents, are the cholinergic fibers from the septum, most of which terminate in a layer immediately above the somata of granule cells, though a few project with the entorhinal afferents. The inner third of the dendritic tree of granule cells (that close to the soma) is occupied by terminals from hippocampal pyramidal cells, i.e., intrinsic associational afferents. Granule-cell axons terminate in pyramidal neurons, closing the circuit. The layered distribution of all afferent terminals and the synapses they form is highly stereotyped, in both location and abundance, facilitating the analysis of the mechanisms regulating axonal growth and synapse formation after lesions.

Unilateral destruction of the entorhinal cortex (entorhinal cortex lesion, ECL) cause the loss of 90% of the synapses in the dentate gyrus outer molecular layer ipsilateral to the lesion. This massive deafferentation is followed by the restitution of lost synapses, leading to a reorganization of hippocampal circuits both ipsilateral and contralateral to the lesion. The laminar distribution of dentate gyrus afferents facilitates the analysis of axonal growth and synapse formation after lesions. Approximately 85% of the synapses lost after unilateral ECL are restored by sprouts from undamaged afferents. Homotypic and heterotypic systems contribute to synapse replacement (Cotman and Nadler, 1978; Steward, 1991; Deller and Frotscher, 1997), suggesting unspecific activation of presynaptic growth. The best-characterized response to dentate gyrus deafferentation is the homotypic sprouting of temporodentate axons after unilateral ECL. The cell bodies of the neurons giving rise to the sprouts are located mainly in layer II of the contralateral EC. Initial studies suggested that collaterals from these neurons grew a relatively long distance thorough the midline to reach the deafferented dentate gyrus. Although this possibility has not been completely ruled out, it is now accepted that sprouting axons are already present in the contralateral dentate gyrus at the time of deafferentation, growing only inside the OML (Steward, 1991; Deller and Frotscher, 1997). The number of synapses formed by these fibers increased 128-fold. Several types of presynaptic growth were observed here, including axonal branching, tangle formation, short axonal extension, terminal hypertrophy, and multiple synapse formation. Although the number of temporodentate synapses increased greatly, this tract was not the major contributor to the new synapses. There are probably intrinsic limits to the number of synapses that one neuron can form and/or strong competition with other afferents for postsynaptic sites. Transplantation of embryonic entorhinal cortex after unilateral ECL enhanced the reinnervation of the OML beyond the normal response, indicating the existence of limits to the number of synapses that an afferent may form, and that these limits do not apply to the postsynaptic neuron.

Reactive synaptogenesis begins 3 or 4 days postlesion, when the first axonal sprouts are detected, reaches maximum rate between 15 and 20 days and it does not conclude until two or three months later. The synapse renewal process is not restricted to the areas directly affected by the lesion: between 2 and 10 days after unilateral ECL 22% of the synapses in the inner molecular layer of the dentate gyrus undergo a renewal cycle (Hoff et al., 1981). A similar but much slower phenomenon is also observed in the opposite cerebral hemisphere. These hippocampal areas do not receive projections from the entorhinal cortex, and synapse turnover seems rather a compensatory response of dentate gyrus granule cells to the loss of afferents in the outer branches of their dendritic tree. The phenomenon extends trans-synaptically to the contacts made by the granule cell axons with hippocampal CA3 pyramidal neurons.

6. CELLULAR AND MOLECULAR STEPS OF SYNAPSE RENEWAL

Reactive synaptogenesis studies, performed in the hippocampus or similar systems, have established that the mechanisms of synapse renewal that operate in the adult are essentially the same as those that acted during development. The greatest difference seems to be that during development, the net number of synapses increases, whereas in the adult renewal is the predominant process.

6.1. Synapse disconnection

The first step in the process of synapse renewal is the disconnection of existing synapses. This step is a formal analog of synapse elimination observed during development. At least two disconnection processes occur in the adult: presynaptic terminal degeneration — a slow process whose intermediate stages can be observed microscopically — and a second, much faster (taking at most a few hours), reversible process that occurs without presynaptic degeneration. In the second process glial cells take part, interposing fine pseudopods between the pre- and postsynaptic elements. Synapse disconnection by the latter mechanism resembles the physiological control of hypothalamic hormone secretion (Hatton, 1985), or the loss of afferents by axotomized neurons (refs. in Cotman et al., 1981). The molecular details of these processes are not known, but intracellular calcium concentration controls depolymerization of microtubules and neurofilaments involved in spontaneous terminal degeneration. Because synaptic activity is associated to considerable variations in intraneuronal Ca^{2+} concentration, and the latter is related to Ca^{2+}-wave propagation by astrocytes, it is tempting to speculate that synaptic activity may control the half life of nerve endings.

6.2. Axon sprouting and growth factors

The process complementary to synapse disconnection is the formation of new neuronal contacts. This involves the growth of axons and/or dendrites and the differentiation of the structures characteristic of mature synapses. The formation of axon sprouts has two essential requirements: the presence of the specific molecules, collectively called *growth factors*, and the existence of an appropriate substrate for the adhesion and growth of the new fibers.

There are several kinds of growth factor: *neurotrophic* factors are substances without which neurons do not survive. Obviously, many substances meet this requirement, e.g., glucose or potassium ions. The distinctive property of neurotrophic factors is that they act at very low concentrations (of the order of 10^{-12} M), and they are usually polypeptides of relatively low molecular weight. There are various families of neurotrophic factors, each of them specific for a defined group of neurons. Thus, NGF (short for *nerve growth factor*) is the best-known member of the neurotrophins, a family of factors essential for the survival of sympathetic and sensory neurons. It was described by Levi-Montalcini and Hamburger in 1951 (reviewed by Levi-Montalcini, 1982) and presents neurotrophic, neuritogenic, and chemotactic activities for sympathetic neurons. There are specific neurotrophic factors for various neuronal types, for example, motoneurons. Until 1982 it was postulated that growth factors were molecules exclusive to the peripheral nervous system and found in the CNS only during embryonic development. However, in 1982 Crutcher and Collins and, independently, Nieto-Sampedro et al., presented convincing evidence that neurotrophic factors were present in the postnatal CNS, and Barde et al. (1982) purified a new trophic factor from porcine brain. Subsequently, Nieto-Sampedro et al. (1983) presented definite evidence that these factors are present in adult CNS, where they probably mediate synaptic plasticity during reactive synaptogenesis. These results were confirmed and extended, and afterwards, with the use of molecular neurobiology techniques, the field expanded explosively. New trophic factors were purified, cloned, sequenced, and grouped in families of similar structure and biological activity.

Neurotrophic factors do not initiate sprouting of neurites (axons or dendrites). They are permissive factors only, allowing neurons to be alive and available to receive instructions from other specific factors, responsible for defined modifications, both structural and functional. Three classes of instructive factors are important for adult synapse renewal: i) *neuritogenic* factors, which cause neurite differentiation; ii) chemotactic, chemotropic or directional factors, which direct the orientation of neurite growth; iii) factors that direct neurotransmitter choice, important for synapse maturation. Some growth factors exhibit more than one of these activities. Thus, NGF is neurotrophic for sympathetic neurons, for some sensory neurons and for CNS cholinergic neurons. For sympathetic neurons, NGF is also neuritogenic and chemotactic. Laminin, a high molecular weight basal membrane protein, is capable, alone or associated to a heparan-sulfate proteoglycan, of initiating neurite growth, both during development and after a lesion. Furthermore, in the case of sensory neurons, NGF associated to laminin has higher neurotrophic and neuritogenic activities than the factor alone. Perhaps the most significant advance in elucidating neuritogenesis has been to connect the mode of action of neuritogenic proteins and trophic factors. Walsh and his team found that intercellular adhesion proteins with neuritogenic activity act through a common domain with the tyrosine-kinase receptor of fibroblast growth factor (Williams et al., 1994; Doherty et al., 1995).

With regard to the mode of action of neurotrophic factors, some details remain to be elucidated, but the general molecular features are known (Russell, 1995). Binding of a factor to its receptor, a tyrosine-kinase, initiates a sequence that begins with the activation of the tyrosine-kinase, followed by a protein phosphorylation cascade that eventually affects Ca^{2+} homeostasis (Schulman, 1995). Intracellular Ca^{2+} levels affect multiple fundamental neural processes (Simpson et al., 1995) and constitute the meeting point of LTP (Malenka, 1995), neurite initiation, and growth and cytoskeletal polymerization and organization (Mattson, 1988; Hoffman, 1995). Physiological levels of neuronal activity induce the production of neurotrophic factors, and treatment with physiological levels of trophic factors increases synaptic efficiency. Many effects of trophic factors depend on the synergy of at least two of them.

The two major sources of growth factors are the postsynaptic cells and glial cells. Innervation and activity regulate the post-synaptic target's production of factors, which decay when innervation is complete and increases after partial or total denervation. This observation explains in part why axon sprouts grow only short distances in the CNS and why, when longer-distance growth is induced, terminals do not penetrate profoundly into the CNS tissue: after a lesion, local axon sprouts would quickly repopulate vacant postsynaptic sites, arresting the production of growth factors by deafferented cells. The other major contributor to growth factor production is glia (Nieto-Sampedro et al., 1983; Rudge, 1993). The temporal course of neuritogenic activity increases in the hippocampus after entorhinal cortex lesion, correlating closely with the time-course of gliosis in that structure and with commissural fiber kinetics of axon sprouting (Nieto-Sampedro, 1988a). Experiments like this, where we examine on one hand the *in vivo* anatomy of the cell, and on the other, assay *in vitro* the production of growth factors, enable indirect data to be obtained on both the cells producing the factors and the cells that are their target. At present, the cloning of many trophic factors, together with *in situ* hybridization techniques, permits directly checking the site of growth factor production.

6.3. Maturation of the new synapses

The final step of synapse renewal is the maturation of the new synaptic contacts. At the neuromuscular junction, the protein *agrin* induces clustering of neurotransmitter receptors in the postsynaptic membrane. The components of the basal membrane of a muscle that had been innervated are capable of defining the precise site where the regenerating terminals of axotomized motoneurons have to form new contacts (Bowe and Fallon, 1995). In such a case, the muscle basal membrane also directs the formation and differentiation of mature neuromuscular junctions, with presynaptic vesicles and postsynaptic foldings. The CNS lacks a basal lamina proper, but agrin and the abundant extracellular proteoglycans play a similar role, organizing new synapse formation. In fact, the brain contains soluble molecules capable of inducing postsynaptic attributes in the muscle basal membrane.

The most-characteristic structures of a mature CNS synapse are presynaptic vesicles in the nerve endings, and postsynaptic densities (PSDs) in dendritic spines of the postsynaptic cell. PSDs are the most-prominent subsynaptic structures when the adult CNS is observed with the electron microscope. During synapse renewal, PSDs appear as dynamic as the synapses to which they belong. Depending on its CNS location, the PSD of a synapse that has lost its presynaptic component may be conserved, or in other cases lost a few hours after deafferentation. Like the synapses themselves, the PSDs seem involved in a renewal cycle in which intermediate stages are observable with the electron microscope. At one of this stages, PSDs with multiple holes and indentations are found, which are probably degraded into small fragments, each of which may give rise to a new dendritic spine (Matus, 2001) and, perhaps, to a new synapse.

7. LONG-TERM POTENTIATION AND SYNAPSE RENEWAL: CELLULAR AND MOLECULAR BASES

The stimuli for synapse renewal that are more natural and frequent in mammals are those that induce learning and memory. Exposure to these stimuli may last a very short time, but perceptions received during seconds or tenths of a second may be remembered many years later. The main questions in studies on learning and memory are, firstly, the mechanism of formation, that is, how a stimulus lasting for a brief time is translated into a long-lasting neural record. The other big question is the nature of that long-lasting neural record or 'engramma'. A widely subscribed hypothesis is that of "synaptic plasticity and memory" (Martin et al., 2000). This was proposed by Cajal, among others, at the beginning of the XX century and is widely shared nowadays (Stryker, 1995; Malenka and Nicoll, 1999; Martin et al., 2000). The hypothesis assumes that the repeated physiological activation of a *hebbian synapse* is translated at a later moment into a morphological change in that synapse. The same processes that increases the efficiency of the synapse, causes a morphological modification that ensures a record of long duration. This field of research is very active and open to controversy. The above hypothesis is a personal preference.

7.1. Long-term potentiation: brief electrical signals drive long-lasting changes

A "Hebbian synapse" is one that behaves as postulated by Donald Hebb (1949): a synapse used repeatedly is 'reinforced', i.e., made more efficient. Once the synapse is reinforced, its stimulation threshold becomes lower, and the synapse can be activated by stimuli of lower intensity than originally necessary, or alternatively, its activation by the same stimulus produces a response of greater amplitude. Bliss and Lømo (1973) electrophysiologically recorded neurons from the rabbit dentate gyrus that was stimulated repeatedly from entorhinal cortex afferents, and found synaptic changes as postulated by Hebb. They called this process *long-term potentiation*, or LTP. Indirect evidence has accumulated indicating that LTP may be the storage device for some types of memory, particularly in the hippocampus (Larkman and Jack, 1995; Malenka, 1995). Other synapses in other CNS locations, as well as inhibitory synapses, may be potentiated (Marty and Llano, 1995). It seems likely that LTP is one of the first steps in synapse renewal.

7.2. Biochemistry of long-term potentiation. Receptors, calcium, kinases and phosphatases

LTP is conveniently induced by simultaneous activation of a synapse population at frequencies between 20 and 200 Hz. This tetanizing stimulation provides the essential requirement of Hebb: concomitant pre- and postsynaptic activities. The strong depolarization of the postsynaptic neuron occurs at a time when the synapse still retains a concentration of neurotransmitter sufficient to act on the postsynaptic receptors. Considering the hippocampus, where the synapses capable of potentiation are excitatory and glutamatergic, presynaptically liberated glutamate acts on two types of ionotropic receptor that coexist in dendritic spines: AMPA receptors (which respond preferentially to the glutamate agonist DL-a-amino-3-hydroxy-5-methylisoxazole-4-propionate, abbreviated as AMPA) and NMDA receptors (which respond preferentially to the glutamate agonist NMDA). During normal synaptic transmission, the arrival of an action potential at the axon terminal causes liberation of glutamate, which, acting on AMPA-type receptors, mediates Na^+-channel opening. AMPA-receptor-gated Na^+-channels support most of the depolarizing postsynaptic current. NMDA-type receptors contribute very little to postsynaptic depolarization, because at the membrane resting potential the channels associated to NMDA-receptors are blocked by Mg^{2+} ions. However, when a train of stimuli depolarizes the postsynaptic membrane, Mg^{2+} dissociates from the NMDA receptor, which, free, is capable of allowing passage of Ca^{2+} and Na^+. NMDA-type receptors may thus be considered voltage-dependent molecular-coincidence detectors, permitting Ca^{2+} entry into the postsynaptic neuron when afferent activity occurs together with postsynaptic depolarization (see, for example, Silva, 2003). There is agreement that LTP can be initiated postsynaptically. However, the relative contribution and importance of postsynaptic changes and pre-synaptic improved efficiency, possibly increasing the probability of neurotransmitter liberation, is still a subject of discussion (Larkman and Jack, 1995).

Depolarization and increase of intracellular postsynaptic Ca^{2+} are both required to reinforce synaptic transmission. Two further processes appear essential to ensure durability of the potentiation. One is the additional regulation of intracellular Ca^{2+} concentration, by activation of so-called "*metabotropic receptors*" (Bartollotto et al., 1994; Pin and Bockaert, 1995; Simpson et al., 1995). The other is the activation of regulators of DNA transcription and protein synthesis (Malenka, 1995; Schulman, 1995). The four families of transcription factors participate in signal amplification and consolidation, in the absence of which synaptic reinforcement lasts for less than 1 hour, i.e., it is relatively short-lived (Hinoi et al., 2002).

Metabotropic receptor activation initiates various intracellular enzymatic cascades, mediated by secondary messengers, including Ca^{2+}, cyclic AMP, cyclic GMP, and phospholipid degradation products, such as inositol phosphates and arachidonic acid. Some of these activities are protein kinases, which catalyze transcription-factor activation by phosphorylation. When LTP is allowed to proceed undisturbed, a prompt consequence of intracellular Ca^{2+} elevation is the activation of protein kinases. Pre-synaptic protein kinase C (PKC) and postsynaptic Ca^{2+} and calmodulin-dependent protein kinase II (CaMKII) are the kinases that have received most attention. The latter is very abundant in dendritic spines. However, cyclic AMP-dependent protein-kinase A is probably important for long-term memory consolidation. It phosphorylates CREB, a transcription factor critical for this process also phosphorylated by CaMK IV in response to growth factors. Finally, a tyrosine kinase phosphorylates NMDA-type glutamate receptors during LTP induction (Schulman, 1995).

An electrophysiological phenomenon equivalent but opposite to LTP was observed initially in cerebellum, then in hippocampus. It is called *long-term depression* or LTD (Malenka, 1995). LTD is induced by low-frequency (1 to 2 Hz) prolonged stimulation (3 to 15 minutes) of a synapse, and can reverse LTP. The mode of LTD induction is remarkably similar to that of LTP, it involves, as for LTP, elevation of intracellular Ca^{2+} concentration. However, the concentration increase is much lower than that after a tetanizing stimulation, and — instead of activating kinases — LTD activates calcineurin, a synaptic phosphatase with high affinity for Ca^{2+} (Lisman, 1989). The existence in neurons of approx. 2-3 MDa multiprotein signaling complexes, or "hebbosomes", specialized in decoding patterns of synaptic activity, has been proposed by Grant and O'Dell (2003). Hebbosomes would co-ordinate the activity of multiple pathways downstream from receptors, translating LTP and LTD into lasting neuronal changes.

7.3. Relationship between synaptic potentiation and morphological changes

Evidence for a relationship between LTP, learning, and changes in dendritic spine number and morphology, has been available for many years but a more direct correlation has been observed recently (Yuste and Bonhöffer, 2001). Postsynaptic morphological changes seem a logical consequence of the postsynaptic cell-protein synthesis required for LTP stabilization. However, the mechanism of communication between pre- and postsynaptic components is less obvious. Two retrograde messengers, nitric oxide and arachidonic acid, seem capable of informing the presynaptic terminal on the state of its postsynaptic counterpart. The first is produced by a Ca^{2+} and calmodulin dependent NO synthase (Bredt and Snyder, 1992). The second is generated by phospholipase A_2, also Ca^{2+}-dependent (Bliss et al., 1990). Both messengers seem to act by increasing the liberation of neurotransmitter.

From this point onward, possible mechanisms for induction of LTP-associated presynaptic morphological changes are more speculative. Neurotrophic-factor synthesis is much increased after stimuli capable of inducing LTP, and this increase is maintained up to 7 days after epileptiform stimulation. If we assume, following our hypothesis, that postsynaptic densities grow by addition of new material, and that after reaching a maximum size they undergo fragmentation, this would be equivalent to dendritic spine division, and would generate vacant PSDs. A vacant or inactive postsynaptic site is an active producer of neurotrophic factors capable of inducing sprouting (Cotman and Nieto-Sampedro, 1982, and references therein). Stimuli capable of inducing LTP would also stimulate synapse renewal and increase the number of vacant spines. The factors liberated at these spines, alone or together with glycosaminoglycans (Ornitz et al., 1995) or adhesion proteins (Williams et al., 1994; Doherty et al., 1995), would have the neuritogenic activity required to initiate axon sprouts destined to occupy the vacant spines. Additionally, synaptotagmin, a synaptic vesicle protein involved in neurotransmitter liberation, promotes the formation of fibroblast filopodia (Feany and Buckley, 1993). These actin-rich structures are typical of presynaptic growth cones; elevation of presynaptic activity produces high levels of both Ca^{2+} and synaptotagmin, inducing ultraterminal sprouting — the kind of axon sprouting most frequent in the CNS. In contrast, low frequency stimulation that induces LTD produces low levels of Ca^{2+}, yet is capable of activating calcineurin, which, in turn, causes axonal growth-cone filopodia retraction (Chang et al., 1995). The tyrosine kinase PYK 2 (Lev et al., 1995) may be a key convergence point of Ca^{2+} levels regulated by electrical activity itself regulated by neurotransmitter-governed ionic channels and by metabolic or growth signals, ruled by metabotropic receptors and growth factors.

REFERENCES

Arenander, A. and de Vellis, J. (1983) Frontiers of Glial Physiology. In *The Clinical Neurosciences* (R. Rosenberg, ed., section V). Churchill Livingstone, New York, pp.53-91.

Barde, Y.A., Edgar, D and Thoenen, H. (1982) Purification of a neurotrophic factor from mammalian brain. *EMBO J.* 1: 549-553.

Bartollotto, Z.A., Bashir, Z.I., Davies, C.H. and Collingidge, G.L. (1994) A molecular switch activated by metabotropic glutamate receptors regulates induction of long-term potentiation. *Nature* 368: 740-743.

Bliss, T. V. P. and Lømo, T. (1973) Long lasting potentiation of synaptic transmission in the dentate area of the anaesthetized rabbit following stimulation of the perforant path. *J. Physiol.* (Lond.) 232: 331-356.

Bliss, T. V. P., Clements, M.P., Errington, M.L., Lynch, M.A. and Williams, J. (1990) Presynaptic changes associated with long-term potentiation in the dentate gyrus. *Seminars in Neurosci.* 2: 345-354.

Bowe, M.A. and Fallon, J.R. (1995) The role of agrin in synapse formation. *Ann. Rev. Neurosci.* 18: 443-462.

Bredt, D.S. and Snyder, S.H. (1992) Nitric oxide, a new neuronal messenger. *Neuron* 8: 3-11.

Chang, H.Y., Takei, K., Sydor, A.M., Born, T., Rusnak, F. and Jay, D.G. (1995) Asymmetric retraction of growth cone filopodia following focal activation of calcineurin. *Nature* 376: 686-690.

Collazos-Castro, J.E. and Nieto-Sampedro, M. (2002). Developmental and reactive growth of dentate gyrus afferents: cellular and molecular interactions. *Restor. Neurol. Neurosci.*, 19: 169-187.

Cornell-Bell, A.H., Finkbeiner, S.M., Cooper, M.S. and Smith, S.J. (1990) Glutamate induces Ca^{2+} waves in cultured astrocytes: long range glial signalling. *Science* 247: 470-473.

Cotman, C. and Nadler, V. (1978): Reactive synaptogenesis in the hippocampus. In: Cotman C. (Ed): *Neuronal Plasticity*. Raven Press, New York. 227 – 271.

Cotman, C.W., Nieto-Sampedro, M. (1982) Brain function, synapse renewal and plasticity. *Annu. Rev. Psychol.* 33: 371-401.

Cotman, C.W., Nieto-Sampedro, M. and Harris, E. W. (1981) Synapse replacement in the nervous system of adult vertebrates. *Physiol. Rev.* 61: 684-784.

Crutcher, K.A. and Collins, F. (1982) In vitro evidence for two distinct hippocampal growth factors: basis of neuronal plasticity? *Science*, 217: 67-68.

Deller, T. and Frotscher, M. (1997): Lesion-induced plasticity of central neurons: sprouting of single fibres in the rat hippocampus after unilateral entorhinal cortex lesion. *Prog. Neurobiol.* 53: 687–727.

Doherty, P., Williams, E. and Walsh, F.S. (1995) A soluble chimeric form of the L1 glycoproteinstimulates neurite outgrowth. *Neuron* 14: 57-66.

Feany, M.B. and Buckley, K.M. (1993) The synaptic vesicle protein synaptotagmin promotes formation of filopodia in fibroblasts. *Nature* 364: 537-540.

Garthwaite, J. (1991). Glutamate, nitric oxide and cell-cell signalling in the nervous system. *Trends in Neurosci.* 14: 60-67.

Grant, S.G.N. and O'Dell, T.J. (2003) The Hebbosome hypothesis of learning: signalling complexes decode synaptic patterns of activity and distribute plasticity. In: *Neurosciences at the Postgenomic Era*, J. Mallet and Y. Christen, eds. Ipsen Foundation, Springer.

Greenough, W.T. and Chang, F.-L. F. (1984) Plasticity of synapse structure and pattern in the cerebral cortex. In: Cerebral Cortex vol. 7 (A. Peters and E.G. Jones, eds.) Plenum Press, New York, pp.391-440.

Hatton, G.I. (1985) Reversible synapse formation and modulation of cellular relationships in the adult hypothalamus under physiological conditions. In *Synaptic Plasticity* (C.W.Cotman, ed.), The Guilford Press, New York, pp.373-403.
Hebb, D. (1949) *The organization of behavior*. John Wiley & Sons, New York.
Hinoi, E., Balcar, V.J, Kuramoto, N. Nakamichi, N., Yoneda, Y.(2002) Nuclear transcription factors in the hippocampus. *Prog. Neurobiol.* 68: 145-165.
Hoff, S., Scheff, S. Kwan, A.Y. and Cotman C. W. (1981). A new type of lesion-induced synaptogenesis: synaptic turnover in non-denervated zones of the dentate gyrus in young adult rats. *Brain Res.* 222: 1-13.
Hoffman, P.N. (1995) The synthesis, axonal transport and phosphorylation of neurofilaments determine axonalcaliber in myelinated nerve fibers. *The Neuroscientist*. 1: 76-83.
James, W (1890) *The Principles of Psychology*. Holt, New York.
Larkman, A.U. and Jack, J.J.B. (1995) Synaptic plasticity: hippocampal LTP. *Current Opinion Neurobiol.* 5: 324-334.
Lee, K.S., Schottler, F., Oliver, M. and Lynch, G. (1981) Electron microscopic studies of brain slices: the effects of high frequency stimulation on dendritic structure. In *"Electrophysiology of isolated CNS preparations"* (Kerkut and Wheal, eds.) Acad. Press, London, pp.189-211.
Lev, S., Moreno, H., Martínez, R., Canoll, P., Peles, E., Musacchio, J.M., Plowman, G.D., Rudy, B. and Schlessinger, J. (1995) Protein tyrosin kinase PYK2 involved in Ca-induced regulation of ion channel and MAP kinase functions. *Nature* 376: 727-729.
Levi-Montalcini, R. (1982) Developmental neurobiology and the natural history of nerve growth factor. *Annu. Rev.Neurosci.* 5: 341-362.
Lisman, J. (1989) A mechanism for the Hebb and anti-Hebb processes underlying learning and memory. *Proc. Nat.Acad. Sci. USA* 86: 9574-9578.
Liu, C.N. and Chambers, W.W. (1958) Intraspinal sprouting of dorsal root axons. *Arch.Neurol.Psychiat.* 79: 46-61.
Malenka, R.C. (1995) LTP and LTD: Dynamic and interactive processes of synaptic plasticity. *The Neuroscientist*. 1: 35-42.
Malenka, R.C. and Nicoll, R.A. (1999) Long-term potentiation- a decade of progress? *Science* 285: 1870-1874.
Marinesco, (1907) Quelque recherches sur la transplantation des ganglion nerveux. *Revue Neurologique* No. 6 (30 mars).
Martin, S.J., Grimwood, P.D. and Morris, R.G. (2000) Synaptic plasticity and memory: an evaluation of the hypothesis. *Ann. Rev. Neurosci.* 23: 649-711.
Marty, A. and Llano, I. (1995) Modulation of inhibitory synapses in the mammalian brain. *Current Opinion Neurobiol.* 5: 335-341.
Mattson, M.P. (1988) Neurotransmitters in the regulation of neuronal cytoarchitecture. *Brain Res. Rev.* 13: 179-212.
Matus A. (2001) Moving molecules make synapses. *Nat Neurosci*. 4: 967-968.
Microglia: Special Issue (1993) *Glia* 7(1): 1-120.
Minea, (1909) Cercetari experimentale asupra Variatiunilor mofologice ab Neuronului sensitiv. Thesis. Bucaresti.
Murphy, S., Simmons, M.L., Agulló, L. (1993) Synthesis of nitric oxide in CNS glial cells. *Trends in Neurosci.* 16: 323-328.
Nedergaard, M. (1994) Direct signalling from astrocytes to neurons in cultures of mammalian brain cells. *Science* 263: 1768-1771.
Newman, E.A. (1986) High potassium conductance in astrocyte end feet. *Science* 233: 453-454.
Nieto-Sampedro, M. (1988) Growth factor induction and order of events in CNS repair. In: *Pharmacological approaches to the treatment of brain and spinal cord injury* (B.A. Sabel and D.G.Stein, eds.) Plenum Press, New York, pp.301-337.
Nieto-Sampedro, M. and Cotman, C.W. (1985) Growth factor induction and temporal order in CNS repair. In *Synaptic plasticity* (C.W. Cotman, ed.). The Guilford Press, New York, pp. 407--455.
Nieto-Sampedro, M., Lewis, E.R., Cotman, C.W., Manthorpe, M., Skaper, S.D., Barbin, G., Longo, F.M. and Varon, S. (1982b) Brain injury causes a time-dependent increase in neuronotrophic activity at the lesion site. *Science* 217: 860-861.

Nieto-Sampedro, M., Manthorpe, M., Barbin, G., Varon, S. and Cotman, C.W. (1983) Injury-induced neuronotrophic activity in adult rat brain: correlation with survival of delayed implants in the wound cavity. *J. Neurosci.* 3: 2219-2229.

Ornitz, D.M., Herr, A.B., Nilsson, M., Westman, J., Svahn, C.-M.and Waksman, G. (1995) FGF binding and FGF receptor activation by synthetic heparan-derived di- and trisaccharides. *Science* 268: 432-436.

Parpura, V., Basarsky, T.A., Liu, F., Jeftinija, K., Jeftinija, S. and Haydon, P.G. (1994) Glutamate-mediated astrocyte-neuron signalling. *Nature* 369: 744-747.

Pin, J.-P. and Bockaert, J. (1995) Get receptive to metabotropic glutamate receptors. *Current Opinion Neurobiol.* 5: 342-349.

Pope, A. (1978) Neuroglia: Quantitative aspects. In *Dynamic properties of glial cells* (E. Schoffeniels and F.G. Tower, eds.) New York, Pergamon Press, pp. 13-20.

Rudge, J.S. (1993) Astrocyte-derived neurotrophic factors. In: *Astrocytes. Pharmacology and function* (S. Murphy, ed.) Acad. Press, San Diego, pp.267-305.

Russell, D.S. (1995) Neurotrophins:mechanisms of action. *The Neuroscientist* 1: 3-6.

Schulman, H. (1995) Protein phosphorylation in neuronal plasticity and gene expression. *Current Opinion Neurobiol.* 5: 375-381.

Silva, A.J. (2003) Molecular and cellular cognitive studies on the role of synaptic plasticity in memory. *J. Neurobiol.* 54: 224-237.

Simpson, P.B., Challiss, R.A.J. and Nahorski, S.R. (1995) Neuronal Ca stores: activation and function. *Trends in Neurosci.* 18: 299-306.

Sontheimer, H. (1995) Glial influences on neuronal signaling. *The Neuroscientist* 1: 123-126.

Speidel, C.C. (1941) Adjustments of nerve endings. *The Harvey Lectures* 36: 126-158.

Sporn, M.B. and Roberts, A.B. (1988) Peptide growth factors are multifunctional. *Nature* 332: 217-218.

Stryker, M.P. (1995) Growth through learning. *Nature* 375: 277-278.

Steward, O. (1991) Synapse replacement of cortical neurons following denervation. In *Cerebral Cortex*, vol. 9, (A. Peters and E. G. Jones, eds.) Plenum Press, New York, pp. 81-131.

Tsukahara (1985) Synaptic plasticity in the red nucleus and its possible behavioral correlates. In: *Synaptic Plasticity* (C.W.Cotman, ed.), The Guilford Press, New York, pp. 201-229.

Virchow, R. (1859) *Cellularpathologie*, Hirschwald, Berlin.

Williams, E.J., Furtness, J., Walsh, F.S. and Doherty, P. (1994) Activation of the FGF receptor underlies neurite outgrowth stimulated by L1, N-CAM and N-cadherin. *Neuron* 13: 583-594.

Yuste, R. and Bonhöffer, T. (2001) Morphological changes in dendritic spines associated with long-term synaptic plasticity. *Annu. Rev. Neurosci.* 24: 1071-1089.

M. NIETO-SAMPEDRO

Institute Cajal, CSIC, Madrid, Spain and Unit of Experimental Neurology, SESCAM, Hospital Nacional de Paraplejicos, Toledo, Spain

21. NEURAL PLASTICITY AND CENTRAL NERVOUS SYSTEM LESION REPAIR

Summary. In the previous chapter, the relationships between neural plasticity and functional processes underlying learning has been described. In this chapter, I will consider the main mechanisms involved in neural tissue repair after its traumatic, or cytotoxic lesion. A particular attention will be paid to available information for reducing secondary neuronal death, and to the molecular and sub-cellular processes involved in glial scar formation and in the inhibition of axonal growth. The possible contribution of aldynoglia, or growth-promoting glia, to axonal regrowth will also be considered. In particular, a detailed presentation will be made of ensheathing glia role in axon section repair in the central and peripheral nervous systems.

1. CENTRAL NERVOUS SYSTEM LESIONS

Brain and spinal chord lesions have an increasing social and economic importance in developed countries. Accidents of all types are the major cause of death in children and young adults and, for all ages combined, surpassed as a cause of death only by heart disease and cancer. However, if we consider the number of potential years of life and work lost, lesions exceed all other problems, because they occur most frequently in people aged under 45. In Spain, about 10 people in a million suffer spinal cord lesions, most of them (81%) traumatic (Collazos-Castro et al., 2003). Long-term survival from such lesions was rare in the near past. Advances in acute treatment of brain and spinal cord lesions have advanced enough to save the lives of many victims, but a reliable cure is not available yet. Present therapies treat acute edema and alleviate symptoms, but no treatment has been described that reliably restores sensorimotor and visceral functions. The best we can do, after vital stabilization, is to teach patients to live as well as possible with the consequences of the lesion, an endeavor made easier by recent advances in rehabilitation.

The aggressions that the central nervous system (CNS) may suffer are of many types and each kind probably evokes a specific physiological response. In neuropathology, however, lesions are classified into two general types, depending on their morphological effects: isomorphic and anisomorphic (Greenfield, 1958). Their effects, at the cellular and molecular levels, differ considerably.

1.1. Anisomorphic lesions

Lesions that alter the gross morphology of the CNS, are termed anisomorphic. They are open lesions, are caused mechanically, and destroy both the boundary between the CNS and the rest of the organism (the *glia limitans*) and the local blood-CNS barrier. Blood-vessel rupture, together with vascular spasm, causes ischemia and its associates, anoxia and hypoglycemia. Blood cells and serum proteins invade the damaged area, and edema, derived from extracellular fluid accumulation as well as from astrocyte swelling, is obvious 24 hours after the injury. Structural and electrophysiological axonal abnormalities in both white and gray matter can be observed immediately after a contusion, with necrosis and degeneration of the myelin of these axons following 8-24 hours later. Removal by blood phagocytes of degenerated myelin and other cell debris occurs 48 hours afterwards.

The degenerative process, however, neither terminates immediately after the lesion nor remains restricted to the injury site. An insidious process extends neuronal death in time and space. The so-called secondary or delayed neuronal death begins 1 or 2 days after the lesion, and provokes the loss of more neurons than the damage directly attributable to the injury or primary death. The neural tissue near the damaged area or connected with it presents depressed electrical activity and will be affected by secondary neuronal death. This area, called the *zone of penumbra* in ischemic lesions, evolves towards secondary death and is possibly responsible for loss of function in the majority of brain and spinal chord traumas.

While blood phagocytes remove cellular remains and secondary neuronal death progresses, astrocytes near the damaged zone proliferate, and their enlarged and fibrous processes form a mesh that recreates the *glia limitans*. Dividing fibroblasts from adjacent connective tissue overlay the fibrous astrocyte mesh and deposit collagen, completing the formation of a new frontier for the CNS, now called "*glial scar*".

1.2. Isomorphic lesions

Isomorphic lesions, typically caused by neurotoxins, Wallerian degeneration, or tumors, do not grossly damage the *glia limitans*. Microglia assume the initiative of the response to isomorphic lesions, abundantly proliferating 3 to 5 days after the lesion, and differentiating into reactive microglia and macrophages. Reactive astrocytes appear following the initial microglial response. They position themselves around reactive microglia, seemingly avoiding them (Fernaud-Espinosa et al., 1993).

2. CENTRAL NERVOUS SYSTEM LESION REPAIR AND THE EVOLUTIONARY LIMITS OF PLASTICITY

The wide and important physiological functions that nervous system plasticity has in mammals present defined limits. Neural plasticity in the peripheral nervous system permits the continuous remodeling demanded by the wear and tear of normal life. Small, rarely incapacitating lesions (generally mechanical) are part of this deterioration. Because injured individuals have the time and opportunity to reproduce, the evolutionary process has been able to select and preserve the neural plasticity mechanisms that permit lesion repair. The presence of a basal lamina and its neuritogenic proteins and the interaction of axon sprouts with Schwann cells and peripheral myelin, permit and stimulate peripheral axon regeneration.

Table 1. Proteoglycans in normal and damaged CNS

Name	Type	MW	Core	Neurites*	Cell source
UMPG	CS/HS	200-220	48	±	Neurons
IMPG	CS/HS	200-220	48	-	Astrocytes
PMPG	CS/HS	200-220	48	-	Neurons
NG2	CS	500	300	-	O2A, oligo progen.
Neurocan	CS		136	-	Neurons/Astroc.
Brevican	CS		145	-	Astroc.
Perlecan	HS/CS/DS		467	±	Blood vessels
Appican	CS	200	90	-	Glia
Glypican-5	HS	220	58	±	Neurons
Phosphacan	CS/KS	400-500	173	±	Glia
DSD-1-PG	Mouse phosphacan homologue			±	Glia/neurons
Syndecans 2-3	HS	40-200	20-120	?	Neurons
Syndecan-4	HS/CS	200	25, 30, 50	±	Glia
Agrin	HS	500	25	-	Neurons/glia
Ryudocan	HS	160	30, 50	syndecan-4	Endothelium
Neuroglycan C	CS	150	56-120	?	Neuroblasts

MW, Mol. Weight x 10^3, CS, chondritin sulphate; HS, heparan sulphate; KS, keratan sulphate; DS, dermatan sulphate.*, +, neurite promotion; -, neurite inhibition; ±, promoting or inhibitory activity, depending on the presence of other growth modifiers; **?**, *data not available.*

In contrast, mammalian CNS plasticity seems aimed mainly at potentiating learning and memory processes. Reactive synaptogenesis mechanisms permit, at most, spontaneous repair of very small lesion sequels — for example, of the rupture or occlusion of blood capillaries. These lesions can be repaired with the help of terminal sprouts and the normal mechanisms that operate in synapse renewal. Larger lesions are incapacitating. They prevent normal life and, in the unlikely case of the injured individual's survival, the prospect of reproduction is practically nil. Thus, effective CNS lesion repair has a minimal probability of selection. Hence, spontaneous lesion repair in the CNS of mammals does not occur as a rule. We will see that, furthermore, the properties of CNS glia lead to axonal growth inhibition than to regeneration.

3. NEURAL RESPONSE TO BRAIN AND SPINAL CHORD TRAUMA

Severe CNS trauma initiates cascades of events, both deleterious and beneficial, that we may wish either to arrest or to potentiate (Nieto-Sampedro and Cotman, 1985; Nieto-Sampedro, 1988a). CNS lesions immediately destroy the local vasculature and the blood-CNS barrier.

In summary, trauma kills both neurons and glia, destroys blood vessels, and damages axon tracts. Because dead neurons are not replaced, and damaged axons do not regenerate spontaneously, permanent functional deficits ensue. To prevent loss of function or to recover the functions lost after injury, experimental approaches have been addressed to i) prevent or reduce secondary neuronal death; ii) regrow injured axons; and iii) restore lost neurons. The last few years have seen significant advances in these topics, and some articles report improvement of function. To date, however, anatomical repair and functional improvement after experimental lesions are small, and the strategies employed are difficult to apply in humans. Furthermore, the evaluation of sensory and motor functions has been particularly confused in experimental models, which frequently do not bear any relation to human pathology and where functional gains cannot be extrapolated to humans.

4. CENTRAL NERVOUS SYSTEM LESION REPAIR: WHAT HAS BEEN ACHIEVED SO FAR?

Ramón y Cajal's (1914) classic "Degeneración and regeneration of the nervous system" includes CNS lesion repair studies of Tello (1911). Tello transplanted fragments of sciatic nerve into various CNS sites and noticed that Schwann cells stimulated central nerve fiber growth. Unfortunately, Cajal's death, the Spanish civil war, and the religio-political opinions of Tello contributed to interrupt these studies. They were taken up more recently by other researchers, with striking though incomplete, results. Aguayo et al. (1981), in Canada, showed that sprouts from damaged central nerves grew long distances through transplants of sciatic nerve, something impossible in the absence of the transplant. The problem of regrowth of the injured axon seemed solved, but functional restoration depends on synapse formation with specific targets, which did not occur. When central fibers exited the transplanted peripheral nerve segment, they were incapable of growing more than a few microns, forming rare and unproductive synaptic contacts. At about the same time, Nieto-Sampedro (1982) showed that lesions induced a significant increase in the production of neurotrophic factors by the injured tissue. Those factors enhanced the survival of transplanted neurons and their partial synaptic integration with the host. Besides providing a synaptic plasticity mechanism, this was an unexpected CNS self-repair response. Although neither the work of Aguayo nor that of Nieto-Sampedro achieved CNS lesion repair, possibly both contributed to changing the pessimistic attitude of researchers: lesion repair had a certain rational basis and seemed possible. A complementary approach was initiated by Caroni and Schwab (1988) and Schwab and Caroni (1988), who found that central myelin expressed axonal growth inhibitors. Antibodies against those inhibitors induced regrowth of corticospinal axons in the injured spinal cord (Schnell and Schwab,1990; 1993).

Corticospinal fibers did not cross the injured area, but rather surrounded the lesion and grew through non-damaged tissue. Therefore, this treatment may be useful for incomplete lesions, and may be synergistic with other therapies. The antibody, besides promoting axonal regeneration of injured nerves, induced collateral sprouts from undamaged fibers, a type of plasticity that might help functional recovery. Human CNS myelin expresses similar axonal-growth-inhibitory proteins, whose activity is also neutralized by the antibody, and which may have clinical potential. While the work on myelin inhibitors was in progress, other investigators, particularly Silver in the USA (Rudge and Silver, 1990) and Nieto-Sampedro in Madrid (Bovolenta et al., 1991, 1992) found that proteoglycans on the surface of reactive astrocytes were the most-general inhibitors of axonal growth to be neutralized, and initiated the characterization of the inhibitors and the preparation of blocking antibodies.

The next significant advance returned to Tello's strategy, and used transplanted intercostal nerves, together with the neurotrophic factor FGF-1, in rats with complete spinal cord section at the T8 level (Cheng et al., 1995, 1996). Olson's group in Sweden showed for the first time partial recovery of hind-limb function correlating with anatomical restitution of the nerve tracts connecting brain and cord, including the corticospinal tract. This type of intervention has been carried out in a few patients, and although they showed no significant functional recovery, the usefulness of the technique cannot yet be ruled out.

Transplantation of olfactory ensheathing cells (ECs) in nerve injuries may be included in Tello's strategy. The special properties of olfactory glia, intermediate between those of astrocytes and Schwann cells, had been described by Doucette (1986, 1990). Ramón-Cueto and Nieto-Sampedro developed culture and purification of ECs from adult olfactory bulb (Ramón-Cueto and Nieto-Sampedro, 1992, 1993). Purified ensheathing glia was transplanted into the spinal cord of adult rats that had received a transection of the central branch of the T8 dorsal root. Unhelped, the sectioned fibers never entered the cord. With the help of EC transplants, bundles of sensory afferents penetrated the cord, crossing the myelin of Lissauer's band and occasionally reaching the contralateral dorsal horn (Nieto-Sampedro and Ramón-Cueto, 1993; Ramón-Cueto and Nieto-Sampedro, 1994). Nieto-Sampedro's group in the Cajal Institute has continued to work on sensory root damage repair (Gudiño-Cabrera et al., 2000; Pascual et al., 1997; 2002, Navarro et al., 1999; Taylor et al.,1999; 2001; Muñeton-Gómez et al, 2003; Verdú et al., 2001), work that has helped us define the scope of the interventions and the analysis of what is lost after spinal cord lesions, what is recovered, and how recovery takes place (Collazos-Castro et al., 2003; Muñetón et al., 2003; Nieto-Sampedro et al., 2003).

The publication of the rhizotomy repair results encouraged the use of EC transplants to repair corticospinal tract lesions (Li et al. 1997), demyelination (Imaizumi et al., 1998), and complete spinal transections (Ramón-Cueto et al., 2000). The conditions for corticospinal tract repair of Li et al. (1997) were optimal as experimental lesions (selective electrolytic lesions) were very small (about 500 µm) and neither bleeding nor fibroblast entry into the cord was allowed. The transplanted cells filled the lesion, directly contacting the damaged axons and joining the proximal and distal borders of the injury. Nevertheless, the number of regenerated axons was small, and the work has been criticized for the possibility that the lesions were incomplete. Several groups have tried to reproduce the results of Ramón-Cueto et al. (2000), so far with little success. In any case, these results have created great expectations of a "cure" for spinal cord lesions, mixed with great confusion as to what can be expected and when.

Over the last hundred years, many surgical transplant strategies have been examined to promote axonal regeneration, including immature astroblasts, microglia, meningeal fibroblasts, neural stem cells, and fragments of embryonic neural tissue. Neurotrophic factors have been supplied, either purified or as implanted cells transfected to produce them. Neuritogenic factors, extracellular matrix molecules, and blocking antibodies against neurite-promoting inhibitors are being assayed. At this time, there are reasons to be guardedly optimistic. We have to define which areas of the CNS and which type of lesions may be repaired, which strategies can be combined, and what results may realistically be expected. Our group is testing potential therapies in models of spinal cord injured by contusion of various degrees of severity, and by specific lesions. Repair is evaluated by electrophysiological, kinematic and kinetic analysis, techniques necessary, though not used systematically in previous studies.

5. REDUCING SECONDARY NEURONAL DEATH

From the biological point of view, the lack of functional recovery after CNS lesions is not a problem of either axonal growth initiation or capacity of long-distance elongation. Rather, it seems related to secondary neuronal death and the formation of a growth-inhibitory environment in the damaged area. Only a relatively small number of neurons die as an immediate consequence of trauma. Acute, or primary, death occurs during the first hours post-lesion. However, a more insidious phenomenon, called secondary neuronal death, extends nerve-cell loss in time and space, and is responsible for the death of more neurons than is the primary lesion. Trauma is followed by an intense glial reaction, accompanied by leukocyte infiltration and liberation of pro-inflammatory cytokines, oxygen and nitrogen free radicals, arachidonic acid metabolites, excitatory aminoacids, quinolinic acid, proteases, and other cytotoxins. All these molecules act synergistically, triggering the neurodegenerative cascade responsible for secondary neuronal death. In the zone of penumbra these neurodegenerative processes are counteracted by the liberation of neuroprotective factors: anti-inflammatory, antioxidant, neurotrophic, and neuritogenic. This is a zone where promoters and inhibitors of neuronal survival, and of axonal degeneration and regeneration, coexist in dynamic balance. The basic strategy to favor functional CNS repair is the modulation and/or modification of that equilibrium. In many cases, the most serious functional deficits after CNS lesions might be limited if we could control secondary neuronal death. Ideally, injured people should be treated soon after trauma, by paramedical personnel, without the need for surgery. That means preferably by parenteral administration of substances capable of blocking or diminishing secondary neuronal death. Excitatory neurotransmitter antagonists, inhibitors of free-radical formation (antioxidants), and neurotrophic factors are the preferred candidates to date. Tested (and abandoned) compounds in these categories include methylprednisolone — a steroid supposed to preserve injured CNS tissue by reducing edema, leukocyte infiltration (Bartholdi and Schwab, 1995), and levels of peroxide radicals (Taoka et al., 2001), and dizocilpine (or compound MK-801) — a promising antagonist of glutamate, acting on NMDA-type receptors, which had limited efficacy and too many undesirable side effects. Other antagonists are subject to clinical trial (e.g., memantine). The current situation is that we do not understand all the complexities of neuronal death secondary to trauma (Herreras and Largo, 2001; Nieto-Sampedro et al., 2002). For a long time we have misled ourselves into believing that the problem of secondary neuronal death was solved, hypothesizing possible neurotoxicities and designing simple models (many *in vitro*) to test these hypotheses. However, no matter how intellectually satisfying it may be to prevent a given type of neurotoxicity in an *ad hoc* model, the problem of post-trauma secondary neuronal death remains unsolved.

The importance of controlling secondary neuronal death cannot be exaggerated. Besides trauma, neuronal death is connected to one of the major clinico-social and economic problems of our age — neurodegenerative diseases that course with progressive neuronal death, e.g. Alzheimer's, Parkinson's and Huntington's diseases.

6. REACTIVE ASTROCYTES: GLIAL SCAR AND INHIBITION OF AXONAL GROWTH

Secondary neuronal death, and the cellular and molecular causes that generate it, also give rise to an axon-growth-hostile environment. The culprits are the cells and molecules of the so-called "glial scar", mainly reactive astrocytes, reactive microglia, fibroblasts, and extracellular matrix. Astrocytes, the main components of the glial scar, are possibly the most-plastic CNS cells, capable of changing in number and morphology in response to perturbations. After a lesion they change shape, becoming "reactive" or "fibrous". The meaning of the word "reactive"referred to astrocytes is by no means precise. Most researchers take it for granted that reactive indicates cells bigger than those that are normal or "resting". Reactive astrocytes show a great increase in expression of intermediate filaments, which gives them the "fibrous" appearance of their alternate name. Astrocyte intermediate filaments are recognized by antibodies against the monomer glial fibrillary acidic protein (GFAP; Bignami and Dahl, 1974), the most characteristic marker of reactive astrocytes. The glial scar essentially consists of a meshwork of *hypertrophic fibrous astrocytes* covering the lesion surface. Fibroblasts from adjacent connective tissue proliferate over the fibrous astrocyte layer, deposit collagen, and complete the formation of a new frontier for the injured CNS — the glial scar. This new frontier frequently separates neurons previously connected, and is a serious obstacle to the re-establishment of new connections. Denervated neurons substitute their original innervation by axon sprouts from undamaged neighboring neurons, which generally does not lead to recovery of the original function. Glial scar formation may represent an attempt by the CNS to isolate itself from uncontrolled influences, reconstituting a new *glia limitans*. However, the glial scar is also a major obstacle to the restitution of damaged connections. Thus, from the clinical point of view, astrocytes are simultaneously beneficial and deleterious. Their roles create a situation where survival (restitution of a glial boundary) and the need to restore lost functions are in conflict. It will be interesting to test whether inhibition of glial scar formation can lead to functional restoration without affecting survival.

Do astrocytes proliferate in adult mammals? Whereas astrocyte number remains stationary in the adult, these cells are able to divide in response to neuronal death or damage. Astrocytes proliferating in the adult mammalian CNS may be mature astrocytes that have undergone de-differentiation, or remaining astrocyte precursors, or new astroblasts arising from stem cells, or all of these possibilities. Although the problem is clearly set out, its definite answer needs further research. Viable astrocytes cannot be cultured from adult mammalian CNS, unless the cultured tissue is adjacent to an injury site (Lindsay, 1986). The astrocytes that then proliferate are cells with morphology and immunological properties similar to those of type 1 neonatal brain astroblasts (Lindsay, 1986). Part of the new astrocytes dividing in the brain and spinal cord after anisomorphic injury definitely arise from stem cells (Holmin et al., 1997; Johansson et al., 1999) attracted to damaged tissue (Johansson et al., 1999; Helmuth, 2000). Astroblasts promote neuritogenesis, are a preferred substrate for neurite extension, and during development are often seen associated to growth cones *in vivo* (Silver, 1984). In contrast, the membranes of fibrous reactive astrocytes induced by Wallerian degeneration or toxin injection contain proteoglycans that cause growth-cone collapse, inhibit neurite outgrowth, and repel growing neurites (Nieto-Sampedro, 1999). Nieto-Sampedro (1999) proposed that the ratio of fibrous astrocytes to astroblasts will determine the overall neurite-promoting or neurite-inhibiting properties of the gliotic tissue. This ratio changes with lesion type and post-lesion time, and gliotic tissue properties change with it. A fraction of the glial scar astrocytes result from proliferation, regulated by well-known mitogens and proliferation inhibitors. Astroblast proliferation is mediated by binding of epidermal growth factor (EGF) to its receptor (EGFR), a well-characterized membrane tyrosine kinase. The natural inhibitor of astrocyte division, neurostatin, is a complex glycolipid, present in both soluble and membrane-bound forms (Abad-Rodríguez et al., 2000). It strongly inhibits EGF-driven astroblast proliferation. Antibodies to EGFR, capable of sequestering neurostatin, induce astrogliosis (Nieto-Sampedro, 1988). It will be very interesting to test whether the reverse is true, i.e., whether neurostatin decreases or inhibits lesion-induced astrogliosis.

Proliferating astrocytes in adult mammalian CNS may originate i) by de-differentiation of mature astrocytes to astroblasts; ii) by differentiation of astrocyte precursors, preserved in the adult; iii) from new astroblasts, derived from stem cells; or iv) from all these sources. The proportion contributed by each of these sources in response to the various perturbations, its regulation and the role of neurostatin in the generation of astroblasts need further investigation. Inflammatory glial reaction, together with leucocyte infiltration and cytotoxine liberation is probably responsible for both, secondary neuronal death and production of neuroprotective factors, anti-inflamatory, antioxydant, neurotrophic and neuritogenic. The penumbra zone is a dynamic area where factors promoting and inhibiting neuronal death and axon growth coexist (reviewed by Tator and Fehling, 1991; Schwab and Bartholdi, 1996; Lipton, 1999). A key point to influence the outcome of trauma is the modulation and/or modification of the balance of these factors to favour neuronal survival and axonal growth

7. THE ASTROCYTE SURFACE: PROMOTER AND INHIBITOR OF AXONAL GROWTH

The dual properties, both beneficial and deleterious, shown by astrocytes are particularly noticeably when considering restitution of the functions lost after a lesion. The origin of *hypertrophic fibrous astrocytes*, the major components of the 'glial scar', and of *astroblasts*, capable of division, can be quite different. Their properties regarding the interaction with neurites are opposite. Hypertrophic fibrous astrocytes are an inhibitory obstacle for regeneration of the damaged axon, whereas astroblasts are an excellent growth substrate for axons and dendrites (neurites). Lesions induce changes in the astrocyte surface: the growth-promoting proteoglycans expressed in astroblasts change to proteoglycans that inhibit neurite initiation, adhesion, and growth in reactive astrocytes (refs. in Nieto-Sampedro, 1999).

Fibrous reactive astrocytes cannot be cultured. The astrocytes growing in culture are epithelioid polygonal cells, similar to astroblasts rather than to mature astrocytes. If these cells are treated with dibutyryl-cyclic AMP (diBcAMP), they become star-shaped, expressing high levels of polymerized GFAP filaments. However, the surface of these cells retains the growth properties of astroblasts (Wandosell et al.,1993). Injury-induced gliosis has been modeled in culture more indirectly. Intraventricular kainic acid injection killed hippocampal neurons and induced strong hippocampal gliosis. Plasma membranes purified from gliotic hippocampi were used as a culture substrate to test their effect on neurite initiation and growth rate in many types of neuron. Gliotic membranes from central tissue of any origin (hippocampus, septum, cortex, striatum, spinal cord) contain both neurite-growth-promoting and neurite-growth-inhibiting proteoglycans in proportions depending on the locus and type of lesion. Antibodies to the inhibitory proteoglycan showed it to be located in neurons in normal tissue, shifting to reactive astrocytes after injury (Nieto-Sampedro, 1999). The reactive astrocyte inhibitor differs from the myelin inhibitors, and probably corresponds to the molecule responsible for the classic 'glial scar' inhibition described by Cajal (1914). One of the monoclonal antibodies against the inhibitory proteoglycan blocked neurite outgrowth inhibition *in vitro*. Its possible *in vivo* action has not been tested yet.

8. ALDYNOGLIA, THE GLIA OF REGENERATIVE SYSTEMS AND CONVENTIONAL MACROGLIA

Two of the few sites of the mammalian CNS where injured axons regenerate spontaneously are the hypothalamic-neurohypophyseal system and the olfactory bulb. Regeneration is possible in these loci because axons are associated to a type of growth-promoting macroglia, similar to peripheral Schwann cells, that we have called aldynoglia or growth-promoting glia. These glial cells include ensheathing glia (ECs) in the olfactory bulb, tanycytes (tan) in the hypothalamus, and pituicytes (pit) in the neurohypophysis. We have cultured aldynoglia, studied their growth properties, characterized their specific surface markers, and described their interaction with diverse axonal types (Gudiño-Cabrera and Nieto-Sampedro, 1999, 2000).

The properties of aldynoglia (ECs, tan, and pit) present similarities and differences with those of CNS macroglia precursors. Their immunophenotypes are similar to those of the O4-positive pro-oligodendrocytes that persist in the adult (refs. in Nieto-Sampedro, 2002). However, from a joint consideration of proliferation properties, growth-promoting properties, interaction with neurons, and immunological markers, the greatest similarity is with Schwann cells. Tanycytes and pituicytes are frequently called astroglia, which they resemble in developmental origin, GFAP, and S-100 protein immunoreactivity. However, both tanycytes and pituicytes show p75-NGF receptor immunoreactivity, in contrast to astrocytes but in common with Schwann cells and ECs (Nieto-Sampedro, 2002). Tanycytes and pituicytes contained the sulfatide and seminolipid antigens recognized by monoclonal O4, in common with cells of oligodendrocytic lineage, Schwann cells, and olfactory bulb ensheathing cells (Gudiño-Cabrera, and Nieto-Sampedro, 2000), but in contrast to astrocytes. Other physiological properties of ECs, tanycytes, and pituicytes, such as their ability to reversibly enfold non-myelinated axons, are also reminiscent of non-myelinating Schwann cells. They are also similar to peripheral glia regarding growth-promoting properties, accounting for the efficient regeneration of peripheral axons, olfactory axons, and hypothalamic fibers (Gudiño-Cabrera and Nieto-Sampedro, 1999, 2000). Furthermore, ECs transplanted into CNS lesions, promote CNS fiber repair (Ramón-Cueto and Nieto-Sampedro, 1994; Li et al, 1997; Pascual et al., 1997; Imaizumi et al, 1998; Navarro et al., 1999).

Adult Schwann cells, ECs, tanycytes, and pituicytes show common features, other than their capability for survival and division in culture, typical of developmentally immature cells. Thus, they have an immature cytoskeleton (strong vimentin immunoreactivity), show markers of motility (polysialylated-neural-cell-adhesion molecule, PSA-NCAM) and migration (p75 NGF receptors. PSA-NCAM is absent or very low in mature astrocytes, but is expressed abundantly throughout adulthood in olfactory bulb glial cells and hypothalamo-neurohypophyseal glia. Antibodies to domains D and F of α-estrogen receptor immunostained the nucleus of hypothalamic neurons, as well as cytoplasmic processes of Schwann cells, ECs, tanycytes and pituicytes, but not astrocytes (Gudiño-Cabrera and Nieto-Sampedro, 1999, 2000).

Neurotrophins and their low-affinity receptor participate in endocrine regulation, particularly that of estrogen (reviewed by Gudiño-Cabrera and Nieto-Sampedro, 1999). Estrogen receptors (ER) are transcription factors and NGF/p75 signaling involves activation of NF-κB. Although p75 NGFR activation by NGF is generally associated to apoptosis, NF-κB is a master switch regulating the expression of many genes, some of which may suppress apoptosis and promote long-term survival. Estrogen is also known to inhibit apoptosis (reviewed by Thompson, 1995). ER-α-immunoreactivity co-localized with that for p75 NGFR. Concomitant expression of ER-α and NF-κB by the glial cells discussed here may confer on them the observed immaturity features and ability to enter the cell division cycle. Another way in which signals from mitogens, estrogen, and neurotrophins may be integrated is through the CREB binding protein (CBP) (Arany et al., 1994). By integrating NGF and estrogen signals through ER and p75 NGFR in Schwann cells and in Schwann-like cells expressing ER and p75 NTR, CBP may regulate the co-ordinated transcription of many genes, leading to the expression of a Schwann-like glial phenotype.

In summary — as regards immunological markers, neurite ensheathing, regeneration-promoting properties, and ability to enter the cell-division cycle — ECs, tanycytes, and pituicytes resemble non-myelinating Schwann cells more closely than CNS macroglia. They may form a unique CNS glial population with enhanced plastic properties. The study of this glia as an *in vivo* regeneration-promoter, will require its systematic transplantation into injured CNS regions where axonal regeneration does not occur, observing what the axonal types, if any, regenerate with its help.

9. FUNCTIONAL CONVERGENCE, LINEAGE AND ONTOGENIC STAGE MARKERS

Cell culture in defined media, and the use of immunological markers have permitted the establishment of progenitor-progeny relationships. Several lineage markers used in the present work would group Schwann cells, olfactory ECs, tanycytes and pituicytes in the same lineage and ontogenic stage. However, Schwann cells originate from the neural crest, olfactory ECs appear to originate from the olfactory placode, and tanycytes and pituicytes arise, like astrocytes and oligodendrocytes, from the subventricular zone. The work reported here, as well as similar observations on perivascular glia, retinal Müller cells, and pineal glia (Gudiño-Cabrera and Nieto-Sampedro, unpublished), indicate that immunological markers do not have a diagnostic value for establishing glial lineage. The above-mentioned cells share specific immunological markers and proliferative and growth-promoting properties, but have lineages that differ from each other. Although O4 antigen is a specific marker of pro-oligodendrocytes at an ontogenic stage subsequent to O2-A precursors, and O4-positive pro-oligodendrocytes persist in the adult and are capable of division, adult O4-positive ECs, tanycytes, and pituicytes are not precursors. Each cell type is quite abundant in the anatomical locus where it is found, and has a precise physiological role in the adult. They seem morphologically, functionally, and immunologically homologous to each other and to non-myelinating Schwann cells and — together with immature Müller and Bergmann cells, as well as other young radial glia types of common phylogenic (but not ontogenic) origin (Reichenbach and Robinson, 1995) — they appear to constitute a distinctive type of central macroglia. In view of their ability to grow and promote growth, we have named them aldynoglia, from the greek αλδαινω (to make grow). Aldynoglia share functional and structural properties, constituting a CNS macroglial type different from astrocytes and oligodendrocytes. It is recognized by its simultaneous expression of p75 NTR, O4 glycolipid, ER-α, vimentin, and GFAP. Though different from astrocytes and oligodendrocytes, aldynoglia are frequently misnamed astroglia. They are involved in naturally occurring promotion of axonal regeneration in the adult, supporting the notion that similar biological needs in different biological systems (need of continuous renewal of olfactory neurons, sensory receptors, or secretory axons) elicit similar gene expression.

10. MOVEMENT: NEURAL DAMAGE AND ITS REPAIR

Muscle contraction and its consequence, movement, underlie all human interactions. Thanks to the co-ordinated action of different muscles it is possible to talk, write, walk, or use tools. Many visceral functions also are possible because of the contraction or relaxation of muscles.

Movements may be classified according to the neural system that induces them. *Reflex movements* are the most elementary form of motor behavior: a sensory signal triggers motor neuron activity, allowing stereotyped, survival-effective acts. Many control systems have been implemented over these basic sensorimotor systems, allowing a greater behavioral range that facilitates appropriate responses in a complex environment. *Voluntary movements* are induced by signals arising from integration and control systems, such as the neocortex and the basal ganglia. *Automatic movements* are stereotyped motor behavior, intermediate between reflex and voluntary movements. Locomotion represents very well these stereotyped movements, the execution of which may be induced either by brain control centers or by peripheral sensory signals. Control systems confer *behavioral significance* on automatic movements. The kinetic chains performed during walking are produced by activation of neuronal networks organized in the spinal cord that generate the patterns required for co-ordinated activation of the muscles involved. The activity of such networks is controlled by sensory information, by the cerebellum, and by brainstem nuclei, which are, in turn, controlled by neocortex and basal ganglia. Following the observations of Grillner (Grillner and Wallén, 1985) it appears that *behaviorally significant locomotion* needs i) a signal to initiate the march, i.e. to assume the adequate posture and begin walking; ii) patterns of activation that generate the movement synergies, i.e. co-ordinated flexion and extension of the limb joints that produce body displacement; iii) a postural control capable of maintaining balance during displacement, which implies controlling the position of the center of gravity while one or more limbs are in the air, and generating enough anti-gravity force with the standing limbs to maintain body height; and iv) the adaptation of the march to the organism objectives in its behavioral context. Now, let us assume a complete spinal cord transection in segment D8 (T8 in animals): in both humans and animals part of the trunk and the hind-limbs will be completely paralyzed. Supraspinal control will be lost, which implies losing the ability to initiate the march voluntarily, to assume an adequate posture, to maintain the balance and to adapt the march to the objectives of the organism. However, because automatic movements also have an important component of sensory control, the appropriate sensory stimulation together with body weight support in animals with complete T8 transection, may induce coordinated flexo-extension joint synergies. Although these are not behaviorally significant movements, they are indicative of the spinal cord's potential for therapeutic intervention.

Unfortunately, to date, we have not been able to repair spinal cord lesions and achieve behaviorally significant locomotion in either experimental animals or in humans. When animals with complete spinal cord lesions and therapeutic interventions are tested on a flat open field, they are as unable to move as are untreated animals. If the rats could not use their forelimbs (with which they are able to lift their own weight), they could not climb. Certainly, they are unable to stand and walk. The best evaluation of functional recovery after therapy is that of Schwab's team (Thallmair et al., 1998), combining kinematic analysis and electromyography. Their animals had incomplete cord lesions, hence residual — though deficient — motor capacity. The administration of the antibody to myelin growth inhibitors, IN-1, promoted motor improvement.

Obviously, using scales where a score is given for task fulfillment is inadequate for measuring motor function. The complexity of movement analysis is compounded by the lesion, the neural response to the lesion, and the effect of whatever therapeutic interventions we have used. Kinematic analysis, combined with kinetic analysis, and electromyography are essential to know what motor functions are lost, what forces are involved, what is recovered, and how recovery takes place. When experimental animals are used, it is essential to distinguish real repair from compensating strategies that mask recovery.

11. LESIONS THAT HAVE BEEN REPAIRED IN RODENTS WITH EC TRANSPLANTS

We proposed EC transplantation as a general strategy for CNS injury repair (Ramón-Cueto and Nieto-Sampedro, 1994; Gudiño-Cabrera and Nieto-Sampedro, 1996), an optimistic proposal that needs reappraisal. EC-promoted repair depends on many variables, and appears limited by, among others, the well-defined migration preferences of ECs in adult CNS (Gudiño-Cabrera and Nieto-Sampedro, 1996). ECs promote fiber regrowth only when the growth direction of the regenerating sprouts coincides with the preferred direction of EC migration, such as in the spinal cord dorsal horn. Other limitations, about which we know less, are sensitivity of ECs to free radicals, the type of their neurotrophic factor output, the type of guiding surface molecules, and the chemiotropic systems, all of which probably contribute to repair variability. Another set of variables that may have a strong influence on the behavior of transplanted ECs are the age of the animal donor and the age and purity of the cells transplanted. The cell types present in EC cultures were characterized by Gudiño-Cabrera and Nieto-Sampedro (2000). Primary cultures of the nerve and glomerular layers of the adult male rat olfactory bulb contained, after 10 days *in vitro*, about 36% p75-NGF receptor-positive cells (ECs with Schwann-like and oligodendroblast-like morphologies), 35% endothelial cells (Factor VIII immunopositive), 15% actin-immunopositive pericytes, 10% OX-42 immunopositive macrophages, and, with proper dissection, less than 1% of fibroblasts (thy 1 and fibronectin immunopositive). ECs used for spinal cord repair could be immunopurified to 92-97% Schwann-like cells, labelled with the fluorescent linker PKH26 (Sigma) and stored at -80 °C for up to 6 months.

The repair of central dorsal rhizotomies with EC transplants has consistently been satisfactory. After rhizotomy, the dorsal root axon forms abundant sprouts that are, however, unable to penetrate again into the spinal cord (Ramón y Cajal, 1914). Dorsal rhizotomy is considered a central lesion and has been used extensively to investigate CNS regeneration because of sensory afferents high intrinsic growth capacity and well-characterized morphological and physiological properties (Fraher, 2000). Reactive astrocytes at the boundary between PNS and CNS (the dorsal root entry zone or DREZ) act as a stop signal for regenerating axons. ECs potent neurite-promoting properties, and their ability to ensheath axons rapidly, to mingle with astrocytes, and to migrate long distances in the adult CNS, make them excellent repair tools. Transplanted ECs modified the spinal environment, causing astrocyte reactivity to diminish or disappear (Verdú et al., 2001), and promoted regeneration and spinal ingrowth of damaged dorsal root fibers (Nieto-Sampedro and Ramón-Cueto, 1993; Ramón-Cueto and Nieto-Sampedro, 1994). Moreover, ECs were as good as or better than Schwann cells in bridging a gap between the proximal and distal stumps of a resected peripheral nerve, a complement frequently required in clinical cases of dorsal root avulsions.

11.1. Sciatic nerve resection repair

The great sciatic nerve carries a heterogeneous population of sensorimotor and autonomic fibers that mediate hind-limb function and body position. The hind-limb paralysis that follows sciatic resection is a challenge for new techniques, which can be compared to the known repair by tubulization combined with transplantation of Schwann cells (Ansselin et al., 1997).

The sciatic nerve of adult female Wistar rats was exposed at the mid-thigh, transected 95 mm from the tip of the third digit, and a distal segment resected. The distal and proximal stumps were fixed by epineural 10-0 suture into each end of a silicone tube, leaving an interstump gap of 12 or 15 mm. In the group with a gap of 12 mm, the tube was pre-filled with collagen gel at a final concentration of 1.28 mg/mL (group C1, n = 10) or with a suspension of about 60,000 ECs in collagen gel (group EC1, n = 14). In the group with the 15 mm gap, the tubes were pre-filled with a laminin-containing gel alone (group C2, n = 6) or with ECs (about 120.000 cells/tube; group EC2, n = 6). Finally, the wound was sutured with 5-0 silk thread in the muscular plane, and the skin was closed with small clips and disinfected. The animals were kept in a warm environment until full recovery from anesthesia and later given amitriptyline (150 µg/mL) in their drinking water to minimize autotomy (Verdú et al., 1999). Functional tests were performed before operation and 60 and 120 days postoperation in groups with a gap of 12 mm, and at 120 and 180 days postoperation in groups with a gap of 15 mm. The sciatic nerve of anesthetized animals (pentobarbital, 30 mg/kg i.p.) was stimulated percutaneously through a pair of needle electrodes near the sciatic notch. Compound muscle action potentials (CMAP) were recorded from the medial gastrocnemius muscle using monopolar needles. Compound nerve action potentials (CNAP) were recorded with needle electrodes inserted near the tibial nerve at the ankle. Compound action potentials were amplified, and displayed on the oscilloscope to measure the amplitude of the negative peak and the latency to the peak. Normal gastrocnemius mean CMAP amplitude was about 45-50 mV and its latency 2.2-2.4 ms, which changed to 1.5 and 3 mV, and a latency of about 4 and 5 ms, by 60 dpo, in groups C1 and G1 of resected animals with an interstump gap 12 mm. After 120 dpo, the CMAP amplitude increased to 11 and 15 mV and the latency shortened to 3 ms. Tibial CNAP could be recorded at 60 dpo in 2 rats of group G1, but in none of group C1. By 120 dpo, the mean tibial CNAP amplitude was 14 and 53 µV in groups C1 and EC1, respectively, and the latency just above 1.5 ms in both groups. No significant differences between groups C1 and EC1 were found at any of the intervals tested. When an interstump gap of 15 mm was left, functional responses could be recorded in 3 rats of group EC2, but in none of the controls. Significant differences were found between these two groups for CMAP and CNAP amplitudes. After functional follow-up, the rats were deeply anesthetized, the operated hindlimb was dissected, and the implanted tube was inspected to verify the presence of a regenerated nerve and its gross quality. Semi-thin sections (1 µm) were cut at the midpoint of the tube and at the distal nerve and stained with toluidine blue. Morphometrical evaluation, including myelinated fiber (MF) counts and area of the nerve cable, was made from photographs of selected fields at X2000 final magnification, with the help of a computer-linked digitizing tablet and specific software. Microdissection showed a regenerating nerve cable inside the silicone tube in 6 of 10 rats in group C1; 11 of 14 in group EC1, and 3 of 6 in group EC2. No nerve cable was seen in rats of group C2. Morphometrical analysis (Verdú et al., 1999) showed that the regenerated cable area and MF number were slightly, though not significantly, higher in group C1 than in group EC1 at both medial and distal levels. Observation of transverse sections at mid-tube revealed that the repaired nerves had a thin perineurium and were organized in multiple fascicles, smaller in group EC2 than in groups EC1 and C1. The number of cell nuclei in transverse sections was higher in EC-transplant nerves than in controls. Macrophages and mast cells were observed in all groups of regenerated nerves more frequently than in the normal sciatic nerve.

11.2. Central dorsal rhizotomy repair with EC transplants

After section of the central branch of a dorsal root, the sensory axons form abundant sprouts that are, however, unable to penetrate again into the spinal cord (Ramón y Cajal, 1914). Cultured and purified ECs transplanted at the point of entry of a centrally sectioned sensory root enabled the navigation of regenerating sensory sprouts through the inhibitory isomorphic gliotic tissue formed after rhizotomy, overcoming the inhibitory properties of reactive glia. Since ECs transplanted into an injured spinal area rich in hypertrophic reactive astrocytes caused astrocyte reactivity to diminish/disappear (Verdú et al., 2001), it seems reasonable that, transplantation of ECs near the DREZ after dorsal rhizotomy will modify the spinal environment to make it less inhibitory, while promoting sprouting and ingrowth of damaged fibers (Nieto-Sampedro and Ramón-Cueto, 1993; Ramón-Cueto and Nieto-Sampedro, 1994). Because of overlap of innervation by adjacent dorsal roots, the damage to a single dorsal root does not cause functional deficits that permit assessing recovery promoted by EC transplants. Accordingly, we performed multiple rhizotomy at the cervical (Taylor et al., 2001), lumbar (Navarro et al., 1999), and lumbosacral (Pascual et al., 2002) levels, and determined the extent of EC-promoted recovery with anatomical, electrophysiological, and behavioral techniques.

Clinical repair of dorsal root avulsions may require in many cases the complement of peripheral nerve repair. As indicated above, ECs have proven as good as Schwann cells in bridging a gap between the proximal and distal stumps of a resected peripheral nerve. As we will see, EC transplantation may not be a general solution for every CNS injury; however, their suppression of astrocyte reactivity and general promotion of neuronal survival and axonal growth represent a step in the right direction.

11.3. Lumbosacral rhizotomy and urinary bladder function

L1 to L3 laminectomy was performed on adult male Wistar rats (2.5 mo. old, 250-300 g), deeply anesthetized with sodium pentobarbital. The bladder was exposed via a lower midline incision, and a polyethylene catheter (1.3 mm outer diameter and 0.8 mm inner diameter) was placed through the bladder dome and secured with a 6-zero polyglycolic acid suture. The abdominal wall was temporarily closed, and abdominal pressure was recorded with a rectal catheter. The dorsal roots innervating the bladder were identified by cystometrography. Responses to buffered saline infused through the catheter (with a peristaltic micro pump) were recorded through a pressure transducer coupled to a multichannel register system, which also directly recorded abdominal pressure. After stabilization, saline solution at room temperature was infused at a rate of 0.16 mL/min for 15 minutes and maximum intravesical pressure and amplitude rate after ventral root (L6 to S3) electrostimulation were recorded. The lumbosacral roots carrying motor fibers to the bladder were identified by the magnitude of the pressure response evoked from the bladder after stimulating the ventral root with monophasic pulses through a hook electrode. Ventral root stimulation increased the intravesical pressure, with the highest response usually at S1. Accordingly, presumptive S1 and adjacent dorsal roots L6, and S2 were bilaterally sectioned close to the spinal cord DREZ, leaving the ventral roots intact. A suspension of purified ECs (Gudiño-Cabrera and Nieto-Sampedro, 1996) was injected into the cord through a glass micropipette by means of repeated air pulses (3-5 msec). Approximately 30,000 ECs in DMEM/F12 per sectioned root were injected 800 µm deep of the entry zone, near the sacral parasympathetic nucleus. Five min from the end of the injection, the micropippete was slowly withdrawn, and transected roots were reapposed to the cord with a drop of bovine fibrin (Pascual et al., 2002).

Cystometrography was performed on the animals one week after surgery, closing the temporary vesicostomy. The bladder was emptied thereafter twice daily by manual expression. Six weeks postoperation, cystometrography was repeated, 5 µL of wheat germ agglutinin/horse radish peroxidase were injected into the body of the bladder wall, and 48 hours later the animals were sacrificed. Preganglionic lumbosacral rhizotomy led to an atonic bladder that retained a large volume of urine. Animals that received EC transplants still had an atonic bladder 6 days postoperation. However, 6 weeks after EC transplantation, dorsal root fibers had grown into the spinal chord and formed synapses in the sacral parasympathetic nucleus. This morphological repair correlated with very significant functional recovery (Pascual et al., 2002).

11.4. Lumbar rhizotomy and hindlimb function

Lumbar roots are the major contributors to the sciatic nerve, and hence to hind-limb function and body position. Extensive lumbar laminectomy was performed on adult female Wistar rats (250-300 g., anesthetized with pentobarbital, 50 mg/kg), the dura was cut, bupivacaine (0.75%) was locally infiltrated to minimize nociceptive input, and L3 to L6 dorsal roots were unilaterally sectioned 1 mm from the entrance into the spinal cord. Immunopurified ECs (30,000 cells in 3 µL DMEM) were injected into the central root stump through a glass micropipette joined to a polythene cannula, by repeated air pulses (20 msec, 10 psi). After transplantation, transected roots were reapposed to their proximal stumps and attached with fibrin. The wound was sutured in the muscular plane and the skin closed with small clips and disinfected. The animals were kept in a warm environment until full recovery from anesthesia, and later given amitriptyline in their drinking water to minimize autotomy.

Multiple preganglionic lumbar rhizotomy abolished both spinal reflexes and the H-wave response evoked by sciatic nerve stimulation and measured in plantar and gastrocnemius muscles. Animals that received EC transplants began to recover the lost neurophysiological properties 14 days post-transplantation, and showed very significant sensory and motor recovery 60 days postoperation (Navarro et al., 1999).

11.5. Cervical rhizotomy and forelimb function

The recovery of function achieved by EC transplantation after multiple rhizotomy at the lumbar and lumbosacral levels encouraged us to examine whether any variations of the EC transplantation technique maight be used to repair dorsal rhizotomies at the cervicothoracic levels. Multiple cervicothoracic rhizotomy in rodents is a good model for brachial plexus avulsion, a comparatively common incapacitating injury after sports and work accidents or complicated child delivery. Our studies show that EC transplantation promotes axonal regrowth with reinnervation of the dorsal horn, activation of postsynaptic dorsal horn neurons, and restitution of polysynaptic reflex activity, leading to recovery of cutaneous, proprioceptive, and autonomic sensory functions. Furthermore, recovery of electrophysiologically functional anatomical connections was translated at the behavioral level into the restoration of withdrawal reflexes and organized supraspinal nocifensor responses to forepaw stimulation (Taylor et al., 2001).

Unilateral C4 to T2 dorsal rhizotomy in the rat led to severe sensory deficits in the forepaw that were significantly corrected by EC transplantation medial to the DREZ of dorsal roots C7/C8. EC transplantation restored *c-fos* protein expression in intrinsic neurons of the superficial dorsal horn laminae after noxious stimuli, such as immersion of the ipsilateral forepaw in water at 52°C (Hunt et al., 1987). The expression of *c-fos* protein was observed in non-operated animals or in sham-operated controls, was lost in the dorsal horn of rats ipsilateral to rhizotomized, and partially restored after EC-transplantation. Therefore, after sensory axon injury, EC transplants help to re-establish functional contacts between repaired axons and intrinsic dorsal horn neurons.

Direct evidence of restitution of functional cutaneous reflex circuits was obtained by observing forepaw withdrawal from radiant heat, mediated by biceps reflex contraction. Biceps activity could be elicited by stimulation of the median nerve ipsilateral to the C4-T2 rhizotomy in rats that received C7/C8 EC transplants. Stimulus-response curves after median nerve activation were obtained for non-operated animals, control rats rhizotomized or rhizotomized and transplanted with cell culture medium, and rats transplanted with ensheathing cells. Activation with pulses of short (0.1 ms) or long (0.5 ms) duration indicated that only small-diameter afferent fibers had regenerated in transplanted animals. The biceps reflexes that were recovered after EC transplantation showed a higher "gain" than in either normal or non-transplanted rhizotomized rats. The biceps muscle discharge evoked by median nerve stimulation in normal adult rats was lost completely after rhizotomy or transplant of DMEM alone, and recovered in EC-transplanted animals. However, the recovered discharge, which was synchronic in normal animals, was still markedly asynchronic in EC transplanted animals six months postoperation (Taylor et al., 2001).

Recovery of sensory input, observed electrophysiologically, occurred in parallel with the restoration of sensory reflexes in the alert animal. Withdrawal reflex and supraspinal nocifensor responses to both radiant heat and mechanical stimuli applied to the ipsilateral forepaw plantar pad were assessed behaviorally before the lesion and up to 2 months after cervical dorsal rhizotomy. Reflex activity in response to mechanical stimuli after brachial rhizotomy was not evoked in either the non-transplanted group or the EC-transplanted group (not shown). In contrast, withdrawal of the ipsilateral forepaw from noxious thermal stimuli recovered significantly in EC-transplanted animals. During the first month following rhizotomy, ipsilateral withdrawal reflex responses were abolished in both experimental groups. However, during the second month after injury, the withdrawal reflex latency in the EC-transplanted group (10-15 s) was significantly better than that in non-transplanted controls (20-22 s). The same behavioral test was used to measure the recovery of organized and unlearned supraspinal responses with a non-parametric scale defined to grade orientation of the head and paw-licking behavior in response to forepaw thermal stimulation. This supraspinal nocifensor score fell from a maximum value in normal animals to '0' after rhizotomy, and remained at '0' in non-transplanted animals. Rhizotomized rats showed no orientation or contact with the ipsilateral forepaw after noxious thermal stimulation. In contrast, the group of EC-transplanted rats recovered a significant supraspinal response during the second month following rhizotomy, i.e. they noticed that the forepaw was stimulated. The development of supraspinal nocifensor responses following EC implantation correlated very significantly with the recovery of flexor withdrawal activity, indicating that both measured responses were evoked by the same regrowing afferents.

12. LESIONS THAT COULD NOT BE REPAIRED WITH EC TRANSPLANTS

EC implant usefulness as an injury repair tool is affected by Ecs' well-defined migration preferences in the adult CNS (Gudiño-Cabrera and Nieto-Sampedro. 1996). This may be one of the reasons why EC transplants could not help repair medial longitudinal fascicle transection, although in the case of corticospinal transection, there may be additional reasons for failure, some of them still unknown. In other CNS injuries and loci, such as cyst formation after photochemical damage in thoracolumbar cord segments, injury repair was partial.

12.1. Oculomotor reinnervation after medial longitudinal fascicle transection

Ocular vergence is mediated by fibers traveling in the medial longitudinal fascicles (MLFs) and is disrupted by MLF transection caudal to the trochlear nucleus. Adult cats were bilaterally implanted with eye coils and a head-restraining system that permits checking eye movements with the search-coil technique (Delgado-García et al., 1986). Two months later, the left abducens nucleus was located by electrical stimulation of the left lateral rectus muscle with hook electrodes and recording of the antidromic field potential. The stereotaxic co-ordinates of the abducens nucleus were used to transect both medial longitudinal fascicles (MLFs) just caudal to the trochlear nucleus, at approximately 0 mm on the rostrocaudal axis. After the lesion, cats were injected bilaterally with a suspension of viable ECs (20,000 cells/µL). Feline ECs were prepared from nerve and glomerular layers of the olfactory bulb of new-born cats (Gudiño-Cabrera and Nieto-Sampedro, 1996), immunopurified, labeled with PKH26 and resuspended in DMEM to a final density of 10,000 cells/µL. Each transected MLF received a total of 2.4 µL of cell suspension, delivered at 6 different ventral to dorsal points (0.4 µL in each site, separated by 300 µm). Starting one day after surgery, cats were checked daily for eye movements until a progressive recovery of eye position in the horizontal plane was observed.

Similar cell-culture and surgical procedures were used for rats, except that, no eye coils were implanted and the MLF was ought stereotaxically. Rats were anesthetized with 4% chloral hydrate (1mL/100g; i.p.) and placed in a stereotaxic device. A small opening was made in the skull, a microknife was advanced through the cerebellum to the brainstem, and unilateral transection of the right MLF was performed stereotaxically at 1-1.3 mm anterior from interaural zero. A suspension of ECs (20.000 cells/µL) was transplanted into the transected MLF, using the same approach and co-ordinates as for the lesion. Cells were injected with a Hamilton syringe at five different points from ventral to dorsal (0.2 µL at each step, separated by 50 µm) so that the total volume injected in the transected MLF was 1 µL (Gudiño-Cabrera et al., 2000).

Host MLF and the neighboring reticular formation did not show any sign of cytoarchitecture disruption, indicating that ECs integrated properly with the surrounding tissue. Although most transplanted ECs remained in place, some of them migrated via the MLF, preferentially in the caudal direction, up to 2.4 mm from the transplant site. A small number of ECs migrated through the rostral MLF or ventrally towards the underlying reticular formation. Migrating ECs formed chains of glial elements lined along *abducens* internuclear neuron axons, an image strongly reminiscent of the association of Schwann cells to regenerating axons in the fiber ensheathment stage called the *promyelinating* stage. Numerous axonal profiles were observed in the MLF, always caudal to the lesion (and transplant) sites (Gudiño-Cabrera et al., 2000). Proximal axons from abducens internuclear neurons branched extensively, apparently unable to cross the lesion site in any of the four cats operated. Cat ECs were prepared specifically for these experiments, and had morphological and immunological properties similar to rat ECs. However, to check that lack of regeneration was not due to variations in EC purity or identity, the same experimental approach was carried out in rats (n=7), using ECs that successfully promoted dorsal root axonal regrowth. As in the case of cats, abducens internuclear neuron axons failed to cross the MLF transection site, although surviving ECs were present.

Abducens axons coursed with abnormal trajectories, typically making U turns or running in parallel to the transection. Their terminals formed large, smooth clubs, suggesting terminal or *en-passant* bouton-like structures or sprouting short, thin collaterals. Parvalbumin and calretinin immunostaining also revealed that sprouts from sectioned MLF axons never crossed the transection site. Calretinin is a good marker of abducens internuclear neurons, while parvalbumin labels other axons coursing along the MLF, some of them probably of vestibular origin. Regenerating axons of both calretinin- and parvalbumin-positive types were observed only caudal to the site of MLF transection and EC transplant. Therefore, the type of fiber that did not regenerate in response to EC transplants was not restricted to abducens internuclear neuron axons, but extended to fibers of other origins.

12.2. Corticospinal tract after mild contusion to the cervical spinal cord

The corticospinal tract is essential for both purposeful locomotion and skilled and delicate finger movements in animals capable of them, such as rats or humans. Damage of corticospinal fibers caused reversible locomotor disturbances in animals moving over irregular surfaces, and hindered skilled forelimb movement.

Mild spinal cord contusion was inflicted on adult male Wistar rats (n=27; 18–24 weeks old, 400–450 g) by dropping a weight (10 g, from 12.5 mm) on to a rectangular impounder (2.5 mm x 3.0 mm) resting on the dura mater of spinal cord segment C7. The contusion destroyed most gray matter in the segment, and completely interrupted the dorsal corticospinal tract (dCTS), while partially sparing the lateral funiculi. One group of rats (n = 14) received 1 million Fast-Blue-labeled ECs, prepared from animals 7-8 weeks old, immediately after the contusion. In one subgroup of these rats (n = 8) the cells were injected into five points of the cord, 0.5 mm apart (about 200,000 cells/2 µL/site), spanning the contusion rostrocaudally. The same total number of ECs was transplanted in another subgroup of animals (n = 6), adding two additional injection points in the dCST, just before the contusion site (Collazos-Castro, 2003). Another group of rats (n = 3) received this same distribution of 1 million ECs into the lesion zone, but eight days after injury. The remaining rats (n=13) were controls transplanted with DMEM or nothing. The dCST was traced 10 weeks after transplantation, by injecting dextran tetramethylrhodamine (mol.weight = 3,000) into the sensorimotor cortex. The animals were perfused two weeks later and immunohistochemistry for p75 NGF receptor and neurofilament was performed to visualize the relation between axonal growth and transplanted ECs.

Spinal contusion led to formation of cysts in the injured area. Survival of ECs in the cystic tissue was variable: sometimes the cells formed a disorganized mass that completely filled the cyst and at other times arranged themselves as string across the cavity. Labelled ECs did not migrate away from the contusion zone, where they could be identified by their strong Fast Blue fluorescence and p75 immunoreactivity. The dCST was completely interrupted in all injured animals, but its regeneration was not promoted by transplanted ECs which, however, survived and proliferated well or very well. In both transplanted and control animals, retraction bulbs from dCST axons were observed ending near the edge of the contused tissue, or at some microns from this border. Neurofilament positive processes indicative of a neurite outgrowth were observed in EC-transplanted animals. However, these sprouts originated from either dorsal root axons or spinal interneurons. Transplanted ECs preserved their neurite-promoting properties, but the growth response was restricted to some neuronal populations. Furthermore, the cells transplanted in the lesion zone were oriented randomly, and the neuritic sprouts precisely followed that random orientation. Growth promotion involved direct contact of axon-EC plasma membrane, but the random orientation of ECs, together with the lack of corticospinal tract regrowth, indicates that EC transplants alone are of limited use for repair of spinal cord lesions arising from mild contusion.

13. LESIONS PARTIALLY REPAIRED WITH EC TRANSPLANTS

Because ECs had neurotrophic, neuritogenic, neurite-ensheathing, and neurite-guiding properties, EC transplants led to partial lesion repair when only some of these activities were expressed. Such was the case in the attempted repair of photochemical damage.

13.1. Cyst formation after photochemical spinal cord damage

Illuminating the spinal cord surface bathed in Rose Bengal solution led to the formation of parenchyma cavities reminiscent of the cysts after contusions. Segments T12-L1 of the spinal cord of female Sprague-Dawley rats (250-300 g; n = 20) anesthetized with sodium pentobarbital (50 mg/kg, i.p.) were exposed by dorsal laminectomy. The dura was cut, and local infiltration with bupivacaine (0.5%) was performed to minimize nociceptive input. Rose Bengal solution (1.5% w/v in saline) was applied directly on the exposed spinal cord for 10 min, excess dye was rinsed off with saline, and the cord was illuminated with two optic fibers 10 mm away from the cord surface and connected to a 150 W halogen source. The cord was illuminated at maximum nominal power for 5 minutes. This photochemical lesion resulted in severe sensorimotor deficits for the first 15 days postoperation. All injured rats dragged their hindlimbs and did not respond to pinprick (Verdú et al., 2001). Then, immunopurified ECs, labeled with PKH26 (60,000 cells in 3 µL of DMEM), were injected bilaterally into each of spinal cord segments T12 to L1 (n = 10, group RB5 + EC) through a glass micropipette, with the help of repeated 20 ms, 10 psi, air pulses (Picospritzer II, General Valve, Fairfield, NJ). The remaining 10 rats were injected with saline (control group, RB5; n = 10). The wound was sutured in the muscular plane with 5-0 silk thread, the skin was closed with small clips, and disinfected, and the animals were kept in a warm environment until full recovery and given amitriptyline (150 µg/mL) in their drinking water to prevent autotomy. Transplantation of olfactory ECs into the damaged spinal cord diminished astrocyte reactivity and parenchyma cavitation. The cystic cavities formed in transplanted rats had a maximal area and volume lower than those in nontransplanted animals. The decrease was not significant at the T12-L1 lesion site, but was significant at the T9-T10 and L4-L6 cord levels. The density of astrocytes in the gray matter segments T12-L1 and L4-L6 was similar in non-transplanted and transplanted rats, but was lower in transplanted rats at the T9-T10 level. Astrocytes in non-transplanted rats showed a hypertrophied appearance, with long robust processes heavily GFAP-positive and overexpressed neuritogenesis-inhibitory proteoglycan. In contrast, most astrocytes in transplanted rats had short thin processes, and only a few were hypertrophied. ECs transplanted into damaged adult spinal cord were neuroprotective, reducing both cystic cavitation and astrocytic gliosis (Verdú et al., 2001).

13.2. Breathing and climbing

Breathing depends on bulbospinal respiratory pathways that carry the descending rhythmic impulses needed to activate the cervical motoneurons originating the phrenic input to the diaphragm. Damage to the upper spinal cord causes loss of ability to breathe, and survival after such injuries is possible only with assisted ventilation. Bulbospinal phrenic motoneuron projection is bilateral in both rats and humans, and bilaterally symmetrical in the rat. High lateral hemisection of the rodent spinal cord destroys part of the respiratory pathway but leaves the opposite side functionally intact. Thus, it spares sufficient supraspinal control to allow survival in normal cage conditions.

The entire spinal gray and white matter of inbred AS strain rats was severed unilaterally by knife cuts at the C2 level. The lesions were reported to cause no cyst formation, persistent macrophage invasion, or heavy glial scarring, although not convincing anatomical evidence was presented. Immediately after the operation, the animals were reported to show unsteady gait and reduced response to pinch of the contralateral hind-paw. The deficits improved spontaneously during the first week postoperation and more rapidly in animals that received implants of about one million cultured ECs embedded in their own extracellular matrix, although by 2 weeks postoperation, both lesion-only and implanted animals were indistinguishable from each other or from unoperated rats (Li et al., 2003). No proper tracing of the connections mediating the recovered functions was reported and the paper is not convincing.

14. CONCLUSIONS AND PERSPECTIVES

The neural plasticity processes involved in learning and lesion repair are formally, and perhaps effectively, similar. Long-term potentiation (LTP) in the brain may mediate memory formation, and in the normal spinal cord very convenient automatisms. However, in the injured spinal cord, LTP equivalents probably mediate very undesirable effects, such as spasticity and neuropathic pain. I have attempted to show that learning and lesion repair are governed by overlapping cell biological events and rules. If this view is correct, it would mean that we could apply what is known about the pharmacology of one of these processes to affect the other.

Aldynoglia is probably the most hopeful CNS lesion repair tool developed in the last two decades. It is capable of normalizing hypertrophic reactive astrocytes when transplanted into a region of heavy gliosis (Verdú et al., 2001), it produces a variety of neurotrophic factors (Ramón-Cueto and Avila, 1998), its surface initiates and guides neurite growth, and the neurites thus initiated are ensheathed by a protective glial membrane fold. Therefore, it seemed reasonable to propose aldynoglia transplantation as a general strategy for CNS injury repair (Ramón-Cueto and Nieto-Sampedro, 1994; Gudiño-Cabrera and Nieto-Sampedro, 1996). However, the situation is not that simple, and similar lesions may be repaired more completely in some loci than in others. Many variables may be involved that limit EC repair capabilities. One of them is EC's well-defined migration preferences in adult CNS. Another is the deleterious effect on free radicals on the survival of ECs, which must be taken into account when transplanting these cells in the free-radical-rich injured tissue. Treatments that combine EC transplantation with exogenous supply of neurotrophic factors, free-radical-destroying molecules, and blocking antibodies to growth-inhibitory proteoglycan have not yet been tested. The newly-created *Experimental Neurology Unit* of the Hospital Nacional de Parapléjicos of Toledo intends to work on these and related topics during the next decade.

REFERENCES

Aguayo, A.J., David, S. and Bray, G.M. (1981) Influences of the glial environment on the elongation of axons after injury: transplantation studies in adult rodents. *J. Exp. Biol*. 95: 231-240.
Ansselin, A.D., Fink T. and Darvey, D.F. (1997) Peripheral nerve regeneration through nerve guides seeded with adult Schwann cells. *Neuropathol. Appl. Neurobiol*. 23: 387-398.
Arany, Z., Sellers,W.R., Livingston, D.M., Eckner, R. (1994) E1A-associated p300 and CREB-associated CBP belong to a conserved family of coactivators. *Cell* 77: 799-800.
Bartholdi, D. and Schwab, M.E. (1995) Methylprednisolone inhibits early inflammatory processes but not ischemic cell death after experimental spinal cord lesion in the rat. *Brain Res*. 672: 177-186.
Bignami A. and Dahl D. (1974) Astrocyte-specific protein and neuroglial differentiation: an immunofluorescence study with antibodies to the glial fibrillary acidic protein. *J. Comp. Neurol*. 153: 27-38.
Bovolenta, P., Wandosell, F. and Nieto-Sampedro, M. (1991) Central Neurite outgrowth over glial scar tissue in vitro. In S.B. Kater, P.C. Letourneau and E.R. Macagno (eds.) *"The Nerve Growth Cone"* Raven Press, pp. 477-488.
Bovolenta, P., Wandosell, F. and Nieto-Sampedro, M. (1992) CNS glial scar tissue: a source of molecules which inhibit central neurite outgrowth. *Progress Brain Res.*, 94: 367-379.
Caroni, P. and Schwab, M. (1988) Two membrane protein fractions from rat central myelin with inhibitory properties for neurite outgrowth and fibroblast spreading. *J. Cell Biol*. 106: 1281-1288.
Cheng, H, Hoffer,B., Strömberg, I., Russell, D., and Olson, L. (1995) The effect of a glial cell line-derived neurotrophic factor in fibrin glue on developing dopamine neurons. *Exp. Neurol*. 104: 199-206.
Cheng, H., Cao, Y. and Olson, L. (1996) Spinal cord repair in adult paraplegic rats: partial restoration of hind limb function. *Science* 273: 510-513.
Cheng, B. and Mattson, M.P. (1991) NGF and bFGF protect rat hippocampal and human cortical neurons against hypoglycemic damage by stabilizing calcium homeostasis. *Neuron* 7: 1031-1041.
Collazos-Castro, J.E. (2003) Ph.D. dissertation. Facultad de Medicina, Universidad Autónoma de Madrid.
Collazos-Castro, J.E., de Castro, F., Gudiño-Cabrera, G., Herreras, O., Insausti-Serrano, R., Navarro, X., Pascual-Piédrola, J.I., Taylor, J.S., Vidal, J. Nieto-Sampedro, M. (2003) Reparación del Trauma Medular. *Boletin SENC* 12: 12-25.
Delgado-García, J.M., del Pozo, F. and Baker, R. (1986) Behavior of neurons in the abducens nucleus of the alert cat. II. Internuclear neurons. *Neuroscience* 17: 953-973.
Doucette, J.R. (1986) Astrocytes in the Olfactory Bulb. In Fedoroff, S. and Vernardakis, A. (eds): *Astrocytes* vol.1, Acad.Press, Orlando, pp.293-310.
Doucette, R. (1990) Glial influences on axonal growth in the primary olfactory system. *Glia*, 3, 433-449.
Fraher, J.P. (2000) The transitional zone and CNS regeneration. *J. Anat*. 196: 137-158.
Greenfield, J.G. (1958) General pathology of nerve cell and neuroglia. In Greenfield, J.G., Blackwood, W., Meyer, A., McMenemey, W.H., Norman, R.M. (eds): *Neuropathology*, London: Ed. Arnold, Ltd. pp. 1-66.
Grillner, S. and Wallén, P. (1985) Central pattern generators for locomotion, with special reference to vertebrates. *Ann. Rev. Neurosci*. 8: 233-261.
Gudiño-Cabrera, G. and Nieto-Sampedro, M. (1996) nsheathing cells: Large scale purification from adult olfactory bulb, freeze-preservation and migration of transplanted cells in adult brain. *Restor. Neurol. Neurosci.*: 10: 25-34.
Gudiño-Cabrera G. and Nieto-Sampedro M. (1999) Estrogen receptor immunoreactivity in Schwann-like brain macroglia. *J. Neurobiol*. 40: 458-470.
Gudiño-Cabrera, G. and Nieto-Sampedro, M. (2000) Schwann-like rowth-promoting macroglia in adult rat brain. *Glia* 30: 49-63.
Gudiño-Cabrera, G., Pastor, A. M., de la Cruz, R. R., Delgado-García, J.M. and Nieto-Sampedro, M. (2000) Limits to the capacity of olfactory ensheathing glia to promote axonal regrowth in the CNS. *NeuroReport* 11: 467-471.
Helmuth, L. (2000) Stem cells hear the call of injured tissue. *Science* 290: 1479-1480.
Herreras, O. and Largo, C. (2002) Las huellas eléctricas en el camino hacia la muerte neuronal isquémica. *Rev. Neurol*. 35: 835-845.
Holmin, S., Almqvist, P., Lendahl, U., and Mathiesen, T. (1997) Adult nestin-expressing subependymal cells differentiate to astrocytes in response to brain injury. *Eur. J. Neurosci*. 9: 65-75.

Hunt, S.P., Pini, A. and Evan, G. (1987). Induction of c-fos-like protein in spinal cord neurons following sensory stimulation. *Nature* 328: 632-634.

Imaizumi, T., Lankford, K.L., Waxman, S.G., Greer, C.A. and Kocsis, J.D. (1998) Transplanted olfactory ensheathing cells remyelinate and enhance axonal conduction in the demyelinated dorsal columns of the rat spinal cord. *J. Neurosci.* 18: 6176-6191.

Johansson, C.B., Momma, S., Clarke, D.L., Risling, M., Lendahl, U., and Frisén, J. (1999) Identification of a neural stem cell in the adult mammalian central nervous system. *Cell* 96: 25-34.

Li, Y., Field, P.M. and Raisman, G. (1997) Repair of adult rat corticospinal tract by transplants of olfactory ensheathing cells. *Science* 277: 2000-2002.

Li, Y., Decherchi, P.. and Raisman, G. (2003) Transplantation of olfactory ensheathing cells into spinal cord lesions restores breathing and climbing. *J. Neurosci.* 23: 727-731.

Lindsay, R.M. (1986). Reactive Gliosis, in: *Astrocytes,* vol.3. (S. Fedoroff and A. Varnadakis, eds.) Acad.Press, New York, pp. 231-262.

Lipton, P. (1999) Ischemic cell death in brain neurons. *Physiol. Rev.* 79: 1431-1568.

Muñetón, V.C., Taylor, J.S, Averill, S.A., Vallejo-Cremades, M.T., King, V.R., Priestley, J.V., Nieto-Sampedro, M. (2003) Transplantation of olfactory ensheathing cells in a model of brachial plexus avulsion increases spinal cord permissivity to dorsal root ingrowth. *J. Neurocytol.*, in press.

Navarro X., Valero A., Gudiño-Cabrera G., Fores J.., Rodriguez F.J., Verdú E., Pascual R., Cuadras J., Nieto-Sampedro M. (1999) Ensheathing glia transplants promote dorsal root regeneration and spinal reflex restitution after multiple lumbar rhizotomy. *Ann. Neurol.* 45: 207-215.

Nieto-Sampedro, M. (1988) Growth factor induction and order of events in CNS repair. In *Pharmacological approaches to the treatment of brain and spinal cord injury* (eds. D.G.Stein and B.A. Sabel). Plenum Press, New York, pp.301-337.

Nieto-Sampedro, M. (1999) Neurite outgrowth inhibitors in gliotic tissue. In *The function of glial cells in health and disease: Dialogue between glia and neurons* (R. Matsas and M. Tsacopoulos, eds.) series *Advances in Exper. Med. Biol.*, Plenum Pub. Corp., New York, 468: 207-224.

Nieto-Sampedro, M.(2002) CNS Schwann-like glia and functional restoration of damaged spinal cord *Prog. Brain Res.* 136: .303-318.

Nieto-Sampedro, M. (2003) CNS lesions that can and those that cannot be repaired with the help of olfactory bulb ensheathing cell transplants. *Neurochem. Res.*, in press.

Nieto-Sampedro, M. and Cotman, C.W. (1985) Growth factor induction and temporal order in CNS repair. In *Synaptic plasticity* (C.W. Cotman, ed.). The Guilford Press, New York, pp. 407-455.

Nieto-Sampedro, M., Collazos-Castro, J.E., Taylor, J.S.,Gudiño-Cabrera, G., Verdú-Navarro, E., Pascual-Piédrola, J.I., Insausti-Serrano, R. (2002) Trauma en el Sistema Nervioso Central y su Reparación. *Rev. Neurol.* 35: 534-552.

Nieto-Sampedro, M., Lewis, E.R., Cotman, C.W., Manthorpe, M., Skaper, S. D., Barbin, G., Longo, F.M. and Varon, S. (1982a) Brain injury causes a time-dependent increase in neuronotrophic activity at the lesion site. *Science* 221: 860-861.

Nieto-Sampedro, M. and Ramón-Cueto, A. (1993) Transplants of ensheathing cells facilitate sensory fiber ingrowth and regeneration into adult spinal cord. *Eur. J. Physiol.* 427, suppl. 1, R51.

Pascual, J.I., Gudiño-Cabrera, G., Insausti, R. and Nieto-Sampedro, M. (1997) Loss and restoration of rat urinary bladder function after lumbosacral rhizotomy and ensheathing glia transplantation. *Soc.Neurosci. Abstr.* 23, 1720.

Pascual, J.I., Gudiño-Cabrera, G., Insausti, R. and Nieto-Sampedro, M. (2002) Spinal implants of olfactory ensheathing cells promote axon regeneration and bladder activity after bilateral lumbosacral dorsal rhizotomy in the adult rat. *J. Urol.* 167: 1522-1526.

Ramón y Cajal, S. (1914) *Estudios sobre la degeneración y regeneración del sistema nervioso.* Imprenta Hijos de Nicolás Moya, Madrid.

Ramón-Cueto, A. and Avila, J. (1998) Olfactory ensheathing glia: properties and function. *Brain Res. Bull.* 46: 175-187.

Ramón-Cueto, A., Cordero, M. I., Santos-Benito, F. F. and Avila, J. (2000) Functional recovery of paraplegic rats and motor axon regeneration in their spinal cords by olfactory ensheathing glia. *Neuron* 25: 425-435.

Ramón-Cueto, A. and Nieto-Sampedro, M. (1992) Glial Cells from Adult Rat Olfactory Bulb: Immunocytochemical Properties of Pure Cultures of Ensheathing Cells. *Neuroscience* 47: 213-220.

Ramón-Cueto, A. and Nieto-Sampedro, M. (1994) Regeneration into the spinal cord of transected dorsal root axons is promoted by ensheathing glia transplants. *Exp Neurol*, 127: 232-244.

Ramón-Cueto, A., Pérez, J. and Nieto-Sampedro, M. (1993) *In vitro* enfolding of olfactory neurites by p75 NGF receptor positive ensheathing cells from adult rat olfactory bulb. *Eur. J. Neurosci.*, 5: 1172-1180.
Reichenbach, A. and Robinson, S.R. (1995) Ependymoglia and ependymoglia-like cells. In: Kettenman, H. & Ransom, B.R. (eds). *Neuroglia*. Oxford Univ.Press, New York, pp. 58-84.
Rudge, J.S. and Silver, J. (1990) Inhibition of neurite outgrowth on astroglial scars in vitro. *J Neurosci.* 10: 3594-3603.
Schnell, L. and Schwab, M.E. (1990) Axonal regeneration in the rat spinal cord produced by an antibody against myelin associated neurite growth inhibitors. *Nature* 343: 269-272.
Schnell, L. and Schwab, M.E. (1993) Sprouting and regeneration of lesioned corticospinal tract fibers in th adult rat spinal cord. *Eur. J. Neurosci.* 5: 1156-1171.
Schwab, M.E. and Bartholdi, D. (1996) Degeneration and regeneration of axons in the lesioned spinal cord. *Physiol. Rev.* 76: 319-370.
Schwab, M.E. and Caroni, P. (1988) Oligodendrocytes and fibroblast spreading in vitro. *J. Neurosci.* 8: 2381-2393.
Silver, J. (1984) Studies on the factors that govern directionality of axonal growth in the embryonic optic nerve and at the chiasm of mice. *J. Comp. Neurol.* 223: 238-251.
Taoka, Y, Okajima, K, Uchiba, M and Johno, M. (2001) Methylprednisolone reduces spinal cord injury in rats without affecting tumor necrosis factor-β production. *J. Neurotrauma.* 18: 533-543.
Tator, C.H. and Fehlings, M.G. (1991). Review of the secondary injury theory of acute spinal cord trauma with emphasis on vascular mechanisms. *J. Neurosurg.* 75: 15-26.
Taylor, J.S., Muñetón-Gómez, V.C., Eguía-Recuero, R. and Nieto-Sampedro, M. (2001) Transplants of olfactory bulb ensheathing cells promote functional repair of multiple dorsal rhizotomy. *Prog. Brain Res.* 132: 651-664.
Tello, J.F. (1911) La influencia del neurotropismo en la regeneración de los centros nerviosos. *Trab. Lab. Invest. Biol.* 9: 123-159.
Thallmair, M., Metz, G.A.S., Graggen, W.J.Z., Raineteau, O., Kartje, G.L. and Schwab, M.E. (1998) Neurite growth inhibitors restrict plasticity and functional recovery following corticospinal tract lesions. *Nature Neurosci.* 1: 124-131.
Thompson, C.B. (1995) Apoptosis in the pathogenesis and treatment of disease. *Science* 267: 1456-1462.
Verdú, E., García-Alías, G., Forés, J., Gudiño-Cabrera, G., Nieto-Sampedro, M., and Navarro, X. (2001) Effects of ensheathing cells transplanted into photochemically damaged spinal cord. *NeuroReport*, 12: 2303-2309.
Verdú, E., Navarro, X., Gudiño-Cabrera, G., Rodríguez, F.J., Ceballos, D., Valero, A., and Nieto-Sampedro, M. (1999) Olfactory bulb ensheathing cells enhance peripheral nerve regeneration. *NeuroReport* 10: 1097-1101.
Wandosell, F., Bovolenta, P. and Nieto-Sampedro, M. (1993) Differences between reactive astrocytes and cultured astrocytes treated with Dibutyryl-cyclic AMP. *J. Neuropathol. Exper. Neurol.* 52: 205-215.

F. ROSSI

Department of Neuroscience, Rita Levi Montalcini Center for Brain Repair, University of Turin, Turin, Italy

22. REGULATION OF THE INTRINSIC GROWTH PROPERTIES IN MAMMALIAN NEURONS

Summary. Successful axon regeneration requires injured neurons upregulate a specific set of growth-associated genes needed to sustain long-distance neuritic elongation. Most of these genes are active during axonogenesis, but they are downregulated at the end of development following the appearance of environmental inhibitory cues. In the mammalian CNS, these cues include molecules issued by target cells or non-neuronal elements localized along the axon, including oligodendrocytes. The extrinsic inhibitory activity can be overcome when the neuronal expression of growth genes is enhanced or following manipulations that shift the balance between attractive and repulsive signalling pathways in the growth cone. Nevertheless, neutralisation of inhibitory molecules in the intact CNS induces aberrant axon growth, indicating that a major physiological function of these molecules is to restrain neuritic plasticity in order to maintain connection specificity.

1. INTRODUCTION

The extremely weak capability of the adult mammalian CNS for spontaneous repair is primarily attributed to adverse environmental conditions that hamper the elongation of injured axons (Bray and Aguayo, 1989; Schwab et al., 1993). However, when confronted with a permissive milieu, distinct populations of central neurons show strongly different growth potentialities, indicating that the outcome of reparative processes is also dependent on the intrinsic properties of the affected nerve cells (Fawcett, 1992). Indeed, it is now well established that the regenerative potential of a neuron is strictly correlated with its ability to activate a specific set of genes required to sustain long-distance elongation of the stem neurite (Fawcett, 2001; Fernandes and Tetzlaff, 2001).

The neuronal potential for axon growth and regeneration declines substantially at the end of development (Skene, 1989). Most mature neurons are not competent to support stem neurite elongation, and retain the capability for only local remodelling of terminal arborisations (Smith and Skene, 1997). Accordingly, at the end of development, growth-associated molecules are downregulated, or expressed at low levels and confined to terminal domains (Kapfhammer and Schwab, 1994). Such a remarkable phenotypic change is related to intrinsic modifications of the maturing neurons (Davies, 1994; Fawcett, 2001) and to the appearance of regulatory molecules in their microenvironment (Schwab et al., 1993; Fawcett, 2001). Although the mechanisms underlying regenerative phenomena in the adult may be partially different from those active during developmental neuritogenesis (Bates and Meyer, 1997; Liu and Snider, 2001; Udvadia et al., 2001; Bonilla et al., 2002), it is clear that a fine regulation of intrinsic neuronal growth properties is crucial both for the proper functioning of the intact CNS and for successful regeneration after injury. Thus, understanding the fine nature and the functional significance of these mechanisms is a prerequisite to designing efficient brain repair strategies.

2. REGULATION OF INTRINSIC GROWTH PROPERTIES IN PNS NEURONS

Contrary to the case of the central nervous system (CNS), vigorous axon regeneration occurs in the peripheral nerovus system (PNS). Axotomised peripheral neurons down-regulate molecules required for signalling functions, while they reactivate growth-associated gene programmes (Hökfelt et al., 1994). The strength of this response is not dependent on the distance of the lesion from the cell body, although it usually builds up faster for proximal than distal injuries (Liabotis and Schreyer, 1995; Kenney and Kocsis, 1998; Fernandes et al., 1999). The cellular changes are typically reversed when the regenerating axons rejoin their targets, but they are maintained for a long time if regeneration is prevented (Bisby, 1988; Woolf et al., 1990).

This behaviour of axotomised PNS neurons indicates that the cell body response may be triggered either by cellular injury or target loss. Indeed, axotomy-like changes can be elicited in intact neurons following either blockade of axonal transport (Woolf et al., 1990; Wu et al., 1993) or target degeneration (Verzé et al., 1996), suggesting that the expression of growth genes is normally suppressed by retrogradely transported inhibitory cues delivered by target cells. Several observations support this conclusion. For instance, cell-body changes in different types of injured neuron can be abolished or attenuated by application of specific target-derived molecules, including NGF (Wong and Oblinger, 1991; Gold et al., 1993; Mohiuddin et al., 1999), BDNF (Fernandes et al., 1998), or FGF-2 (Blottner and Herdegen, 1998). Conversely, upregulation of injury/growth-related genes can be elicited in intact neurons by means of anti-NGF antibodies (Gold et al., 1993; Shadiack et al., 2001).

Additional insights on the mechanisms that regulate axonal growth properties in peripheral neurons come from studies on dorsal root ganglion (DRG) cells. In these neuron a strong cell-body response is elicited after section of the peripheral axon, whereas a much weaker reaction occurs when the central branch is severed (Wong and Oblinger, 1990; Schreyer and Skene, 1993). Accordingly, the peripheral neurite is able to regenerate vigorously, whereas the central one shows poor growth capabilities. However, regeneration of the central branch into peripheral nerve implants (Richardson and Issa, 1984; Oudega et al., 1994) or along the dorsal columns of the spinal cord (Neumann and Woolf, 1999) can be considerably enhanced if a conditioning lesion of the peripheral neurite is made several days before axotomy of the central branch. Thus, the activation of growth-related genes in DRG neurons is primarily regulated by inhibitory signals flowing through the peripheral branch. A lesion of the central neurite is not sufficient to overcome this control.

The axons of acutely isolated, and hence axotomised, DRG cells typically grow in an "arborising" mode, a reflection of the constitutive ability for terminal arborisation plasticity. Within several hours, however, the growth pattern changes substantially into an "elongating" mode, which is required for long-distance regeneration. (Smith and Skene, 1997). Such a switch of growth mode depends on the transcription of new genes, since it can be prevented by adding inhibitors of protein synthesis to the culture medium. In addition, it is regulated by retrogradely-transported cues issued by target cells. In fact, neurons previously axotomised or treated with axon flow blockers *in vivo* are immediately competent to sustain elongating neuritic growth *in vitro*. Thus, target-derived cues set the growth status of DRG neurons by actively suppressing the gene program needed to sustain stem neurite elongation.

Although these data indicate that the regenerative potential of peripheral neurons is regulated primarily by target-derived inhibitory cues, there is also evidence that the cell-body response may be influenced by "positive", lesion-induced signals. One of the best-characterised of such signals is Leukaemia Inhibitory Factor (LIF). Following axotomy, LIF is expressed by Schwann cells and satellite cells in peripheral ganglia (Banner and Patterson, 1994; Sun et al., 1994). In addition, this substance is taken up by injured axons and retrogradely transported to the cell body (Curtis et al., 1994). Axotomy-induced changes of peptide expression in sympathetic ganglia are considerably attenuated in LIF-deficient mice (Rao et al., 1993; Sun and Zigmond 1996). Application of LIF to intact sympathetic neurons does not induce significant changes of peptide expression, indicating that, by itself, it is not sufficient to trigger a cell body reaction. However, concomitant administration of anti-NGF antibodies produces a particularly strong response, suggestive of additive interactions between these two molecules (Shadiack et al., 1998).

The intrinsic growth potential of peripheral neurons appears to be regulated by a fine balance between target-derived negative cues and lesion-induced positive signals (Fig. 1). Under normal conditions, retrograde cues maintain the signalling phenotype of peripheral neurons, and restrict neurite growth processes. Following axotomy, the removal of target-derived inhibition, together with the appearance of lesion-induced growth-promoting factors, allows the affected neurons to build up a strong compensatory response. After successful regeneration, the flow of target-derived signalling is restored, and growth-associated molecules are again suppressed. By this mechanism, the neurite growth status of peripheral neurons can be finely regulated in order to secure adaptive function in normal conditions and allow efficient repair after injury.

3. REGULATION OF INTRINSIC GROWTH PROPERTIES IN CNS NEURONS

In the CNS, both the basal expression of growth-related genes and the strength of the cell-body response to injury vary extremely between distinct neuron populations (Lieberman, 1970; Herdegen et al., 1997; Zagrebelsky et al., 1998). Accordingly, when permissive conditions are created, different neuron populations show different regenerative capabilities (Rossi et al., 1995, 2001; Bravin et al., 1997) — the growth potential always being strictly correlated with the intensity of cell-body changes (Herdegen et al., 1993; Schaden et al., 1994; Vaudano et al., 1995; Zagrebelsky et al., 1998). Hence, repair processes in the CNS are characterised by the great variability of the cellular reaction to injury. This fact is a reflection of distinctive features of different central neuron phenotypes, which are likely related to their specific functional role.

The strength of the cellular response to injury, and the success of the ensuing regeneration, are also dependent on lesion conditions, the most important of which is the distance of the injury from the cell body. For instance, retinal ganglion cells (Vidal-Sanz et al., 1987; Doster et al., 1991; Hüll and Bähr, 1994) or rubrospinal neurons (Tetzlaff et al., 1994; Fernandes et al., 1999) upregulate growth-associated genes and regenerate into peripheral nerve implants only when their axons are severed within a few millimetres of the cell body. Although there are exceptions to this rule, it is clear that the regulation of intrinsic growth properties of central neurons is more complex than that of their peripheral counterparts.

As in the PNS, growth-associated genes can be activated in central neurons by blocking axonal transport (Leah et al., 1993; Zagrebelsky et al., 1998), suggesting that retrogradely transported inhibitory cues are also effective in the CNS. Among such cues, inhibitory molecules delivered by target cells have been shown to regulate the expression of growth-associated genes in several populations of adult CNS neurons (Hughes et al., 1997; Bormann et al., 1998; Haas et al., 1998). Additional support to this conclusion comes from the observation that the cell-body response to injury is conspicuously attenuated, if not completely abolished, when collateral branches are spared (Leah et al., 1993). In many instances, however, even a complete target disconnection is not sufficient to induce cell-body changes, indicating that other cues contribute to this regulation. On the basis of the above-mentioned observation that the response to injury depends on the distance between the injury site and the cell body, it has been proposed that in the CNS retrograde inhibitory signals are also issued by glial cells located along the axon. Accordingly, the cell-body reaction can be initiated only when most of this inhibition has been removed (Skene, 1989, 1992).

CNS myelin or derived factors are likely candidates to exert such function. Indeed, myelin maturation is temporally correlated to the decline of growth potential in many CNS pathways (Kapfhammer and Schwab, 1994; Z'Graggen et al., 2000; Gianola and Rossi, 2001). In addition, the observation that retinal ganglion cells react equally well to an intraorbital optic nerve transection, irrespective of their position relative to the optic disk, indicates that the crucial factor is not the actual distance of axon injury from the soma, but, rather, the length of the remaining myelinated axon segment in the optic nerve (Doster et al., 1991; Meyer et al., 1994). Candidate molecules for this constitutive regulatory function are several myelin-associated neurite-growth-associated proteins, including Nogo-A (former NI-250, Huber and Schwab, 2000), myelin-associated glycoprotein (MAG, McKerracher, 1994; Mukhopadhyay, 1994), and oligodendrocyte-myelin glycoprotein (OMgp, Wang et al., 2002a). These proteins are constitutively expressed in the central white matter, suggesting that they exert some important function in the intact brain. Application of neutralising antibodies against Nogo-A induces the expression of growth-associated genes in intact Purkinje cells (Zagrebelsky et al., 1998; Buffo et al., 2000) or spinal neurons (Bareyre et al., 2002). The cell-body changes are accompanied by aberrant sprouting from uninjured axons, which grow out of their normal innervation territories (Buffo et al., 2000; Bareyre et al., 2002). Finally, focal demyelination in the spinal cord induces local sprouting of transected rubrospinal neurons, but not upregulation of growth-associated genes, confirming that removal of inhibitory influences along most of the axon length is required to trigger cell-body changes (Hiebert et al., 2001). Taken together, these observations indicate that Nogo-A, and possibly other myelin-associated proteins, constitutively inhibit growth-associated gene programs and restrict structural plasticity in adult CNS neurons.

The intrinsic growth potential of adult CNS neurons is thus subjected to a dual inhibition exerted by retrogradely-transported signals issued from target cells and non-neuronal elements along the axon (Fig. 1). The functional significance of these regulatory cues is, however, probably different. As described in the previous section, evidence from the peripheral nervous system indicates that target-derived signals arrest axon elongation and induce the genes related to the signalling/transmitter phenotype of the neuron (Hökfelt et al., 1994). A similar function has been proposed for NGF on cholinergic neurons of the basal forebrain (Higgins et al., 1989; Vantini et al., 1989). In contrast, the inhibitory action of myelin-derived cues would help to secure connection specificity, by preventing unwanted neuritic growth and confining structural plasticity within well-defined territories. During development, this control would be important to prevent aberrant branching (Colello and Schwab, 1994) and to channel outgrowing axons along proper pathways (Schwab and Schnell, 1991). In the adult, it may be required to regulate the expression of neuronal growth genes (Zagrebelsky et al., 1998, Bareyre et al., 2002) and their targeting to terminal domains (Kapfhammer and Schwab, 1994), and to confine structural remodelling of terminal arborisations (Buffo et al., 2000; Bareyre et al., 2002).

In addition to these inhibitory molecules, the expression of growth-associated genes in central neurons may also be influenced by positive environmental cues. For instance, in several types of central neuron cell-body changes are enhanced by growth-promoting tissues transplanted at the lesion site (Hüll and Bähr, 1994; Vaudano, 1995; Broude et al., 1997). The molecular mechanisms underlying this effect are still unclear, although trophic factors may be important. It has been also proposed that inflammation-related molecules enhance the neuronal response to injury (Fernandes and Tetztlaff, 2001).

Altogether, these data suggest that multiple positive and negative environmental cues interact to regulate the intrinsic regenerative potential of central neurons (Fig. 1). In the CNS, however, the outcome of these interactions is strongly conditioned by the specific phenotypic features of each neuron population, leading to a differential sensitivity to extrinsic signals. As a consequence, the features of the response to injury, and the ensuing regenerative potential, vary extremely between different types of neuron, so that the various neuronal populations will differ in their response to treatments and manipulations aimed at promoting regeneration.

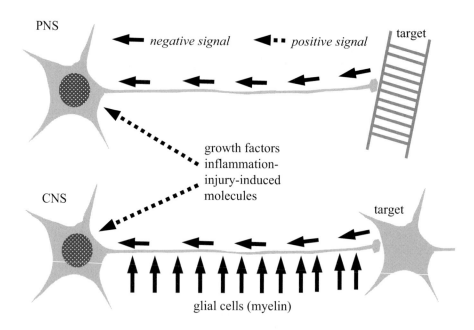

Figure 1. Regulation of growth potential in PNS and CNS neurons. Expression of neuronal growth-associated genes depends on the balance between positive signals (dashed arrows), mostly induced by injury, and negative cues (arrows) delivered from target cells or, in the CNS, by non-neuronal elements located along the axon.

4. WORKING WITHIN: BOOSTING THE REGENERATIVE RESPONSE OF NEURONS

To date, most of the approaches designed to improve brain repair have been directed at reducing environmental inhibition. However, as mentioned above, this may be effective only for certain categories of central neurons, which are inclined to spontaneous regeneration. In addition, the recent observation that the activation of neuronal growth genes allows neuritic elongation into non-permissive territories, and most notably CNS myelin, indicates that repair processes can be effectively promoted by strengthening the intrinsic growth potential of injured neurons. The main strategies designed to achieve this goal have hitherto been directed at potentiating the cell body response to injury or interfering with signalling pathways in the growth cone, so as to overcome environmental inhibition.

Implantation of a peripheral nerve segment into the vitreous body promotes axon regeneration following optic nerve crush (Berry et al., 1996). This effect has been attributed to trophic factors acting on the cell body of axotomised retinal ganglion cells. Indeed, direct application of neurotrophins (BDNF) to the somata of injured rubrospinal neurons enhances the expression of growth-associated genes, and improves axon regeneration into peripheral nerve implants (Kobayashi et al., 1997). The treatment is also effective in long-term axotomised cells (Kwon et al., 2002). A similar increase in the expression of growth-associated genes has been observed in corticospinal neurons (Giehl and Tetzlaff, 1996). In this case, however, BDNF does promote sprouting of the injured axons, but not regeneration into peripheral nerves (Hiebert et al., 2002), again highlighting the differential responsiveness of each CNS neuron population.

Expression of growth-related genes has been potentiated by application of the purine nucleoside inosine in goldfish neurons (Petrausch et al., 2000). Administration of this molecule to adult rats following unilateral injury of the corticospinal tract (Benowitz et al., 1999) or cortical ischaemia (Chen et al., 2002) results in vigorous compensatory sprouting of intact axons, accompanied by some functional recovery. It is still unclear whether inosine might also promote regeneration of injured neurites. Enhanced regeneration, together with increased expression of GAP-43, has also been reported following administration of the immunosuppressant drug tacrolimus (FK506), both in peripheral (Gold et al., 1995, 1998) and in central neurons (Madsen et al., 1998).

Expression of growth-associated genes can also be induced by producing transgenic animals or by virus-mediated gene transfer. Potentiation of neuritic growth capabilities has been obtained *in vitro* by inducing the expression of different integrins (Condic, 2001) or the small proline-rich repeat protein 1A (Bonilla et al., 2002). Overexpression of major growth-cone proteins, such GAP-43 or CAP-23, induces spontaneous sprouting in different populations of PNS and CNS neurons, and promotes collateral reinnervation at the neuromuscular junction (Aigner et al., 1995; Caroni et al., 1997). When the GAP-43 gene is specifically targeted at Purkinje cells, which are distinguished for their poor regenerative potential, profuse neuritic sprouting occurs after injury, but these neurons remain unable to regenerate into growth-permissive transplants (Buffo et al., 1997). Similar results have been obtained with thalamic neurons (Mason et al., 2000), showing that overexpression of individual growth genes may promote axon growth, but it is not sufficient to sustain long-distance neurite elongation. Indeed, significant regeneration of the central branch of DRG neurons into peripheral nerve implants has been observed in double-transgenic mice overexpressing both GAP-43 and CAP-23 (Bomze et al., 2001). On the whole, these results indicate that the switch from the constitutive "arborising" growth mode to regenerative elongation of the stem neurite is probably a multi-step process, requiring the coordinated expression of a whole set of genes.

Most-promising results have been obtained recently by approaches that reduce the sensitivity of growth cones to environmental inhibitory cues. For instance, developmental studies indicate that Rho GTPase may be involved in translating environmental inhibitory signals into cytoskeletal reorganisation, leading to growth-cone collapse. This protein can also be activated by MAG through the neurotrophin p75 receptor, and hence, may contribute to transducing environmental inhibition in the adult CNS. Indeed, inactivation of Rho GTPase by treatment with C3 transferase, or transfection with a dominant negative form of Rho, allows cultured neurons to grow on inhibitory substrates, whereas a similar treatment *in vivo* results in enhanced sprouting of retinal axons following optic nerve crush (Lehmann et al., 1999).

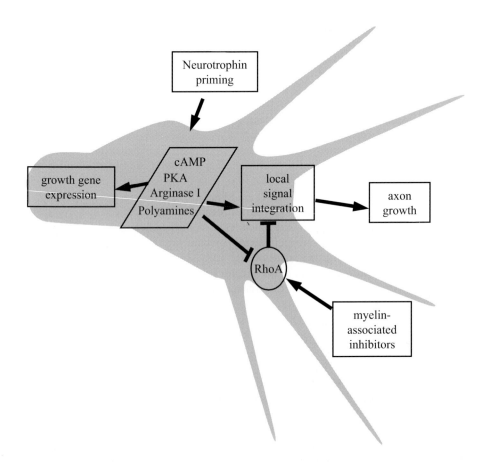

Figure. 2. Interaction between growth-promoting and inhibitory mechanisms at the growth cone. The activation of the cAMP pathway elicits the upregulation of growth-associated genes and allows local signal integration at the growth cone to overcome environmental inhibition mediated by RhoA.

Cerebellar or DRG neurons exposed *in vitro* to neurotrophins also become insensitive to MAG inhibition (Cai et al., 1999). Such priming effect of neurotrophins is associated with intracellular elevation of cAMP and is blocked by PKA inhibitors (Cai et al., 1999). Injection of cAMP into DRG neurons mimics the effect of peripheral axotomy, and promotes significant regeneration of the central branch into the dorsal columns of the spinal cord (Neumann et al., 2002; Qiu et al.,

2002). A recent study shows that cAMP induces upregulation of arginase I, which in turn produces several polyamines (Cai et al. 2002). Application of these molecules to cultured neurons replicates the effects produced by cAMP. Interestingly, arginase I is expressed in developing neurons, which are insensitive to the inhibitory activity of myelin, but it is spontaneously downregulated at the end of development (Cai et al., 2002). Thus, the sensitivity of a neuron to environmental signals appears to be controlled by the level of certain metabolites in the growth cone. For instance, elevation of cAMP in the growth cone may suppress the local inhibitory action of Rho GTPase, and also trigger gene expression in the cell body (Fig. 2). Although our present understanding of these mechanisms is still limited, it is clear that the dissection of the molecular pathways that regulate axon growth in response to specific extrinsic signals will be crucial for effective promotion of repair processes.

5. WHY IS REGENERATION INHIBITED?

Over the last few years, the study of the mechanisms underlying inhibition of axon growth in the mammalian CNS has progressed substantially, and numerous key molecules have been identified. Some of these molecules, mostly associated with glial scarring or inflammation, are component of the tissue response to injury, and are not present in the intact CNS (Fawcett and Asher, 1999). Other factors, including the above-mentioned myelin-associated proteins or target-derived neurotrophic factors, are constitutively expressed in the mature CNS, and, in many cases, their levels are not significantly modified after injury (Huber et al., 2002; Wang et al., 2002b). Thus, although most studies about these factors are focussed particularly on their role in the injured brain, it is clear that their main function is to regulate growth rather than hamper regeneration.

As proposed above, in the mature CNS, target-derived factors and myelin-associated proteins co-operate to suppress neuronal growth potential in order to induce specific signalling phenotypes and maintain connection specificity. This suggests that the highly plastic mammalian CNS neurons have to be kept at bay to prevent progressive disorganisation of the normal wiring of neural circuits. However, if such function is important in the intact brain, it may also make sense after injury: repair processes may conceivably lead to the formation of aberrant circuitries, with detrimental effects on the recovery of function. Paradoxically, regeneration could be actively suppressed because it is maladaptive.

Several reports showing that properly patterned connections can be re-established in the adult mammalian brain, argue against this conclusion. For instance, positional cues are re-expressed in the superior colliculus after deafferentation (Wizenmann et al., 1993, Knöll et al., 2001), and retinal axons regenerated through a peripheral nerve implant do restore a rough topographic organisation (Sauvé et al., 2001). Severed olivocerebellar axons that regenerate into a cerebellar transplant recognise neurochemically distinct subsets of Purkinje cells (Rossi et al., 2002) and form topographic-specific connections by collateral reinnervation in the adult cerebellar cortex (Zagrebelsky et al., 1996). Similarly, properly patterned projections can be re-established by sprouting axons in the corticopontine (Thallmair et al., 1998) and corticospinal systems (Bareyre et al., 2002). Although these reports indicate that it is possible to repair specific connections in the adult CNS, the best evidence still comes from collateral reinnervation. To date, the number of severed regenerating axons that succeed in rejoining their target regions in the CNS is usually too small to draw any definite conclusion about the accuracy of the newly formed synapses.

Studies on the peripheral nervous system, where numerous axons regenerate, have obtained controversial results concerning the specificity of reinnervation and the recovery of adaptive functions (Carlstedt, 2000). However, it has recently been shown that extremely precise innervation patterns can be restored after nerve crush, but not after axotomy (Nguyen et al., 2002). This indicates that specific repair can be achieved if the normal tissue architecture is maintained, whereas aberrant growth occurs when the microenvironment is disrupted. This may hold true also for the CNS. The regeneration of properly patterned connections has been convincingly demonstrated only for compensatory sprouting from intact axons, which grow into a largely preserved tissue. In contrast, the growth of severed axons through injured CNS regions is often erratic and disorganised. Thus, although the adult CNS may be endowed with previously unsuspected potentialities for self-repair, inhibition of axon growth after injury may still be needed to prevent aberrant growth through a disrupted environment. Accordingly, efficient repair strategies should not only neutralise environmental inhibition and strengthen intrinsic neuronal growth properties, but also provide the regenerating axons with appropriate guidance cues.

REFERENCES

Aigner, L., Arber, S., Kapfhammer. J.P., Laux, T., Schneider, C., Botteri. F. et al. (1995) Overexpression of the neural growth-associated protein GAP-43 induces nerve sprouting in the adult nervous system of transgenic mice. *Cell, 83*, 269-78.

Banner. L.R., & Patterson, P.H. (1994) Major changes in the expression of the mRNAs for cholinergic differentiation factor and its receptor after injury to the adult peripheral nerves and ganglia. *Proceedings of the National Academy of Science (USA), 91*, 7109-7113.

Bareyre, F.M., Haudenshild, B., & Schwab, M.E. (2002) Long-lasting sprouting and gene expression changes induced by the monoclonal antibody IN-1 in the adult spinal cord. *Journal of Neuroscience, 22*, 7097-7110.

Bates, C.A., & Meyer, R.L. (1997) The neurite-promoting effect of laminin is mediated by different mechanisms in embryonic and adult regenerating mouse optic axons *in vitro. Developmental Biology, 181*, 91-101.

Benowitz, L.I., Goldberg, D.E., Madsen, J.R., Soni, D., & Irwin, N. (1999) Inosine stimulates extensive axon collateral growth in the rat corticospinal tract after injury *Proceedings of the National Academy of Science (USA), 96*, 13486-13490.

Berry, M., Carlile, J., & Hunter, A. (1996) Peripheral nerve explants grafted into the vitreous body of the eye promote the regeneration of retinal ganglion cell axons severed in the optic nerve. *Journal of Neurocytology, 25*, 147-170.

Bisby, M.A. (1988) Dependence of GAP-43 (B50, F1) transport on axonal regeneration in rat dorsal root ganglion neurons. *Brain Research, 458*, 157-161.

Blottner, D., & Herdegen, T. (1998) Neuroprotective Fibroblast Growth Factor type-2 down-regulates the c-Jun transcription factor in axotomized sympathetic preganglionic neurons of adult rat. *Neuroscience, 82*, 283-292.

Bravin, M., Savio, T., Strata, P., & Rossi, F. (1997) Olivocerebellar axon regeneration and target reinnervation following dissociated Schwann cell grafts in surgically injured cerebella of adult rats. *European Journal of Neuroscience, 9*, 2634-2649.

Bray, G.M., & Aguayo, A.J. (1989) Exploring the capacity of CNS neurons to survive injury, regrow axons and form new synapses in adult mammals. In F.J. Seil (Ed.) *Neural Regeneration and Transplantation* (pp. 67-78). New York: Alan Liss.

Bomze, H.M., Bulsara, K.R., Iskandar, B.J., Caroni, P., & Skene, J.H.P. (2001) Spinal axon regeneration evoked by replacing two growth cone proteins in adult neurons. *Nature Neuroscience, 4,* 38-43.

Bonilla, I.E., Tanabe, K., & Strittmatter, S.M. (2002) Small proline-rich repeat protein 1A is expressed by axotomized neurons and promotes axonal outgrowth. *Journal of Neuroscience, 22,* 1303-1315.

Bormann, P., Zumsteg, V.M., Roth, L.W., & Reinhard, E. (1998) Target contact regulates GAP-43 and alpha-tubulin mRNA levels in regenerating retinal ganglion cells. *Journal of Neuroscience Research, 52,* 405–419.

Broude, E., McAtee, M., Kelley, M.S., & Bregman, B.S. (1997) c-Jun expression in adult rat dorsal root ganglion neurons: differential response after central or peripheral axotomy. *Experimental Neurology, 148,* 367-377.

Buffo, A., Holtmaat, A.J.D.G., Savio, T., Verbeek, S., Oberdick, J., Oestreicher, A.B. et al. (1997) Targeted overexpression of the neurite growth-associated protein B-50/GAP-43 in cerebellar Purkinje cells induces sprouting in response to axotomy, but does not allow axon regeneration into growth permissive transplants. *Journal of Neuroscience, 17,* 8778-8791.

Buffo, A., Zagrebelsky, M., Huber, A.B., Skerra, A., Schwab, M.E., Strata, P., et al. (2000). Application of neutralising antibodies against NI-35/250 myelin-associated neurite growth inhibitory proteins to the adult rat cerebellum induces sprouting of uninjured Purkinje cell axons. *Journal of Neuroscience, 20,* 2275-2286.

Cai, D., Shen, Y., De Bellard, M.E., Tang, S., & Filbin, M.T. (1999) Prior exposure to neurotrophins blocks inhibition of axonal regeneration by MAG and myelin via a cAMP-dependent mechanism. *Neuron, 22,* 89–101.

Cai, D., Deng, K., Mellado, W., Lee, J., Ratan, R.R., & Filbin, M.T. (2002) Arginase I and polyamines act downstream from cyclic AMP in overcoming inhibition of axonal growth in vitro. *Neuron, 35,* 711–719.

Carlstedt, T. (2000) Approaches permitting and enhancing motor neuron regeneration after spinal cord, ventral root, plexus and peripheral nerve injuries. *Current Opinion in Neurology, 13,* 683-686.

Caroni, P., Aigner, L., & Schneider, C. (1997) Intrinsic neuronal determinants locally regulate extrasynaptic and synaptic growth at the adult neuromuscular junction. *Journal of Cell Biology, 136,* 679-692

Chen, P., Goldberg, D.E., Kolb, B., Lanser, M., & Benowitz, L.I. (2002) Inosine induces axonal rewiring and improves behavioral outcome after stroke. *Proceedings of the National Academy of Science (USA), 99,* 9031-9036.

Colello, R.J., & Schwab, M.E. (1994) A role for oligodendrocytes in the stabilization of optic axon numbers. *Journal of Neuroscience, 14,* 6446-6452.

Condic, M.L. (2001). Adult neuronal regeneration induced by transgenic integrin expression. *Journal of Neuroscience, 21,* 4782–4788.

Curtis, R., Scherer, S.S., Somogyi, R., Adrian, K.M, Ip. Y.Y., Zhu, Y. et al. (1994) Retrograde axonal transport of LIF is increased by peripheral nerve injury: correlation with increased LIF expression in distal nerve. *Neuron, 12,* 191-204.

Davies, A.M. (1994) Intrinsic programmes of growth and survival in developing vertebrate neurons. *Trends in Neurosciences, 17,* 195-199.

Doster, K.S., Lozano, A.M., Aguayo, A.J., & Willard, M.B. (1991) Expression of the growth-associated protein GAP-43 in adult rat retinal ganglion cells following injury. *Neuron, 6,* 635-647.

Fawcett, J.W. (1992) Intrinsic neuronal determinants of regeneration. *Trends in Neurosciences, 15,* 5-8.

Fawcett, J.W. (2001) Intrinsic control of regeneration and the loss of regenerative ability in development. In N.A. Ingoglia & M. Murray (Eds.). *Axonal Regeneration in the Central Nervous System* (pp. 161-183). New York, Basel: Marcel Dekker Inc.

Fawcett, J.W., & Asher, R.A. (1999) The glial scar and central nervous system repair. *Brain Research Bulletin, 49,* 377-391.

Fernandes, K.J., & Tetzlaff, W.G. (2001) Gene expression in axotomized neurons: identifying intrinsic determinants of axonal growth. In N.A. Ingoglia & M. Murray (Eds.). *Axonal Regeneration in the Central Nervous System* (pp. 219-266). New York, Basel: Marcel Dekker Inc.

Fernandes, K.J., Fan, D.-P., Tsui, B.J., Cassar, S.L., & Tetzlaff, W.G. (1999) Influence of the axotomy to cell body distance in rat rubrospinal and spinal motoneurons: different regulation of GAP-43, tubulins, and neurofilament-M. *Journal of Comparative Neurology, 414,* 495-510.

Fernandes, K.J., Kobayashi, N.R., Jasmin, B.J., & Tetzlaff, W.G. (1998) Acetylcholinesterase gene expression in axotomised rat facial motoneurons is differentially regulated by neurotrophins: correlation with trkB and trkC mRNA levels and isoforms. *Journal of Neuroscience, 18,* 9936-9947.

Gianola, S., & Rossi, F. (2001) Evolution of the Purkinje cell response to injury and regenerative potential during postnatal development of the rat cerebellum. *Journal of Comparative Neurology, 430,* 101-117.

Giehl, K.M., & Tetzlaff, W.G. (1996) BDNF and NT-3, but not NGF, prevent axotomy-induced death of rat corticospinal neurons in vivo. *European Journal of Neuroscience, 8,* 1167-1175.

Gold, B.G., Storm-Dickerson, T., & Austin, D.R. (1993) Regulation of the transcription factor c-Jun by nerve growth factor in adult sensory neurons. *Neuroscience Letters, 154,* 129-133.

Gold, B.G., Katoh, K., & Storm-Dickerson, T. (1995) The immunosuppressant FK506 increases the rate of axonal regeneration in rat sciatic nerve. *Journal of Neuroscience, 15,* 7509-7516.

Gold, B.G., Yew, J.Y., & Zeleny-Pooley, M. (1998) The immunosuppressant FK506 increases GAP-43 mRNA levels in axotomized sensory neurons. *Neuroscience Letters, 241,* 25-28.

Haas, C.A. & Frotscher, M. (1998) The role of NGF in axotomy-induced c-Jun expression in medial septal neurons. *International Journal of Developmental Neuroscience, 16,* 691-703.

Herdegen, T., Skene, J.H.P., & Bähr, M. (1997) The c-Jun transcription factor - bipotential mediator of neuronal death, survival and regeneration. *Trends in Neurosciences, 20,* 227-231.

Herdegen, T., Brecht, S., Mayer, B., Leah, J., Kummer, W., Bravo. R., et al. (1993) Long-lasting expression of JUN and KROX transcription factors and nitric oxide synthase in intrinsic neurons of the brain following axotomy. *Journal of Neuroscience, 13,* 4130-4145.

Hiebert, G.W., Dyer, J.K., Tetzlaff, W., & Stevees, J.D. (2000) Immunological myelin disruption does not alter expression of regeneration-associated genes in intact or axotomized rubro-spinal neurons. *Experimental Neurology, 163,* 149-156.

Hiebert, G.W., Khodarahmi, K., McGraw, J., Steeves, J.D., & Tetzlaff W. (2002) Brain-derived neurotrophic factor applied to the motor cortex promotes sprouting of corticospinal fibers but not regeneration into a peripheral nerve transplant. *Journal of Neuroscience Research, 69,* 160-168.

Higgins, G.A., Koh, S., Chen, K.S., & Gage, F.H. (1989) NGF induction of NGF receptor gene expression and cholinergic neuronal hypertrophy within the basal forebrain of the adult rat. *Neuron, 3,* 247-256.

Hökfelt, T., Zhang, X., & Wiesefeld-Hallin, Z. (1994) Messenger plasticity in primary sensory neurons following axotomy and its functional implications. *Trends in Neurosciences, 17,* 22-30.

Huber, A.B., & Schwab, M.E. (2000) Nogo-A, a potent inhibitor of neurite growth and regeneration. *Biological Chemistry, 381,* 407-419.

Huber, A.B., Weinmann, O., Brösamle, C., Oertle, T., & Schwab, M.E. (2002). Patterns of Nogo mRNA and protein expression in the developing and adult rat and after CNS lesions. *Journal of Neuroscience, 22,* 3553–3567

Hughes, P.E., Alexi, T., Hefti, F., & Knusel, B. (1997) Axotomized septal cholinergic neurons rescued by nerve growth factor or neurotrophin-4/5 fail to express the inducible transcription factor c-Jun. *Neuroscience, 78,* 1037-1049.

Hüll, M., & Bähr, M. (1994) Regulation of immediate early gene expression in retinal ganglion cells following axotomy and during regeneration through a peripheral nerve graft. *Journal of Neurobiology, 25,* 92-105.

Kapfhammer, J.P., & Schwab, M.E. (1994) Inverse patterns of myelination and GAP-43 expression in the adult CNS: neurite growth inhibitors as regulators of neuronal plasticity? *Journal of Comparative Neurology, 340,* 194-206.

Kenney, A.M., & Kocsis, J.D. (1998) Peripheral axotomy induces long-term c-Jun amino-terminal kinase-1 activation and activator protein-1 binding activity by c-Jun and JunD in adult dorsal root ganglia *in vivo. Journal of Neuroscience, 18,* 1318-1328.

Knöll, B., Isenmann, S., Kilic, E., Walkenhorst, J., Engel, S., Wehinger, J., et al. (2001) Graded expression patterns of ephrin-As in the superior colliculus after lesion of the adult mouse optic nerve. *Mechanisms of Development, 106,* 119-127.

Kobayashi, N.R., Fan, D.-P., Giehl, K.M., Bedard, A.M., Wiegand, S.J., & Tetzlaff, W. (1997) BDNF and NT4/5 prevent atrophy of rat rubrospinal neurons after cervical axotomy, stimulate GAP-43 and

Tα1-tubulin mRNA expression, and promote axonal regeneration. *Journal of Neuroscience, 17,* 9583-9595.

Kwon, B.K., Liu, J., Messerer, C., Kobayashi, N.R., McGraw, J., Oschipok, L., & Tetzlaff, W. (2002) Survival and regeneration of rubrospinal neurons 1 year after spinal cord injury. *Proceedings of the National Academy of Science (USA), 99,* 3246-3251.

Leah, J., Herdegen, T., Murashov, A., Dragunow, M., & Bravo, R. (1993). Expression of immediate early gene proteins following axotomy and inhibition of axonal transport in the rat central nervous system. *Neuroscience, 57,* 53-66.

Lehmann, M., Fournier, A., Selles-Navarro, I., Dergham, P., Sebok, A., Leclerc, N., et al. (1999). Inactivation of Rho signaling pathway promotes CNS axon regeneration. *Journal of Neuroscience, 19,* 7537-7547.

Liabotis, S., & Schreyer, D.J. (1995) Magnitude of GAP-43 induction following peripheral axotomy of adult rat dorsal root ganglion neurons is independent of lesion distance, *Experimental Neurology, 135,* 28-35.

Lieberman, A.R. (1971) The axon reaction: a review of the principal features of perikaryal response to axon injury. *International Reviews of Neurobiology, 24,* 49-124.

Liu, R.-Y., & Snider, W.D. (2001) Different signaling pathways mediate regenerative versus developmental sensory axon growth. *Journal of Neuroscience, 21,* RC164

Masden, J.R., MacDonald, P., Irwin, N., Goldberg, D.E., Yao, G.-L., Meiri, K.F., et al. (1998) Tacrolimus (FK506) increases neuronal expression of GAP-43 and improves functional recovery after spinal cord injury in rats. *Experimental Neurology, 154,* 673-683.

Mason, M.R., Campbell, G., Caroni, P., Anderson, P.N., & Lieberman, A.R. (2000) Overexpression of GAP-43 in thalamic projection neurons of transgenic mice does not enable them to regenerate axons through peripheral nerve grafts. *Experimental Neurology, 165,* 143-152.

McKerracher, L., David, S., Jackson, D.L., Kottis, V., Dunn. R.J., Braun, P.E. (1994) Identification of myelin-associated glycoprotein as a major myelin-derived inhibitor of neurite growth. *Neuron, 13,* 805-811.

Meyer, R.L., Miotke, J.A., & Benowitz, L.I. (1994) Injury induced expression of growth-associated protein-43 in adult mouse retinal ganglion cells *in vitro*. *Neuroscience, 63,* 591-602.

Mohiuddin, L., Delcroix, J.D., Fernyhough, P., & Tomlinson, R.D. (1999) Focally administered Nerve Growth Factor suppresses molecular regenerative responses of axotomized peripheral afferents in rats. *Neuroscience, 91,* 265-271.

Mukhopadhyay, G., Doherty, P., Walsh, F.S., Crocker, P.R., & Filbin, M.T. (1994) A novel role of myelin-associated glycoprotein as an inhibitor of axonal regeneration. *Neuron, 13,* 1-20.

Neumann, S., & Woolf, C.J. (1999) Regeneration of dorsal column fibers into and beyond the lesion site following adult spinal cord injury. *Neuron, 23,* 83-91.

Neumann, S., Bradke, F., Tessier-Lavigne, M., & Basbaum, A.I. (2002) Regeneration of sensory axons within the injured spinal cord induced by intraganglionic cAMP elevation. *Neuron, 34,* 885–893.

Nguyen, D.T., Sanes, J.R., & Lichtman, J.L. (2002). Pre-existing pathways promote precise projection patterns. *Nature Neuroscience 5,* 861-867.

Oudega, M., Varon, S., & Hagg, T. (1994) Regeneration of adult rat sensory axons into intraspinal nerve grafts: promoting effects of conditioning lesion and graft predegeneration. *Experimental Neurology, 129,* 194-206

Petrausch, B., Tabibiazar, R., Roser, T., Jing, Y., Goldman, D., Stuermer, C.A.O., et al. (2000) A Purine-Sensitive Pathway Regulates Multiple Genes Involved in Axon Regeneration in Goldfish Retinal Ganglion Cells. *Journal of Neuroscience, 20,* 8031–8041.

Qiu, J., Cai, D., Dai, H., McAtee, M., Hoffman, P., Bregman, B.S, et al. (2002) Spinal axon regeneration induced by elevation of cyclic AMP. *Neuron, 34,* 895–903.

Rao, M.S., Sun, Y., Escary, J.L., Perreau, J., Tresser, S., Patterson, P.H., et al. (1993) Leukemia inhibitory factor mediates an injury response but not a target-directed developmental transmitter switch in sympathetic neurons. *Neuron, 11,* 1175-85

Richardson, P.M., & Issa, V.M.K. (1984) Peripheral injury enhances regeneration of spinal axons. *Nature, 284,* 264–265.

Rossi, F., Jankovski, A., & Sotelo, C. (1995) Differential regenerative response of Purkinje cell and inferior olivary axons confronted with embryonic grafts: environmental cues versus intrinsic neuronal determinants. *Journal of Comparative Neurology, 359,* 663-677.

Rossi, F., Buffo, A., & Strata, P. (2001). Regulation of intrinsic regenerative properties and axonal plasticity in cerebellar Purkinje cells. *Restorative Neurology and Neuroscience, 19*, 85-94.

Rossi, F., Saggiorato, C., & Strata, P. (2002) Target-specific innervation of embryonic cerebellar transplants by regenerating olivocerebellar axons in the adult rat. *Experimental Neurology, 173*, 205-212.

Sauvé, Y., Sawai, H., & Rasminsky, M. (2001) Topological specificity in reinnervation of the superior colliculus by regenerated retinal ganglion cell axons in adult hamsters. *Journal of Neuroscience, 21*, 951-960.

Schaden, H., Stürmer, C.A.O., & Bähr, M. (1994) GAP-43 immunoreactivity and axon regeneration in retinal ganglion cells of the rat. *Journal of Neurobiology, 25*, 1570-1578.

Schreyer, D.J., & Skene, J.H.P. (1993) Injury-associated induction of GAP-43 expression displays axon branch specificity in rat dorsal root ganglion neurons. *Journal of Neurobiology, 24*, 959-70.

Schwab, M.E., & Schnell, L. (1991) Channeling of developing corticospinal tract axons by myelin-associated neurite growth inhibitors. *Journal of Neuroscience, 11*, 709-721.

Schwab, M.E., Kapfhammer, J.P., & Bandtlow, C.E. (1993) Inhibitors of neurite growth. *Annual Reviews of Neuroscience, 16*, 565-595.

Shadiack, A.M., Vaccariello, S.A., Sun, Y., & Zigmond, R.E. (1998) Nerve growth factor inhibits sympathetic neurons' response to an injury cytokine. *Proceedings of the National Academy of Science (USA), 95*, 7727-7730.

Shadiack, A.M., Sun, Y., & Zigmond, R.E. (2001) Nerve Growth Factor antiserum induces axotomy-like changes in neuropeptide expression in intact sympathetic and sensory neurons. *Journal of Neuroscience, 21*, 363–371.

Skene, J.H.P. (1989) Axonal growth-associated proteins. *Annual Reviews of Neuroscience, 12*, 127-156.

Skene, J.H.P. (1992) Retrograde pathways controlling expression of a major growth cone component in the adult CNS. In P.C. Letourneau, S.B. Kater & E.R. Macagno (Eds.), *The Nerve Growth Cone* (pp. 463-475). New York: Raven Press.

Smith, D.S., & Skene, J.H.P. (1997) A transcription-dependent switch controls competence of adult neurons for distinct modes of axon growth. *Journal of Neuroscience 15*, 646-658.

Sun, Y., & Zigmond, R. (1996) Involvement of leukemia inhibitory factor in the increase of galanin and vasoactive intestinal peptide mRNA and the decreases in neuropeptide Y and tyrosine hydroxylase mRNS after axotomy of sympathetic neurons. *Journal of Neurochemistry, 67*, 1751-60.

Sun, Y., Rao, M., Zigmond, R.E., & Landis, S.C. (1994) Regulation of vasoactive intestinal peptide expression in sympathetic neurons in culture and after axotomy: the role of cholinergic differentiation factor/leukemia inhibitory factor. *Journal of Neurobiology, 25*, 415-430.

Tetzlaff, W.G., Kobayashi, N.R., Giehl, K.M.G., Tsui, B.J., Cassar, S.L., & Bedard, A.M. (1994) Response of rubrospinal and corticospinal neurons to injury and neurotrophins. In F.J. Seil (Ed.). *Neural Regeneration. Progress in Brain Research, Vol. 103* (pp 271-286). Amsterdam: Elsevier.

Thallmair, M., Metz, G.A.S., Z'Graggen, W.J., Raineteau, O., Kartje, G.L., & Schwab, M.E. (1998). Neurite growth inhibitors restrict plasticity and functional recovery following corticospinal tract lesions. *Nature Neuroscience, 1*, 124-131.

Udvadia, A.J., Köster, R.W., & Skene, J.H.P. (2001) GAP-43 promoter elements in transgenic zebrafish reveal a difference in signals for axon growth during CNS development and regeneration. *Development 128*, 1175-1182.

Vantini, G., Schiavo, N., Di Martino, A., Polato, P., Triban, C., Callegaro, L., et al., (1989). evidence for a physiological role of nerve growth factor in the central nervous system of neonatal rats. *Neuron, 3*, 267-273.

Vaudano, E., Campbell, G., Anderson, P.N., Davies, A.P., Woolhead, C., Schreyer, D.J., et al. (1995) The effects of a lesion or a peripheral nerve graft on GAP-43 upregulation in the adult brain: an *in situ* hybridisation and immunocytochemical study., *Journal of Neuroscience, 15*, 3594-3611.

Verzè, L., Buffo, A., Rossi, F., Oestreicher, A.B., Gispen W.H., & Strata, P. (1996) Increase of B-50/GAP-43 immunoreactivity in uninjured muscle nerves of *mdx* mice. *Neuroscience, 70*, 807-815.

Vidal-Sanz, M., Bray, G.M., Villegas-Perez, M.P., Thanos, S., Aguayo, A.J. (1987). Axonal regeneration and synapse formation in the superior colliculus by retinal ganglion cells in the adult rat. *Journal of Neuroscience 7*, 2894-909

Wang, K.C., Koprivica, V., Kim, J.A., Sivasankaran, R., Guo, Y., Neve, R.L., et al. (2002) Oligodendrocyte-myelin glycoprotein is a Nogo receptor ligand that inhibits neurite outgrowth. *Nature, 417*, 941-944.

Wang, X., Chun, S.-J., Treloar, H., Vartanian, T., Greer, C.A., Strittmatter, S.M. (2002a). Localization of Nogo-A and Nogo-66 receptor proteins at sites of axon–myelin and synaptic contact. *Journal of Neuroscience, 22,* 5505–5515.

Wang, X., Chun, S.-J., Treloar, H., Vartanian, T., Greer, C.A., Strittmatter, S.M. (2002b) Localization of Nogo-A and Nogo-66 receptor proteins at sites of axon–myelin and synaptic contact. *Journal of Neuroscience, 22,* 5505–5515.

Wizenmann, A., Thies, E., Klostermann, S., Bonhoeffer, F., & Bähr, M. (1993). Appearance of target-specific guidance information for regenerating axons after CNS lesions. *Neuron, 11,* 975-983.

Wong, J., & Oblinger, M.M. (1990) A comparison of peripheral and central axotomy effects on neurofilament and tubulin gene expression in rat dorsal root ganglion neurons. *Journal of Neuroscience, 10,* 2215–2222.

Wong, J., & Oblinger, M.M. (1991) NGF rescues substance P expression but not neurofilament or tubulin gene expression in axotomized sensory neurons. *Journal of Neuroscience, 11,* 543–552.

Woolf, C.J., Molander, C., Reynolds, M., & Benowitz, L.I. (1990) GAP-43 appears in the rat dorsal horn following peripheral nerve injury. *Neuroscience, 34,* 465-478.

Wu, W., Mathew, T.C., & Miller, F.D. (1993) Evidence that the loss of homeostatic signals induces regeneration–associated alterations in neuronal gene expression, *Developmental Biology, 158,* 456-466.

Zagrebelsky, M., Rossi, F., Hawkes, R.,& Strata P. (1996) Topographically arranged climbing fibre sprouting in the adult rat cerebellum. *European Journal of Neuroscience, 8,* 1051-1054.

Zagrebelsky, M., Buffo, A., Skerra, A., Schwab, M.E., Strata, P., & Rossi, F. (1998) Retrograde regulation of growth-associated gene expression in adult rat Purkinje cells by myelin-associated neurite growth inhibitory proteins. *Journal of Neuroscience, 18,* 7912-7929.

Z'Graggen, W.J., Fouad, K., Raineteau, O., Metz, G.A., Schwab, M.E., Kartje, G.L. (2000) Compensatory sprouting and impulse rerouting after unilateral pyramidal tract lesion in neonatal rats. *Journal of Neuroscience, 20,* 6561-6569.

F. BERGMANN AND B.U. KELLER

Center of Physiology, University of Göttingen, Germany

23. GLUTAMATE, CALCIUM AND NEURODEGENERATIVE DISEASE: IMPACT OF CYTOSOLIC CALCIUM BUFFERS AND THEIR POTENTIAL ROLE FOR NEUROPROTECTIVE STRATEGIES

Summary. Disruptions of glutamate- and Ca^{2+}-dependent signal cascades are critical steps underlying "excitotoxic" cell damage in different forms of neurodegenerative disease, including Alzheimer disease, Parkinson disease, Huntington disease and amyotrophic lateral sclerosis (ALS). Recent investigations have provided increasing evidence that a highly interactive network of proteins and organelles precisely controls the spatio-temporal profile of $[Ca^{2+}]i$ and – accordingly – the activation pattern of "downstream" excitotoxic cascades. This network displays a substantial cell-to-cell variability, which might partially explain the profound heterogeneity in neuronal damage induced by a given insult. The cellular basis of this "selective neuronal vulnerability" has been studied in great detail in mouse models of ALS, which are characterized by a highly selective degeneration of motoneurons in the brain stem and spinal cord. Comparative studies between vulnerable and resistant cells demonstrated the critical importance of Ca^{2+} buffering, diffusion and uptake for neuronal vulnerability. As more detailed information becomes available, it might become feasible to develop novel neuroprotective strategies based on a targeted "stabilisation" of Ca^{2+} homeostasis in selectively vulnerable systems. The future potential of such strategies will be discussed.

1. INTRODUCTION

Research of the last years has created a wealth of information about the disruptions of cellular signal cascades associated with neurodegenerative disease, and molecular genetics has provided a valuable set of animal models that closely resemble human pathologies. For example, mouse models of Alzheimer's disease (AD), Parkinson's disease (PD), Huntington's disease (HD) and amyotrophic lateral sclerosis (ALS) have been created by manipulations of genes encoding presenilin, synuclein, huntingtin and SOD1, respectively. This chapter will concentrate on recent insights about the role of glutamate- and Ca^{2+}-related signal cascades, and will focus on the potential role of Ca^{2+} buffer and stores for specific neuronal pathologies. It first examines elementary properties of Ca^{2+} handling in neurons, with particular emphasis on mechanisms that might become useful target sites for neuroprotective strategies. We then evaluate the impact of Ca^{2+} buffers and organelles in models of AD, PD, HD, ALS and ischemia. Finally, we discuss a framework for neuroprotective strategies aiming at stabilisation of Ca^{2+} buffers and stores in vulnerable cells.

2. ELEMENTARY PROPERTIES OF CA^{2+} BUFFERS AND STORES IN NEURONS

In the most basic scheme, a Ca^{2+} ion entering through a channel diffuses for about a millisecond before it is captured by a binding partner (buffer) in the cytosol. Suitable partners include proteins, lipids, organelles or other negatively charged cytosolic components of the cell. The underlying mechanisms have both been investigated by detailed experimental and theoretical approaches as summarized in previous reviews (Neher, 1995; Zucker, 1999; Augustine, 2001). In short, Ca^{2+}-binding to buffers is thought to occur within several nanometers of the channel pore, where the volume defined by the "free diffusion" zone around an open channel is called a "nanodomain". Model calculations suggest that Ca^{2+} levels reach concentrations up to 100µM, and signal cascades that are located in the direct vicinity (< 100nm) of an open channel are regulated by Ca^{2+} levels in this domain. A significantly larger area is defined by the local accumulation of $[Ca^{2+}]i$ in "microdomains" (Fig. 1), which also form around open Ca^{2+} channels and are characterized by a steady-state, dynamic equilibrium with the local buffers of the cell. Critical parameters are binding rates, affinities and diffusional mobilities of protein buffers like calbindin-D28k, calretinin and parvalbumin. Peak $[Ca^{2+}]i$ levels in microdomains reach values up to 10µM, depending on the presence of localized diffusion barriers such as the cytoskeleton, organelles and local clusters of proteins. Nano- and microdomains rapidly collapse after channel closure, where lifetimes (usually several milliseconds) depend on the diffusional mobility of local buffers. Lateral dispersion of microdomains eventually leads to a cumulative rise in global cytosolic $[Ca^{2+}]i$ with a corresponding activation of downstream signal cascades.

2.1. Protein buffers

The overwhelming fraction of Ca^{2+} ions (>98%) in a cell is rapidly bound by either fixed or mobile Ca^{2+}-binding proteins like calbindin-D28k, calretinin or parvalbumin. For example, calbindin-D28k represents a "BAPTA-like" buffer that binds an incoming Ca^{2+} ion immediately after cell entry according to its fast on-rate $k_{on} = 10^8$ $M^{-1}s^{-1}$. Provided that they display sufficient diffusional mobility and their affinity is in the concentration range of interest (K_d around 300nM for calbindin-D28k), protein buffers (PBs) may accelerate the lateral dispersion of Ca^{2+} transients in local domains, a mechanism referred to as "buffered diffusion". It is interesting to note, however, that quite different effects result from *immobile* protein buffers that are firmly attached to fixed local structures around a channel. In this case, a high density of PBs significantly *prolongs* local accumulations of $[Ca^{2+}]i$, in particular if a fast binding rate for Ca^{2+} occurs in combination with slow off-kinetics of the buffer.

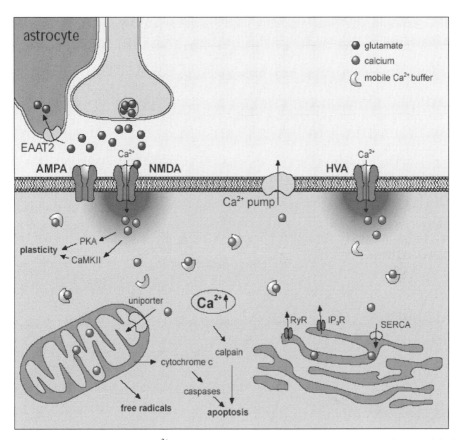

Figure 1. Basic scheme of Ca^{2+} regulation in neurons and excitotoxic cascades. Simplified scheme of elements involved in regulation of cellular Ca^{2+} homeostasis and excitotoxic cascades. Ca^{2+} ions, which enter the cell via different influx pathways (AMPA- , NMDA-type glutamate receptors, high voltage-activated Ca^{2+} channels (HVA)), are rapidly captured by mobile buffers, transported into organelle stores or extruded over the cell membrane. Ca^{2+} transport into the endoplasmic reticulum (ER) is driven by action of the sarco-endoplasmic reticulum Ca^{2+} ATPase (SERCA), transport into mitochondria occurs via the mitochondrial Ca^{2+} uniporter. Ca^{2+} is released from the ER via ryanodine (RyR) or inositol triphosphat (IP_3) receptors. Local Ca^{2+} elevations trigger processes like synaptic plasticity, which involves Ca^{2+}-dependent enzymes protein kinas A (PKA) and Ca^{2+}/calmodulin-dependent protein kinase II (CaMKII). Extensive glutamatergic stimulation under pathological conditions can appear, when i.e. glutamate uptake into astrocytes via the glutamate transporter EAAT2 is impaired. This increases cytosolic Ca^{2+} levels, enhances mitochondrial Ca^{2+} loading, triggers free radical generation and may induce cellular cascades leading to cell death.

Compared to calbindin-D28k, "EDTA-like" protein buffers like parvalbumin (K_d for Ca^{2+} around $50 \cdot 10^{-9}M$) display a different functional profile, primarily characterized by a slower on-rate and a significant affinity for Mg^{2+} ($K_d = 10^{-4}M$).

These differences have important functional consequences, as the unbinding process of Mg^{2+} controls the availability of Ca^{2+} binding sites and thus accounts for larger nanodomains. Similarly, increased Ca^{2+} binding affinities provide a smaller Ca^{2+} binding capacity in areas with high levels of $[Ca^{2+}]i$ like those found in microdomains (> 1µM), where high affinity binding sites are thought to be saturated. By considering these elementary consequences of "buffered diffusion", it is becoming clear that parvalbumin shapes the spatio/temporal profile of cytosolic $[Ca^{2+}]i$ in a way that is substantially different from that imposed by an equivalent concentration of calbindin-D28k. With respect to potential neuroprotective strategies, these examples illustrate the versatile impact of buffer systems and thus underlines the necessity to specify the physical/chemical environment of Ca^{2+}-dependent signal cascades before any conclusions about the beneficial value of buffer modifications can be drawn. In this context, it is interesting to note that concentrations of protein buffers vary substantially between different cell types. One way to quantitatively determine protein buffers is given by the "added buffer" approach (Neher, 1995, 1998).

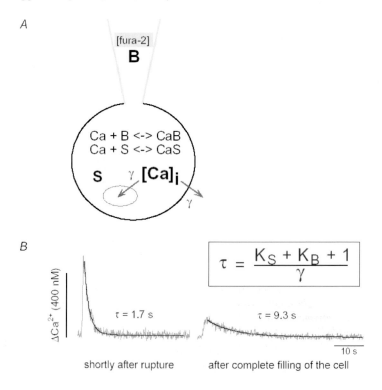

Figure 2. The added buffer approach. A In a linear model, the cytosolic $[Ca^{2+}]$ is dependent on the amount of endogenous buffer (S), which binds Ca^{2+} reversibly, and the transport rate (γ) of Ca^{2+} across cellular membranes. A gradual introduction of an exogenous buffer (B) such as fura-2 via the patch pipette permits the determination of the cytosolic Ca^{2+} buffer

capacity (K_S), which reflects the relative fraction of bound vs. free Ca^{2+} ions in the cell. B K_S can be calculated by determining the recovery time constant τ of evoked Ca^{2+} transients during filling of a patch-clamped cell with exogenous buffer (fura-2), when the buffering capacity of the added buffer (K_B) is known. Essential for the calculation of K_S is the ability of the exogenous buffer to prolong the recovery time of Ca^{2+} transients and to reduce the transient amplitude.

One way to quantitatively determine protein buffers is given by the "added buffer" approach (Neher, 1995, 1998). As illustrated in Fig. 2, a gradual introduction of an exogenous buffer like fura-2 in patch-clamped neurons permits the determination of the cytosolic buffer capacity K_S, which reflects the relative fraction of bound vs. free Ca^{2+} ions in the cell. By utilizing this approach, exceptionally low endogenous Ca^{2+} binding ratios were found in chromaffin cells as well as in hypoglossal and spinal motoneurons ($K_S = 40$, $K_S = 41$, $K_S = 50$ respectively), whereas Purkinje cells display exceptionally high buffer capacities ($K_S = 900$; 6 day old mice), presumably resulting from a powerful expression of calbindin-D28k (Fig. 3). The "added buffer" approach also allows to calculate Ca^{2+} transport rates (γ), which reflect the combined action of slow uptake into organelles and Ca^{2+} transport across the plasma membrane. Similar to buffering capacities, transport rates show a considerable variation between different cell types. They were found to be relatively slow in hypoglossal, spinal and oculomotor neurons ($\gamma = 60$ s^{-1}, $\gamma = 140$ s^{-1} and $\gamma = 156$ s^{-1} respectively) compared to those in the calyx of Held (400 s^{-1}, 21°C), but were several-fold faster than in adrenal chromaffin cells ($\gamma = 13$ s^{-1}).

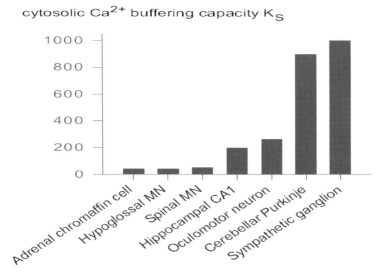

Figure 3. Comparison of buffering capacities. Cytosolic Ca^{2+} buffering capacities (K_S) are compared among different cell types. Note that K_S varies substantially between cells and that neurons that are damaged in ALS - hypoglossal and spinal motoneurons (MNs) - display several fold lower K_S values compared to cells usually resistant in ALS.

2.2. Organelles as Ca^{2+} stores

A different type of cytosolic Ca^{2+} regulation is achieved by organelles such as the endoplasmic reticulum and mitochondria. As transport across their membranes is slow compared to Ca^{2+} binding kinetics of calbindin-D28k, organelle-related mechanisms are thought to shape $[Ca^{2+}]i$ on the time scale of seconds rather than milliseconds. Organelle uptake is primarily based on active transport mechanisms like ATP-dependent SERCA (sarco endoplasmic reticulum calcium ATPase) pumps that dominate Ca^{2+} transport into the endoplasmic reticulum (ER). Ca^{2+} transport into mitochondria is primarily achieved by uniporters that utilize the electrochemical potential across the inner mitochondrial membrane as energy source for Ca^{2+} transport. Accordingly, disruptions of mitochondrial electrochemical potential by mitochondrial "uncouplers" like CCCP represent a commonly used strategy to characterize mitochondrial Ca^{2+} transport mechanisms. It should also be noted that not only Ca^{2+} uptake but also Ca^{2+} release from the ER via ryanodine or inositol triphosphat (IP_3) receptors may shape local Ca^{2+} signals and contribute to spreading of Ca^{2+} signalling across the cell.

2.3. Ca^{2+} transport across plasma membranes

Active Ca^{2+} transport across the plasma membrane operates on a time scale that is comparable to uptake mechanisms into organelles and is thought to regulate recovery of global Ca^{2+} transients like those observed during persistent, high frequency action potential activity of neurons. Suitable molecular entities that achieve such transport include ATP-dependent Ca^{2+} ATPases and Na^+/Ca^{2+} exchangers that utilize the Na^+ gradient across the plasma membrane to extrude elevated levels of $[Ca^{2+}]i$. Another protein that achieves Ca^{2+} transport across these membranes is the Ca^{2+}/H^+-exchanger that may account for pH-dependent regulation of $[Ca^{2+}]i$.

2.4. Functional impact of local Ca^{2+} profiles

Not only a global rise in $[Ca^{2+}]i$, but also very local changes in $[Ca^{2+}]i$ are important determinants of Ca^{2+} controlled signal cascades. In particular, microdomains are thought to control a number of highly localized signal cascades including synaptic modulation, Ca^{2+}-dependent afterhyperpolarizations (AHPs) and neurotoxic pathways that seem to represent critical determinants of cell death. A particular prominent example for local, Ca^{2+} controlled signalling is long term potentiation (LTP), which is thought to underlay synaptic plasticity and therefore processes like learning, memory and development. Synaptic Ca^{2+} elevations following influx through NMDA receptors induce LTP by a mechanism involving activation of Ca^{2+} dependent enzymes calcium/calmodulin-dependent protein kinase II (CaMKII) and protein kinase A (PKA). Interestingly, the absence of calbindin-D28k severely impairs the ability to induce LTP in the hippocampus. This can be explained by alteration of the local Ca^{2+} profile following influx through NMDA receptors, which is predicted to exert higher amplitude, larger extension and faster decay in the absence of this Ca^{2+} buffering protein. This clearly influences the level of Ca^{2+} that is "seen" by enzymes dependent on their localization to the Ca^{2+} influx site. In conclusion, this example strongly indicates the necessity of an exact temporal and spatial pattern of change in $[Ca^{2+}]i$ to allow defined control of signal cascades by Ca^{2+} (Franks and Sejnowski, 2002).

It is known for many years that a substantial rise in cytosolic $[Ca^{2+}]$ may trigger secondary Ca^{2+} dependent phenomena, which promote neurotoxicity and eventually result in cell death. This includes activation of the Ca^{2+} dependent protease calpain and endonucleases, which promote cytoskeletal breakdown and DNA degradation. More recently is became clear that activation of neurotoxic pathways may be dependent on local accumulation of Ca^{2+} and even more be linked to specific routes of Ca^{2+} influx, close to which neurotoxic signalling mechanisms are located.

Experiments revealed i.e. that Ca^{2+} influx through NMDA receptor channels is more harmful than Ca^{2+} influx through voltage-activated Ca^{2+} channels. Apparently, glutamate toxicity is triggered by $[Ca^{2+}]$ increases in subcellular microdomains near the NMDA receptor, as only fast Ca^{2+} chelators were able to protect against NMDA-receptor mediated neurotoxicity (Sattler and Tymianski, 2000). Another mechanism of Ca^{2+} mediated neurotoxicity involves mitochondrial Ca^{2+} overload and release of cytochrome *c* into the cytosol resulting in activation of various caspases (in particular caspase 3, 7 and 9), key enzymes in induction of apoptotic cell death. As the Ca^{2+} levels reached in microdomains can exceed the global cytosolic $[Ca^{2+}]$ by several fold, local accumulation of Ca^{2+} due to influx or release may have important impact on mitochondrial Ca^{2+} uptake and activation of apoptosis-related cascades.

3. CELL-SPECIFIC CA^{2+} HOMEOSTASIS AND SELECTIVE NEURONAL VULNERABILITY

Research of the last few years has clearly demonstrated that the network of proteins and organelles that controls Ca^{2+} levels and associated excitotoxic cascades in neurons is not uniform, but shows a substantial variability between different neuron types. More detailed studies have indeed suggested that specific profiles of Ca^{2+} handling in neurons might correlate with their selective vulnerability. Table I gives an overview of those animal models of neurodegeneration that have recently been investigated, including models of Alzheimer's disease (AD), Parkinson's disease (PD), Huntington's disease (HD), amyotrophic lateral sclerosis (ALS) and ischemia.

3.1. Amyotrophic lateral sclerosis and selective degeneration of motoneurons

In amyotrophic lateral sclerosis (ALS), a neurodegenerative disorder characterized by a loss of motor neurons in the motor cortex, brainstem and spinal cord, there is a body of evidence implicating glutamate toxicity as a contributory factor in selective neuronal injury (Heath and Shaw, 2002). Probably the best accepted evidence arises from the fact that interference with glutamate transmission is so far the only neuroprotective therapeutic strategy that has shown benefit in terms of slowing disease progression in ALS patients. Biochemical studies have shown decreased glutamate levels in central nervous system (CNS) tissue and increased levels in the cerebrospinal fluid (CSF) of ALS patients. Consistently, there is evidence for altered expression and function of glial glutamate transporters, particularly excitatory amino acid transporter 2 (EAAT2).

Table I

Neurodegen. Disease	Targeted Cells	Cellular Feature/Pathology	Potential Neuroprotection
ALS	vulnerable MN	low PB disrupt of mitochondria	elevate PB stabilize mitochondria
HD	striatum (spiny neurons)	disrupt of mitochondria	stabilize mitochondria elevate PB ?
PD	substantia nigra	low PB oxidative stress	elevate PB ?
AD	hippocampus neocortex	medium PB disrupt of ER	elevate PB reduce ER-Ca release
Ischemia	hippocampus	medium PB	lower PB increase VACC inact

Studies on cultured motoneurons have shown that glutamate toxicity is apparently mediated via calcium permeable alpha-amino-3-hydroxy-5-methyl-4-isoxazole propionic acid (AMPA)-subtype glutamate receptors. Mitochondrial dysfunction implicating cellular energy deficits and impaired Ca^{2+} handling may also contribute to excitotoxicity. Whereas the exact pathomechanism still remains to be elucidated, there is overwhelming evidence that disruption of Ca^{2+}-related signal cascades represents a critical element in motoneuron degeneration. Elevated Ca^{2+} concentrations in motoneurons have been found in tissue from ALS-patients and transgenic mouse models of the disease as well as in cultured motoneurons after transferring CSF from ALS patients. Moreover, an in vitro model of motoneuron degeneration based on prolonged glutamate exposure of dissociated cells revealed that Ca^{2+} ions play a central role in motoneuron injury.

So far, several parameters have been associated with the selective vulnerability of motoneurons in ALS and corresponding murine models. This includes highly Ca^{2+}-permeable AMPA receptor channels, a high neurofilament content and low endogenous concentrations of PBs (Shaw and Eggett, 2000). The potential role of

Ca^{2+} buffers in selective vulnerability of motoneurons was also suggested by determining the Ca^{2+} buffering capacity K_S using the "added buffer" approach described above (Fig. 2). K_S values in vulnerable hypoglossal and spinal motoneurons (K_S = 41 and K_S = 50 respectively) were found to be 5-6 times lower compared to the binding ratios in resistant oculomotor (K_S = 264) or hippocampal CA1 neurons (K_S = 160-207) (Vanselow and Keller, 2000) (Fig. 3). Recent work indicates that a prominent role of mitochondria, which take up approximately 50 % of cytosolic Ca^{2+} loads, might also contribute to the vulnerability of motoneurons to ALS-associated injury.

3.2. Huntington's disease and striatal degeneration

Glutamate-mediated excitotoxicity has also been implicated in the pathogenesis of Huntington's disease (HD), a neurodegenerative disorder, in which GABAergic projecting cells (spiny neurons) in the striatum are preferentially lost due to mutations in the huntingtin protein. Early evidence arose from animal experiments showing that intrastriatal injection of glutamate receptor agonists mimicked many of the pathological and neurochemical features of human HD. More recent studies on transgenic mouse models of HD demonstrated a decreased expression level of a metabotropic glutamate receptor, mGluR2, presynaptically localized on corticostriatal terminals, which is thought to serve as negative feedback control for glutamate release. Deficiency of this type of receptor might therefore result in disinhibition of glutamate release. This scenario might be enhanced by decreased glial glutamate transport, which has been demonstrated in HD transgenic mice (Behrens et al., 2002). Very recently, the importance of glutamate transmission in HD has been confirmed by a study showing that riluzole similar as in ALS prolongs survival time in transgenic HD mice. Mitochondrial dysfunction leading to disruption of mitochondrial Ca^{2+} handling is also believed to contribute to neuronal degeneration in HD (Panov et al., 2002). Whereas glutamate toxicity in ALS seems to be mediated by AMPA receptors, in HD, it apparently acts via activation of N-methyl-D-aspartate (NMDA) receptors. Glutamate excitotoxicity and mitochondrial dysfunction both promote increase of the intracellular Ca^{2+} levels, which is thought to be crucial in mediating neuronal cell death in HD.

The understanding of the selective vulnerability of striatal neurons in HD has remarkably increased during recent years (Calabresi et al., 2000). Electrophysiological data showed that vulnerable neurons are readily and reversibly depolarized by glutamate whereas cholinergic interneurons that are typically spared in HD are much less sensitive and depolarize to a lesser extent. Differential expression of NMDA receptor subtypes and a positive interplay between NMDA and group I mGluR in vulnerable cells have been proposed to underlay the differential sensitivity to excitotoxic events. Additionally, recent work demonstrated that inhibition of mitochondrial metabolism, as observed in HD, renders spiny neurons sensitive to NMDA-mediated synaptic excitation, whereas mitochondrial inhibition had no effect on synaptic transmission in cells not affected in HD. It has also been hypothesized that an early loss of calbindin-D28k in HD could render neurons more vulnerable to pathological or even physiological Ca^{2+} fluxes.

3.3. Alzheimer's disease and degeneration of hippocampal neurons

Alzheimer's disease and degeneration of hippocampal neurons. In Alzheimer's disease (AD), neuronal death occurs primarily in the neocortex and hippocampus, where a subset of pyramidal cells and their projections are particularly vulnerable. In some cases, injury results from mutations in the β-amyloid precursor protein (β-APP) or presenilin genes. Association between pathological hallmarks of AD (neurofibrillary tangles and amyloid plaques) and perturbed Ca^{2+} homeostasis has been established in studies of patients as well as in animal and cell culture models of AD (Mattson and Chan, 2001). Central to the neurodegenerative process is the inability of neurons to properly regulate intracellular Ca^{2+} levels. Presinilin and β-APP mutations perturb Ca^{2+} homeostasis in a way that sensitizes neurons to apoptosis and excitotoxicity. Environmental factors for AD, i.e. high calorie diets or low level of intellectual activity, are being identified that may also impact on neuronal Ca^{2+} homeostasis.

As the disruption of intracellular Ca^{2+} homeostasis is thought to involve enhanced Ca^{2+} release from the endoplasmic reticulum (ER), the presence of relatively large pools of ER Ca^{2+} release sites in hippocampal and cortical neurons may predispose these cells to degeneration in AD. Additionally, a high expression level of presenilin and β-APP in vulnerable neurons and the presence of neurofilament might contribute to their selective vulnerability in AD.

3.4. Degeneration of substantia nigra neurons in Parkinson's Disease

In Parkinson's disease (PD), neurons in the substantia nigra selectively degenerate leading to a loss of dopamine and a disruption in the neuronal circuitry controlling movement. Oxidative stress and mitochondrial dysfunction are implicated to play a major role in PD-related neurodegeneration (Jenner and Olanow, 1998). The potential involvement of environmental factors is suggested by the observation that the toxin 1-methyl-4-phenyl-1,2,3,6-tetrahydropyridine (MPTP) can induce PD-like pathology and symptomology in rodents. Animal studies revealed that further factors possibly contributing to neurodegeneration in PD include overactivation of glutamate receptors. Local application of glutamate receptor agonists could produce parkinsonism in rats and antagonising excitatory neurotransmission in the substantia nigra has proven benefit in experimental animals leasoned with MPTP (Klockgether and Turski, 1993). Hence, PD stands in the row of neurodegenerative disorders, where excitotoxicity and resulting disruption of neuronal Ca^{2+} homeostasis seems to play a role.

The factors determining the selective vulnerability of striatal neurons in PD are still little understood. As mitochondrial dysfunction and oxidative damage seem to play important roles, a high basal level of oxidative stress due to dopamine metabolism has been implicated to render substantia nigra neurons susceptible to further injury. That cell specific Ca^{2+} buffer systems might impact on the selective vulnerability of striatal neurons has arisen from the observation that calbindin-D28k is primarily expressed in midbrain dopaminergic neurons reported to be relatively resistant to degeneration in PD and certain of its animal models (Damier et al., 1999). Also calretinin expression has been reported to correlate with neuronal resistance in PD and its animal models to a certain extent. However, experiments failed to show a causal role of Ca^{2+} binding proteins for selective vulnerability of neurons in PD.

3.5. Acute neurodegeneration during ischemia

Neuronal injury caused by occlusion of cerebral arteries is believed to be mediated by excessive activation of glutamate receptors. Extracellular glutamate levels increase due to release of glutamate and reduction of glutamate uptake following neuronal and glial depolarization as a result of energy depletion. Extensive studies have been undertaken to elucidate the molecular cascades following glutamate stimulation in animal and cell culture models of ischemia. Whereas the early phase is characterized by Na^+ influx through NMDA and AMPA receptor channels leading to cell swelling, profound neuronal damage is promoted by a substantial elevation in cytoplasmic $[Ca^{2+}]$, which derives from Ca^{2+} influx through NMDA receptors and further amplification by voltage activated Ca^{2+} influx and Ca^{2+} release from intracellular stores. The massively increased intracellular second messenger Ca^{2+} triggers numerous deleterious processes, including free radial formation, membrane degradation, mitochondrial dysfunction, inflammation, DNA-damage and apoptosis (Dirnagl et al., 1999).

In the case of global cerebral ischemia, certain neuronal cell types have been shown to be injured preferentially. This increased vulnerability to ischemia is mainly attributable to hippocampal CA1 and cortical pyramidal neurons. The molecular mechanisms underlying the cell-specific pattern of global ischemia-induced neuronal death are little understood, but a role for gap junctional coupling has been proposed.

4. CA^{2+} BUFFERS AND STORES AS POTENTIAL TARGETS FOR FUTURE NEUROPROTECTIVE STRATEGIES

The finding that cell-specific adaptations of buffers and stores are elementary determinants of excitotoxic neuronal damage in several forms of neurodegenerative disease has triggered a series of studies aiming at a targeted stabilisation of vulnerable cells by changes in cytosolic Ca^{2+} handling. Interestingly, such studies have indeed underlined the significance of cell-specific heterogeneities in neuronal Ca^{2+} regulation as indicated by the following neuroprotective strategies. (see also Table I)

4.1. Neuroprotection by elevated protein buffers - ALS

Several studies investigating neuronal Ca^{2+} homeostasis in ALS indicated that neuronal populations with *high* buffering capacities are best protected against ALS-related excitotoxic insults (Vanselow and Keller, 2000). Although the molecular mechanisms underlying this selective vulnerability are not completely understood, several studies were performed to stabilize vulnerable motoneurons by elevated PBs. Indeed, several groups (Beers et al., 2001; Van Den Bosch et al., 2002) could show that elevated concentrations of parvalbumin were able to protect vulnerable cells in a cell culture and mouse model of ALS. In general, this result can be explained by several different mechanisms.

One interpretation is that elevated protein buffers limit maximum amplitudes of cytosolic Ca^{2+} responses, thus reducing the probability for activation of apoptosis-related signal cascades. A second interpretation is that in motoneurons with low endogenous PB concentrations elevated parvalbumin prevents local accumulations of $[Ca^{2+}]i$ around open Ca^{2+} channels due to its high diffusional mobility. For example, if mitochondria were located in close vicinity of voltage dependent Ca^{2+} channels, rapid dispersion of local Ca^{2+} gradients could reduce the risk for mitochondrial Ca^{2+} overload and associated production of reactive oxygen species (ROS). In earlier studies, both mechanisms have been shown to contribute to selective motoneuron degeneration during excess Ca^{2+} influx (Carriedo et al., 2000).

4.2. Elevating protein buffers in AD

Immunohistochemical analysis of post mortem human brain tissue has suggested that neurons expressing high levels of calbindin are relatively resistant to death in AD (Iritani et al., 2001). Therefore the attempt was made to elevate calbindin in vulnerable neurons assessing the neuroprotective potential of this Ca^{2+} binding protein during AD-related injury. Indeed, over expression of calbindin protected neurons against Aβ- and mutant presenilin-mediated injury *in vitro*. The protective effect of calbindin was explained by prevention of sustained elevations in $[Ca^{2+}]i$, which are known to damage mitochondrial electron transport resulting in oxidative stress and increased vulnerability to apoptosis. Consistently, the Ca^{2+} buffer BAPTA protected synaptosomes against the adverse effect of presenilin-1 mutations on mitochondrial function and reduced activation of apoptosis-related signal cascades.

4.3. Altering protein buffers in HD and PD

Both in Huntington's disease and Parkinson's disease, the causal involvement of PB systems in selective vulnerability of neurons is still controversial. However, as excitotoxic disturbance in neuronal Ca^{2+} homeostasis occurs during disease progression, the strategy of altering PBs in vulnerable neurons might have neuroprotective potential. Indirect evidence for a neuroprotective role of increased PBs comes from experiments showing that neurotrophin-4/5, which is known to elevate calbindin levels in cells, was able to protect striatal projection neurons as well as substantia nigra neurons against injury in animal models of HD and PD (Alexi et al., 2000).

4.4. Neuroprotection by reduced protein buffers

In mouse models of ischemia and stroke, several studies have provided evidence that *reduced* concentrations of PBs can protect neurons against Ca^{2+}-mediated damage. In the hippocampus, reduced concentrations of calbindin protected cells against ischemia-induced degeneration, presumably by retarding the dispersion of local gradients around open Ca^{2+} channels. As these channels are characterized by rapid, Ca^{2+}-dependent inactivation, enhanced local accumulation of $[Ca^{2+}]i$ provides a strong, negative feedback mechanism that reduces Ca^{2+} influx during prolonged depolarisations. This interpretation is supported by studies on transgenic animals (Klapstein et al., 1998), where "knock-out" of the Ca^{2+}-binding protein calbindin protected hippocampal cells during ischemia.

4.5. Neuroprotection by modifications of organelle Ca^{2+} stores

An example, where modification of organelle buffers contributes to neuroprotection is provided by the administration of creatine in ALS and HD. In both diseases, mitochondrial dysfunction has been found to play a key role. Dietary creatine supplementation increases survival and delays motor symptoms in transgenic murine models of ALS and HD. The beneficial effect of creatine is explained by an increase in brain phosphocreatine levels, which compensate for the bioenergetic deficit that results from mitochondrial impairment. Additionally, creatine reduces oxidative stress and stabilizes mitochondrial permeability transition (Tarnopolsky and Beal, 2001).

Considering Alzheimer's disease, there is evidence that stabilization of the ER can promote neuroprotection. Under pathological conditions, mutant presenilin, which is located in the ER membrane, enhances Ca^{2+} release from $Ins(1,4,5)P_3$- and ryanodine sensitive stores. Drugs that prevent Ca^{2+} release from the ER such as dantrolene protect against the adverse effects of the presenilin mutations and stabilize cellular Ca^{2+} homeostasis *in vitro*.

5. CONCLUDING REMARKS

An increasing number of investigations indicates that cytosolic Ca^{2+} handling is critical for the pathogenesis of neurodegenerative disease, and neuron-specific variations in Ca^{2+} buffers, organelles and transport mechanisms may partially explain the selective vulnerability of defined neuronal populations in AD, PD, HD and ALS. By investigating in detail the underlying, highly interactive network of Ca^{2+}-regulating proteins and organelles, it is possible to identify a number of risk factors that expose specific neuronal populations to exceptional risks during pathophysiological conditions and glutamate-related excitotoxic insults. Based on encouraging results from animal models of human neurodegeneration, it is reasonable to assume that a targeted stabilisation of Ca^{2+} buffers and stores might also display beneficial effects in human neurodegenerative disorders.

REFERENCES

Alexi T, Borlongan CV, Faull RL, Williams CE, Clark RG, Gluckman PD, Hughes PE (2000) Neuroprotective strategies for basal ganglia degeneration: Parkinson's and Huntington's diseases. Prog Neurobiol 60:409-470.

Augustine GJ (2001) How does calcium trigger neurotransmitter release? Curr Opin Neurobiol 11:320-326.

Beers DR, Ho BK, Siklos L, Alexianu ME, Mosier DR, Mohamed AH, Otsuka Y, Kozovska ME, McAlhany RE, Smith RG, Appel SH (2001) Parvalbumin overexpression alters immune-mediated increases in intracellular calcium, and delays disease onset in a transgenic model of familial amyotrophic lateral sclerosis. J Neurochem 79:499-509.

Behrens PF, Franz P, Woodman B, Lindenberg KS, Landwehrmeyer GB (2002) Impaired glutamate transport and glutamate-glutamine cycling: downstream effects of the Huntington mutation. Brain 125:1908-1922.

Calabresi P, Centonze D, Bernardi G (2000) Cellular factors controlling neuronal vulnerability in the brain: a lesson from the striatum. Neurology 55:1249-1255.

Carriedo SG, Sensi SL, Yin HZ, Weiss JH (2000) AMPA exposures induce mitochondrial Ca(2+) overload and ROS generation in spinal motor neurons in vitro. J Neurosci 20:240-250.

Damier P, Hirsch EC, Agid Y, Graybiel AM (1999) The substantia nigra of the human brain. II. Patterns of loss of dopamine-containing neurons in Parkinson's disease. Brain 122 (Pt 8):1437-1448.

Dirnagl U, Iadecola C, Moskowitz MA (1999) Pathobiology of ischaemic stroke: an integrated view. Trends Neurosci 22:391-397.

Franks KM, Sejnowski TJ (2002) Complexity of calcium signaling in synaptic spines. Bioessays 24:1130-1144.

Heath PR, Shaw PJ (2002) Update on the glutamatergic neurotransmitter system and the role of excitotoxicity in amyotrophic lateral sclerosis. Muscle Nerve 26:438-458.

Iritani S, Niizato K, Emson PC (2001) Relationship of calbindin D28K-immunoreactive cells and neuropathological changes in the hippocampal formation of Alzheimer's disease. Neuropathology 21:162-167.

Jenner P, Olanow CW (1998) Understanding cell death in Parkinson's disease. Ann Neurol 44:S72-84.

Klapstein GJ, Vietla S, Lieberman DN, Gray PA, Airaksinen MS, Thoenen H, Meyer M, Mody I (1998) Calbindin-D28k fails to protect hippocampal neurons against ischemia in spite of its cytoplasmic calcium buffering properties: evidence from calbindin-D28k knockout mice. Neuroscience 85:361-373.

Klockgether T, Turski L (1993) Toward an understanding of the role of glutamate in experimental parkinsonism: agonist-sensitive sites in the basal ganglia. Ann Neurol 34:585-593.

Mattson MP, Chan SL (2001) Dysregulation of cellular calcium homeostasis in Alzheimer's disease: bad genes and bad habits. J Mol Neurosci 17:205-224.

Neher E (1995) The use of fura-2 for estimating Ca buffers and Ca fluxes. Neuropharmacology 34:1423-1442.

Neher E (1998) Usefulness and limitations of linear approximations to the understanding of Ca++ signals. Cell Calcium 24:345-357.

Panov AV, Gutekunst CA, Leavitt BR, Hayden MR, Burke JR, Strittmatter WJ, Greenamyre JT (2002) Early mitochondrial calcium defects in Huntington's disease are a direct effect of polyglutamines. Nat Neurosci 5:731-736.

Sattler R, Tymianski M (2000) Molecular mechanisms of calcium-dependent excitotoxicity. J Mol Med 78:3-13.

Shaw PJ, Eggett CJ (2000) Molecular factors underlying selective vulnerability of motor neurons to neurodegeneration in amyotrophic lateral sclerosis. J Neurol 247 Suppl 1:I17-27.

Tarnopolsky MA, Beal MF (2001) Potential for creatine and other therapies targeting cellular energy dysfunction in neurological disorders. Ann Neurol 49:561-574.

Van Den Bosch L, Schwaller B, Vleminckx V, Meijers B, Stork S, Ruehlicke T, Van Houtte E, Klaassen H, Celio MR, Missiaen L, Robberecht W, Berchtold MW (2002) Protective effect of parvalbumin on excitotoxic motor neuron death. Exp Neurol 174:150-161.

Vanselow BK, Keller BU (2000) Calcium dynamics and buffering in oculomotor neurones from mouse that are particularly resistant during amyotrophic lateral sclerosis (ALS)-related motoneurone disease. J Physiol 525 Pt 2:433-445.

Zucker RS (1999) Calcium- and activity-dependent synaptic plasticity. Curr Opin Neurobiol 9:305-313.

C. LÓPEZ-GARCÍA AND J. NACHER

Cellular Neurobiology, University of Valencia, Spain

24. POSTNATAL NEUROGENESIS AND NEURONAL REGENERATION

Summary. The discovery of neurogenesis in the adult brain has challenged one of the central dogmas of neuroscience. Pioneer reports in rodents seed the ground for a detailed description in birds and reptiles, which was finally confirmed in discrete regions of several mammalian species including humans. This neurogenetical capability may serve as the basis for neuronal regeneration, as has already been described in the reptilian brain, and thus may represent a promising therapeutic approach. Consequently, in the last years there has been an important effort to deepen our knowledge of the biology, the functional significance and the regulation of adult neurogenesis.

1. INTRODUCTION

In the middle of the last century, the application of radioactive DNA precursors (i.e., tritiated thymidine) and autoradiographic techniques in laboratory animals led to the identification of mitotically active nervous cells, which give rise to granule neurons in the olfactory bulbs, the dentate gyrus of the hippocampus, and the cerebellum. These findings opened new vistas, evidencing that some nervous centers increase their neuronal populations during postnatal life. Subsequently, they were assessed in non-mammalian species, especially in lower vertebrates, in which postnatal neurogenetic activity was also evidenced in the spinal cord, retina, cochlea, etc., and in other particular centers of reptiles and birds. Postnatally generated neurons were incorporated into the nervous tissue and contributed to the continuous growth of some centers (retina, cochlea); they substituted killed neurons after programmed cell death (canary singing center), or were even the motor for regenerative neo-histogenetic processes after a lesion (in the lizard medial cortex). Postnatal neurogenetic activity and regenerative potential are linked phenomena, and both are residual ontogenetic characters, which appear robust in lower vertebrates but become progressively weaker in mammals. Experimental modulation of postnatal neurogenesis may help the nervous system to repair itself is a possibility.

2. POSTNATAL NEUROGENESIS

Production and incorporation of new neurons during postnatal life has been evidenced in a series of nervous centers. In the next sections, we will briefly describe the current knowledge of this phenomenon throughout the evolutionary scale.

2.1. Olfactory mucosa and olfactory bulbs

The sensory neurons in the olfactory mucosa have been shown to be subjected to a continuous removal and regeneration during adulthood. This phenomenon is clearly related with the postnatal neurogenetic activity and regenerative potential of the olfactory bulbs. The precursors of the olfactory bulb neurons generated during adulthood reside in the subventricular zone surrounding the lateral ventricle. Once generated, the neuroblasts migrate tangentially towards the olfactory bulb through the rostral migratory stream, where they maintain proliferative activity. Once in the bulbs, they undergo radial migration and differentiate into granule and periglomerular interneurons.

2.2. Retina

In lower vertebrates, the pigmented epithelium at the retinal margin acts as a retinal progenitor center that persists throughout life and gives rise to all neuronal types of the retina. It contributes to the continuous increase in size of the eye and the retina of lower vertebrates (fishes and amphibians), and its cells are the motor of regenerative potential after massive retinal damage. The amniote chick retina also shows postnatal neurogenesis in its periphery. However, its potentiality is reduced as the progenitor cells give rise only to bipolar and amacrine neurons, failing to produce photoreceptor, horizontal, or ganglion cells. Unlike fish and amphibian progenitors, chick retina progenitors do not increase their rate of proliferation in response to acute damage and, consequently, do not enable retinal regeneration. The proliferation of progenitor cells in lower vertebrate and avian retina is increased by different molecules, including insulin, insulin growth factor-I, and epidermal growth factor.

2.3. Cochlea

In sharks, a large number of inner hair cells are generated postnatally. A similar phenomenon was found in the amphibian cochlea and in that of amniotes (birds and mammals), and these auditory cells were also demonstrated to have a considerable regenerative potential. In birds, traumatic loss of hair cells induces a dramatic increase in the proliferation rate of supporting cells, which give rise to new hair cells.

2.4. Hippocampal dentate gyrus

During embryonic development, some neuroepithelial cells leave their ependymal location at the caudal edge of the lateral ventricles and reach an intraparenchymal location in the hippocampus, forming a secondary proliferative matrix just beneath the dentate gyrus granular layer: the subgranular zone. During early postnatal life, this region produces thousands of neurons that are recruited into the adjacent granule layer following an outside-to-inside pattern. In rodents, this highly proliferative period, during which nearly 85% of dentate granule cells are generated, extend over the first three weeks after birth; in primates, it lasts until the third postnatal month. Afterwards, in both rodents and primates (including humans), as well as in many other mammals, hippocampal neurogenesis persists, although reduced, during adult life. This neurogenetic activity is mainly restricted to the subgranular zone (neuronal progenitor cells have also been found scattered throughout the adult hippocampus). However, the rate of proliferation and addition of new granule neurons in primates appears to be lower than the very high rate in rodents.

2.5. Canary singing center

The male canary's singing pattern changes every year as the result of annual replacement of the striatal high vocal center neurons, which project to the robust nucleus of the archistriatum. Seasonally, under hormonal control, ependymal cells proliferate and give rise to immature neurons that migrate along radial glia until they reach the high vocal center. There, they mature and send new axons to the robust nucleus, substituting dead projection neurons generated in a previous year.

2.6. Cerebellar cortex

During early embryonic stages, certain neuroepithelial cells located in the roof of the fourth ventricle move to a subpial location, forming the transitory "outer granular layer" of the cerebellar cortex, which is also considered a secondary proliferative matrix. During early postnatal life (similarly to what happens in the hippocampal subgranular zone), progenitor cells proliferate and give rise to immature granule neurons, which migrate through the molecular and Purkinje layers and then differentiate, forming the granule layer of the cerebellar cortex. This postnatal neurogenesis is transitory, and ceases soon after birth.

2.7. Spinal Cord

In adult fishes and amphibian larvae, the cells lining the spinal cord ependyma, especially those displaying a cytoplasmic process that reach the outer parenchymal (meningeal) surface (namely, tanycyte-radial glia), retain proliferative activity. This proliferation may be enhanced after crush or mechanical injury, thus leading to functional neuronal regeneration. In fact, fishes can regenerate their spinal cord, although this phenomenon is age-dependent. Spinal cord transection in larval frogs leads to regenerative events that restore the control of initial movements, but in adult frogs results in impairment of movement. In chick embryos, spinal cord transection leads to regeneration, provided that the procedure is performed before day E15.

2.8. Other structures with discrete postnatal neurogenesis

The hypothalamus also incorporates new neurons during the first postnatal week, but not later, and this may also happen in adult dorsal root ganglia. Recent work has shown incorporation of new neurons in the neocortex of adult primates, although it has also been reported that these neurons have a transitory existence. It should be noted, however, that some other authors were not able to replicate these findings on adult neocortical neurogenesis; consequently, this is still a matter of debate. Neurogenesis has also been described in some cortical areas and the amygdala of the young adult rat, and (only after specific lesion of neocortical neurons) in mice. However, in rodents, the rate of neuronal incorporation into these areas seems to be very discrete when compared with that into the adult hippocampus or olfactory bulb. It has to be mentioned that the usual DNA labeling agents (tritiated thymidine, 5-bromodeoxyuridine) are not true markers for some proliferating cells (e.g., in regenerating planarians), as they are not incorporated into these mitotic cells, but probably synthesized by them. The contrary possibility may also happen: some non-dividing cells (e.g., Purkinje neurons) can take up the DNA marker as a consequence of DNA repair activity, leading to false positives.

2.9. Fate of postnatally generated neurons

In lower vertebrates, postnatally generated neurons usually incorporate into the nervous tissue. They are recruited in the neural circuitry and contribute to the continuous growth of these particular centers (e.g., fish retina), where they survive for undeterminate periods of time. In other centers (e.g., the canary singing center), the postnatally generated neurons migrate, differentiate, and become functional during defined periods of time which are under strict hormonal control. Finally, in mammalian centers with postnatal neurogenesis, we can detect a differential fate for the postnatally generated neurons. Those generated in the very early postnatal periods are incorporated into the brain parenchyma and survive for undeterminate periods of time (e.g., those of the dentate gyrus, the olfactory bulbs, and the cerebellum). In contrast, most of the neurons generated in later periods of life undergo programmed cell death. The reason for this neuronal death, is possibly that these new cells have to overcome critical obstacles in order to be recruited into the adult nervous parenchyma, which appears more stabilized (with less plasticity, stable circuits and great difficulty in delivering growth factors to the incoming immature neurons). In fact, all regions retaining neuronal incorporation/production in the adult CNS also show abundant programmed cell death. However, many adult-generated neurons do survive, at least transitorily, and incorporate into the circuitry of the olfactory bulb, the hippocampus, or even the cerebral cortex.

3. NEURONAL REGENERATION

In some favorable cases, severed peripheral nerve or axonal fiber bundles regrow until they re-innervate the nerve targets and/or proper muscles, restituting functional performance. We then speak of "nervous" or "axonal" regeneration, which may, of course, occur preferentially when the lesion is close to the terminal synaptic end, usually far away from the motoneuron or sensory cell body.

The concept of "neuronal regeneration" is reserved for those cases in which particular groups of lesioned/destroyed neurons are replaced by recently generated ones. This phenomenon has been well documented in some structures of the CNS of lower vertebrates, and in some "peripheral" nervous centers (i.e., neurons in the olfactory mucosa and auditory hair cells of the macula) of mammals. However, the only case of true neuronal regeneration of an amniote central nervous center is that of the lizard medial cortex.

The lizard medial cortex is a center homologous to the hippocampal dentate gyrus, which exhibits robust postnatal neurogenetic activity, and thus may regenerate. Below we will describe the postnatal neurogenetic characteristics of the mammalian dentate gyrus, their experimental manipulation, and the expectancies raised by this neuronal re-generation and/or repair.

3.1. The case of the lizard medial cortex

The lacertilian cerebral cortex may be regarded as an archicortex or "reptilian hippocampus" and its medial region as a "lizard fascia dentata" on grounds of their anatomy, cyto-chemo-architectonics, ontogenesis, and postnatal development. In normal conditions, cells in the ependyma subjacent to the medial cortex of adult lizards proliferate and give rise to immature neurons, which migrate through the inner plexiform layer until they reach the medial cortex cell layer. Finally, these recruited neurons differentiate and give rise to zinc-containing axons directed to the rest of the cortical areas, thereby resulting in continuous growth of the medial cortex and its zinc-enriched axonal projection. This happens in adult lizards, in which the ependyma subjacent to the medial cortex remains as a residual neuroepithelium, the sulcus septoarchicorticalis.

The biological meaning of the continuous growth of the lizard medial cortex and its zinc-rich axonal projection system is an enigma. The lizard medial cortex is involved in spatial memory performance, like the mammalian dentate gyrus and its hippocampal mossy fibers. The continuous growth of the lizard medial cortex may enable a parallel increase of spatial memory performance, which fits well with the age-related increase of territorial domains observed in lizards.

Specific lesion of the medial cortex granule neurons with the neurotoxin 3-acetylpyridine (which prevents ATP synthesis and thus damages the neurons most active during the drug delivery period) triggers a burst of mitotic activity in the subjacent ependymal sulcus septoarchicorticalis. Newly generated immature neurons migrate through the inner plexiform layer until reaching the medial cortex cell layer, using the vertical shafts of radial glia as guide and support, aided by the expression of the polysialylated form of the neural cell adhesion molecule, which confers antiadhesive properties to migratory cells. Once they are recruited in the cell layer, they mature, emit axons and substitute the previously killed neuronal somata, which are being removed simultaneously by glial cells. Very soon after the neurotoxin injection, the resident microglia transitorily disappear from the medial cortex, thereby creating a permissive milieu for the survival and migration of immature neurons. As microglia are not present in the very first moments after lesion, cell debris removal is initially undertaken by radial ependymoglia. Microglia re-appear later in large numbers and also participate in the cleaning process, both in the granule-cell layer and in the plexiform layers, where incoming axons presumably re-establish new functional synaptic contacts. Finally, several weeks later, the lesioned lizard recovers normal spatial cognitive performance (i.e., finding a hole for escape — a capacity lost after neurotoxin lesion), and the histological appearance of its medial cortex is normal (i.e., indistinguishable from that of a non-lesioned lizard). Consequently, the lizard cerebral cortex is a good model for the study of neuronal regeneration and the complex factors that regulate its neurogenetic, migratory, and neo-synaptogenetic events.

Nevertheless, there are certain limits for neuronal regeneration in lizards. For instance, it does not occur in winter, because low temperature prevents migration of newly generated neurons, and short daylight periods decrease the proliferating activity of ependymal stem cells.

3.2. The case of the mammalian dentate gyrus

The lizard medial cortex and the dentate gyrus of the mammalian hippocampus share common ontogenetic, structural, and phylogenetic features, which may be predictory of similar properties concerning neuronal regeneration.

Neuronal progenitors in the adult dentate gyrus are located in the subgranular zone, and it has recently been demonstrated that these cells express the intermediate filament GFAP and have the characteristics of astrocytes. Radial glia-like cells persist in the subgranular zone of young adult rats, where they proliferate and express nestin, a neural precursor marker, suggesting that these astroglial cells may be the progenitors of recently generated granule neurons.

An intense mitotic response, similar to that observed in lizards, was found in the rodent hippocampus after treatments resulting in neuronal degeneration. Traumatic brain injury leads to increased neurogenesis in the adult dentate gyrus, and specific mechanical and excitotoxic lesions of the adult dentate gyrus proper also induce the proliferation of progenitors in the subgranular zone, especially near the lesion site. Deafferentation of the rat dentate gyrus by transection of the perforant path also induces granule neurogenesis. Human temporal lobe epilepsy is frequently associated with a marked loss of hippocampal neurons. Chemoconvulsant models of temporal lobe epilepsy, such as pilocarpine-induced or kainic-acid-induced seizures, increase both cell proliferation in the subgranular zone and dentate granule neurogenesis. Similarly, electrical models such as amygdala kindling increase neurogenesis in the hippocampus of adult rats. Consequently with the rodent models described above, an increase in nestin immunoreactive cells has been found in the hippocampus of young humans suffering temporal lobe epilepsy.

Ischemic stroke also induces neuronal damage in the hippocampus, and consequently, different models of transient global ischemia in rodents lead to increased neurogenesis in the adult dentate gyrus. The N-methyl-D-aspartate (NMDA) and alpha-amino-3-hydroxy-5-methylisoxazole-4-propionate (AMPA) types of glutamate receptor, and basic fibroblast growth factor mediate this enhancement of granule neurogenesis. Moreover, it has recently been demonstrated that transient ischemia increases neurogenesis in the subventricular zone, and leads to incorporation of new neurons to the rat neocortex.

Removal of adrenal steroids by adrenalectomy leads to massive apoptotic neuronal cell death in the granular layer of the dentate gyrus and to increased granule neurogenesis, in both the adult and the aged hippocampus.

Experimental induction of diabetes with streptozotocin in rats induces deleterious changes in the adult hippocampus, and increases the number of proliferating cells in the subgranular zone.

4. MODULATION OF POSTNATAL NEUROGENESIS IN THE MAMMALIAN DENTATE GYRUS

The possibility that some "neuronal regeneration" could take place in the brain and spinal cord of higher vertebrates has an obvious importance. The possibility of manipulating and enhancing hidden subjacent regenerative potentialities gives some optimism for neural repair strategies. Transplantation of neural stem cells might also be a great contribution, provided that the newly generated neurons reach the damaged areas, survive, are recruited into the proper places, extend axons to the correct targets and thus incorporate into the correct circuitry.

For all that, the events following a massive lesion of the dentate granule cells do not seem to lead to significant neuronal regeneration in mammals. Recently published reports do indicate, however, that neuronal regeneration may exist in the neocortex of mice after selective degeneration of corticothalamic projection neurons, and in the striatum and the CA1 hippocampal layer after ischemia. Moreover, the incorporation of new neurons into the damaged CA1 layer increases significantly after treatment with growth factors, which also leads to improvements in spatial memory performance. Many of these new neurons participating in the regenerative events extend axonal projections to their usual targets, although no direct evidence has been provided yet that these "regenerated", or even "normal" postnatally generated ones, establish functional synaptic contacts.

The ability to modulate this adult neurogenesis has implications for therapeutic approaches. Several intrinsic molecules, suc as growth factors, neurotransmitters, and hormones influence adult hippocampal neurogenesis, many of them in an interconnected way.

4.1. Pharmacological and other extrinsic factors

The proliferation of precursor cells and the incorporation of newly generated granule neurons can be modulated by the NMDA-type glutamate receptor. Blockade of this receptor by treatment with competitive or non-competitive receptor antagonist increases the number of newly generated cells in the dentate gyrus. A similar effect is obtained by lesioning the entorhinal cortex, the main excitatory input to the granule neurons. Blockade of AMPA/kainate receptors increases, although to a lesser extent, cell proliferation in the hippocampus, while the neuroleptic haloperidol, a potent dopamine D2 receptor antagonist, increases cell proliferation in the subgranular zone. The neurotransmitter serotonin, via the 5HT1A receptor, also promotes hippocampal neurogenesis.

Glucocorticoids inhibit the production of granule neurons in the adult dentate gyrus; this effect is achieved through an NMDA-receptor-dependent mechanism, probably through the mineralocorticoid receptor. In contrast, estrogen promotes an increase in the number of granule neurons; consequently, the reproductive status influences hippocampal neurogenesis. Recent findings indicate that serotonin mediates these effects of estrogen in neurogenesis.

Several growth factors influence the production of new neurons in the adult dentate gyrus. Insulin-like growth factor-1 promotes adult hippocampal neurogenesis. Peripheral injection of fibroblast growth factor stimulates cell proliferation and progenitors isolated from the adult hippocampus can be induced to produce neurons after exposure to this growth factor. The neurotrophin brain-derived neurotrophic factor (BDNF) enhances neuronal incorporation into the olfactory bulb and other areas of the CNS, and a neurogenic role has been suggested for this molecule in the adult hippocampus, as it is elevated concomitantly with treatments that enhance neurogenesis, such as diet restriction or administration of the sodium-channel blocker riluzole. Interestingly, these actions of hormones, neurotransmitters, and neurotrophic factors on adult neurogenesis are probably mediated through the cAMP-CREB cascade. Similarly to what happens with growth factors, some cytokines — such as interleukin-6 — are able to modulate adult neurogenesis in the dentate gyrus.

As mentioned above, regulation of adult hippocampal neurogenesis involves several molecules, and it seems that at least some of these effects are inheritable, as dentate neurogenesis in rodents is influenced by strain and even gender.

Extrinsic factors also modulate adult hippocampal neurogenesis. Dietary restriction increases neurogenesis and induces BDNF expression in the adult dentate gyrus. Deficiency in vitamin E also enhances neuronal production in the granule-cell layer. Different types of stressor, such as psychosocial stress, exposure to a predator's odor, or prenatal restraint stress, have been shown to decrease adult hippocampal neurogenesis. Seasonal changes, especially photoperiod length, also affect cell proliferation and incorporation of new neurons into the adult hippocampus of some mammalian species, similarly to the already described case of lizard.

Some pharmacological treatments have also been found to affect adult hippocampal neurogenesis. Substances commonly used in the treatment of mood disorders, such as lithium or enhancers of serotoninergic transmission, also increase adult hippocampal neurogenesis, suggesting a link between adult hippocampal neurogenesis and depression. Even widely used compound, such as acetylsalicylic acid, or procedures, such as acupuncture, have been found to modulate hippocampal neurogenesis after experimental ischemia. Additionally, several drugs of abuse have been found to interfere with adult neurogenesis in the dentate gyrus. Opiates and nicotine decrease the production of new granule neurons. Chronic ethanol abuse, in contrast, leads to increased proliferation in the adult dentate gyrus.

4.2. Enhanced postnatal neurogenesis in some active states

Recent studies indicate a strong link between hippocampal neurogenesis and learning. Exposure to an enriched environment increases adult hippocampal neurogenesis and improves spatial memory. Another line of support to the link between neurogenesis and memory/learning is that training in associative learning tasks that require the hippocampus increases the number of newly generated granule neurons, while a reduction in the number of newly generated neurons, using a toxin against proliferating cells, impairs this associative hippocampal-dependent learning. Simpler activities can also induce increases in the production and incorporation of new granule neurons into the adult hippocampus: habitual running not only increases hippocampal neurogenesis in adults but also induces LTP and improves learning. Endogenous insulin-like growth factor 1 and probably BDNF mediate these effects.

4.3. Postnatal neurogenesis varies with age

The production of new neurons persists, although at very low levels, in the dentate gyrus of middle-aged and aged mammals. The reduction in neurogenesis during aging may be due to decreased proliferation of neuronal precursors, because cell proliferation is also strongly diminished in the middle-aged and aged rat hippocampus. On the other hand, this reduction may be promoted by a decrease in the number of neuronal progenitor cells. Recent work demonstrates that nestin-immunoreactive radial-glia-like cells can also be found in the aged rat hippocampus, although their number is extremely reduced.

Recent reports indicate that, as in the young adult brain, hippocampal neurogenesis can be enhanced in the aged brain, thereby reversing aging-induced deficits. The production of new granule cells increases in aged mice living in an enriched environment, a condition that also improves performance in a hippocampal dependent memory task. Granule-cell production can also be enhanced in aged rats by removing adrenal steroids. In fact, it has been postulated that the age-related decline in hippocampal neurogenesis is due to the increase in circulating levels of glucocorticoids that occurs with aging. Administration of insulin-like growth factor 1 also increases granule-neuron production and improves memory in the aged brain. Similarly, treatment with an NMDA-receptor antagonist increases cell proliferation and neurogenesis, as well as the number of radial-glia-like cells, in the aging hippocampus.

REFERENCES

Altman J (1963) Autoradiographic investigation of cell proliferation in the brains of rats and cats. Anat Rec 145: 573-592.
Alvarez-Buylla A, Temple S (1998) Stem cells in the developing and adult nervous system. J Neurobiol 36: 105-110.
Bayer SA, Altman J, Russo RJ, Zhang X (1993) Timetables of neurogenesis in the human brain based on experimentally determined patterns in the rat. Neurotoxicol 14: 83-144.
Benraiss A, Arsanto JP, Coulon J, Thouveny Y (1999) Neurogenesis during caudal spinal cord regeneration in adult newts. Dev Genes Evol 209: 363-369.
Cameron HA, Tanapat P, Gould E (1998) Adrenal steroids and N-methyl-D-aspartate receptor activation regulate neurogenesis in the dentate gyrus of adult rats through a common pathway. Neuroscience 82: 349-354.
Corwin JT, Warchol ME (1991) Auditory hair cells: Structure, function, development, and regeneration. Annu Rev Neurosci 14: 301-333.
Cotanche DA, Lee KH, Stone JS, Picard DA (1994) Hair cell regeneration in the bird cochlea following noise damage or ototoxic drug damage. Anat Embryol (Berl) 189: 1-18.
Eriksson PS, Perfilieva E, Björk-Eriksson T, Alborn AM, Nordborg C, Peterson DA, Gage FH (1998) Neurogenesis in the adult human hippocampus. Nature Med 4: 1313-1317.
Gould E, Gross CG (2002) Neurogenesis in adult mammals: some progress and problems. J Neurosci 22: 619-623.
Graziadei PP, Monti-Graziadei GA (1985) Neurogenesis and plasticity of the olfactory sensory neurons. Ann N Y Acad Sci 457: 127-142.
Gueneau G, Privat A, Drouet J, Court L (1982) Subgranular zone of the dentate gyrus of young rabbits as a secondary matrix. A high-resolution autoradiographic study. Dev Neurosci 5: 345-358.
Holder N, Clarke JDW (1988) Is there a correlation between continuous neurogenesis and directed axon regeneration in the vertebrate nervous system? TINS 11: 94-99.
Horner PJ, Power AE, Kempermann G, Kuhn HG, Palmer TD, Winkler J, Thal LJ, Gage FH (2000) Proliferation and differentiation of progenitor cells throughout the intact adult rat spinal cord. J Neurosci 20: 2218-2228.
Johns PR, Easter SS (1977) Growth of the adult goldfish eye. II. Increase in retinal cell number. J Comp Neurol 176: 331-342.
Kornack DR, Rakic P (1999) Continuation of neurogenesis in the hippocampus of the adult macaque monkey. Proceedings of the National Academy of Sciences of the United States of America 96: 5768-5773.
Lopez-Garcia C, Molowny A, Nacher J, Ponsoda X, Sancho-Bielsa F, Alonso-Llosa G (2002) The lizard cerebral cortex as a model to study neuronal regeneration. An Acad Bras Cienc 74: 85-104.
Molowny A, Nacher J, Lopez-Garcia C (1995) Reactive neurogenesis during regeneration of the lesioned medial cerebral cortex of lizards. Neuroscience 68: 823-836.
Nottebohm F (2002) Why are some neurons replaced in adult brain? J Neurosci 22: 624-628.
Rakic P (2002) Neurogenesis in adult primate neocortex: an evaluation of the evidence. Nature Reviews Neuroscience 3: 65-71.
Reh TA, Levine EM (1998) Multipotential stem cells and progenitors in the vertebrate retina. J Neurobiol 36: 206-220.
Reznikov KY (1991) Cell proliferation and cytogenesis in the mouse hippocampus. Adv Anat Embryol Cell Biol 122: 1-74.
Seki T, Arai Y (1995) Age-related production of new granule cells in the adult dentate gyrus. NeuroReport 6: 2479-2482.
Zupanc GKH (2001) Adult neurogenesis and neuronal regeneration in the central nervous system of teleost fish. Brain Behavior and Evolution 58: 250-275.

C.A. HAAS AND M. FROTSCHER

Institute of Anatomy and Cell Biology, University of Freiburg, Germany

25. MIGRATION DISORDERS AND EPILEPSY

Summary. Normal development of the cerebral cortex requires radial migration of cortical neurons from the ventricular zone towards the pial surface. During recent years new insights into the different steps of cortical layer formation have been gained from the study of genetic disorders in humans and from the investigation of spontaneous or engineered mouse mutants. From these studies individual molecules have been identified which are crucially involved in the different steps of migration, i.e. migration onset, the actual migration along radial glial fibers and neuronal positioning. Here we will review current knowledge of the molecular pathways governing correct cortical layer formation which is a prerequisite for normal brain function.

1. INTRODUCTION

The central nervous system (CNS) of adult mammals is composed of billions of neurons which are precisely located and interconnected in a highly specific fashion. Fully differentiated nerve cells are highly specialized and post-mitotic. As a consequence, any kind of trauma or damage will result in a severe imbalance or interruption of the complicated brain network. In humans, neurological disorders are the reflection of an unbalanced function of the nervous system. The cause for neurological diseases can be neurodegenerative events leading to selective cell death of groups of specialized neurons, as it is the case in Parkinson disease, or the progressive cell death in a variety of brain regions as it is observed in Alzheimer's disease. Not only neuronal cell death can cause severe dysfunction of the nervous system, also changes in connectivity may affect proper brain function.

Epileptic disorders are accompanied by neuronal cell death and changes in neuronal connections, which can be observed in different parts of the brain. Epilepsy is the most common neurological disease in humans. It is characterized by the occurrence of repetitive seizures which are due to an altered brain function. Large groups of neurons are activated repeatedly during a seizure and GABAergic inhibition does not control these excitatory events, as it is the case in normal brain function. Epilepsy may develop from brain damage, stroke or an infection of the brain. But there are also developmental malformations which result in epilepsy, especially defects in neuronal migration are often associated with epileptic disorders. In this chapter we will review the different steps of cortical development and we will outline the current knowlegde about migration disorders which lead to epileptic syndroms.

The mammalian neocortex is composed of six layers which are interconnected in a complex fashion with each other and with other brain regions. The development of this intricate network has to be tightly regulated in order to guarantee proper cortical function. Forebrain neurons are born in the ventricular zone and migrate from their site of origin towards the pial surface to accumulate below the marginal zone and to form the cortical plate. In order to find their way properly, the postmitotic neurons travel along guiding structures, the radial glial cells, which are orientated perpendicular to the pial surface. After reaching the cortex neurons position themselves in layers that finally will become the adult cortex.

The sequence of events leading to cortical lamination is rather complicated and involves several steps: earliest born neurons leave the ventricular zone and form the so-called preplate. Later arriving neurons split the preplate into the outer marginal zone (beneath the pial surface) and a deeper layer, called the subplate. Neurons generated later must pass through the subplate and past previously deposited neurons to form the cortical plate. This mode of migration implies that newly generated neurons bypass older, already positioned neurons so that the youngest neurons settle themselves directly below the marginal zone, thus building the six-layered cortex in an *inside-out* fashion (Angevine and Sidman, 1961).

Recently, several genes have been identified which control different steps of cortical development and which are crucial for the correct formation of cortical layers. If one of these genes is mutated, cortical structure cannot form properly and disturbed brain function is the consequence in affected individuals, often leading to severe epilepsy and mental retardation. These new insights have come from studies of genetic disorders in humans as well as from experiments in mouse mutants, both lines leading to a better understanding of cortical development.

2. NEURONAL MIGRATION ONSET

As soon as neurons have become post-mitotic in the ventricular zone of the telencephalon, they have to leave the germinal matrix and start their migration towards the cortical plate. In the human disorder periventricular heterotopia (PH) a fraction of newly postmitotic neurons appears incapable of leaving the ventricular zone. In adult individuals with PH, one population of differentiated neurons accumulates along the lateral ventricle, whereas another population migrates normally indicating that the abnormal neurons in PH appear to have a defect in migration onset (Eksioglu et al., 1996). Periventricular heterotopia is an X-chromosome-linked disease which results in females being mosaics due to random X-chromosome inactivation while affected males appear not to survive gestation (Eksioglu et al., 1996). The major neurological manifestation in females with PH is epilepsy, which begins in the second and third decade of life, and is most likely caused by the heterotopic accumulation of neurons along the lateral ventricle. The gene responsible for PH has been cloned and identified as *filamin 1* gene (Fox et al., 1998), which encodes filamin 1 (FLN1), an actin-binding protein, known to play a critical role in the control of cell shape, migration, filipodia formation and chemotaxis (for review see Stossel et al., 2001). The available data suggest that the role of FLN1 in neuronal migration is mediated by interaction with the cytoskeleton. Very recently, it was shown that interaction with the Filamin-interacting protein (FILIP) regulates cell migration out of the ventricular zone (Nagano et al., 2002). FILIP is an intracellular protein, which is associated with Filamin 1 and regulates its interaction with F-actin. During early development FILIP induces degradation of Filamin 1 thereby tethering post-mitotic neurons to the ventricular zone, as soon as FILIP activity decreases, Filamin 1 can associate with the cytoskleton and neurons start to migrate towards the pial surface (Nagano et al., 2002).

3. THE MIGRATION PROCESS

After departing from the ventricular zone neurons must migrate long distances towards the cortical plate. This process of migration has been found to be abnormal in two other disorders in which neurons leave the ventricular zone and migrate for some distance before they stop their migration. One of these disorders in humans is "type I" lissencephaly (literally translated this means *smooth brain*) which is characterized macroscopically by the absence of gyri and an atypical, four-layered cortex. This lamination pattern, however, does not have anything in common with the normal six-layered cortex, except that the marginal zone is preserved. In lissencephaly, the majority of cortical neurons are situated in the fourth layer. Mutations of at least two genes lead to a lissencephaly phenotype, *LIS1* and doublecortin (*DCX*). The *LIS 1* gene was mapped to chromosome 17, and already one mutated allele, resulting in a 50% decrease in LIS1 protein, is sufficient to cause the lissencephaly phenotype (Reiner et al., 1993). The *DCX* locus has been localized to the X-chromosome causing a lissencephaly phenotype in males, but a less severe disorder in heterozygous females called double cortex (DC) or subcortical band heterotopia (Pinard et al., 1994; des Portes et al., 1998). In DC, there is a normal six-layered outer cortex and an additional accumulation of neurons in the subcortical white matter. Like PH, DC is generally regarded as a mosaic phenotype in females owing to random inactivation of the X-chromosome. Individuals with lissencephaly suffer from intractable epilepsy and from severe mental retardation, whereas females with DC are affected similarly, but much less severe.

The precise role of LIS1 and doublecortin in neuronal migration is only partially understood. The *LIS1* gene product is the regulatory subunit of the brain-specific, intracellular enzyme platelet-activating factor acetylhydrolase (PAF-AH), which deacylates and inactivates PAF, a potent proinflammatory phospholipid (Hattori et al., 1994). Lis1 colocalizes with microtubles and regulates their dynamics (Sapir et al., 1997) suggesting a role in neuronal migration by interaction with the cytoskeleton. Mutated mice with graded reduction of Lis1 activity display dose-dependent migration defects in laminated brain structures such as the neocortex, hippocampus and cerebellum (Hirotsune et al., 1998). More specifically, it was shown recently that lack of Lis 1 results in a malformation of the subplate and reduced neuroblast proliferation in the ventricular zone (Gambello et al., 2003), indicating a role of Lis1 in neuronal migration and neurogenesis during brain development.

Doublecortin is a microtubule-associated protein which is exclusively expressed by neurons during the phase of migration (Gleeson et al., 1999). It stimulates microtubule polymerization suggesting that its function in neuronal migration might also involve regulation of the microtubule skeleton. Possibly DCX and LIS1 together influence an essential microtubule-based process in neuronal migration, but this remains to be elucidated.

4. PENETRATION OF THE SUBPLATE: THE REELIN SIGNALING PATHWAY

As soon as neurons complete their migration they become organized into the cortical plate and find their final positions. This last step of migration has been the focus of recent research efforts. A naturally occurring mouse mutant, the *reeler* mouse, displays an inversion of cortical layers with an accumulation of neurons in the normally cell-sparse marginal zone (layer I). The primary defect in *reeler* seems to be that cortical neurons do not penetrate the *pre-plate*, but rather accumulate beneath it resulting in an inversion of cortical layers. In addition, the *pre-plate* neurons are pushed upward towards the outer margin of the cortex, leading to the formation of a new layer, the superplate (Figure 1; Caviness and Sidman, 1973).

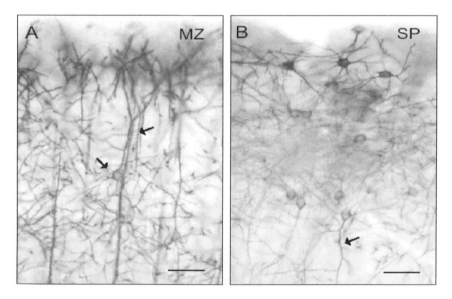

Figure 1. Pyramidal cells in wild type and in reeler cortex visualized by Golgi impregnation. A. Wild type mouse. Pyramidal cells with their apical dendrites (arrow) are orientated in strict order perpendicular to the pial surface. The apical dendrites form a terminal tuft in the marginal zone, which is almost devoid of neuronal cell bodies. B. Reeler mouse. Altered organization of neurons in the reeler cortex. An accumulation of neurons (the superplate) is visible in the marginal zone. Individual pyramidal cells with an inverted orientation (arrow) can be distinguished. Scale bars: 80 μm. MZ, marginal zone; SP, superplate

The mutated gene in *reeler* mice was found to encode a large extracellular matrix protein named reelin (d'Arcangelo et al., 1995; Hirotsune et al., 1995) that is synthesized and secreted by Cajal-Retzius cells, early generated horizontal neurons situated in the marginal zone of the neocortex and hippocampus (Figure 2; d'Arcangelo et al., 1997; Haas et al., 2000) suggesting that signals coming from the marginal zone are crucial for the formation of cortical lamination.

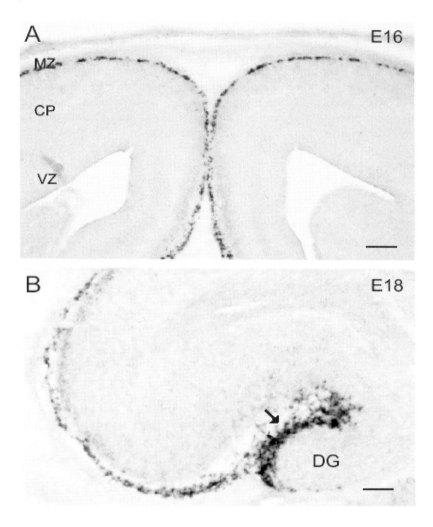

Figure 2. Localization of Cajal-Retzius cells in the marginal zones of mouse cortex and hippocampus by in situ hybrization for reelin mRNA. A. Embryonic day 16. Note the exclusive labeling of Cajal-Retzius cells in the marginal zone of the cortex. B. Embryonic day 18. Reelin mRNA-positive cells are accumulated along the hippocampal fissure (arrow). Scale bars: 250 μm. MZ, marginal zone; CP, cortical plate; VZ, ventricular zone; DG, dentate gyrus.

Several other mouse mutations have been instrumental in elucidating the signaling pathway mediating reelin function. The mouse mutants yotari and scrambler are phenocopies of reeler and have led to the discovery of disabled 1 (dab1), an intracellular adaptor protein which is phosphorylated upon reelin binding (Sheldon et al., 1997; Rice et al., 1998). In the search for the missing link, the reelin receptor, two lipoprotein receptors, apolipoprotein E receptor 2 (ApoER2) and very low density lipoprotein receptor (VLDLR), were identified in mouse mutations. When knocked out together, the reeler/scrambler phenotype is recapitulated, showing severe migration defects in the neocortex, hippocampus and cerebellum (Trommsdorf et al., 1999; d'Arcangelo et al., 1999). Reelin binds to the transmembrane receptors, ApoER2 and VLDLR, present on migrating cortical neurons, thereby activating the phosphorylation of dab1, the intracellular binding protein, which couples the reelin effects to intracellular signaling cascades. The detailed steps downstream of dab1 are only poorly understood. It is clear, however, that reelin deficiency leads to hyperphosphorylation of the tau protein (Hiesberger et al., 1999), thus providing a link of reelin function to the cytoskeleton.

The precise mechanism, how the reelin signaling pathway affects neuronal positioning during development, is only partially understood. Different models have been proposed: Reelin may act as a stop signal for migrating neurons leading to a detachment of these neurons from radial glial fibers. In the absence of reelin, new neurons would accumulate on radial glial fibers and invert the normal cortical layering. Alternatively, reelin could act as a chemoattractant for migrating or a chemorepellant for early-born neurons (for review see Frotscher, 1998; Rice and Curran, 2001). The situation is more complex and context-dependent as it became clear from recent studies. When reelin was ectopically expressed under the control of a *nestin*-promotor, which targets reelin expression to the ventricular zone, no effect was noted, since endogenous reelin was still present in the marginal zone. However, when these transgenic mice were crossed with *reeler* mice lacking reelin, a partial rescue of the *reeler* phenotype was achieved, although the expression of reelin occurred in the ventricular zone and not in the marginal zone, as it is the case in normal brain development (Magdaleno et al., 2002). These observations indicate that reelin does not simply act as a positional cue for migrating neurons. Recent experiments in hippocampal slice cultures propose an indirect effect of reelin by acting on the radial glial scaffold necessary for correct migration of neurons during development (Förster et al., 2002).

5. GRANULE CELL DISPERSION AND REELIN

Recent studies suggest that reelin seems to be involved in brain disorders such as temporal lobe epilepsy (TLE). TLE is one of the most common neurological disorders in humans, which is often accompanied by Ammon's horn sclerosis (AHS).

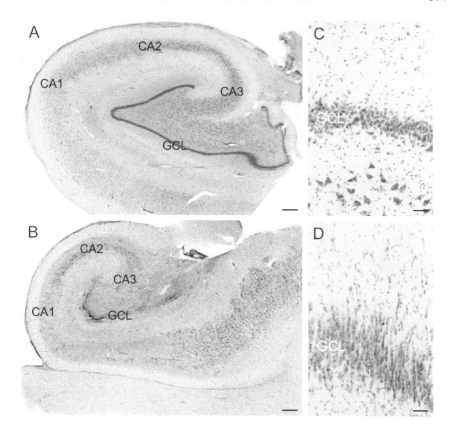

Figure 3. Morphology of normal and epileptic human hippocampus stained with cresyl violet. A. Human control hippocampus with normal distribution of neurons in hippocampal subfields and in the dentate gyrus, where the granule cells are located in a dense layer. B. Epileptic human hippocampus with characteristic features of Ammon's horn sclerosis. Selective cell loss is obvious in hippocampal subfields CA1 and CA3. The granule cell layer is dispersed. C. Granule cell layer of a normal human dentate gyrus. The granule cells are arranged in a densely packed layer. D. Loss of dense granule cell packing (granule cell dispersion) in temporal lobe epilepsy. Scale bars A, B: 600 μm; C, D: 75 μm. CA1, CA2, CA3, hippocampal subfields; GCL, granule cell layer.

This pathology is characterized by a selective neuronal loss in the hippocampal subfields CA1 and CA3 and in the hilus of the dentate gyrus (Armstrong, 1993). In addition, there is an enlargement of the granule cell layer, termed granule cell dispersion (GCD; Figure 3; see Houser, 1990). It is assumed that surviving neurons sprout following hippocampal cell loss and that new axon collaterals reinnervate the partially denervated hippocampus. The sprouting fibers may form a new hippocampal circuitry that is thought to contribute to seizure pathogenesis as well as seizure propagation (Mathern et al., 1997; Sloviter, 1994). Recently it was shown that also dispersed granule cells contribute to a reorganization of the hippocampal circuit by the formation of new collaterals into the molecular layer. These sprouting fibers establish recurrent synapses with granule dendrites (Freiman et al., 2002).

Little is known how granule cell dispersion develops in TLE patients. Since GCD was found to be independent of the etiology of the disease, Houser (1990) postulated that a local migration defect could be the cause of GCD. Interestingly, mice with mutations in the reelin signal transduction pathway show a disturbed lamination pattern of granule cells in the dentate gyrus. In *reeler* mice granule cell lamination is completely lost (Figure 4), whereas VLDLR and ApoER2 knockout mice show milder forms of dispersion (Rakic and Caviness, 1995; Gebhardt et al., 2002; Drakew et al., 2002) indicating that the reelin signaling pathway plays a role in correct positioning of granule cells. Along this line, increased numbers of Cajal-Retzius (CR) cells were found both in the hippocampus of *reeler* mice (Coulin et al., 2001) and in tissue samples from patients with temporal lobe epilepsy (Blümcke et al., 1999), raising the possibility that alterations in CR cells and in the reelin pathway underlie neuronal migration defects in reeler mutants and in humans with TLE. Because TLE frequently is drug-resistant, hippocampal surgery is often necessary to achieve seizure control. In a recent study, hippocampal tissue removed from TLE patients for therapeutic reasons was analyzed for changes in the expression of the reelin signaling cascade. Interestingly, an inverse correlation of the degree of GCD and the number of reelin mRNA-expressing CR cells and the relative amount of reelin mRNA, respectively, was found in the dentate gyrus by *in situ* hybridization and by quantitative real-time RT-PCR. In other words, in cases with strong dispersion little reelin mRNA could be detected, whereas in autopsy controls and in cases with mild dispersion a high reelin expression was monitored (Haas et al., 2002; Frotscher et al., 2003). In addition, the components of the reelin signal transduction pathway were expressed in human hippocampi obtained from TLE patients indicating that reduced levels of reelin could contribute to the development of GCD in TLE patients (Haas et al., 2002). This assumption is supported by the fact that neurogenesis of dentate granule cells continues into adulthood and can even be stimulated by seizures (Parent et al., 1997). Moreover, radial glial cells, the guiding structures for migrating neurons during development, re-appear in epileptic hippocampi with granule cell dispersion (Crespel et al., 2002). Together, these findings strongly suggest that GCD may be caused by mal-positioning of seizure-induced, newly born granule cells which migrate along radial glial fibers and fail to find their right position in the granule cell layer due to a local reelin deficiency.

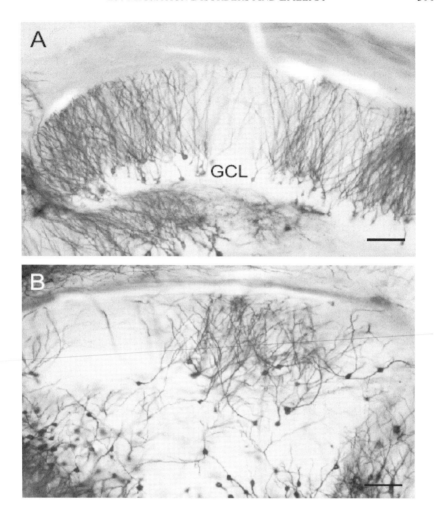

Figure 4. Granule cells in the dentate gyrus of wild type and reeler mouse. Individual neurons are visualized by Golgi impregnation. A. Laminar distribution of granule cells in the dentate gyrus of a wild type mouse. The dendrites extend into the molecular layer. B. Granule cells in the reeler mouse. Note the loss of laminar distribution of granule cells. Dendrites are oriented in different directions. Scale bars: 60 μm. GCL, granule cell layer.

6. CONCLUSIONS

The genetic analysis of migration defects in man and mouse has provided new insights into the molecular events governing cortical lamination (see Figure 5). PH with the associated mutation in FLN1 as well as the function of FILIP indicate that the onset of migration represents an important intial step. The interaction of FLN1 and FILIP with actin suggests that this process depends on an actin-mediated mechanism. Lissencephaly and double cortex appear to be caused by defects in the actual migration process. Since Lis1 as well as doublecortin strongly interact with microtubules, it is assumed that the effect of both molecules on neuronal migration involves the cytoskeleton. Mutations in the genes of the reelin pathway seem to be involved in positioning of neurons at the end of their migration. Moreover, a reelin deficiency appears to be critically involved in pathological changes such as granule cell dispersion in Ammon's horn sclerosis. The precise role of reelin in these pathological processes remains to be elucidated.

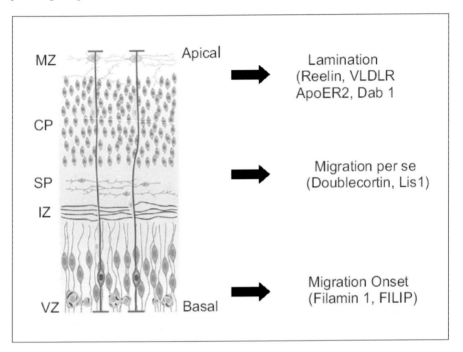

Figure 5. Summary diagram of essential steps in cortical lamination and the molecules involved. MZ, marginal zone; CP, cortical plate; SP, subplate; IZ, intermediate zone; VZ, ventricular zone.

ACKNOWLEDGEMENTS

The authors thank all those who contributed with their time and talents to the studies reviewed in this chapter. In especially, we thank Dr. Alexander Drakew for the Golgi studies. This work was supported by the Deutsche Forschungsgemeinschaft (SFB 505; TR-3).

REFERENCES

Angevine, J. B., & Sidman, R. L. (1961) Autoradiographic study of cell migration during histogenesis of the cerebral cortex in the mouse. Nature, 192, 766-768.

Armstrong, D. D. (1993) The neuropathology of temporal lobe epilepsy. J Neuropath Exp Neurol, 52, 433-443.

Blümcke, I., Beck, H., Suter, B., Hoffmann, D., Fodisch, H. J., Wolf, H. K., Schramm, J., Elger, C.E., & Wiestler, O. D. (1999) An increase of hippocampal calretinin-immunoreactive neurons correlates with early febrile seizures in temporal lobe epilepsy. Acta Neuropathol, 97, 31-39.

Caviness, V. S., Jr., & Sidman, R. L. (1973) Time of origin of corresponding cell classes in the cerebral cortex of normal and reeler mutant mice: an autoradiographic analysis. J Comp Neurol, 148, 141-151.

Coulin, C., Drakew, A., Frotscher, M., & Deller, T. (2001) Stereological estimates of total neuron numbers in the hippocampus of adult reeler mutant mice: Evidence for an increased survival of Cajal-Retzius cells. J Comp Neurol, 439, 19-31.

Crespel, A., Coubes, P., Rousset, M., Alonso, G., Bockaert, J., Baldy-Moulinier, M., & Lerner-Natoli, M. (2002) Immature-like astrocytes are associated with dentate granule cell migration in human temporal lobe epilepsy. Neurosci Lett, 330, 114-118.

D'Arcangelo, G., Miao, G. G., Chen, S. C., Soares, H. D., Morgan, J. I., & Curran, T. (1995) A protein related to extracellular matrix proteins deleted in the mouse mutant reeler. Nature, 374, 719-723.

D'Arcangelo, G., Nakajima, K., Miyata, T., Ogawa, M., Mikoshiba, K., & Curran, T. (1997) Reelin is a secreted glycoprotein recognized by the CR-50 monoclonal antibody. J Neurosci, 17, 23-31.

D'Arcangelo, G., Homayouni, R., Keshvara, L., Rice, D. S., Sheldon, M., & Curran, T. (1999) Reelin is a ligand for lipoprotein receptors. Neuron, 24, 471-479.

Des Portes, V., Pinard, J. M., Billuart, P., Vinet, M. C., Koulakoff, A., Carrie, A., Gelot, A., Dupuis, E., Motte, J., Berwald-Netter, Y., Catala, M., Kahn, A., Beldjord, C., & Chelly, J. (1998) A novel CNS gene required for neuronal migration and involved in X-linked subcortical laminar heterotopia and lissencephaly syndrome. Cell 92, 51-61.

Drakew, A., Deller, T., Heimrich, B., Gebhardt, C., Del Turco, D., Tielsch, A., Förster, E., Herz, J., & Frotscher, M. (2002) Dentate granule cells in reeler mutants and VLDLR and ApoER2 knockout mice. Exp Neurol, 176, 12-24.

Eksioglu, Y.Z., Scheffer, I. E., Cardenas, P., Knoll, J., DiMario, F., Ramsby, G., Berg, M., Kamuro, K., Berkovic, S. F., Duyk, G. M., Parisis, J., Huttenlocher, P. R., & Walsh, C. A. (1996) Periventricular heterotopia: An X-linked dominant locus causing aberrant cerebral cortical development. Neuron, 16, 77-87.

Förster, E., Tielsch, A., Saum, B., Weiss, K. H., Johanssen, C., Graus-Porta, D., Müller, U., & Frotscher, M. (2002) Reelin, disabled 1, and β_1 integrins are required for the formation of the radial glial scaffold in the hippocampus. Proc Natl Acad Sci USA, 99, 13178-13183.

Fox, J. W., Lamberti, E. D., Eksioglu, Y. Z., Hong, S. E., Feng, Y., Graham, D. A., Scheffer, I. E., Dobyns, W. B., Hirsch, B. A., Radtke, R. A., Berkovic, S. F., Huttenlocher, P. R., & Walsh, C. A. (1998) Mutations in filamin 1 prevent migration of cerebral cortical neurons in human periventricular heterotopia. Neuron, 21, 1315-1325.

Freiman, T. M., Gimbel, K., Honegger, J., Volk, B., Zentner, J., Frotscher, M., & Deller, T. (2002) Anterograde tracing of human hippocampus in vitro- a neuroanatomical tract tracing technique for the analysis of local fiber tracts in human brain. J Neurosci Methods, 120, 95-10.

Frotscher, M. (1998) Cajal-Retzius cells, Reelin and the formation of layers. Curr Opin Neurobiol, 8, 570-575.

Frotscher, M., Haas, C. A., & Förster, E. (2003) Reelin controls granule cell migration in the dentate gyrus by acting on the radial scaffold. Cereb Cortex, 13, 634-640.

Gambello, M. J., Darling, D. L., Yingling, J., Tanaka, T., Gleeson, J. G., & Wynshaw-Boris, A. (2003) Multiple does-dependent effects of Lis1 on cerebral cortical development. J Neurosci, 23, 1719-1929.

Gebhardt, C., del Turco, D., Drakew, A., Tielsch, A., Herz, J., Frotscher, M., & Deller, T. (2002) Abnormal positioning of granule cells alters afferent fiber distribution in the mouse fascia dentata: Morphologic evidence from reeler, apolipoprotein E receptor 2-, and very low density lipoprotein receptor knockout mice. J Comp Neurol, 445, 278-292.

Gleeson, J. G., Lin, P. T., Flanagan, L. A., & Walsh, C. A. (1999) Doublecortin is a microtubule-associated protein and is expressed widely by migrating neurons. Neuron, 23, 257-271.

Haas, C. A., Deller, T., Krsnik, Z., Tielsch, A., Woods, A., & Frotscher, M. (2000) Entorhinal cortex lesion does not alter reelin mRNA expression in the dentate gyrus of young and adult rats. Neuroscience, 97, 25-31.

Haas, C. A., Dudeck, O., Kirsch, M., Huszka, C., Kann, G., Pollak, S., Zentner, J., & Frotscher, M. (2002) Role for reelin in the development of granule cell dispersion in temporal lobe epilepsy. J Neurosci, 22, 5797-5802.

Hattori, M., Adachi, H., Tsujimoto, M., Arai, H., & Inoue, K. (1994) Miller-Dieker lissencephaly gene encodes a subunit of brain platelet-activating factor acetylhydrolase. Nature, 370, 216-218.

Hiesberger, T., Trommsdorf, M., Howell, B. W., Goffinet, A., Mumby, M. C., Cooper, J. A., & Herz, J. (1999) Direct binding of Reelin to VLDL receptor and ApoE receptor 2 induces tyrosine phosphorylation of disabled-1 and modulates tau phosphorylation. Neuron, 24, 481-89.

Hirotsune, S., Fleck, M., Gambello, M., Bix, G., Chen, A., Clark, G., Ledbetter, D., McBain, C., & Wynshaw-Boris, A. (1998) Graded reduction of Pafah1b1 (Lis1) activitiy results in neuronal migration defects and early embryonic lethality. Nature Genet, 19, 333-339.

Hirotsune, S., Takahare, T., Sasaki, N., Hirosè, K., Yoshiki, A., Ohashi, T., Kusakabe, M., Murakami, Y., Muramatsu, M., Watanabe, S., Nakao, K., Katsuki, M., & Hayashizaki, Y. (1995) The reeler gene encodes a protein with an EGF-like motif expressed by pioneer neurons. Nature Genet, 10, 77-83.

Houser, C. R. (1990) Granule cell dispersion in the dentate gyrus of humans with temporal lobe epilepsy. Brain Res, 535, 195-204.

Magdaleno, S., Keshvara, L., & Curran, T. (2002) Rescue of ataxia and preplate splitting by ectopic expression of reelin in reeler mice. Neuron, 33, 573-586.

Mathern, G. W., Babb, T. L., & Armstrong, D. L. (1997) Hippocampal sclerosis. Engel, J., Pedley, T. A. (Eds.) Epilepsy: A comprehensive textbook. Philadelphia: Lippincott-Raven, p. 133-155.

Nagano, T., Yoneda, T., Hatanaka, Y., Kubota, C., Murakami, F., & Sato, M. (2002) Filamin A-interacting protein (FILIP) regulates cortical cell migration out of the ventricular zone. Nature Cell Biol, 4, 495-501.

Parent, J. M., Yu, T. W., Leibowitz, R. T., Geschwind, D. H., Sloviter, R. S., & Lowenstein, D. H. (1997) Dentate granule cell neurogenesis is increased by seizures and contributes to aberrant network reorganization in the adult rat hippocampus. J Neurosci, 17, 3727-3738.

Pinard, J. M., Motte, J., Chiron, C., Brian, R., Andermann, E., & Dulac, O. (1994) Subcortical laminar heterotopia and lissencephaly in two families: a single X linked dominant gene. J Neurol Neurosurg Psychiatr, 57, 914-920.

Rakic, P., & Caviness, V. S., Jr. (1995) Cortical development: view from neurological mutants two decades later. Neuron, 14, 1101-1104.

Reiner, O., Carrozzo, R., Shen, Y., Wehnert, M., Faustinella, F., Dobyns, W. B., Caskey, C. T., & Ledbetter, D. H. (1993) Isolation of a Müller-Dieker lissencephaly gene containing G protein beta-subunit-like repeats. Nature, 364, 714-721.

Rice, D. S., Sheldon, M., D'Arcangelo, G., Nakajima, K., Goldowitz, D., & Curran, T. (1998) Disabled-1 acts downstream of reelin in a signaling pathway that controls laminar organization in the mammalian brain. Development, 125, 3119-3729.

Rice, D. S., & Curran, T. (2001) Role of reelin signaling pathway in the central nervous system development. Ann Rev Neurosci, 24, 1005-1039.

Sapir, T., Elbaum, M., & Reiner, O. (1997) Reduction of microtubule catastrophe events by LIS1, platelet-activating factor acetylhydrolase subunit. EMBO J, 16, 6977-6984.

Sheldon, M., Rice, D. S., d'Arcangelo, G., Yoneshima, H., Nakajima, M., Mikoshiba, K., Howell, B. W., Cooper, J. A., Goldowitz, D., & Curran, T. (1997) Scrambler and yotari disrupt the disabled gene and produce a reeler-like phenotype in mice. Nature, 389, 730-733.

Sloviter, R. S. (1994) The functional organization of the hippocampal dentate gyrus and its relevance to the pathogenesis of temporal lobe epilepsy. Ann Neurol, 35, 640-654.

Stossel, T. P., Condeelis, J., Cooley, L., Hartwig, J. H., Noegel, A., Schleicher, M., & Shapiro, S. S. (2001) Filamins as integrators of cell mechanics and signalling. Nature Rev Mol Cell Biol, 2, 138-145.

Trommsdorff, M., Gotthardt, T., Hiesberger, T., Shelton, J., Stockinger, W., Nimpf, J., Hammer, R.E., Richardson, J. A., & Herz, J. (1999) Reeler/disabled-like disruption of neuronal migration in knockout mice lacking the VLDL receptor and ApoE receptor 2. Cell 97, 689-701.

J. A. ARMENGOL

Department of Human Anatomy and Embryology, University of Sevilla and Division of Neuroscience, CABD, CSIC-UPO, Sevilla, Spain

PART D: INTRODUCTION

26. INVASIVE STRATEGIES AS THERAPEUTIC APPROACHES FOR CENTRAL NERVOUS SYSTEM DISEASES

Summary. The main goal of experimental therapies for the treatment of degenerative neurological diseases ranges from the prevention of molecular events underlying neuronal death to the replacement of concrete damaged neuronal populations. Invasive therapeutic strategies such as grafting of embryonic neural cells or of adult bone marrow stem cells, the induction of trophic factor synthesis in brain cells by virus-mediated genetic transfer, and the electrical stimulation of the brain by the placement of electrodes into concrete brain circuits have accrued in the past decades. However, successful results obtained in experimental animals do not necessarily correlated with similar clinical advances. The rationale of these different invasive therapeutic strategies to treat brain damage and neurodegenerative processes is reviewed in this chapter. The discussion is focused on some noticeable discrepancies between experimental advances represented by the proposed therapies and their actual clinical benefits.

1. INTRODUCTION

Paleontologic and anthropologic studies demonstrate that invasive manipulations of the skull, and hence of the brain, under mystic, therapeutic or cannibalistic rituals, existed from Paleolithic times. Thus, ancient-Egyptian descriptions of the morphology, hypothetical roles, and living and post-mortem manipulations of human brain (as an example see The Edwin Smith Papyrus) are a consistent example of the intellectual attraction that the brain exerted in man. The French neurologist Paul Broca, with his comments on the first pre-Columbian Peruvian trepanned skull in 1867 and following studies, contributed to the analysis of the evidence and possible explanations of old trepanations from a modern neuroscientific point of view.

Technical advances from the 19th century to date have yielded therapeutic strategies devoted to repairing some of the damage caused by central neurological disorders or diseases, but were confronted with the prevailing axiomatic concept: *brain tissue is unable to regenerate*. Therefore, all neurosurgical treatments were aimed at ameliorating uncomfortable or side-effects that other therapeutic strategies did not solve. However, at the same time, the knowledge of mechanisms underlying brain functions was greatly furthered by accidental or directed invasive approaches. As an example, the reader will readily remember the Phineas P. Gage case, or the elegant studies of Penfield and Rasmussen (1950) on the somatotopic representation of the body surface in the neocortex.

The seminal works of Ramon y Cajal (1913-1914) established the basis of the research line centered on the possibility that the brain can change through adult life. In fact, one of current neuroscience's most exciting discoveries on brain functions was the experimental demonstration of its plastic properties, and the capacity of the nervous system to respond and regenerate after injury. This completely changed the rigid and hieratic concept that defined adult neurons as immutable after their ultimate differentiation.

Invasive treatments of neurological diseases range from the ablation of specific central nervous system (CNS) areas (e.g., temporal lobe, globus pallidus, etc.) to refined strategies of engineered-cell grafting to restore anatomical and functional losses. The aim of the present chapter is to analyze the advance of invasive strategies from laboratory animal models to their application in patients affected by CNS disorders. Particular attention will be given to the main procedures currently used, namely transplants of embryonic neurons or modified stem cells, gene therapy and electric brain stimulation. The reason why strategies performed successfully in experimental animals have unfortunately, not yet attained the same degree of efficacy in human patients will also be discussed.

2. INVASIVE TREATMENTS. GENERAL CONCEPTS

Invasive strategies for brain treatment can be classified in four categories: (i) technical approaches aimed at improving traumatic, ventricular or vascular problems, whose principal objective is to avoid or to palliate the acute and massive loss of brain parenchyma; (ii) invasive strategies to eliminate tumors or control their development; (iii) techniques to restore damaged circuitries where the loss of parenchyma is secondary to the disruption of normal functional trophic phenomena which maintain neuronal homeostasis. The decrease or loss of neurotransmitters, the progressive degeneration of specific neuronal cell types, or the degeneration elicited by the abnormal accumulation of substances with deleterious effects within the adult brain are the best examples of such disorders. Pathologies generalized under the term neurodegenerative diseases include those whose genetic causes are now well known (e.g., Huntington's disease, spinocerebellar ataxias, etc.), those of idiopathic nature or caused by multiple etiologies (including genetics) not yet completely understood (e.g., Alzheimer's or Parkinson's diseases), and long-term deleterious effects of CNS trauma, such as the complete or partial section of the spinal cord; and finally (iv) strategies aimed at restoring the congenital, acute, or progressive loss of glial cells that produce myelin sheaths.

The main goal of therapeutic strategies for CNS pathologies, which irrespective of their etiology impede normal brain function by neuronal cell loss, resides in the addition of embryonic neurons, neural crest-derived cells, or genetically engineered stem cells, to restore damaged neural circuitries. Such circuitries, apart from the neuroendocrine system, can be broadly classified into two main groups: (i) *global* or *modulatory* systems, whose role depends of the correct balance between neurotransmitters and their receptors rather than of a highly refined connectivity, and (ii) *point-to-point circuit* systems, whose normal functions essentially depend on the integrity of the specific and refined topographic arrangement of the synaptic connections (Sotelo and Alvarado-Mallart, 1991). In the former, the pharmacological recovery of critical threshold levels of neurotransmitter within the extracellular space will greatly ameliorate the symptoms. In the second, the restoration of the function involves the anatomical rebuilding of the appropriate input-output synaptic contacts of damaged circuitries. In both cases, the results of invasive methods, such as the implanting of exogenous biological or non-biological material, depend on the reaction of the host tissue, the adaptation and half-life of the implanted material and, finally, on the interplay between the reaction of the nervous system reaction and the nature of the implanted material.

3. GRAFTING PROCEDURES

G. Thompson (1890) was the pioneer in performing grafts of nervous tissue, and described them as a useful tool which required further research. However, the routine use of grafts of embryonic central nervous system tissues begun in the 70's, followed rapidly by their use in clinical trials on Parkinsonian patients from 1987 on (for a historical rev., see Björklund, 1991). Transplant experiments have generated a great amount of morphological, physiological, and molecular information about (i) how the adult central nervous system reacts after a lesion; (ii) the degree of functional integration between the host and the grafted neurons; and (iii) the temporal evolution and long-term benefits of grafting procedures versus pharmacological or other classical therapeutics on animal models and patients with neurodegenerative diseases (see this book's section E).

3.1. Transplants in global systems

From the seventies, functional host-graft integration has been well documented in animal transplant experiments (Döbrössy and Dunnet, 2001). The idea that degenerative diseases could be treated by transplants in which cells act as biological pumps of trophic factors or specific neurotransmitters was strongly supported by experimental data. Early works of M.J. Perlow, and A. Björklund and U. Stenevi (see Björklund, 1991), reporting the reversion of Parkinsonian symptoms in animal models after the grafting of fetal dopaminergic neurons, was an important step in the history of transplant procedures. Transplants of fetal dopaminergic neurons have frequently been used in Parkinsonian patient therapy. In successful cases, the amelioration of motor symptoms is directly correlated with transplanted cell survival within the host's brain. Furthermore, positron emission topography analysis demonstrates that surviving grafted dopaminergic cells are functionally active within the host brain one year or more after transplantation (for a rev., see Dunnet et al., 2001).

Besides fetal dopaminergic neurons, almost all adult dopamine-synthesizing cells have been used in the search for a stable source of neurotransmitter-producing grafts without immune problems. Carotid body glomus cells are one of these. The graft of glomus cells reverses motor symptoms in both rat and monkey models of Parkinson's disease. Grafted glomus cells survive for long periods within the host brain, and their axons invade the host parenchyma (Figure 1 D-E).

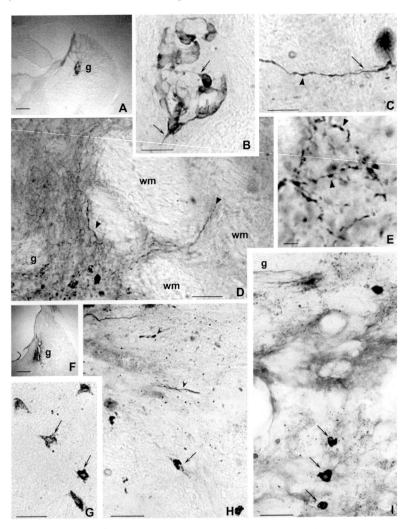

Fig. 1. Tyrosine hydroxylase (TH) immunohistochemistry of the striatum of hemiparkinsonian rats transplanted with glomus carotid body cells, three (A-E, F-I) and one month (G-H) after graft. Cell bodies of young adult donors developed fusiform or complex morphologies (B), from which arose neurite-like prolongations (B-C, arrows). Fibers from the graft, with

beaded morphology (C-E, arrowheads) resembling dopaminergic axons, invaded the gray matter of the host striatum (D, arrows), mimicking the nigro-striatal afferents. Cells of aged-donor rats developed irregular morphologies (G-I, arrows), while thick TH-immunoreactive fibers located near the graft were interrupted, and possessed enlargements similar to small axonal torpedoes (H, arrowheads). A and F illustrate the striatal placement of the graft (g). Bars: (A) 1 mm; (B-C) 20 μm; (D) 50 μm; (E) 10 μm; (F) 750 μm; (G-H) 20 μm; (I) 25 μm. Figures A to E reproduced from Espejo et al., Neuron, 20:197-206 (1998), with permission from Elsevier Science.

The smooth round round glomus cell somate differentiates to adopt bi- or multipolar neuron-like morphology (Figure 1, B-C). This differentiation process is indicative that neural crest-derived glomus cells are able to recognize and respond to environmental host signals that induce them to re-differentiate. Adult host cells are also able to react under the influence of grafted cells. Thus, some dopaminergic axons observed early in glomus experiments do not belong to the grafted cells, but arise from host dopaminergic axons, surviving after the experimental Parkinsonism, and sprouting under the effect of GDNF and TGF secreted by grafted neural crest-derived cells (Espejo et al., 2001). These inductive relationships are not exclusive to the interaction between grafted neural crest-derived cells and the adult striatum of Parkinsonian animals; but are a generalized phenomenon occurring in the host brain after neural grafts (see part 3.2). Furthermore, in grafted animals housed in an enriched environment, the hypothetical increase in the number of synaptic inputs received by the graft, results from a positive effect on the best behavioral recovery of transplanted animals (Döbrössy and Dunnet, 2001).

Experiments of adult cell autotransplants, for instance of glomus cells, address a relevant question: Is the age of the graft a conditioning factor for the successful evolution of grafted cells, and could it hence influence the long-term restoration of the symptoms? In fact, a recent autograft study describes the progressive loss of the synthetic capacities of donor cells with the age of the patient (V. Arjona et al., personal communication, Neurosurgery 2002, in press). The experiments required to analyze this question are simple: it is sufficient to use glomus cells from old donor rats whose age corresponds to the average age of Parkinsonian patients and, thereafter, to analyze the motor behavior and the graft's morphological evolution in the same animal model. Our preliminary results show that after a graft of glomus cells taken from two-year-old donor rats, Parkinsonian rats manifest a moderate recovery of their motor abnormalities. However, motor asymmetries re-appear from one month post-graft. Morphological analysis of these grafts has shown that grafted cells retain the expression of tyrosine hydroxylase (Figure 1, G-I). Therefore, these cells are theoretically still able to synthesize dopamine. However, their aberrant morphology (Figure 1 G-I) and the presence of broken axons (Figure 1, H), indicate that these grafted cells are progressively degenerating. Further studies are needed to elucidate whether such cells are able or not to secrete GDNF, and how long a sustained secretion of this trophic factor lasts. Our present investigation (using detection of apoptotic enzymes such as the caspase-3-active form) of the mechanisms underlying the degeneration of these cells, provided us with a preliminary impression that, contrary to results of previous experiments, aged-carotid-body glomus cells are unable to survive within the adult host striatum.

3.2. Transplants in point-to-point circuits systems

Various systems organized in a point-to-point manner (e.g., retino-geniculate, hippocampal, thalamo-cortical, cortical-cortical, cortico-spinal, etc.) have been widely used to elucidate the capacities of restoration after brain damage caused by different pathologies (e.g., stroke, trauma, neurodegenerative diseases)(see Björklund, 1991; Döbrössy and Dunnet, 2001). The main goal of these experiments is to replace lost neurons, and anatomically restore their input-output connections. Experiments using neurotoxically lesioned or mutant cerebella have provided key data on the morphological and functional recovery capacities of damaged point-to-point circuits after grafting procedures.

Embryonic cerebellar transplant studies demonstrated (i) the specific neurotropic effect exerted by the adult host brain on specific grafted embryonic neuronal cell types of which it is devoid; (ii) the tropic and trophic influences of embryonic grafts on adult target-deprived afferent fibers; and, (iii) the ability of target-deprived adult fibers to follow specific cues and to establish correct topographic and synaptic connections with grafted neurons. Purkinje cells of embryonic cerebellar anlagen transplanted into adult Purkinje cell degeneration (*pcd*) mutant mice exit from the graft and actively colonize the Purkinje cell deprived molecular layer of the host. Furthermore, when a *pcd* phenotype is partially performed by the injection of kainate into the cerebellar cortex of adult rats, grafted Purkinje cells colonize only the zone of the host molecular layer devoid of Purkinje cells, while undamaged host Purkinje cells that remain exert a stop signal on young Purkinje cells, impeding their progression through the rest of the host molecular layer.

Climbing fibers are one of the two types of extra-cerebellar input to Purkinje cells. After invasion of the *pcd* molecular layer by grafted Purkinje cells, host climbing fibers re-grow, synapse with the grafted Purkinje cells, and re-establish the morphological characteristics of climbing fiber-Purkinje cell dendrite synapses of the normal adult cerebellum (Figure 2, A). This morphological restoration is also correlated with the functional recovery of the typical all-or-none climbing fiber-Purkinje cell excitatory postsynaptic potentials (Sotelo and Alvarado-Mallart, 1991).

In kainate-lesioned adult cerebellum, the grafted embryonic cerebellar anlage integrates into a host cavity and develops into a small cerebellar structure (Figure 3, A). The adult climbing fibers, devoid of their Purkinje cell partners by the effect of kainate, grow inside the graft, climb on grafted Purkinje cells, and delineate their dendritic trees (Figure 3, C). Incoming climbing fibers distribute throughout the molecular layer of the graft in a patchy manner, resembling the striped pattern of the adult olivocerebellar projection (Figure 3, D-E). In contrast, the other extra-cerebellar input to the cerebellar cortex - the mossy fibers - rarely penetrate into these types of graft, and do not synapse with the granule cells of the mini-cerebellar structure. Therefore, under these experimental conditions, grafted Purkinje cells receive only one of the two extracerebellar cortical inputs, and consequently the lesioned circuit remains only partially restored.

26. INVASIVE STRATEGIES FOR CENTRAL NERVOUS SYSTEM DISEASES

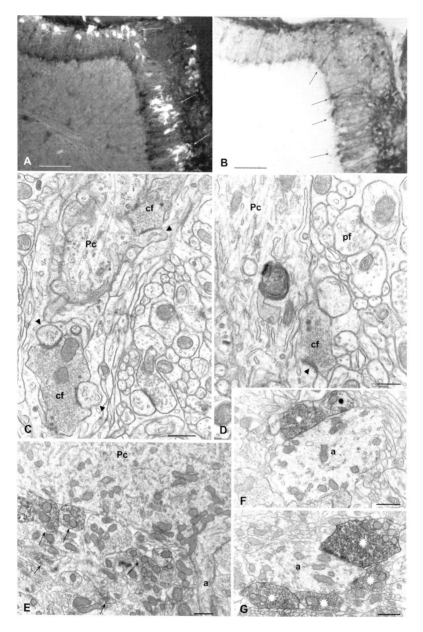

Fig. 2. Embryonic cerebellar grafts into pcd adult mouse cerebellum, seven days (A-B) and one month (C-G) after transplantation. Young grafted Purkinje cells [revealed by calbindin immunohistochemistry (A, arrows)] migrate through the host molecular layer over the Bergmann fibers. During this migratory process, adult host Bergmann fibers re-express nestin (B, arrows). Host climbing fibers (C-D, cf) synapse with the dendritic spines (C-D,

arrowheads) of grafted Purkinje cells (C-D, Pc), located within the host molecular layer. GABAergic axonal profiles of host basket cell -detected by GAD65 immunocytochemistry- surround the body (E, arrows) and the initial segment of the axon (F-G, stars) of grafted Purkinje cells (Pc). The distribution of labeled terminals resembles the typical basket pinceau. A, Purkinje cell axon. pf, parallel fiber. Bars: (A-B) 100 µm; (C-G) 1 µm. Courtesy of Professor C. Sotelo. Institute of Neuroscience. CSIC-UMH. Alicante. Spain.

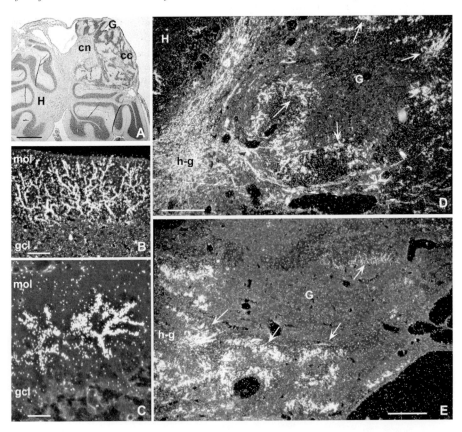

Fig. 3. Embryonic cerebellar grafts into adult kainate lesioned cerebellum. The graft develops into a mini-cerebellum with cerebellar cortex (A, cc) and cerebellar nuclei (A, cn). Host climbing fibers labeled by the injection of tritiated leucine in the host inferior olive delineate the host Purkinje cell dendritic trees (B). Olivocerebellar fibers entering the graft also climb on the dendritic tree of the graft Purkinje cells (C). The arrows in D and E show labeled host climbing fibers which distribute through the molecular layer of the graft in a patchy manner. G, graft. H, host. h-g, host-graft interface. gcl, granule cell layer. mol, molecular layer. Bars: (A) 1 mm; (B) 150 µm; (C) 50 µm; (D-E) 300 µm.

In embryonic cerebellar grafts into *pcd* adult cerebellum, the small dendritic spines of the Purkinje cells integrated within the molecular host layer receive parallel fiber synapses. Furthermore, the host's basket cells send their axons to engrafted Purkinje cell somata and initial axon segments, forming the typical inhibitory "pinceaux" (Figure 2). The two excitatory and the main inhibitory inputs of the Purkinje cells have therefore been successfully restored. However, with few exceptions the axon of grafted Purkinje cells did not attain the output gateway of the cerebellum: that is, the cerebellar nuclei (Sotelo and Alvarado-Mallart, 1991). Thus, despite the high level of morphological and electrophysiological integration of grafted Purkinje cells within the cerebellar cortical circuit, the absence of an efficient output makes the "restoration" by the transplant cerebellar circuit physiologically non-functional regarding co-ordination with the rest of the motor system.

In addition to the specific integration between graft and host neurons, the embryonic graft of cerebellar anlage molecularly influences adult glial cells. Purkinje cells exit from the graft and follow a migratory route. After crossing the graft-host interface, they use the Bergmann glia to reach the host molecular layer. Young postmitotic Purkinje cells induce in host Bergmann fibers the transient re-expression of the intermediate filament nestin -an intermediate filament expressed by radial glial cells (in this case immature Bergmann glia) - during their migration along these glial fibers (Figure 2). Nestin expression is suppressed when grafted Purkinje cells attain their final location within the host molecular layer. These results indicate that embryonic PCs can trigger in adult cerebellum the molecular changes necessary for their own migration and ultimate synaptic integration in the host cortical circuitry (Sotelo et al., 1994). Further molecular analysis of the interactions between grafted embryonic neurons and adult glial cells, or even those between embryonic and adult neurons, would help to elucidate the regulatory interplay between graft and host cells -for instance the secretion of growth factors indispensable for neuronal survival during normal development- to enhance the survival of grafted neurons within adult host brain parenchyma.

The spinal cord regeneration obtained after different strategies, and measured by the level of restoration of the motor behavior in paraplegic animal models, enkindled the dream that paralyzed patients could walk again. However, the path was bedeviled with disappointments. The experimental spinal cord lesion before the therapeutic approach, the strategies used to promote regeneration and the tests used to measure spinal cord recovery are quite different between the various research groups (for a reflection see Pearson, 2003). The experiments of Ramon-Cueto have obtained stable bridges between the two ends of the hemisectioned spinal cord, using grafts of olfactory ensheathing glial cells. Furthermore, neuronal tracers injected into one spinal stump are detected in the other side –an unequivocal demonstration that damaged axons regrow across the glial bridge and distribute through the distal segment of the damaged spinal cord (Figure 4, B-C). This axonal growth is functionally correlated with the improvement of lower extremities movements of grafted rats (Figure 4, D). Experimental evidence is accruing on the long extensions run by regenerated axons after crossing the glial bridge. However, data are still not available on either the identity of the regenerated axons or the distribution of synaptic endings of regrown axons on the motor neurons and/or on lamina VII interneurons of the spinal cord.

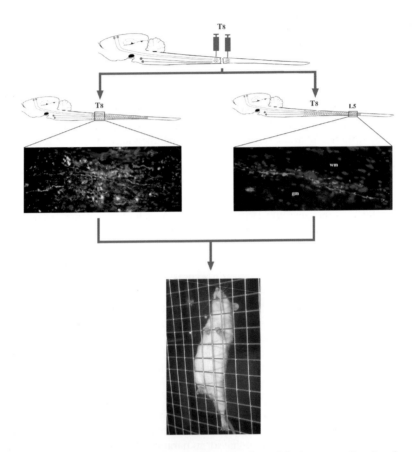

Fig. 4. Olfactory ensheathing glial cell (OEG) transplants (blue) promote functional recovery of adult paraplegic rats and long-distance axonal regeneration of sectioned corticospinal and brainstem axons (green and red, respectively) in their spinal cords. After transplantation of pure OEG suspensions into both spinal cord stumps, following complete transection at thoracic level 8 (T8)(A). Damaged corticospinal (green) and brain stem (red) axons grew through the glial scar created at the transection site (B). Corticospinal (red) regenerated axons elongated through the distal spinal cord stump for the longest distances (3 cm, L5 segment) (C). Photograph D shows one transplanted paraplegic rat presenting voluntary hindlimb movement seven months after surgery. Reproduced from Santos-Benito and Ramon-Cueto, Anat. Rec., 271B :77-85 (2003), with permission from Willey Interscience.

Detailed morphological and electrophysiological studies devoted to analyzing the synaptic distribution of these axons are required to demonstrate that motor recovery is due to restoration of the point-to-point corticospinal circuit rather than a broader and unspecific neural input. In fact, discrete degrees of motor recovery have been obtained in lesioned spinal cords that received the transplant of global system neurons such as serotoninergic embryonic neurons. In any case, I hope that in the next few years, experimental transplantation of ensheathing glia, together with molecular analysis on the causes of axonal growth permissivity by host glia, will yield fruitful results and improve the prospects of paralyzed patients.

3.3. Transplants of stem cells

The explosive increase in understanding of stem cells properties gives great expectation for their putative use as graft donors in neurodegenerative diseases. Neuronal proliferation in the adult brain was discovered early in the sixties by J. Altman. After a long period of silence, the origin, cell cycle, and migratory behaviors of these neural stem cells have been extensively investigated from the nineties to date, (for a rev., see Alvarez-Buylla et al., 2001). Neuronal stem cells have been found in several adult CNS territories, and have been postulated as putative mechanisms for self-renewal of dead neurons. Nevertheless, the neurogenetic role of adult stem cells in normal conditions has been questioned by studies that find mitotically active neural stem cells without a clear increase in neuronal cell number (for a reflection, see Rakic, 2002). Furthermore, the low number of stem cells occurring in the adult brain and their low rate of division make it difficult to imagine that their mere trophic stimulation would be enough to replace the neuronal cell loss taking place in neurodegenerative diseases. Thus, the strategy again becomes the grafting of the damaged host brain with exogenous engineered neural stem cells able to synthesize the lost neurotransmitter, and to provide complementary trophic factors preventing the progressive nigral neuronal cell death in Parkinson's disease, or the complex degenerative process involved in aging and/or Alzheimer's disease.

Recent experimental evidence has created expectation for the possible use of the patient's adult bone marrow stem cells for autotransplantation. Various experiments have demonstrated that daughter cells of adult neural stem cells share properties of hematopoietic cell lines. In turn, adult bone marrow stem cells contain a population of highly undifferentiated, multipotent cells whose progeny can originate cells expressing glial and neuronal phenotypes (see Bonilla et al., 2002). In fact, adult bone marrow stem cells grafted into postnatal host brain differentiate in all neural cell lineages (Figure 5). A number of grafted cells integrate within the host's subventricular zone before differentiation (Figure 5, A). Thus, the subventricular zone, one of the stem-cell-containing regions of the adult nervous systems (Alvarez-Buylla et al., 2001), seems to possess permissive cues for transplanted bone marrow stem cells. Whether contact with this zone is a prerequisite for further neural differentiation of bone marrow stem cells remains unknown. Experiments of the Martinez research team have demonstrated for the first time that adult bone marrow cells can differentiate into oligodendrocytes (Figure 5, B-C) which migrate towards white-matter brain areas (Figure 5, D). Further analysis should elucidate how these cells acquire the oligodendroglial fate. Such experiments open up the possible therapeutic use of adult bone marrow stem cells for the treatment of demyelinating diseases, such as multiple sclerosis.

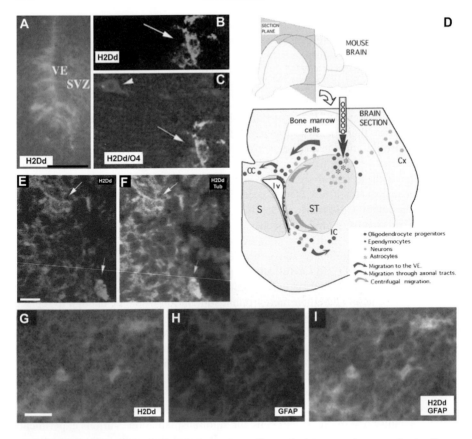

Fig. 5. Adult bone marrow hematopoietic stem cells grafted into newborn rat brain. Some grafted cells integrated within the subventricular zone (A), and differentiated into ependymocytes (A), astrocytes (G-I), oligodendrocytes (B-C) and neurons (E-F). Confocal microphotographs illustrate the matching of donor cell marker (H2Dd) with oligodendroglial (C, O4), neuronal [F, β-III-Tubulin (Tub)], and astroglial (I, GFAP) markers. D illustrates the migratory routes and fate of grafted cells. CC, corpus callosum. Cx, cerebral cortex. IC, internal capsule. S, septum. ST, striatum. Bars: (A) 100 μm; (B-C) 40 μm; (E-I) 20 μm. Reproduced from Bonilla et al., Eur. J. Neurosci., 15:575-582 (2002), with permission from Blackwell Science.

Bone marrow stem cells grafted into adult hosts seem able to colonize the CNS and differentiate into highly complex neurons (e.g., Purkinje cells), which survive for more than one year after transplantation (Figure 6). It is also remarkable that transplanted bone marrow stem cells reach the host brain after their intravascular delivery, crossing the blood-brain barrier. These results raise the possibility of future treatments in which donor cells could be administered peripherally, thereby avoiding any kind of damage to the patient's brain. However, recent observations that grafted cells fuse with remaining host cells rather than differentiate in new host cell populations, raise new discouraging doubts about the short-term application of this therapeutic possibility for clinical use in humans.

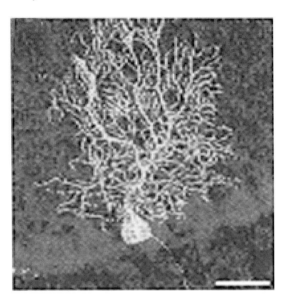

Fig. 6. Adult bone marrow stem cells differentiated within host cerebellum in Purkinje cells one year after intravascular administration of the bone marrow stem cells. Confocal analysis illustrates the matching between the GFP-marker of the stem cells (green) and GAD (red) marker of the Purkinje cell. Reproduced from Priller et al., J. Cell Biol., 26:733-738 (2001), with permission from The Rockefeller University Press.

Biotechnological advances open up an ultimate and exciting strategy for the possibility of designing artificial cells, effectively synthesizing defective neurotransmitters or trophic factors which could enhance the recovery of damaged global systems. In fact, the artificial production of individual cell components such as cell membranes, DNA and RNA is nanotechnologically feasible, and only their definitive integration into a functional artificial cell able to secrete its products is lacking (Pohorille and Deamer, 2002). This would be a promising line of future research to solve two of the main problems of transplant sources: (i) the number and reproducibility of donor cells, and (ii) the host immune response.

4. GENE THERAPY

Advancement in the knowledge of the genetic and molecular basis of the physiopathology of neurodegenerative diseases (e.g., the complete sequencing of the human genome) opens up a broad new field of research. The detection of the intimate basis and mechanisms involved in inherited neurodegenerative diseases enables strategies for family planning to prevent the passing down of the disease. In other cases, unfortunately still the majority, genetic planning is not yet possible.

Further experimentally and conceptually refined studies on the molecular causes of each disease will enable attempts to restore the neurotransmitters lost (or their synthetic pathway) or to block the expression of the mutated genes.

Molecular experiments are encouraging for the possible treatment of degenerative disease affecting global systems. The use of viral vectors for cell gene transfer permits redirecting the expression of neurotransmitter enzymes by glial cells or remaining undamaged neurons located within damaged circuits. The re-expression of enzymes involved in the synthesis of dopamine, γ-aminobutyric acid (GABA), GDNF, or CNTF promises reasonably good results in palliating the symptoms and the degenerative neuronal loss of Parkinson's and Huntington's diseases.

Gene therapy has three main drawbacks. The first is the toxicity of the intracerebral delivery of a virus and its activation of the immune system. Toxicity is always a risk after viral administration. Recent studies have related toxicity with the virus concentration. Therefore, the most suitable solution to this problem is to use low titers of vector, while obtaining a reasonable degree of transgene expression (Mittoux et al., 2002). The second issue is that the absence of viral dispersion through the brain parenchyma outside the placement site could limit the extent of gene transfer. However, some types of virus, such as adenoviruses, possess higher rates of neuronal transport (in both retrograde and anterograde directions). This means that fibers passing through or ending within the placement site can take up the virus. Thus, if synthetic and secretor machineries of these neurons function normally, the final amount of transgenic product could be increased, and, more important, released locally at the degenerated site (Mitoux et al., 2002). The third problem is the duration of gene expression after delivery. Recent findings in rats and monkeys have shown acceptably long-term GDNF-transgene expression after the administration of adenovirus-associated virus and lentivirus, and these have been postulated as candidates for further human clinical trials (Björklund et al., 2000).

The use of viral vectors for cell transfer of neurotrophic factors (e.g., CNTF or GDNF) represents an elegant substitute for the earlier administration of neurotrophic factors by intracerebral pump implants. Administration of viral vectors avoids the need to replace empty pumps, the obstruction of the delivery device, and other problems inherent in the periodic manipulation of the system. However, there is one more problem to be overcome for the neuroprotective treatment to be successful. The disease must be diagnosed at its onset, when neuronal damage and cell loss can still be stopped. Unfortunately, in most cases the evident clinical symptoms of neurodegenerative diseases appear too late, when damage has already affected the majority of neurons involved. Thus, diagnostic procedures aimed at the early detection of the disease, even before the appearance of clinical signs, is a priority that should be tackled urgently.

5. ELECTRICAL STIMULATION OF THE CENTRAL NERVOUS SYSTEM

At the end of the 19th century, electrical stimulation of the brain was used clinically in human patients as a diagnostic tool prior to ablation of cerebral areas by researchers as Krause, Sherrington, Keen or Horsley. It has been used for experimental and behavioral analyses from the sixties, and in the last two decades, the electrical stimulation of deep brain structures has become an accepted therapy for several neurological disorders. Placement of electrodes to stimulate periaqueductal gray-matter projecting areas has been used to relieve chronic central neurogenic pain. In this case, electrical stimulation allows the excitatory or positive stimulation of cerebral areas, increasing the release of enkephalin in the spinal cord by stimulation of the mesencephalic central gray. Similarly, the stimulation of sensory nuclei or fiber tracts, such as the ventrocaudal nucleus of the thalamus or the internal capsule, produces a positive effect that is evidenced by the appearance of parasthesia or muscle contractions, respectively. Dural cerebral cortex stimulation inhibits intractable central neurogenic pain by the stimulation of sensory pathways that modulate pain transmission.

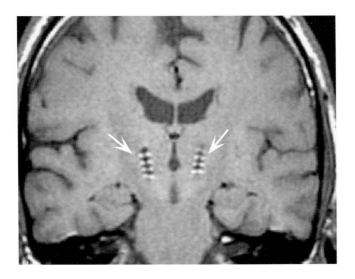

Fig. 7. Nuclear magnetic resonance. Coronal section of a Parkinsonian patient bilaterally implanted with high frequency stimulation electrodes in the subthalamic nuclei (arrows). Courtesy of Dr. Arjona. Head of the Department of Neurosurgery. H.U. Virgen de las Nieves. Granada. Spain.

The positive effects of electric stimulation have also been used in several brain areas (e.g., cerebellum, thalamus or basal ganglia) to regulate the uncontrolled triggers of epileptic foci. However, when electrodes are placed within the motor control nuclei of the motor cortex → basal ganglia → thalamus → motor cortex circuit, the effect of electrical stimulation is inhibitory, mimicking that after removal of the nuclei obtained in the old-fashioned ablation therapies. Empiric clinical evidence has demonstrated that dystonias, idiopathic tremors, and the tremor and akinesia of Parkinson's disease cease or ameliorate under high-frequency stimulation of basal ganglia. According to clinical results, of the different nuclei stimulated (the medial globus pallidus, the ventral intermedius nucleus of the thalamus, the substantia nigra pars reticulate, and the subthalamic nucleus), the subthalamic nucleus is one of the most suitable targets for treatment of Parkinsonian disorders (Figure 7). However, it remains unclear how high-frequency stimulation acts to inhibit basal ganglia. Furthermore, in addition to the paradoxical inhibitory response after electrical stimulation, basal ganglia respond differently depending on the frequency of stimulation. Thus, motor symptoms disappear under high-frequency stimulation (up to 80 Hz; the range of frequency clinically used is between 130 and 200 Hz), while low frequencies (40-50 Hz) have opposite effects (see Dostrovsky and Lozano, 2002).

Mechanisms underlying the inhibitory effect of high-frequency stimulation are poorly understood. Three mechanisms are currently under discussion. The first, motor control, involves a simple modulator feedback loop: periphery (muscle receptors)→thalamus→cortex→periphery (motor unit). In deregulated circuits, the burst firing rhythm of thalamic neurons could perpetuate the entry of stimuli from the periphery, explaining the tremor. High-frequency stimulation blocks or "jams" the periodic entry of inputs to the loop, thereby arresting the tremor (Benabid et al., 2002). Another possible mechanism is that long-lasting high-frequency stimulation leads to a reduction in the excitability of subthalamic neurons. In fact, the inactivation of calcium and sodium channels which seems to follow high-frequency stimulation could be explained by the inactivation of those channels by the current itself, or by the increase of extracellular potassium concentrations, changing the neuronal membrane voltage, and inactivating these voltage-dependent channels (Benabid et al., 2002; Dostrovsky and Lozano, 2002). The third possibility suggested is the high-frequency stimulation of striatal and external globus pallidus afferent GABAergic fibers to the medial globus pallidus. The increase of GABA release in these fiber terminals triggers the repolarization of medial globus pallidus neurons (Dostrovsky and Lozano, 2002), which secondarily increases the activity of the thalamic neurons. This last mechanism seems to be supported by viral-vector-mediated cell-transfer experiments in adult rats. The enhancement of GABAergic activity within the medial globus pallidus, by the directed expression of glutamic acid decarboxylase (GAD) in the subthalamic excitatory neurons, results in a substantial neuroprotection of the neurons of the substantia nigra pars compacta, accompanied by the recovery of motor behavior (Luo et al., 2002). The complex neurotransmitter balance underlying the normal functioning of basal ganglia hinders experimental approaches to solving this question. Indeed, no experiments clearly demonstrating the validity of any of these hypotheses have been reported. Thus, it is possible to infer that the eventual inhibition of the subthalamic nuclei, and the recovery of motor symptoms obtained under high-frequency stimulation, could be the result of the conjoint action of more than one (perhaps all) of the proposed mechanisms (Benabid et al., 2002).

6. INVASIVE TREATMENT. HOW AND WHEN? THE REAL STATUS OF THE QUESTION

In all cases, invasive strategies involve the access to the CNS from outside. Hence, the focal disruption of the glial barriers (i.e., blood-brain barrier), elicits a response of the host brain to the surgical treatment like that to an anisomorphic lesion. Immune responses to implanted biological material (e.g., embryonic, chromaffin, or engineered stem-cells grafts) have been evident from the first experiments. Three strategies have been designed to impede the eventual rejection of the graft. (i) The earliest, explored from the beginning of human grafting procedures, was the use of autotransplants of adrenal or neural crest-derived cells. The possibility of obtaining hematopoietic stem cells, which can be engineered and re-directed before graft, opens up new prospects in this now-old-question. (ii) As in other types of transplant, a second strategy was the use of immunosuppressive therapies, whose disadvantages -such as the increased risk of infectious diseases, etc.,- are well known (the reader is directed to general surgery and immunology books). (iii) Finally, the creation of a protective environment at the host-graft interface, using Sertoli's cells to attenuate the host's rejection reaction and permit the graft integration (Senut et al., 1996), could also help to solve this problem.

The best clinical results were obtained after grafting young dopaminergic neurons obtained from legal abortions in Parkinsonian patients. In the majority of cases, however, the amelioration of motor impairments and psychological symptoms of the disease after grafting required the adjuvant of classical pharmacological treatment. Furthermore, in spite of the good results attained by this therapy, two major disadvantages have consistently been described: (i) even in the best cases, fewer than 10% of grafted neurons survive, restricting the number of graft patients who show evident amelioration of their symptoms, and (ii) thirty percent of successfully transplanted patients presented undesirable side-effects, such as tardive dyskinesia (Freed et al., 2001). Side-effects in pharmacologically treated patients can be reversed by stopping the administration of dopamine agonists, but in transplanted patients clinical correction is already impracticable, and further discomfort is added to the symptoms of the original disease. It therefore seems reasonable to employ all the pharmacological and related therapies before deciding on the use of transplants as the routine protocol. Recent data indicate that the implanting of engineered stem cells, or viral gene transfer, aimed at increasing the levels of neurotrophic factors, could solve this problem, and appear today as the most-suitable invasive strategies for Parkinson's and Alzheimer's diseases. Their main disadvantage resides in the need to have a reasonable number of surviving neurons to rescue. That is, these procedures imply the early clinical diagnosis of the disease, which is to date not always feasible.

Huntington's disease is one of the most dramatic neurodegenerative diseases. The Huntington's patient suffers considerable life impairments before the conclusion of the disease. Recently, Peschanski's group reported substantial beneficial effects after the graft of striatal neuroblasts (Bachoud et al., 2002). Although further studies are needed to refine grafting procedures and reduce their problems, the clinical benefits obtained in grafted Huntington's patients indicate the use of this as a symptomatically effective method.

Finally, high-frequency electrical stimulation, although quasi-empirical, appears an elective symptomatic method for chronic pain, some kinds of epilepsy, and those motor disorders resistant to pharmacological treatments or in which they cannot be applied.

7. CONCLUSION

Apart from ethical issues, which are treated in Chapter 55, there is still a long experimental road to travel before routine invasive strategies for the effective total treatment of neurodegenerative diseases are achieved. High-frequency stimulation alleviates the symptoms of some neurological diseases. Despite technical problems, engineered stem-cell grafts and virally mediated gene-transfer methods are beginning to appear as suitable strategies in neurodegenerative disorders involving neuronal cells in the global system or myelin sheath loss. In contrast, the effective restoration of damaged point-to-point circuits by the synaptic integration of grafted embryonic neurons remains today a dream which only the full understanding of the molecular basis of the mechanisms of cell-cell recognition and synaptic plasticity could begin to make real.

REFERENCES

Alvarez-Buylla A., Garcia-Verdugo J.M. and Tramontin A.D. (2001) A unified hypothesis on the lineage of neural stem cells *Nat. Rev. Neurosci.* 2, 287-293.

Bachoud-Levi A.C., Hantraye P. and Peschanski M. (2002) Fetal neural grafts for Huntington's disease: a prospective view. *Mov. Disord.* 17, 439-444.

Benabid A.L., Benazzous A. and Pollak P.(2002) Mechanisms of deep brain stimulation. *Mov. Disord.* 17, Suppl. 3: 73-74.

Björklund A. (1991) Neural transplantation - an experimental tool with clinical possibilities. *TINS* 14, 319-322.

Björklund A., Kirik D., Rosenblad C., Georgievska B., Lundberg C. and Mandel R.J. (2000) Towards neuroprotective gene therapy for Parkinson's disease: use of adenovirus, AAV and lentivirus vectors for gene transfer of GDNF to the nigrostriatal system in the rat Parkinson model. *Brain Res.* 886, 82-98.

Döbrössy M.D. and Dunnet S.B. (2001) The influence of environment and experience on neural grafts. *Nature Rev. Neurosci.* 2, 871-879.

Dostrovsky J.O. and Lozano A.M. (2002) Mechanisms of deep brain stimulation. *Mov. Disord.* 17, Suppl. 3: 63-68.

Dunnet S.B., Björklund A., and Lindvall O. Cell therapy in Parkinson's disease – stop or go? *Nature Rev. Neurosci.* 2, 365-369.

Espejo E.F., Gonzalez-Albo M.C., Moraes J.-P., El Banoua F., Flores J.A. and Caraballo I. (2001) Functional regeneration in a rat Parkinson's model after intrastriatal grafts of glial cell line-derived neurotrophic factor and transforming growth factor β_1-expressing extra-adrenal chromaffin cells of the Zuckerandl's organ. *J. Neurosci.* 21, 9888-9895.

Freed C.R., Green P.E., Breeze R.E., Tsai W.-Y., DuMouchel W., Kao R., Dillon S., Winfield H., Culver S., Trojanowski J.Q., Eidelberg D. and Fahn S. (2001) Transplantation of embryonic dopamine neurons for severe Parkinson's disease. *New Engl. J. Med.* 344, 710-719.

Lou J., Kaplitt M.G., Fitzsimons H.L., Zuzga D.S., Liu Y., Oshinsky M.L. and During M.J. (2002) Subthalamic GAD gene therapy in a Parkinson's disease rat model. *Science* 298, 425-429.

Mittoux V., Ouary S., Monville C., Losivoski F., Poyot T. Conde F., Escartin C., Robichon R., Brouillet E., Peschanski M. and Hantraye P. (2002) Corticostriatopallidal neuroprotection by adenovirus-mediated ciliary neurotrophic factor gene transfer in a rat model of progressive striatal degeneration. *J. Neurosci.* 22, 4478-4486.

Pearson H. (2003) In search of a miracle. *Nature* 423, 112-113.

Pohorille A. and Deamer D. (2002) Artificial cells: Prospects for biotechnology. *TINS* 20, 123-128.

Rakic P. (2002) Adult neurogenesis in mammals: An identity crisis. *J. Neurosci.* 22, 614-618.

Ramon y Cajal S. (1913-1914) *Estudios Sobre la Degeneración y Regeneración del Sistema Nervioso.* Imprenta Moya: Madrid. Translated into English as *Degeneration and Regeneration of the Nervous System* (May R.M. trans. and ed.). London: Oxford University Press, 1928. Reprinted and edited with additional translations by J. DeFelipe and E. G. Jones (1991), *Cajal's Degeneration and Regeneration of the Nervous System.* New York: Oxford University Press.

Senut M-C., Suhr S.T. and Gage F.H. (1996) Intracerebral xenografts: Sertoli cells to the rescue? *Nat. Biotechnol.* 14, 1650-1651.

Sotelo C. and Alvarado-Mallart R.M. (1991) The reconstruction of cerebellar circuits. *TINS* 14, 350-355.

Sotelo C., Alvarado-Mallart R.M., Frain M. and Vernet M. (1994) Molecular plasticity of adult Bergmann fibers is associated with radial migration of grafted Purkinje cells. *J. Neurosci.* 14, 124-33.

A. RIBEIRO

Laboratory of Neuroscience's, Faculty of Medicine, University of Lisbon, Portugal

27. GENERAL ASPECTS OF NEUROPHARMACOLOGY IN RELATION TO BRAIN REPAIR FOLLOWING TRAUMA

Summary. Trauma is the main cause of death in individuals between the ages of 1 and 44 years. However, there has been little progress in the development of effective pharmacological agents to protect brain of injured patients. There is still little information on the mechanisms involved in neuronal cellular insult after severe head injury, especially in humans. Glutamate acts both as a primary excitatory neurotransmitter and a potential neurotoxin within the mammalian brain. Experimental evidence suggests that hyperactivity of the glutamate NMDA currents system contributes to neuronal death in brain trauma. Also, in animal models of neurotrauma, this neural injury is followed by gliosis which has been linked to the severity of brain injury. In the context of traumatic brain injury, investigation of substances that can selectively inhibit the glutamate NMDA receptor subtype in humans while also therapeutically controlling glial cell responses following brain trauma would be of great help in brain damage. This has led to the application of pharmacological strategies to limit secondary injury and subsequent neurological deficits. The research on pharmacological agents will help to promote functional rewiring of brain and spinal cord after injury in order to obtain neural repair and neurological rehabilitation. Beta-blockers, e.g. propanolol; benzodiazepines; haloperidol; chlorpromazine; tricyclic antidepressants, e.g. amitriptyline; methotrimeprazine; buspirone; lithium carbonate; dopamine agonists; adenosine based therapies; endocanabinoids; bradykinin b2; galanin; NMDA antagonists; cholinomimetic agents; sodium channel blockers; neurotrophic factors; interleukin-1 receptor antagonists; are part of the considerable number of substances tested experimentally. Many of these therapies have progressed into human clinical trials in severe traumatic brain injury.

1. INTRODUCTION

It is difficult to accurately determine the number of people affected annually by the devastating effects of traumatic brain injury (see e.g. Regner et al, 2001). It is clear, however, that the impact of traumatic brain injury exceeds the financial cost of acute health care. Traumatic brain injury is a major cause of disability and death in most Western countries. The long-term outcome of patients with traumatic brain injury has been targeted specifically for improvement during the last decade. The initial brain injury (known as the primary injury) may occur in one area of the brain (focal injury) or may affect the entire brain (diffuse injury). The outcome depends upon many factors, including the intensity of the brain insult and the effectiveness of the interventions. In the last twenty five years, the management of traumatic brain injury has progressed dramatically, in consequence of a better comprehension of the physiologic events that lead to cause the secondary neuronal injury as well as to great advances in the care of critically ill patients. Trauma is the leading cause of death in individuals between the ages of 1 and 44 years. However, there has been little progress in the development of effective pharmacological agents to protect brain-injured patients. To date, there is still little information on the mechanisms involved in neuronal cellular insult after severe head injury, especially in humans. Glutamate acts both as a primary excitatory neurotransmitter and a potential neurotoxin within the mammalian brain. Experimental evidence suggests that hyperactivity of the glutamate NMDA currents system contributes to neuronal death in brain trauma. Also, in animal models of neurotrauma, this neural injury is followed by gliosis, which has been linked to the severity of brain injury. In the context of traumatic brain injury, investigation of substances that can selectively inhibit the glutamate NMDA receptor subtype in humans while also therapeutically controlling glial cell responses following brain trauma would be of great help in brain damage. This has led to the application of pharmacological strategies to limit secondary injury and subsequent neurological deficits. The research on pharmacological agents will help to promote functional rewiring of brain and spinal cord after injury in order to obtain neural repair and neurological rehabilitation. (see e.g. Faden, 1993; Zuccarelli, 2000).

2. DEFINITION OF TRAUMATIC BRAIN INJURY

Application of a physical force suddenly to the head can cause injury to the brain. There are several types of brain injury following trauma: First, non-penetrating injuries, these injuries may cause loss of consciousness, and may result in gross and widespread brain damage; Second, penetrating injuries, via high velocity missiles penetrating the skull and brain, or less frequently the skull is subject to a crush injury. The patient may suffer serious or even fatal injury without previous loss of consciousness; Third, concussion, violent shaking of the brain with transient functional impairment with no definition of underlying nerve cells. The existence of this condition has not been confirmed by animal experiments or clinical studies. Finally, contusion, means bruising of the brain and is a rather loose term in clinical use. Headache, fatigue, dizziness, reduced libido, and irritability are some of the symptoms that make part of post-concussional syndrome or 'mild head injury'. Attention, language and memory impairment and in many occasions, coma followed by disorientation are together with a number of psychiatric symptoms namely anxiety part of the clinical picture of traumatic brain injury. Neuropharmacology has therefore to be aware of the causes and symptoms in order to use its therapeutic armamentarium to help in these different symptoms.

3. ANATOMY AND PHYSIOLOGY OF TRAUMATIC BRAIN INJURY AND SPINAL CORD INJURY

Traumatic injuries to the brain or spinal cord cause irreversible tissue damage by at least three mechanisms: First, through consequences of mechanical disruption of neurons or their projections; secondly, through biochemical or metabolic changes that are initiated by the trauma; and finally through inflammatory reactions or gliotic changes. The cellular elements and the chemical neuro-mediators in secondary injury following traumatic brain injury or spinal cord injury act via interconnections between the cellular elements and their secretions; the immune system and the nervous system are highly regulated in normal physiology, which benefits the organism. When these cells suffer traumatic insult in the CNS, the connections between the systems become more than tight; these systems act together to strangulate the tissue, depriving it of the local control over microcirculation and necessary oxygen, rendering membrane potentials useless to modulate neuronal function. Surgical interventions during the acute stages of traumatic brain injury and spinal cord injuries continue to progress as do biochemical and bioelectric therapeutics during the chronic and rehabilitation stages. There is some hope, too, for effective neuropharmacological intervention at the initial stages, before secondary injury starts to occur. The fact that chemical mediators of secondary injury are already part of normal physiology, whether during development or adulthood, means that their activity can be modified by specific agonists and antagonists to restore homeostasis or to promote the safe pathways that can lead to regeneration. This is the orientation of much of current basic and clinical research. During the past decade, considerable experimental and clinical data have been accumulated regarding cellular and biochemical events associated with posttraumatic tissue.

4. BRAIN PLASTICITY AND TRAUMA

In rats traumatic brain injury significantly inhibits expression of long-term potentiation in hippocampal slices in vitro, and changes in functional synaptic plasticity in the hippocampus may contribute to cognitive disorders associated with traumatic brain injury (Sick et al., 1998).

5. CEREBRAL ISCHEMIA AND TRAUMA

Cerebral ischemia leads to brain damage caused by pathogenetic mechanisms that are also activated by neurotrauma. These mechanisms include among others excitotoxicity, overproduction of free radicals, inflammation and apoptosis. Furthermore, cerebral ischemia and trauma both trigger similar auto-protective mechanisms including the production of heat shock proteins, anti-inflammatory cytokines and endogenous antioxidants. Neuroprotective therapy aims at minimizing the activation of toxic pathways and at enhancing the activity of endogenous neuroprotective mechanisms. According to Leker (2002) the similarities in the damage-producing changes in endogenous substances in the brain may imply that neuroprotective compounds found to be active against one of these conditions may indeed be also protective in the other.

6. NEUROPHARMACOLOGY

Multiple neuropharmacologic agents have been used early in the treatment of posttraumatic brain injury:

6.1. Beta-Blockers

After traumatic brain injury agitation is frequent and disruptive for patient care, distressing and difficult to treat. The use of propranolol has been advocated to control agitation after brain injury. It reportedly lacks some of the deleterious cognitive and emotional effects of other medications and physical restraint. The effectiveness of propranolol in reducing the intensity of agitation during the initial hospitalisation after closed-head injury (e.g. Brooke et al, 1992) recommends its use in these situations.

6.2. Benzodiazepines and neuroleptics

Agitation is one of the most dramatic behaviour changes caused by head injury. When there is an imminent danger of harm to the patient himself or to others, or when aggressive behaviour makes medical management difficult, the benzodiazepines have been found very useful (e.g. Sandel et al., 1993). Antipsychotic drugs are only indicated in head injury when the agitation is the cause of a clinical emergency, and in such a case potent drugs such as haloperidol are better choice, since they have less sedative effects. They are also effective when the clinical features are similar to those observed in the classical schizophrenia. Antiepileptic drugs have been used successfully to treat agitation-aggressiveness, especially in paroxyistic behaviour disorders. When using neuroleptics in agitated patients with organic brain disorders, caution should be taken, since they have the potential to precipitate seizures.

6.3. Tricyclic antidepressants

Amitriptyline or other tricyclic antidepressants may be helpful in diminishing agitation associated with frontal lobe damage without impeding cognitive recovery (Jackson et al., 1985).

6.4. Methotrimeprazine

Medical management of the agitation associated with traumatic brain injury has been problematic. Methotrimeprazine has come to be the preferred drug and used routinely for effective treatment of agitation.

6.5. Buspirone

Recovery from head injury is a long process, with agitation being a well-known stage in the recovery. Several medications have been used in the treatment of this stuations namely buspirone. (Levine 1988)

6.6. Lithium carbonate

Lithium carbonate can be a useful medication in the treatment of aggressive behaviour and affective instability after brain injury, but that is of particular significance for the neurotoxicity in the patients, especially if used together with neuroleptic agents (Glenn et al., 1989).

6.7. Antiepileptics

In recent years the number of new antiepileptic agents has been growing, as have their potential applications (Kennedy et al., 2001). Traumatic brain injury patients often show behavioural disturbances that can be treated with antiepilectic drugs. These conditions include bipolar disorders, as well as post-traumatic seizures and agitation.

6.8. Dopamine agonists

Bromocriptine. Cognitive impairments are pervasive and persistent sequelae of human traumatic brain injury. Chronic treatment with the D2 receptor agonist, bromocriptine attenuates both working memory and spatial learning acquisition deficits. Cortical impact produces working memory deficits in rats and chronic bromocriptine treatment regime conferred cognitive and neural protection after traumatic brain injury (see e.g. Kline et el., 2002).

6.9. Adenosine

Robertson et al. (2001) described that adenosine concentration is markedly increased in cerebrospinal fluid of children which suffered traumatic brain injury. There is a clear association between increases in adenosine in the cerebrospinal fluid and glutamate concentrations in the case of brain injury. Cerebrospinal fluid adenosine concentration is increased in a time-and severity-dependent manner in infants and children after severe head injury. The association between cerebrospinal fluid adenosine and glutamate concentrations may reflect an endogenous attempt to neuroprotect against excitotoxicity after severe traumatic brain injury. (for reviews on the potential neuroprotective actions of adenosine see de Mendonça et al., 2000; Ribeiro et al., 2003).

6.10. Endocanabinoids

Cannabinoid compounds are potential candidates for therapeutic development after brain injury. They are acting through receptor-dependent and receptor-independent mechanisms. These properties have been known since antiquity. However, only in the last decade important advances in the understanding of the physiology, pharmacology, and molecular biology of the cannabinoid system have given this field of research fresh impetus and have pointed out the interest in the possible therapeutic use of these compounds. Cannabinoids may act as neuroprotective agents (prevention of excitotoxicity by inhibition of glutamate release, antioxidant effects, anti-inflammatory actions) in disease states such as cerebral ischemia, brain trauma, and multiple sclerosis (Grundy, 2001). Traumatic brain injury releases harmful mediators that lead to secondary damage. On the other hand neuroprotective mediators are also released, and the balance between these classes of mediators determines the final outcome after injury. Recently, it was shown that the endogenous brain cannabinoids, anandamide and 2-Arachidonoyl glycerol are also produced after traumatic brain injury in rat and mouse respectively, and when administered after traumatic brain injury, they reduce brain damage. In the case of 2-Arachidonoyl glycerol, better results are seen when it is administered together with related fatty acid glycerol esters. Significant reduction of brain edema, better clinical recovery, and reduced infarct volume and hippocampal cell death are noted. This new neuroprotective mechanism may involve inhibition of transmitter release and of inflammatory response. 2- Arachidonoyl glycerol is also a potent modulator of vascular tone, and counteracts the endothelin-induced vasoconstriction that aggravates brain damage; it may thus help to restore blood supply to the injured brain (Mechoulam et al., 2002).

6.11. BradykininB2 – antagonists

Brain injury after cold lesion is partially mediated by bradykinin and can be successfully treated with bradyknin B2 antagonists (Gorlach et al., 2001).

6.12. Galanin

The intensity of behavioural deficits after traumatic brain injury has been shown to be related in part to alterations in the balance between the release of excitatory and inhibitory neurotransmitters. Release of extracellular excitatory neurotransmitters dramatically increases following experimental traumatic brain injury. The neuromodulatory action of thr peptide, galanin, has been shown on behavioural morbidity, as measured by sensory motor and memory performance tasks, associated with experimental traumatic brain injury in the rat. Physiologically galanin may reduce certain components of traumatic brain injury morbidity, possibly by modulating neuronal excitability (Liu et al., 1994)

6.13. Novel pharmacotherapies for use in traumatic brain injury

6.13.1. Magnesium

Previous experimental studies have demonstrated that brain intracellular free magnesium significantly declines following traumatic brain injury and that the administration of magnesium salts attenuates the post-traumatic neurological deficits. More recent studies have established that magnesium salts administered after trauma enter the brain intracellular space reduce the size of lesion volume. Such protection could be obtained through attenuation of both necrotic and apoptotic induced cell death. Magnesium salts are currently on clinical trial in traumatic brain injury. The neuroprotective effect of magnesium chloride has been demonstrated by improving behavioural and neurochemical outcome in several models of experimental brain injury, Administration of magnesium chloride significantly reduces the injury-induced damage in the cortex but does not alter posttraumatic cell loss in the CA3 region of the ipsilateral hippocampus. In addition to its beneficial effects on behavioural outcome, magnesium chloride treatment attenuates cortical histological damage when administered following traumatic brain injury (see Bareyre et al., 2000).

6.13.2. Cyclosporine-A

Cyclosporine-A is known to inhibit opening of the mitochondrial permeability transition pore. Administration of cyclosporine-A after traumatic brain injury has been shown to attenuate axonal injury and decrease the resultant lesion volume. Therefore, inhibitors of mitochondrial transition pore opening and resultant attenuation of apoptosis show some promise as potential neuroprotective agents.

6.13.3. Substance P antagonists

Recent evidence has shown that these antagonists may decrease lesion volume and improve neurological outcome after ischemia. Similar findings have also been reported in relation to traumatic brain injury. The fact that substance P antagonists are known to reduce neurogenic inflammation, oedema formation and are clinically being trailed as both antidepressants and antinociceptive agents, suggests that these substances warrant deeper investigation as therapeutic agents following traumatic brain injury.

6.13.4. Oestrogen and progesterone

There are numerous contradictions in the literature concerning the potential neuroprotective effects of the hormones oestrogen and progesterone. Recent studies suggest that both hormones are protective in traumatic brain injury and further studies are required to ascertain the mechanisms associated with this protection as well as their potential for clinical application (Vink et al., 2001).

7. EXPERIMENTAL APPROACHES

7.1. Metabotropic Receptors

Glutamate toxicity, mediated via ion channel-linked receptors, plays a key role in traumatic brain injury pathophysiology. Excessive glutamate release after traumatic brain injury also activates protein G-linked metabotropic glutamate receptors (mGluRs). Traumatic brain injury produces an alteration in metabotropic glutamate receptor protein expression that spontaneously recovers by 15 days after injury (Gong et al., 1999).

7.1.1. Group I Metabotropic Glutamate Antagonists

Injury-induced acute activation of mGluR1 receptors contributes to both the cellular pathology and the behavioural morbidity associated with traumatic brain injury (Lyeth et al., 2001). In vitro studies suggest that activation of Groups II and III metabotropic glutamate receptors may provide some degree of neuroprotection and may be potential targets for the development of therapeutic strategies. Administration of selective Group II metabotropic glutamate receptor agonists protects neurons against in vivo traumatic brain injury. These receptors may thus be a promising target future neuroprotection. (Zwienenberg et al., 2001).

7.2. NMDA Antagonists

Cognitive deficits produced by head trauma involving both neuroexcitation and deafferentation can be attenuated with chronic application of glutamatergic antagonists during the period of deafferentation injury and this attenuation is correlated with axo-dendritic integrity. (Phillips et el., 1998). Although several noncompetitive N-methyl-d-aspartate (NMDA) receptor antagonists have been shown to be substantially efficacious in experimental models of brain trauma, side effects associated with this class of compounds have limited their clinical application. Therefore, new non-competitive NMDA receptor antagonists have been developed, being non-toxic but retaining in efficacy. Since memory dysfunction and hippocampal damage are common and potentially related to consequences of brain trauma in humans, NMDA antagonist treatment may have clinical utility. (Leoni et al., 2000) The hippocampus is selectively vulnerable to experimental traumatic brain injury, and finding beneficial effects with glutamate receptor antagonists or decreasing extracellular levels of glutamate point out that a glutamate-mediated excitotoxicity may be responsible for this selective damage. Thus, excitotoxicity does not significantly contribute to hippocampal neuronal at loss after traumatic brain injury and, in contrast to classic studies of excitotoxicity in vivo, the pattern of hippocampal cell death after traumatic brain injury is extremely acute. (Carbonell ewt al., 1999).

7.3. Kynureanate

Inhibition of excitatory amino acid neurotransmission with pharmacological agents improves physiologic, metabolic, and neurobehavioral outcome following experimental brain trauma. However, pharmacological compounds which are known to have beneficial effects on neurobehavioral and physiological recovery following brain injury may significantly attenuate post-traumatic neuronal cell loss. Recent data from Hicks et al. (1999) demonstrated that pharmacological intervention with excitatory amino acid receptor antagonists may be of therapeutic value in the treatment of brain injury.

7.4. Cholinomimetic Agents

Novel cholinomimetic therapeutics (e.g. with tetrahydroaminoacridine) improve cognitive outcome following traumatic brain injury in rats, (Pike et al., 1997).

7.4.1. Muscarinic Receptors

Attenuation of lactate accumulation in scopolamine-treated rats suggests that traumatic brain injury induced muscarinic receptor activation also contributes to increased glycolytic metabolism an/or ionic imbalances, (Lyeth et al., 1996). Changes in spatial memory following experimental traumatic brain injury include long-term changes that are 'overt': detected by routine behavioural assessments, or 'covert': undetected in the absence of a secondary pharmacological challenge, such as by the cholinergic antagonist, scopolamine. It therefore, seems that there are three distinct stages of functional recovery: an initial period when overt deficits are present; a second period following recovery from overt deficits within which covert deficits can be reinstated by a pharmacological challenge, and finally a third period following recovery from both overt end covert deficits. Covert deficits apparently persist long after the recovery of overt deficits and, like other neurological deficits, the rate of recovery is dependent upon the magnitude of traumatic brain injury. Finally, spatial memory deficits can occur in the absence of histological evidence of cell death in the hippocampus (Dixon et al., 1995). Mild to moderate traumatic brain injury is associated with enduring impairments of cognitive function either in humans or animals. However, few studies have been done to optimize the role of post injury pharmacological strategies for attenuating the observed cognitive impairment after traumatic brain injury. (see Pike et al., 1995).

7.5. Sodium Channel Blockers

Release of the excitatory amino acid neurotransmitter glutamate has been implicated in secondary tissue damage following central nervous system trauma and ischemia.

Sodium channel blockers inhibit ischemia-induced glutamate release on traumatic brain injury in rats (Sun and Faden 1995). Treatment significantly attenuated behavioural deficits at 24h and 1 week. At 2 weeks, neuronal loss in the CA1 and CA3 pyramidal cell layers of the hippocampus was significantly decreased by administration of sodium channel blockers.

7.6. Lactate

In experiments carried out by Ros et al. (2001) to test the hypothesis that lactate reduces the neurotoxicity induced by glutamate in vivo, when L-lactate is perfused together with glutamate there was a significant reduction in the size of the lesion and there was no greater reduction in dialysate lactate than after glutamate alone. The neuroprotective role of L-lactate is attributed to its ability to meet the increased energy demands of neurones exposed to high concentrations of glutamate.

7.7. Neurotrophic Factors

Neurotophic factors have been shown to have potential therapeutic applications in neurodegenerative diseases, and nerve growth factor (NGF) is neuroprotective in models of excitotoxicity. NGF administration, in the acute, posttraumatic period following fluid-percussion brain injury, apparently improves post-traumatic cognitive deficits (see e.g. Sinson et al., 1995) and a neurotrophic factor treatment following traumatic brain injury is neuroprotective (Kim et al., 2001). Brain-derived neurotrophic factor has been shown to be neuroprotective in models of excitotoxicity, axotomy and cerebral ischemia. Blaha et al. (2000) demonstrated the therapeutic potential of brain-derived neurotrophic factor following traumatic brain injury in the rat. However, post-traumatic brain-derived neurotrophic factor infusion does not significantly affect neuromotor function, learning, memory or neuronal loss in the hippocampus, cortex or thalamus when compared to vehicle infusion in brain-injured animals. In contrast brain-derived neurotrophic factor is not protective against behavioral or histological deficits caused by experimental traumatic brains injuries:

7.8. Interleukin-1 Receptor Antagonist

The effect of systemic administration of human recombinant interleukin-1 receptor antagonists on behavioral outcome and histopathologic damage after lateral fluid-percussion brain injury of moderate severity suggest that inhibitors of cytokine pathways may be therapeutically useful for the treatment of brain trauma (see e.g. Sanderson et al., 1999).

8. ROLE OF APOPTOSIS

The initial mechanical tissue disruption of brain injury is always followed by a period of a secondary injury that increases the size of the lesion. This secondary injury has long been thought to be due to the continuation of cellular destruction through necrotic (or passive) cell death. Recent evidence from brain injury and ischemia suggested that in this circumstance occurs apoptosis, an active form of programmed cell death seen during development. This process could play an important role in the central nervous system during injuries in adulthood. There is now strong morphological and biochemical evidence from a number of laboratories demonstrating the presence of apoptosis in the case of spinal cord injury (see e.g. Beattie et al., 2000). Apoptosis occurs in populations of neurons, oligodendrocytes, microglia, and perhaps, astrocytes. The death of oligodendrocytes in white matter tracts proceeds for many weeks after injury. It appears that, there is a close relationship between microglia and oligodendrocytes dying. This suggests that microglial activation may be involved in this process. There is also evidence for the activation of important intracellular pathways known to be involved in apoptosis in other cells and systems. For example, some members of the caspase family of cysteine proteases are activated after spinal cord injury. It appears that the evolution of the lesion after brain injury can involve both necrosis and apoptosis.

9. ROLE OF TEMPERATURE

Hypothermia: The use of hypothermia to treat various neurological emergencies, initially introduced into clinical practice in the 1940s and 1950s, had been considered without interest by the 1980s. However in the early 1990s, there was a revival of its use in the treatment of severe traumatic brain injury. The success of mild hypothermia led to the broadening of its application to many other neurological emergencies. Mild hypothermia has been applied with varying degrees of success in many neurological emergencies, including traumatic brain injury, spinal cord injury, ischemic stroke, subarachnoid hemorrhage, out-of-hospital cardiopulmonary arrest, hepatic encephalopathy, perinatal asphyxia (hypoxic-anoxic encephalopathy), and infantile viral encephalopaty. According to Inamasu and Ichikizaki (2002) there is no total consensus on the efficacy and safety of mild hypothermia. though preliminary clinical studies have shown that mild hypothermia can be a feasible and relatively safe treatment. Recently it has been shown in vitro patch clamp studies that in conditions of 32°C, adenosine inhibits NMDA receptor currents in hypoxia suggesting that adenosine might be the substance involved in neuroprotection during hypothermia (see Sebastião et al., 2001).

10. ROLE OF NITRIC OXIDE

Glutamate toxicity has been implicated in almost all aspects related to lesions is brain injury including traumatic, ischemic, and hemorrhagic damage. In vitro as well as in vivo methods to measure nitric oxide production showed a direct correlation with brain tissue oxygen tension in subarachnoid hemorrhage patients. These types of result point toward the importance of nitric oxide production in the response of brain to injury (Khaldi et al., 2002).

11. ROLE OF POTASIUM HOMEOSTASIS

Failure of ionic homeostasis can result in neuronal hyperexcitability and abnormal synchronization. Glial cells play a crucial role in the homeostasis of the brain microenvironment, the effects of traumatic brain injury on rat hippocampal glia were investigated by D'Ambrosio et al., (1999), using a fluid percussion injury model and patch-clamp experiments in hippocampal slices. They found after fluid percussion injury impaired glial physiology there is a reduction in transient outward and inward K (+) currents, as well as an abnormal extracellular K(+) accumulation in the post-traumatic hippocampal slices, accompanied by the appearance of afterdischarges. After pharmacological blockade of excitatory synapses and of K(+) inward currents, uninjured slices showed the same altered K(+) accumulation in the absence of abnormal neuronal activity. So, it appears that traumatic brain injury can cause the failure of glial K(+) homeostasis, which in turn can promote abnormal neuronal function. These authors proposed a new potential mechanistic link between traumatic brain injury and subsequent development of disorders such as memory loss, cognitive decline, seizures, and epilepsy.

12. EPILEPSY INDUCED BY TRAUMATIC BRAIN INJURY

Traumatic brain injury is a main cause of symptomatic epilepsy in young adults. Coulter et al., (1996) studied the physiological and anatomical epileptogenic effects of a prior incident of traumatic brain injury in rats subjected to a fluid percussion brain injury, as well as they performed electrophysiological studies using combined hippocampal-entorthinal cortical slices. These slices showed greater disinhibition in the dentate gyrus than in other regions. In the traumatic brain injury hippocampal-entorhinal cortical slices it was caused a self-sustaining epileptic activity. This type of activity was never obtained in control slices. Hippocampal-entorhinal cortical slices prepared from pilocarpine treated animals generated self-sustaining epileptic activity with fewer stimulus trains than did traumatic brain injury slices. In anatomical studies, both traumatic brain injury and pilocarpine hippocampi revealed significant loss of neurons within the hilar region. Traumatic brain injury induces a series of changes within the limbic system of rats, which are qualitatively similar in many aspects but quantitatively less severe than changes seen in rats with chronic temporal lobe epilepsy. These changes in physiology and anatomy after traumatic brain injury observed in the limbic system, may contribute to the development of epilepsy following head trauma.

13. OVEREXPRESSION OF BCL-2 IS NEUROPROTECTIVE

It has been suggested that the cell death regulatory protein, Bcl-2, participates in the pathophysiology of various neurological disorders, including traumatic brain injury. Evaluating the cognitive function and histopathologic sequelae after controlled cortical impact brain injury allowed Nakamura et al. (1999) to conclude that overexpression of Bcl-2 protein may play a protective role in neuropathologic sequelae following traumatic brain injury.

14. CONCLUDING REMARKS

To summarize as Bullock et al. (1999), laboratory studies have identified numerous potential therapeutic interventions that might have clinical application for the treatment of human traumatic brain injury. Many of these therapies have progressed into human clinical trials in severe traumatic brain injury. Numerous trials have been completed, and many others have been prematurely terminated or are currently in various phases of testing. The results of the completed Phase III trials have been generally disappointing, compared with the expectations produced by the successes of these interventions in animal laboratory studies.

REFERENCES

Bareyre FM, Saatman KE, Raghupathi R, Mcintosh TK. Positinjury treatment with magnesium chloride attenuates cortical damage after traumatic brain injury in rats. J Neurotrauma 2002; 17: 1029-1039

Beatti MS, Farooqui AA, Bresnahan JC. Review of current evidence for apoptosis after spinal cord injury. J Neurotrauma 2000; 17: 915-925.

Blaha GR, Raghupathi R, Saatman KE, McIntosh TK, Brain-derived neurotrophic factor administration after traumatic brain injury in the rat does not protect against behavioral or histological deficits. Neurocience 2000; 99: 483-493

Bullock MR, Lyeth BG, Muizelaar JP. Current status of neuroprotection trials for traumatic brain injury: lessons from animal models and clinical studies. Neurosurgery 1999, 45: 207-217

Carbonell WS, Grady MS. Evidence disputing th importance of excitatoxicity in hippocampal neuron death after experimental traumatic brain injury. Acad Sci 1999; 890: 287-98

Coulter DA, Rafiq A, Shumate M, Gong QZ, DeLorenzo RJ, Lyeth BG. Brain injury-induced enchanced limbic epileptogenesis: anatomical and physiological parallels to an animal model of temporal lobe epilepsy. Epileppsy Res 1996; 26: 81-91

D'Ambrosio R, Maris DO, Grady MS, Winn HR, Janigro D. Impaired K(+) homeotasis and altered lectrophysiological properties of post-traumatic hippocampal glia. J Neurosci 1999; 19: 8152-8162

De Mendonça A, Sebastião AM, Ribeiro JÁ. Adenosine: does it have a neuroprotective role after all? Brain Res Rev 2000; 33: 258-274

Dixon CE, Liu SJ, Jenkins LW, Bhattachargee M, Whitson JS, Yang K, Hayes RL. Time course of increased vulnerability of cholinergic neurotransmission following traumatic brain injury in the rat. Brain Res 1995; 70: 125-131

Faden AI. Neuroprotection and traumatic brain injury: theoretical option or realistic proposition. Curr Opin Neurol 2002; 15: 707-712

Glenn MB, Wroblewski B, Parziale J, Levenie L, Whyte J, Rosenthal M. Am J Phys Med Rehabil 1989; 68: 221-226

Gong QZ, Philips LL, Lyeth BG. Metabotropic glutamate receptor protein alterations aftaer TBI in rats J Neurotrauma 1999; 16: 893-902

Gorlach C, Hortobagyi T, Hortobagyi S, Benyo Z, Relton J, Whalley ET, Wahl M. Bradykinin B2, but not B1, receptor antagonism has a neuroprotective affect after brain injury. J Neurotrauma 2001; 18: 833-838

Grundy RI, Rabuffetti M, Beltramo M. Cannabionoids and neuroprotection. Mol Neurobiol 2001; 24: 29-51

Hicks RR, Smith DH, Gennarelli TA, McIntosh T. Kynurenate is neuroprotective following experimental brain injury in the rat. Brain Res 1994; 655: 91-96

Inamasu J, Ichikizaki K. Mild hypothermia in neurologic emergency: an update. Ann Emerg Med 2002; 40: 220-2230

Jackson RD, Corrigan JD, Arnett JA. Amitriptyline for agitation in head injury. Arch Phys Med Rehabil 1985; 66: 180-181

Kennedy R, Burnett DM, Greenwald BD. Use of antiepileptics in traumatic brain injury: a review for psychiatrists. Ann Clin Psychiatry 2001; 13: 163-171

Khaldi A, Chiueh CC, Bullock MR, Woodward JJ. The significance of nitric oxide production in the brain after injury. Ann N Y Acad Sci 2002; 962: 53-59

Kim BT, Rao VL, Sailor KA, Bowen KK, Dempsey RJ. Protective effects of glial cell line-derived neurotrophic factor on hippocampal neurons after traumatic brain injury in rats. J Neurosurg 2001; 95: 674-679

Kline AE, Massuci JL, Marion DW, Dixon CE. Attenuation of working memory and spatial acquisition deficits after a delayed and chronic treatment regimen in rats subjected to traumatic brain injury by controlled cortical impact. J Neurotrauma 2002; 19: 415-425

Leker RR, Shohami E. Cerebral ischemia and trauma-different etiologies yet similar mechanisms: neuroprotective opportunities. Brain Res Rev 2002; 39: 55-73

Leoni MJ, Chen XH, Mueller Al, Cheney J, McIntosh TK, Smith DH. NPS 1506 attenuates cognitive dysfunction and hippocampal neuron death following brain trauma in the rat. Exp Neurol 2000; 166: 442-449

Levine AM. Buspirone and agitation in head injury. Brain Inj 1988; 2 165-167

Liu A, Lyeth BG, Hamm RJ. Protective affect of galanin on behavioral deficits in experimental traumatic brain injury. J Neurotrauma 1994; 11: 73-82

Lyeth BG, Gong QZ, Dhillon HS, Prasad MR. Effects of muscarinic receptor antagonism on the phosphatidylinositol bisphosphate signal transduction pathway after experimental brain injury. Brain Res 1996, 742:63-70.

Lyeth BG, Gong QZ, Shields S, Muizelaar JP, Berman RF. Group I metabotropic glutamate antagonist reduces acute neuronal degeneration and behavioral deficits after traumatic brain injury in rats. Neurol 2001; 169:191-199.

Mechoulam R, Spatz M, Shohami E. Endocannabionoids and neuroprotection. Sci STKE 2002 23; RE5.

Nakamura M, Raghupathi R, Merry DE, Scherbel U, Saatman KE, Mcintosh TK. Overexpression of Bcl-2 is neuroprotective after experimental brain injury in transgenic mice. J Comp Neurol 1999, 412: 681-692

Philips LL, Lyeth BG, Hamm RJ, Reeves TM, Povlishock JT Glutamate antagonism during secondary deafferentation enhances cognition and axo-dentritic integrity after traumatic brain injury. Hippocampus 1998; 8: 390-401

Pike BR, Hamm RJ, Temple MD, Buck DL, Lyeth BG. Tetrahydroaminoacridine, a cholinesterase inhibitor, on cognitive performance following experimental brain injury. J Neurotrauma 1997; 14: 897-905

Pike BR, Hamm RJ, Temple MD, Buck DL, Lyeth BG. Tetrahydroaminoacridine, a cholinesterase inhibitor, on cognitive performance following experimental brain injury. J Neurotrauma 1997; 14: 897-905

Regner A, Alves LB, Chemale I, Costa MS, Friedman G, Achaval M, Leal L, Emanuelli T. Neurochemical characterization of traumatic brain injury in humans. Neurotrauma 2001; 18: 783-792

Ribeiro JA, Sebastião AM, de Mendonça A. Adenosine receptors in the nervous system: pathophysiological implications. Prog Neurobiol 2003; 68: 377-392

Robertson CL, Bell MJ, Kochanek PM, Adelson PD, Ruppel RA; Carcillo JÁ, Wisniewski SR, Mi Z, Janesko KL, Clark RS, Marion DW, Graham SH, Jackson EK. Increased adenosine in cerebrospinal fluid after severe traumatic brain injury in infants and children: association with severity of injury and excitotoxicity. Crit Care Med 2001 29: 2287-2293

Ros J, Pecinska N, Alessandri B, Landolt H, Fillenz M. Lactate reduces glutamate-induced neurotoxicity in rat cortex. J. Neurosci Res. 2001 ; 66 : 790-794

Salzman C. Treatment of the agitation of late-life psychosis and Alzheimer's disease. Neuroleptics Eur Psychiatry 2001; 16 Suppl 1: 25s-28s

Sandel ME, Olive DA, Rader MA. Chlorpromazine-induced psychosis after brain injury. Brain Inj 1993; 7: 77-83

Sanderson KL, Raghupathi R, Saatman KE, Martin D, Miller G, Mcintosh TK. Interleukin-1 receptor antagonist attenuates regional neuronal cell death and cognitive dysfunction after experimental brain injury. Blod Flow Metab 1999: 1118-1125

Sebastião AM, de Mendonça A Moreira, T, Ribeiro JA, Activation of synaptic NMDA receptors by action potential-dependent release of transmitter during hypoxia impairs recovery of synaptic transmission on reoxygenation. J Neurosci 2001: 21: 8564-8571.

Sick TJ, Perez-Pinzon MA, Feng ZZ. Impaired expression of long-term potentiation in hippocampal slices 4 and 48h following mild fluid-percussion brain injury in vivo. Brain Res 1998; 785: 287-292

Sinson G, Voddi M, Mcintosh TK. Nerve growth factor administration attenuates cognitive but not neurobehavioral motor dysfunction or hippocampal cell loss following fluid-percussion brain injury in rats. J. Neurochem 1995; 65: 2209-2216

Sun FY, Faden Ai. Neuroprotective effects of 619C89, a use-dependent sodium channel blocker, in rat traumatic brain injury. Res 1995 27; 673: 133-140

Vink R, Nimmo Aj, Cernak I. Na overview of new and novel pharmacotherapies for use in traumatic brain injury. Clin exp Pharmacol Physiol 2001; 28: 919-921

Zuccarelli LA. Altered cellular anatomy and aphysiology of acute brain injury and spinal cord injury. Crit care Nurs Clin North Am. 2000; 12: 403-411

Zwienenberg M, Gong QZ, Berman RF, Muizelaar Jp, Lyeth BG. The effect of groups II and III metabotropic glutamate receptor activation on neuronal injury in a rodent model of TBI. Neurosurgery 2001; 48: 1119-1126.

I. LISTE AND A. MARTÍNEZ-SERRANO

Centre for Molecular Biology "Severo Ochoa". Autonomous University of Madrid.Campus Cantoblanco, Madrid, Spain

28. STEM CELLS AND NERVOUS TISSUE ENGINEERING

Summary. Initial steps to the use of neural stem cells for brain repair dates from more than a decade. In this chapter, we will review the historical background of neural stem cell applications in experimental models of human neurodegenerative processes. A detailed profile of *in vitro* properties of different isolates of human neural stem cells will be presented. We will also concentrate on common aspects, as well as in some controversial issues related to the transplantation of forebrain human neural stem cells and precursor cells, mainly regarding the generation and transplantation of human dopaminergic neurons for treatment of Parkinson's disease. Future perspectives in this extremely active field of research will also be commented.

1. INTRODUCTION

The idea (and experimental demonstration) that neural stem or precursor cells could be used for brain repair dates back more than a decade, when rodent neural stem/precursor cell lines were obtained and successfully transplanted into the rodent brain. The purpose of those initial experiments was to study basic developmental mechanisms, and also to explore the cells' potential for cell and gene therapy (see Martínez-Serrano and Björklund, 1997 and Martínez-Serrano et al., 2001, as well as studies quoted therein, for a review). In those original, seminal studies, principles such as the capacity of stem and precursor cells to survive, integrate, migrate, repopulate the damaged brain, and also to serve as vehicles for gene transfer, were validated. Partially on this basis was built the whole new and exciting field of gene and cell therapy for central nervous system (CNS) diseases. Because the initial work done with rodent cells has been extensively reviewed, in this chapter we will focus primarily on neural stem/precursor cell isolates of *human* origin. We will provide a critical view of their properties, the multiple sources of neural human stem/precursor cells currently available, and the *in vivo* performance, after implantation, of these neural sources, focusing on their putative clinical application. Although numerous aspects of these new technologies are still not well understood, in this chapter we will try not only to identify problems or drawbacks, but also to provide or suggest ideas on how to make progress in so promising a field as cell therapy.

Rigorously evaluation of the potential of neural stem-cell grafts in models of human neurodegenerative disorders/diseases requires good, solid animal models for pre-clinical experimentation. Otherwise, interpretation of experimental transplantation results will be difficult. Rodent and primate models of Parkinson's disease offer such opportunity. In these animal models, some of the anatomical features and symptoms of the disease are recreated, and experimental therapies can be tested. Furthermore, some of these models have validated concepts that in the past were extrapolated from experimental animals to the human setting (Björlund and Lindvall, 2000; Dunnet and Björlund, 2000). Therefore, part of the discussion in the present chapter will use the scenario of current efforts aimed at generating dopaminergic neurons from human stem cells (hSC) for the treatment of Parkinson's disease.

Focusing now on *human* neural stem cells (hNSC), precursors, and progenitors, these different cell types have been successfully isolated from a wide variety of sources (not all of them covered in detail in this chapter), and best characterized when derived from i) hES cells, ii) fetal neural tissue, and iii) neonatal and adult nervous system, from biopsies or autopsies (Martínez-Serrano et al., 2001; Navarro et al., 2003). Moreover, numerous methods have been applied for the generation of cultures highly enriched in hNSCs, through either the selective expansion of the cells of interest with appropriate mitogens (EGF, bFGF and LIF), or their FACS-based purification/isolation (on the basis of surface antigen expression or fluorescent reporter genes expressed from cell-type specific gene promoters; see Martínez-Serrano et al., 2001; Navarro et al., 2003). Once a neural fate is induced in hES cells, or NSCs are enriched/isolated from neural tissue, these neural cells are normally expanded in culture. For culturing, two main strategies have been devised: *i)* the propagation of *cell strains*: mortal cells that can be serially passaged for a limited period of time, usually in the form of suspension aggregates (the so-called neurospheres) and which proliferate under the stimuli provided by epigenetic factors (mitogens added to the culture medium, like EGF, bFGF, LIF); or *ii)* the establishment of non-transformed, immortal *cell lines* of hNSCs, through the combined action of mitogens and a cell cycle regulatory gene, normally *avian myc (v-myc)*.

In both cases, removal of mitogens (and the optional addition of serum, retinoic acid or other factors) results in growth arrest and differentiation of most (but not all) of the cells (Martínez-Serrano et al., 2001; Navarro et al., 2003). For the sake of clarity, and for rigor with the cell biology nomenclature, these two types of cultures — neurospheres and established, immortal cells — will be referred to here as *cell strains* and *cell lines*, respectively.

2. *IN VITRO* PROPERTIES OF DIFFERENT ISOLATES OF HUMAN NEURAL STEM CELLS

In culture, different isolates of hNSCs behave differently, with regard to cell division rate, proliferation limit, heterogeneity, stability of their capacity to generate differentiated cell types and cell cycle parameters, telomere maintenance, and expression of senescent phenotype. To make the discussion of all these hNSC properties productive, we will focus primarily on fetal, forebrain-derived hNSCs

(Carpenter et al., 1999; Flax et al., 1998; Ostenfeld et al., 2000; Vescovi et al., 1999; Villa et al., 2000). However, it is also important to mention other studies where different properties have been noted among cultures derived from different brain regions of the fetal or adult human brain, indicating that some regional specification is preserved in culture (Keyoung et al., 2000; Ostenfeld et al., 2002; Roy et al., 1999; 2000; Skogh et al., 2001; Wright et al., 2002). For medical applications, a checklist of the requisites of putative therapeutic cell preparation shows that various cell properties merit careful, detailed consideration, as explained below.

2.1. Homogeneity vs. heterogeneity

Homogeneity is a desirable property of a therapeutic cell preparation. This concept proceeds from the fact that complete control over the cell-biological properties of a cell sample to be used for transplantation seems to be a gold standard for researchers. However, we would like to emphasize that, as a standing principle, heterogeneity should not be considered *a priori* a negative property. Human whole-organ transplantation and fetal-tissue grafting in Psrkinson's patients represent two good examples of naturally generated heterogeneous transplant sources having a clear therapeutic impact on the host. If cell preparations (cultured hNSCs) are considered from a scientific standpoint, however, spontaneous generation of heterogeneity in culture represents a far from optimal situation.

Compelling data have been reported in the literature illustrating that hNSC cell strains are heterogeneous at the molecular, ultra-structural, cellular, and functional (viability, proliferation and lineage marker gene expression before and after differentiation) levels (Navarro et al., 2003). Human neurospheres have been found to differ one from the other even within a single clone of cells, or in samples derived from embryos of the same gestation. Heterogeneity in the neurospheres could be caused, at least in part, by the absence of selective pressure in culture (cells can accumulate mutations freely, or drift easily when cultured in bulk, as in any cell strain not subjected to selection). Cell lines, in contrast, are clonal and homogeneous cell populations at all levels studied so far, and are usually derived after drug-selection of parental cultures. As mentioned above, this should not be understood as an advantage of cell lines over strains in absolute terms. However, clonality and homogeneity clearly facilitate molecular and cellular study and manipulation of the cells for grafting, and for understanding the properties and characterization of the final cell therapeutic product. Nevertheless, much further research is needed to be able to evaluate whether homogeneity is, or is not, a must.

2.2. Stability

Stability in cell division rate and other cell properties (such as capacity to generate differentiated cell types) is also of importance for any cell therapeutic product. For a clinical product to be derived from a given cell source, a large source of cells with known and predictable properties needs to be available at any given moment. Strains of hNSCs continuously evolve in culture and change their properties with time, for instance, in terms of proliferation rate, neuron generation, in particular that of specific neurotransmitter neuronal cell types (such as DA neurons), and also get their capacity to produce oligodendrocytes reduced with time in culture (see Navarro et al., 2003). Established cell lines are not completely free of these drawbacks, although they seem to be more stable in culture, at least in some respects. For instance, growth rate, clonogenic potential, and rate of neuron generation have been determined to be stable properties of cells of the hNS1 line (formerly called HNSC.100; Villa et al., 2000), for up to four years in continuous culture (Navarro et al., 2003). However, oligodendrocyte progenitor cell generation capacity was lost from this cell line over years in culture, as commonly happens for cell strains (Martinez-Serrano A., unpublished). In more-recent studies, we have also shown, in other human ventral mesencephalic established cell lines, that their capacity for dopaminergic neuron generation is reduced with passaging (data from the first one-two months in culture, the longest time studied so far, since their very recent successful derivation; Liste I., Villa A. and Martinez-Serrano A., unpublished).

2.3. Senescence

Senescence of hNSC cell strains, regardless of their origin (hES cells, fetal, neonatal or adult CNS tissue samples) has often been documented in the scientific literature (Navarro et al., 2003; Villa et al., 2000). Reported population-doubling limits range from 10 to 90 (in the presence of LIF), but, as expected for any somatic cell, the cultures finally come non-productive and senesce (Wright et al., 2002). Senescence was attributed to the loss of telomerase activity and concomitant telomere shortening (Ostenfeld et al., 2000). However, recent Q-FISH (quantitative fluorescent in situ hybridization) studies from our group (Navarro et al., 2003) indicate that telomere erosion is not occurring in hNSCs strains (obtained from different sources and laboratories). Moreover, no numerical or structural chromosomal abnormalities are present in hNSC strains. This indicates that their telomeres are functional, insofar as telomere functionality is currently understood. However, up to 60% of the cells in these strains were found to express a cellular senescent phenotype (SA-β-gal$^+$). The conclusion drawn from these studies should be that hNSC strains senesce in culture, not because of telomere shortening and a genetic catastrophe, but rather because of an unspecific cell culture shock phenomenon (or stress-induced premature senescence, SIPS), due simply to inappropriate culture conditions. Whilst a limitation in the proliferation capacity of hNSC strains is not per se a negative property (indeed, it might be a positive one), the presence of more than half the cells in a given culture showing clear signs of senescence, and thus altered biology, should be a concern for the clinical use of such cell strains (Maser and DePinho, 2002). Established cell lines, besides their long proliferation record, show no signs of senescence when analyzed with the same methods (Navarro et al., 2003).

3. WHAT IS THE REAL IDENTITY OF AN *IN VITRO* PROPAGATED, CULTURED NEURAL STEM CELLS?

Most people in the field of cell therapy would agree with the view that a valid operational definition of a neural stem cell describes a cell which i) has been instructed/specified as belonging to the nervous system, ii) self-renews, and iii) is able to generate multiple types of cell in that tissue (multi-potential for the generation of all the cell types present in the nervous system). Present knowledge on nervous system development is still too limited to provide real definition for the NSC (or the various NSCs that may exist in different developmental moments and nervous system regions). Most likely, and as recently suggested for rodent NSCs, the nature and properties of the in vitro propagated cells are the result of the culture conditions imposed by the investigator (an *in vitro* environment and a strong mitogen cocktail). However, the identity of cultured cells the relationships between them and their *in vivo* counterparts are far from clear.

The question of how the above-mentioned attributes of stem cells match with the properties of in-culture propagated cells remains unanswered. For instance, the capacity for self-renewal (a gold standard for a stem cell) is a property, assigned usually to cultures that proliferate only in response to mitogens. But how cell proliferation in culture relates to the homeostatic self-renewal of stem-cell population (preventing its self-extinction) has not yet been determined.

As another example, the capacity for terminal, functional differentiation of *in vivo* stem cells is not a matter of discussion. Such capacity should also be a requirement for cultured cells to be used in CNS repair. However, the literature abounds in reports describing the expression of only immature markers in NSC derivatives (such as β-III-tubulin or GFAP), together with the presence of undifferentiated cells. In fact, clear data demonstrating terminal differentiation are rarely sought, except in a few reports describing mature electrophysiological properties of neurons (see below). The fact is that in most cases, only morphology or expression of certain markers are taken as signs of differentiation, and the actual capacity for terminal differentiation is almost never demonstrated. In this situation, it is reasonable to ask whether or not many of the reported cultures of hNSCs are able to achieve terminal differentiation.

The two examples above are illustrative of a situation commonly found in the field, when properties of *in vivo* cells are assigned to *in vitro* cells, without having complete certainty that such extrapolations can be made. It often happens that first cells are isolated and then, instead of their actual properties being scrutinized, the assumed stem-cell-like properties are sought in that preparation, in order to be allowed to call that particular cell preparation "a neural stem or precursor-cell culture". The result is a large collection of published "NSC" cultures of widely differing properties, and which surely do not represent the same cell — that is, the true NSC. The alternative, and possibly more correct, prospective isolation of NSCs on the basis of their known *in vivo* properties, is a field much less explored up to now (Vescovi et al., 1999). Prospective isolation and study of non-cultured cells will surely help to define the nature of the NSC present *in vivo*, at least at the level of basic scientific studies.

In spite of present uncertainties about the precise nature of *in vitro* cultured NSCs, at some point, ideas must be practical, and work goal-oriented. When it comes to the therapeutic application of cells manipulated and cultured *in vitro*, one must use the tools (cells) available, and focus on the properties making the cells at hand easy to work with, particularly those relating to propagation in culture, even if these are not the perfect cells.

Prospective alternatives, such as the above-mentioned direct purification of cells with given properties (expression of surface antigens, or promoters driving reporter gene expression; Martínez-Serrano et al., 2001; Navarro et al., 2003) from nervous system samples, are of great interest from a basic research point of view. They may also become of clinical interest if sources containing large numbers of the desired cells could be identified.

4. HUMAN NEURAL STEM CELLS, CULTURED *IN VITRO* AND REGARDLESS OF THE PROPAGATION METHOD USED, ARE NEITHER TRANSFORMED NOR COMPLETELY NORMAL CELLS

One of the concern-raising aspects in stem-cell biotechnology relates to the potential changes and transformations the cells may undergo in culture, both in terms of time-dependent alterations of their intrinsic properties (such as identity and growth and differentiation potential), and in terms of the cells' acquiring or accumulating mutations that would render them transformed or malignant. The effort to find alternative methods for the direct isolation of the desired cells from bulk cell samples originates in part from this concern.

It is worth mentioning at this point that human cells (as opposed to mouse cells) are highly resistant to transformation and to becoming genetically unstable in culture, due to enhanced, stringent protection and repair mechanisms against DNA damage (Wright and Shay, 2000). However, the question then is: "Are any of the present propagation systems for hNSCs free of these mutation-and-transformation-related concerns?" In spite of the current interest in stem cells (at both scientific and social levels), the answer from the perspective of the molecular and cell biologist should be "No". Any cell in culture, particularly when it is not subjected to a selection procedure other than its self-capacity for proliferation, can freely accumulate mutations at a speed that results in an accelerated evolution in the culture dish. Considering the spontaneous mutation rate and the genome size in humans, calculated mutation rates would account for a mutation every third division, for any human cell in culture (Rocanova et al., 2003). Since continuous passaging of the cells self-selects for cells with enhanced growth properties, altered cells have all the opportunities they need to overgrow and take over the culture. The evolving properties of cultured hNSCs may be easily understood on such grounds.

These risks (undesired properties, transformation) should, however, be weighed against the experimental evidence opposing the acquisition of transformed properties. There are no reports in the literature describing spontaneous transformation of hNSCs in culture. [As far as our own experience goes, we have never seen any sign of transformation since we started to work with hNSCs in culture (in 1997), even for cell lines continuously cultured for up to four years; Navarro et al., 2003]. The lack of transformation of hNSCs in culture may be due to specific protective mechanisms as noted above, and also because human cell transformation requires the progressive acquisition of new properties and multiple gene mutations that bring them close to a transformed phenotype (Chang et al., 2003; Hanahan and Weinberg, 2000).

However, and in spite of the considerations against transformation presented above, we need to face the situation that hNSCs may become altered in culture. In addition, premature senescence or cell culture induced stress may be linked to events triggering transformation (Maser and DePenho, 2002).

The issue of transformation of *v-myc* cell lines merits deeper discussion, since it is commonly seen in the scientific literature that *v-myc* immortalized cells are often regarded as transformed cells. In theoretical terms, *v-myc* expression *might* represent a concern, since the actual mechanism by which this avian protein establishes hNSCs in culture is poorly understood. However, the sole expression of *v-myc* does not allow calling an established cell line a transformed cell line. *v-Myc*, the only cell-cycle-controlling gene so far reported to be efficient in establishing hNSC cell lines, is a rather peculiar gene, and its functions/activities are not well understood. *Avian-myc* (*v-myc*, p110 $^{gag\text{-}myc}$) is a fusion protein that combines the *gag* portion of an avian retroviral genome with the coding exons 2 and 3 of the chicken *c-myc* gene (mRNA version). There have been no studies carried out in hNSC lines aimed at elucidating the putative roles of the gag and myc portions. Therefore, when discussing *v-myc* established cell lines of hNSCs, we have to realize that we are dealing with a new entity in biology. Properties of hNSCs established by *v-myc* expression have been reported, reviewed, and discussed thoroughly elsewhere (Martínez-Serrano et al., 2001; Navarro et al., 2003). This evidence can be summarized as i) *v-myc* lines share most of their properties with cell strains, except for an increased stability, homogeneity, and a lack of senescence; and ii) there is solid, published evidence demonstrating that *v-myc* established lines are not transformed. It is also worth mentioning that overexpression of *c-myc* is unable to establish cell lines of hNSCs (Villa et al. 2000) and is not transforming *in vivo* (Fults et al., 2002).

From a consideration of the above arguments (and although established cell lines are not transformed), it is fair to state that these *v-myc* cell lines are not free of the risk of accumulating mutations in culture, as cell strains or any other cultured cell would.

In the near future, all of these concerns about changes occurring in cultured cells will need to be weighed against the (promised) therapeutic benefit that an implant of suitable cells may confer on a patient. Sometimes, molecular and cell biologists exaggerate the theoretical drawbacks of cultured cells, as a means to stimulate the further discovery of better cells. However, at present, many medical procedures — surgery, for instance — incur a high risk level (1-3% lethality), but are, by no means, rejected either by the medical community or by society as a whole. Therefore, and to conclude this section, future discussions should aim to clarify whether or not a cultured cell is a product suitable for human therapy. Regulatory agencies must play a fundamental role in this matter. An important point in these discussions is the long record of laboratory animals that have so far been transplanted with cultured NSCs of rodent or human origin, and in which tumors have never been described.

5. TRANSPLANTATION OF FOREBRAIN HUMAN NEURAL STEM CELLS AND PRECURSOR CELLS

The field of transplantation of forebrain-derived hNSCs and precursor cells has been extensively and recently reviewed (Martínez-Serrano et al., 2001; Navarro et al., 2003). Therefore, we would like to concentrate here on the common aspects found in some recent studies, and on the controversial points they collectively raise. Three main discussion aspects are identified, which seem highly relevant from the pre-clinical standpoint.

First, not all the grafted hNSCs (both strains and cell lines) in a given transplant terminally differentiate. This observation has been made for the whole range of possible host developmental ages, from the fetal to the adult mammalian brain, mainly in rodents, but also in primates (Blackshaw and Cepko, 2002; Martínez-Serrano et al., 2001; Navarro et al., 2003). From a basic scientific point of view, the presence of pools of undifferentiated cells in vivo may be understood as if hNSCs were be able to self-control their population in order to avoid self-extinction (true self-renewal, possibly, see comments above). This may be seen as a manifestation of an (putative) underlying homeostatic principle inherent in their nature as stem cells. Although direct experimental evidence for *in vitro* or *in vivo* homeostasis of grafted hNSCs hardly exists, these observations should foster further research in this line. From a pre-clinical standpoint, however, the incapacity for terminal differentiation may be seen as a hurdle for cell therapy development. Those *in vivo* quiescent cells might become reactivated if correct stimuli are provided (for instance by ongoing inflammatory processes commonly seen in many neurological diseases). Perhaps those cells unable to differentiate and becoming quiescent would be excellent vehicles for gene therapy: answer should await clear experimental demonstration indicating so, and a complete scrutiny of their in vivo properties.

Second, long-range migration is commonly seen in transplanted hNSCs. Whereas in some of the older studies (reviewed in Martínez-Serrano et al., 2001; Navarro et al., 2003), labeled (BrdU or ^{3}H-thymidine) grafted cells were not usually found far from the implantation site, when genuine human markers are used to detect the cells, investigators have found human cells in the grafted rodent brain many millimeters away from the implantation site (in the frontal cortex and midbrain from intrastriatal implants, as an example). This feature applies to the fetal or neonatal brain, which can be highly permissive for cell migration, but is also observed in adult hosts.

Third, an experimental twist partially aimed at resolving the two points above has been developed, based on the pre-differentiation of the cells prior to transplantation (see Wu et al., 2002). Pre-differentiation has the consequence, in some settings, of limiting grafted cell differentiation and migration, but may offer ways to enhance the differentiation of the grafted cells, particularly at discrete, selected locations. This technical trick has recently been used in a report that *in vitro* pre-differentiated ("primed") cells preferentially differentiate into cholinergic neurons at cholinergic locations (Wu et al., 2002). Several other types of phenotypically defined neurons were also present accompanying cholinergic cells. It is not clear at this point whether cholinergic neuron generation is the result of instructive or selective (survival) mechanisms (trophic factors for selected neurons provided by the appropriate regions *in vivo*). In any case, this strategy may provide a way to implant specific neuronal types (in this case, cholinergic) at given locations.

To conclude this section, two questions may be worth raising: first, "What can be learnt from transplants of forebrain-derived hNSCs or precursor cells into the intact developing or adult brain, that could be relevant for the treatment of neurodegenerative diseases?". Seemingly, there is no immediate application of the results from these studies for the development of cell therapies, but the hope is that through the understanding of developmental and basic aspects of the performance of the grafted cells, the field should be able progressively to shape up more application–oriented procedures for cell replacement.

Secondly, "What about the function of grafted cells?". Graft-induced effects on the host will be more fully covered in the next section. Here we would like to point out the issue of the functional development of the grafted cells at the cellular and physiological levels. Two recent reports have addressed this aspect, using rodent cells in both cases (Navarro et al., 2003). Thus, an immortalized precursor cell line (RN33B, ^{ts}Tag perpetuated) or neurons derived from mouse ES cells were implanted into the neonatal or the adult lesioned brain, respectively. Neurons showing the typical morphology and correct electrophysiological properties were found *in vivo*.

6. GENERATION AND TRANSPLANTATION OF HUMAN DOPAMINERGIC NEURONS

The generation of human dopaminergic neurons for the treatment of Parkinson's disease is at a central position in the goal list of stem-cell biotechnologists (Arenas, 2002; Isacson, 2002; Brundin and Hagell, 2001; also, Kokaia, Lindvall and Martínez-Serrano, unpublished results). Cell replacement in Parkinson's disease is conceptually simple, as long as the implantation of a regulated source of dopamine (DA) in the striatum (caudate/putamen) is achieved (Björklund and Lindvall, 2000; Dunnet and Björklund, 2000). Currently, available sources of human DA neurons for transplantation are logically, very limited, and include fresh human ventral mesencephalic (vm) tissue, or DA neurons generated from cultured cells (which promise to become at some point an unlimited, predictable and controlled source of cells for therapies), such as hES cells, forebrain hNSCs (strains or lines), and also vm DA neuron progenitors (Arenas, 2002; Björklund and Lindvall, 2000; Brundin and Hagell, 2001; Dunnet and Björklund, 2000; Isacson, 2002; also, Kokaia, Lindvall and Martínez-Serrano, unpublished results).

Although all these tissue/cell sources have the positive and common characteristic of (putatively) being able to serve as a source of DA neurons which might exert a therapeutic effect, they are not free of potential drawbacks, which can be briefly described as follows.

First, fresh human vm tissue is possibly the best but the most limited source of DA neurons for transplantation, and the outcome of these transplants depends very much on practical issues deriving from tissue collection and implantation procedures (Isacson, 2002; Brundin and Hagell, 2001; also, Kokaia, Lindvall and Martínez-Serrano, unpublished results).

Secondly, mouse ES cells have not been transplanted as such, or after manipulations aimed at their genetic modification (for instance for the expression of Nurr1, as discovered using mouse NSCs (Wagner et al., 1999), or after an instruction/patterning/pre-differentiation scheme in culture (Kim et al 2002). The variation in *in vivo* results obtained using mouse ES cells seem to depend on ES line used, on the laboratory, and on the pre-differentiation or inductive procedures applied to the cells in culture. In some cases, good integration, neuronal maturation, and functional recovery have been reported (Kokaia, Lindvall and Martínez-Serrano, unpublished results). However, in one case, teratoma formation was reported in up to 20% of the studied animals (implanted with non pre-differentiated ES cells). A common feature of all studies (and a consistent one of ES differentiation studies *in vitro*), has been the finding of mixed neurotransmitter neuronal populations in the analyzed brains (usually containing TH, serotonin and GABA neurons). Primate ES cells induced to become dopaminergic by SDIA (stromal cell-derived dopaminergic inducing activity) also result in the generation of a mixed population of dopaminergic, noradrenergic, GABA-ergic, cholinergic, and serotoninergic neurons.

Thirdly, human forebrain NSCs have also been converted into DA neurons by the addition of inductive cocktails, with varying efficacy, and also found to be contaminated with other neuronal cell types (Kokaia, Lindvall and Martínez-Serrano, unpublished results). Therefore, a common feature of all systems reported so far where DA neurons have been generated *in vitro* is the presence of numerous other neurotransmitter phenotypes co-existing with the desired DA neurons. Whether this is a positive or negative aspect of these technologies remains to be clarified, but, in any case, it should be considered in order to ensure the correct interpretation of functional effects in the recipient experimental brain.

Lastly, another aspect where research should be focused relates to the limited survival of stem-cell-derived human DA neurons *in vivo* (Areans, 2002; Brundin and Hagell, 2001; also, Kokaia, Lindvall and Martínez-Serrano, unpublished results). An alternative to ES and NSCs for DA neuron generation is mesencephalic, pre-specified neuronal progenitors. Human VM progenitors grow in culture more slowly than forebrain human precursors do, and for limited periods of time (Ostenfeld et al., 2002; Kokaia, Lindvall and Martínez-Serrano, unpublished results). In the best case, propagation for up to eleven months has been achieved using low-oxygen culture conditions, although total yield and TH^+ neuron generation rate for the longest-expanded cells was not reported in detail. In general, reports have concentrated on short-time experiments (1-3 weeks expansion), in which acceptable rates of TH^+ neuron generation were achieved (10-60% of the total cells). In contrast, longer term expanded progenitors yield much lower (<1%), or almost no TH^+ neuron generation (Osterfeld et al., 2002; Storch et al., 2001). Common features in these reports are the limited proliferation capacity of the human vm progenitors, the diminishing capacity for human DA neuron generation with time in culture, and poor survival *in vivo*, when this was tested (ranging from occasional TH^+ cells to 0.2% of the grafted cells; Kokaia, Lindvall and Martínez-Serrano, unpublished results).

Recalling the arguments presented above for forebrain transplants, it will also be essential to carry out detailed migration and differentiation studies of the transplanted human DA neurons.

7. CONCLUDING REMARKS AND FUTURE PERSPECTIVES

The field of stem-cell biology is currently in vigorous growth, generating an increasing amount of evidence justifying the implementation of these new technologies in real therapies for neurodegenerative disorders. Some highly positive achievements have been secured in the past and also recently using rodent stem/precursor cells, but little is known about the functional impact of human NSC transplants in models of neurodegeneration. A number of issues need to be further studied, regarding the survival, integration, migration and differentiation of human stem/precursor cells *in vivo*. Those studies, conducted first in laboratory animals (rodents, and then primates), will hopefully guide this research field to its ultimate goal: cell replacement of specific neuronal types in the brain. For this, cells will need to be endowed with functional properties, and should exert functional effects in the host, resulting in behavioral recovery, symptom amelioration, and, optimally, a cure.

At a first sight, this goal may look distant, but the tremendous research effort currently under way will hopefully bring it close to reality within a few years.

Finally, we would like to point out that these research efforts should be aimed at the development of cell therapeutic products of world-wide application, affordable for all public-health systems, in order to ensure that the emerging therapies will be available to every single patient needing them.

REFERENCES

Arenas E (2002) Stem cells in the treatment of Parkinson's Disease. Brain Res. Bull., 57, 795-808.
Björklund A and Lindvall O (2000) Cell replacement therapies for central nervous system disorders. Nat. Neuroscience, 3, 537-544.
Blackshaw S and Cepko CL (2002) Stem cells that know their place. Nat Neurosci 5, 1251-1252.
Brundin P and Hagell P (2001) The neurobiology of cell transplantation in Parkinson's disease. Clinical Neuroscience Research, 1, 507-520.
Carpenter MK, Cui X, Hu Z, Jackson J, Sherman S, Seiger A and Wahlberg L (1999) In vitro expansion of a multipotent population of human neural progenitor cells. Exp. Neurol., 265-278.
Chang S, Khoo CM, Naylor ML, Maser RS, DePinho RA (2003) Telomere-based crisis: functional differences between telomerase activation and ALT in tumor progression. Genes Dev. 17, 88-100.
Dunnett SB, Björklund A (2000) Prospects for new restorative and neuroprotective treatments in Parkinson's disease. Nature. 399(6738 Suppl):A32-9.
Flax JD, Aurora S, Yang C, Simonin C, Wills AM, Billinghurst LL, Jendoubi MJ, Sidman RL, Wolfe JH, Kim SU and Snyder EY (1998) Engraftable human neural stem cells respond to developmental cues, replace neurons, and express foreign genes. Nat. Biotech., 16, 1033-1039.
Fults D, Pedone C, Dai C and Holland EC (2002) MYC expression promotes the proliferation of neural progenitor cells in culture and in vivo. Neoplasia, 4, 32-39.

Hanahan D and Weinberg RA (2000) The hallmarks of cancer. Cell 100, 57-70.

Isacson O (2002) Models of repair mechanisms for future treatment modalities of Parkinson's disease. Brain Res. Bull. 57, 839-846.

Keyoung HM, Benraiss A, Roy NS, Louissant A, Wang S, Rashbaum WK, Okano H and Goldman SA (2000) The nestin and mushashi promoters identify and select two distinct pools of neural stem cells from fetal human brain. Soc. Neurosci. Abstr, 312.12.

Kim JH, Auerbach JM, Rodriguez-Gomez JA, Velasco I, Gavin D, Lumelsky N, Lee SH, Nguyen J, Sanchez-Pernaute R, Bankiewicz K, McKay R (2002) Dopamine neurons derived from embryonic stem cells function in an animal model of Parkinson's disease. Nature. 418,:50-56.

Martínez-Serrano A and Björklund A (1997) Immortalized neural progenitor cells for CNS gene transfer and repair. Trends Neurosci., 20, 530-538.

Martínez-Serrano A, Rubio FJ, Navarro B, Bueno C and Villa A (2001) Human Neural Stem and Progenitor Cells: In vitro and in vivo properties, and their potential for gene therapy and cell replacement in the CNS. Curr. Gene Ther., 1, 279-299.

Maser RS, DePinho RA (2002) Connecting chromosomes, crisis, and cancer. Science 297, 565-569.

Navarro B, Villa A, Liste I, Bueno C and Martinez-Serrano A (2003) Isolation, proliferation, survival and differentiation of human neural stem/precursor cells in vitro and in vivo. In Neural Stem Cells, Kluwer Eds. Jane Bottenstein, Ed. Kluwer.

Ostenfeld T, Caldwell MA, Prowse KR, Linskens MH, Jauniaux E and Svendsen CN (2000) Human neural precursor cells express low levels of telomerase in vitro and show diminishing cell proliferation with extensive axonal outgrowth following transplantation. Exp. Neurol., 164, 215-226.

Ostenfeld T, Joly E , Tai YT , Peters A , Caldwell M, Jauniaux E and Svendsen CN (2002) Regional specification of rodent and human neurospheres. Dev. Brain Res., 134, 43–55.

Roccanova L, Ramphal P, Rappa P 3rd (2003) Mutation in embryonic stem cells. Science 292, 438-440.

Roy NS, Wang S, Harrison-Restelli C, Benraiss A, Fraser RA, Gravel M, Braun PE, Goldman SA (1999) Identification, isolation, and promoter-defined separation of mitotic oligodendrocyte progenitor cells from the adult human subcortical white matter. J Neurosci 19, 9986-9995.

Roy NS, Wang S, Jiang L, Kang J, Benraiss A, Harrison-Restelli C, Fraser RAR, Couldwell WT, Kawaguchi A, Okano H, Nedergaard M and Goldman SA (2000) In vitro neurogenesis by progenitor cells isolated from the adult human hippocampus. Nat. Medicine, 6, 271-277.

Skogh C, Eriksson C, Kokaia M et al (2001) Generation of regionally specified neurons in expanded glial cultures derived from the mouse and human lateral ganglionic eminence. Mol. Cell. Neurosci., 17, 811-820.

Storch A, Paul G, Csete M, Boehm BO, Carvey PM, Kupsch A, Schwarz J (2001) Long-term proliferation and dopaminergic differentiation of human mesencephalic neural precursor cells. Exp Neurol. 170,:317-25.

Vescovi, A. L., Parati, E. A., Gritti, A., Poulin, P., Ferrario, M., Wanke, E., Frölischthal-Schoeller, P., Cova, L., Arcellana-Panlilio, M., Colombo, A., and Galli, R. (1999) Isolation and cloning of multipotential stem cells from the embryonic human CNS and establishment of transplantable human neural stem cell lines by epigenetic stimulation. Exp. Neurol. 156: 7–83.

Villa A, Snyder EY, Vescovi, A and Martínez-Serrano A (2000) Establishment and properties of a growth factor dependent, perpetual neural stem cell line from the human CNS. Exp. Neurol., 161, 67-84.

Wagner J, Akerud P, Castro DS, Holm PC, Canals JM, Snyder EY, Perlmann T, Arenas E (1999) Induction of a midbrain dopaminergic phenotype in Nurr1-overexpressing neural stem cells by type 1 astrocytes. Nat Biotechnol. 17, 653-9.

Wright LS, Li J, Klein S, Wallace K, Johnson JA, Svendsen CN (2002) Increased neurogenesis and GFAP expression in human neurospheres following LIF treatment: A micro-array study. Program No. 329.3. 2002 Abstract Viewer/Itinerary Planner. Washington, DC: Society for Neuroscience, 2002. Online.

Wright WE and Shay JW (2000) Telomere dynamics in cancer progression and prevention: fundamental differences in human and mouse telomere biology. Nat. Med. 6, 849-851

Wu P, Tarasenko YI, Gu Y, Huamg LYM, Coggeshall RE and Yu Y (2002) Region-specific generation of cholinergic neurons from fetal human neural stem cells grafted in adult rat. Nat. Neurosci., 5, 1271-1278.

X. NAVARRO AND E. VERDÚ

Group of Neuroplasticity and Regeneration, Institute of Neuroscience and Department of Cell Biology, Physiology and Immunology, University Autònoma de Barcelona, Bellaterra, Spain

29. CELL TRANSPLANTS AND ARTIFICIAL GUIDES FOR NERVE REPAIR

Summary. Recovery from peripheral nerve injury depends on a variety of factors, both intrinsic and extrinsic to neurons. Wallerian degeneration, neuronal reaction, axonal growth and target reinnervation are sequential processes that occur after a peripheral nerve injury. The capability of severed axons to regenerate and recover functional connections is dependent on the site and type of lesion, and the distance over which axons must re-grow to span the injury. The management of most nerve injuries is presently limited to surgical repair, by direct suture of the stumps or by interposition of a nerve graft to provide axons growing from the proximal stump with a substrate supporting regeneration. The development of artificial nerve grafts, composed of a guide filled with molecular and cellular elements that promote axonal regeneration is a strategy to facilitate regeneration and to solve the secondary problems of autograft and allograft repair.

1. INTRODUCTION

The peripheral nervous system is constituted by groups of neurons whose cell bodies are located in the spinal cord or within spinal ganglia, their intrinsic central connections, and their axons, which extend through peripheral nerves to reach target organs. Functionally, the peripheral nerves contain several types of nerve fibers. Afferent sensory fibers can be unmyelinated or myelinated (the latter ranging from 2 to 20 µm in diameter) and terminate at the periphery either as free endings or in a variety of specialized sensory receptors in the skin and deep tissues. Efferent motor fibers originate from motoneurons in the anterior horn of the spinal cord and end in neuromuscular junctions in skeletal muscles; the majority can be divided into two types: alpha-motor fibers, with diameters between 10 and 17 µm, which synapse with skeletal muscle fibers, and gamma-motor fibers, 3 to 8 µm in diameter, innervating muscle spindles. Efferent autonomic fibers in somatic peripheral nerves are constituted by postganglionic sympathetic fibers, generally unmyelinated, which innervate smooth muscle and glandular structures. The number and type of nerve fibers vary markedly depending on the nerve and the anatomical location. In somatic nerves, such as those of the limbs, the number of unmyelinated fibers is approximately twice that of myelinated ones. Most peripheral nerves are mixed, providing motor, sensory, and autonomic innervation to the corresponding projection territory.

Nerve fibers, both afferent and efferent, are grouped in fascicles surrounded by connective tissue in the peripheral nerve. The fascicular architecture of a peripheral nerve changes throughout its length, with a higher number of smaller fascicles in distal than in proximal segments. In addition to bundles of nerve fibers, the peripheral nerves are composed of three supportive sheaths: epineurium, perineurium and endoneurium. The epineurium is the outermost layer, continuous with the mesoneurium and the connective sheaths of surrounding tissues. It is a loose connective tissue and carries the blood vessels that supply the nerve. The perineurium is the sheath surrounding each fascicle in the nerve. It consists of inner layers of flat perineurial cells and an outer layer of collagen fibers organized in longitudinal, circumferential, and oblique bundles. The perineurium contains a blood vessel network connected with epineurial vessels and with endoneurial capillaries. The perineurium is the main contributor to the tensile strength of the nerve, acts as a diffusion barrier and maintains the endoneurial fluid pressure. The endoneurium is composed of fibroblasts, collagen and reticular fibers, and extracellular matrix, occupying the space between nerve fibers within the fascicles. The endoneurial collagen fibrils are packed around each nerve fiber to form the walls of the endoneurial tubules. Inside these tubules, axons are accompanied by Schwann cells, which either myelinate or just surround the axons. The basal laminae produced by Schwann cells are arranged in continuous tubes around the axon/Schwann-cell units.

Following injuries to peripheral nerves, the motor, sensory, and autonomic functions conveyed by the involved nerves will be partially or totally lost in the denervated segments of the body, due to the interruption of axon continuity, degeneration of nerve fibers distal to the lesion and eventual death of axotomized neurons. These deficits can be compensated by reinnervation of denervated targets following two compensatory mechanisms: regeneration of injured axons and collateral branching of undamaged axons in the vicinity. However, clinical and experimental evidence usually shows that these mechanisms do not enable for a satisfactory functional recovery, especially after severe injuries.

2. NERVE INJURIES

2.1. Classification of nerve injuries

Axonal damage by crush, transection, ischemia, or inflammation leads to interruption of axonal continuity with ensuing degeneration of nerve fibers distal to the lesion site. The nerve trunk injured, and the type and severity of the lesion, determine the need for surgical repair and the technique to apply. The prognosis for functional return is correlated with the degree of intraneural disruption. This is why the most popular classifications of mechanical nerve lesions (those of Seddon and of Sunderland) are based upon the morphology of the lesion and the nerve sheaths damaged. The most detailed classification, that of Sunderland, differentiates five degrees of injury. The *first degree* (*neurapraxia* of Seddon) corresponds to focal blocking of impulse conduction, usually due to compression. Large nerve fibers are more readily damaged than small ones. Histologically there is damage of the myelin sheaths, but axonal continuity is maintained and, therefore, distal nerve fibers do not degenerate. This lesion reverts within a short time, once the cause is eliminated. In *second degree* injuries (*axonotmesis* of Seddon) the axons are interrupted, but the connective sheaths remain intact. Nerve fibers distal to the lesion degenerate, and recovery requires regeneration of the severed axons. Recovery is usually good because the maintenance of endoneurial tubules and basal lamina of Schwann cells gives support to regenerative axons and allows the appropriate reinnervation of target organs by the original neurons. The *third degree* of nerve lesion implies disruption of axons and of endoneurial tubules, although not of the perineurium. This results in distal degeneration and in internal disorganization of the nerve fascicles. The ensuing intrafascicular fibrosis constitutes an obstacle for regeneration, and the misalignment of endoneurial tubules leads to deficiencies in tissue reinnervation. In *fourth degree* injury the perineurium is also damaged, only the epineurium remaining intact — resulting in disorganization of the fascicular architecture, intraneural scar formation, and loss of directional guidance for regenerating axons. Formation of a neuroma as the next step is common. Finally, the *fifth degree* of lesion (*neurotmesis* of Seddon) implies the complete transection of the nerve, with degeneration and destructuration of the distal segment. In all cases there is separation of the nerve stumps due to elastic forces; the length of the gap created depends upon the size of the nerve and the adherences to surrounding tissue. When the damage affects a segment of the peripheral nerve, as in severe trauma or in *resection* lesion, the gap between the proximal and the remaining distal stump is correspondingly longer.

From the third degree of injury, the prognosis for recovery is usually poor, and surgical repair indicated. The structural and fascicular alterations impede an appropriate spontaneous regeneration and reconstitution of the nerve. Therefore, treatment will attempt to re-establish the gross continuity of the nerve trunk, preventing the formation of scar tissue at the lesion site, and providing the injured axons with a proper terrain for regrowth, i.e. the distal degenerating nerve.

2.2. Mechanisms of nerve injury

Nerve injuries can result from a variety of causes, including mechanical lesions, ischemia, immunologic attack, metabolic disorders, toxic agents, or radiations. Injuries due to metabolic or general diseases, such as diabetes mellitus, uremia or intoxications, are generally more widespread (polyneuropathies) and their prognosis and treatment depend mainly on the underlying cause. From the point of view of experimental axonal regeneration, the paradigmatic types of injury are mechanical.

2.2.1. Compression

Exogenous pressure affecting a peripheral nerve causes decreased blood flow and structural deformation of nerve fibers. Thinning of myelin sheaths and focal demyelination are the main consequences, leading to conduction block. The large myelinated fibers are more susceptible than the small or unmyelinated ones, so loss of muscle strength and of touch sensation are predominant symptoms. Long periods of intense compression can progress to produce interruption of the axonal membrane and endoneurial and perineurial sheaths. Chronic mild compression is commonly seen in entrapment neuropathies, such as compression of the median nerve at the carpal tunnel or of the ulnar nerve at the elbow.

2.2.2. Section

Laceration with sharp instruments results in either total or partial section of the nerve. Due to elastic recoil, proximal and distal nerve stumps retract, creating a gap. Surgical repair is needed to reconstitute the continuity of nerve sheaths and make regeneration feasible. The blood flow is almost unimpaired after laceration, although some degree of crush damage usually appears in segments close to the section.

2.2.3. Stretch

The structure and elastic components of peripheral nerves allow them to adapt to a certain degree of stretch, such as changes in joint angle during limb movement. When longitudinal traction is applied, the undulations of the nerve disappear before the nerve suffers load, which is supported mainly by the perineurial connective layers. Under traction, the epineurium is the first structure to rupture, then the perineurium is stretched and the axons may start to break. Axonal injury due to stretch is not limited to one segment, but may cause disruption of axons at different sites of the nerve. In addition, stretch may also induce nerve damage due to compression within the fascicles, and to vessel occlusion and ischemia.

2.2.4. Ischemia

The peripheral nerve is quite resistant to ischemia, due to the extensive anastomotic epineurial and perineurial network of blood vessels. Chronic ischemia, resulting from vessel occlusion, arteriolar angiopathy, or capillary disease, can cause nerve injury ranging from first to fourth degree in Sunderland's classification. During an ischemic period, first there is a decrease of the membrane resting potential, and the appearance of paresthesiae, followed by a decrease in action potential amplitude and finally conduction block, with anesthesia and paralysis. Unmyelinated fibers are more resistant than myelinated ones to ischemia, due to the lower metabolic need of the former. Upon a long-lasting period of ischemia, damage to Schwann cells, focal demyelination and axonal degeneration may occur.

3. CELLULAR AND MOLECULAR BASIS OF NERVE REGENERATION

After injuries that cause rupture of peripheral nerve fibers, axons and myelin sheaths distal to the lesion site are degraded by Wallerian degeneration. The degenerative end products are eliminated by the co-operative action of Schwann cells and infiltrating macrophages (Fig. 1). The first signs of Wallerian degeneration are observed within 24 hours after nerve injury, and they last about two weeks. By 48 hours, all axons show complete disruption of their internal structure with disintegration of the cytoskeleton. Denervated Schwann cells phagocytose myelin debris to some extent; however, the main pathway for phagocyting of myelin and axonal debris is the recruitment of hematogenous macrophages. From 2-3 days after injury there is a significant infiltration of macrophages into the degenerating nerve, attracted by chemotactic and inflammatory cytokines (such as leukemia inhibitory factor (LIF), interleukin (IL)-1α and IL-1β), secreted by reactive Schwann cells.

Schwann cells of the distal nerve segment are stimulated, by the loss of axonal contact and by cytokines secreted by macrophages, to proliferate after injury. The de-differentiated cells line up within the endoneurial tubes to form the bands of Büngner, which later provide support for regenerating axons. The highest rate of Schwann cell multiplication is reached by 3 days after lesion, and then continues with decreasing frequency for 2-3 weeks, reaching a more than three-fold increase in number. There is also an increase in the collagen content of the distal nerve segment resulting in reduction of the lumen size of endoneurial tubules.

If Wallerian degeneration serves to create a microenvironment distal to the injury that is favorable for the axonal regrowth of surviving neurons, retrograde reaction and chromatolysis represent the metabolic changes necessary for regeneration. The most consistent morphological changes in the neuronal body after axotomy are dissolution of the Nissl bodies, nuclear eccentricity, nucleolar enlargement, cell swelling, and retraction of dendrites. The inflow of calcium ions and the suppression of retrograde transport of neurotrophic factors due to the lesion induce the expression of immediate early genes (IEG) in the neuron soma. The axotomized neurons shift from a "neurotransmitter" state to a "regenerative" state, with a decrease in the synthesis of neurotransmission-related products and an increase in that of growth-associated proteins and structural components of cytoskeleton and membrane. Chromatolysis may sometimes lead to death of the axotomized neurons, a phenomenon that eliminates any possibility of regeneration. Neuronal survival depends on several factors, including age, severity of injury and proximity of injury to the cell body. Neurons in the adult are less likely to die than are immature neurons, whereas lesions near the cell bodies induce higher proportion of neuronal death than distal lesions. Proximal to the lesion, growth cones emerge from the severed axons and elongate if they find a favorable terrain in their microenvironment. In the absence of a guiding structure, such as the distal nerve stump, regenerating axons take a tortuous course and form a neuroma — that is an enlargement composed of immature nerve fibers and connective tissue. If the regenerating axons gain the distal nerve, then elongate within the distal endoneurial tubes, in association with the Schwann cell membrane and the basal lamina. The materials for axonal growth are provided by the cell body via axonal transport. The rate of axonal regeneration is initially very slow, and reaches a constant value by 3-4 days after injury: about 3-4 mm/day after a crush injury, and slower, 2-3 mm/day after nerve transection and repair.

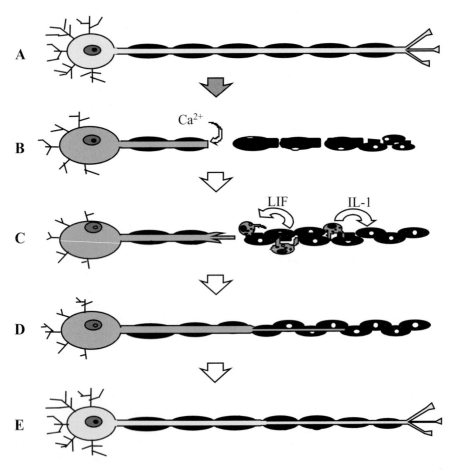

Figure 1. Degeneration and regeneration after peripheral nerve injury. A) Normal nerve fiber. B) Transection of the fiber results in distal fragmentation of axon and myelin sheaths. C) Fine sprouts emerge from the proximal axonal end. Schwann cells proliferate, and macrophages invade the distal segment. Macrophages and Schwann cells phagocytose degrading materials. D) Schwann cells in the distal segment line up in bands of Büngner. Advancing axons are embedded in Schwann cell cytoplasm, attracted by a gradient of neurotrophic factors. E) Axonal reconnection with end organs and maturation of the nerve fiber

The factors that stimulate and control axonal regeneration originate from multiple sources, but the most important influences derive from the local environment of the lesion. Axonal elongation requires a suitable substrate of trophic and tropic factors, provided by reactive Schwann cells and the extracellular matrix within the degenerated nerve stump. Schwann cells play a key role in nerve regeneration because they constitute a favorable substrate for axonal growth, and synthesize a number of trophic factors to support neuronal survival and regeneration.

Several sprouts emerge from each parent axon and may advance in the distal nerve; thereby the total number of axons in the distal segment may exceed the number of axons in the proximal nerve, even for a long time. When axons reach synaptic loci in peripheral tissue, usually all but one axon sprout are withdrawn gradually. Nerve fibers that regenerate through erroneous distal pathways to targets that can not reinnervate, such as motor axons to the skin, are preferentially eliminated.

The functional significance of regeneration is to replace the distal nerve segment lost during degeneration, allowing reinnervation of target organs and restitution of their corresponding functions. However, the regenerative process usually can not reconstitute a normal nerve structure or enable a normal function, especially when the lesion is severe. After nerve injury and repair, the diameter of regenerated axons, their conduction velocity, and excitability remain below normal levels for a long time; consequently, recovery of reinnervated organs is incomplete and often inappropriate. The limitation of nerve regeneration is more marked when the lesion creates loss of continuity in the nerve and the outcome is dependent upon the interstump gap length.

Although motor and sensory axons regenerate through the injury site towards distal territories, reinnervation of peripheral targets does not always lead to recovery of complex functions. The selectivity of axon-target reconnection plays an important role in the impairment of function after nerve injury and regeneration. Tissue specificity, i.e. preferential growth of axons towards a distal nerve stump rather than to other tissues, has been well documented. Fascicular specificity — the preferential regeneration through the original nerve fascicle — was suggested in some early studies but not confirmed by most investigations. Target organ specificity — the appropriate reinnervation of each type of end organ (muscle, sensory receptor, etc.) by axons that originally served that organ — is less than perfect, although preferential motor reinnervation has been observed. Cell adhesion molecules differentially expressed by sensory or motor Schwann cell endoneurial tubules seem to play a role in preferential guidance, but later improvement in target specificity is the result of progressive withdrawal or pruning of misdirected axons, which, without appropriate distal reconnection suffer retrograde atrophy and degeneration.

Various factors may contribute to poor long-term functional effects after nerve injuries: 1) Irreversible damage to the neuronal cell body due to axotomy and retrograde degeneration, excluding the possibility of regeneration. 2) Inability for the growth of axons due to the nerve lesion or to underlying diseases. Laceration of the nerve with an associated tissue gap, or distal degeneration due to generalized neuropathy may both impede regeneration. Absence of regeneration and reinnervation precludes functional recovery, and is associated with atrophy of denervated target organs. 3) Poor specificity of reinnervation by regenerating axons, when target organs become reinnervated by nerve fibers with different function. Aberrant reinnervation is pronounced during regeneration over long distances. Plasticity of central connections may compensate functionally for this lack of specificity, although in humans, plasticity has only a limited effect on disturbed sensory localization or fine motor control after injuries.

4. REPAIR OF PERIPHERAL NERVE INJURIES

After peripheral nerve injuries, the capability of severed axons to regenerate and recover functional connections depends on the site and type of lesion, and the distance over which axons must regrow to span the injury. After nerve crush regeneration is usually successful because the continuity of the endoneurial tubes is preserved. This is in contrast to the limited growth across a gap imposed after complete transection or resection of a nerve. The use of biochemical or metabolic factors to support and enhance regeneration remains speculative and, thus, the management of most peripheral nerve injuries is limited to surgical repair. The objective is to provide axons growing from the proximal stump with an appropriate substrate, such as the distal degenerating nerve, to support regeneration. If the injury is left unrepaired, the regenerative sprouting of fibers in the proximal stump forms a neuroma, as the growing axons do not find a favorable terrain to elongate through the extraneural milieu to reach the distal stump.

4.1. Suture and autograft repair

When the lesion is a clean cut, the classical method of repair is mobilization of the nerve followed by *epineurial suturing* of proximal and distal stumps. However, this does not ensure correct matching of the fascicular organization of the nerve trunk. With a microsurgical approach, it is possible to identify and coapt individual fascicles by means of perineurial sutures, thus improving the chances for appropriate regeneration. Identification of sensory and motor fascicles is improved by the use of neurophysiological stimulation and of histochemical stainings. With respect to timing, most evidence indicates that the sooner the repair of a transected nerve, the better the functional recovery. A shortly delayed repair (2-3 weeks) has also been advocated, based on a conditioning effect of the injured nerve. Longer periods from axotomy to repair result in progressive reduction of the number of neurons that successfully regenerate their axons and reinnervate denervated targets.

When the gap created by tissue destruction and nerve retraction is too long to allow apposition and suture without tension, a *nerve graft* is usually interposed. The purpose of introducing a graft between the stumps of a transected nerve is to offer mechanical guidance as well as a stimulating environment for the advancing axons. The Schwann cells of the graft and their basal lamina play an essential role in promoting neurite growth. It is generally agreed that gaps in injured nerves are most successfully bridged by the use of autologous nerve grafts, which behave in the same way as the distal segment of the severed nerve itself. Wallerian degeneration occurs somewhat more slowly in the graft than in the distal stump, but autografts retain their neural structure for considerable periods and do not show immunoreaction. Regenerating axons have to cross two suture lines, with increasing chances for misdirection. Experimental studies have shown that autograft repair enables similar number of regenerated axons and functional recovery as in direct suture repair. However, autograft repair involves some problems, such as the need of a second surgical step, elimination of the donor nerve function, a limited supply of donor nerves, and the mismatch between nerve and graft dimensions. Donor nerves are limited to those with pure sensory function, such as sural and antebrachial cutaneous nerves. The harvested nerve can be divided in segments to allow for a fascicular grafting in larger injured nerves.

4.2. Alternative grafts

The use of *nerve allografts* has not been encouraging, because the specific tissue immunity rejection, directed mainly against Schwann cells and myelin sheaths of the graft, precludes axonal regeneration. Immunosuppressive therapy is needed for reducing graft rejection, but its secondary complications frequently surpass the benefits. With evolving principles for immunosuppression, particularly with the use of cyclosporin A and FK506, allografts are coming into clinical application for selected cases in which the nerve gaps exceed the length that can be reconstructed with available autograft material. In order to reduce the risks of long-term immunosuppression, temporary treatments have been assayed in experimental models. Under immunosuppression, axons regenerate through the allograft ensheathed by the donor Schwann cells, but rejection of the donor cells follows a few days after treatment is withdrawn. Regenerated axons lose their function, become demyelinated and may degenerate. Eventually if the regenerated axons survive and have succeeded in reinnervating a target organ, host Schwann cells repopulate the allograft and nerve function may be recovered.

Immunogenicity of nerve allografts can be reduced by making them acellular. Different procedures have been applied to eliminate non-neuronal cells while maintaining the connective sheaths of the graft. Experimental studies have shown that such allograft scaffold is repopulated by migrating host Schwann cells and consequently supports axonal regeneration. The distance that Schwann cells are able to migrate along an acellular graft is, however, limited.

Any other tissue containing a basal lamina can be used as a bridge for nerve repair, in a similar fashion to that of acellular nerve grafts. This option has been used primarily with segments of skeletal muscle, made acellular by freezing-thawing or chemical extraction. Regenerative axons grow into the empty cylinders of muscle basal lamina that contain neuritotropic molecules, such as laminin and fibronectin, following the migration of Schwann cells. There appears to be a limiting length for *muscle grafts*, due to the limited migration of Schwann cells. Myelination and maturation of regenerated axons have been found to take place more slowly in the muscle graft than in the distal nerve segment.

The use of blood vessels as a conduit to close a gap in peripheral nerves has also received considerable attention. Vein grafts are easily available from autologous origin, and have been used successfully for bridging nerve gaps in animal studies and in human patients. Vascular grafts act as a conduit, but also provide molecular and cellular elements from their own wall. Invaginated vein grafts that expose regenerating axons directly to the collagen-rich adventitia resulted in greater vascularity and earlier regeneration than with standard vein conduits. However, veins tend to collapse, hampering the regenerating nerve fibers; therefore they are considered for the repair of only short gaps.

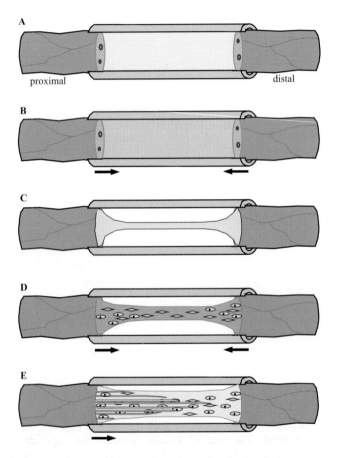

Figure 2. Phases of regeneration within a guide used to bridge the nerve gap. A) The nerve stumps are introduced and sutured to the ends of the guide. B) During the first days, the guide is filled by fluid from the nerve stumps. C) A fibroconnective cable bridges the nerve stumps. D) Non-neuronal cells (fibroblasts, endothelial cells, and Schwann cells) migrate from both stumps, enriching the connective cable with neurotropic and neurotrophic molecules. E) Axons grow from the proximal stump in association with the migrating Schwann cells.

4.3. Tubulization repair

The development of an artificial nerve graft, comprising a conduit filled with exogenous elements that promote axonal regeneration, has been pursued as an effective alternative to solve the secondary problems of autograft and allograft repairs. Tubulization — the implantation of a tube as guide to bridge a nerve gap — has provided a useful model for studying the cellular and biochemical events during peripheral nerve regeneration, and it has been successfully applied for repairing injured nerves in different animal models and in human patients. Proximal and distal nerve stumps are introduced a few millimeters into the ends of a tube, and held in place by means of one or a few epineurial sutures, leaving a gap between the nerve stumps. Such nerve guides thus offer a closed space where neurotrophic elements provided by the injured nerve stumps accumulate and support axonal growth. Accordingly, tubulization allows successful regeneration over longer gaps than in unrepaired nerves. However, regeneration by nerve guides also has a limit, depending upon the length of the gap.

Tubulization has several theoretical advantages over classical nerve repair methods: it is easier from the surgical point of view, avoids tension, limits excessive collagen and scar formation at the suture lines, provides directional guidance to the regenerating axons and prevents axonal escape into the surrounding tissues. In addition, nerve guides offer a controlled microenvironment, where different adjuncts to promote axonal regeneration may be applied in situ. Nerve guides may be produced in different sizes (diameter and length) and stored for use when needed.

Initial attempts to find a guide for nerve regeneration included the use of decalcified bone conduits, cellulose wrapping, metal (tantalum, stainless steel) cuffs, and arteries or veins, without much success. In the early 1980s, the silicone tube model was introduced as an important tool for studying basic biological mechanisms of nerve regeneration. In the last two decades many materials have been employed to make nerve guides and applied in experimental studies. Materials of biological origin include mesothelial conduits, dura mater, fascia, epineurium and collagen guides, while of the synthetic ones the most important are silicone, polyethylene, acrylic copolymers, and polyglycolic and polylactide derivatives.

In clinical practice, tube repair has been assessed as an alternative to direct nerve suture or to nerve autografts when a short gap (less than 2 cm) has to be bridged. In a prospective randomized study comparing primary suture versus silicone tube repair of transected median or ulnar nerves in the forearm, Lundborg and colleagues reported no differences between the two techniques for up to 1 year after surgery. In another randomized trial a bioresorbable polyglycolic acid conduit was compared with standard repair, either end-to-end or with a short nerve graft, of digital nerve transection in the hand. The overall results revealed no significant difference between the two groups, although the group with a conduit showed improved recovery of sensory discrimination.

5. NERVE GUIDES: FROM BASICS TO CLINICAL APPLICATION

Nerve regeneration through a nerve guide requires the reconstitution of a new neural structure. During the first days after a tube has been implanted, its lumen is filled with fluid, derived mainly from blood extravasation. Then, a loose fibrin matrix is formed within the tube, bridging both nerve stumps. Non-neuronal cells (fibroblasts, endothelial cells, Schwann cells) migrate from both stumps along the connective cable, providing neovascularization, connective strands, and basal lamina, and secreting a variety of neurotrophic and extracellular matrix compounds that allow the growth of axonal sprouts from the cut nerve proximal stump (Fig. 2).

The fluid surrounding the regenerative cable in the tubes contains macromolecular factors, which display neurotrophic activity and promote Schwann cell migration and proliferation. By 3 weeks, non-neuronal cells completely bridge the chamber, and unmyelinated axons are found distally, while myelinated axons arrive at the midpoint of the tube. By 4 weeks, myelinated axons have regenerated along the whole tube. With longer time, the regenerated nerve increases in caliber. The newly regenerated nerve is usually centered in the tube lumen, and is composed of small fascicles with myelinated and unmyelinated axons, surrounded by a thin perineurium (Fig. 3).

The success of intratubular regeneration depends upon the capability of the injured nerve to provide enough of the humoral and cellular elements that constitute the initial regenerative cable. The main limitation of tubulization is the gap length to which it may be applied. The limiting or critical gap length for nerve regeneration varies in different animal species, suggesting a mass-distance relationship. Experimental reports have shown that within silicone or other synthetic guides, regeneration is always successful with a gap of up to 4 mm in the mouse, up to 10 mm in the rat, and up to 30 mm in large primates, but normally fails for longer gaps. The physico-chemical characteristics of the tube, mainly in terms of size, permeability, durability and wall composition, affect the chances of regeneration; further advances in biomaterials will likely improve the effectiveness of tubulization repair.

5.1. Physical parameters of nerve guides

Various parameters of the guide used for nerve repair, such as internal diameter, microgeometry of inner surfaces, thickness, permeability and chemical composition affect the degree of nerve regeneration. The most suitable characteristics of a nerve guide, from a review of the different studies, are summarized in Table 1. In fact, as indicated by Fields and co-workers, the ideal material for tubulization has not been obtained yet, but the best materials are inert (biocompatible), thin and flexible, translucent, bioresorbable, inhibitory to processes causing pathology, and beneficial to processes contributing to healing and regeneration.

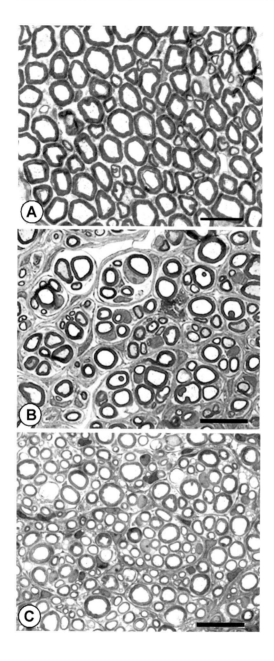

Figure 3. Transverse sections of mouse sciatic nerves. A) Intact nerve. B) Regenerated nerve after a 6 mm resection and autograft repair. C) Regenerated nerve after a 6 mm resection and repair with an artificial graft comprising a guide seeded with autologous Schwann cells. Bar = 16 μm

Table 1. Ideal characteristics of a nerve guide for the repair of nerve gaps

- Interstump gap: ≤ 4 mm in mice, ≤ 10 mm in rats, ≤ 30 mm in primates
- Tube lumen area: 2.5 times the cross-sectional area of the nerve to repair
- Wall: flexible, thin, non-collapsible, translucent
- Inner surface: smooth, homogeneous
- Permeability: impermeable or highly permeable
- Material: biocompatible, resorbable, available

The length of the nerve guide is determined by the gap to be bridged. The increase in gap length progressively reduces the degree of functional recovery and the number of regenerated axons, and increases the time until reinnervation of target organs. However, regeneration does not decrease linearly with increasing length of the gap — there is a critical length above which regeneration fails (see above). The first stage of regeneration in nerve guides is the formation of a fibrin cable that acts as a guiding surface for cell migration. Matrix diameter within silicone tubes decreases when increasing the gap length. Failure of regeneration across long gaps occurs if the nerve stumps are not able to provide a proper cable with enough regeneration-promoting elements inside the tube. In fact, regeneration through guides is significantly reduced or even absent if the distal nerve stump is excluded. Prefilling the chambers with extracellular matrix compounds promotes the generation of a larger-diameter cable that supports axonal growth across longer gaps than when the chamber is empty. No such length relationship is sustained by experimental studies for nerve graft repair, because the graft provides basal lamina and reactive Schwann cells throughout its length as in the degenerating distal stump.

The formation of the initial fibrin matrix is also dependent on the guide's internal diameter. A large internal diameter allows a thicker regenerating cable with lower fibrin density and faster migration of Schwann cells and axons than in tubes fitted to the diameter of the nerve. However, if the diameter of the tube is much greater than that of the nerve, vascularization, Schwann cell migration and axonal advance are drastically retarded, probably due to a critical reduction in the concentration of neurotrophic factors inside the chamber. When the tube diameter is near that of the nerve fibrosis and neuroma develop at the ends of the tube and, in the long term, ischemia and degeneration appear inside the tube due to compression of the enlarging regenerating nerve. Best results have been obtained with tubes having an inner cross-section area about 2.5 times that of the nerve. Moreover, regeneration is improved when the tube wall is the thinnest possible, but non-collapsible, and flexible.

The microgeometry of the guide's internal surface has also been shown to affect nerve regeneration. Tubes with a rough surface induced the formation of a loose connective tissue stroma that contained only a few regenerated axons and adherences to the tube wall, in contrast with the well-formed nerve cable of numerous axons found centered in smooth-walled tubes of the same material.

5.2. Resorption and permeability of nerve guides

Many different materials have been used for tubulization (Table 2). Of them, silicone and other plastic tubes have been the most frequently used experimentally. Durable tubes have the disadvantage that they remain in the body, where they may induce a chronic inflammatory response and eventual compression of the regenerated nerve, and a second operation might be necessary for removal. The use of nerve guides made of biodegradable materials could avoid this problem if, after allowing the outgrowth and maturation of the nerve, they were gradually resorbed without significant deformation. Experimental comparisons have shown that resorbable guides, particularly those made of collagen, polyglycolic acid, and polylactide-caprolactone polymer, promoted regeneration over long gaps in a higher proportion of animals than did permanent tubes, such as those of silicone, Teflon, and polysulfone. It has been hypothesized that resorbable tubes allow a better nutrient supply to the regenerative nerve, enhance the constitution of the initial matrix and the subsequent cellular cable, and increase their flexibility as they degrade.

Table 2. Materials most frequently used for nerve guides

Biological materials	Non-resorbable polymers	Bioresorbable polymers
artery	silicone	polyester
vein	polyethylene	polyglycolic acid
mesothelial tubes	polytetrafluoroethylene	polylactide
skeletal muscle	polyvinyl chloride	polylactide-caprolactone
epineurium	ethylene-vinyl acetate	
collagen	polyacrylonitrile-polyvinyl	
hyaluronate	polysulfone	
	cellulose	

Maintenance of stable wall configuration and lumen caliber is difficult to attain with polymer compositions that are also bioresorbable. Biodegradable tubes tend to swell or bend as the wall material becomes degraded. Such distortions, when sufficient to compromise luminal integrity, can interfere with nerve regeneration, even when repairing short gaps. Thick-walled guides induce compression of the initially regenerated nerve because of swelling after in vivo implantation. The degradation rate may thus, influence the effectiveness of bioresorbable guides if swelling and fragmentation of their wall occur, especially in the short term, when the regenerating nerve fibers are more susceptible to compression and extraneural factors. The process of nerve regeneration requires good coordination between the rate of tube reabsorption and the time necessary for growth and maturation of the nerve within the tube. Bioresorbable nerve guides should degrade slowly (over several months) in accordance with the rate of axonal growth and maturation, maintaining mechanical continuity and lumen stability for longer time than the axons require to cross the gap.

The degree of permeability of the nerve guide may also affect nerve regeneration. For bridging short or medium gaps, such as 4 mm in the mouse and 4-8 mm in the rat, tubulization is usually successful, and semipermeable tubes behave similarly to or slightly worse than impermeable silicone tubes. However, when the gap is longer than the maximal length that can be regenerated within silicone tubes, i.e. 6 mm in the mouse and 15 mm in the rat, or the distal end is left without a nerve insert, semipermeable tubes (with molecular weight cut-off range of 30-100 kDa) improve the regeneration success rate. The favorable effects of permeable tubes may be attributed to different reasons, such as metabolic exchange across the tube wall, diffusion into the guide lumen of growth-promoting factors generated in the external environment, retention of trophic factors secreted by the nerve stumps, or a combination of all these. The best results have been found with highly permeable guides which, in addition, allow the infiltration of reparative cells (macrophages and fibroblasts) that contribute to the formation of the initial regenerative cable. In contrast, semipermeable guides with permeability cut-off between 100 and 1000 kDa behave worse than less and more-permeable guides. Hence, the tube wall stability over time and the pore size seem to be important factors in determining the flow of different constituents that may promote or inhibit regeneration.

6. INTRODUCTION OF EXOGENOUS FACTORS IN NERVE GUIDES

6.1. Extracellular matrix components

As effective regeneration in tubes depends on accumulation of neurotrophic agents in the fluid filling the guide and on the formation of an initial connective matrix between the nerve stumps across the gap, manipulation of the local environment within the tube lumen has been used in attempts to secure this initial step and thereby improve and accelerate the growth of axons regenerating through the guide. Prefilling the lumen with components of the naturally formed intratubular matrix, such as fibrin, collagen, fibronectin and laminin-containing gels, has been reported to enhance nerve regeneration by promoting formation of a large-diameter matrix that supports axonal growth across longer gaps than when the chamber is empty. However, gel substrates, even if containing neuritotropic factors, may impair regeneration by physically impeding the migration of regenerating axons and non-neuronal cells if they are too dense or provide a network of pores that are too narrow. Gels composed of collagen or laminin, at the maximal dilution that still allows gelification, were the most successful in enhancing regeneration, but the improvement was limited and results were still worse than those found with an autograft.

A suitable exogenous matrix within a nerve guide should have neuritotropic activity, be diluted in order to provide pores wide enough for cellular and axonal migration, and longitudinally oriented pathways to mimic endoneurial tubules of the nerve. By using an aligned matrix, the rate and direction of axonal elongation should improve due to contact guidance with the fibrils aligned along the tube axis. This option has been reported for tubes prefilled with a fibrin matrix from dialyzed plasma, and more recently, with magnetically aligned collagen- or laminin-containing gels, and with a collagen-glycosaminoglycan matrix copolymer. Thus, the composition, density, and structural organization of the exogenous intratubular matrix may each play a role in determining the fate of axonal regeneration over tubulized gaps.

6.2. Neurotrophic factors

The effects of growth factors on neurite outgrowth, and their ability to prevent neuronal death after injury, suggest that they may be useful in promoting regeneration after injury. Experimental studies have reported beneficial effects of several neurotrophic factors, administered either locally or systemically, in enhancing nerve regeneration. The addition of nerve growth factor (NGF) inside a tubular guide partially prevents neuronal loss in dorsal root ganglia following nerve transection and tube repair and facilitates regeneration of rodent peripheral nerves. The spectrum of neurotrophic factors reported to stimulate axonal regrowth include the other neurotrophin family members (brain-derived neurotrophic factor, neurotrophin-3, and neurotrophin-4/5), insulin-like growth factors (IGF-I and IGF-II), ciliary neurotrophic factor (CNTF), basic fibroblast growth factor (FGF-2), and glial-cell-line-derived neurotrophic factor (GDNF). In general, these strategies to promote nerve regeneration based on tubulization repair have tested the introduction of only one neurotrophic factor, within a limited range of concentrations and time windows, and have been performed with different animal models and gap lengths. Application of neurotrophic factors in tubuled long gaps has limited success, while in short or mid-length gaps, where regeneration occurs normally, it is less relevant. Local delivery of trophic factors over a long time (days to weeks), by means of miniosmotic pumps connected to the nerve guide or by guidance channels designed for slow release of the factor incorporated in their wall, is of greater interest, and resulted in increased regeneration in long gaps. However, it has been suggested that a complex mixture of different neurotrophic factors and extracellular matrix components may potentiate peripheral nerve regeneration better than any one individual component.

7. SCHWANN CELL TRANSPLANTS

Two key points determine the success of tubulization repair. First, the initial formation of a suitable acellular matrix bridging the proximal and distal stumps, and second, the repopulation of this matrix by migrating Schwann cells, which will stimulate and guide the regenerating axonal sprouts.

Over the past decade a series of studies has pursued the design of an artificial nerve graft that enhances nerve regeneration and may become an alternative to the classical autograft repair. Such graft should mimic the main components of a natural graft: a biocompatible nerve guide that encloses the gap to be repaired, prefilled with an extracellular matrix that provides physical support and neuritotropic cues, and seeded with competent Schwann cells that replace the host cells and secrete a variety of neurotrophic factors. Because of their importance in creating a suitable environment for nerve regeneration, Schwann cells have been widely considered the ideal cells for transplantation in the peripheral nervous system.

Strategies toward developing an artificial cellular graft should include at least: 1) the ability to isolate Schwann cells from peripheral nerve segments within a short time period, 2) the proof that transplanted cells survive in the host and promote axonal regeneration, and 3) the availability of tubulization materials with suitable properties in that they are biocompatible, bioresorbable and permeable. Schwann cells can be isolated and expanded in primary cultures from fresh neonatal nerves or from predegenerated adult nerves, with or without addition of glial growth factors and mitogenic agents. Within a degenerating nerve Schwann cells deprived of axonal contact proliferate and upregulate the synthesis and release of neurotrophic and neurotropic factors. Thus, in vivo pre-degenerated nerves are an excellent source of adult Schwann cells, easily expandable in vitro. Furthermore, preactivation of the Schwann cells increases their regeneration-promoting effect substantially.

Any attempt to seed a nerve guide with glial cells, silicone and plastic tubes are not the most suitable since they impede cell adhesion and interchange of nutrients with the extraneural milieu. Semipermeable and resorbable materials, including collagen, acrylic copolymer, and polylactide-caprolactone polymer, have been preferentially used as nerve guides to contain Schwann cell transplants. Alternatively, scaffolds of acellular nerve, frozen muscle, or tendon sheaths have been populated with Schwann cells previously isolated in culture and implanted.

The ideal medium in which to suspend transplanted Schwann cells within the guide has not yet been defined. Direct injection of a cell suspension in culture medium has been used, but it may be better to embed the cells in a gel that facilitates cell attachment and prevents leaking of the cells through the guide ends. Laminin-containing gels seem the most suitable matrix for seeding Schwann cells, based on evidence that laminin is a good substrate for Schwann cell adhesion and has neuritotropic action.

Several studies (Table 3) have shown that syngeneic Schwann cells transplanted in nerve guides enhance axonal regeneration with respect to control guides with saline solution or extracellular matrix components. Although syngeneic Schwann cells have been shown to survive after implantation, a number of the transplanted cells die because of host rejection. Prelabeled transplanted Schwann cells survive within the guide several months after implantation, but in a larger number if their origin is autologous rather than syngeneic. A concentration-dependent effect has been reported for Schwann cells transplanted in vivo. Thus, a high number of syngeneic cells have to be transplanted initially to ensure that enough remain viable after implantation to promote regeneration, whereas a lower number of autologous cells may provide better results. In comparison of cell transplants from syngeneic, isogeneic, or autologous donors in the same animal model, the best functional and morphological regeneration was obtained with autologous Schwann cells, followed by isogeneic and finally syngeneic cells. These data indicate that Schwann cell transplants enhance axonal regeneration, but also that immune compatibility between donor and host is an important factor that affects their capability to survive and promote regeneration. Alternatively, the results of allogeneic cell transplants may be improved by antigen matching between donor and host or by immunosuppression.

Table 3. Artificial nerve graft assays: nerve guides seeded with Schwann cells

Autors	Guide	Matrix	Schwann cells	density
Feltri et al., 1992	Silicone	medium	syngeneic	1×10^6
Guénard et al., 1992	PAN/PVC	laminin	syngeneic heterologous	8×10^5 12×10^5
Keeley et al., 1993	Aquavene	collagen	syngeneic	1×10^4
Kim et al., 1994	Collagen	collagen	syngeneic	4×10^5
Levi et al., 1994	PAN/PVC	laminin	autologous	3×10^6
Ansselin et al., 1997	Collagen/K	medium	syngeneic	5×10^5
Rodríguez et al., 2000	PLC	laminin	autologous syngeneic	1.5×10^5 1.5×10^5
Dumont et al., 1997	extracted nerve	medium	syngeneic	1×10^5
Fansa et al., 2001	vein acellular muscle	medium medium	isogeneic isogenic	6×10^5 6×10^5

The demonstration that human Schwann cells can be isolated in culture, and that they survive, enhance axonal regrowth, and myelinate regenerated axons after transplantation in the peripheral nervous system of immune-deficient rodents, is relevant for clinical expectations. For clinical application, small nerve pieces resected from the injured nerve stumps during an exploratory intervention may be a source of autologous Schwann cells to construct such artificial cellular graft for implantation in a short-delay repair.

Approaches toward treating injuries to the nervous system using glial cell transplants that are genetically modified to secrete certain growth factors are under intensive investigation. For peripheral nerve repair the most suitable vehicle is the Schwann cell. Schwann cells upregulate the expression and secretion of a variety of neurotrophic factors during Wallerian degeneration, but less is known about their capabilities after culturing and in vivo transplantation. The combination of an increased production of certain neurotrophic factors at the site of the lesion with the permissive pathway provided by exogenously transplanted Schwann cells is likely to enhance axonal regeneration when the local environment is poor, as in long nerve gaps.

Further in vitro and in vivo experiments are needed to engineer an ideal artificial nerve graft, requiring multivariable and combined research in new materials for nerve guides, in extracellular matrix components, and in glial cell transplantation. New developments have to be compared, in terms of morphological regeneration and of functional restoration with what has up to now been considered the gold standard for repair of nerve gaps: the autologous nerve graft.

REFERENCES

Bunge, R.P. (1993) Expanding roles for the Schwann cell: ensheathment, myelination, trophism and regeneration. *Current Opinion in Neurobiology*, 3, 805-809.

Doolabh, V., Hertl, M.C., & Mackinnon, S.E. (1996) The role of conduits in nerve repair: a review. *Reviews in Neuroscience*, 7, 47-84.

Evans, P.J., Midha, R., & Mackinnon, S.E. (1994) The peripheral nerve allograft: a comprehensive review of regeneration and neuroimmunology. *Progress in Neurobiology*, 43, 187-233.

Fawcett, J. (Ed.). (1993) Repairing the damaged nervous system. *Seminars in Neuroscience*, 5, 383-459.

Fawcett, J.W., & Keynes, R.J. (1990) Peripheral nerve regeneration. *Annual Review of Neuroscience*, 13, 43-60.

Fields, R.D., LeBeau, J.M., Longo, F.M., & Ellisman, M.H. (1989) Nerve regeneration through artificial tubular implants. *Progress in Neurobiology*, 33, 87-134.

Fu, S.Y., & Gordon, T. (1997) The cellular and molecular basis of peripheral nerve regeneration. *Molecular Neurobiology*, 14, 67-116.

Hall, S. (2001) Nerve repair: a neurobiologist's view. *Journal of Hand Surgery*, 26B, 129-136.

Kline, D.G. (2000) Nerve surgery as it is now and as it may be. *Neurosurgery*, 46, 1285-1293.

Kline, D.G., & Hudson, A.R. (1995) Nerve injuries. Operative results for major nerve injuries, entrapments, and tumors. Philadelphia: Saunders.

Liu, H.M. (1996) Growth factors and extracellular matrix in peripheral nerve regeneration, studied with a nerve chamber. *Journal of the Peripheral Nervous System*, 1, 97-110.

Lundborg, G. (1988) *Nerve injury and repair*. London: Churchill Livingstone.

Lundborg, G. (2000) A 25-year perspective of peripheral nerve surgery: evolving neuroscientific concepts and clinical significance. Journal of Hand Surgery, 25A, 391-414.

Mirsky, R., & Jessen, K.R. (1996) Schwann cell development, differentiation and myelination. *Current Opinion in Neurobiology*, 6, 89-96.

Rodríguez, F.J., Verdú, E., Ceballos, D., & Navarro, X. (2000) Neural guides seeded with autologous Schwann cells improve nerve regeneration. *Experimental Neurology*, 161, 571-584.

Stoll, G., & Müller, H.W. (1999) Nerve injury, axonal degeneration and neural regeneration: basic insights. *Brain Pathology*, 9, 313-325.

Sunderland, S. (1991) *Nerve injuries and their repair. A critical appraisal*. Edinburgh: Churchill Livingstone.

Terenghi, G. (1999) Peripheral nerve regeneration and neurotrophic factors. *Journal of Anatomy*, 194, 1-14.

Verdú, E., & Navarro, X. (1998) The role of Schwann cell in nerve regeneration. In B. Castellano, B. González, & M. Nieto-Sampedro (Eds), *Understanding Glial Cells* (pp. 319-359). Boston: Kluwer Academic.

Yannas, I.V. (2001) *Tissue and organ regeneration in adults*. New York: Springer-Verlag.

P. BACH-Y-RITA

Departments of Orthopaedics and Rehabilitation Medicine, and Biomedical Engineering, University of Wisconsin, Madison, U.S.A.

30. MODERN LATE NEUROLOGIC REHABILITATION: NEUROSCIENCE AND MOTIVATING FUNCTIONAL REHABILITATION

Summary. The potential for late recovery of function following brain damage exists, but rehabilitation programs are usually designed for the early stages of recovery. Neuroscience issues related to late brain reorganization include brain plasticity (including receptor plasticity, the up- or downregulation of synaptic and non-synaptic neurotransmitter receptors), and volume transmission (non-synaptic diffusion neurotransmission in the brain). Special purpose computer-assisted motivating rehabilitation devices aid in the process of late functional recovery. Cost-effective rehabilitation is facilitated by these devices, which allow reduction of hands-on therapist hours, and the ability to carry out rehabilitation programs at home and over the Internet.

1. INTRODUCTION

Neurological rehabilitation has traditionally had little relation to scientific findings, since the conceptual framework of the neurosciences was not conducive to brain plasticity and functional recovery by means of brain reorganization. However, these are now fully accepted in both the neurosciences and the field of rehabilitation, and so major advances are occurring. In addition to the neuroscience bases, motivation and other psychosocial factors are essential for late neurologic rehabilitation.

This chapter includes material from other publications of the author (e.g., Bach-y-Rita, 1995; 2000; 2001a; 2002; 2003).

2. BRAIN PLASTICITY

In normal learning, and following damage, the brain responds to functional need. With learning, cortical representation can change, such as the greatly increased representation of a fingertip area in monkeys following training in haptic exploration (Jenkins, et al, c.f. Bach-y-Rita, 1995). Brain plasticity also is the basis for recovery from direct brain damage such as due to a stroke or head injury, or indirect brain damage such as in blindness. In the later case, the brain must reorganize following the loss of a major sensory input. As an example, Pascual-Leone and Torres (1993) demonstrated sensorimotor cortex changes in Braille readers.

Broca's description of cerebral localization (Broca 1861) led to a century of domination of the neurosciences by concepts of connectionism and strict localization (Bach-y-Rita, 1995). This had a negative effect on the field of rehabilitation. In the absence of a concept of brain plasticity, clinicians could not expect reorganization. However, even during that century, plasticity was recognized by scientists such as Bethe, Goldstein, Foerster, and others (reviewed in Bach-y-Rita, 1972; 1995) and Franz (Colotla and Bach-y-Rita, 2002). Some of the rehabilitation approaches now becoming popular where started by them.

The scientific literature regarding specific mechanisms of brain plasticity (e.g. Bach-y-Rita, 1995; 2000) has been reviewed. One that has not been recognized as being important in brain plasticity is nonsynaptic diffusion neurotransmission, which is generally called volume transmission.

2.1. Volume Transmission

Volume transmission includes the diffusion, through the extracellular fluid of neurotransmitters released at points that may be remote from the target cells, with the resulting activation of extrasynaptic receptors. Relevant reviews include the following: (Agnati, et al, 2000; Bach-y-Rita, 2001; 1995; Fuxe and Agnati, 1991).

Volume transmission appears to play a important role in normal functions such as vision, and may be particularly important in mass-sustained functions such as mood and pain (Bach-y-Rita, 1991). Among the range of functions in which volume transmission plays a role are many of those mediated by noradrenaline (c.f., Bach-y-Rita, 1995). Volume transmission may be the primary information transmission mechanism in several abnormal functions, such as mood disorders, spinal shock, spasticity, shoulder-hand and autonomic dysreflexia syndromes, pain, and drug addiction.

Routtenberg's studies (reviewed by him in Fuxe and Agnati, 1991) showed that transmitters could readily move in the extracellular space and travel over distances by directional flow. In recent years, a large number of receptor subtypes have been identified; these provide a mechanism for selective activation even if the volume transmission neurotransmitter is massively diffused over an area (Bach-y-Rita, 2003). High affinity receptors on cells distant from the transmitter release can bind the neurotransmitter and cause a neuron response, while closer neurons without those specific receptor sub-types will not respond. Volume transmission appears to be common in certain brain regions; nonsynaptic interneuronal communication is very common in the greater limbic system and may play an important role in the organization and regulation of behavior by the core and para-core regions of the brain (Nieuwenhuys, c.f. Agnati, et al, 2000). Bach-y-Rita and Aiello (2001) have suggested that the parts of the human brain that show the greatest size increase over other animals, such as pre-frontal cortex, may be exactly those parts in which highly non synaptic-based functions have their neuronal representation.

2.2. Receptor plasticity

Receptor plasticity, both at the synapses and on the cell membrane away from synapses (reached by volume transmission) may play a major role in the reorganization of function following brain damage. Noradrenaline-mediated neurotransmission offers an example of the differences between synaptic and volume transmission. With volume transmission, the effect of non-junctional noradrenaline is likely to be longer lasting than a similar quantity of noradrenaline released at synapses. In the absence of a junction, inactivation of the noradrenaline is slowed; while the synapse has a full panoply of degradation enzymes and re-uptake mechanisms, non-junctional receptor sites have few if any of these inactivating device. Due to the slow activation and inactivation, there is a reduced need for nerve action potentials and for neurotransmitter production, transport and heat generation. The brain is efficient in the use of energy; thus the widespread sustained activation of groups of neurons via diffusion of neuroactive substances through the extracellular fluid should be more energy efficient than synaptic activation of those same neurons (Bach-y-Rita, 1995). Our brain energetics (Aiello and Bach-y-Rita, 2000) and modeling studies (Aielo and Bach-y-Rita, 2001) support this interpretation.

The effects of brain damage are not uniform in the various neurotransmitter systems. In a rat model, Westerberg, et al (1989) studied the effects of transient cerebral ischemia in the rat hippocampal subfields on excitatory amino acid receptor ligand binding. They noted that their results demonstrate a lack of correlation between receptor changes in the early recovery period following ischemia and the development of neuronal necrosis in different hippocampal regions, and also note long-lasting receptor changes in areas considered resistant to an ischemic insult. Some receptors are down-regulated while others appear to be up-regulated on the surviving cells. Volume transmission may contribute to the survival of partially denervated neurons and to brain reorganization after brain damage by selective up- and down-regulation of receptors. These issues have been discussed elsewhere (e.g., Bach-y-Rita, 1995; 2001), in regard to both brain and spinal cord injury.

De Keyser, Ebinger and Vauquelin (1989) demonstrated an up-regulation of D1 Dopamine receptors in the neocortex of persons who's death was due to a recent unilateral infarct of the ventral midbrain, producing a unilateral relative dopamine depletion. In autopsies of patients 9, 19 and 27 days post stroke, there was a 27 to 37% increase of receptors on the lesioned side. Since the massive brainstem lesion unilaterally destroyed ascending dopamine fibers, it is very likely that the receptor up-regulation was not related to dopamine synapses; rather it is likely that it represents extrasynaptic volume transmission-related receptors.

3. LATE REHABILITATION

Among the rehabilitation programs that have been specifically designed to obtain late recovery from stroke is the program of Tangeman, Banaitis and Williams (1990). They studied a group of 40 patients from 1 to 23 years post-stroke. Comments from their patients indicated that the stress of the acute stroke phase prevented them from benefiting completely from their acute inpatient rehabilitation. A period at home provided the patients with the opportunity to directly experience how the stroke affected their daily lives, and they had renewed interest in improving their skills and level of independence. The authors noted the appropriateness of home therapy, but chose a clinic-based program for the study for practical reasons. They demonstrated significant gains in functional measures, and concluded that "...functional improvement after intensive rehabilitation therapy would appear to be possible for chronic stroke patients who are at least one year poststroke." Other late rehabilitation programs include our NMRC program in which most of the rehabilitation takes place at home without the need for a skilled therapist (Bach-y-Rita, 1995).

Functional rehabilitation programs that are of interest to the individual patient make maximum use of the patient's motivation and participation. Home programs are effective for late rehabilitation and are also appropriate for some cases of early rehabilitation. We explored the feasibility of home stroke rehabilitation in Sweden (de Pedro-Cuesta, Widén-Holmqvist and Bach-y-Rita, 1992, c.f., Bach-y-Rita, 1995). A Pilot study was completed that suggested the feasibility and cost-effectiveness of home stroke which led to a successful prospective randomized study (Widén-Holmqvist, et al, 1998, c.f., Bach-y-Rita, 2000).

An innovative approach to late rehabilitation is the delivery of services within an educational model. This is particularly pertinent for late rehabilitation, where the disabled person is not sick, but merely disabled. As we discussed elsewhere (Bach-y-Rita, et al, 2002), for more than 20 years, a Community College in California (Cabrillo College) has been offering a course in which the students are persons who have had a stroke. This cost-effective program offers classes in independent living skills, mobility and fitness, speech and language development and counseling in a small group setting, in which students choose their own functional goals.

3.1 Facial Paralysis

A facial paralysis home rehabilitation program started as a research project (Balliet, Shinn and Bach-y-Rita, 1982) and developed into a clinical service (Bach-y-Rita, 1995). In patients with persisting facial nerve dysfunction following surgery for the removal of an acoustic neuroma, a VI-XII craneal nerve anastomosis is a surgical option to restore innervation to the denervated facial muscles. Although persons with VII-XII anastomoses do not have brain damage comparable to stroke or traumatic brain injured persons, they do present a situation in which brain reorganization can be explored and documented. The subjects have had total VII nerve sections, followed by the anastomosis of all or a portion of the ipsilateral XII nerve to the peripheral stump of the VII nerve. The new innervation to the ipsilateral facial muscles will be from the XII nerve, and thus from the XII nerve nucleus and its cortical representation and other inputs. Even many years after the neural damage, a large increase in function is possible in patients with various etiologies. Movements such as eye closure, automatic bilateral (voluntary or emotional) smile, and the ability to eat in public without synkinesic movements; this requires reorganization of the brain, such that the XII craneal nerve nucleus and nerve, genetically programmed to produce tongue movements, instead control facial muscles and movements.

4. SENSORY SUBSTITUTION

Tactile vision substitution studies were initiated as models of brain plasticity; persons with congenital vision loss presented a Jacksonian model [Hughlings Jackson emphasized the opportunities for discovery offered by the experiments made on the brain by disease (excerpts in Clarke and O'Malley, c.f., Bach-y-Rita, 1995). A major source of afferent information had been eliminated before they had the opportunity to develop the mechanisms for the analysis of the information from that sensory system. Thus, a thorough study of persons learning to use a sensory substitution system, with the information from an artificial receptor delivered to the brain through sensory systems (e.g., tactile) that have remained intact, offered unique opportunities to evaluate mechanisms of brain plasticity.

A person who has suffered the total loss of a sensory modality has, indirectly, suffered a brain lesion. In blind persons, about 2 million fibres from the optic nerves are absent. In the absence of a modality such as sight, behaviour and neural function must be reorganized. However, blind persons have not necessarily lost the capacity to see (Bach-y-Rita, 1972), since we do not see with the eyes, but with the brain. In normal sight, the optical image does not get beyond the retina. From the retina to the central perceptual structures, the image, now transformed into nerve pulses, is carried over nerve fibres. It is in the central nervous system that pulse-coded information is interpreted, and the subjective visual experience results. It appears to be possible for the same subjective experience that is produced by a visual image on the retina to be produced by an optical image captured by an artificial eye (a TV camera), when a way is found to deliver the image from the camera to a sensory system that can carry it to the brain. Optical images picked up by a TV camera are transduced into a form of energy (vibratory or direct electrical stimulation) that can be mediated by the skin receptors. The visual information reaches the perceptual levels for analysis and interpretation via somatosensory pathways and structures. Our studies with the tactile vision substitution systems (TVSS) have been extensively described (e.g., Bach-y-Rita, 1972; 1995; Bach-y-Rita et al., 1969; Sampaio, Maris and Bach-y-Rita, 2001). We developed TVSS to deliver visual information to the brain via arrays of stimulators in contact with the skin of one of several parts of the body (abdomen, back, thigh).

Our principal work has been with congenitally blind persons. There are psychological issues that must be addressed before a system can be practical. Thus, exploring the face of one's loved-one can be very disappointing, since the emotional messages that the long experience with vision have provided have not been perceived with our TVSS (similar problems are confronted by congenitally blind persons who acquire sight through surgical correction of the cause, such as congenital cataracts, that had prevented vision).

In the Bach-y-Rita, and Hughes, 1985 report (c.f., Bach-y-Rita, 1995), we noted that "An understanding of the functional equivalence between visual and vibrotactile processing would have both basic scientific and practical implications, the former because it would bear on whether information for the various perceptual systems ought to be considered modality specific or amodal, and the latter because the data would suggest the possibilities and constraints for vision substitution and other prosthetic developments...Although the early system was termed a tactile vision substitution system, we have been reluctant to suggest that blind users of the device are actually seeing. Others ... have not been so reluctant, claiming that since blind subjects are being given similar information to that which causes the sighted to see and are capable of giving similar responses, one is left with little alternative but to admit that they are seeing (and not merely "seeing")".

4.1. Vestibular substitution

Persons who have bilateral vestibular damage (BVD), such as from the ototoxicity of Gentamicin, experience functional difficulties that include postural "wobbling" (both sitting and standing), unsteady gait, and oscillopsia. Supported by a National Institutes of Health SBIR grant, we developed a vestibular substitution system using a head-mounted accelerometer and an electrotactile brain-machine interface (BMI) through the tongue (TVS). The use of TVS produces a strong stabilization effect on head-body co-ordination in BVD subjects. Under these conditions, we identified three characteristic and unique head motion features (d.c. drift, sway, and long-period perturbations) that consistently appear in the head-postural behaviour of BVD subjects, but are greatly reduced or eliminated with use of TVS (Tyler, Danilov and Bach-y-Rita, 2003). BVD subjects using the TVS reported feeling "normal" "stable" or having reduced perceptual "noise" while using the TVS and for short periods after removing the TVS.

5. COMPUTER ASSISTED MOTIVATING REHABILITATION (CAMR)

In the middle 70's, when newly-introduced electronic pong games could be connected to a home TV and played by 2 persons controlling individual joy-sticks, we adapted a pong game for the functional training of upper extremities of persons with a hemiparetic limb (Cogan, et al, c.f., Bach-y-Rita, 1995). Three different hand pieces were machined to coincide with the varying grips of the stroke patients for use with a modified exercise track. A patient had to reach forward to 90 degrees of shoulder flexion with full elbow extension in order to reach balls at the top of the screen. With practice, and during the emotionally involving game, the patient ceases to think of arm movements and begins to think in terms of accomplishing the goal. We noted that "The game concept helps to maintain a high level of interest, enhances motivation, and adds enjoyment to the hard work of rehabilitation.... during a pong game, the patient has an immediate goal for every movement of the arm. The patient also receives immediate visual feedback as to the accuracy of the movement patients quickly find themselves absorbed in playing the game. Some patients prefer playing against the built-in computer, or against their unaffected arm. Others enjoy the socialization which develops during a game with an aide or another patient." This may have been the first clinical application of what today would be called non-immersive Virtual Reality.

More than 20 years after the publication of the Cogan, et al paper, the second version was designed and built by the students of an engineering design class in Cuernavaca, Mexico. In addition to controlling the game, the system can record the movements of the patients' hemiparetic limb and these objective data are intended to be used to track the patients' rehabilitation progress. Using this system, subjects practiced with the "palanca" (as version 2 has been called), controlling a computer video game, daily for 25-40 minutes for a total of thirteen days. All subjects attended the entire thirteen days, except two, who missed two sessions each. Funding was available to subjects who were unable to afford transportation to the center daily to ensure maximum participation. All subjects participated enthusiastically in the video game activity. However, one subject with severe cognitive deficits due to frontal lobe damage, with poor attention to task as well as a severe left hemianopsia, required constant positive feedback and encouragement, as well as a quiet environment to participate in the activity. Patients got very involved and showed a high level of concentration. One hemiparetic aged man initially refused to use the system because he did not think his arm could perform the task, but after a short session in which he performed well, he asked if he could take it home. He later demonstrated, with surprise, his ability to extend his wrist, which he had been unable to do for 22 years (Bach-y-Rita, et al 2002).

6. REHABILITATION IN THE FUTURE

Prospective, randomized clinical studies, such as the home stroke rehabilitation study described above, offer the opportunity to obtain scientifically valid evidence by which rehabilitation methods can be evaluated. Such clinical trials, together with the objective-quantified studies, should lead to scientifically-validated rehabilitation.

Traditional neurologic rehabilitation is costly, inefficient, labor intensive, and artificially fractionated into multiple specialities. There is little rationale for the timing and intensity of presently practiced rehabilitation therapy. Greater attention to theory and research will lead to radical changes in the delivery of rehabilitation services that will virtually eliminate Rehabilitation as practised today. Demands for demonstrated efficacy at a reasonable cost will bring to the fore approaches such as the Constraint-Induced (CI) Movement Therapy developed by Taub for late rehabilitation (c.f., Taub and Crago, 1995) and the interesting and motivating real-life activity therapy that we have been emphasizing over the last 20 years (c.f., Bach-y-Rita 1995; Bach-y-Rita, et al, 2002).

Among the major changes that will occur, the rehabilitation team will virtually disappear. In any case, there is no evidence that it is better than other form of delivery of rehabilitation services (Keith, c.f., Bach-y-Rita, 1995), and it is very expensive in terms of cost, time and efficiency. In a preliminary study (never completed because of his fatal illness) of the efficiency of parcelling rehabilitation therapy into numerous sub- disciplines such as Physical Occupational, recreational speech psychology and others.

If intense therapy is appropriate, a single cross-trained therapist, who can spend more hours with a patient than several separate therapists, may provide therapy more efficiently. A Neurologic Rehabilitation Therapist, combining aspects of each of the present therapies, will emerge in the reorganization of rehabilitation services. For home rehabilitation for those patients for which it is appropriate, a single therapist is practical and efficient, as demonstrated in our Stockholm studies (see above). In-patient rehabilitation will still be required for many patients, and for parts of the programs (in some cases several short hospital stays) of even those patients rehabilitated principally at home. Stroke patients and their families are least stressed, and happier in general, when rehabilitation can be provided at home in a familiar and supportive environment, in the cases of patients who have such support systems. Furthermore, outside of the hospital, patients are less likely to acquire hospital infections. Many home rehabilitation tasks can be related to real-life activities, such as washing the dishes or sweeping the floor, and so carry-over is less difficult when the rehabilitation is already based on functional real-life activities. Attitude alone has a dramatic effect (c.f., Bach-y-Rita 1995), and the environment and psychological factors are of enormous importance.

For neurologic rehabilitation, cost-effective rehabilitation programs that are well-documented and validated should include the packaging of programs, possibly on compact disks or on the internet. These could be provided as a library of programs from which not only specialists (Physiatrists and Neurologists) but also primary care physicians could select to prescribe the rehabilitation of their patients and monitor their progress.

Neuropharmacology offers the possibility of major advances: not only for specific problems such as agitation, motivation and spasticity, but for correcting lesion-induced neurotransmitter imbalances, for facilitating pathways of the various sorts discussed above (neural as well as those related to the microenvironment of the brain cells).

Motivating therapy will dominate. An example is the ingenious approach taken by Gauthier and his colleagues (Gauthier, Hofferer and Martin, 1978) to obtaining eye movement control in children with cerebral palsy who had eye co-ordination deficits. They had noted, as had others before them, that watching a pendulum aided in the training, but found that the children refused to watch because they found it too boring. They developed a fascinating functional pendulum, by projecting children's movies (Snow White, Lassie) at a galvanometer-controlled mirror, which reflected the image to the back-side of a projection screen. The children sat in front of the screen with their heads fixed, so that to follow the pendular movements of the image, they had to use eye movements. They had 6 hours a week of intense therapy (3 movies) and within a month improved to the point that they could learn to read.

And finally, the timing of rehabilitation service delivery will change. No longer will intense therapy be forced on sick patients, many not ready for it. Various forms of rehabilitation therapy will be spread more judiciously across the full course of the disability.

For patients who have the possibility of home rehabilitation, close monitoring will occur over the Internet, and rehabilitation will be electronic and computation-based, again with simple, inexpensive mechanical patient interfaces. The computer-based devices will also facilitate data collection. A new class of therapy aid will be developed for home visits; they will interact closely, personally, and electronically with the therapy staff, and will monitor both the rehabilitation program and the psychosocial factors, helping to create a positive, hopeful and motivating environment. Equipment such as walkers and parallel bars will be simple and inexpensive, and wheelchairs will be plastic, easily assembled from molded parts. Motivating video-game based therapy appropriate for the age and interests of the patient will be available as a library of programs, so that the therapy staff will modify the rehabilitation, based on progress, by means of changes in the programs in the electronic library. The Internet, and virtual reality rehabilitation will substitute for presently used expensive devices. There will be no separation of acute and post-acute rehabilitation; rather, there will be a continuum, lasting as long as is needed - usually several years.

REFERENCES

Agnati, L., Fuxe, K., Nicholson, C., & Syková, E. (2000). Volume transmission revisited. *Prog Brain Res, 125*.

Aiello, G. L., & Bach-y-Rita, P. (2000). The cost of an action potential. *J Neurosci Methods, 103*, 145-149.

Aiello, G. L., & Bach-y-Rita, P. (2001). Hebbian brain cell-assemblies: nonsynaptic neurotransmission, space and energy considerations. In D. Cihan & A.L. Buczak & M. J. Embrechts & O. Ersoy & J. Ghosh & S. W. Kercel (Eds.), *Intelligent Engineering Systems through Artificial Neural Networks* (Vol. 11, pp. 441-447). New York: ASME Press.

Bach-y-Rita, P. (1972). *Brain Mechanisms in Sensory Substitution*. New York: Academic Press.

Bach-y-Rita, P. (1991). Thoughts on the role of volume transmission in normal and abnormal mass sustained functions. In K. Fuxe & L. F. Agnati (Eds.), *Volume Transmission in the Brain* (pp. 489-496). New York: Raven Press, Ltd.

Bach-y-Rita, P. (1995). Nonsynaptic Diffusion Neurotransmission and Late Brain Reorganization. New York: Demos-Vermande.

Bach-y-Rita, P. (2000). Conceptual issues relevant to present and future neurologic rehabilitation. In H. Levin & J. Grafman (Eds.), *Neuroplasticity and reorganization of function after brain injury* (pp. 357-379). New York: Oxford University Press.

Bach-y-Rita, P. (2001a). Theoretical and practical considerations in the restoration of functions following stroke. *Topics in Stroke Rehab., 8*, 1-15.

Bach-y-Rita, P. (2001b). Nonsynaptic diffusion neurotransmission in the brain: Functional considerations. *Neurochem. Res., 26*, 871-873.

Bach-y-Rita, P. (2002). Brain Damage, Recovery From. In V. S. Ramachandran (Ed.), *ENCYCLOPEDIA OF THE HUMAN BRAIN* (Vol. 1, pp. 481-491). New York: Academic Press.

Bach-y-Rita, P. (2003). Late post-acute neurologic rehabilitation: neuroscience, engineering and clinical programs. *Arch Phys. Med. Rehab., 84*, 1100-1108

Balliet, R., Shinn, J. B., & Bach-y-Rita, P. (1982). Facial paralysis rehabilitation: Retraining selective muscle control. *Intern. J. Rehabil. Med., 4*, 67-74.

Bach-y-Rita, P., Collins, C. C., Saunders, F., White, B., & Scadden, L. (1969). Vision substitution by tactile image projection. *Nature, 221*, 963-964.

Bach-y-Rita, P., Wood, S., Leder, R., Paredes, O., Bahr, D., Wicab-Bach-y-Rita, E., & Murillo, N. (2002). Computer-assisted motivating rehabilitation (CAMR) for institutional, home, and educational late stroke programs. *Topics Stroke Rehab., 8*, 1-10.

Broca, P. (1861). Remarques sur le siege de la faculte du langage articule; suivies d'une observation d'a phemie (perte de la parole). *Bull. Soc. Anat. Paris, 6*, 330-357.

Colotla, V. A., & Bach-y-Rita, P. (2002). Shepherd Ivory Franz: His Contributions to Neuropsychology and Rehabilitation. *Cognitive, Affective, & Behavioral Neuroscience, 2*, 141-148.

De Keyser, J. D., Ebinger, G., & Vauquelin, G. (1989). Evidence for a widespread dopaminergic innervation of the human cerebral cortex. *Neuroscience Letters, 104*, 281-285.

De Pedro-Cuesta, J., Widen-Holmqvist, L., & Bach-y-Rita, P. (1992). Evaluation of stroke rehabilitation by randomized controlled studies: a review. *Acta Neurol Scand, 86*, 433-439.

Fuxe, K., & Agnati, L. F. (1991). *Volume transmission in the Brain.* New York: Raven Press.

Gauthier, G. M., Hofferer, J. M., & Martin, B. (1978). Film projecting system as a diagnostic and training technique for eye movements of cerebral palsied children. *Electroencephalography and Clinical Neurophysiology, 45*, 122-127.

Pascual-Leone, A., & Torres, F. (1993). Plasticity of the sensorimotor cortex representation of the reading finger in Braille readers. *Brain, 116*, 39-52.

Sampaio, E., Maris, S., & Bach-y-Rita, P. (2001). Brain Plasticity: "Visual" Acuity of Blind Persons via the Tongue. *Brain Research, 908*, 204-207.

Tangeman, P. T., Banaitis, D. A., & Williams, A. K. (1990). Rehabilitation of chronic stroke patients: changes in functional performance. *Arch. Phys. Med. Rehabil., 71*, 876-880.

Taub, E., & Crago, J. E. (1995). Behavioral plasticity following central nervous system damage in monkeys and man. In B. Julesz & I. Kovacs (Eds.), *Maturational Windows and Adult Cortical Plasticity.* Redwood City, CA: Addison-Wesley.

Tyler, M.E., Danilov, Y. et al. (2003). Closing and open-loop control system: vestibular substitution through the tongue. J. Integrative Neurosci 2: 1-6.

Westerberg, E., Monaghan, D. T., Kalimo, H., Cotman, C. W., & Wieloch, T. W. (1989). Dynamic changes of excitatory amino acid receptors in the rat hippocampus. *J. Neurosci., 9*, 798-805.

Widén-Holmqvist, L., von Koch, L., Kostulas, V., Holm, M., Widsell, G., Tegle, R. H., Johansson, K., Almazan, J., & de Pedro Cuesta, J. (1998). A randomized controlled trial of rehabilitation at home after stroke in south-west Stockholm. *Stroke, 29*, 591-597.

S. ISENMANN

Clinic and Policlinic of Neurology, Friedrich-Schiller-University, Jena, Germany

31. VIRAL GENE DELIVERY

Summary. Somatic gene therapy aims at increasing the expression of deficient or protective factors, or at the suppression of deleterious factors. This goal will probably be achieved best by the use of viral vectors. Currently, vectors are being optimised with respect to transduction properties, levels of transgene expression, and reduction of toxicity and immunogenicity. While a multitude of gene transfer approaches have been completed in experimental models, gene therapy is now slowly moving into the clinic, and promising results can be hoped for in several cases. Caution is required, however, not to overstrain hope in this emerging therapeutic modality, and drawbacks in gene therapy studies outside the nervous system warrant meticulous caution for safety concerns.

1. INTRODUCTION

For many neurological diseases still no satisfactory treatments are available. Molecular genetics has in the past decade helped to identify causative genes and mutations for a number of familial neurological diseases, and progress in our understanding of pathogenic events led to the hope that it may soon be possible to specifically and causally interfere with such mechanisms. Here, we will introduce the concept and basic science of gene therapy approaches, illustrate some examples of experimental and clinical gene therapy studies, and discuss future perspectives of this emanating treatment modality that may help ameliorate the course of neurological disorders, and change the lives of affected patients.

2. WHAT IS GENE THERAPY?

2.1. The concept of gene therapy

Classical pharmacotherapy usually aims at relief of disease symptoms (e.g., pain, impairment of movement, altered muscle tone) rather than targeting of underlying (causative) molecular genetic alterations. Typically, drugs have short half lives, necessitating administration on a regular scheme, often several times daily. In addition, pharmacotherapy for CNS conditions is inevitably administered systemically, and hence prone to cause systemic side effects.

Gene therapy was originally envisaged as a concept to overcome many of the restrictions and side effects of classical pharmacological therapies. The original concept involved the replacement of gene products (enzymes) deficient in autosomal recessive inborn errors of metabolism. Thereby, following somatic gene transfer, these enzymes will be produced by the very cell population that express them under physiological conditions. Gene delivery would be a single event, followed by long-term – ideally, permanent – gene expression and, hence, definite phenotypic compensation. If this was feasible, gene therapy would be much more effective than classical pharmacotherapy and protein delivery – (facilitated) administration of recombinant proteins. While this concept still holds, current experience shows that gene therapy is still an ambitious goal. On the other hand, with ongoing research on the molecular biology and pathophysiology of nervous system diseases, and progress in gene transfer techniques, possible applications of gene delivery to the nervous system are getting even broader. At the same time, scenarios emerge where gene transfer or – with view to future clinical application – gene therapy can become an important therapeutic modality in diseases as diverse as metabolic diseases, acute neuronal demise in stroke, brain trauma and spinal cord injury, chronic neurodegenerative conditions such as Alzheimer's and Parkinson's disease, and brain tumours, and many of these concepts are now aimed at more or less symptomatic relief rather than causative cure. Therapeutic strategies for neuronal rescue usually follow one of three basic principles: to replace missing enzyme or protein functions in recessive diseases; to provide general trophic support to neurons at risk in acute or chronic neurodegeneration, or to decrease the extent of protein dysfunction in dominant negative gain of function mutations.

The basic principle of gene therapy, however remained the same: to deliver genetic blueprints of desired factors specifically and effectively to target tissues or target cells or, especially in the CNS, to target regions, and express the proteins of interest. Still, due to specificity of the approach only minor – if any – side effects, and no systemic drawbacks should be encountered.

2.2. Gene transfer: principles and basic science

In gene therapy, it is not the remedy (i.e. protein or active compound) itself that is administered, but rather the nucleic acid (i.e., DNA or, less frequently, mRNA) that is delivered to target cells. Therefore, the cellular "biomachinery" is utilized to produce beneficial factors, for an extended period of time. In many instances, gene transfer will be *additive*, i.e., genes for factors that are deficient, or that would be favourable, are introduced by means of gene transfer, and an additional function is introduced. However, the contrary is also conceivable: to eliminate by *ablative* gene transfer gene products that may be deleterious, e.g., in dominant gain of function mutations, or to abolish genes associated with tumour growth (such as constitutively active oncogenes or growth factor receptors, such as ErbB2, for example). The latter can be achieved by dominant negative approaches, by antisense oligonucleotides, antisense RNA, catalytic ribozymes, RNA "decoy" and small inhibitory RNA (Xia et al., 2002), and will not be discussed to much detail in this chapter.

Among the factors that can be added are genes for deficient enzymes in metabolic diseases. Here, it is important to note that many such diseases are inherited in an autosomal recessive manner indicating that expression of even a minor amount of the respective protein may suffice to maintain a normal phenotype.

In chronic neurodegeneration, neurons of a particular phenotype, located in specified anatomical locations or compartments (such as midbrain neurons in the substantia nigra in Parkinson's disease), using a particular neurotransmitter (e.g., dopamine in Parkinson's disease, or acetyl choline in Alzheimer's disease) are afflicted by the premature degeneration process. Possible gene transfer strategies, therefore, may aim at replacing the respective neurotransmitter, e.g., by gene transfer of synthesizing enzymes. Alternatively, neurotrophic factors such as BDNF or GDNF may target the degeneration process and provide neurotrophic support; anti-apoptotic factors such as Bcl-2, or caspase inhibitors such as XIAP may protect lesioned neurons (Reed, 2002). Many neurodegenerative diseases share as a pathologic feature insoluble extra- or intracellular protein aggregates. While it is not yet entirely clear if these protein aggregates contribute to the pathophysiology and neuronal dysfunction, or if they are a mere by-product of the pathophysiological changes, evidence from immunization trials in experimental Alzheimer's models and in patients suggests that clearing of protein aggregates may halt the degeneration process, and even improve function. Therefore, expression of chaperone proteins that can help resolve protein aggregates, and activation of the ubiquitine proteasome pathway may be further promising therapeutic principles.

In more acute neuronal degeneration such as cerebral ischemia or hemorrhage, brain trauma, or spinal cord injury, anti-excitotoxic factors (e.g., glutamate receptor antagonists), free radical scavengers, and inhibitors of mitochondrial dysfunction may be effective (Sapolsky, 2003). For ensuing axonal regeneration, that is a prerequisite for functional recovery, in addition outgrowth and regeneration-promoting, anti-scarring, and anti-inflammatory factors may be useful. Table 1 gives an overview on principles and factors that may be effective in promoting neuronal function and regeneration, and inhibiting degeneration.

2.3. Gene therapy targets

Other chapters in this section cover to some detail the pathophysiological cascades relevant to neuronal degeneration in various CNS pathologies, from which promising gene therapy targets may be devised. While it was held until some ten years ago that neurons die almost exclusively by excitotoxicity and necrosis in acute insults, and apoptotic mechanisms started to emerge as important contributors to neuronal demise in chronic neurodegenerative diseases, it appears now that the two types of neuronal degeneration may – to varying extent – contribute to neuronal damage in most of the conditions discussed here. It should be stressed, however, that the mere naming of the prevailing cell death mechanism does by no means imply that the mode of neuronal degeneration – and, therefore, the targets for successful treatment options – are understood to every detail. Rather, we should be aware that in spite of the discovery of a number of deleterious molecules, most interactions are probably not yet sufficiently clear. These mechanisms will contribute to cell damage in various conditions to differing extent, and there are several facets of excitotoxic and apoptotic cascades that involve different molecules in different cells, and – depending on the primary pathology – in various brain regions (*selective vulnerability*). Therefore, in spite of similar molecules involved, refined strategies both for the choice of targets and for the vector, and route of administration will have to be devised. Bearing this in mind, we may still list some of the more important players – and hence attractive molecular targets – in the degeneration process. For excitotoxic damage, these include excess glutamate receptor activation, calcium influx, ion channels, energy deprivation, oxygen radicals, nitric oxide, proteases, lipid peroxidation, inflammatory processes (Sapolsky, 2003). In apoptotic neuronal damage, proteins of the Bcl-2 family, caspases, IAPs, mitochondrial dysfunction, DNases, transcription factors are involved (Reed, 2002). Neurotrophic factors such as BDNF or GDNF; SOD; antiapoptotic Bcl-2 homologues and IAPs may be protective in a wide range of pathologies.

2.4. Gene structure, promoters, regulatory elements

Effective gene delivery to the brain requires – in addition to the nucleic acid sequence of interest – a promotor to drive gene expression. Initial experimental studies often used viral promotors (typically, CMV [cytomegalovirus] or RSV [rouse sarcoma virus]) in order to achieve rapid, high-level transgene expression in a broad host cell range, and such studies were important to show in principle the feasibility of CNS gene transfer in animal models. With increasing experience, however, more specific and adjustable expression was desired. Therefore cell-type specific promoters were tested. Neuronal expression can now be achieved using neuron specific synapsin, neuron specific enolase (NSE), or tubulin $\alpha 1$ promoters (Kügler et al., 2001), or, alternatively, glial specific silencer elements. Conversely, transgene expression can be directed to astrocytes or oligodendrocytes by the glial fibrillary acidic protein (GFAP) or myelin basic protein (MBP) promoters, respectively.

Generally, the genomic gene structure is characterized by the presence of exons (gene sequences that are translated to protein) with intervening introns (untranslated sequences that have regulatory functions). In contrast – owing to limitations of vector capacity (see below) – for gene transfer often cDNAs are used that contain mere exon structures. Increasing evidence now indicates that regulatory elements such as introns and, for example, the woodchuck hepatitis virus posttranscriptional regulatory element (WPRE) can increase and stabilize transgene expression. Generally, regulation of transgene expression is desired. To this end, promoters that can be efficiently turned down are desirable for security aspects, e.g. when unexpected side effects or immunological reactions may prevail, or transgene expression is no longer needed. Regulation (in- and decreasing levels of transgene production as well as switching it on and off) may be achieved efficiently by taking advantage of heterologous transcription factors that can be regulated by administration or withdrawal of low molecular weight ligands such as tetracycline or steroids.

3. GENE TRANSFER TECHNIQUES

3.1. In vivo and ex vivo gene transfer

The original concept of CNS gene therapy involves an *in vivo* gene transfer approach, whereby a vector is introduced directly into the CNS, and consecutively the therapeutic transgenes are expressed by targeted host cells. An alternative approach includes *ex vivo* gene transfer. Here cells (of autologous or heterologous origin) are transfected or transduced *in vitro*, and transgene expressing cells are then transplanted into the CNS (the brain, spinal cord, CSF, or eye, for example). In this scenario, transplanted cells are often designed to act as biological "minipumps" that continuously produce and release therapeutic compounds (Aebischer & Ridet, 2001; Isenmann & Bähr, 1997).

With increasing knowledge on the molecular pathological and physiological processes involved in neurological insults and neurodegeneration, and genetic findings in familial neurodegenerative diseases, an increasing number of promising candidate genes for gene transfer is emerging. Delivery to the CNS, however, remains a major challenge that has not been solved satisfactorily to date. Administration of recombinant proteins may in experimental models be useful for *proof of principle* demonstration that certain factors are effective in defined pathologies. However, in many pathologies (except, perhaps, cerebral ischemia and the acute phase of spinal cord trauma) long-term delivery of therapeutic factors will be desired, and will only be achieved by delivery of respective coding sequences, i.e., DNA (or mRNA). Brain targeting by systemic delivery is difficult due to the blood-brain barrier (BBB; Pardridge, 2002). Although strategies have been devised to circumvent this barrier, e.g., by transiently opening the BBB or conjugation of proteins, DNA, or vectors to transferring receptor (OX26) antibodies, systemic delivery may still not be desired to prevent systemic side effects. Local routes of administration may include neurosurgical procedures to the brain parenchyma or the CSF, and may utilize transport properties of vectors in the brain along axonal projection trajectories.

3.2. Viral and non viral DNA transfer

In vitro, gene transfer to cells and tissues may be achieved by direct incubation with recombinant DNA, without further modification (*naked DNA*). In principle, naked DNA can also be injected into the brain. Delivered DNA is expressed both locally and in distant regions following axonal transport. More refined gene delivery strategies to the brain include bio-ballistic methods using a *gene gun*, *in vivo* electroporation, and complexing of nucleic acids with cationic lipids (*lipoplexes*), or polymers (*polyplexes*). Major obstacles to these approaches include limited cellular uptake, inefficient nuclear transport (and hence persistence in the cytosol, with ensuing degradation), low level transcription, and limited duration of transgene expression. Therefore, delivery of naked or complexed DNA may not be fundamentally superior to delivery of recombinant (stabilized) proteins.

Viruses have specialized to infect cells, and deliver their genetic material to the nucleus, where introduced nucleic acids are transcribed. Therefore, viral vectors have emerged as attractive tools for therapeutic gene delivery. Technically speaking, two principal prerequisites must be met in order to utilize viruses for this purpose: the potential to replicate must be abandoned (while infectivity, i.e., the potential to effectively deliver the genetic material contained within the particle must be retained), and the nucleic acids must be modified such that pathogenic virus genes are deleted and replaced by the desired therapeutic genes (while retaining the capacity to package the modified nucleic acids into virions). In order to produce gene transfer vectors, as a general rule, genes coding for essential viral genes are deleted (and the respective proteins provided *in trans* for production of vector particles), and a nucleic acid containing therapeutic gene sequences instead of viral genes, and the viral packaging signal, is packed by producer cells to obtain vector particles in which the therapeutic DNA is contained in virion particles (for details see Isenmann & Bähr, 1997; Costantini et al., 2000).

3.3. Viral vectors for CNS gene transfer

Several vector systems have been devised that meet these basic requirements, and have been utilized with some success in experimental gene therapy studies. More specifically, prerequisites for clinical gene transfer trials using viral gene transfer include the use of replication-deficient, safe vectors (with the exception of several cancer gene therapy protocols that use replication-competent, or replication-conditional vectors that may be toxic to tumour cells); targeted and restricted transgene expression, e.g. through – in addition to targeted local delivery – surface receptors, cell-type specific promoters and, importantly, minimal vector toxicity and immunogenity (see Table 2; Aebischer & Ridet, 2001; Costantini et al., 2000; Isenmann & Bähr, 1997).

3.3.1. Properties of viral gene transfer vectors

The first vectors devised this way were derived from retroviruses, more specifically **oncoviruses** such as Moloney murine leukaemia virus (MoMLV). These enveloped RNA viruses can transduce a wide range of host cells. Their use for gene transfer to the CNS *in vivo* is limited by the requirement of cell division for integration and transgene expression. They have been used, however, for a multitude of *ex vivo* gene transfer approaches, where cells as diverse as (autologous) fibroblasts, hematopoietic precursors or neural precursors, or (heterologous) cell lines and tumours have been transduced *in vitro*, followed by transplantation. This approach has also been used for a few clinical trials in neurodegenerative disorders (see below). However, the dosage of protein is difficult to determine, and stable and regulated expression is still a challenge. When used for *in vivo* gene transfer, oncovirus vectors bear a risk of tumour formation by insertional mutagenesis. Recently, this scenario has become reality in two children patients who were otherwise successfully treated for adenosine deaminase (ADA) deficiency causing severe combined immune deficiency (SCID). **Lentivirus** vectors, derived from HIV, SIV, or FIV viruses belong to another subfamily of retroviruses. They have the advantage of effectively transducing also non-dividing cells. The host range has been broadened to include many cell types including neurons by pseudotyping with the vesicular stomatitis virus G glycoprotein (VSV-G). For safety aspects, three- and four- plasmid systems have been devised to preclude recombination of replication competent virus particles. To increase safety further, self-inactivating vectors have been generated by manipulations of the U3 regions of the 5' and 3' LTR. Lentiviral gene transfer to the CNS yields high-level, long term neuronal transgene expression, with retrograde transport, and no significant immune response.

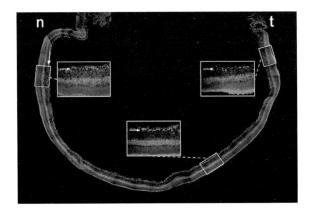

Fig. 1. Adenoviral gene transfer to retinal ganglion cells in the adult rat retina. GFP fused to HSV VP22 is expressed from a CMV promoter following adenoviral gene transfer. Many retinal ganglion cells (arrows) are transduced throughout the retinal surface (n, nasal; t, temporal part of the retina; see Kretz et al., 2003, for details).

Adenoviral gene transfer to the brain was first described some ten years ago. Wild-type adenoviruses are dsDNA viruses with a genome of approximately 36 kb and have a natural tropism for epithelia. Anti-adenoviral antibodies occur frequently in the population. The first generation of replication-defective vectors was constructed by deleting E1 and/or E3 genes. These vectors transduce many cell types and typically confer transgene expression for several weeks, although in some reports significantly longer periods of expression were noted. In many paradigms, first generation vectors provoke a significant immune reaction in experimental animals. Ensuing vector generations with constructs containing additional deletions in the E2 and/or E4 genes were less immunogenic. These vectors typically accommodate just under 10 kb of foreign DNA. The gene therapy field was shaken in 1999 when a 19-years old volunteer died from a severe immunological reaction to the vector and ensuing multi organ failure following administration of a high dose of adenoviral vectors for liver gene therapy of ornithine transcarbamylase (OTC) deficiency. A major breakthrough in adenoviral vector technology came with the advent of **"gutless" vectors** that contain only the sequences necessary for packaging and replication of the genome, but lack all structural genes, thereby increasing transgene capacity up to 37 kb, while reducing immunogenity and prolonging the duration of expression.

Adeno-associated virus (AAV) is a non-pathogenic ssDNA virus with a genome (and vector capacity) of approx. 4.8 kb. The ITRs promote both extrachromosomal replication and genomic integration that in the presence of the *rep* gene products may occur preferentially site-specific into human chromosome 19q. Long-term transgene expression is facilitated both in the episomal state, and following integration. Typically, replication and packaging of AAV vectors requires an (adenoviral) helper virus, that is later heat inactivated; an adenovirus-free method has, however, also been described. AAV vectors show little toxicity and circulating antibodies, although frequently present, are rarely neutralizing.

Herpes simplex virus (HSV-1) is a common pathogen in adults, with more than 90% of the population carrying circulating antibodies. HSV-1 usually causes cold sores, but occasionally life-threatening HSV encephalitis can occur. The genome organisation is complex, with a 152 dsDNA molecule coding for more than 80 genes. Infection with wild-type HSV-1 consists of two cycles, with virus replication occurring in the lytic cycle, and a latent cycle in which no replication takes place, and only one promotor (the latency active promotor, LAP) is active. Two types of vectors are derived from HSV-1: **Recombinant HSV-1 vectors** essentially contain the entire viral genome with mutations in at least one viral gene to reduce toxicity, and provide space for 30-50 kb of transgene DNA. Current vectors carry deletions in multiple immediate-early genes that encode transactivating factors, thereby essentially eliminating expression of most viral genes. Replication conditional recombinant HSV-1 vectors can replicate specifically in tumour cells, and may specifically kill brain tumours (see below). **HSV amplicon vectors** consist of a plasmid containing the HSV origin of replication and the packaging signal, allowing for packaging as concatenate into HSV virions in the presence of HSV helper functions that are provided by a set of cosmids or a bacterial artificial chromosome (BAC) plasmid. Amplicon vectors are not toxic or antigenic, and have a large transgene capacity (up to 150 kb).

As would be expected, the virus-based vector systems introduced have different properties with respect to cell-type specificity and transduction; transgene capacity; cellular state and location of introduced nucleic acids; kinetics and duration of transgene expression; and toxicity and immunogenity (Table 2). It emerges that, although various vector systems have been employed, none of the currently available ones is perfect in a sense that it would without restrictions be utilized for clinical gene therapy for nervous system diseases. In addition, different pathologies (e.g., acute neuronal demise in stroke; acute axonal damage in spinal cord trauma; widespread chronic neuronal degeneration in Alzheimer's and localized degeneration in Parkinson's disease; life-long persisting enzyme deficiency in lysosomal storage diseases; brain tumours) will – in addition to transfer of different therapeutic genes, and use of specified vectors – also call for different application strategies. The development and optimisation of **hybrid vectors** that are aimed at combining specific advantages of various vector system while eliminating their specific disadvantages is an active field of *vectorology* (vector development/optimisation technology). To this end, HSV/AAV, adenovirus/AAV, and adenovirus/retrovirus chimeric, or hybrid vectors have been devised. Other vector systems under development and evaluation include **foamy virus** vectors (retrovirus, spumaviridae), **sendai virus** (hemagglutinating virus of Japan) vectors, **Ebstein Barr virus** (EBV) vectors, and **alphavirus** (Sindbis Virus, Semliki Forest virus) vectors. Interestingly, it emerges that vector-derived gene transfer can be modified and enhanced by combination with nonviral techniques, such as liposomes, polyethylenimine, or protein transduction domains such as HIV TAT, HSV-1 VP22 (cp. Fig. 1), and *antennapedia*.

Importantly, transduction and expression characteristics are not only determined by the gene transfer vector, gene construct and promotor used, but also by the route of administration, site of action, local diffusion (that in the brain is generally limited), and transport characteristics of the vector and gene product, and these variables necessitate careful consideration and evaluation in a specific context, rather than a uniform protocol for conditions as diverse as acute hemispheric cerebral ischemia (stroke), or chronic degeneration of cholinergic neurons in the neocortex, hippocampus and basal nucleus over decades (Alzheimer's disease). In the long term, gene therapy vectors will likely be different from the vectors used at present, and the use of vectors that are tailored for specific applications, organs, tissues and cell types, and, possibly, even for individual patients may be envisaged.

4. PREREQUISITES FOR CLINICAL GENE THERAPY TRIALS

4.1. Disease models, pathophysiology, and safety aspects

While the concept of gene therapy for neurological diseases is appealing, the obstacles to successful genetic therapies are manifold, and drawbacks have occurred – just as with conventional pharmacological therapies. In general, two major prerequisites have to be fulfilled before any novel therapeutic regimen can be brought to clinical application: it should be both effective and safe. For gene therapy, effectivity will depend on the understanding of biochemical pathways and pathophysiological principles, and the ability to effectively interfere with pathological processes. This can be tested on a molecular basis in cell or tissue culture, and in animal models. With respect to stroke, for example, a number of animal models are available that allow to model the pathology of cerebral ischemia, and examine putative protective compounds in rodents. Caution is warranted, however, with respect to translation of such work to clinical application in humans. Thus, a vast number of promising studies on successful pharmacological neuroprotection for cerebral ischemia, have – in spite of major efforts – not been successfully translated to clinical trials: Possible reasons for this failure have been intensly discussed in the past years, and include the notion that standardized cerebral ischemia in young, healthy, lissencephalic rodents may differ considerably from etiologically and locally heterogeneous stroke in elderly humans bearing a number of cardiovascular risk factors. Keeping this in mind, promising results obtained with experimental gene transfer approaches must also be interpreted with caution.

Chronic neurodegeneration as occurs in the system degenerations, e.g. Parkinson's, Alzheimer's, Huntingtons's diseases, or ALS and the spinocerebellar ataxias, may be more reliably modelled, since for all of the mentioned conditions, familial forms have been described. In the past ten years, disease genes, and mutant human alleles have been identified, and used to model the conditions, on a molecular genetic basis, in transgenic rodents. While still relevant differences exist between the genome, cellular metabolism, and CNS anatomy of mice and men, such genetic models may be particularly relevant for modelling human pathology in laboratory animals. With respect to safety of gene transfer, several aspects have to be considered. The vector as such must be safe, i.e. it must not evoke any major adverse reaction (e.g., immune reaction, toxicity, insertion mutagenesis); the route of administration (e.g., neurosurgical procedures, temporary opening of the BBB) must be safe, and, finally, the transgene (i.e., the "drug" proper) must be well tolerated.

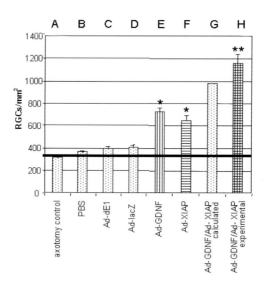

Fig. 2. Experimental gene transfer for neuroprotection. Rescue of retinal ganglion cell neurons in vivo following transection of the optic nerve and administration of adenoviral vectors. Following axotomy alone (A), or with control treatments (injection of buffer [B], Ad.dE1 containing no transgene cassette [C], or Ad.lacZ expressing E. coli β-galactosidase [D]), around 350 – 400 RGCs/mm^2 survive, with no significant difference between groups. Ad.GDNF (E) and Ad.XIAP (F) both protect a significant number of RGCs from axotomy-induced degeneration. Double treatment (H) rescues significantly more RGCs than either treatment alone. This number is higher that would be expected if the two treatments acted only in an additive fashion (G), indicating synergistic neuroprotection. For experimental and statistical details, see (Straten et al., 2002).

In addition, safety concerns include the caution not to cause germ line mutations in the case of integrating vectors, and the absence of recombination events. While basic aspects of these requirements can be tested in rodents, they must also be carefully examined in larger mammals, i.e. dogs and/or monkeys, before transition to the clinic is warranted.

Along these lines, to date, a number of gene therapy studies have been carried out in laboratory animals in models of cerebral ischemia, Alzheimer's and Parkinson's disease and in ALS, as well as in several genetic models of metabolic diseases affecting the brain.

5. PRECLINICAL GENE THERAPY STUDIES AND CLINICAL TRIALS

It would by far exceed the scope of this chapter to enumerate the multitude of animal studies that have reported to some extent successful gene therapy applications in animal models of acute or chronic neuronal damage, metabolic defects, or brain tumours. Suffice it to note that hundreds of reports have over the past 15 years or so described experimental *ex vivo* and *in vivo* gene transfer, and considerable progress has been made over the past years owing to improved vectors, and more refined insight into pathophysiological processes, and cell death cascades. Here, we will focus on some of the more remarkable and promising examples, and highlight several instances where clinical trials have already been initiated or may be initiated soon, on the basis of experimental studies.

5.1. Metabolic diseases

A number of metabolic diseases have severe CNS phenotypes, often leading to mental retardation in infants, and premature death. Among these conditions are lysosomal storage diseases, and defects in amino acids metabolism. Current treatment options include for some of these disorders dietary restrictions, substitution of the deficient enzyme, or bone marrow transplantation, but for most there is no satisfactory treatment available. Gene therapy may be particularly useful here.

Sly syndrome (mucopolysaccharidosis [MPS] Type VII) is an autosomal dominantly inherited lysosomal storage disease caused by a defect in β-glucuronidase. In a mouse model, successful phenotypic correction has recently been shown by FIV-mediated *in vivo* gene transfer of the *β-glucuronidase* gene (Brooks et al., 2002). In this study it was particularly remarkable that behavioural abnormalities were not only halted, but even reverted. Similar approaches are conceivable for many of the more than 40 known storage diseases, and such approaches may be translated to the clinic soon, provided safe vectors and application procedures are available.

Canavan's disease is a leukodystrophy caused by a defect in amino acid metabolism due to aspartoacyclase (ASPA) deficiency. AAV-mediated ASPA delivery to the brain has been achieved and was well tolerated in two affected children according to a pilot study (Leone et al., 2000). Functional effects of such treatment have been evaluated in similarly treated *aspa* knockout mice, and improvement has been shown histologically, and by magnetic resonance (MR) spectroscopy, and MR imaging of the brain. Meanwhile, a clinical gene therapy trial employing neurosurgically administered ASPA expressing AAV has been initiated (Janson et al., 2002). This is particularly encouraging in the CNS gene therapy field, since – although most metabolic diseases are rather rare – correction of disease phenotypes in autosomal recessive metabolic diseases by somatic gene transfer was the initial goal when the concept of gene therapy emerged.

5.2. Neurodegenerative diseases

Parkinson's disease is characterized mainly by motor symptoms caused by dysfunction and degeneration of dopaminergic neurons in the substantia nigra of the ventral midbrain, leading to lack of the neurotransmitter dopamine in the striatum. Pharmacological dopaminergic treatment options are available and often lead to symptomatic relief for years, before side effects and fluctuations prevail. Both rodent and monkey models have been established that mimic the pathophysiological and histological hallmarks of the human disease, and can be utilized for experimental treatment strategies. Gene therapy approaches have aimed at either replacement of dopamine synthesizing enzymes, i.e., tyrosin hydroxylase (TH), the rate limiting enzyme with or without additional replacement of aromatic L-amino acid decarboxylase (AADC), GTP cyclohydrolase 1 (GTP CH1) and vesicular monoamine transporter-2 (VMAT-2), or at protecting the neurons at risk by transfer of vectors expressing the trophic factor, GDNF. A number of animal studies have reported beneficial results for either approach using various vectors. Interestingly, combination treatment with an adenoviral vector expressing the caspase inhibitor, XIAP, that rescued nigral neurons from MPTP-mediated toxicity, and a second vector expressing GDNF, that preserved axonal terminals in the striatum, had synergistic protective effects in rodents (Eberhardt et al., 2000). In monkeys, lentiviral GDNF transfer also rescued nigral neurons and preserved dopaminergic function (Kordower et al., 2000). These findings may soon prompt clinical trials using lentivirus-mediated GDNF transfer for Parkinsonian patients. Deep brain stimulation of the subthalamic nucleus (STN) following implantation of electrodes is another treatment option for Parkinson's patients. In a recent report, "functional subthalamectomy" was achieved by AAV-mediated glutamic acid decarboyxlase (GAD) gene transfer to the STN in rats, and this conversion of the STN to a predominantly inhibitory, GABA releasing system protected nigra neurons (Luo et al., 2002). Again, a clinical gene therapy trial is underway making use of this strategy.

In **motor neuron disease**, or amyotrophic lateral sclerosis (ALS), dysfunction and degeneration of motor neurons in the primary motor cortex and the ventral horn of the spinal cord causes progressing weakness of the limb muscles. Typically, death occurs few years after onset from neuromuscular respiratory failure. To date, the disease course can be slowed slightly by riluzole, a glutamate-blocking drug, but no causative treatment or cure is available. In a fraction of familial cases, mutations in the gene for superoxide dismutase (SOD)-1 have been found, and transgenic mice were generated as disease models and to evaluate therapeutic strategies. Transgenic overexpression of *bcl-2*, or AAV-mediated *bcl-2* gene transfer slowed motor neuron degeneration, with the latter strategy obviously only being effective in the treated segments. Other promising candidate therapeutic factors include calbindin, or glial glutamate transporters. Owing to the wide distribution of motor neurons along the entire CNS axis, however, the affected cell population will hardly be specifically accessible to gene transfer, and strategies mainly aim at supporting the neurons at risk. Based on experimental findings that ciliary neurotrophic factor (CNTF) is a trophic factor for motor neurons, in the mid 1990ies two large clinical trials were initiated with a total of some 1,300 patients that received subcutaneous CNTF injections. Overall, there was no indication of efficacy, while many patients experienced side effects. About the same number of patients was treated with recombinant BDNF, with similarly disappointing results except, perhaps, the notion of a subgroup analysis that patients in the highest dose group that experienced severe diarrhoea as side effect also appeared to have a higher survival rate. One of two studies (with a total of some 450 patients) using insulin-like growth factor (IGF)-1 appeared to show some beneficial effect, but given the negative results of the second study, this was not followed further (Apfel, 2001; Thoenen & Sendtner, 2002). It emerged from these studies that were conducted in the early days of neurotrophic factor research, that delivery of these putative protective agents was a major problem. This led to experimental studies using *ex vivo* gene transfer of neurotrophic factors to various cell types, followed by transplantation to the CNS, or the CSF. Intrathecal CNTF infusion, or transplantation of CNTF secreting encapsulated xenogenic cells in a total of 10 patients was well tolerated, yet without obvious beneficial effects (Aebischer et al., 1996). It will have to be seen if refined administration techniques and, possibly, *in vivo* gene transfer strategies may have the potency to significantly alter the course of motor neuron disease.

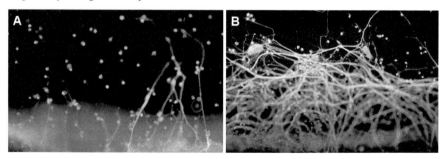

Fig. 3. Experimental gene transfer for axonal regeneration. Adenovirally delivered Bcl-X_L promotes neurite outgrowth of adult RGCs in vitro. A, sparse intrinsic neurite outgrowth from adult retinal explants. B, intense neurite outgrowth in retinal explants overexpressing Bxl-X_L following adenoviral transduction (see Isenmann et al., 2003, Kretz et al, 2004).

In **Alzheimer's disease**, cholinergic neurons in the hippocampus, cerebral cortex, and basal forebrain nuclei dysfunction and degenerate, leading to the known progressive neuropsychological syndrome. The disease process is accompanied by the deposition of insoluble amyloid plaques, which may be a treatment target as suggested by recent studies employing antibody treatments. Cholinesterase inhibitors and memantine, a non-competitive NMDA receptor antagonist, are currently used for mild to moderate cognitive symptoms, but no causative treatment option is available. Before mutations in the *APP*, *PS-1* and *PS-2* genes were identified and transgenic mouse models of the disease were available, fimbria fornix transection was used as a model for forebrain cholinergic neuron degeneration. A wealth of studies indicated that NGF when infused into the brain or delivered by genetically engineered cells could rescue cholinergic neurons, and these findings were then translated to Alzheimer's disease patients. Following feasibility studies with intraventricular NGF infusions in several patients, eventually a phase I clinical study was initiated. Autologous fibroblasts were transduced with a retrovirus vector to express NGF, and transplanted to the nucleus basalis of several Alzheimer's patients. Ongoing evaluation includes neuropsychological testing and MRI scans (Tuszynski, 2002). However, it should be noted that Alzheimer's disease involves large parts of the brain, and therefore local protective treatment will likely only result in partial and temporary symptomatic relief.

Based on experimental findings, neurotrophic factors are promising treatment options in a number of neurological conditions (Aebischer & Ridet, 2001; Apfel, 2001; Thoenen & Sendtner, 2002; Tuszynski, 2002). However, systemic administration has consistently lead to severe side effects. *In vivo* or *ex vivo* gene therapy strategies may help to avoid side effects while targeting the remedies to the desired location and CNS cell populations at risk. While this is beyond the scope of the initial gene therapy concept – to introduce causal therapies – this may still represent a significant progress given the lack of long-term efficient therapies for these conditions to date.

5.3. Brain tumors

Of a total of several hundred clinical gene therapy trials in clinical medicine, approximately 2/3 have been initiated for cancer gene therapy. This is largely due to the poor prognosis of a number of tumour entities. Glioblastoma patients, for example, have a mean survival of approx. one year in spite of current treatment modalities (surgery, radiation, chemotherapy). Unlike in neurodegenerative or metabolic diseases, the aim of cancer gene therapy is to efficiently eliminate malignant cells. This may be achieved by intracellular generation of toxic compounds, enzymatic activation of pro-drugs, expression of tumour suppressor or apoptosis-inducing proteins, inhibition of activated oncogenes, inhibition of angiogenesis, or enhancement of immune responses to tumour antigens by replication-deficient vectors, or by selectively replication-competent virus vectors that replicate in and kill exclusively dividing cells, i.e., in the CNS by and large tumour cells (Kirn et al., 2001; Rampling et al., 2000). In order to safely eliminate tumour cells, two major challenges have to be met: Since glioblastoma cells migrate and hence are distributed widely in the brain, therapeutic agents must act also distant of the tumour bulk to target these scattered cells; and since CNS neurons cannot be replaced to significant extent, these cells must be spared from lethal insults targeted at tumour cells. It emerges that many of the animal models used to date (typically, heterologous tumour cells are transplanted into the brains of rodents to form tumour bulks) are not optimal owing to the origin and nature of the tumour cells, tumour bulk formation without migratory phenotype, and immunological phenomena.

A multicenter, randomised clinical gene therapy study comprising 248 patients was conducted in the late 1990ies. In the gene therapy group, fibroblasts expressing a HSV-TK producing retrovirus vector were injected into the tumour bed following surgical resection. While the treatment was tolerated and appeared to be safe, there was no beneficial effect of this gene therapy approach, probably largely owing to technical shortcomings (Rainov, 2000). Currently, some 15 clinical brain tumour gene therapy trials are carried out (most of them phase I, and most located in the USA). While none reported unequivocal favourable effects as yet, this number indicates the great hope posed on such approaches, and may reflect the progress made with respect to vector technology, and application strategies.

6. PERSPECTIVES

Besides the examples selected, gene therapy may in the future be a treatment option for a multitude of neurologic diseases, and only part of them will be genetically determined. Thus, polyneuropathies, (chronic) pain, muscle disorders, inflammatory and infectious diseases, and seizure diseases may in the future also efficiently be addressed by gene therapy, and acute CNS lesions such as stroke, brain trauma, and spinal cord injury will likely become gene therapy targets (Sapolsky, 2003).

On the other hand, it is evident that gene therapy is still in its infancy, and much work is yet to be done in order to overcome some of the hurdles that have emerged. Much work is to be completed on the basic principles of genetic medicine – although it has to be mentioned with caution that the identification of disease genes and pathogenetic mechanisms has so far not led to straightforward therapeutic options in neurodegenerative diseases. Rather, "symptomatic", protective strategies are being pursued in many instances. Like with approaches towards pharmacological neuroprotection, however, it remains to be seen how experimental results translate to the clinic, and if sparing neurons, or structures, equals sparing function.

Technical requirements for viral vectors that may be used in the clinical setting include high transduction capacity, and possibly targeted transduction; sufficient transgene capacity to accommodate not only cDNAs of interest, but also appropriate (cell-type specific, regulatable) promoters, with the facility to switch off expression if necessary; transcription regulatory elements; appropriate levels and duration of transgene expression; lack of toxicity and immune response; appropriate and safe routes of administration. Detailed understanding of the pharmacokinetics and pharmacodynamics of gene transfer systems is another prerequisite for safe and ultimately successful future trials. It is important to note that within few years, considerable progress has been made in this area, and it seems fair to predict that gene therapy will in the future improve the outcome of a number of diseases of the nervous system. In order to implement gene therapy for CNS conditions, continued integration of the expertise of different specialities including virology, molecular and medical genetics, neuroscience, neuropathology, clinical neurology, and neurosurgery will be essential.

Fortunately, so far no severe adverse reactions have been reported in neurological gene transfer studies that are to be attributed to the genetic therapies, although several hundreds of patients have been treated for glioblastoma, and several dozends for metabolic and degenerative diseases, mainly in phase I studies to test safety and tolerability. We are currently in a process where we get used to the notion that gene therapy may well be looked at as a specialized form of drug development – and clinical applications should follow the same rigorous standards as clinical trials in drug development (Carter, 2000).

Gene therapy will likely progress into the clinic soon in the following areas: malignant brain tumours, and glioblastoma in particular (for the bad prognosis with current treatments), neurodegenerative diseases such as Parkinson's (for relatively easy accessibility of circumscribed neuronal circuits to intervention, and targeted delivery), and eye diseases including retinal degenerations and glaucoma (for easy accessibility, and the fact that the eye is a separate compartment, and offers some degree of immune privilege). In the near future primarily patients with progressed disease will receive gene therapies within clinical study protocols. With increasing experience, however, patients with progressive debilitating disease may in the future be treated relatively early in the disease course to prevent or slow disease progression. Hints that early treatment may be advantageous also come from transplantation studies in Parkinson's disease, where it appears that patients receiving transplants at a progressed stage of the disease may have virtually no benefit, while those treated at early disease stages clearly do (Freed et al., 2001). This will as a further prerequisite demand for reliable clinical diagnosis of such conditions, e.g., by genetic and biochemical analyses, and refined imaging techniques, while at present chronic neurodegenerative can only be diagnosed with high enough certainty at relatively progressed stages, when a large number of neurons are lost.

Combination of (local, cell-type specific, regulated) gene therapy with systemic pharmacological therapies may show additional, or even synergistic protective effects, but this has not been explored to date, and may be addressed particularly when gene therapy strategies have been established. Alternative novel strategies such as activation of endogenous stem/progenitor cells, or neuron replacement with progenitor or stem cell derived cells may in the future become realistic, and may be combined with gene therapy.

REFERENCES

Aebischer, P., & Ridet, J. (2001). Recombinant proteins for neurodegenerative diseases: the delivery issue. *Trends Neurosci, 24*(9), 533-540.

Aebischer, P., Schluep, M., Deglon, et al. (1996). Intrathecal delivery of CNTF using encapsulated genetically modified xenogeneic cells in amyotrophic lateral sclerosis patients. *Nat Med, 2*(6), 696-699.

Apfel, S. C. (2001). Neurotrophic factor therapy--prospects and problems. *Clin Chem Lab Med, 39*(4), 351-355.

Brooks, A. I., Stein, C. S., Hughes, et al. (2002). Functional correction of established central nervous system deficits in an animal model of lysosomal storage disease with feline immunodeficiency virus-based vectors. *Proc Natl Acad Sci U S A, 99*(9), 6216-6221.

Carter, B. J. (2000). Gene therapy as drug development. *Mol Ther, 1*(3), 211-212.

Costantini, L. C., Bakowska, J. C., Breakefield, X. O., & Isacson, O. (2000). Gene therapy in the CNS. *Gene Ther, 7*(2), 93-109.

Eberhardt, O., Coelln, R. V., Kügler, S., et al. (2000). Protection by synergistic effects of adenovirus-mediated X-chromosome- linked inhibitor of apoptosis and glial cell line-derived neurotrophic factor gene transfer in the 1-methyl-4-phenyl-1,2,3,6- tetrahydropyridine model of Parkinson's disease. *J Neurosci, 20*(24), 9126-9134.

Freed, C. R., Greene, P. E., Breeze, R. E., Tsai, W. Y., DuMouchel, W., Kao, R., Dillon, S., Winfield, H., Culver, S., Trojanowski, J. Q., Eidelberg, D., & Fahn, S. (2001). Transplantation of embryonic dopamine neurons for severe Parkinson's disease. *N Engl J Med, 344*(10), 710-719.

Isenmann, S., & Bähr, M. (1997). Ohne Umweg ins Gehirn: Gentherapie im Nervensystem. In T. Herdegen & M. Bähr & T. R. Tölle (Eds.), *Klinische Neurobiologie* (pp. 357-408). Heidelberg: Spektrum.

Isenmann, S., Kretz, A., & Cellerino, A. (2003). Molecular determinants of retinal ganglion cell development, survival, and regeneration. *Prog Retin Eye Res, 22*(4), 483-543.

Janson, C., McPhee, S., Bilaniuk, L., et al. (2002). Clinical protocol. Gene therapy of Canavan disease: AAV-2 vector for neurosurgical delivery of aspartoacylase gene (ASPA) to the human brain. *Hum Gene Ther, 13*(11), 1391-1412.

Kirn, D., Martuza, R. L., & Zwiebel, J. (2001). Replication-selective virotherapy for cancer: Biological principles, risk management and future directions. *Nat Med, 7*(7), 781-787.

Kordower, J. H., Emborg, M. E., Bloch, J., et al. (2000). Neurodegeneration prevented by lentiviral vector delivery of GDNF in primate models of Parkinson's disease. *Science, 290*(5492), 767-773.

Kretz, A., Wybranietz, W. A., Hermening, S., Lauer, U. M., & Isenmann, S. (2003). HSV-1 VP22 augments adenoviral gene transfer to CNS neurons in the retina and striatum in vivo. *Mol Ther, 7*(5), 659-669.

Kretz A, Kugler S, Happold C, Bahr M, Isenmann S. (2004). Excess Bcl-XL increases the intrinsic growth potential of adult CNS neurons in vitro. Mol Cell Neurosci. 26(1), 63-74.

Kügler, S., Meyn, L., Holzmüller, H., et al. (2001). Neuron-specific expression of therapeutic proteins: evaluation of different cellular promoters in recombinant adenoviral vectors. *Mol Cell Neurosci, 17*(1), 78-96.

Leone, P., Janson, C. G., Bilaniuk, L., et al. (2000). Aspartoacylase gene transfer to the mammalian central nervous system with therapeutic implications for Canavan disease. *Ann Neurol, 48*(1), 27-38.

Luo, J., Kaplitt, M. G., Fitzsimons, H. L., Zuzga, D. S., Liu, Y., Oshinsky, M. L., & During, M. J. (2002). Subthalamic GAD gene therapy in a Parkinson's disease rat model. *Science, 298*(5592), 425-429.

Pardridge, W. M. (2002). Drug and gene delivery to the brain: the vascular route. *Neuron, 36*(4), 555-558.

Rainov, N. G. (2000). A phase III clinical evaluation of herpes simplex virus type 1 thymidine kinase and ganciclovir gene therapy as an adjuvant to surgical resection and radiation in adults with previously untreated glioblastoma multiforme. *Hum Gene Ther, 11*(17), 2389-2401.

Rampling, R., Cruickshank, G., Papanastassiou, V., et al. (2000). Toxicity evaluation of replication-competent herpes simplex virus (ICP 34.5 null mutant 1716) in patients with recurrent malignant glioma. *Gene Ther, 7*(10), 859-866.

Reed, J. C. (2002). Apoptosis-based therapies. *Nat Rev Drug Discov, 1*(2), 111-121.

Sapolsky, R. M. (2003). Neuroprotective gene therapy against acute neurological insults. *Nat Rev Neurosci, 4*(1), 61-69.

Straten, G., Schmeer, C., Kretz, A., Gerhardt, E., Kügler, S., Schulz, J. B., Gravel, C., Bähr, M., & Isenmann, S. (2002). Potential synergistic protection of retinal ganglion cells from axotomy- induced apoptosis by adenoviral administration of glial cell line- derived neurotrophic factor and x-chromosome-linked inhibitor of apoptosis. *Neurobiol Dis, 11*(1), 123-133Thoenen, H., & Sendtner, M. (2002). Neurotrophins: from enthusiastic expectations through sobering experiences to rational therapeutic approaches. *Nat Neurosci, 5 Suppl*, 1046-1050.

Tuszynski, M. H. (2002). Growth-factor gene therapy for neurodegenerative disorders. *Lancet Neurology, 1*, 51-57.

Xia, H., Mao, Q., Paulson, H. L., & Davidson, B. L. (2002). siRNA-mediated gene silencing in vitro and in vivo. *Nat Biotechnol, 20*(10), 1006-1010.

Table 1. Gene therapy targets

therapeutic principles	example: factors	example: disease
additive gene transfer (add [replace] missing function)		
replace deficient enzyme in inborn errors of metabolism (loss of function)	Aspartoacyclase	Canavan's disease
replace neurotransmitter synthesising enzymes in neurodegeneration	TH, AADC, GTP CH 1	Parkinson's disease
	AChE	Alzheimer's disease
add neurotrophic support for degenerating neurons	GDNF	Parkinson's disease
	NGF	Alzheimer's disease
	GDNF, CNTF	motoneuron disease
introduce antiapoptotic factors for neurons at risk of degeneration	Bcl-2, Bcl-X_L,	stroke, trauma,
	p35, XIAP	chronic neurodegeneration
remove insoluble protein aggregates	HSP-70	trinucleotide repeat diseases
	neprilysin	Alzheimer's disease
anti-excitotoxic	free radical scavenger	SOD-1
counteract hyperexcitation by GABA activation	GAD	excititoxicity, Parkinson's disease
counteract cellular calcium overload	calbindin D28K	stroke, excitotoxicity
counteract cellular energy depletion in acute insults	glucose transporter	stroke, excitotoxicity
reduce inflammation in stroke and neurodegeneration	IL-1 receptor antagonist	stroke, Alzheimer's disease
gene or cell ablation/inactivation strategies (abrogate deleterious function)		
counteract excitotoxicity through inhibiting NMDA (over-)activation	NMDA-R-Antagonist	antisense in stroke
gene ablation/inactivation	pathological gain of function	e.g., ALS, SCA
remove/inactivate deleterious genes/factors	tumour growth factors, oncogenes	tumours (glioma)
prodrug activation of toxic metabolites	HSV-TK, CDA	tumours (glioma)
induce tumour cell apoptosis	p53, Fas	tumours (glioma)
combination strategies		
remove deleterious factor (dominant gain of function) and introduce deficient factor, either by two (unrelated) approaches, or by gene replacement (targeted *knock in*)	possibly in trinucleotide repeat diseases	SCA, Huntington's disease
	presenilin mutations	familial Alzheimer's disease

Table 2. Viral gene transfer vectors

	Retrovirus (Oncovirus)	Retrovirus (Lentivirus: HIV, SIV)	"conventional" Adenovirus	"gutless" Adenovirus	AAV	HSV / Amplicon	naked DNA / liposomes
properties	enveloped	enveloped	non-enveloped	non-enveloped	non-enveloped	enveloped	---
nucleic acid	RNA	RNA	dsDNA	dsDNA	ssDNA	dsDNA	DNA (RNA)
max. insert size	8 kb	8 kb	8 kb	30 kb	4.8 kb	40 / 150 kb	unlimited
Titer (per ml)	10^8	$>10^8$	10^{12}	10^{11}	10^{12}	10^{10}	unlimited
route of gene transfer	ex vivo	ex vivo / in vivo	ex vivo / in vivo	ex vivo / in vivo	ex vivo / in vivo	ex vivo / in vivo	ex vivo / in vivo
genomic integration	yes	yes	no	no	possible	no	(rare)
duration of gene expression	short-term	long-term	short-term	short-term / long-term	long-term	long-term	short-term
stability	good	(not tested)	good	good	good	good	good
production	rel. easy	complicated	easy	easy	large scale production difficult	good	easy
immunogenicity	(--)	(--)	++	+	(--)	++	--
preexisting immunity	unlikely	unlikely, possibly in HIV infected individuals	yes	yes	yes	yes	no
security concerns	insertion mutagenesis	insertion mutagenesis?	inflammation toxicity	inflammation toxicity	inflammation toxicity	inflammation toxicity	no
major advantages	efficient cell entry, low toxicity, no viral genes expressed, suited for transduction of stem cells, tumor cells	persistent gene transfer, retrograde transport	efficient cell entry, high titres and expression levels, retrograde transport	efficient cell entry, high transgene capacity, reduced immunogenicity	low toxicity, (partly site specific) integration	esp. amplicon: large packaging capacity high titres, neurotropism, retrograde transport	easy to produce, no size restriction, no toxicity, may be coupled to proteins to facilitate/direct cell entry
major disadvantages	low titres, transduces only dividing cells, insertion mutagenesis	possibly insertion mutagenesis	immunogenity, viral gene expression, possibly recombination	difficult to construct, helper virus contamination	low titres, need helper (adeno-) virus	toxicity, immunogenity	low transfection efficacy, degradation

P. KERMER AND M. BÄHR

Department of Neurology, University Hospital Göttingen (UKG), Göttingen, Germany

PART E: INTRODUCTION

32. GENERAL REMARKS ON THE BRIDGING BETWEEN BASIC RESEARCH AND CLINICS.

Summary. Transfer of basic research results to clinical application with generation of novel treatment options for neurodegenerative diseases is an important issue but appeared to be extremely difficult in the past. This introductory chapter addresses these difficulties embracing health care expenses, basic research as well as therapeutic options in the hospital with their limitations and possibilities.

1. INTRODUCTION

Both acute or chronic brain damage is caused by a wide range of different disorders like stroke, trauma or neurodegenerative diseases and does not only account for a high percentage of long-term disability in the affected patients of all ages but also puts a substantial financial burden on society. Although the gain of knowledge on the molecular mechanisms underlying neuronal loss after brain damage over the past decade has been impressive, new therapeutic strategies that entered the clinic have been very limited. Thus, an optimized translation of basic research results into possible and feasible treatment options at the bedside by recruiting basic researchers, physician-scientists and clinicians into a team is of outstanding importance.

Summarizing some epidemiologic data and annual costs of different neurological disorders involving acute or chronic brain damage as well as selected new treatment options and clinical trials that evolved from research results, this chapter will discuss substantial problems arising when translating results from bench to bedside. Moreover, the authors will try to provide an outlook on how past disappointments applying new drugs in the clinic may give rise to better therapeutic strategies trying to cure, halt or just delay the severe consequences of acute or chronic brain damage in the future.

2. ABOUT EPIDEMIOLOGY AND HEALTH CARE EXPENSES

The diversity of disorders that result in acute or chronic brain damage among others includes stroke, trauma (see Chapter 39/40), tumors and the entity of so-called neurodegenerative diseases like Alzheimer's, Parkinson's, or Huntington's disease, amyotrophic lateral sclerosis (ALS), etc. While these disorders mostly affect neuronal cells, Multiple Sclerosis and many other inflammatory diseases also compromise myelinating cells in the nervous system (see also Chapter 41). Regardless of the cell type or the region in the nervous system that is affected first, brain damage accounts for a high percentage of long-term disability. Moreover, it hits people of all ages, thereby causing tremendous expenses for society. For example, with an incidence of approximately 250-400 per 100000 and a mortality rate of around 30 % stroke following heart disease and cancer is the number three cause of mortality in the Western hemisphere (Dirnagl.et al., 1999). Studies report total long-term per patient costs (over 4 million disabled survivors in the US alone) of over $200000 (Payne et al., 2002) summing up to billions in expenses for local health systems and social security. Moreover, with life expectancy on the rise neurodegenerative diseases like Alzheimer's cause increasing problems for society. This dementia with degeneration of neurons in the basal forebrain, hippocampus and cortex affects around 10 % of the population over 65 years of age. By the age of 85 this percentage has reached up to 50 % (Antuono and Beyer, 1999). Other neurodegenerative disorders like ALS (see Chapter 47), Chorea Huntington or Parkinson's disease, while being less frequent, not only occur in older patients but usually have an age of onset that peaks between the age of 40 to 50, thereby affecting people during employment and incriminating retirement plans as well as social security.

3. MOLECULAR BIOLOGY AND ANIMAL MODELS

Over the last two decades awareness of the severe effects of acute and chronic brain damage has risen substantially. 'Brain awareness weeks', public relations of a diversity of foundations, hospitals, nursing homes, the media and many more have contributed significantly to public education and at last prevention. Not surprisingly, this increased public interest was initiated and accompanied by improved and extended funding opportunities as well as grant approvals for neuroscience. Intensified research activity has lead to a great gain of knowledge about the molecular mechanisms underlying cell death in the brain. Uncovering apoptosis as one important contributor to neuronal loss with the discovery of caspases (cysteine aspartyl-specific proteases) as death proteases (Alnemri et al., 1996) and the significance of mitochondria, the 'cellular power houses', which are located at the crossroads between apoptosis and necrosis as extreme opposites of cellular death pathways (Kroemer and Reed, 2000), have had a big impact on research efforts to find new therapeutic strategies not only for brain damage but a multitude of disorders where programmed cell death is involved. Moreover, genetic studies have broadened our horizon regarding the pathogenesis of several neurodegenerative diseases. For example, there are familiar forms of Parkinson's disease with mutations in proteins which are believed to be involved in protein degradation (see Chapter 33/34). Chorea Huntington is an autosomal-dominant hereditary disease, which is caused by an erranous elongation of CAG-trinucleotid repeats encoding for glutamate, and represents the most frequent member of the so-called trinucleotide-repeat diseases (Lindblad and Schalling, 1999). While the function of the affected protein - called Huntingtin - still needs to be elucidated, it is believed to interfere with the apoptotic cascade (Sanchez Mejia and Friedlander, 2002; see Chapter 46). In another hereditary disease affecting motoneurons and being referred to as spinal muscular atrophy, the genetic defect could be linked to the neuron-specific anti-apoptotic protein NAIP (neuronal inhibitor of apoptosis protein; Roy et al., 1995). In addition, concepts dealing with the toxicity of excitatory amino acids like glutamate (excitotoxicity; Choi, 1992; see Chapter 48), or the findings which lead to the neurotrophic factor hypothesis (Barde, 1989) laid the foundation for a multitude of compounds that were tested for neuroprotective effects. Animal models have been developed which mimic the symptoms and characteristics of human diseases causing brain damage. For example, various stroke models in rodents do exist in which different characteristics of the clinical situation during stroke in humans can be studied (see Chapter 42). Moreover, mice are available that have been genetically engineered to mimic symptoms observed in hereditary neurodegenerative diseases like Parkinson's and Alzheimer's disease as well as ALS. These models are supplemented by investigations after lesion or viral gene transfer in knock-out, transgenic and wild-type animals to verify in vivo functions of new genes and to test potential neuroprotective genes and drugs that had promising results in vitro. Among those compounds that were successful in animal models of brain damage are antioxidants, caspase inhibitors, neurotrophic factors, supplements, glutamate antagonists and many more. More recently, synthetic drugs are being developed to improve the results and effects of their natural counterparts.

4. TRANSFER OF KNOWLEDGE INTO THE HOSPITAL

1. Problems and Limitations

Unfortunately, the transfer of new therapeutic strategies from bench to bedside has been very disappointing thus far. Up to date, new neuroprotective compounds that have been approved for clinical use are limited. Akatinol-Memantine®, for example, was originally approved for treatment of Parkinson's disease and dementia more than 10 years ago (e.g. Gortelmeyer and Erbler, 1992). Meanwhile, the mechanism of action for this neuroprotective substance has been unravelled and it could be classified as non-competitive inhibitor of the NMDA type of glutamate receptors. Today, memantine experiences a kind of renaissance after a large phase III trial on treatment of Alzheimer's disease (see Chapter 45) was finished with positive results. However, in the patient suffering from brain damage most other drugs with neuroprotective effects as revealed by basic research models in vitro and in vivo, failed to exert their protective activity or were simply not applicable on a daily basis in the clinic (see Chapter 49).

Of course, we have to ask for the reasons of such failure. Several points have to be considered when talking about clinical drug application. Side effects of a treatment are one such point that has to be mentioned. If a new drug causes adverse effects in patients, it is hardly applicable for daily use in the clinic and will not be accepted for treatment. Though clinicians have to balance the positive and negative therapeutic effects every time they apply a medication the negative ones will also cause the problem of low compliance by the patients themselves. As a consequence adverse effects for several reasons made many neuroprotective agents already fail during phase I and II trials which therefore could not even enter the hospital for large scale testing. For example, many glutamate receptor antagonists like dizlocipine (MK-801) proved to be very effective in various animal models of neuronal damage. However, for MK-801 clinical development had to be discontinued as a consequence of safety concerns, particularly psychotic side effects. Doubtless, the profile of positive and negative effects is closely related to the specificity of the compound, the applied dosage and the way of its application. The drug might be cleaved or inactivated in the periphery of the body if injected intravenously or given as a pill. A famous example is levodopa as treatment for Parkinson's disease. Given without decarboxylase inhibitors, the drug does not reach therapeutic levels in the brain. In addition, negative effects far away from the site of damage limit the therapeutic success because it is not possible to administer the drug directly at the target site or to achieve the levels of concentration of the medication in the brain without overloading and intoxicating the rest of the body. In the case of nimodipine, an antagonist of L-type calcium channels that had been evaluated in controlled randomized trials for neuroprotection, clinical testing during stroke had to be terminated because of worsened outcome and increased mortality, possibly due to hypotension as a side effect. The problem of getting the drug into the brain appears to be even bigger in neurodegenerative diseases where the blood brain barrier is usually intact and can hardly be penetrated by larger molecules. In contrast, in animal models, several drugs were applied by stereotaxic injection close to the region of the lesion avoiding the problems discussed above all at once. Needless to say, this is not routinely possible in patients.

Aside from artificial drug administration, animal models have to be standardized in time, extent of the lesion and therapeutic regimen. Additionally, those studies are usually performed in otherwise healthy animals. While being necessary for correct statistics and clear interpretation of the results, the reality in the hospital is different. Especially when dealing with older patients, individuals are often multi-morbid and treated with a list of medication, thereby narrowing the therapeutic options for physicians dramatically. More important though seems to be the therapeutic time window, in particular when dealing with acute brain damage. For example, attempts to resolve blood clots in stroke patients only proved to be successful when applied within a short time from onset of symptoms. That requires a rapid admission of patients to the hospital once they suffer from signs of ischemia. However, daily work on stroke units tells us that only a small percentage of patients can be included in such therapeutic strategies. Applying such thoughts on animal models where neuroprotective substances are tested, uncovers their limits immediately. Usually, the drugs in such models are applied early after lesion, at the time of damage or sometimes even before the insult occurs. Strategies where treatment starts hours after the cell death stimulus reflecting the reality in emergency rooms are rare and often not successful. In the case of chronic neurodegeneration, the situation is even worse. While the available animal models are essential, contribute a lot to the understanding of the molecular mechanisms of cell death during brain damage, provide a general basis for therapeutic strategies and facilitate the optimization of therapies, they are often less helpful regarding drug application in a clinical setting. If we look at Parkinson's disease, for instance, patients usually develop symptoms when the number of dopaminergic neurons in the substantia nigra has decreased by 50-60 with the slow process of neuronal degeneration already lasting for years (Andersen, 2001). Thus, as physicians we can not achieve the rescue of the affected neurons but we rather have to aim at the protection of the remaining neuronal population. Staying with Parkinson's disease we realize another problem that has to be solved for all the diseases mentioned so far. There are hereditary forms of the disease with different mutations in several genes, there is the so-called idiopathic form where we do not know the cause, a toxic form from long-lasting poisoning and a form of the disease that is embedded in other syndromes where neuronal degeneration not only occurs in the substantia nigra but in other independent brain regions at the same time (multiple system atrophy etc, Oertel and Bandmann, 1999). Moreover, from histological point of view, we distinguish the disease with the formation of cellular inclusion bodies (Lewy-bodies) and without. It seems reasonable to speculate that, while sharing the same or at least similar symptoms, all these forms of Parkinson's disease represent different entities, following different pathophysiological pathways and consequently requiring different treatment.

Last not least, it should be kept in mind that studies involving patients usually use clinical parameters (scales, mortality) as outcome measures, which are far from being as accurate as the parameters used in animal model studies. Thus, definition of new surrogate-markers for disease progression and monitoring therapy effects poses one of the most important challenges for future clinical trials.

2. Possible Solutions

Keeping all that in mind, we have to ask ourselves how the current approach to scientific and clinical handling of acute and chronic brain damage can be optimized and focused, thereby facilitating the transfer from what we learn at the bench to what we do at the bedside. First and foremost, we think that the interaction of basic researchers and their counterparts in the hospital should be improved. Recruiting basic researchers together with physician scientist and clinicians into a team should enable us to work out strategies for better experimental designs in the laboratory thus accelerating the transfer of knowledge. At round tables and seminars where scientist and clinicians come together, experimental problems can be discussed in a broader way covering the spectrum from what answers are needed to what solutions are possible. Another improvement that appears essential is public relations. Presenting results from basic bench-work under consideration of the clinical setting to a broad public not related to the field will not only improve education and generate independent input supporting prevention of disease but also raise interest in the subject, increase private funding as well as motivate people to participate in clinical studies and in the validation of new therapeutic options.

In the laboratory, the problems could be approached from several directions simultaneously. Instead of testing yet another drug that acts on a cytoprotective pathway already known and analyzed in detail, we have to think about new strategies based on what was uncovered during recent years. The easiest thing that comes to mind is the combination of different neuroprotective agents acting on different pathways or at different levels of the same signal transduction pathway. While requiring a thoroughly designed study protocol to allow a validation of results obtained by such studies, on the long run we might be able to reduce adverse effects by reducing the dosage of a single drug while still benefiting from additive effects of the drug combination. Moreover, drug specificity has to be improved. Molecular biologists and protein biochemists are already working on synthetic small molecule drugs that can mimic protective effects of natural proteins but avoid the negative ones. That might be achieved by modulation of binding characteristics of a protein deleting or adding domains, mutating binding sites etc. Another important point is drug administration. An intact blood brain barrier prevents proteins from entering the brain. While surgical approaches to circumvent this barrier will only be feasible for a small number of patients in the future, the design of compounds bearing moieties which facilitate the transport into the central nervous system (e.g. TAT-proteins; Becker-Hapak et al., 2001; Ford et al., 2001) represents a more promising way. Alternatively, vehicles like genetically modified viral vectors are already explored for their therapeutic use (Baekelandt et al., 1999). In addition, antisense strategies lowering the expression levels of disease-driving genes, if specific enough, appear as an interesting approach for the treatment of hereditary diseases where mutated proteins are held responsible for pathophysiology (Estibeiro and Goodfray, 2001). Finally, new animal models have to be developed, which get even closer to the situation in humans and at the same time, lesion models and neuroprotective strategies should be adjusted to realistic time frames and procedures reflecting the situation in the patient. In such a way, results obtained from in vivo models will have a much bigger impact on the possibilities at the bedside.

On the other hand, in the clinic, current classifications of certain diseases have to be revised and maybe renewed. Like it has been discussed above for Parkinson's disease, we should revisit the different processes that result in acute and chronic brain damage and define accurate markers/surrogate-markers for disease progression and therapy effects. Intensifying research on genetic mapping and potential marker proteins, we might be able to come up with new entities regrouping diseases according to their pathophysiological mechanism in addition to histology or symptoms. Doing so, with some subgroups we might even get closer to observations made in specific animal models. In addition, new imaging techniques may provide us with new tools that allow demonstration of cellular and axonal integrity/lesions in vivo. It is exciting to hypothesize that we will be able in the future to provide every patient suffering from brain damage with a more specified and diversified therapy that considers individual symptoms as well as pathophysiology and that we may be able to non-invasively follow up these patient groups.

5. PERSPECTIVE

Altogether, in recent years basic research significantly extended our understanding of the processes involved in cell death during acute and chronic brain damage. While the gap between gained knowledge and therapeutic strategies we can revert to in the hospital still seems huge, we can be optimistic that, if efforts in different disciplines are coordinated and optimized, this gap will get smaller in the near future. Using our scientific resources appropriately we should be able to improve the existing and increase the number of weapons we carry to fight the severe consequences of brain damage.

REFERENCES

Alnemri ES, Livingston DJ, Nicholson DW, Salvesen G, Thornberry NA, Wong WW, Yuan J. (1996) Human ICE/CED-3 protease nomenclature. Cell 87(2):171.
Andersen JK. (2001) Does neuronal loss in Parkinson's disease involve programmed cell death? Bioessays 23(7):640-6.
Antuono P, Beyer J. (1999) The burden of dementia. A medical and research perspective. Theor Med Bioeth. 20(1):3-13.
Baekelandt V, De Strooper B, Nuttin B, Debyser Z. (2000) Gene therapeutic strategies for neurodegenerative diseases. Curr Opin Mol Ther. 2(5):540-54.
Barde YA. (1989) Trophic factors and neuronal survival. Neuron. 2(6):1525-34.
Becker-Hapak M, McAllister SS, Dowdy SF. (2001) TAT-mediated protein transduction into mammalian cells. Methods. 24(3):247-56.
Choi DW. (1992) Excitotoxic cell death. J Neurobiol. 23(9):1261-76.
Dirnagl U, Iadecola C, Moskowitz MA. (1999) Pathobiology of ischaemic stroke: an integrated view. Trends Neurosci. 22(9):391-7.
Estibeiro P, Godfray J. (2001) Antisense as a neuroscience tool and therapeutic agent. Trends Neurosci. 24(11 Suppl):S56-62.
Ford KG, Souberbielle BE, Darling D, Farzaneh F. (2001) Protein transduction: an alternative to genetic intervention? Gene Ther. 8(1):1-4.
Gortelmeyer R, Erbler H. (1992) Memantine in the treatment of mild to moderate dementia syndrome. A double-blind placebo-controlled study. Arzneimittelforschung. 42(7):904-13.
Kroemer G, Reed JC. (2000) Mitochondrial control of cell death. Nat Med. 6(5):513-9.
Lindblad K, Schalling M. (1999) Expanded repeat sequences and disease. Semin Neurol. 19(3):289-99.
Oertel WH, Bandmann O. (1999) Multiple system atrophy. J Neural Transm Suppl. 56:155-64.
Payne KA, Huybrechts KF, Caro J, Craig Green TJ, Klittich WS. (2002) Long Term Cost-of-Illness in Stroke: An International Review. Pharmacoeconomics. 20(12):813-25.
Roy N, Mahadevan MS, McLean M, Shutler G, Yaraghi Z, Farahani R, Baird S, Besner-Johnston A, Lefebvre C, Kang X, et al. (1995) The gene for neuronal apoptosis inhibitory protein is partially deleted in individuals with spinal muscular atrophy. Cell. 80(1):167-78.
Sanchez Mejia RO, Friedlander RM. (2001) Caspases in Huntington's disease. Neuroscientist. 7(6):480-9

C. KRARUP

Department of Clinical Neurophysiology, Rigshospitalet, Copenhagen, Denmark

33. MECHANICAL LESIONS OF THE PERIPHERAL NERVOUS SYSTEM

Summary. Mechanical injury to peripheral nerve is associated with changes in conduction properties and safety in mild lesion that cause demyelination or Wallerian degeneration in severe lesions that cause loss of axonal continuity. In the PNS the microenvironment is permissive for axonal regrowth and this forms the only basis for functional recovery. However, for regeneration to be of clinical benefit, other conditions must be fulfilled including reinnervation of target organs and reestablishment of function of both the regenerated axons and the denervated target organs. The different requirements for clinical recovery in animal models and in patients are discussed in this chapter.

1. INTRODUCTION

Degeneration of neurons or axons is the final result of harmful effects of metabolic, ischemic, hereditary, inflammatory, toxic, or mechanical injuries. Major efforts are aimed at reducing the degree of irreversible cellular breakdown since nervous tissue does not posses the ability to form new neurons in the adult organism. In contrast, axonal degeneration in the peripheral nervous system may be repaired by regrowth from the proximal stump, and this provides the basis for reinnervation of target organs. However, functional recovery is dependent on a number of factors including correct reinnervation of sensory receptors, muscle and other organs, and reestablishment of axonal function. The outgrowth of axons is directed by adoptive changes in the neuronal cell body, and in particular by interactions between the growth cone and the interstitial tissue of the distal nerve stump that provides the framework for regenerative processes. The clinical results of nerve regeneration are often unsatisfactory with persistent deficits or to symptoms that may be due to misdirection of axonal growth. Thus, patients often show poor control of fine movement due to activation of muscle incorrectly innervated, complain of reduced sensation, pain and demonstrate poor localization of sensory stimuli. There is therefore a continuing need for improved understanding of the processes that determine regeneration of axons.

It is the aim of this chapter to review clinical and experimental studies that seek to provide information regarding factors that influence functional recovery after mechanical injury of peripheral nerve fibers.

2. CONSEQUENCES OF MECHANICAL INJURY

The effects of mechanical injury may be classified as axonal lesions in continuity or as loss of axonal continuity. Though simplified, this distinction has clinical implications since recovery after lesions in continuity associated with changes in myelin usually is rapid and complete, whereas recovery after loss of axonal continuity depends on several factors such as the type of injury, the site of the lesion and the time interval between injury and surgical treatment. In the clinical setting recovery is often incomplete, as for example in patients with facial nerve injury, which is of particular relevance considering the precision necessary to communicate specific mimic expressions (Fig. 1).

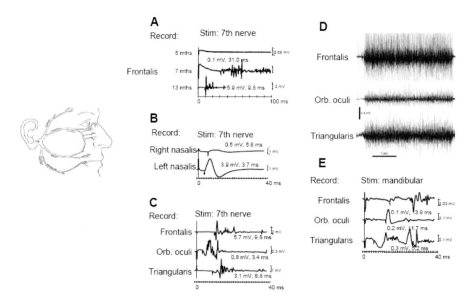

Figure 1. Patient with right facial palsy following facial nerve lesion at a middle ear operation, examined 5, 7 and 13 months after surgery. A, at 5 months there was no motor response from facial muscles when stimulating the nerve at the stylomastoid foramen (left), at 7 months a small long latency response was present, and at 13 months the response had increased in amplitude. B, even at 13 months, however, the CMAP from the nasalis muscle was only about 10-15% of that at the control side in agreement with regeneration of only a limited number of nerve fibers. C, CMAPs could be recorded from all facial branches but had low amplitudes and long latencies indicating that fibres were immature. D, at attempt at eye closure there was strong activity also in the frontalis and triangularis muscles associated with synkinesias in these muscles. E, the aberrant motor activity was associated with motor responses in the muscles innervated by the upper facial branch when stimulating the mandibular branch with an electrical pulse. A direct response was recorded from the triangularis muscles whereas recurrent responses were recorded from the frontalis and orbicularis oculi muscles consistent with aberrant branching.

Even though regeneration usually is substantial with reinnervation of muscles supplied by the temporal, zygomatic and mandibular branches, most patients have poor control of facial expression due to synkinesias of other muscles not involved in a particular facial movement; for example eye closure may occur when wrinkling of the forehead or pursing of the lips is attempted. Thus, in spite of regeneration patients demonstrate the following problems associated with peripheral nerve lesions causing Wallerian degeneration (Fig. 1A): 1) insufficient number of axons growing to the target (Fig. 1B), 2) partial but incomplete reestablishment of conduction properties of regenerated axons (Fig. 1C), 3) disturbed control of voluntary activity (Fig. 1D), and 4) presence of aberrant branching as indicated by the presence of axon reflexes (Fig. 1E).

3. DEMYELINATION VERSUS WALLERIAN DEGENERATION

Lesions in continuity may result from acute compression, crushing or chronic entrapment of nerve (Gilliatt, 1980a; Gilliatt, 1980b), and the resulting neuropathological changes depend on the severity and duration of the physical impact, which may occur as a consequence of external factors or anatomical structures. Acute compression is associated with demyelination of mainly large fibers which if severe or prolonged may result in Wallerian degeneration. Chronic entrapment neuropathy is usually associated with combined demyelination and axonal degeneration, and this type of lesion is extremely common in clinical practice (carpal tunnel syndrome, tardy ulnar nerve palsy). Recovery after demyelination occurs over days to weeks (Trojaborg, 1970). Loss of axonal continuity results in Wallerian degeneration associated with break down of both the axonal components and myelin distal to the site of the lesion, which occurs over hours to several days in different species, and during which the axon is still able to conduct an action potential. Failure of conduction begins distal and then involve more proximal segments (Gilliatt & Hjorth, 1972). This delayed failure of conduction may lead to the erroneous conclusion that the clinical and electrophysiological findings are caused by conduction block, *i.e.* demyelination with axonal continuity, rather than by loss of axonal continuity. Distinction between conduction block and axonal failure is definite only after loss of distal excitability, and repeated conduction studies are therefore required.

3.1 Wallerian degeneration

The degeneration of the distal nerve stump after loss of axonal continuity was already known to occur by Waller (Müller & Stoll, 1998; Stoll *et al.*, 2002), and the neuropathological and temporal development of Wallerian degeneration was delineated by Cajal (Ramon y Cajal, 1928). It has over the last several years become clear that Wallerian degeneration occurs as a reaction to disturbed axon-glial interaction, and that break down of the axon and myelin are dependent on both diffusible factors and macrophages (Brück, 1997). Axonal breakdown occurs as a consequence of Ca^{++} influx in the distal axon mediated through specific ion-channels (George *et al.*, 1995), and is dependent on intrinsic axonal factors as shown in the mouse mutant with delayed Wallerian degeneration (Wld^s) (Gillingwater & Ribchester, 2001). These mice show poor recruitment of macrophages during Wallerian degeneration possibly due to deficient proliferative response by Schwann cells. Wallerian degeneration has been shown to be closely connected with upgrading of cytokines in the distal nerve stump (Stoll & Müller, 1999; Stoll *et al.*, 2002). During Wallerian degeneration Schwann cells dedifferentiate and proliferate within their basal lamina (Schwann cell tubes, band of Büngner) stimulated by growth factors produced in connection with loss of axonal contact; subsequently they enter a quiescent phase followed by renewed proliferation and myelination when contacted by growing axons (Pellegrino & Spencer, 1985). If axonal regeneration is prevented Schwann cells and their basal lamina tubes undergo atrophy which hinder regeneration (Sunderland & Bradley, 1950; Weinberg & Spencer, 1978).

Regeneration is strongly dependent on the cellular, structural and humoral changes that occur during Wallerian degeneration (Brown *et al.*, 1994). Thus prevention of macrophage recruitment is associated with poor axonal regeneration (Dahlin, 1995; Fugleholm *et al.*, 1998; Sørensen *et al.*, 2001).

3.2 Axonal regeneration

Axons send growth cones from the proximal stump into the distal denervated stump during regeneration guided by cytokines and growth factors produced by Schwann cells and macrophages (Liu & Snider, 2001; Politis *et al.*, 1982). Regeneration of axons, however, also entails reconnecting to target organs, recovery of function of these organs, and finally recovery of the normal axon-glial relationship, axonal caliber, and ion-channel functions are required for proper function to be regained.

4. PHASES OF RECOVERY

4.1. Elongation

Assessment of the rate of regeneration is usually carried by measuring the time of reinnervation of muscle supplied by the nerve but this does not allow evaluation of elongation rate, and axonal growth in vivo requires either localization of the growth cone or other means of measuring the advancement of the nerve fiber. In patients the presence of a Tinel's is often used to gauge the presence of immature mechanically sensitive axonal sprouts, and hence regeneration of nerve fibers, although this must be considered an uncertain sign (Fig. 2). In experimental animals this mechanical sensitivity has been used in the so-called pinch-test (Danielsen *et al.*, 1995). We have, however, developed a method using chronic implanted electrodes that allow recording of action potentials from individual myelinated fibers to measure the rate of elongation (Fugleholm *et al.*, 1994; Krarup *et al.*, 1988; Krarup & Loeb, 1988). In the cat this method showed a rate of 3-4 mm/day after crushing of the nerve whereas the rate was lower at 2.5 mm/day after sectioning and suture.

33. MECHANICAL LESIONS OF THE PERIPHERAL NERVOUS SYSTEM

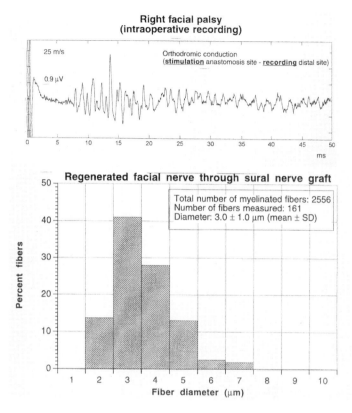

Figure 2. Intraoperative recording of nerve action potential from a sural nerve autograft inserted about 12 months before to treat a right sided long standing facial palsy (top graph). The graft was anastomosed to the normal left facial nerve and passed subcutaneously to the right side, and the nerve action potential was obtained by stimulating the graft through hook electrodes placed over the left side and recording through hook electrodes placed over the nerve on the right side. A small piece of the distal end of the nerve was examined histologically (lower graph), and showed that there were about 2500 regenerated fibers, most with small diameters of less than 7 μm, corresponding to the low conduction velocity.

This difference may be related to the greater branching of axons in the distal stump after sectioning than after crushing (Fugleholm *et al.*, 2000). The influence of Schwann cells on elongation was assessed by freezing of the nerve distal to the site of crushing or sectioning, a process which destroyed these cells. Whereas freezing of the nerve for a distance of 20 mm distal to crushing had no effect on elongation (Fig. 3A), it slowed elongation after sectioning (Fig. 3B). Distal to the frozen nerve segment, elongation accelerated as axons entered the distal stump with normal endoneurium (Fugleholm *et al.*, 1994).

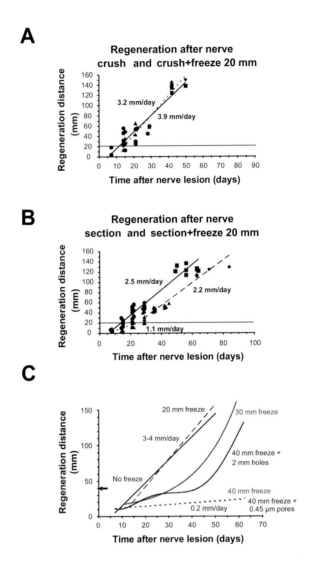

Figure 3. Regeneration of tibial nerve in cat to show effect of freezing on elongation. A, the rate of elongation after crushing was 3.2 mm/day and similar at 3.9 mm/day after crushing and freezing over 20 mm. B, the rate of elongation after section was lower at 2.5 mm/day, and in contrast to crushing, freezing over 20 mm slowed elongation to 1.1 mm/day through the frozen segment; distal to the frozen segment the elongation rate increased to 2.2 mm/day. C, in this composite the elongation rate was reduced to 0.2 mm/day when freezing was extended to 40 mm distal to crushing, and this was not affected by allowing contact to the environment through 0.45 μm pores. By contrast, when contact was provided through 2 mm pores elongation could take place in connection with revascularisation. At freezing over a distance

of 30 mm, elongation was delayed through the frozen segment but then pick up speed distal to it. Modified from Fugleholm et al., 1994, 1998 and from Sørensen et al., 2001.

These findings indicate that axons may regenerate for limited distances without Schwann cells in agreement with other studies (Ide *et al.*, 1990), but at elimination of cells over longer distances of 30 mm, elongation after crushing was delayed, and at 40 mm it was almost completely stopped (Fig. 3C) (Fugleholm *et al.*, 1998). Outgrowth of axons has been considerably improved by implantation of short grafts in a 'stepping stone' arrangement (Maeda *et al.*, 1993), and we also found that allowing revascularisation markedly improved elongation (Fig. 3C) (Sørensen *et al.*, 2001), possibly associated with an endothelial growth factor stimulating effect. In contrast to effects of semipermeable tubes as growth conduits, we could not demonstrate regeneration after freezing when contact occurred through a 45 μm millipore filter (Fig. 3C), indicating that access by humoral factors or oxygen was insufficient to influence axonal growth (Sørensen *et al.*, 2001).

4.2 Reinnervation of target organs

Reinnervation of target organs is a necessary but not sufficient condition for recovery after Wallerian degeneration, since aberrant connections between the CNS and the organ may prevent useful function (Sunderland, 1991). Following nerve crush the maintenance of nerve fiber basal lamina ensures faster and more precise regeneration than after nerve section where axons must reestablish contact to the basal lamina tubes guiding it to its proper target (Lundborg, 1988). The fundamental effect of nerve section on the disruption of the relation between the growing axon and the Schwann cell tube has major impact on both the number of fibers and the specificity of reinnervation. Thus the amplitudes of the muscle and nerve responses recovered to a much greater extent after nerve crush than after nerve section (Fugleholm *et al.*, 2000).

Table 1: Recovery of motor and sensory conduction in the median nerve 2-15 years after repair at wrist or elbow in 7 patients

Conduction studies	Amplitude (% of normal)	Number of motor units (% of normal)
CMAP (APB)	7-63	4-34
CSAP (median n. wrist)	3-32	

CMAP, compound muscle action potential. APB, abductor pollicis brevis muscle. CSAP, compound sensory action potential recorded from the median nerve at wrist

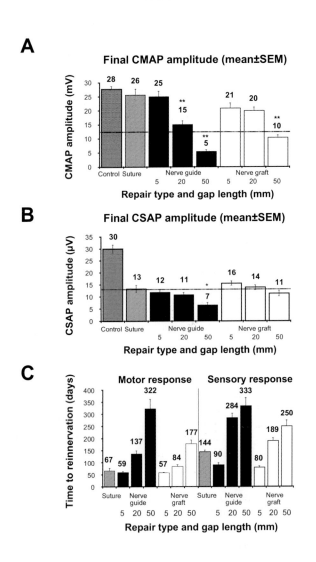

Figure 4. Recovery of the muscle response (A) and the sensory response (B) in monkey median nerve after section and repair with direct suture and repair of gap lengths of 5, 20, or 50 mm with nerve guide or nerve graft. Normal amplitudes of the CMAP were obtained at 5 mm gap lengths repaired with both guide and graft and at 20 mm gap length when repaired with a graft but not with a guide. At 50 mm gap length normal amplitudes were obtained with neither graft nor guide. The CSAP recovered to a far lesser extent than the CMAP at all types of repair. C, the time to first recording of motor and sensory responses at different repair procedures showed longer regeneration time when using guide than graft for distances of 20 and 50 mm, whereas there was no difference for short gap length. Modified from Krarup et al., 2002.

It is well established that the degree of recovery in patients with peripheral nerve lesions is variable; thus, in seven patients with complete median nerve lesions examined 2-15 years after surgical repair, motor and sensory responses had recovered by widely varying amplitudes of the compound muscle and sensory action potentials, and the number of motor units evaluated by means of "motor unit number estimation" (MUNE) (McComas, 1995) was small, indicating that the number of regenerating axons was less than 10% even in some patients with lesions at the wrist (Table 1).

In order to ascertain the factors that influence the extent of recovery after different types of nerve lesions, muscle and sensory responses were recorded from monkey median nerves with 0, 5, 20, 50 mm nerve tissue gap lesions repaired with direct suture, nerve graft or a collagen nerve guide (Krarup et al., 2002). These studies showed that the CMAP and CSAP amplitudes were influenced by the type of repair as well as the length of the nerve gap (Fig. 4A,B): the CMAP reached normal values after direct suture and gap length of 5 mm repaired by nerve graft or guide as well as after repair of the 20-mm nerve gap repaired by graft, while it markedly decreased at longer distances repaired by graft or guide.

The amplitude of the CSAP in general recovered to a lesser extent than the CMAP. This difference reflects that the motor response is the result of summation of motor unit potentials, dependent on the size of the motor units, while the CSAP is the result of single nerve fiber potentials. The size of motor unit responses increased markedly during reinnervation (see below). Thus there was no evidence of greater regenerative capacity of motor than sensory fibers since the number of recovered motor units corresponded to the size of the CSAP, and in cat nerve there was no difference in the rate of elongation between motor and sensory fibers (Moldovan, Sørensen, Krarup, unpublished). A longer time to reinnervation was a striking difference between nerve guide compared with graft repair for comparable nerve gap distances (Fig. 4C). When analyzed as a function of reinnervation time the CMAP amplitude (Fig. 5A) as well as the number of motor units (Fig. 5C) decreased strongly, indicating that axonal outgrowth is time dependent.

Furthermore, the size of motor unit potentials decreased at delayed reinnervation indicating reduced branching of regenerated axons. Thus the main determining factor for successful reinnervation in monkeys appeared to be the rate of regeneration (Fig. 6). In other experimental models prolonged denervation significantly reduced reinnervation through effects on the distal nerve stump as well as directly on motor neurons (Fu & Gordon, 1995a; Fu & Gordon, 1995b), and these studies further support that regeneration must take place through a narrow time window.

It is well established that the distal nerve stump attracts axons growing from the proximal stump and that this most likely occurs through diffusible factors (Lundborg, 2000; Ramon y Cajal, 1928). However, it is much less clear to what extent and by what mechanisms reinnervation of target organs is a guided or a random process. In rat double labeling studies have shown that motor fibers preferentially regrow into motor pathways, termed preferential motor reinnervation, PMR (Brushart, 1993), and that sensory fibers, though to a lesser extent, reinnervate sensory pathways (Madison et al., 1996). PMR has been suggested to be related to molecular recognition between the axon and Schwann cells in bands of Büngner (Martini *et al.*, 1994), and this interaction may be further improved by electrical stimulation (Al-Majed *et al.*, 2000b; Al-Majed *et al.*, 2000a). The mechanism of PMR is considered to occur as a consequence of random growth of axon sprouts into the distal nerve stump followed by pruning of erroneously projecting branches (Brushart *et al.*, 1998; Brushart, 1991), and hence that branching of the growing axon may present a compensatory mechanism for recovery of specific contact between the CNS and the target organ.

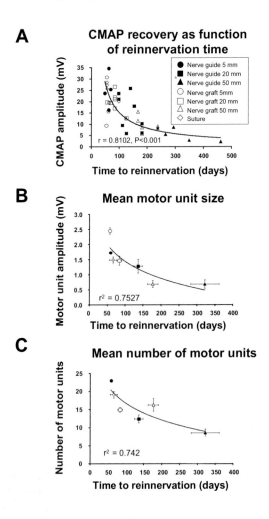

Figure 5. There were marked declines in the amplitudes of the CMAPs (A), the sizes of motor unit potentials (B), and the number of motor units (C) with increasing reinnervation time. Modified from Krarup et al., 2002.

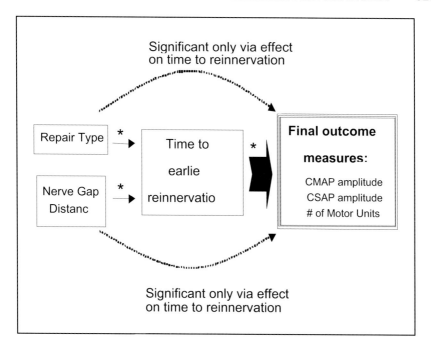

Figure 6. The repair procedures and the nerve gap lengths affected recovery after nerve lesions through a strong effect on reinnervation time. The outcome after nerve lesions was strongly dependent on the time of reinnervation. Modified from Krarup et al., 2002.

It is unknown to what degree PMR may take place in primates, and we have therefore examined this question in monkeys with median nerve lesions (see above) (Madison et al., 1999). PMR was studied by counting motor units in the reinnervated APB using MUNE and statistical methods were applied to calculate if the number could be explained by a random process. The number of motor units could not be explained by chance alone after repair with direct suture or over short gap lengths whereas a random process was sufficient in long gap lesions. Axonal branching and aberrant growth of motor branches into sensory projection territories could be studied as recurrent motor responses evoked by electrical stimulation digital nerve fibers that have also been found in humans with peripheral nerve lesions (Krarup et al., 1990; Montserrat & Benito, 1990). By following the temporal development of these responses, it appeared that the number of motor units that could be elicited by digital stimulation decreased with time in nerves repaired with direct suture or over short gap lengths whereas this did not take place in the long gap lesions, suggesting that PMR occurred as a consequence of pruning of aberrant motor branches (Fig. 7).

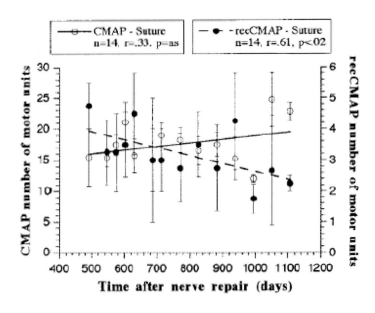

Figure 7. Number of motor units recorded from the abductor pollicis brevis muscle evoked by direct stimulation of the median nerve (CMAP, filled circles, full line) or by stimulation of digital nerve fibers (recCMAP, open circles, dashed line) in monkeys related to time after section and suture of the median nerve at the wrist. The CMAP showed a weak increase in the number of reinnervated motor units, whereas the aberrant response showed a gradual decline in the number of motor units suggesting pruning of motor axon branches that had grown into the sensory territory. From Madison et al., 1999.

Even though PMR therefore appeared to occur in the non-human primate it was markedly dependent on the severity of the lesion as expressed by the length of the gap between the proximal and the distal nerve stumps, and a similar limitation has also been found in studies in the rat (Brushart et al., 1995), where regeneration across long nerve gaps occurred according to fascicular arrangement. In human median nerve lesions we used a similar statistical method as in monkey and found that 7% of about 200 motor units could be recovered by random reinnervation. The number of motor units counted in seven patients ranged from 4 to 34% of the control side indicating that the reinnervation was guided in the three hands with the largest recovered motor responses whereas the poor recovery reinnervation in the other four hands occurred as a random process (Krarup, Boeckstyns, Moldovan, unpublished). The likely presence of PMR in humans has motivated the use nerve guide silicone tubes in the treatment of patients with short nerve gap length lesions (Lundborg, 2000; Lundborg et al., 1997) as an alternative to epineural or fascicular suture, and we have started a study with a similar design using a biodegradable collagen nerve guide (Intgra LifeSciences, N.J., USA) in patients with median or ulnar nerve lesions.

That branching of regenerating nerve fibers may, however, also be a source of aberrant reinnervation as suggested by the poor functional recovery in rat facial nerve lesions associated with increased sprouting (Angelov *et al.*, 1996; Dohm *et al.*, 2000).

These dual effects of axonal sprouting probably explain why recovery after nerve lesions in spite of several attempts of new treatment strategies remains at best moderate.

4.3. Maturation of nerve fibers

It is well established that newly regenerated axons lack myelin and have small diameters and that maturation involves enlargement of caliber and remyelination, though the diameter and the conduction velocity only attain values of 80-90% and the internodal length only about one-third of control nerve (Cragg & Thomas, 1964; Devor & Govrin-Lippmann, 1979; Fugleholm *et al.*, 2000; Hiscoe, 1947; Vizoso & Young, 1948). The axonal caliber increase during growth and maturation is presumably controlled by neurofilament transport (Hoffman *et al.*, 1987).

Though axons acquire normal characteristics during maturation ion channel distribution remains abnormal. Thus fast K^+-channels, normally confined to the para- and internode (Baker *et al.*, 1987; Scherer & Arroyo, 2002), are found in nodal axolemma after regeneration and may render the membrane unstable and assume immature traits (Kocsis & Waxman, 1987). Using threshold electrotonus (Bostock *et al.*, 1998; Burke *et al.*, 2001) in regenerated cat nerve we have found persistent abnormalities during hyperpolarization that suggest persistent hyperpolarization of the axon membrane (Moldovan & Krarup, in press), further supporting that regenerated axons may never assume entirely normal properties.

5. CONCLUSIONS

It has been found that, though peripheral axons may regenerate after Wallerian degeneration, a number of factors and processes must be integrated for useful function to be reestablished. Thus *both* the neuronal cell body metabolic machinery must be switched to provide proper building material to extend axonal processes and possibly to take part in directional guidance, *and* the interaction between the growth cone and structural, humoral and cellular environment must be properly modified. These prerequisites may be fulfilled in a limited time window during which the distal stump of the denervated nerve provides an optimal environment for outgrowth of axons, reinnervation of target organs and recovery of axonal function. Such a window of opportunity could include proper balancing of attractive and repulsive factors, and suggests that treatment strategies should be aimed at acceleration of axonal growth.

REFERENCE

Al-Majed, A. A., Brushart, T. M., & Gordon, T. (2000a). Electrical stimulation accelerates and increases expression of BDNF and trkB mRNA in regenerating rat femoral motoneurons. *European Journal of Neuroscience* 12, 4381-4390.

Al-Majed, A. A., Neumann, C. M., Brushart, T. M., & Gordon, T. (2000b). Brief electrical stimulation promotes the speed and accuracy of motor axonal regeneration. *Journal of Neuroscience* 20, 2602-2608.

Angelov, D. N., Neiss, W. F., Streppel, M., Andermahr, J., Mader, K., & Stennert, E. (1996). Nimodipine accelerates axonal sprouting after surgical repair of rat facial nerve. *Journal of Neuroscience* 16, 1041-1048.

Baker, M., Bostock, H., Grafe, P., & Martius, P. (1987). Function and distribution of three types of rectifying channel in rat spinal root myelinated axons. *Journal of Physiology* 383, 45-67.

Bostock, H., Cikurel, K., & Burke, D. (1998). Threshold tracking techniques in the study of human peripheral nerve. *Muscle & Nerve* 21, 137-158.

Brown, M. C., Perry, V. H., Hunt, S. P., & Lapper, S. R. (1994). Further studies on motor and sensory nerve regeneration in mice with delayed Wallerian degeneration. *European Journal of Neuroscience* 6, 420-428.

Brück, W. (1997). The role of macrophages in Wallerian degeneration. *Brain Pathology* 7, 741-752.

Brushart, T. M., Gerber, J., Kessens, P., Chen, Y. G., & Royall, R. M. (1998). Contributions of pathway and neuron to preferential motor reinnervation. *Journal of Neuroscience* 18, 8674-8681.

Brushart, T. M., Mathur, V., Sood, R., & Koschorke, G. M. (1995). Dispersion of regenerating axons across enclosed neural gaps. *Journal of Hand Surgery.American Edition* 20, 557-564.

Brushart, T. M. E. (1991). The mechanical and humoral control of specificity in nerve repair. In *Operative nerve repair and reconstruction*, ed. Gelberman, R., pp. 215-230. Lippincott, Philadelphia.

Brushart, T. M. E. (1993). Motor axons preferentially reinnervate motor pathways. *Journal of Neuroscience* 13, 2730-2738.

Burke, D., Kiernan, M. C., & Bostock, H. (2001). Excitability of human axons. *Clinical Neurophysiology* 112, 1575-1585.

Cragg, B. G. & Thomas, P. K. (1964). The conduction verlocity of regenerated peripheral nerve fibres. *Journal of Physiology* 171, 164-175.

Dahlin, L. B. (1995). Prevention of macrophage invasion impairs regeneration in nerve grafts. *Brain Research* 679, 274-280.

Danielsen, N., Kerns, J. M., Holmquist, B., Zhao, Q., Lundborg, G., & Kanje, M. (1995). Predegeneration enhances regeneration into acellular nerve grafts. *Brain Research* 681, 105-108.

Devor, M. & Govrin-Lippmann, R. (1979). Maturation of axonal sprouts after nerve crush. *Experimental Neurology* 64, 260-270.

Dohm, S., Streppel, M., Guntinas-Lichius, O., Pesheva, P., Probstmeier, R., Walther, M., Neiss, W. F., Stennert, E., & Angelov, D. N. (2000). Local application of extracellular matrix proteins fails to reduce the number of axonal branches after varying reconstructive surgery on rat facial nerve. *Restorative Neurology and Neuroscience* 16, 117-126.

Fu, S. Y. & Gordon, T. (1995a). Contributing factors to poor functional recovery after delayed nerve repair: prolonged axotomy. *Journal of Neuroscience* 15, 3876-3885.

Fu, S. Y. & Gordon, T. (1995b). Contributing factors to poor functional recovery after delayed nerve repair: prolonged denervation. *Journal of Neuroscience* 15, 3886-3895.

Fugleholm, K., Schmalbruch, H., & Krarup, C. (1994). Early peripheral nerve regeneration after crushing, sectioning, and freeze studied by implanted electrodes in the cat. *Journal of Neuroscience* 14, 2659-2673.

Fugleholm, K., Schmalbruch, H., & Krarup, C. (2000). Post reinnervation maturation of myelinated nerve fibers in the cat tibial nerve: chronic electrophysiological and morphometric studies. *Journal of the Peripheral Nervous System* 5, 82-95.

Fugleholm, K., Sørensen, J., Schmalbruch, H., & Krarup, C. (1998). Axonal elongation through acellular nerve segments of the cat tibial nerve: importance of the near-nerve environment. *Brain Research* 792, 309-318.

George, E. B., Glass, J. D., & Griffin, J. W. (1995). Axotomy-induced axonal degeneration is mediated by calcium influx through ion-specific channels. *Journal of Neuroscience* 15, 6445-6452.

Gilliatt, R. W. (1980a). Acute compression block. In *The physiology of peripheral nerve disease*, ed. Sumner, A. J., pp. 287-315. Saunders, Philadelphia.

Gilliatt, R. W. (1980b). Chronic nerve compression and entrapment. In *The physiology of peripheral nerve disease*, ed. Sumner, A. J., pp. 316-339. Saunders, Philadelphia.

Gilliatt, R. W. & Hjorth, R. J. (1972). Nerve conduction during wallerian degeneration in the baboon. *Journal of Neurology, Neurosurgery and Psychiatry* 35, 335-341.

Gillingwater, T. H. & Ribchester, R. R. (2001). Compartmental neurodegeneration and synaptic plasticity in the Wld[s] mutant mouse. *Journal of Physiology* 534, 627-639.

Hiscoe, H. B. (1947). Distribution of nodes and incisures in normal and regenerated nerve fibers. *The Anatomical Record* 99, 447-475.

Hoffman, P. N., Cleveland, D. W., Griffin, J. W., Landes, P. W., Cowan, N. J., & Price, D. L. (1987). Neurofilament gene expression: a major determinant of axonal caliber. *Proceedings of the National Academy of Sciences of the United States of America* 84, 3472-3476.

Ide, C., Osawa, T., & Tohyama, K. (1990). Nerve regeneration through allogeneic nerve grafts, with special reference to the role of the Schwann cell basal lamina. *Progress in Neurobiology* 34, 1-38.

Kocsis, J. D. & Waxman, S. G. (1987). Ionic channel organization of normal and regenerating mammalian axons. In *Progress in Brain Research*, eds. Seil, F. J., Herbert, E., & Carlson, B., pp. 89-101. Elsevier Science Publishers B.V., Amsterdam.

Krarup, C., Archibald, S. J., & Madison, R. D. (2002). Factors that influence peripheral nerve regeneration: an electrophysiological study of the monkey median nerve. *Annals of Neurology* 51, 69-81.

Krarup, C. & Loeb, G. E. (1988). Conduction studies in peripheral cat nerve using implanted electrodes: I. Methods and findings in controls. *Muscle & Nerve* 11, 922-932.

Krarup, C., Loeb, G. E., & Pezeshkpour, G. H. (1988). Conduction studies in peripheral cat nerve using implanted electrodes: II. The effects of prolonged constriction on regeneration of crushed nerve fibers. *Muscle & Nerve* 11, 933-944.

Krarup, C., Upton, J., & Creager, M. A. (1990). Nerve regeneration and reinnervation after limb amputation and replantation: clinical and physiological findings. *Muscle & Nerve* 13, 291-304.

Liu, R. Y. & Snider, W. D. (2001). Different signaling pathways mediate regenerative versus developmental sensory axon growth. *Journal of Neuroscience* 21, RC164.

Lundborg, G. (1988). *Nerve injury and repair*, pp. 1-222. Churchill Livingstone, Edinburgh, London, Melbourne, New York.

Lundborg, G. (2000). A 25-year perspective of peripheral nerve surgery: evolving neuroscientific concepts and clinical significance. *Journal of Hand Surgery.American Edition* 25, 391-414.

Lundborg, G., Rosen, B., Dahlin, L., Danielsen, N., & Holmberg, J. (1997). Tubular versus conventional repair of median and ulnar nerves in the human forearm: early results from a prospective, randomized, clinical study. *Journal of Hand Surgery.American Edition* 22, 99-106.

Madison, R. D., Archibald, S. J., & Brushart, T. M. (1996). Reinnervation accuracy of the rat femoral nerve by motor and sensory neurons. *Journal of Neuroscience* 16, 5698-5703.

Madison, R. D., Archibald, S. J., Lacin, R., & Krarup, C. (1999). Factors contributing to preferential motor reinnervation in the primate peripheral nervous system. *Journal of Neuroscience* 19, 11007-11016.

Maeda, T., Mackinnon, S. E., Best, T. J., Evans, P. J., Hunter, D. A., & Midha, R. T. R. (1993). Regeneration across 'stepping-stone' nerve grafts. *Brain Research* 618, 196-202.

Martini, R., Schachner, M., & Brushart, T. M. (1994). The L2/HNK-1 carbohydrate is preferentially expressed by previously motor axon-associated Schwann cells in reinnervated peripheral nerves. *Journal of Neuroscience* 14, 7180-7191.

McComas, A. J. (1995). Motor unit estimation: Anxieties and achievements. *Muscle & Nerve* 18, 369-379.

Montserrat, L. & Benito, M. (1990). Motor reflex responses elicited by cutaneous stimulation in the regenerating nerve of man: axon reflex or ephaptic response? *Muscle & Nerve* 13, 501-507.

Müller, H. W. & Stoll, G. (1998). Nerve injury and regeneration: basic insights and therapeutic interventions. *Current Opinion in Neurology* 11, 557-562.

Pellegrino, R. G. & Spencer, P. S. (1985). Schwann cell mitosis in response to regenerating peripheral axons in vivo. *Brain Research* 341, 16-25.

Politis, M. J., Ederle, K., & Spencer, P. S. (1982). Tropism in nerve regeneration in vivo. Attraction of regenerating axons by diffusible factors derived from cells in distal nerve stumps of transected peripheral nerves. *Brain Research* 253, 1-12.

Ramon y Cajal, S. (1928). *Degeneration and regeneration of the nervous system*, pp. 1-769. Oxford University Press, London.

Scherer, S. S. & Arroyo, E. J. (2002). Recent progress on the molecular organization of myelinated axons. *Journal of the Peripheral Nervous System* 7, 1-12.

Sørensen, J., Fugleholm, K., Moldovan, M., Schmalbruch, H., & Krarup, C. (2001). Axonal elongation through long acellular nerve segments depends on recruitment of phagocytic cells from the near-nerve environment. Electrophysiological and morphological studies in the cat. *Brain Research* 903, 185-197.

Stoll, G., Jander, S., & Myers, R. R. (2002). Degeneration and regeneration of the peripheral nervous system: from Augustus Waller's observations to neuroinflammation. *Journal of the Peripheral Nervous System* 7, 13-27.

Stoll, G. & Müller, H. W. (1999). Nerve injury, axonal degeneration and neural regeneration: basic insights. *Brain Pathology* 9, 313-325.

Sunderland, S. (1991). *Nerve Injuries and their Repair. A Critical Appraisal*, pp. 1-538. Churchill Livingstone, Edinburgh, London, Melbourne, New York.

Sunderland, S. & Bradley, K. C. (1950). Endoneurial tube shrinkage in the distal segment of a severed nerve. *Journal of Comparative Neurology* 93, 411-420.

Trojaborg, W. (1970). Rate of recovery in motor and sensory fibres of the radial nerve: clinical and electrophysiological aspects. *Journal of Neurology, Neurosurgery and Psychiatry* 33, 625-638.

Vizoso, A. D. & Young, J. Z. (1948). Internode length and fibre diameter in developing and regenerating nerves. *Journal of Anatomy* 82, 110-134.

Weinberg, H. J. & Spencer, P. S. (1978). The fate of Schwann cells isolated from axonal contact. *Journal of Neurocytology* 7, 555-569.

V. CEÑA, M. FERNÁNDEZ, C. GONZÁLEZ-GARCÍA AND
J. JORDÁN

Regional Center of Biomedical Investigations, University of Castilla La Mancha, Albacete, Spain

34. STROKE AND ISCHEMIC INSULTS

Summary. Brain ischemia is a common disease that is responsible for about 120 deaths per 100,000 habitants in Western countries. When cerebral blood flow decreases, there is an area, named penumbra, surrounding the core of ischemia, where neurons are perfused through collateral vessels. In this area, several molecular mechanisms involving excitatory amino acid release, increases in intracellular calcium, mitochondrial dysfunction, free radical production, release of mitochondrial proteins, proteases (mainly caspases) activation and neuronal death. These mechanisms activated in the penumbra area are responsible for the final fate of neurons and the degree of neuorological damage that is produced in the patient.

1. INTRODUCTION

Stroke is the third leading cause of death in western countries being mortality rate about 120 per 100,000 persons per year. The probability of suffering a stroke is higher than 95% when cerebral blood flow decreases to about 25% of control levels. Under these conditions, mitochondrial function fails and ATP concentration in the neurons drops within minutes after cessation of blood supply. At the core of the infarct, this causes osmotic cell lysis and necrotic death. Surrounding this necrotic core is a region, called penumbra, where perfusion from collateral vessels is kept between 25-50 % of control values. This would allow enough cerebral blood flow to maintain mitochondrial ATP production. However, mitochondrial function gradually fails over the ensuing days resulting in secondary cell death occurring mainly via apoptosis. Regardless of whether neuronal death is acute or secondary, ischemia is accompanied by increased efflux of excitatory amino acids, bioenergetical failure causing massive cell depolarisation, with efflux of K^+ and uptake of Na^+, Cl^-, and Ca^{2+} (for a review, see Kristian and Siesjo, 1997), disruption of energy production, increased free radical generation, intracellular Ca^{2+} dyshomeostasis and induction of the apoptotic cascade.

2. EXCITATORY AMINO ACID RELEASE

Excitatory amino acids, glutamate and aspartate, are endogenous compounds acting as neurotransmitters in brain, through the activation of three types of ionotropic receptors named after the initially described agonist activating them: N-methyl-D-aspartate (NMDA), alpha-amino-3-hydroxy-5-methyl-4-isoxazole-propionic acid (AMPA) and kainic acid receptors. NMDA receptor-associated channels are permeable to Na^+, K^+ and Ca^{2+} in a voltage-dependent manner, whereas AMPA and kainic acid receptors are linked to Na^+ permeable channels (Lerma, 2003). In addition, glutamate might also activate metabotropic receptors that induce G protein-mediated changes in second messengers. During brain ischemia there is a marked release of glutamate from the brain that can be monitored in plasma and cerebrospinal fluid from patients suffering ischemic stroke. Consistently with these clinical findings, pre-treatment of rats with the NMDA receptor antagonist, MK-801, decreases by 30 % infarct size following middle cerebral arterial occlusion showing that excessive stimulation of these receptors take place during cerebral ischemia, leading to neuronal degeneration.

Following glutamate release, there is a marked stimulation of glutamate receptors leading to neuronal degeneration that shows two components: an acute Na^+/Cl^--dependent neuronal swelling and delayed Ca^{2+}-mediated cell death where a massive and prolonged Ca^{2+} influx to the cytoplasm and reactive oxygen production has been observed.

3. CALCIUM

Calcium ions play a key role as regulators of numerous cellular functions. Therefore, cells tightly control free intracellular calcium concentration ($[Ca^{2+}]_i$; see Berridge et al., 2003). Since the extracellular calcium concentration ($[Ca^{2+}]_o$) is several orders of magnitude higher (1 mM) than the intracellular one (about 100 nM), even a small increase in the permeability of cell membranes to Ca^{2+} ions lead to a significant rise in $[Ca^{2+}]_i$. Calcium can enter cells via voltage- and agonist-operated Ca^{2+} channels and can be released from intracellular stores, i.e. the endoplasmic reticulum or the so called calciosomes. The mitochondria also represent a potential Ca^{2+} source; however, during resting physiological conditions mitochondrial Ca^{2+} content is low (about 200 nM).

During the last two decades it has become widely accepted that neuronal damage following brain ischemia is due to a perturbation of cellular Ca^{2+} metabolism. Brain ischemia, which compromises the cellular bioenergetic status, leads to cell depolarisation and to a rise in $[Ca^{2+}]i$. When ischemia ends, by reperfusion, the bioenergetic potential usually recovers and ion gradients are restored. However, neuronal damage can be observed following hours or days of reperfusion.

A marked and prolonged increase in $[Ca^{2+}]i$ is harmful to cells because it leads to activation of Ca^{2+}-dependent enzymes. Calcium might activate phopholipase A_2 increasing arachidonic acid production that is metabolised by lipoxygenases or cyclooxygenases producing reactive oxygen species. On the other hand, endonucleases (enzymes that degrade DNA) play an important role in apoptosis generation and might also be activated by Ca^{2+} in absence of caspase-3. Also phosphatases, like calcineurin, or proteases, like calpain, might be activated by Ca^{2+}.

4. REACTIVE OXYGEN SPECIES

On the other hand, evidence has accumulated showing reactive oxygen species generation under excitotoxic or ischemic conditions. It was shown that NMDA toxicity involves both nitric oxide (•NO) and reactive oxygen species, *i.e.* superoxide anion radical (•O_2^-), hydrogen peroxide (H_2O_2), and their highly cytotoxic by-product hydroxyl radical (•OH). An •OH generation also occurred *in vivo* in rat striatum under NMDA or glutamate exposure and during focal or global cerebral ischemia (Globus et al., 1995).

An increasing amount of evidence has led different authors to propose free radical production as a potential mechanism of brain injury in stroke. An •OH generation was actually reported in models of brain ischemia (Globus et al., 1995), where increased extracellular concentrations of excitatory amino acids might be the cause of this formation of reactive oxygen species, leading to oxidative stress and subsequent neuronal loss. In fact, evidence has accumulated showing that, when neurones are directly exposed to excitatory amino acids, a reactive oxygen species production occurs possibly through mechanisms involving phospholipase A_2, nitric oxide synthase or xanthine oxidase. For instance, an •O_2^- generation was reported in neuronal cultures exposed to glutamate or NMDA. Mepacrine, a phospholipase A_2 inhibitor, could prevent this production, suggesting that polyunsaturated fatty acids released from membranes by this enzyme might be potential precursors of reactive oxygen species *in vivo*. On the other hand, NG-nitro-L-arginine methyl ester, a nitric oxide synthase inhibitor, could reduce •OH efflux in rat striatum following NMDA perfusion. It was suggested that, when •NO is produced in the presence of •O_2^-, both compounds can interact and form a peroxynitrite anion intermediate (•$ONOO^-$).

The free radical superoxide anion is a product of normal cellular metabolism, produced mainly in mitochondria due to a switch from the normal four-electron reduction of O_2 to a one-electron reduction. Superoxide is harmful to the cell because it reduces iron-III to iron-II, and interacts with nitric oxide radical forming the strong oxidant, peroxynitrite. Superoxide is scavenged by superoxide dismutases generating hydrogen peroxide that interacts with transition metal ions to form the hydroxyl radical, the strongest oxidant formed in biological systems. Destructive actions of the hydroxyl radical include alterations of DNA and initiation of chain reactions of lipid peroxidations.

Lipid peroxides can interact with either iron-II or iron-III and give rise to alkoxyl and peroxyl radicals, respectively, each of which can initiate new chains of lipid peroxidation. Peroxidized plasmalemmal, endoplasmic reticulum, mitochondrial membrane lipids greatly interfere with cell function. It is vital for the cell to scavenge the peroxides produced and it does this mainly using the selenoenzyme glutathione peroxidase, which, in turn, uses, as cofactor, reduced-glutathione (GSH). The efficiency by which glutathione peroxidase can scavenge peroxides increases with increasing GSH concentration. In other words, relatively small increases in GSH concentration have a marked effect on the ability of glutathione peroxidase to scavenge peroxides. Indeed, increasing intracellular GSH has been demonstrated to increase the ability of cells to scavenge strong oxidants. Conversely, decreasing intracellular GSH results in greater damage following oxidative stress. GSH is also important in the regeneration of ascorbate, which has been used to reduce the vitamin E radical back to vitamin E. Hence, GSH plays a very central role in the ability of cells to manage oxidative stress.

5. MITOCHONDRIA

Changes in Ca^{2+} influx and O_2^- production within the mitochondria prior to cell death are events shared by cells treated with neurotoxic agents such as NMDA, veratridine or β-amyloid.

In addition, cell Ca^{2+} overload can also cause mitochondrial failure, that might lead to cell death. Recently, the mitochondria have became the main focus of interest in apoptotic cell death pathways (Kroemer and Reed, 2000) and probably also play a key role in delayed post-ischemic cell death (Kristian and Siesjo, 1998). This is because, during recovery, re-energized mitochondria take up most of the Ca^{2+} that has entered the cell during the insult. This Ca^{2+} uptake activates mitochondrial phospholipases, and seems to trigger increased production of free radicals as well as release of mitochondrial proteins, some of which are proapoptogenic (Susin et al., 1998). These factors seem to be released when the mitochondrial membrane permeability is increased by the opening of a large conductance channel: the mitochondrial permeability transition pore (MPTP). Thus, gradually raising $[Ca^{2+}]_i$ followed by mitochondrial Ca^{2+} accumulation and increased free radical production finally leads to irreversible damage to mitochondria, and to the triggering of a cell death program. We will focus now on some aspects of ischemia-induced and Ca^{2+}-dependent mitochondrial dysfunction leading to the assembling of a mitochondrial transition pore.

The mitochondria participate, together with endoplasmic reticulum, ion pumps and channels located in the cellular membrane, in the regulation of the $[Ca^{2+}]_i$ (Carafoli, 2003). Mitochondria might modulate the $[Ca^{2+}]_i$ during and following intense activation of Ca^{2+} conductance in plasma membranes. At steady state, there is a balance between influx and efflux of Ca^{2+} across mitochondrial membrane. Mitochondria start to accumulate Ca^{2+} when the cytosolic Ca^{2+} concentration rises over a "set point" (about 500 nM).

At the moment, three transport systems that control the entrance and exit of this cation into the mitochondria have been described: the Ca^{2+} uniporter, the $2Na^+/Ca^{2+}$ antiporter and the $Ca^{2+}/2H^+$ antiporter (Figure 1). The Ca^{2+} uniporter is an electrogenic transporter that shows a low affinity for Ca^{2+} and can be inhibited by Ruthenium Red and Mg^{2+}. On the other hand, the Na^+/Ca^{2+} antiporter, which can be inhibited by L-type Ca^{2+} channels blockers such as diltiazen, and the energy dependent $Ca^{2+}/2H^+$ antiporter system, located in the mitochondria of some cells, carry out Ca^{2+} release in response to specific stimuli. In some pathological situations, an alternative mechanism can trigger Ca^{2+} extrusion from mitochondria: the MPTP formation. This pathway is outlined below, and is believed to play a crucial role in cellular death programs. Its formation results in electric potential dissipation and release of substances from the mitochondrial matrix and the intermembrane space. Taken together, these systems allow the mitochondria to respond to $[Ca^{2+}]_c$ fluctuations with a slow variation in $[Ca^{2+}]_m$ levels. Following an increase in $[Ca^{2+}]_m$ there is a stimulation of Ca^{2+}-sensitive dehydrogenase producing an increasing in NAD(P)H and stimulation of respiratory channel. However, in pathological conditions, where the $[Ca^{2+}]_c$ is maintained hat high levels, Ca^{2+} enters mitochondria in excess. This results in several different effects, such as activation of the mitochondrial Ca^{2+} uniporter system, saturation of the Ca^{2+} efflux systems, and eventually drastic changes in the mitochondrial interior.

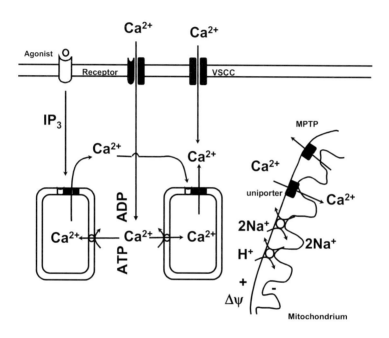

Figure 1. Mechanisms of calcium handling by the neuron

Mitochondria have been proposed to be the most important source of reactive oxygen species in cells of the nervous system. In the mitochondrial electronic transport chain (ETC), a non-enzymatic reduction of O_2 generating $\bullet O_2^-$ may occur, parallel to the enzymatic reduction of O_2 to H_2O by cytochrome oxidase. In some cases, up to a 2% of the total consumed O_2 by ETC can be transformed to O_2^- by the coenzyme Q. This reaction occurs due to the fact that electron transport is generally a very reactive process and involves components with a negative redox potential, such as flavine (complex I) and ubiquinone (complex III). Moreover, under conditions where ADP levels have been diminished, such as during treatment with different ETC complex inhibitors, an increased $\bullet O_2^-$ production has been observed.

The MPTP is a multiprotein complex formed in the contact sites among the inner and outer mitochondria membranes. Cytoplasmic (hexokinase), outer membrane (VDAC), inner membrane (ANT) and mitochondrial matrix (cyclophilin D; Cyp D) proteins participate in its structure (Halestrap et al., 2002).

Under physiological conditions, the different components of the MPTP are found disaggregated (Crompton, 2000). VDAC contributes to outer membrane permeabilization, ANT controls, in a specific way, the influx through the inner membrane of phosphorylated and non-phosphorylated adenine nucleotides, and cyclophilin D exhibits a peptidyl propyl isomerase activity, which is crucial for protein folding (Andreeva et al., 1999). Some of these components can be also bound to other proteins. VDAC associates with the mitochondrial benzodiazepine receptor and, in this manner, regulates extramitochondrial cholesterol transfer to the intermembrane space. Moreover, at the contact sites, creatine phophokinase association facilitates the energy transport through the creatine/phosphocreatine system.

When an apoptotical stimulus reaches the mitochondria, the different MPTP protein components come together to form a pore of about 1.0 to 1.3 nm ratio. This pore triggers the flux of molecules smaller than 1500 Da in a non-selective manner. Its opening produces an inner mitochondrial membrane permeabilization. It appears that this pore can act at least at two different levels of conductance and reversibility. At low level of conductance, the MPT pore opening is reversible and does not entail a large amplitude swelling of mitochondrial matrix, although it does cause a collapse of the mitochondrial potential ($\Delta\varphi m$; see Skulachev, 1999). However, at high level of conductance, the MPT pore opening is irreversible and leads to large amplitude swelling of the mitochondrial matrix. This may result in i) release of proteins and other solutes from the mitochondrial matrix into the cytosol; ii) a decrease in the electric transmembrane potential, with the consequence of ATP level depletion; and iii) mitochondrial swelling, due to entrance of water, that eventually will produce the outer membrane breaking, releasing several intermembrane space components into the cytoplasm.

There are many factors that regulate formation and opening of the MPTP. Calcium, reactive oxygen species, and members of the Bcl-2 family proteins, which have been related to neuronal death processes, modulate MPTP formation. Under conditions where $[Ca^{2+}]_i$ is sustained high like in neurodegenerative processes, ischemia and excitotoxicity, uptake uniporter systems are activated, allowing Ca^{2+} entry into mitochondria, while the responsible systems for the mitochondrial Ca^{2+} extrusion are saturated. Under these situations, mitochondria are not able to increase $[Ca^{2+}]_m$ indefinitely, and, eventually, release their Ca^{2+}. The mechanism responsible for this release implies the increase of the inner mitochondria membrane permeability though MPTP formation. Oxidative stress and substrates such as phosphate, acetoacetate and oxalacetate, are factors that stimulate Ca^{2+} release from mitochondria through MPTP formation. Reactive oxygen species are able to induce MPTP opening (Galindo et al., 2003), a process that is prevented by antioxidant agents such as vitamin E or glutathione.

Blockade of MPTP formation protects cell cultures from apoptotic stimuli. As it has been pointed previously, during brain ischemia an excessive glutamate receptor activation occurs that yields an increase in $[Ca^{2+}]_i$ inducing its accumulation in the mitochondria and subsequent MPTP opening. In cell cultures, where accumulation is prevented by mean of drugs able to dissipate the electric potential, such as the protonophore FCCP, cellular death processes are also prevented. Hence, MPTP formation may be crucial for the activation of signalling cascades that result in cell death. In addition, the inhibition of reactive oxygen species production by means of drugs with catalase or superoxide dismutase activities has been shown to efficiently prevent MPTP formation, both in isolated mitochondria, and in cell culture. Finally, MPTP blockers, such as cyclosporin A (CsA) and bongkrekic acid, also exhibit neuroprotective effects. Although CsA has many other actions, it is likely that its capacity to bind to Cyp D is responsible for its cytoprotective effects. Other drugs, that, like CsA, are able to inhibit calcineurin (such as the immunophilin FK506), do not show comparable protective effects. CsA prevents decreases in cellular viability in hippocampus following ischemic treatment. Bongkrekic acid that binds to ANT, blocks the interaction of the translocator with Bax and has been shown to inhibit caspase-9 release from brain mitochondria. This drug offers neuroprotection against ischemic injury in the brain, and protects neurons against NMDA-induced excitotoxic injury. In the latter setting, it also protects against the drop in $\Delta\Psi m$ and ATP levels.

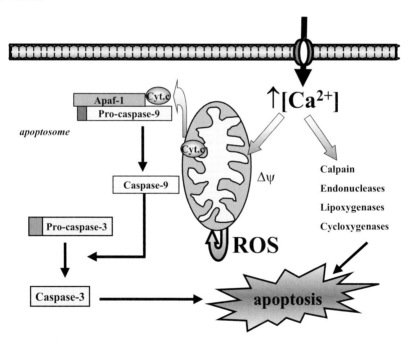

Figure 2. Intrinsic apoptotic pathway showing the role of mitochondria.

6. CASPASES

Protease and more specifically caspases are considered the executioners of ischemia-induced neuronal death. As illustrated in Figure 2, following ischemia, it is believed that caspases are activated through the so-called intrinsic mechanism that involves mitochondria (Blomgren, 2003). The decrease in $\Delta\varphi m$ can trigger a transient MPT of low conductance, which causes release of cytochrome c, activation of caspase 3 and apoptotic cell death. Caspase-3 activation may feed back to the mitochondria and cause further depolarisation and dysfunction. If the bioenergetic status is compromised, or massive amount of cytochrome c is released, the electrochemical potential for H^+ is collapsed and ATP production ceases being the result necrotic cell death.

At least 79 peptides that are released during MPTP opening have been characterized. These include catabolic enzymes, endozepine, cytochrome c, Smac/Diablo, apoptosis inductor factor (AIF) and some members of the caspases family, such as caspase-2, capase-3 and caspase-9. Once they have been released into the cytoplasm, these agents could activate different apoptotic pathways (Figure 2). Cytochrome c is a cofactor for the cytosolic multiprotein complex apoptosome. AIF release causes chromatin condensation and DNA fragmentation, and can act, independently of caspases, to induce apoptosis. The apoptosome initiates a cascade of caspase activation that involves different members of the family (caspase-9, caspase-3) leading to activation of caspase-3 that is believed to be the final executioner. Once activated, caspase-3 can cleave different proteins including poly-(ADP-ribose) polymerase (PARP), cytoskeletal proteins and endonucleases that contribute to cell death. In summary, brain ischemia initiates a complex mechanism including excitatory amino acid release, calcium dyshomeostasis, mitochondrial dysfunction and caspase activation that leads to neuronal death.

REFERENCES

Andreeva L, Heads R and Green CJ (1999) Cyclophilins and their possible role in the stress response. Int J Exp Pathol. 80: 305-315.

Berridge MJ, Bootman MD and Roderick HL (2003) Calcium signalling: dynamics, homeostasis and remodelling. Nat Rev Mol Cell Biol. 4: 517-29.1

Blomgren K, Zhu C, Hallin U and Hagberg H (2003) Mitochondria and ischemic reperfusion damage in the adult and in the developing brain. Biochem Biophys Res Commun. 304: 551-559.

Carafoli E (2003) Historical review: Mitochondria and calcium: ups and downs of an unusual relationship. Trends Biochem Sci. 28: 175-181.

Crompton M. (2000) Mitochondrial intermembrane junctional complexes and their role in cell death. J Physiol. (Lond.) 529: 11-21.

Galindo MF, Jordan J, Gonzalez-Garcia C and Ceña V (2003) Reactive oxygen species induce swelling and cytochrome c release but not transmembrane depolarization in isolated rat brain mitochondria. Br J Pharmacol. 139: 797-804.

Globus MY, Alonso O, Dietrich WD, Busto R and Ginsberg MD (1995) Glutamate release and free radical production following brain injury: effects of posttraumatic hypothermia. J Neurochem. 65: 1704-1711.

Halestrap AP, McStay GP and Clarke SJ (2002) The permeability transition pore complex: another view. Biochimie. 84: 153-166.

Kristian T and Siesjo BK (1997) Changes in ionic fluxes during cerebral ischaemia. Int Rev Neurobiol. 40: 27-45.
Kristian T and Siesjo BK (1998) Calcium in ischemic cell death. Stroke 29: 705-718.
Kroemer G and Reed JC (2000) Mitochondrial control of cell death. Nat Med. 6: 513-519.
Lerma J. (2003) Roles and rules of kainate receptors in synaptic transmission. Nat Rev Neurosci. 4: 481-495.
Skulachev VP (1999) Mitochondrial physiology and pathology; concepts of programmed death of organelles, cells and organisms. Mol Aspects Med. 20: 139-184.
Susin SA, Zamzami N and Kroemer G (1998) Mitochondria as regulators of apoptosis: doubt no more. Biochim Biophys Acta. 1366: 151-165.

H. WILMS AND G. DEUSCHL

Department of Neurology, University Hospital of Schleswig-Holstein, Campus Kiel, Germany

35. PARKINSON'S DISEASE I: DEGENERATION AND DYSFUNCTION OF DOPAMINERGIC NEURONS

Summary. Parkinson disease (PD) is the second most common progressive neurodegenerative disorder, with a prevalence of approximately 1-2% at the age of 65. PD is due to the relatively selective loss of dopaminergic neurons in the substantia nigra pars compacta, which leads to a profound reduction in striatal dopamine (DA). There is growing recognition that (PD) is likely to arise from the combined effects of genetic predisposition as well as largely unidentified environmental factors. Although responsible for only a small minority of cases of PD, recently identified genetic mutations have provided tremendous insights into the basis for neurodegeneration and have led to growing recognition of the importance of abnormal protein handling in Parkinson's as well as other neurodegenerative disorders. The mutant gene products all cause dysfunction of the ubiquitin-proteosome system, identifying protein modification and degradation as critical for pathogenesis. A number of environmental factors are known to be toxic to the substantia nigra. We review the various genetic and environmental factors thought to be involved in PD, as well as the mechanisms that contribute to degeneration of dopaminergic neurons in PD.

1. INTRODUCTION

Idiopathic Parkinson's disease (PD), first described by James Parkinson in 1817, is the second most common neurodegenerative disorder after Alzheimer's disease. It affects 1- 2 % of the general population over the age of 65. The disease predominantly affects dopaminergic neurons of the substantia nigra pars compacta, culminating in their demise with subsequent depletion of dopamine in the striatum. After approx. 50% of the dopamine neurons and 75-80% of striatal dopamine is lost, patients start to exhibit the classical symptoms of PD including bradykinesia, postural reflex impairment, resting tremor, and rigidity. The present chapter will summarise the physiological function of dopaminergic neurons in motor control, explain the different aetiologies currently discussed for their degeneration in PD and explain the resulting pathophysiology of PD.

2. THE BASAL GANGLIA MODEL AND PARKINSON´S DISEASE

Research since the late 1970s has added considerably to our knowledge of basal ganglia structure and function. A convergence of clinical findings, data from human post-mortem studies, and animal models of Parkinson´s disease contributed to the development of a model of the pathophysiology of Parkinson's disease with considerable explanatory power (DeLong 1990) regarding the physiological control of motor function, and the pathophysiology of PD and the motor complications of its medical treatment.

The basal ganglia are a group of subcortical nuclei that form a functional system. The primate basal ganglia are composed of the striatal complex, the two segments of the globus pallidus, the two portions of the substantia nigra and the subthalamic nucleus. The primary afferent structure of the basal ganglia is the striatal complex, which receives innervation from the neocortex including the precentral motor areas and postcentral sensory fields in a somatotopically organised fashion. The corticostriate projections are excitatory and use glutamate as their primary neurotransmitter. In primates the dorsal striatum has two components, the medial caudate and lateral putamen, separated by fibres of the internal capsule. Detailed anatomical studies have shown that putaminal output reaches the internal segment of the globus pallidus (GPi) and the substantia nigra pars reticulata (SNr) through two different pathways, which arise from two different putaminal populations: A monosynaptic "direct" pathway innervates the GPi / SNr, an "indirect" pathway projects to the external segments of the globus pallidus, which innervates the GPi / SNr, through the subthalamic nucleus too. The neurotransmitter of both striatal projection neurons and Gpi / SNr neurons is the inhibitory amino acid χ-aminobutyric acid (GABA).

The Gpi and SNr are the output stations of the basal ganglia: These structures project via the ventral anterior / ventral lateral (VA / VL) and mediodorsal (MD) thalamic nuclei to the primary motor cortex, premotor cortices and prefrontal cortex, with both Gpi and SNr possessing an additional projection to the pedunculopontine nucleus. Moreover, the SNR sends its efferents to the superior colliculus and mesopontine tegmentum. These brainstem nuclei project back to the basal ganglia predominantly on the subthalamic nucleus, these brainstem projections may play an important role in postural and other motor control mechanisms via reticulospinal neurons. Thalamic nuclei are considered to be the major route through which the basal ganglia modulate motor performance. The GPe projects primarily to the subthalamic nucleus (STN). The latter exerts an excitatory / glutaminergic drive to the GPi, SNr, SNc and GPe. The SNc is reciprocally connected with the striatum, giving rise to the massive striatal dopaminergic connection. The subthalamic nucleus (STN) is a regulator of basal ganglia output to the thalamus. By virtue of its major excitatory inputs to the GPi and SNr, the STN is a major determinant of GPi/ SNr neuron activity and basal ganglia output. Thalamocortical neurons are excitatory and glutaminergic, and regulation of the activity of thalamocortical activation is suggested to be the means by which basal ganglia output influences motor performance: In the normal state corticostriatal activation of the direct pathway will produce a GABAergic inhibition of GPi / SNr neurons, which in turn will disinhibit their thalamic target, thus facilitating the thalamic projection to he precentral motor fields. It is hypothesized that voluntary movements are generated by the cerebral cortex and the basal ganglia inhibit competing motor programs, that would interfere with the desired motor action (Mink, 1996).

Studies using the model of MPTP-induced parkinsonism in monkeys show that loss of dopaminergic neurons in the SNc leads to loss of striatal dopaminergic innervation resulting in a reduction of the normal inhibition of the nigrostriatal pathway on GABAergic neurons. They increase their activity, thus overinhibiting the GPe. Decreased activity of GABAergic pallidosubthalamic neurons disinhibits the STN. Hyperactivity of the STN provides an additional excitatory input to the GPi / SNr. This results in an excessive activity of the GPi / SNr and increased inhibition of the thalamic nuclei receiving basal ganglia input. The decreased thalamocortical neuron activity results in deficient excitatory motor input to motor, premotor, and prefrontal cortices and may cause the motor deficits of PD. The important role of the STN in the pathogenesis of parkinsonism was determined by lesioning the STN in MPTP-monkeys to ameliorate symptoms (Bergman, Wichmann et al. 1990).

Efforts have been made to explain the treatment related dyskinesias, particularly peak dose dyskinesias, which occur commonly in PD patients: Animal models of peak dose dyskinesias have been made in an effort to explore the hypothesis that these dyskinesias result from diminished basal ganglia output to the thalamus. Results derived from experiments with either unilaterally 6-OHDA-lesioned rats or MPTP-lesioned primates indicate that chronic intermittent L-dopa or dopamine agonist therapy results in relative normalisation of striato-GPi / SNr projection neuron activity. When chronic intermittent treated animals are then challenged with L-dopa or dopamine agonist therapy, both the striato-GPe and striato-GPi / SNr respond to stimulation of their dopamine receptors, with the striato-GPe neurons decreasing in activity and striato-GPi / SNr neurons increasing in activity. The effect of the decrease in striato-GPe activity is to disinhibit the GPe and suppression of STN activity. In subjects who never received chronic intermittent L-dopa or dopamine agonist therapy, the result is relative normalisation of basal ganglia output to the thalamus. In the chronic intermittent-treated subjects, the restoration of striato-GPi / SNr activity by treatment results in diminished GPi / SNr output as striato-GPi / SNr output is driven above levels seen normally in unlesioned subjects. A model of basal ganglia function in health and Parkinson's disease is depicted in Fig. 1.

3. AETIOLOGIES OF DEGENERATION OF THE SUBSTANTIA NIGRA IN PARKINSON'S DISEASE

Despite many years of focused research, the causes of the degeneration of dopaminergic neurons remains to be elucidated. Understanding the cause of PD is critical as that knowledge could lead to directed research that will develop new and potent therapies. The relative contributions of environmental versus genetic factors regarding the cause of PD have been hotly debated.

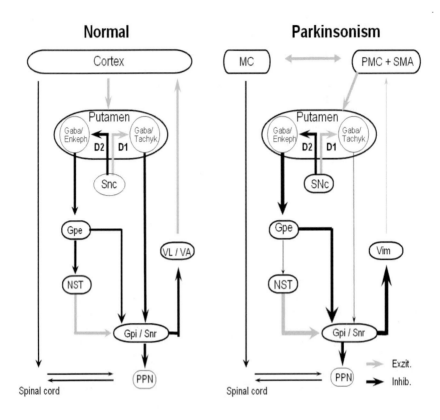

Fig.1: Model of basal ganglia function in health and Parkinson's disease. Black arrows indicate an inhibitory, gray arrows indicate an excitatory function of the pathway. Physiologically, STN and Gpi maintain a reciprocal control upon the STN activity. In Parkinson's disease increased excitatory output from the STN leads to augmented activation of the Gpi and SNr, which leads to increased inhibition of the VL / VA and PPN. Abbreviations: MC = motor cortex, pmc = premotor cortex, sma = supplementary motor cortex, SNc = substantia nigra pars compacta, SNc = substantia nigra pars reticulata, Gpe = external segment of the globus pallidus, Gpi = internal segment of the globus pallidus, NST = N. subthalamicus, VL / VA = Ventralis lateralis and ventralis anterior of the thalamus, PPN = pedunculopontine nucleus.

3.1. Neurotoxicity

Epidemiological studies have for a long time suggested that pesticide exposure is associated with an increased risk of developing PD (Gorell, Johnson et al. 1998). Compelling evidence in favour of an environmental agent came from the discovery that inadvertent injections of the protoxin n-methyl-4-phenyl-1,2,3,6-

tetrahydropyridine (MPTP) causes a parkinsonian syndrome in humans that is virtually indistinguishable from PD (Langston, Ballard et al. 1983). Researchers have capitalised on this discovery to develop an animal model of PD. After administration MPTP crosses the blood-brain barrier and is converted in astrocytes by monoxides B to its active metabolite 1-methyl-4-pyridinium (MPP+). MPP+ was found to be a mitochondria poison that inhibits mitochondrial respiration at complex I of the electron transport system. The selectivity of MPP+ for dopaminergic neurons is due to the fact that it is an excellent substrate for the dopaminergic transporter and is thereby accumulated preferentially in dopaminergic neurons. Following recognition of MPTP's neurotoxicity and its mechanism of action, several laboratories reported a selective defect in complex I of the electron transport chain in PD.

Other agricultural chemicals have been shown to induce a parkinsonian state as well: One of them is rotenone, a common pesticide and naturally occurring compound derived from the roots of certain plant species. Rotenone is commonly used as an insecticide in vegetable gardens, and is also used to kill or sample fish populations in lake and reservoirs. Exposure to the herbicide 1, 1´dimethyl-4´-bipyridinium, or paraquat has emerged as another putative risk factor for PD on the basis of its structural similarity to MPP+. Systemic injection of paraquat into mice causes a dose-dependant decrease in dopaminergic nigral neurons and striatal dopaminergic neurons followed by reduced ambulatory movements (Betarbet, Sherer et al. 2000). Manganese which is used in overlapping geographical areas with paraquat, has been shown to decrease locomotor activity and potentate MPTP effects, suggesting that exposure to mixtures of chemicals may also be relevant etiologically. These results highlight the possibility that environmental toxins, including pesticides that inhibit mitochondrial function, may contribute to degeneration of dopaminergic neurons in PD.

Influence of the immune system and oxidative stress on the progression of dopaminergic neurodegeneration in Parkinson´s disease

While different mechanisms including environmental toxins and genetic factors initiate neuronal damage in the substantia nigra and striatum, there is now strong unequivocal evidence that activation of neuroinflammatory cells, especially microglia, with subsequent production of pro-inflammatory cytokines and molecules aggravates this neurodegenerative process.

In order to understand the course of idiopathic PD and secondary Parkinsonian syndromes the factors perpetuating the degenerative process need to be identified in addition to its initial trigger factors. It was shown that following an acute exposure to the neurotoxin 1-methyl-4-phenyl-1,2,3,6-tetrahydropyridine (MPTP), manganese and other toxins the degenerative process continues for years in absence of the toxin. Large numbers of reactive microglia have been observed in the substantia nigra pars compacta in human post-mortem tissue from patients with PD as well as patients suffering from MPTP-induced parkinsonism, indicating that this inflammatory process might aggravate neurodegeneration. By releasing various kinds of noxious factors such as cytokines or proinflammatory molecules (e.g. proteolytic enzymes, reactive oxygen intermediates or NO) microglia may damage CNS cells. Nitric oxide (NO) is neurotoxic due to inhibition of complex 1 and 2 of the respiratory chain. Moreover it reacts with superoxide anion to generate peroxynitrite, a highly reactive molecule capable of oxidising proteins, lipids and DNA, which causes striatal neurodegeneration in a mouse model of PD in vivo. The cytokine tumor necrosis factor-α (TNF-α) is an important factor in the regulation of inflammatory processes and apoptotic cell death: In brains of PD patients, TNF-α producing glial cells have been detected in the substantia nigra and immunoreactivity for TNF-α receptors was found in cell bodies and processes of most dopaminergic neurons. TNF-α- and interleukin-6 (IL-6) levels are elevated in the striatum and cerebrospinal fluid of PD patients possibly indicating an involvement of these cytokines in the pathology of the disease. In transgenic mice overexpression of Il-6 causes encephalopathy with astrogliosis, neuronal loss, demyelinisation and edema. The stimuli triggering microgliosis in Parkinsonian syndromes are unknown so far: However, analysis of neuronal loss in the mesencephalon of PD patients shows that it is not uniform but that neurons containing neuromelanin (NM) are predominantly involved. Moreover, NM was found in patients suffering from juvenile, idiopathic and (MPTP)-induced parkinsonism. In an autopsy study of a patient with (MPTP)-induced parkinsonism extraneuronal NM was found in close vicinity to activated microglial cells even 12 years after exposure to the neurotoxin (Langston, Forno et al., 1999). We have hypothesised that extraneuronal melanin might trigger microgliosis, microglial chemotaxis and microglial activation in PD with subsequent release of proinflammatory and possible neurotoxic mediators like NO, TNF-α and IL-6. The addition of human NM to microglial cell cultures induced positive chemotactic effects, activated the pro-inflammatory transcription factor nuclear factor kappa B (NF-κB) via phosphorylation and degradation of the inhibitor protein κB (IκB), and led to an upregulation of TNF-α, IL-6 and NO. The impairment of NF-κB function by the IκB-Kinase inhibitor sulfasalazine was paralleled by a decline in neurotoxic mediators. NM also activated p38 mitogen-activated protein kinase, the inhibition of this pathway by SB203580 diminished phosphorylation of the transactivation domain of the p65 subunit of NF-κB. These findings demonstrate a crucial role of NM in the pathogenesis of Parkinson´s disease by augmentation of microglial activation, leading to a vicious cycle of neuronal death, exposure of additional neuromelanin and chronification of inflammation (Wilms, Rosenstiel et al. 2003). Another potential source of increased oxidative stress is iron. In the SN iron is normally bound either by ferritin or NM, and is associated with Lewy bodies in Parkinson´s disease (Jellinger, Kienzl et al. 1992). Since iron partakes only in its unbound form in the generation of free radicals, it is noteworthy that several molecules including dopamine and 6-OHDA release ferritin-bound iron in vitro (Double, Maywald et al. 1998). Moreover intranigral iron injection reduces striatal dopamine levels, conversely iron levels in the substantia nigra were elevated after experimental lesions of the nigrostriatal tract.

Dopamine metabolism is another potential source of increased oxidative stress in PD: Levodopa is toxic to dopaminergic neurons in vitro (Spencer, Jenner et al., 1994) and in the 6-hydroxydopamine rodent model of PD (Blunt, Jenner et al., 1993). Enzymatic metabolism of dopamine by monoaminooxidase generates hydrogen peroxide, which is inactivated by glutathione peroxidase and its cofactor gluthatione. In rodents increased presynaptic turnover of dopamine evoked by injection of reserpine, which interferes with the storage of dopamine in synaptic vesicles, induced a significant rise in the level of oxidized glutathione (Spina and Cohen, 1989), however this has not been observed in human PD and in several animal models of altered dopaminergic function.

Anti-inflammatory drugs may be one of the new approaches in the treatment of PD. Indometacine and dexamethasone were found to be neuroprotective in MPTP-animal models of PD. In our study the antagonization of microglial activation by a pharmacological intervention targeting microglial NF-κB or p38 MAPK could point to additional venues in the treatment of Parkinson's disease, too (Wilms, Rosenstiel et al. 2003).

3.2. Genetic factors

The majority of PD cases are sporadic and do not result from obvious genetic defects. However, a small percentage of approx. 5% have a familial form of PD, usually marked by earlier disease onset. The discovery of these inherited forms of PD shifted the emphasis back to genetic factors. The genetic anomalies that cause PD are varied and complex. These conditions are most likely caused by abnormalities in the transport, degradation and aggregation of proteins that lead to cell-specific changes and ultimately to degeneration of dopaminergic neurons: Lewy bodies, rounded eosinophilic inclusions predominantly found in the soma and neurites of melanin-containing neurons in the SN, are a common pathological hallmark of sporadic PD. They are composed of α-synuclein, ubiquitin and proteosomal subunits. Mutations in genes encoding for these products are hypothesised to lead to aberrations of the proteolytic pathway and to aggregation of proteins as Lewy bodies. While Lewy-bodies might be biologically inert, they are space-occupying entities that are bound to infer with nerve cell function.

The first gene to be linked to PD (PARKIN 1) encodes α–synuclein, a presynaptic protein of unknown function (George 2002) and major component of Lewy bodies. Two missense mutations (single base-pair changes) in this gene, leading to early-onset autosomal-dominant PD, were discovered in a few European unrelated families. Linkage data identified a disease locus on chromosome 4q21-23. Affected individuals have typical idiopathic PD, including levodopa responsiveness and Lewy bodies, although the age of onset is somewhat lower and progression appears to be more rapid. These observations have inspired studies to explain how mutations of α-synuclein can cause degeneration of dopaminergic cells in the substantia nigra focusing on the structural properties of this protein. Unfolded or misfolded proteins represent a major stress to cells in general, and inefficient clearance of such proteins is a hallmark of aging cells. Such proteins with abnormal conformation, as well as otherwise damaged or oxidised proteins that are not degraded, tend to aggregate as inclusions. In vitro biochemical analyses suggest that the random coil structure of -synuclein and its tendency to aggregate into misfolded structures under certain circumstances might be sufficient to confer toxic properties to this protein. Native α-synuclein in vitro is an unfolded protein with no ordered secondary structure. In the presence of lipid-containing membranes α-synuclein displays an alpha helical structure. In higher concentrations it forms oligomers and undergoes another transformation turning into a beta sheet typical of amyloid fibrils. α-synuclein expression in cultured neuroblastoma cells was found to induce apoptosis. Since the fibrillary, β-pleated sheet conformation of α-synuclein was observed in the cells, the authors concluded that this form was responsible for neuronal damage.

Reactive oxygen species as well as ane of the missense mutations (A53T), but not the other (A30P) increases the rate of fibril formation. It is not known for sure whether the fibrils or some oligomeric intermediates promote neurodegeneration. It has been demonstrated that A30P and A53T protofibrils may produce toxicity through the formation of pore-like structures that can permeabilize cell membranes (Lashue, Hartley et al, 2002). Circular and elliptical rings corresponding to protofibrils were visualized by atomic force microscopy in a mixture of A53T and wild type -synuclein (Conway, Harper et al., 2000). These annular species resembled pore-forming toxins of bacteria such as Clostridium perfringens. This finding suggests that toxicity may arise from a common structural feature of fibrilization intermediates. The pore- or channel-like activity of protofibrils could be responsible for, or at least contribute to, disease pathology by permeabilizing cell membranes and causing leakage of cellular contents that could lead to cell death.

Catecholamines including dopamine inhibit the conversion of toxic α-synuclein protofibrils to the stable α-synuclein filaments which may explain the selective susceptibility of the dopaminergic system in PD.

Overexpression of wild-type or PD-causing α-synuclein mutants in transgenic animals recapitulates many of the behavioural, pathologic, and biochemical features of human PD: Mice developed numerous α–synuclein-immunoreactive nerve cell bodies and processes. One study reported the presence of α–synuclein immunoreactive inclusions that were granular and as such distinct from the filaments found in human disease. There was no loss of dopaminergic nerve cells in the substantia nigra, but a reduction in dopamine nerve terminals in the striatum. Another model of PD, which indicates that α-synuclein overexpression alone can cause nigrostriatal degeneration, has been developed in the fly. The transgenic flies display Lewy-body-like inclusions and a loss of dopaminergic neurons with advancing age.

A mutation in PARK2, a gene encoding for parkin, which is an E3 ubiquitin ligase has been associated with juvenile onset autosomal-recessive parkinsonism (Kitada, Asakawa et al. 1998). Various mutations are associated with PD, including deletions and point mutations. Clinically, the disease usually begins when the patient is in his or her 20s, is prominently associated with dystonia and diurnal fluctuations, and progresses slowly but has early and severe levodopa-induced dyskinesias, but no dementia. Pathologically, degeneration of neurons in the substantia nigra pars compacta without formation of Lewy bodies has been found. The ubiquitin targeted pathway of intracellular protein catabolism utilises initially three classes of proteins, corresponding to the three steps in the first phase of the pathway—ubiquitin activation, ubiquitin conjugation, and target protein ligation—to attach the polyubiquitin tail to the protein to be destroyed. Ubiquitin C-terminal hydroxylases in the final step release ubiquitin from this polyubiquitin tail for re-use after digestion of the target protein by the 26S proteasome complex. An inability to release ubiquitin from the polyubiquitin tail, as a result of deficient ubiquitin C-terminal hydrolase activity, with the resultant absent or incomplete digestion of the target protein, could allow the accumulation of the building blocks for neurotoxic fibrils. Mutations in the parkin gene cause the enzyme to lose its activity, leading to the abnormal accumulation of its non-ubiquinated form with failure to form Lewy bodies, leading to earlier selective neural cell deaths. This observation brings into question the relevance of Lewy bodies for the pathogenesis of PD. Lack of these inclusions in the brains of patients with *parkin* mutations who develop the disease at a much younger age than individuals with other forms of PD might even support a protective role of Lewy bodies.

A missense mutation, isoleucine to methionine at residue 93 (I93M), in the ubiquitin C-terminal hydrolase L1 (UchL1) gene on chromosome 4p has been found in two siblings with typical PD in a small German pedigree in which the disease is transmitted as an autosomal dominant disorder with incomplete penetrance (Leroy, Boyer et al. 1998). UchL1 is one of the most abundant proteins in the brain, also present in Lewy bodies, and belongs to a family of enzymes (terminal hydrolase family) that is responsible for degrading polyubiquitin chains back to the ubiquitin monomer, which can then be recycled and used to clear other proteins.

Diminished enzymatic activity of the ^{mutant} form of UCH-L1 may disturb recycling of ubiquitin monomers with subsequent dysfunction of the proteosomal-proteolytic pathway. Since the I93M mutation in this gene has yet to be discovered in other families with PD, it is still the subject of debate whether this mutation is necessarily a cause of PD or is merely a harmless polymorphism occurring in a single family.

Additional loci have been linked to inherited PD, but the genes associated with these affected families are not known. A locus located on chromosome 2p13 has been described in a subset of families with autosomal dominant inheritance and typical Lewy body pathology (PARK3), the gene responsible for this disease has not been identified yet (Gasser, Muller-Myhsok et al. 1998). Affected family members show signs of dementia; correspondingly in addition to neuronal loss in the substantia nigra and typical brainstem Lewy bodies, neurofibrillary tangles and Alzheimer plaques were found post-mortem.

A further locus on chromosome 4p has been found to segregate with disease in another family with autosomal dominant Lewy body parkinsonism- the 4p 14-15 haplotype (PARK4) (Farrer, Gwinn-Hardy et al. 1999). The PARK4 locus appears to segregate with both PD and postural tremor, furthermore, affected family members with PD have several atypical features, including early weight loss, dysautonomia, and dementia. Further families with multiple cases of young onset PD all occurring within the same generation have been investigated for autosomal-recessive forms of parkinsonism. An autosomal recessive locus on chromosome 1, PARK6, has recently been described in a large Sicilian family and is linked to chromosome 1 (1p35-p36). Very close to the PARK6 locus is another recently discovered PD locus (PARK7, 1p36), which is significantly linked to a Dutch family with early onset autosomal recessive PD. PARK7 results from mutations in *DJ-1*, a ubiquitous protein of unknown function. There has been a further description of another locus on chromosome 12 (12p11.2-q13) linked to the development of PD in a Japanese family with autosomal dominant disease with low disease penetrance (PARK8). Nigral degeneration without Lewy bodies was found in the neuropathologic examination in some patients. Kufor-Rakeb syndrome, an autosomal recessive nigro-striatal-pallidal-pyramidal neurodegenerative disorder, has been mapped to a 9-cM region of chromosome 1p36 and designated PARK9. Loci and genes linked to familial PD have been summarised in Table 1.

Table 1: Loci and genes linked to familial PD

Locus	Chromosomal location	Gene	Mode of inheritance
Park 1	4q21.3	α-synuclein	Autosomal dominant
Park 2	6q25.2-27	Parkin	Autosomal recessive
Park 3	2p13	unknown	Autosomal dominant
Park 4	4p15	unknown	Autosomal dominant
Park 5	4p14	UCH-L1	Autosomal dominant
Park 6	1q35-36	unknown	Autosomal recessive
Park 7	1p36	DJ-1	Autosomal recessive
Park 8	12p11.2-q13	unknown	Autosomal dominant
Park 9	1p36	unknown	Autosomal recessive, (Kufor-Rakeb syndrome)

The discovered mutations point to the critical role of protein folding and degradation through the ubiquitin proteasome pathway as a central common mechanism leading to neuronal cell death. Accumulation of oxidatively damaged proteins and impairment of proteasome activity in the SN of patients with sporadic PD substantiate the conclusions drawn from the genetic data that altered proteolysis is a key pathogenic process in this disorder. Interestingly, oxidative stress increases the tendency of α-synuclein to aggregate presumably into toxic oligomers, while overexpression of α-synuclein in vitro leads to the generation of excessive amounts of reactive oxygen species thereby creating a vicious cycle. Individuals are variably exposed to numerous environmental factors that lead to increased oxidant stress through food and drinking water including the complex I inhibitors mentioned above, which might be one possible link between the role of environmental factors and genetic predisposition in dopaminergic neurodegeneration (Fig. 2).

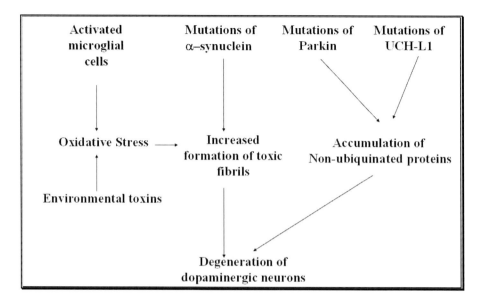

Fig.2: Proposed pathogenic mechanisms leading to degeneration in dopaminergic neurons in PD via environmental toxins or gene mutations.

REFERENCES

Bergman, H., T. Wichmann, et al. (1990). "Reversal of experimental parkinsonism by lesions of the subthalamic nucleus." Science 249(4975): 1436-8.

Betarbet, R., T.B. Sherer et al. (2000). "Chronic systemic pesticide exposure reproduces features of Parkinson's disease" Nat Neurosci 3(12): 1301-6.

Blunt, S. B., P. Jenner, et al. (1993). "Suppressive effect of L-dopa on dopamine cells remaining in the ventral tegmental area of rats previously exposed to the neurotoxin 6-hydroxydopamine." Mov Disord 8(2): 129-133.

Conway, K. A., J. D. Harper et al. (2000). "Fibrils formed in vitro from alpha-synuclein and two mutant forms linked to Parkinson's disease are typical amyloid. " Biochemistry 39(10): 2552-63.

DeLong, M. R. (1990). "Primate models of movement disorders of basal ganglia origin." Trends Neurosci 13(7): 281-5.

Double, K. L., M. Maywald, et al. (1998). "In vitro studies of ferritin iron release and neurotoxicity." J Neurochem 70(6): 2492-9.

Farrer, M., K. Gwinn-Hardy, et al. (1999). "A chromosome 4p haplotype segregating with Parkinson's disease and postural tremor." Hum Mol Genet 8(1): 81-5.

Gasser, T., B. Muller-Myhsok, et al. (1998). "A susceptibility locus for Parkinson's disease maps to chromosome 2p13." Nat Genet 18(3): 262-5.

George, J. M. (2002). "The synucleins." Genome Biol 3(1).

Gorell, J. M., C. C. Johnson, et al. (1998). "The risk of Parkinson's disease with exposure to pesticides, farming, well water, and rural living." Neurology 50(5): 1346-50.

Jellinger, K., E. Kienzl, et al. (1992). "Iron-melanin complex in substantia nigra of parkinsonian brains: an x- ray microanalysis." J Neurochem 59(3): 1168-71.

Kitada, T., S. Asakawa, et al. (1998). "Mutations in the parkin gene cause autosomal recessive juvenile parkinsonism." Nature 392(6676): 605-8.

Langston, J. W., P. Ballard, et al. (1983). "Chronic Parkinsonism in humans due to a product of meperidine-analog synthesis." Science 219(4587): 979-80.

Langston, J. W., L. S. Forno, et al. (1999). "Evidence of active nerve cell degeneration in the substantia nigra of humans years after 1-methyl-4-phenyl-1,2,3,6-tetrahydropyridine exposure." Ann Neurol 46(4): 598-605.

Lashuel, H. A., D. Hartley, D et al. (2002). "Neurodegenerative disease: amyloid pores from pathogenic mutations. " Nature 418(6895): 291.

Leroy, E., R. Boyer, et al. (1998). "The ubiquitin pathway in Parkinson's disease." Nature 395(6701): 451-2.

Mink, J. W. (1996). "The basal ganglia: focused selection and inhibition of competing motor programs" Prog Neurobiol 50(4): 381-425

Spencer, J. P., A., Jenner et al. (1994) "Intense oxidative DNA damage promoted by L-dopa and its metabolites. Implications for neurodegenerative disease." FEBS Lett 353(3): 246-50.

Spina, M. B. and G. Cohen (1989). "Dopamine turnover and glutathione oxidation: implications for Parkinson disease" Proc Natl Acad Sci U S A 86(4): 1398-400.

Wilms, H., P. Rosenstiel, et al. (2003). "Activation of microglia by human neuromelanin is NF-kB dependent and involves p38 mitogen-activated protein kinase: implications for Parkinson's disease." FASEB J 17(3): 500-2.

C. WINKLER[1] AND D. KIRIK[2]

[1]*Department of Neurology, Hannover Medical School, Hannover, Germany*

[2]*Department of Physiological Sciences, Lund University, Lund, Sweden*

36. PARKINSON'S DISEASE II: REPLACEMENT OF DOPAMINE AND RESTORATION OF STRIATAL FUNCTION

Summary. Motor dysfunction in Parkinson's disease is primarily due to the loss of dopamine within the striatum. While pharmacotherapy is rather straightforward during the first years of the disease, treatment during later stages of the disease poses a considerable challenge to the clinician. In this chapter new therapeutic approaches are discussed which aim at restoring striatal dopamine function in Parkinson's disease. During early stages of the disease growth factors may be thus used to prevent the protracted dopamine neuron degeneration, whereas during later stages of the disease intracerebral transplants of dopamine-producing cells or gene therapy approaches may be used to replace dopamine within the striatum.

1. INTRODUCTION

Parkinson's disease (PD) is characterized by a progressive degeneration of the dopamine (DA) neurons in the substantia nigra (SN), and the cardinal symptoms of PD, i.e. bradykinesia, rigidity, tremor and postural instability, usually develop over many years. The pharmacological gold standard for therapy of PD is the replacement of DA using the DA precursor l-3,4-dihydroxyphenylalanine (L-DOPA). Initially, the response to treatment is excellent but over a course of several years most patients develop severe therapy-related side effects such as motor fluctuations, abnormal involuntary movements or psychiatric disturbances. The search for alternative treatment strategies is therefore of considerable interest. During recent years new therapeutical approaches have been developed that aim to restore the striatal DA content. Within the following sections we will first focus on restoration of the striatal DA content by replacement of the lost DA neurons using intracerebral transplantation of fetal DA cells. Then we will discuss application of the neurotrophic factor GDNF (glial cell line-derived neurotrophic factor) in order to prevent neurodegeneration and to induce regeneration of the DA system. Finally, we will discuss recent gene therapy approaches for PD using viral vectors expressing the tyrosine hydroxylase (TH) gene, the rate limiting enzyme for production of DA, in order to restore the striatal DA content, or viral vectors expressing GDNF in order to prevent neurodegeneration.

2. INTRACEREBRAL TRANSPLANTATION OF DOPAMINERGIC NEURONS IN ORDER TO RESTORE STRIATAL FUNCTION

Following a decade of experimental work in animal models, clinical neurotransplantation in PD was initiated based on a straightforward concept and well-defined biological mechanisms, i.e. those neurons that die in PD are to be replaced by new DA neurons. The best cells currently available for such transplantation purposes are DA neurons that are harvested from aborted fetal human ventral mesencephalon. Such grafts are usually placed within the caudate-putamen (CPU; striatum in the rat), because there is experimental evidence that DA cells implanted into the SN are not able to grow axons to the CPU, where DA is primarily needed (see also Fig 1). In animal models of PD, transplants of fetal DA tissue have been shown to survive, to reinstate a new DA innervation in the previously denervated striatum, to release DA in an autoregulated fashion, and to establish afferent and efferent synaptic connections with the host brain.

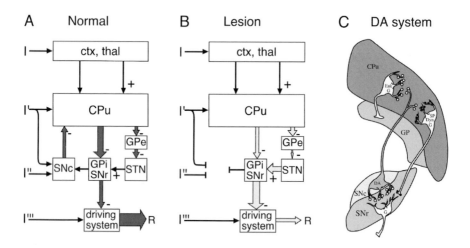

Figure 1: Model of basal ganglia function. (A) Motor responses are elicited from a set of subcortical driving systems following activating stimuli (I', I'', I''') controlled by inhibitory projections from the output stations of the basal ganglia, i.e. the globus pallidus internus (GPi) and the substantia nigra pars reticulata (SNr). These output nuclei are themselves under control of the caudate-putamen (CPU) via the direct striatopallidal (CPU -> GPi) and striatonigral (CPU -> SNr) pathways, and via an indirect pathway involving the GP externus (GPe) and the subthalamic nucleus (STN). (B) A lesion of DA system in the SN pars compacta (SNc) increases the threshold for activation of the driving system due to change of activity in the direct in indirect projection pathways. (C) The DA neurons of the SNc modulate information flow through the basal ganglia. D1-receptors are depicted as dark circles, D2-receptors as light circles. In addition to GABA (G), projection neurons in the CPU express either enkephalin (Enk), or substance P (SP) plus dynorphin (Dyn).

Within the last 15 years several hundred PD patients have thus received intracerebral transplants of fetal DA neurons. Survival of the grafts in the human brain can be demonstrated using imaging techniques, primarily [^{18}F]-dopa positron emission tomography, and in a number of patients who have come to autopsy neuropathological evidence for graft survival and reinnervation of the CPU has been provided. Clinical benefit does not occur directly after transplantation but usually develops over a period of 1 – 2 years following transplantation, which is thought to reflect the maturation process of the grafted DA neurons within the host brain. Pronounced improvements are usually observed on bradykinesia and rigidity, and motor fluctuations and time spent in "off" are also clearly reduced. Some of the patients have so far been followed up for more than 10 years. In these patients the clinical benefit appears to be stable over time and there is currently no indication that the grafted cells may undergo the same protracted degeneration as the host DA neurons.

These data show that transplantation of human fetal DA tissue provides long-term benefit for patients with PD. However, transplant-induced behavioural improvement is not uniform in PD patients. Thus, while some patients show dramatic reductions of motor disabilities following transplantation, which may even allow complete withdrawal of all oral medication therapy, most patients are still clearly Parkinsonian and dependent on their oral medication. Indeed, within some of the most extensive transplantation studies, mean symptomatic relief defined as improvements in PD rating scales was only 30 – 40%. In these studies putaminal [^{18}F]-dopa uptake was increased by 55 – 69% above the pregraft baseline, but this value still represents only half of what would be expected in age matched controls.

Several reasons have been discussed as probable causes for the incomplete functional recovery after grafting. There has been a debate whether transplant-induced functional recovery may be further improved with better graft survival and with a dense reinnervation within all parts of the CPU. Currently, DA neuron graft survival may be as low as 5 – 10% and increasing the number of fetuses implanted in each patient can only partly compensate this. Recent neuropathological analysis in PD patients, however, has shown that despite 80.000 – 130.000 surviving DA neurons per putamen, reinnervation was obtained in only 24 – 78% of the postcommissural putamen, while other regions such as the ventral striatum did not receive any transplant-derived reinnervation at all. Insufficient spread of the grafted cells within the target structure rather than overall poor graft survival may thus be a major determinant for incomplete behavioural recovery. Furthermore, it has been suggested that the ectopic location of the implants in the CPU may prevent complete behavioural normalization, because the transplanted DA cells would not receive the afferent input they would normally receive in the SN; and it has been argued that other transplant locations may have to be considered since DA denervation in PD is not restricted to the CPU but also involves the mesolimbic and mesocortical DA systems.

During recent years animal experiments have been designed to answer some of the questions highlighted in the previous paragraph. Most of these experiments were performed in the rat 6-hydroxydopamine (6-OHDA) lesion model, which is the most extensively studied animal model of PD. Injection of the neurotoxin 6-OHDA into the medial forebrain bundle induces a near-complete destruction of the nigral DA neurons and their striatal terminals. Alternatively, 6-OHDA is injected into the striatum, which induces a progressive degeneration of part of the nigral neurons. Both the complete lesion (considered to reflect late stage PD) and the partial lesion (considered to reflect early or mid-stage PD) induce behavioural deficits, which resemble the symptoms observed in PD patients.

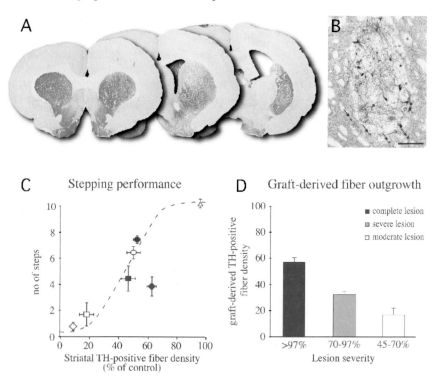

Figure 2: Neurotransplantation for PD. (A) Multiple intrastriatal grafts of fetal DA cells induce a dense DA striatal reinnervation in rats with complete unilateral 6-OHDA lesions. (B) High power view from A, showing an intrastriatal graft deposit with numerous DA cells which extend processes into the host striatum. Asterisks denote the internal capsule, arrows show some of the grafted cells which are integrated within the host striatum. (C) Performance in the stepping test, which measures forelimb akinesia, in relation to the striatal TH-positive fibre density. While unlesioned animals (open triangle) can perform normally in this test, rats with partial lesions (open circle, open square) and complete lesions (open diamond) reduce their performance according to an S-shaped curve (dashed line). While transplants in rats with partial lesions of the nigrostriatal system (filled circle, filled square) induce behavioural recovery along the S-shaped curve, behavioural recovery in rats with complete lesions (which include lesions of the nonstriatal DA system; filled diamond) and transplants is less pronounced suggesting that spared portions of the host DA system within nonstriatal brain regions may be necessary for the grafts to exert their optimal functional effects. (D) Graft-derived DA fibre outgrowth is dependent on the lesion severity. For details, see also Winkler et al., 2000.

Similar to transplantation in PD patients, in these animals transplantation of DA cells using the standard transplantation approach usually induces reinnervation within up to 75% of the head of the CPU while more rostral or caudal parts of the CPU and the ventral striatum do not receive a graft-derived DA innervation. Even within reinnervated areas the density of DA fibres is not normalized but amounts to only 30 – 50% of normal. Functional analysis in such transplanted animals shows that some simple spontaneous motor behaviours may be normalized whereas more complex motor behaviours are either unaffected or only partially normalized by the grafts.

These limitations can in part be overcome by distributing the fetal DA cells over multiple implantation sites, which will provide a more extensive and evenly distributed DA innervation over larger areas of the striatum. Indeed, in some studies graft-derived innervation has been demonstrated within the whole rostro-caudal extent of the striatum including the ventral striatum, thus increasing the density of DA fibres to a mean of 60% (varying between 40 – 80%) of normal (see Figs. 2A, B). Behavioural recovery in such animals is further enhanced as compared to animals with fewer or smaller grafts, but still normalization of all behavioural deficits is not obtained. In Fig. 2C, this is illustrated for the stepping test, which measures forelimb akinesia. Performance in this test has been shown to be DA-dependent. Thus, DA-denervating lesions induce a reduction of stepping performance, which is more pronounced with increasing size of the lesion. Indeed, behavioural performance in this test follows an S-shaped curve (dashed line in Fig. 2C). In animals with complete lesions and extensive intrastriatal transplants (filled diamond in Fig. 2C), behavioural recovery should be near to normal in case it occurred along the S-shaped curve. However, behavioural recovery is far less pronounced.

These data thus speak against a simple relationship between graft-induced striatal DA reinnervation and functional recovery in the rat PD model. It has been suggested that intrastriatally placed DA neurons may not fully integrate within their ectopic graft location but rather retain immature properties. Indeed, incomplete behavioural recovery is paralleled by an incomplete normalization of functional cellular parameters: Electrophysiological studies have shown that neuronal discharge rates or threshold currents required to evoke activation of striatal neurons, are not completely normalized by the grafts. Furthermore, the lesion-induced down regulation of substance P in host striatal neurons projecting directly to the globus pallidus (GP) internus and SN pars reticulata, or cytochrome oxidase activity and drug-induced immediate early gene expression within basal ganglia nuclei of the so called indirect projection pathway, is only partially normalized by the grafts (see Fig. 1 for details of basal ganglia circuitry). These data may thus indicate persistent dysfunction of all striatal projection neurons following intrastriatal grafting of fetal DA neurons. Since the host afferent connections with the grafted DA neurons are rather sparse, it has been suggested that the grafts may be able to act as a biological minipump to release DA tonically, but that regulated DA release via appropriate synaptic connections is not sufficiently obtained in order to normalize complex cellular or motor function.

During recent years attempts have been made to place the DA cells in the SN hoping that the full range of afferent connections, e.g. from hypothalamus or brain stem regions, may be more easily established in the homotopic graft location. While such grafts do not grow axons to the striatum, they still induce some minor behavioural improvements. The role of DA release within the SN has thus been considered of such importance that a pilot study has been initiated where PD patients have received intranigral DA grafts in addition to intraputaminal grafts. Although the behavioural impact of additional intranigral grafts in patients with intraputaminal grafts is still unclear, the data suggest that fetal DA cells survive following transplantation into the SN, and that the intranigral grafting procedure can be performed without serious side effects.

Patients with PD have so far been considered for transplantation studies after the failure of the standard therapy regimens. Such PD patients are usually in advance stages of their disease so that, similarly to rats with complete DA-denervating lesions, DA denervation is not restricted to the striatum but also involves limbic or cortical forebrain areas. The functional impairments may therefore reflect a combination of striatal and non-striatal dysfunction. The incomplete functional recovery observed after transplantation of fetal DA cells in both PD patients and in rats with complete DA-denervating lesions may thus in part be explained by the failure of the grafts to restore DA innervation in non-striatal forebrain areas.

The role of host-derived DA innervation within non-striatal forebrain regions for performance in motor behaviours has been investigated in rats with partial intrastriatal DA-denervating lesions, which leaves the DA innervation in most non-striatal forebrain areas intact. As illustrated in Fig. 2C, behavioural improvement in partially lesioned animals with intrastriatal DA grafts occurs along the dashed line (filled circle and filled square), thus indicating that spared portions of the host DA system may be necessary for the DA grafts to exert their optimal functional effects. However, even in these animals behavioural recovery was not complete. Rather, graft-derived DA fibre-outgrowth did not exceed a threshold value of approximately 70% of normal irrespectively of the lesion severity. Thus, in animals with complete DA-denervating lesions graft-induced fibre outgrowth was 3-fold or 1.5-fold larger as compared to animals with severe or moderate intrastriatal lesions (Fig 2D). In agreement with other studies the denervated striatum seems thus to provide a positive stimulating effect on the outgrowth of axons from the grafted cells, and this stimulatory effect is reduced in animals where part of the intrinsic DA system is left intact.

These experimental data may have implications for transplantation in PD patients. First, it is suggested that a certain number of surviving DA neurons is required to obtain functional recovery in spontaneous motor behaviours, and that the functional outcome may be further improved with increasing numbers of surviving DA cells. In agreement with this, behavioural recovery after transplantation in PD patients is more pronounced when more tissue is grafted. Furthermore, since the CPU is heterogeneously organized it is suggested that function may be further improved by evenly distributing the cells and thus, obtaining an even DA reinnervation within the target structure. However, it is so far unclear whether complete normalization of all Parkinsonian symptoms may be obtained simply by normalizing the DA content within the CPU. Recent positron emission tomography data indeed suggest that although basal and drug-induced DA release may be restored to normal levels following transplantation into the CPU, PD symptoms such as postural imbalance or gait disturbances may not be sufficiently improved by the grafts. In agreement with previous data demonstrating profound DA deficiency in non-striatal regions in PD patients, reinnervation of the CPU alone, therefore, in the absence of a functional meso-limbocortical DA pathway, may thus be insufficient to completely normalize behaviour in patients with PD.

PD patients in advanced stages of the disease involving non-striatal areas as well may thus be less suitable candidates to receive intracerebral DA transplants because functional impairments related to DA denervation in non-striatal areas may have an impact on the overall functional efficacy of the transplants as seen in animal experiments. On the other hand, transplant-derived fibre outgrowth and functional efficacy may be reduced in patients in earlier stages of the disease, i.e. when the host-derived DA innervation in the CPU is only partly reduced in comparison to late stage disease patients.

Recently, two randomized placebo-controlled double blind clinical trials have been completed in order to assess investigator bias and placebo effects in clinical neurotransplantation. However, the primary outcome measures of neither study were met, which has in part been attributed to changes of otherwise established methods such as the preparation and storage of tissue prior to transplantation or the immunosuppressive regimen. Further analysis of the data has shown that fetal DA transplants can induce significant behavioural recovery in patients with PD. However, the transplant-induced benefits were less pronounced as expected from the previous open labeled studies. Furthermore, severe transplant-induced dyskinesia and dystonia was observed in more than 15% of the patients in both studies. The reasons for the occurrence of these dyskinesias are still unclear, but some critical issues will have to be addressed in the future such as the treatment of the tissue prior to transplantation or the number of implantation tracts. Furthermore, it has been argued whether an excessive graft-derived production of DA within specific striatal regions may be critical for the development of dyskinesia. In conclusion, intracerebral transplantation of fetal DA cells is a powerful tool to induce functional benefit in patients with PD. However, more experimental and clinical studies are required before this method may be applied to larger numbers of patients.

3. APPLICATION OF RECOMBINANT GDNF PROTEIN IN PARKINSON'S DISEASE IN ORDER TO PREVENT NEURODEGENERATION AND TO INDUCE REGENERATION OF THE DOPAMINERGIC SYSTEM

Since neurodegeneration in PD progresses over many years therapeutic approaches that aim at preventing the further nigral DA neuron degeneration and thus halt motor deterioration are of considerable interest. A number of substances such as antioxidants, antiapoptotic agents or growth factors have been identified to promote the survival of DA neurons. As one of the most potent survival factors on DA neurons in experimental studies a neurotrophic factor was purified from a rat glial cell line in 1993 and hereafter termed glial cell line-derived neurotrophic factor (GDNF). This factor has been considered as particularly interesting for therapeutic approaches in PD, since it plays an important role as a local and target-derived survival factor for DA neurons during ontogeny and for the formation of the terminal DA fields in the striatum.

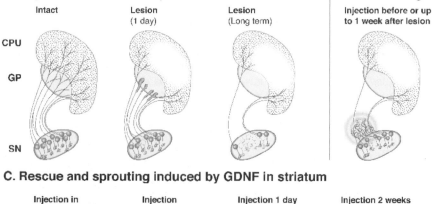

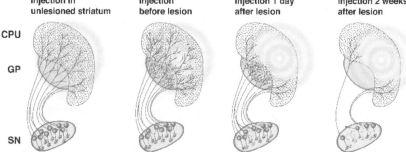

Figure 3: Schematic illustrations of the effects of GDNF protein injections in the intrastriatal 6-OHDA lesion model. (A) Intrastriatal 6-OHDA injections induce an acute destruction of the striatal DA fibre terminals and, over a course of several weeks, DA neuron cell loss in the SN. (B) Administration of GDNF into the SN can effectively prevent the delayed DA neuron cell loss, but there is no protection of the axonal connection to the striatum. (C) GDNF injections into the unlesioned striatum will increase the turnover of DA, but there is no evidence of DA fibre sprouting in the striatum. In contrast, protection and sprouting of DA fibre terminals may occur following an intrastriatal 6-OHDA lesion. This effect, however, is dependent on the time of administration in relation to the lesion. Abbreviations: CPU: caudate-putamen unit; GP: globus pallidus; SN: substantia nigra.

Repeated intracerebral injections or pump infusions of recombinant GDNF protein have been shown to exert robust effects on both, the intact and the lesioned DA system in a number of different animal models of PD. In the intact DA system, administration of GDNF has been shown to increase DA tissue content and turnover of DA, elevate the activity of tyrosine hydroxylase, and furthermore, increase overall motor activity. In the lesion models, there is evidence that GDNF is able to promote the rescue of lesioned nigral DA neurons, to prevent the progressive DA axon degeneration and to induce regeneration of striatal DA fibre terminals. These effects are proposed to be dependent on both, the time-point of GDNF administration with respect to the lesion, and the place of GDNF application, as illustrated in Fig. 3 for the intrastriatal 6-OHDA lesion model. As illustrated in Fig. 3A, injection of 6-OHDA into the striatum induces an acute lesion of the terminal DA fields within the striatum, followed by a progressive degeneration of the nigral DA neurons and their axons over several weeks. This model thus allows to investigate the protective effects of GDNF on DA neurons that are either acutely lesioned or undergoing a protracted degeneration. Furthermore, GDNF-induced regeneration and sprouting of spared DA neurons can be studied using this model.

Thus, intranigral GDNF administration prior to the lesion may effectively prevent nigral DA cell death, but the nigrostriatal DA pathway will not be preserved and thus, regeneration within the denervated striatal target will not occur (Fig. 3B). In contrast, administration of GDNF in the striatum prior to the lesion will not only protect the cell bodies but also induce significant preservation of the nigrostriatal DA pathway and moreover, induce sprouting of DA fibres within the striatum and globus pallius (Fig. 3C). Interestingly, such a sprouting effect is not observed when GDNF is injected into the unlesioned striatum (Fig. 3C) suggesting that the lesion itself renders the nigral neurons more receptive to the effects of GDNF. Since a substantial part of the DA system has already undergone neurodegenerative changes at the time of diagnosis of PD, experimental designs in which the neuroprotective factors are supplied after the onset of neurodegeneration are specifically relevant for the exploration of their clinical potential. Thus, intranigral administration of GDNF starting on the day of the lesion can protect the cell bodies without preventing the axonal degeneration and this effect, though less pronounced, is still observed when GDNF is injected into the SN within one week after the intrastriatal 6-OHDA lesion (Fig. 3B). Intrastriatal administration of GDNF starting on the day after the lesion is also effective in protecting the nigral DA neurons, but the striatal DA fibre density is not preserved because the nigrostriatal DA axons have already retracted back to the level of the globus pallidus at the onset of GDNF treatment (Fig. 3C). Nonetheless, the axon degeneration may be halted at the level of the globus pallidus, and within this structure sprouting of DA fibres is observed. If GDNF is repeatedly injected into the striatum several weeks after the intrastriatal 6-OHDA lesion, rescue of DA neurons will be negligible since most nigral DA neurons will already have died, and significant regeneration of the striatal DA innervation and normalization of behavioural deficits will not be observed.

Experimental data thus suggest that there is a critical time window for application of GDNF and that the functional benefits after administration of GDNF are most pronounced when GDNF is given early after induction of the neurodegenerative process. Although GDNF is also capable to induce sprouting within the remaining DA system when administration of GDNF is delayed after the lesion, functional benefits will be less pronounced. In the clinical setting, in order to effectively prevent DA neuron degeneration and to enhance function within the remaining DA system, GDNF should thus be applied relatively early during the disease process when a substantial part of the nigral DA neurons is still alive. However, it has to be emphasized that none of the animal models of PD replicates the human disease and thus, the clinical outcome following GDNF administration in PD cannot be easily predicted.

The place to which GDNF should be administered has to be carefully considered. Since animal data provide evidence that GDNF may be protective also when injected intracerebroventricularly, a subsequent multicentre trial was initiated, in which PD patients received intraventricular injections of GDNF at monthly intervals. In one patient, who came to autopsy, there was no evidence of striatal DA reinnervation and no GDNF-immunoreactivity was observed in the striatum. It is thus unclear whether GDNF was able to penetrate in sufficient concentration from the ventricles into the striatum. PD symptoms were not improved in this patient either, but side effects such as hallucinations, inappropriate sexual conduct or nausea were observed, some of which were temporally related to the GDNF injections. Since GDNF receptors are widely distributed within the brain, some of the side effects may thus have been related to GDNF function outside the nigrostriatal system.

In order to reduce the probability of adverse events due to so far unknown GDNF-induced effects outside the DA system in the brain, local delivery of the neurotrophic factor into the caudate-putamen, where GDNF has shown to exert its most pronounced functional effects in animal studies, is thus currently favoured for further clinical trials. In a recent phase I clinical trial five PD patients received continuous delivery of recombinant GDNF into the postcommissural putamen for more than one year without serious side effects. In these patients intraparenchymal catheters were placed in the putamen and connected to programmable pumps, which were regularly refilled with GDNF. Motor performance as measured in PD rating scales was improved by approximately 50% in these patients, while [^{18}F]-dopa uptake was increased by approximately 20%. Although it is currently unclear whether these effects are due to direct stimulation of the remaining DA system, i.e. increase in DA turnover and DA tissue content, or whether there is a prevention of the ongoing neurodegenerative process or even fibre sprouting, these promising results clearly justify further investigations in more patients. Overall, GDNF is to date a promising substance that may offer a new route for therapeutic intervention in PD.

4. GENE THERAPY FOR PARKINSON'S DISEASE

Since neurodegeneration in PD progresses over many years and decades, it is likely that GDNF should be administered over long periods in order to preserve or restore function within the nigrostriatal DA system. Local production of the neurotrophic factor may thus offer distinct advantages over repeated GDNF injections. During recent years viral vectors have been designed to locally deliver the GDNF gene into the brain. So far, adenovirus, adeno-associated virus or lentivirus vectors designed to express GDNF have been effectively used to induce long-term production of GDNF in the animal brain. Such GDNF-expressing viral vectors have been as effective in preventing the progressive DA neurodegeneration and inducing functional recovery in different animal models of PD as repeated injections or infusions of recombinant GDNF protein. Each of these viruses has specific characteristics with regard to the size of the genes that can be inserted into them, but all of them are capable of infecting resident non-dividing brain cells. Host-derived immune reactions or even toxic effects related to the virus injections are negligible, primarily when adeno-associated virus or lentivirus vectors are used. Furthermore, tumor formation following viral vector gene therapy has not been reported.

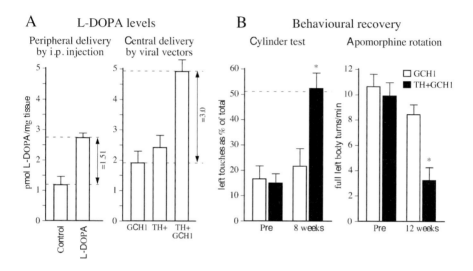

Figure 4: Effects of TH gene transfer in the rat PD model. (A) Following i.p. injection of 6 mg/kg L-DOPA and blockade of peripheral and central decarboxylase activity accumulation of 1.5 pmol L-DOPA/mg tissue above that seen following saline injection is observed. L-DOPA levels following TH plus GCH1 gene transfer clearly exceed the threshold value obtained with peripheral L-DOPA injection. (B) The cylinder test measures exploratory paw use when the animals are placed in a glass cylinder. Equal use of both paws is considered as normal behaviour (dashed line). Left paw use, which is severely impaired following the lesion, is normalized following combined TH plus GCH1 gene transfer. Similarly, apomorphine-induced rotational behaviour is reduced following TH plus GCH1 gene transfer. Data from Kirik et al., 2002.

Serious side effects of viral vector-induced GDNF expression have so far not been reported in animal models of PD. However, there is now growing evidence that prolonged expression of GDNF in the brain can induce down-regulation of the L-DOPA-synthesizing enzyme tyrosine hydroxylase (TH) and thus may prevent functional effects that would otherwise be expected following GDNF therapy. Furthermore, since currently available viral vectors are highly efficient in transducing neurons, GDNF release is not limited to the striatum but also occurs in the striatal projection regions, such as globus pallidus and SN pars reticulata, via anterograde transport within transduced neurons. Within these striatal target regions aberrant GDNF-induced DA fibre sprouting has been observed, and although there is currently no proof that such aberrant fibre sprouting has negative side effects, it is not unlikely that normal basal ganglia function may be affected. There are thus two requirements that may have to be fulfilled before GDNF gene therapy for PD may enter the clinic, and this may also hold true for other gene therapy approaches: First, regulation of gene transfer-derived protein expression in the brain, e.g. using more refined viral vector systems, has to be achieved in order to switch transgene production on or off, and second, transgene expression may have to be targeted to specific cells, e.g. glial cells, in order to limit transgene expression to the site of injection.

Another promising but still experimental approach to promote functional recovery in PD, is gene transfer of the L-DOPA-synthesizing enzyme TH into the brain. Since side effects following long-term oral L-DOPA therapy are at least in part due to the intermittent supply of L-DOPA, transfer of the TH gene into the striatum may provide continuous production of L-DOPA and consequently, stimulate striatal DA receptors in a more physiological way. Furthermore, since L-DOPA release would be limited to the striatum, other L-DOPA-related side effects such as nausea, circulations problems or halluzinations may be reduced. There is now evidence that L-DOPA production within the striatum using a gene therapy approach can exceed the amount of L-DOPA that reaches the striatum following peripheral L-DOPA administration. However, in order to produce sufficient amounts of L-DOPA in the brain, expression of TH has to be combined with coexpression of the rate-limiting enzyme for production of the cofactor tetrahydrobiopterin, GTP-cyclohydrolase-1 (GCH1). This is exemplified in Fig. 4 for gene transfer using adeno-associated viral vectors. Thus, following combined TH plus GCH1 gene transfer to the striatum in a rat model of PD, sufficient amounts of L-DOPA are produced to normalize motor performance in PD-relevant behaviours. Furthermore, reduction of apomorphine-induced rotational behaviour indicates that the lesion-induced upregulation of striatal DA receptors is reversed due to the TH gene therapy approach.

In conclusion, gene therapy approaches for PD, whether aimed at preventing neurodegeneration of the host DA system or at restoring L-DOPA production in the striatum, provide an interesting tool to study DA function in the brain and may in the future develop into a clinical concept for treatment of PD.

REFERENCES

Arenas, E. (2002). Stem cells in the treatment of Parkinson's disease. *Brain Res Bull, 57*(6), 795-808.

Björklund, A., Kirik, D., Rosenblad, C., Georgievska, B., Lundberg, C., & Mandel, R. J. (2000). Towards a neuroprotective gene therapy for Parkinson's disease: use of adenovirus, AAV and lentivirus vectors for gene transfer of GDNF to the nigrostriatal system in the rat Parkinson model. *Brain Res, 886*(1-2), 82-98.

Björklund, A., & Lindvall, O. (2000). Cell replacement therapies for central nervous system disorders. *Nat Neurosci, 3*(6), 537-544.

Castilho, R. F., Hansson, O., & Brundin, P. (2000). Improving the survival of grafted embryonic dopamine neurons in rodent models of Parkinson's disease. *Prog Brain Res, 127*, 203-231.

Cenci, M. A., Whishaw, I. Q., & Schallert, T. (2002). Animal models of neurological deficits: how relevant is the rat? *Nat Rev Neurosci, 3*(7), 574-579.

Freed, C. R., Greene, P. E., Breeze, R. E., Tsai, W. Y., DuMouchel, W., Kao, R., et al. (2001). Transplantation of embryonic dopamine neurons for severe Parkinson's disease. *N Engl J Med, 344*(10), 710-719.

Gill, S. S., Patel, N. K., Hotton, G. R., O'Sullivan, K., McCarter, R., Bunnage, M., et al. (2003). Direct infusion of glial cell line-derived neurotrophic factor in Parkinson's disease. *Nat Med, 9*(5), 589-595.

Hurelbrink, C. B., & Barker, R. A. (2001). Prospects for the treatment of Parkinson's disease using neurotrophic factors. *Expert Opin Pharmacother, 2*(10), 1531-1543.

Kirik, D., Georgievska, B., Burger, C., Winkler, C., Muzyczka, N., Mandel, R. J., et al. (2002). Reversal of motor impairments in parkinsonian rats by continuous intrastriatal delivery of L-dopa using rAAV-mediated gene transfer. *Proc Natl Acad Sci U S A, 99*(7), 4708-4713.

Kordower, J. H., Palfi, S., Chen, E. Y., Ma, S. Y., Sendera, T., Cochran, E. J., et al. (1999). Clinicopathological findings following intraventricular glial-derived neurotrophic factor treatment in a patient with Parkinson's disease. *Ann Neurol, 46*(3), 419-424.

Lang, A. E., & Lozano, A. M. (1998). Parkinson's disease. Second of two parts. *N Engl J Med, 339*(16), 1130-1143.

Lindvall, O., & Hagell, P. (2000). Clinical observations after neural transplantation in Parkinson's disease. *Prog Brain Res, 127*, 299-320.

Winkler, C., Kirik, D., Björklund, A., & Dunnett, S. B. (2000). Transplantation in the rat model of Parkinson's disease: ectopic versus homotopic graft placement. *Prog Brain Res, 127*, 233-265.

Zurn, A. D., Widmer, H. R., & Aebischer, P. (2001). Sustained delivery of GDNF: towards a treatment for Parkinson's disease. *Brain Res Brain Res Rev, 36*(2-3), 222-229.

R. ALBERCA AND E. MONTES

Department of Neurology, Alzheimer's Disease Unit. H.U. Virgen del Rocío, Sevilla, Spain

37. ALZHEIMER'S DISEASE

Summary. Alzheimer's disease (AD) is a very frequent, chronic progressive disorder characterized pathologically by extracellular (amyloid deposits) and intracellular neurofibrillary tangles aggregates. There are two main forms of the disease: the very uncommon autosomal dominant form, and the non-hereditary form, which is related to different risk factors, such as age and Apo E genotype, and afflicts 5-10% of people over 65 years of age. The disease leads to dysfunction of specific populations of neurons, disrupts cerebral neurotransmission pathways, and finally causes neuronal death and cerebral atrophy. Clinical signs are the consequence of the selective and hierarchical degeneration of neurons. Disordered neurotransmission is thought to be responsible for the memory loss and other cognitive and behavioural alterations leading to dementia, which is the clinical hallmark of the disease. Both etiologically and pathogenically based treatments are lacking, but treatment strategies are being developed that inhibit formation of Aβ or improve its solubility, reduce tau phosphorylation, suppress inflammation and regulate apoptosis. Symptomatic treatments are available, but the disease progresses inexorably, and patients finally die from intercurrent illnesses. AD is a devastating disorder to sufferers and their families, and provokes enormous social costs. The only way to cope with the disease is to improve early diagnosis and to introduce etiopathogenically related treatments for its cure, or to stop its progression and diminish its frequency.

1. INTRODUCTION

Alzheimer's disease (AD) is an anatomo-clinical syndrome due to different etiopathogenic mechanisms. The disease can be either sporadic, or inherited as an autosomal dominant disorder due to mutations in three different genes [preseniline 2 (PS2) on chromosome1; preseniline 1 (PS1) on chromosome 14, and amyloid precursor protein (APP) on chromosome 21]. The prevalence of the inherited AD is 5.3 per 100,000 persons at risk (Campion et al. 1999) and the disorder is characterized by early-onset dementia which follows an autosomal dominant mode of inheritance. Such families should be referred to a genetic clinic for counselling. The inherited form of AD lacks clinical significance because it is an infrequent disorder, but the identification of mutations has enabled the creation of genetically engineered animal models, for investigating the *in vivo* consequences of mutations and to explore new therapeutic strategies.

The sporadic form of AD afflicts 5-10% of persons older than 65 years. Women clearly predominate, mainly because of their higher life expectancy. There are between 200,000 and 400,000 AD patients in our country (Spain), and a very conservative estimate has calculated the median annual direct cost per patient to be about 15,100 euros. However, both social and personal costs are enormous but difficult to measure. The sporadic form of AD is related to several risk factors. The most important is age: prevalence of the disease doubles every five years after age 60, and increases from 1% at 60-65 years of age to around 40% at 85 years and older (Von Strauss et al. 1999). Another important risk factor is possession of the Apo e4 allele (located on chromosome 19q13.2), which is associated with more than half of the susceptibility to AD: the lifetime risk of AD for an individual without e4 is about 9%; whereas for persons with e4 is 29% (Seshadri et al. 1995). Epidemiological studies have disclosed other possible risk factors, such as other susceptibility genes, gender, head trauma, number of years of education, positive family history, neurotoxins, low levels of vitamin B12, and Down's syndrome. Moreover, both the incidence and intensity of the dementia seem to be related to hypercholesterolemia, diabetes and systolic arterial hypertension: treatment of this latter condition has reduced the incidence of the dementia (Guo et al. 1999). AD is also associated to low blood levels of vitamin B12 and folic acid, and this may have important consequences in the prevention and treatment of the disease. Estrogen replacement and anti-inflammatory medications can reduce the risk of AD development, as we will see later.

2. ETIOPATHOGENESIS AND RELATED TREATMENT STRATEGIES

Most of the etiopathogenic discoveries in inherited form of AD have been applied to the sporadic form of AD. Hereditary forms lead to accumulation of amyloid, and many workers in this field have accepted the idea that the main pathological hallmark of AD, both hereditary and non-hereditary, is the abnormal accumulation of $A\beta$ peptide, a pleated-sheet amyloid peptide derived by β- and γ-secretase cleavages of the amyloid precursor protein (APP). $A\beta$ peptide production is prevented by α-secretase, which cleaves APP within the $A\beta$ domain. In normal individuals, the peptides ($A\beta$ 41, $A\beta$ 42, and $A\beta$ 43) are both produced and cleared, but it is hypothesised that in AD toxic $A\beta$ peptides accumulate pathologically. All mutations cause increased production of the rapidly aggregating $A\beta$ 42 (but in the sporadic form there is no evidence for the increased production); thus, the clearance of amyloid may be altered. $A\beta$ peptides eventually assemble in fibrils, which are deposited as the core of the senile plaques, and there is evidence that $A\beta$ accumulation, which occurs near synapses and impairs neurotransmission, correlates with clinical decline. Proteases are good targets to be inhibited (Sinha et al. 1999), and γ- and β-secretase inhibitors are being developed to limit the formation of $A\beta$ peptides, even though these secretases are difficult to access by inhibitors because they have a subcellular location, in the lumen of the Golgi body and endosomes. Moreover, zinc and copper induce β-amyloid aggregation (and this combination of metals and amyloid produces enormous amounts of hydrogen peroxide in the brain); substances that chelate these ions, such as clioquinol, could improve the condition. Other compounds to inhibit production, deposition, aggregation, toxicity, and clearance of $A\beta$ are under study.

Clearance of "debris" and toxins is in part carried out by activated microglia. In fact, microglia are clustered around brain amyloid deposits, but the phagocytosis of the Aβ peptides by these cells seems to be blocked in AD (Frackowiack et al. 1992). A therapeutic and preventive approach is immunisation against β amyloid and stimulation of the glial clearance of Aβ peptides. Schenk et al. (1999) experimentally introduced vaccination with pre-aggregated Aβ 42; the vaccine both prevented and reduced amyloid deposits in PDAPP transgenic mice. The behavioural deficits were alleviated if the mouse was vaccinated prior to the deposition of amyloid, providing a framework for investigation of these therapeutic strategies. However, clearance of Aβ peptide may accelerate vascular amyloid deposits or provoke other undesirable side-effects (such as complement activation) in cases with abundant cerebral deposits. In fact, adverse events have caused clinical investigation of amyloid vaccination in humans to be temporarily suspended. An alternative might be a vaccine that stimulates the beneficial immune response —but not the more dangerous T-cell immune response — as well as passive transfer of Aβ peptide antibodies, which also reduced cerebral levels of Aβ.

A "cascade" of unwanted effects seems to follow amyloid deposition. Deposition of Aβ peptides elicit local neurotoxic effects, inducing dystrophic neuritis and an infiltration of microglia that express acute phase reactants and pro-inflammatory cytokines. Anti-inflammatory agents could suppress inflammation, slow the rate of AD pathology, and protect the brain from the reactive glial responses associated with Aβ deposition. It was found that a six-month course of ibuprofen caused a 50% reduction in Aβ peptide deposition in mice (Lim et al. 2000), an effect that could be due either to a reduction of pro-amyloidogenic proteins or to increased Aβ peptide clearance. It is possible that NSAID treatment could result in stimulation of phagocytosis and inhibition of neurotoxins. In fact, according to a meta-analysis (McGeer et al. 1996), NSAID consumption is associated with a 50% reduction of risk in AD, but we need confirmatory studies and more-precise information about possible preventive treatments regarding effectiveness, dosing, duration of treatment, and adverse effects, especially with modern NSAIDs.

There is an increasing evidence that AD causes severe oxidative stress as a result of either Aβ mediated oxyradical generation or perturbed ionic calcium balances within neurons and mitochondria. Dietary intake of vitamin E from food may be associated with a reduced risk of AD (Morris et al. 2002), and drugs with strong antioxidant properties, such as vitamin E and selegiline, have shown a slight efficacy in AD (Sano et al. 1997). New antioxidants are under clinical development. On the other hand, serum homocysteine is increased in AD, and homocysteine could potentiate endothelial and neuronal oxidative injury. McCaddon et al. (2002) have suggested that cerebral oxidative stress could impair the metabolism of homocysteine and compromise the reduction of vitamin B12 to its active state. Accordingly, treatment of AD patients with increased homocysteine should include glutathionylcobalamin (McCaddon et al. 2002), but the efficacy of such treatment has not yet been proved. Moreover, neurons can be damaged through other mechanisms. Conversely, glutamatergic overreactivation of the NMDA receptors can cause cell damage, and seems to be involved in pathogenesis of AD. Certain antagonists of these receptors, such as memantine, could diminish this pathological mechanism, and, in fact, this product has been shown to be clinically effective in AD patients with moderate and severe dementia (Reisberg et al. 2000).

Estrogens are believed to have trophic, anti-inflammatory, antioxidant and anti-Aβ effects. Postmenopausal women, who are chronically hypoestrogenic, may be more vulnerable to neurological diseases and may suffer increasingly from cognitive dysfunction as they age. High dosage of estradiol can improve cognition in AD postmenopausal women (Asthana et al. 2001), and hormonal replacement therapy has been associated with a 29% reduction in the risk of the disease (Yafee et al. 1998). We are waiting for the results of other, well designed, prospective studies to ascertain the capacity of estrogens to diminish the frequency of the disorder and to improve the condition, especially with the new products under development which should include compounds without the adverse effect of estrogenic feminisation.

Neurofibrillary tangles are formed from poorly-soluble paired helical filaments (PHF) composed principally of hyperphosphorylated tau isoforms and ubiquitin. They are characteristically located within neuronal bodies, dendrites, and dystrophic neurites seen in proximity to Aβ deposits, and are probably associated with altered intracellular transport. Several kinases have been reported to act on tau to induce abnormalities associated with PHFs. Thus, another therapeutic avenue could be to inhibit these kinases or to act on hyperphosphorylation of tau protein.

Finally, cell alterations lead to disordered neurotransmission at multiple levels and systems. In particular, there is a substantial presynaptic cholinergic deficit in AD, as shown by loss of neurons in the nucleus basalis of Meynert and reduced neocortical activity of choline acetyltransferase and acetylcholinesterase. Some experimental evidence indicates that correction of the deficit (acetylcholinesterase inhibitors, muscarinic receptor agonists) might decrease amyloid deposition in the brain. However, acetylcholinesterase inhibitors are the main line of the current symptomatic treatment and this is discussed later at length. On the other hand, neuronal growth factor can experimentally enhance the size and the activity of residual cholinergic cells. Autologous transplants of modified skin fibroblasts can supply neuronal growth factor to the brain, and this approach, and others with neurotrophic molecules, able to penetrate the blood-brain barrier, are under development.

3. PATHOLOGY AND CLINICAL EXPRESSION OF THE DISORDER

The pathological prevalence of the disease in people aged 85 years and older is 33%, a number which double the estimates of clinical prevalence (Polvikoski et al. 2001). Pathological diagnosis remains the gold standard for recognition of the disease: according to modern criteria, it should be based on the presence of both senile plaques, and neurofibrillary tangles, following CERAD guidelines (Mirra et al. 1991) and recommendations of the National Institute of Aging (1997). AD selectively affects cerebral neuronal populations, and progression of the lesions follows the stages described by Braak and Braak (1991). Stages I and II, which are clinically silent, are characterised by neurofibrillary tangles confined to the transenthorrinal region. During stages III and IV, both neurofibrillary tangles and senile plaques spread to entorhinal and transentorhinal regions, which become severely affected: these lesions are probably related to early memory loss in AD. Stages V and VI are characterized by isocortical alterations and are clinically expressed as a fully developed dementia. Primary cortical areas, especially primary visual cortex, are the last to be involved. Neuronal and synaptic loss and neurotransmission deficits accompany amyloid deposition and tau pathology, and all are related to the dementia. The disease eventually causes a diffuse cerebral atrophy, especially marked on posterior, associative cortex.

The clinical phenotype of AD centres around an amnesic disorder related to early mesial temporal lobe degeneration. The disorder has a long, pre-dementia phase with isolated memory loss, and strategies are being developed to pick up these "pre-symptomatic" AD cases. Amnesic "Mild Cognitive Impairment, (MCI)" (Petersen et al. 2001) is a clinical entity, which includes non-demented patients with isolated memory loss (Table I). Most of MCI cases are really "pre-symptomatic" AD, and have an increased risk of developing full clinical symptoms of AD in the following years, at an annual rate of 12% (instead of 1% in controls); consequently, they should be longitudinally monitored. Apo E4 carrier status, atrophy of mesial temporal structures on MR, and increased tau and diminished Aβ 42 peptide in CSF all increase the likelihood that these MCI patients to eventually develop AD. Various clinical trials (vitamin E, anticholinesterase, COX-2 inhibitors) to improve cognition in MCI and to stop its progression to dementia are in progress.

Table 1

Diagnostic criteria for "Mild Cognitive Impairment" (Petersen et al. 2001)

1. Memory complaint, preferably corroborated by an informant
2. Objective memory impairment
3. Normal general cognitive function
4. Intact activities of daily living
5. Not demented

Dementia is the clinical hallmark of AD. The typical clinical syndrome of AD includes a memory defect with difficulty in learning and recalling new information together with anomia progressing to fluent aphasia, alterations in visuospatial skills with environmental disorientation, apraxia, and a dysexecutive syndrome with defective planning, judgement, and insight. Apathy and depression usually appear early. Psychosis and agitation become apparent when the dementia is moderate to severe, and can precipitate institutionalisation. Parkinsonian signs, gait abnormalities, primitive reflexes, and epileptic seizures all appear late, and paratonia and myoclonus closely predict the final months of the disorder. Patients remain bedridden with flexion hypertonia, and eventually die from intercurrent infections, five to ten years after the first clinical manifestations. The disease is clinically heterogeneous and may appear with intense early language, visual, or praxic manifestations. The so-called presenile and senile forms arise before and after 65 years of age. Presenile AD usually runs a more rapid course, and shows intense cortical deficits.

Minimental State Examination (MMSE) is widely used for detecting dementia. Specificity of MMSE is high, but sensitivity is poor. Clinical suspicion can be corroborated by asking relatives about the patient's difficulties in learning and retaining information or in handling complex tasks, and also about behavioural alterations. Neuropsychological test batteries, including functional assessment, abnormal behaviour scales, and global staging scales, are useful to determine the cognitive/functional status of the patient and the stage of the disorder. Formal clinical diagnosis of AD is based on certain clinical criteria (DSM IV, NINCDS) that imply both the existence of a dementia and the exclusion of other causes of the disorder (Table II). Routine blood analysis — including sedimentation rate, and B12, folic acid, and thyroid studies — are mandatory to exclude symptomatic dementias, and CT should be used at least once to exclude structural brain lesions, especially those amenable to treatment. CSF examination, structural (MRI) and functional (SPECT and PET) neuroimaging, and other more sophisticated studies, should be carried out only in some particular circumstances. Dementia can be confused with delirium, amnesia, aphasia, psychosis, conversion states and depression. AD must be differentiated from reversible dementias (overuse of medication, toxins, hypothyroidism and vitamin-deficiency states, normotensive hydrocephalus, some medical conditions, subdural hematomas, and tumours), subcortical dementias (progressive supranuclear palsy, Huntington's disease, Parkinson's disease, corticobasal degeneration), other cortical, diffuse degenerative dementias (Lewy bodies dementia), degenerative dementias with focal onset (frontal dementia, non-fluent progressive aphasia, semantic dementia, and subcortical gliosis), vascular dementias (multi-infarct dementia, dementia due to strategically situated lesions), and mixed dementias. The diagnostic accuracy of this exclusionary approach is limited. Specificity and sensitivity of clinical criteria depend on several factors and are said to be about 80%. However, in a large, multicentre series of clinicopathological cases, the specificity of NINCDS criteria was much lower, and diagnosis of AD appeared to be extremely difficult in old patients or when the disorder coexisted with other pathologies (Mayeux et al. 1998). A more positive, "inclusionary" approach would improve our diagnostic skills. The existence of the Apo e4 allele notably increases both specificity and positive predictive value of the clinical diagnosis (Mayeux et al. 1999). Medial temporal lobe atrophy is seen on cerebral MRI in AD patients, but the ability of the technique to differentiate AD from other dementias is uncertain (Knopman et al. 2001). AD frequently shows biparieto-temporal hypoperfusion of regional cerebral blood-flow studied by single-positron emission computed tomography (SPECT), but the specificity of this finding is low.

Table 2

Diagnostic criteria of Alzheimer's disease simplified from NINCDS criteria

A. Probable
   1. Dementia
   2. Deficits in two or more cognitive areas
   3. Progressive course of cognitive deficits
   4. No disturbance of consciousness
   5. Impairment of activities of daily living
   6. Onset between 40 and 90 years of age
   7. Absence of other systemic and neurological disorders

B. Definite
   1. The clinical criteria for probable Alzheimer's disease
   2. Histopathological evidence obtained from a biopsy or autopsy

Positron emission tomography (PET) study of the cerebral metabolic state with [18F]fluorodeoxyglucose was 94% sensitive and 73% specific in a large anatomo-clinical series (Silverman et al. 2001). Moreover, the authors found that cognitive impairment was unlikely to occur during the next 3 years in patients with negative PET scans (Silverman et al. 2001). In a community-based sample, increased CSF-tau was found very early in AD: the finding had a sensitivity of 93%, a specificity of 86%, and an area of 0.95 on ROC to discriminate AD from controls and depression (Andreassen et al. 1999). A low level of aβ42 peptide together with the increase of tau, in CSF seems to be more specific than either circumstance alone, and the same can be said for tau isoforms (phosphorylated tau). The exact contribution to AD diagnosis of these techniques has not been definitively established in large, prospective, anatomo-clinical series, but the Apo E genotype, MRI and PET, and tau (or phosphorylated tau) together with aβ42 peptide determination in CSF, are probably useful diagnostic markers for the "positive" differentiation of early AD from depression and aging, and for the exclusion of other degenerative dementias.

4. SYMPTOMATIC TREATMENT OF THE DISEASE

Despite the encouraging investigations and partial results in etio-pathogenically related treatments, no current therapy has been shown to halt or reverse the underlying disease process. However, a nihilistic approach to the disease is unjustified because pharmacological cognitive enhancement therapies, cognitive behavioural interventions, psychotropic agents, symptomatic behavioural approaches for non-cognitive symptoms, psychosocial interventions and care-giver support, have all been shown to be at least modestly effective. These therapeutic strategies should be planned according to a longitudinal model of the disorder, because clinical manifestations and needs differ according to the stage of AD, and some of them are short-lived (see Table III).

Table 3

Treatment strategies

A. – *Etiopathogenic*
1. Inhibit formation of toxic amyloid (inhibitors of secretases?), promote its clearance (vaccine and other procedures), and affects its levels (statins?, NAIDS)
2. Suppress inflammation (COX inhibition)
3. Suppress oxidative injury (vitamin E from nutrients, treatment with vitamin E and other antioxidants)
4. Ameliorate synaptic and axonal degeneration (?)
5. Attenuate tau abnormalities (target kinases?)
6. Prevent cell death (caspase inhibitors) and promote cell viability (trophic factors)
7. Replace neurons (stem cells)

B. – *Symptomatic*
1. Improve cognition with teatments that are pharmacological (cholinergic receptor agonists-antagonists), act on other neurotransmission systems, or are neuropsychological
2. Improve non-cognitive manifestations with behavioural and pharmacological interventions (atypical antipsychotics, antidepressants, anxiolytics, others)

4.1. Cognitive-enhancing pharmacological treatment

The cholinergic theory implies that cholinergic deficit is related to cognitive dysfunction of AD, even if other additional factors are likely to participate in the mental decline. Accordingly, drugs that potentiate cholinergic function should improve cognitive performance in these patients. There have been several approaches to treating the cholinergic deficit in AD increasing the synthesis of acetylcholine, decreasing its hydrolysis, or acting on nicotinic and muscarinic receptors — but the only current strategy leading to meaningful clinical effect has been to reduce the hydrolysis of acetylcholine through inhibition of acetylcholinesterase: the increase of acetylcholine in central synapses improves both, cholinergic transmission and cognition (and could possibly affect Aβ deposition).

Several acetylcholinesterase inhibitors, such as donepezil (up to 10 mg once a day), rivastigmine (up to 6 mg twice a day), and galantamine (up to 12 mg twice a day) have been developed and are the main pharmacological daily treatment in AD

(Takiron is used only in exceptional circumstances). The pharmacological activities of the various acetylcholinesterase inhibitors differ, as they can inhibit AChE activity via non-competitive (blockade of deacetylation process) or competitive mechanisms. The most important clinical effects of their action relate to the duration of the effect, and perhaps to the selectivity of the enzyme inhibition (with regard to both butyrylcholinesterase and acetylcholinesterase peripheral activity), the activity on nicotinic receptors and their action on other neurotransmission systems.

Improvement related with acetylcholinesterase inhibitors has been demonstrated by comparing scores on various cognitive scales — of patients receiving the active product with those of patients given placebo. The positive effect seems similar for all drugs (between 3 to 5 points in ADAS-cog); and treatment benefits include improvement in cognitive measurements, temporary stabilization of the disorder and amelioration of the rate of decline (the duration of the effect is equivalent to about 6 to 12 months' worth of delay in symptom progression). These agents have been shown not only to improve overall functioning and cognition, but also to delay institutionalization and to reduce demands on care giving. Adverse effects are usually slight and include nausea, vomiting, gastric pain, and anorexia.

Acetylcholinesterase inhibitors are indicated in people with AD with slight to moderate dementia (patients scoring between 24 and 10 points in MMSE), but a recent study has suggested that patients with severe dementia should probably also be a target for this treatment (Davis et al. 1999). Not all patients do well with this treatment, but those not responding to one anticholinesterase may respond to another. The most common effect appears to be a halt in the expected decline, and the differences between acetylcholinesterase inhibitors are mainly in their capacity to cause adverse events and in the rate of their introduction. The duration of treatment is questioned, but incomplete data suggest that therapy should not be stopped until very late, because the response seems to persist for several years, and an abrupt early withdrawal can lead to a precipitous worsening. If deterioration follows withdrawal, the medication should be reinstated.

4.2. Symptomatic treatment of non-cognitive manifestations

Cognitive and non-cognitive manifestations of AD are interrelated, probably because they are to some extent the consequence in shared neurotransmission systems. For example, cholinergic deficit correlates with impaired cognition and low cholineacetyltransferase activity in both frontal and temporal cortices, and has also been found to correlate with clinical overactivity (Minger et al. 2000). In fact, acetylcholinesterase inhibitors partially control behavioural symptoms and diminish the need for psychotropic medication (Cummings et al. 2000). Consequently, these should be added only when the cholinergic treatment has failed to improve non-cognitive manifestations of such intensity that they do not respond to behavioural psychological interventions.

Depression is found in 10 to 20% of AD patients, and usually presents early during the course of the disease. This problem can either be reactive or appear secondary to medications or medical illnesses, but most cases are related to AD pathology, probably a reflection of impaired of noradrenergic and serotonergic neurotransmission. Different scales (Neuropsychiatric Inventory, Geriatric Depression Scale, and others) can assess the existence and intensity of depression. Treatment is mandatory in major depression, and should avoid antidepressants with anticholinergic and cardiovascular side effects. There is no convincing evidence that any particular antidepressant is superior to any other, and differences in potential side-effects determine the choice. Serotonin re-uptake inhibitors (fluoxetine, paroxetine, etc., in their usual doses) are considered the first-line antidepressants in these patients.

Psychotic symptoms (delusions, hallucinations, and misidentifications) usually appear when dementia is moderate or severe, at an annual incidence rate up to 47% (Ballard et al. 1997). They are thought to be related to a loss of balance between mesolimbic dopaminergic and cholinergic systems, leading to dopaminergic hyperactivity, but the literature is inconsistent, and the physiopathology of delusions and hallucinations can differ. The existence and intensity of psychotic symptoms can be assessed by scales such as the Neuropsychiatric Inventory of Cummings and the BEHAVE-AD of Reisberg. Psychosis should be treated if symptoms are disturbing or if they promote the institutionalisation of the patient. Classical neuroleptics (haloperidol, thioridazine) are being substituted by atypical antipsychotics, either risperidone (up to 2 mg/day) or olanzapine (up to 10 mg/day), because they cause lesser extrapyramidal, cognitive and cardiovascular adverse effects. The need for antipsychotics should be periodically monitored because the annual resolution rate for psychotic symptoms is 43% (Ballard et al. 1997), half of them spontaneously resolving within a three month period.

Anxiety and agitation are frequently treated with short-half-life benzodiazepines, but these drugs can worsen the condition, and atypical antipsychotics should be considered if symptoms are long-lasting. Carbamazepine, propranolol, and trazodone have also been used to control agitation. Usual therapies are employed for other non-cognitive manifestations such as insomnia, urinary incontinence, sexual disturbances and epileptic seizures.

REFERENCES

Andreasen N, Minthon L, Clarberg A, et al. Sensitivity, specificity, and stability of CSF-tau in AD in a community-based patient sample. Neurology 1995;53:1488-1494

Asthana S, Baker LD, Craft S, et al. High-dose estradiol improves cognition for women with AD. Results of a randomized study. Neurology 2001;57:605-612

Ballard C, O'Brien J, Coope B, Fairbain A, Abid F, Wilcock G. A prospective study of psychotic symptoms in dementia sufferers: psychosis in dementia. International Psycogeriatrics 1997;9:57-64

Braak H, Braak E. Neuropathological staging of Alzheimer related changes. Acta Neuropathol (Berl.):1991;82:239-259

Campion D, Dumanchin C, Hannequin D, et al. Early-onset autosomal dominant Alzheimer disease: prevalence, genetic heterogeneity, and mutation spectrum. Am J Hum Genet 1999;65:664-670

Cummings JL. Cholinesterase inhibitors: a new class of psychotropic compounds. Am J Psychiatry 2000;157:4-15

Frackowiak J, Wisnieski HM, Wegiel J, et al. Ultrastructure of the microglia that phagocytose amyloid and the microglia that produce β-amyloid fibrils. Acta Neuropathol 1992;84:225-233

Guo Z, Fratiglioni L, Zhu L, Fastbom J, Winbland B, Viitanen M. Occurrence and progression of dementia in a community population aged 75 years and older. Relationship of antihypertensive medication use. Arch Neurol 1999;56:991-996

Lim GP, Yang F, Chu T, et al. Ibuprofen suppresses plaque pathology and inflammation in a mouse model for Alzheimer's disease. J Neurosci 2000;20:5709-5714

McCaddon A, Regland B, Hudson P, Davies G. Functional vitamin B_{12} deficiency and Alzheimer's disease. Neurology 2002;58:1395-1399

McGeer PL, Schulzer M, McGeer EG. Arthritis and anti-inflammatory agents as possible protective factors for Alzheimer's disease: a review of 17 epidemiological studies. Neurology 1996;47:425-432

Minger SL, Esiri MM, McDonald B, et al. Cholinergic deficits contribute to behavioral disturbance in patients with dementia. Neurology 2000;55:1450-1457

Mirra SS, Heyman A, McKeal D, et al. The Consortium to Establish a Registry for Alzheimer's disease (CERAD). II. Standardization of the neuropathological assessment of Alzheimer's disease. Neurology 1991;41:479-486

Morris MC, Bienias JL, Tanguey CC, Bennet DA, Aggarwal N, Wilson RS, Scherr PA. Dietary intake of antioxidant nutrients and the risk of Alzheimer disease in a biracial community study. Lance 2002;287:3230-3237

Petersen RC, Stevens JC, Ganguli M, Tangalos EG, Cummings JL, DeKosky ST. Practice parameter: Early detection of dementia: Mild cognitive impairment (an evidence-based review). Report of the quality standards subcommittee of the American Academy of Neurology. Neurology 2001;56:1133-1142

Polvikovski T, Sulkava R., Myllykangas L, et al. Prevalence of Alzheimer's disease in very elderly people. A prospective neuropathological study. Neurology 2001;56:1690-1696

Reisberg B, Windscheif U, Ferris SH, et al. Memantine in moderately severe to severe Alzheimer's disease (AD): results of a placebo-controlled 6-month trial. Neurobiol Aging 2000;21(suppl. 1):S275

Sano M, Ernesto C, Thomas RG, et al. A controlled trial of selegiline , alphatocopherol, or both as treatment for Alzheimer 's disease. N Engl J Med 1997;336:1216-1222

Schenk D, Barbour R, Dunn W, et al. Immunisation with amyloid-β attenuates Alzheimer-disease-like pathology in the PDAPP mouse. Nature 1999;400:173-177

Seshadri S, Drachman DA, Lippa C. Apolipoprotein E4 allele and the lifetime risk of Alzheimer's disease. Arch Neurol 1995;52:1074-1079

Silverman DHS, Small GW, Chang CY, et al. Positron emission tomography in evaluation of dementia. JAMA 2001;286:2120-2127

Sinha S, Anderson JP, Barbour R, et al. Purification and cloning of amyloid precursor protein beta-secretase from human brain. Nature 1999;40:537-540

The National Institute for Aging and Reagan Institute working group on diagnostic criteria for the neuropathological assessment of Alzheimer's disease. Consensus recommendations for the postmortem diagnosis of Alzheimer's disease. Neurobiol Aging 1997;18:S1-S2

Von Strauss EM, Viitanen D, De Ronchi D, et al. Aging and the occurrence of dementia. Arch Neurol 1999;56:587-592

Yaffee K, Sawaya G, Lieberburg I, Grady D. Estrogen therapy in postmenopausal women: effects on cognitive function and dementia. JAMA 1998;279-688-695

R. DENGLER AND J. BUFLER

Department of Neurology, Medical School Hannover, Germany

38. AMYOTROPHIC LATERAL SCLEROSIS AND RELATED DISORDERS

Summary. Amyotrophic lateral sclerosis (ALS) is a degenerative disorder involving the upper and lower motor neuron sparing the extraocular and sphincter muscles. The course of the disease is relentlessly progressive with a mean survival time of 3 years. ALS occurs mainly sporadic (sALS) and is in about 10 % of cases familial (fALS). Although the etiopathogenesis of ALS is still not known, the information on the pathomechanisms of motor neuron death has been mounting and a first drug has been shown to prolong survival. The important milestone in ALS research in the last decade was the finding that about 5 to 10% of the familial cases, i.e. 2 % of all patients, carry mutations of the SOD 1 gene.

1. INTRODUCTION

The first comprehensive description of the disease dates back to Charcot in 1873 who also coined the term amyotrophic lateral sclerosis (sclerose lateral amyotrophique) (10). His precise and systematic clinical observations still hold. He also described the involvement of the upper and lower motor neuron in post mortem studies. The term motor neuron disease frequently used in English speaking countries has been introduced by Gowers. Famous people dying from ALS were the US Baseball player Lou Gehrig, the movie actor David Niven and probably also Mao Tse Tung. Still alive, although on the respirator, is the astrophysicist Stephen Hawkings who has a very slow variant of the disease

2. EPIDEMIOLOGY

ALS has a worldwide incidence of 1 to 2.5 cases per 100.000 and a prevalence of 3 to 8 per 100.000. These numbers appear to increase slightly over the last decades although it is unclear if improved diagnosis and prolonged survival or other variants such as environmental changes play the dominant role. Approximately 90% of the cases are sporadic (sALS) and the other 10% familial (fALS). The age of onset is 50 to 70 years peaking at 60 years although younger and older patients are common. The mean age of onset of fALS seems to be somewhat lower. There is a slight male preponderance (ratio male to female 1.6 to1) in sALS which is not seen in fALS. The mean survival time after disease onset is 3 years with considerably shorter and longer (up to 20 years) time courses. The life expectancy has probably slightly improved because of modern treatment.

An endemic form of the disease is found in the western Pacific (Guam, New Guinea, Japanese Kii Peninsula) clinically presenting as Parkinson-ALS-dementia complex with predominant involvement of the upper motor neuron. At the end of world war II there was a prevalence in Guam of about 100 per 100.000. It is still debated whether the incidence has declined after the occupation of the island and whether genetic or environmental factors are the major pathogenetic players.

3. CLASSIFICATION

Sporadic ALS can be clinically classified as follows:

- Classical ALS: clinical involvement of upper and lower motor neuron
- Progressive muscle atrophy (PMA): clinical involvement of the lower motor neuron only; upper motor neuron signs may occur with time
- Progresive bubar palsy: begins mostly as pseudobulbar dysfunction and progresses usually to classical ALS
- Primary lateral sclerosis: pure upper motor neuron involvement over years; mild lower motor neuron signs may occur with time.

The recently revised "**El Escorial Criteria**" (14) define different levels of certainty of the diagnosis which may be of relevance for drug trials and other scientific purposes: The revised Criteria distinguish on clinical grounds:

- suspected ALS,
- possible ALS,
- probable ALS,
- definite ALS

and list as new categories

1. **definite familial ALS laboratory supported**:
   clinical condition which may fit with beginning motor neuron disease associated with molecular genetic testing positive for SOD 1 mutation

2. **laboratory supported probable ALS:**
   clinical condition fitting with possible ALS which reveals signs of active denervation in electromyographic studies in, at least, two limbs

Patients can usually be enrolled in controlled treatment studies only when they fulfil the criteria of probable ALS. The application of these new categories makes it possible to include patients earlier than in the past, i.e in clinical stages of the disease when treatment is more promising.

4. ETIOLOGY AND PATHOMECHANISMS

The etiology of ALS is still largely unknown. Although there was much progress in molecular genetics of ALS and in the understanding of biochemical disturbances, the pathomechanisms of disease development are not well understood. The following section firstly describes the known and putative genetic defects and secondly possible molecular pathomechanisms.

4.1. Genetics

A major breakthrough in ALS research was the detection of mutations in the SOD 1 gene in 10 to 20 % of fALS cases in 1993 (24). A description of the actual state of the art of the genetics in ALS can be found in (2).

4.1.1. Copper/zinc superoxide dismutase (SOD 1) mutations

About 10 % of ALS cases are familial and a total of 2 % are caused by a mutation in the SOD 1 gene (24). More than 70 different mutations in all exons have been described. All are dominant except for the recessive D90A mutation which is the only one causing a fairly uniform phenotype (predominant upper motor neuron signs and leg involvement in conjunction with slow disease progression). It is now accepted that the SOD1 mutations cause a gain of toxic function rather than a loss of dismutase activity. The nature of this toxic function is still a matter of debate and will be discussed later. Meanwhile, several human SOD1 mutations have been used to produce transgenic mice which develop motor neuron disease and serve as valuable animal models of ALS, especially for drug treatment studies

4.1.2. Other genes

Approximately 1 % of ALS patients have deletions in the gene for the heavy neurofilament subunit (16). The role of this mutation for developing the disease is unclear. Type 3 of autosomal recesive ALS, a disease of juvenile onset and slow progression mainly involving the upper motor neuron, could now be linked to 2q33. Type 1 which is more common and affects lower motor neurons first has recently been linked to 15q15-q22. A form of autosomal dominant juvenile ALS showing slow progression and upper and lower motor neuron involvement has been linked to 9q34. Mutations or polymorphisms in the gene for apurinic/apyrimidinic exonuclease (APEX) have been associated with sALS in one study which requires further confirmation. The genes associated with spinal muscular atrophy (survival motor neuron gene on chromosome 5) do apparently not play a role in development of ALS.

4.2. Pathomechanisms

Possible mechanisms causing disease in the SOD1 mutations have been intensively studied, in particular, as they could also play a role in the more frequent sALS.

4.2.1. Nitrosylation of proteins

Mutant SOD 1 can enhance the generation of peroxynitrite (ONOO⁻) resulting in increased tyrosin and protein nitrosalytion. Nitrotyrosin mimics phosphotyrosine and may block protein phosphorylation. In contrast to phosphotyrosin, nitrotyrosin is negatively charged and may cause altered protein conformation (25) and, as a consequence, protein aggregation. This could especially interfere with normal neurofilament production (see later). On the other hand, studies in transgenic mice with the G93A mutation do not favour the concept of abnormal nitrosylation.

4.2.2. Copper toxicity

Conformational changes of mutant SOD1 can result in abnormal exposure of the copper atom at the active site and can cause increased peroxidase activity as well as a tendency of the enzyme to lose the copper atom. The copper : zinc ratio has been shown to be decreased in the mutant molecules. This could be toxic in various ways. The role of copper has been further stressed by the finding that copper chelators can prolong survival time of spinal motor neurons of G93A transgenic mice (**2**).

4.2.3. Abnormalities of neurofilaments (NFL)

Characteristically, motor neurons in sALS and SOD 1 mutant fALS reveal an abnormal accumulation of NFLs in proximal axons and cell bodies (1,16). These changes could be primary as well as secondary in the disease process. Studies in animals mutant for both NFL and SOD 1 have shown that normal NFL structure is not necessary for motor neuron death suggesting that NFL changes in ALS may be secondary. Other studies, however, demonstrated that NFL aggregation can cause motor neuron death (4) and, therefore may play a role in ALS development. NFL damage in ALS may also be caused by oxidative stress resulting from abnormal SOD 1 function as is indicated by an association of accumulated NFLs and lipofuscin.

4.2.4. Protein aggregation

Motor neurons of ALS patients show various intracellular inclusions containing abnormal protein aggregates. Accumulation of SOD 1 protein associated with ubiquitin and neurofilaments could be demonstrated in sALS as well as in SOD1 mutant transgenic mice (9,11). The role of these inclusions in the disease process is unclear (see also pathology).

4.2.5. Mitochondrial changes

Mitochondrial changes are found in motor neurons of mutant SOD 1 transgenic mice and in sALS in man (**2**). Cytochrome C oxidase activity in spinal cords of ALS patients may be reduced or absent similarly to patients with primary mitochondrial DNA deletions. The role of these mitochondrial changes in the disease process is unclear.

4.2.6. Excitotoxicity

The concept of excitoxicity in neurodegenerative disorders and, in particular in ALS, postulates that death of neurons can result from overstimulation by excitatory amino acids such as glutamate, the excitatory transmitter in the central motor system. Motor neuron overstimulation in conjunction with increased Calcium permeability of AMPA type glutamate receptors could result in intracellular Calcium accumulation which could trigger abnormal enzymatic activities causing cell death. This scenario may be accentuated by the fact that motoneurons express very small amounts of Calcium buffering proteins such as parvalbumin. Several arguments favour a major role of this pathomechanism in ALS.

Cerebrospinal fluid levels of glutamate have been found to be increased in ALS. The level of the astrocytic glutamate transporter EAAT2 has been reported to be decreased in motor cortex and spinal cord although other reports were unable to confirm these results. Oxidative products in mutant SOD 1 transgenic mice seem to inactivate EAAT2 in contrast to the wild type suggesting that SOD 1 mutations may exert toxicity via excitotoxicity. In addition, spinal motor neurons of transgenic mice with the G93A mutation exhibit increased sensitivity to glutamate toxicity. In man, the expression of GluR2, a specific subunit of AMPA-type glutamate receptors associated with low calcium permeability, is reduced in spinal motor neurons of ALS patients (26). This might be an additional factor explaining the selective vulnerability of motor neurons of ALS patients. Further strong support for the concept of excitotoxicity comes from the therapeutic efficacy of the glutamate antagonist Riluzole which has been shown to prolong survival of ALS patients by several months (6,19).

4.2.7. Virus etiology

Viral infections have frequently been disussed as possible cause of ALS. A long debate on the persistence of polio virus infection has been finished without convincing results. A recent study (7) describes enterovirus nucleic acid in cells of the spinal gray matter of ALS patients. There seems to be a close homology to DNA sequences of ECHO viruses 6 and 7. Indirect evidence for abnormal retrovirus presence in the serum of ALS patients has also been described. The significance of these exciting findings has to be determined in further studies.

4.2.8. Pathogenetic similarities to other neurodegenerative disorders

A common feature of neurodegenerative disorders such as ALS, Parkinson´s or Alzheimers´s disease, is the selective vulnerability of distinct populations of neurons in the disease process. Anatomically this means degeneration of the nigrostriatal system in Parkinson´s disease, preferential loss of cholinergic and corticocortical projection neurons in Alzheimer´s disease and loss of upper and lower motor neurons in ALS. Some of the pathomechanisms of ALS, e.g. oxidative stress (3), disturbances of phosphorylation and pathological intracellular aggregation of neurofilaments (22), mitochondrial dysfunction (17) and glutamate induced excitotoxiticty (5) may also play an important role in Parkinson´s and Alzheimer´s disease, and probably also other degenerative CNS disorders. These mechanisms may rather form a common final path for neuronal death and may be used for designing pharmacological therapies. Selective vulnerability, however, appears to be the central key to the understanding of the etiopathogenesis of the different neurodegenerative disorders.

5. ANATOMICAL AND MOLECULAR PATHOLOGY

5.1. Anatomical pathology

5.1.1 Upper motor neuron

Macroscopic inspection of the brain may show a selective atrophy of the motor cortex . Microscopically there is a loss and reduction of sizes of the Betz cells in the cortical layer V and their dendritic trees in conjunction with a reactive astrocytic gliosis. Furthermore there is a degeneration of the Betz cell axons with Wallerian degeneration and myelin pallor at various levels of the corticospinal tract. The frequent finding that cortical changes can be marginal as well as myelin pallor above the level of the medulla in cases with clinically typical upper motor neuron signs may point to a primary axonopathy of the dying-back type in ALS.

5.1.2. Lower motor neuron

Typically, there is reduction of numbers and sizes of motor neurons in the ventral horn and the bulbar motor nuclei. The oculomotor nuclei and the sacral motor nucleus of Onufrowitz for the sphincter muscles are spared possibly because their motor neurons lack monosynaptic contact with the corticospinal system and express calcium buffering proteins. The cell somata of affected motor neurons may show various inclusion bodies which will be described later and dendritic degeneration. Areas of cell degeneration reveal reactive astrocytic gliosis. Motor axons may be swollen in the most proximal portions and show signs of retraction and degeneration in the distal ones.

5.1.3. Non-motor neuropathology

Pallor of the dorsal columns of the spinal cord is not rare although most frequent and, in part, typical in fALS. The ascending spinocerebellar tracts can also exhibit myelin pallor. There may be cell loss in the subtantia nigra although its significance is unclear.

5.1.4. Pathology outside the nervous system

Affected muscles reveal groups of angulated atrophic fibres and typical fibre-type grouping as unspecific signs of denervation and reinnervation from collateral sprouting of surviving motor neurons.

Changes of collagen have been described in the skin and recently also in the spinal cord. Mitochondrial changes in liver cells have recently been reported and require confirmation.

5.2. Molecular pathology

The molecular pathology of ALS is less well defined than that of other neurodegenerative diseases such as Alzheimer's or the various Parkinson syndromes. ALS is not associated with abnormalities of the metabolism of alpha-synuclein or tau protein except for the endemic form of Guam which shows the pathology of a tauopathy with intracellular neurofibrillary tangles. There is, however, a large body of literature describing various inclusion bodies some of which contain neurofilamentous material.

5.2.1. Ubiquinated inclusions

Ubiquitin, a 76 amino acid polypeptide, is physiologically involved in protein degradation. Ubiquinated inclusions (UBIs) occur in many neurodegenerative disorders associated with accumulation of abnormal proteins although there is no evidence for a primary abnormality of ubiquitin metabolism. UBIs have also been called basophilic inclusions. In sALS, Lewy body-like, densely packed UBIs and skein-like UBIs, both highly ubiquinated, are found in degenerating motor neurons as well as in seemingly normal ones (18). So far no specific protein targets for ubiquitination could be identified in ALS. UBIs do not show co-localization with tau, neurofilaments or alpha-synuclein although smaller filaments found in these inclusions may be neurofilaments. UBIs are, however, fairly characteristic and can be demonstrated in up to 80% and more of sALS cases if looked for systematically (18). They are mainly found in lower motor neurons and infrequently in the brain, especially in demented cases. Their relation to the disease process, however, still awaits clarification.

5.2.2. Bunina bodies and Hirano bodies

Bunina bodies appear to be specific for both sALS and fALS although they can be demonstrated less frequently than UBIs. They are found in the soma of anterior horn cells as small eosinophilic inclusions and seem to be of lysosomal origin. Recently they were shown to be immunoreactive for cystatin C. Their role in the degeneration of motor neurons is not well understood. Hirano bodies are smaller, probably unspecific eosinophilic inclusions found in the cell soma, dendrites and axons of motor neurons in ALS.

5.2.3. Hyaline bodies

Hyaline inclusions or conglomerates contain as a major component phosphorylated and non-phosphorylated neurofilament proteins in contrast to the above described UBIs. They may rarely occur in other neurological diseases and in normal subjects and are apparently less specific for ALS than the above described inclusions. Hyaline bodies may mainly occur in fALS with SOD 1 mutation (18). Their pathogenesis, however, is unclear.

5.2.4. Spheroids and globules

Neurofilaments form a major class of cytoskeletal proteins and are especially important for slow axonal transport. Consequently they play a major role in large projection neurons. Axonal swellings in the anterior horn have been described in ALS containing the larger spheroids more proximally and the smaller globules more peripherally both composed of conglomerations of highly phosporylated neurofilaments. Spheroids may also be seen in dendrites. Spheroids and globules are unspecific although their numbers are certainly increased in ALS. Spheroids like the above described hyaline bodies arise from abnormal neurofilament metabolism and may slightly differ in the underlying molecular mechanisms.

6. CLINICAL SIGNS AND DIAGNOSIS

6.1. Symptoms

The disease onset is usually focal, most frequently distal in an arm or leg. Painless loss of skill and weakness of a hand are typical first symptoms. Initial involvement of the upper extremities is found in about 40 to 50 % of cases, of the lower extremities and of the bulbar region each in about 25 %. Many patients report a liability to cramps preceding first symptoms by months or even years. Symptoms progress slowly from distal to proximal, to the other side, to the other extremity and to the bulbar region sparing the extraocular and sphincter muscles.

In addition, brisk deep tendon reflexes are typically found whereas a Babinski sign is seen less frequently (30 %). Spasticity may be missing and is rarely predominant, mostly in the legs. Clinically involved muscles as well as unremarkable ones show fasciculations which are rarely missing in the examination and are a hallmark of the disease.

Bulbar signs are dysarthria and dysphonia, later also dysphagia and an atrophy of the tongue. Weakness of neck and trunk muscles are usually less impressive. By definition, there are no sensory deficits although intermittent paresthesiae may be reported. Weight loss is typical and results from both muscle wasting and dysphagia. Autonomic dysfunction is not prevalent although there is hypersalivation due to dysphagia. Respiratory deficiency usually heralds the terminal phase.

Cognitive functions are mostly clinically normal. Subtle psychological tests, however, may demonstrate impairment of frontal functions. Although some patients may show what Charcot called "belle indifference" most patients develop a mild to moderate depression in the course of the disease. Suicides, however, are very rare. Pathological loughing, crying or yawning do not indicate affective lability but are pseudobulbar signs pointing to frontal disinhibition.

Characteristically, the course of the disease is relentlessly progressive. The rate of impairment, however, differs between patients. Ten % survive 5 years and 5 % 10 years. Death is mostly caused by respiratory failure resulting from the weakness of respiratory muscles.

6.2. Diagnosis

There are no laboratory markers except for molecular genetic tests in fALS for the various SOD1 mutations. Serum and cerebrospinal fluid show normal results or unspecific changes such as a mild increase of CK. Diagnosis is still delayed, usually by several months, also in developed countries and may be made sooner when neurologists are involved early (12).

Imaging is mainly important for exclusion of other diseases. Cerebral MRI is mostly normal. Occasionally, hyperintensity of the pyramidal tract is seen in T2 weighted images or an atrophy of the motor cortex or the bulbar region. MRI spectroscopy can demonstrate signs of neurodegeneration in the motor cortex although still inconsistently. Diffusion tensor imaging may have the potential to prove early degeneration of the central motor pathways and is currently under study (15). The role of functional MRI is open. PET can detect activation of microglia which, however, may be unspecific.

The most important technical approach is still electromyography (13) which reveals signs of acute and chronic denervation and reinnervation in several muscles in different body regions. Pathological results can be found early in seemingly unaffected muscles indicating a disseminated process extending beyond the areas of single nerves, nerve roots or plexus. Motor unit number estimate (MUNE), existing in methodologically different variants, is an elegant electromyographic approach to monitor the decline of motor neuron numbers in selected muscles in the natural course of the disease and during treatment studies. Recently, transcranial magnetic stimulation has shown its capability to detect upper motor neuron involvement. Its sensitivity may be enhanced by recording contralateral and ipsilateral activation simultaneously, testing intracortical inhibition by paired pulse stimulation and studying pyramidal tract conduction impairment by the new technique of triple stimulation.

Table 1: Differential Diagnosis of ALS:

1.	**Combination of upper and lower motor neuron lesion**
1.1	structural
	cervical myelopathy and malformations
1.2	toxic
	heavy metals: Pb, Hg, Mn; organic solvents
1.3	metabolic
	Hyperparathyroidism, Hexosaminidase-Deficiency
1.4	infectious
	HTLV 1, Jakob-Creutzfeld Disease,
1.5	immunologic
	paraproteinemic
2.	**Pure nuclear lesions**
2.1	structural
	entrapment syndromes, cauda tumor
2.2	neurodegenerative, hereditary
	spinal muscular atrophy, HMSN type 2, Kennedy syndrome
2.3	toxic-metabolic
	various motor neuropathies
2.4	infectious
	Polio and Post-Polio-Syndrome, other Enteroviruses
2.5	immunologic
	multifocal motor neuropathy, paraproteinemic neuropathy
2.6	other
	crampus-fasciculation syndrome, inclusion body myositis
3.	**Pure central lesion**
3.1	structural
	craniocervical or spinal ccompression
3.2	toxic-metabolic
	Lathyrism, Cassava
3.3	infectious
	HTLV 1
3.4	immunologic
	encephalomyelitis disseminata
3.5	neurodegenerative, hereditary
	familial spastic paraparesis, hereditary ataxias
3.5	other
	paraneoplastic, ischemic pseudobulbar palsy

6.3. Differential diagnosis

The clinically most important conditions mimicking ALS are listed in table 1. Three conditions shall be discussed in more detail.

Cervical myelopathy is occasionally confused with ALS resulting in unnecessary neck surgery. This could be avoided by involving an experienced neurologist and by applying strict clinical and electrophysiological criteria.

Kennedy syndrom (x-chromosomal-recessive, bulbospinal muscular atrophy) is a motor neuron disease caused by a CAG triple-repeat mutation in the androgen receptor gen. As it is inherited in an x-linked recessive manner it affects only men. Leading clinical symptoms are bulbar signs, mild and mostly symmetric atrophy and weakness of extremity muscles, weak or lacking reflexes, tremor of the perioral region and of upper extremities and gynecomasty. The latter occasionally results in surgery, mostly before the disease is identified. Upper motor neuron signs are generally missing. Life expectancy is much longer than that of ALS and disability of patients remains mostly mild to moderate. Kennedy syndrome is frequently confused with ALS of bulbar onset.

Multifocal motor neuropathy (MMN) can clinically resemble beginning ALS. It is characterized by focal muscle atrophy and weakness, mostly involving the distal upper extremities. Usually there is no sensory deficit. Patients show cramps and fasciculations and tendon reflexes are mostly preserved or may even be brisk. The rate of progression, however, is much slower than that of ALS and the distribution of symptoms remains limited. MMN is immunogenic and is mostly associated with high serum titers of anti GM1-ganglioside antibodies. The electrophysiological demonstration of focal motor conduction blocks is essential for diagnosis. It is important to make the diagnosis early as MMN can be treated by intravenous immunoglobulins.

7. TREATMENT

7.1. Pharmacologic treatment

In the past drugs with various pharmacological activity such as antibiotics, hormones, cytostatics, imunosuppressants, antiinflammatory agents and many others have been tried in ALS without success. Recently. the antiglutamatergic drug riluzole was the first to show a significant influence on the course of the disease in two large controlled randomised and double blind studies (6,19). Patients taking 100 mg of the substance daily had a significantly greater chance to be alive at all times over the study period of 18 months than those in the placebo group. Briefly, this means prolongation of survival by several months. It appears that young patient age and treatment start in an early stage of the disease are favourable and may be associated with longer extra survival time. Although the positive effects of riluzole have been used as argument in favour of the excitoxic concept, the exact mode of action of the drug is not known. In addition to its unspecific antiglutamatergic and antiexcitotoxic action, it probably has other pharmacologic properties i.e. sodium channel blocking activity (20). The effects of other tested antiglutamatergic drugs, e.g. gabapentin, did not reach significance (21).

Controlled trials with different nerve growth factors using various routes of application were so far unsuccessful although these substances can increase the survival of motoneurons in vitro and in animal models. Subcutaneous application of CNTF was associated with weight loss and significant faster disease progression in the verum group (23). It was speculated that activation of interleukin 6 was responsible for these negative effects. Clinical trials with IGF-1 exhibited non-significant effects (8) and a double blind randomised clinical trial with intrathekal application of BDNF was stopped because there was no difference between the placebo and verum group. Similarly, a clinical trial with a substance (SR57745A) stimulating the synthesis of BDNF was not successful. A major and still unsolved problem is by which route nerve growth factors can be administered to act as selectively as possible on motoneurons.

The efficacy of substances such as high-dose Vitamin E, carnitin, the antibiotic minocyclin, melatonin, the serotonin precursor 5-hydroxy-tryptophane and others proven effective in mutant SOD 1 transgenic mice is so far undetermined in man.

Among others, there are currently two larger studies on the way testing the TNF alpha-inhibitor pentoxifyllin and another antiapoptotic substance.

7.2. Symptomatic treatment

Symptomatic treatment has been improved considerably in the last decade. Larger neuromuscular centers are now well experienced with treatment of cramps, spasticity, various forms of pain, pathological laughing and crying, anxiety and depression.

Nutrition becomes a critical problem with increasing dysphagia and risk of aspiration. It is best managed by applying percutaneous endoscopic gastrostomy (PEG) before the patient has lost too much weight and has relevant respiratory problems. PEG doubtlessly improves quality of life although it is not yet clear whether it prolongs survival.

Hypersalivation and drooling is socially disabling and may be treated by administration of anticholinergic drugs and rarely by irradiation of the salivary glands. A new and elegant mode of treatment is focal injection of botulinum toxin type A or B into the salivary glands which decreases saliva production for several months.

The main therapeutic problem in the later stages of the disease is respiratory insufficiency. For some time, non-invasive ventilation via a mask is an appropriate help for many patients and is used increasingly. It can be easily performed at home supported by family members or other aids. In our experience, most patients do not want tracheostomy and subsequent permanent invasive ventilation although there are great differences from one country to another due to socio-ethnic backround. An example is Japan where large numbers of patients are invasively ventilated for long times.

Palliative treatment in the terminal phase requires experience and aims mainly at alleviation of anxiety and respiratory discomfort. In Europe and in the US, most patients still die at home, or with steadily increasing frequency in an appropriate environment such as a hospice.

REFERENCES

1. Al-Chalabi A, Andersen PM, Nilsson P et al.: Deletions of the heavy neurofilament subunit tail in amyotrophic lateral sclerosis. Hum Mol Genet 1999; 8:157-164
2. Al Chalabi A, Leigh PN: Recent advances in amyotrophic lateral sclerosis. Curr Opin Neurol 2000; 13:397-405
3. Bains JS Shaw CA: Neurodegenerative disorders in humans: the role of glutathione in oxidative stress-mediated neuronal death. Brain Res Brain Res Rev 1997; 25: 335-358
4. Beaulieu JM, Nguyen MD, Julien JP: Late onset death of motor neurons in mice overexpressing wild type peripherin. J Cell Biol 1999; 147:531-544
5. Beal MF: Aging, energy, and oxidative stress in neurodegenerative diseases. Ann Neurol 1995; 38:357-366
6. Bensimon G., L. Lacomblez, V. Meininger: A controlled trial of riluzole in amyotrophic lateral sclerosis. ALS/Riluzole Study Group. N. Engl. J. Med. 330 (1994) 585-591
7. Berger MM, Kopp N, Vital C et al: Detection and cellular localization of enterovirus RNA sequences in spinal cord of patients with ALS. Neurology 2000; 54:20-25
8. Borasio GD, Robberecht W, Leigh PN, Emile J, Guiloff RJ, Jerusalem F. Silani V, Vos PE, Wokke JH, Dobbins T: A placebo-controlled trial of insulin-like growth factor-I in amyotrphic lateral sclerosis. European ALS/IGF-I Study Group. Neurology 1998; 51: 583 – 586
9. Bruijn LI, Houseweart MK, Kato S et al.: Aggregation and motor neuron toxicity of an ALS-linked SOD 1 mutant independent from wild type SOD 1, Science 1998; 281:1851-1854
10. Charcot JM: Lecons sur les maladies du systeme nerveux. Delahaye, Paris, 1873
11. Chou SM, Wang HS, Taniguchi A, Bucala R: Advanced glycation endproducts in neurofilament conglomeration of motoneurons in familial and sporadic amyotrophic lateral sclerosis. Mol Ned 1998; 4:324.332
12. Dengler R: Current treatment pathways in ALS: a European perspective. Neurology 1999; 53:S4-10
13. Dengler R, Ludolph A, Zierz S (Eds.): Amyotrophe Lateralsklerose. Thieme, Stuttgart, 2000
14. EL Escorial Revisited. Revised criteria for the diagnosis of amyotrophic lateral sclerosis: Reqirements for the diagnosis of ALS. 1998; http://www.wfnals.org/Articles/elescorial1998criteria.htm.
15. Ellis CM, Simmons A, Jones DK et al.: Diffusion tensor MRI assesses corticospinal tract damage in ALS. Neurology 1999;53:1051-1058.
16. Figlewicz DA, Krizius A, Martinoli MG et al.: Variants of the heavy neurofilament subunit are associated with the development of amyotrophic lateral sclerosis. Hum Mol Genet 1994; 3:1757-1761
17. Heales SJ, Bolanos JP, Stewart VC, Brookes PS, Land JM, Clark JB: Nitric oxide, mitochondria and neurological disease. Biochim Biophys Acta 1999; 1410: 215-228
18. Ince P, Neuropathology. In Brown RH, Meiniger V, Swash M (Eds.). Amyotrophic Lateral Sclerosis. Martin Dunitz, London, 2000, pp 83-112
19. Lacomblez L, Bensimon G, Leigh PN et al.: Dose ranging study of riluzole in amyotrophic lateral sclerosis. Amyotrophic Lateral Sclerosis/ Riluzole Study Group II. Lancet 1996; 347:1425-1431
20. Mohammadi B., Lang N., Dengler R., Bufler J.: Interaction of high concentrations of Riluzole with recombinant skeletal muscle sodium channels and adult-type nicotinic receptor channels. Muscle Nerve 2002; 26:539-545
21. Miller RG, Moore LA, Young C et al.: Placebocontrolled trial of gabapentin in patients with amyotrophic lateral sclerosis. WALS Study Group. Neurology 1996; 74: 1383-1388
22. Miller RG, Petajan JH, Bryan WW, Armon C, Barohn RJ, Goodpasture JC, Hoagland RJ, Parry GJ, Ross MA, Stromatt SC : A placebo-controlled trial of recombinant human ciliary neurotrophic (rhDNTF) factor in amyotrophic lateral sclerosis. RhCNTF ALS Study Group. Ann Neurol 1996; 39: 256 – 260

23. Morrison BM, Hof PR, Morrison JH: Determinants of neuronal vulberability in neurodegenerative diseases. Ann Neurol 1998; 44: S32-44
24. Rosen DR, Siddique T, Patterson D et al.: Mutations in CU/Zn superoxide dismutase gene are associated with familial amyotrophic lateral sclerosis. Nature 1993; 362:59-62
25. Torreilles F, Salman-Tabcheh S, Gerin M, Torreilles J: Neurodegenerative disordes. The role of peroxynitrite. Brai Res-Brain Res Rev 1999;30:153-163
26. Williams TL, Day NC, Ince PG et al.: Calcium permeable alpha-amino-3-hydroxy-5-methyl-4-isoxazole propionic acid receptors: a molecular determinant of selective vulnerability in amyotrophic lateral sclerosis. Ann Neurol 1997; 42:200-207.

L. LEY AND T. HERDEGEN

Institute of Pharmacology, University Hospital of Schleswig-Holstein, Campus Kiel, Germany

39. PHARMACOLOGICAL STRATEGIES FOR NEURODEGENERATION AND OVERVIEW OF CLINICAL TRIALS

Summary. The search for neuroprotective drugs has consumed enormous resources over the past decade. A broad spectrum of compounds with diverse mechanisms of action has been investigated for the treatment of neurodegenerative diseases. Numerous drugs which appeared to be promising in preclinical investigations had been withdrawn in patients because of the lack of efficacy or an unfavourable risk benefit ratio. The difficulties assigning results from animal studies to patients are considered the main reason for this failure. For future drug development it is important to pay more attention to the establishment of animal models that reflect the situation in the patient and to diagnostic procedure for an optimal therapeutic regimen. Nevertheless, novel biotechnical applications such as viral gene transfer, fusion peptides, transfer of stem cells or locally implanted cell factories might overcome the pharmacokinetic obstacle of the blood-brain-barrier and offer promising therapeutic approaches (either as mono-therapy or as 'cocktails') within the next decade.

1. INTRODUCTION

There is hardly any research area for the treatment of human diseases with such as a tremendous discrepancy between investments and clinically relevant outcome as the therapy of neurodegenerative disorders. The "decade of the brain" propagated in the early 1990's, is still continued by comprehensive investments of man power, financial resources and administrative programs. But none of the novel strategies derived from neurobiological research has reached the approval as recommended therapy.

The therapeutic strategies for the treatment of stroke provide the best example for the discrepancy between scientific efforts and clinical failure. More than 200 clinical studies with a similar number of innovative (bio-)pharmaceutical compounds were performed for the treatment of stroke. Due to complete failure of these trails, the therapy of stroke in 2003 is restricted to the old (symptomatic) armament of lysis, steroids and (pseudo-effective) rheologic drugs. In principle, the same holds true for other neurodegenerative disorders such as Alzheimer's (AD), Parkinson's (PD) or Huntington's disease (HD).

A most critical issue is the range of actions of neuroprotective drugs. Neuordegeneration comprise a mixture of inflammation, excititotoxicity, apoptosis, necrosis, withdrawal of trophic molecules and extracerebral pathologies which ideally demands the application of therapeutic "cocktails". This strategy, however, is a pharmaceutical nightmare in terms of clinical testing (what are the appropriate controls?), side effects and drug interactions. Nevertheless, the combination of different therapeutic principles might be a promising strategy.

On the other hand, numerous data from basic neurobiological research indicate that starting from different origine, neurodegenerative diseases can run into a common end route such as activation stress kinases, caspases, calcium-dependent proteases, generation of ROS, induction of pro-apoptotic effector proteins, frustrate enter of the cell cycle, activation of death-inducing ligands, synthesis of pro-degenerative extracellular matrix molecules, disintegration of the cytoskeleton or block of axonal transport. In consequence to these common executive end routes, the same neuroprotective molecules can be applied in different neurodegenerative diseases.

The novel 'hightech' options of biotechnology provide new hopes in spite of the failure of the first applications, but the resolution of critical issues has raised the perspective of successful implementation into the clinics. Thus, embryonic cell transfer, stem cell therapy, cell factory for the delivery of protective compounds, viral gene transfer, transfer of fusion-peptides, vaccination or, as recent option, the application of erythropoetin have been improved in terms of pharmacokinetics, mode of application, benefit-risk-ratio, efficiency or precise definition of the appropriate sub-population of patients. Of course, these tools are far from being transformed into a daily easy-to-handle regimen for everyone including the patients with movement disorders or mental illness, the leading symptoms of the targeted neurodegenerative diseases.

The global efforts and the new research area of *clinical neurobiology* has provided profound insights into the molecular pathophysiology and pathogenesis of neurodegenerative disorders, the indispensable fundament for a successful clinical treatment. It has become clear and must not considered as mere optimism, that the modern biotechnology has set the stage for successful realisation of molecular approaches within the next decade. The first gentherapy for PD was performed in August 2003.

2. THE PROBLEM OF DELIVERY - LIMITATIONS FOR THE ACCESS OF DRUGS

The skull encapsuled brain is separated from the circulation by a specific and complex filter system, the blood-brain-barrier (BBB). The membranes, gap junctions, endothelial cells and astrocytes are the structural guardians which control the transport of trophic molecules to the neuropil, and which restrict the access of small and/or lipophilic molecules. In consequence, the majority of hydrophilic and/or large sized drugs do not enter the nervous system. In addition, multiple drug resistance proteins and transporters effectively clear the drugs from the cerebrum into the circulation.

Several strategies try to overcome the pharmacokinetic challenge of the BBB. (a) The viral gene transfer (see chapter 31). This route of application offers the chance of a locally increased synthesis of (lacking) molecules. Critical issues concern the risk of invasive application of the viral construct and the difficulty to control or modulate the rate and location of product synthesis. (b) The embryonic and stem cell transfer aim at the replacement of dying or already lost neuronal populations (see chapters 28 and 36).The first few successfully transplanted patients represent the holy promising fire at the end of the dark tunnel. Some patients suffering from severe Parkinson's disease do not reveal the characteristic progress of the disease even more than 10 years following the transplantation of embryonic dopaminergic neurons. (c) The implantation of 'cell factories'. Xenogenic cell lines which produced neurotrophic molecules were inserted in close proximity of the spinal cord of patients suffering from amyotrophic lateral sclerosis (ALS) (Aebischer et al. 1996); (Aebischer et al. 1999). (d) TAT-fused peptides as small molecules do not demand invasive intracerebral application but might reach the brain following systemic application. The fusion with the so-called TAT sequence which derives from HIV, enables the penetration not only of the BBB, but also of intracellular compartments such as nucleus (Bonny et al. 2001).

It is important to realise that the disease itself might prevent the successful use of drugs. This is obvious in stroke when the causative occlusion of the blood vessels blocks the access of drugs to the damaged area. Similar, destruction or atrophy of the neuroparenchyma and subsequent replacement by glial scar might enlarge the distance of drug perfusion from the capillaries to the damaged area. Moreover, the retrograde transport is essentially limited due to axonal degeneration in PD and ALS, excluding the synaptic uptake of drugs from circulation or the target area.

3. THE ISSUE OF 'LOCAL ACTION' AND TARGET SPECIFICITY

The liquor system envelops and internets the entire brain including the spinal cord. Thus, drugs are dissipated from the cerebral spinal fluid throughout the brain. Similarly, the arterial circulation does not offer a compartment specific delivery of drugs in spite of the cascade-like distribution of the (end-) arterioles with restricted circuits.

As alternative, local actions have to be achieved through the interaction with compartment-specific biochemical or morphological properties such as the targeting of compartment-specific protein subtypes. Moreover, the combined recognition/co-stimulation of two receptive structures by one therapeutic molecule will substantially increase the drug specificity and limit the unspecific activation of unwanted neuronal networks.

4. THE ISSUE OF PROPER EXPERIMENTAL MODELS AND THE THERAPEUTIC WINDOW

What do we really know about the intraneuronal, glial and intercellular processes (in the physical and metaphysical meaning of the word *processes*) which underlie the initiation and development of neurodegenerative diseases? Most of the data from *in vitro* systems and animal research derive from *acute* degenerative experiments, and - for statistical purposes – from the use of mono- or oligo-causal set ups. In Alzheimer's disease, for example, degenerated but nevertheless alive neurons survive in an atrophic state which can protract over months with features of disturbed cell cycle. Only few experiments have reflected these aspects of chronic pathophysiology (Green et al. 2003).

Besides the issue of experimental models, the success of therapy essentially depends on the precise diagnosis including the early onset of therapy in time. At present, anti-degenerative treatments start after the appearance of clinically relevant symptoms, i.e. after destruction of 70-80% neurons in AD or PD, or after several hours following the onset of stroke, i.e. when the majority of ischemic neurons have died. Moreover, the pharmacological or biotechnological interventions encounter the attenuated resistance of the surviving neurons which might also be more sensitive against side effects of the regimen. In contradistinction, time-point and routes of application are optimised under experimental conditions resulting in substantial and promising neuroprotection.

The issue of the therapeutic window, i.e. the onset of an effective therapy for reversible processes, is particularly relevant following acute injury i.e. following stroke. Under experimental conditions, the majority of drugs which are neuroprotective when applied before the occlusion of the blood vessels, loose their curing effects when given few hours after ischemia. First reports on the application of erythropoetin indicated that this drug might be effective even several hours after the stroke (Ehrenreich et al. 2002).

5. INADEQUATE THERAPY MIGHT WORSEN THE PATHOLOGY

The majority of current experimental concepts propagate neuroprotection by inhibition of apoptosis and inflammation– this conceivable therapeutic approach, however, is compromised by the risk of an inadequate, incomplete or counterproductive interference with the pathological processes. For example, apoptosis includes the presentation of phosphatidylserine residues at the surface of injured cells such as neurons. These residues act as signals for microglia which reinforce or finalise the (neuronal) death by phagocytosis or generation of immune responses. Provided that the efficiency of anti-apoptotic or anti-inflammatory strategies obey a dose-response relationship, sub-optimal doses could result in lingering of alive but heavily damaged neurons with persisting phosphatidylserine residues at their surface which provoke an ongoing signalling for activation of microglia with inflammatory or autoimmune reactions.

Recently, the concept of the "protective auto-immunity against the enemy within" (Schwartz et al. 2003) respects the existence of beneficial and protective autoimmune responses following injury. Abrogation of these responses handicap the regenerative and self-protective power of the brain. In principle, these beneficial autoimmune responses do not differ from disease-provoking autoimmune reactions. The decisive triggers are not yet fully understood which determine the outcome of basically similar reactions. In consequence, we have to consider that therapeutic (anti-inflammatory) efforts against auto-immune responses carry the risk to attenuate the endogenous healing power of the nervous system.

6. CLINICAL TRIALS IN NEURODEGENERATIVE DISEASES

In the last decade many new pharmacological strategies for neurodegenerative diseases came to clinical development. However, the majority of these strategies was either not effective, displayed severe side-effects, were toxic or showed unfavourable benefit-risk ratio. Moreover, the current therapeutic concepts improve the clinical symptoms, but do not or only moderately reduce the progression of pathology (Stocchi 2003). This section summarises the therapeutic strategies for the most relevant neurodegenerative diseases i.e. Alzheimer's disease, Parkinson's disease, amyothrophic lateral sclerosis, Huntington`s disease and degeneration following cerebral ischemia that have been tested in clinical trials, i.e. came to clinical development.

6.1. Alzheimer's disease (AD)

AD is characterised by a progressive loss of cholinergic neurones. The actual therapy aims to improve the symptomatic clinical outcome by reinforcement of the cholinergic transmission for instance by inhibition of the acetylcholine-esterase (Trinh et al. 2003). The acetyl-cholinesterase inhibitors donezepil, rivastimine and galantamine have been licensed for treatment of Alzheimer's disease. However, no other class of drugs has reached the clinic so far. Metrifonate, another acetyl-cholinesterase inhibitor, was even withdrawn from clinical development due to side effects (Dubois et al. 1999).

Several non-cholinergic therapeutic strategies have been or are being investigated in clinical trials that have a potential for disease modification. Promising therapeutic strategies such as the antioxidants α-tocopherol and idebenone, the cyclooxygenase inhibitors flurbiprofen, rofecoxib and naproxen (Eriksen et al. 2003; Etminan et al. 2003), the nootrophic drugs gingko biloba extract, nicergoline and piracetam, modulators of NMDA and AMPA-receptors (glutaminergic pathway) like memantine, the glial cells stabiliser propentofylline and stimulators of the nerve growth factors (NGF) are under investigation and partly showed beneficial effects on the progression of AD (Farlow 2002). Due to their anti-inflammatory potential statins might also be beneficial in the treatment of AD and are currently investigated (Scott and Laake 2001). However, inflammatory reactions may be part of the auto-immune response which represents the cerebral effort to fight against the neurodegeneration. Thus, the concept of anti-inflammation remains to be proven.

A further promising approach is the modification of amyloid and tau-protein formations, e.g. by vaccination with Aß proteins or treatment with amyloid and tau-protein modifying agents (Janus 2003). However, a clinical trial with the A-β-42 amyloid vaccine AN-1792 has been premature terminated due to association with meningoencephalitis (Imbimbo 2002). Improvement of the vaccination regimen may overcome this handicap. In the future gene therapy and manipulation of gene products might be a possible approach to cure AD. So far no therapeutic strategy has been established or is being tested in clinical trials.

6.2. Parkinson disease (PD)

Novel concepts for the treatment of Parkinson's disease (PD) intend to slow down the degeneration of the dopaminergic neurones in the substantia nigra i.e. the degeneration of the presynaptic structures of the dopaminergic neurones in the striatum and the subsequent "dying back". Similar as with AD, the current therapy is symptomatic, i.e. the reinforcement and/or maintenance of the dopaminergic transmission and the reduction of the relatively enhanced cholinergic transmission (Stocchi and Olanow 2003). Recent data indicate a central role of mitochondrial pathology (with subsequent initiation of apoptosis) for the pathogenesis of PD. However, the evidence is still missing that the preservation of mitochondrial function results in neuroprotection of dopaminergic neurons.

The MAO-B antagonists selegeline and lazabemide were shown to delay onset of requiring L-dopa for about 6 to 12 month (1996). Apparently, dopaminergic neurones in the substantia nigra are particularly sensitive to oxygen radicals. As with other degenerative diseases of the brain, unspecific therapeutic approaches use free radical scavengers like α-tocopherol, but there is no proof of evidence for a real beneficial effect.

Several anti-excitotoxic agents were tested in clinical trials. Modulation of glutaminergic pathways by the NMDA-receptor antagonists amantadine and remacemide showed no beneficial effects (Schachter and Tarsy 2000; Crosby et al. 2003). Inhibition of the voltage dependent sodium channel by riluzole to prevent the release of glutamate was also ineffective in PD (Jankovic and Hunter 2002).

Bioenergetic agents like creatine, coenzym Q10, gingko biloba extract, nicotinamide, riboflavin, acetyl-carnitine, or lipoic acid might have beneficial effects in PD. Indeed, creatine and coenzmy Q have been shown to protect dopamine neurones in MPTP-treated rodents, and coenzmy Q showed reduction in the disease progression in PD patients (Shults et al. 2002). Furthermore, there has been considerable interest in the potential of dopamine agonists such as L-dopa for neuroprotection and for attenuation of PD (Clarke and Deane 2001; Whone et al. 2003). Dopamine agonists have been shown to protect dopamine neurones from toxins in in-vitro and in-vivo models.

The dopamine agonists pramipexole, ropinirole and cabergoline were shown to slow down progression of PD, but display considerable side effects like dyskinesia and confusion.

Another therapeutic approach is the inhibition of the apoptotic pathway by anti-apoptotic agents. Metabolites of the MAO-B antagonist selegiline have been shown to alter expression of several genes involved in apoptosis including SOD1 and 2, glutathione peroxidase, c-Jun, glyceraldehyde-3-phosphate dehydrogenase (GAPDH), BCL-2 and BAX (Stocchi and Olanow 2003). The development of CEP-1347, an inhibitor of the mixed lineage kinases (MLK) which are upstream of the c-Jun N-terminal kinases (JNK) has raised expectations for protection of nigral neurones. Inhibition of MLKs and JNKs by CEP-1347 was most effective in animal models of various neurodegenerative diseases (Saporito et al. 1999). Meanwhile a phase III clinical trial with CEP-1347 in PD is in progress.

Beyond drug application, the transplantation of embryonic dopaminergic neurones or stem cells is a promising therapeutic strategy of future treatment of PD as discussed elsewhere in the book. However, to date, no drug has been established to be neuroprotective in PD.

6.3. Amyotrophic lateral sclerosis (ALS)

The degeneration of spinal cord neurones is the characteristic feature of amyotrophic lateral sclerosis (ALS). The current etiologic theories implicate glutamate excitotoxicity and oxidative stress as main factors in the pathogenesis of ALS. Furthermore, patho-physiology of ALS involve the activation of microglia. Tumor necrosis factor (TNF)-α expression appears to be induced prior to disease onset. Neuro-inflammation also seems to play a putative role in the pathogenesis of ALS. Recent studies demonstrate cyclooxygenase-2 (COX-2) up-regulation in the CNS tissue of ALS patients. COX-2 controls the production of the inflammatory prostaglandin PGE2 that is significantly elevated in ALS. Anti-inflammatory compounds such as the COX-2 inhibitor celecoxib and minocycline are currently in early clinical stages for treatment of ALS (Carter et al. 2003).

The NMDA-antagonist riluzole was shown to be modestly effective in prolonging survival of ALS patients for about 2 years (Miller et al. 2000). Gabapentin, that reduces glutamate activity by an unknown mechanism, demonstrates modest reduction in the loss of muscle strength (Miller et al. 1996).

In 1996, treatment of ALS by the implantation of polymer encapsulated xenogenic cell line marked a novel milestone in the neurobiotechnology; the cells were engineered to secrete hCNTF neuronal chimeras producing (Aebischer et al. 1996); (Aebischer et al. 1999). This encounter of biotechnology, neurobiology and neurological diseases provided a transient improvement in some patients, whereas a minority of patients suffered from severe side effects. This was also true for the systemical application of neurotrophins CNTF and BDNF (Miller et al. 1996; Bradley 2003). However, trials of subcutaneous ciliary neurotrophic factor (CNTF), subcutaneous and intrathecal application of brain-derived neurotrophic factor (BDNF), and intracerebroventricular glial-derived neurotrophic factor (GDNF) have all failed to show efficacy. Of all growth factors, recombinant human insulin-like growth factor-1 (myotrophin) has shown the most promising effects (Borasio et al. 1998).

Antioxidants like α-tocopherol, vitamin C, coenzmy Q10, B-carotene and N-acetylcysteine were partly tested in clinical trials. No clear efficacy could be established (Desnuelle et al. 2001). Selegiline, a MAO-B inhibitor with anti-oxidative properties, has also failed to demonstrate clinical improvement of ALS (Lange et al. 1998).

Creatine is believed to be neuroprotective by blocking the mitochondria transposition pore, which is involved in both excitotoxic and apoptotic cell death. Clinical trials of creatine in ALS are in progress (Klivenyi et al. 1999).

6.4. Huntington's disease (HD)

The pathogenic feature of Huntington's disease (HD) is the generation of long trinucleotide repeats and the subsequent degeneration of neurones in the cerebellum. The actual treatments comprise non specific drugs which are also used for other neurodegenerative diseases (2001; Lucetti et al. 2002) demonstrating the lack of specific regimens for HD.

The NMDA-antagonist remacemide and the antioxidant coenzmy Q10 were shown to have no beneficial effects in HD (Schilling et al. 2001). However, the NMDA-antagonist amantadine exerts significant improvement of dyskinesia but no improvement of psychiatric symptoms (Lucetti et al. 2002).

The tetracycline antibiotic minocycline, an inhibitor of apoptotic pathways with anti-inflammatory characteristics, is currently under clinical evaluation.

6.5. Cerebral ischemia

In most cases, cerebral ischemia is the result of a thrombotic event in the circulation outside the brain. In consequence, the current therapeutic approaches aim to improve the thrombotic pathology and the underlying cardiovascular defects. Following occlusion of cerebral blood vessels, thrombolysis within a very narrow time window is by now the only approved and effective intervention.

Acute cerebral ischemia is the perfect example for the discrepancy between experimental splendour and clinical failure of novel neurobiological procedures. Numerous reasons account for this discrepancy. For example, animals undergoing cerebral artery occlusion are healthy in the cardiovascular system; the experimental intervention occurs before the occlusion or within a narrow time frame; substances can be applied intrathecally; many experimental drugs are not monitored for putative side effects, and finally it remains to be elucidated which experimental model reflects the clinical neuropathology following stroke. In addition, the application of one drug might not be sufficient to rescue the ischemic neuroparenchym which suffers from multifactorial damage such as coincident inflammation, apoptosis, necrosis, withdrawal of trophic factors, excitotoxicity or breakdown of membrane potentials. In consequence, the optimal therapeutic regimen might comprise a 'cocktail' of different compounds. The test of such a 'cocktail', however, is a nightmare for pharmaceutical companies and clinical stroke units.

Numerous scientific and experimental studies have been performed to rescue neuronal death following ischemia and or to improve the neurological outcome. Many drugs which act on specific targets involved in the presumed ischemic cascade have been tested in clinical trials in the last decade. So far, none of these drugs was shown to be effective in improving outcome after acute cerebral ischemia. The voltage-gated calcium channel antagonists nimodipine and flunarizine were investigated in several phase III clinical trials in acute cerebral ischemia, but failed (Horn and Limburg 2001). Inhibitors of voltage-gated sodium channel like fosphenytoin and lubeluzole preventing presynaptic glutamate release were also ineffective in clinical trials (Gandolfo et al. 2002; Sareen 2002). Studies with the non-competitive NMDA-receptor antagonist selfotel and cerestat were prematurely discontinued or showed an unfavourable risk benefit ratio (Lees 1997; Davis et al. 2000). Eliprodil, that acts as an antagonist at the polyamide-site of the NMDA-receptor and the GABA-A receptor agonist clomethiazol was also ineffective in acute cerebral ischemia (Akins and Atkinson 2002) (Mohandas et al. 2002). Magnesium blocks the voltage gated ion channel of the NMDA receptor non – competitive and is just being tested in a clinical trial. Data are expected in 2003/2004.

Free radicals play a major role in cell damaging. Thus, a variety of compounds acting as free radical scavengers and antioxidants such as tirilazad mesylate, ebselen and citicholine were investigated in acute cerebral ischemia with disappointing results (Yamaguchi et al. 1998; Bath et al. 2001; Davalos et al. 2002).

The nootrophic compound piracetam that acts at cell membranes and elevates cAMP levels showed some efficacy in dementia but had no beneficial effects in acute cerebral ischemia and might even have unfavourable effects on early death (Ricci et al. 2002).

Enlimomab, a monoclonal antibody against ICAM involved in the apoptotic cascade, had an unfavourable risk-benefit ratio in acute ischemia (2001).

A promising strategy is the application of the endogenous glycoprotein erythropoetin. Intravenous high-dose of recombinant erythropoetin (rhEPO) in patients is well tolerated in acute ischemic stroke and associated with an improvement in clinical outcome at 1 month (Ehrenreich et al. 2002).

Inhibitors of the inducible nitric oxide synthase (iNOS), calpain (calcium dependent cysteine protease) inhibitors, opioid and serotonin antagonist as far as kainate inhibitors and antisense-oligonucleotides that modify gene expression are currently in preclinical and clinical early stages that might present encouraging strategies to prevent neurodegeneration. (Kidwell et al. 2001).

REFERENCES

(1996). "Effect of lazabemide on the progression of disability in early Parkinson's disease. The Parkinson Study Group." Ann Neurol 40(1): 99-107.

(2001). "A randomized, placebo-controlled trial of coenzyme Q10 and remacemide in Huntington's disease." Neurology 57(3): 397-404.

(2001). "Use of anti-ICAM-1 therapy in ischemic stroke: results of the Enlimomab Acute Stroke Trial." Neurology 57(8): 1428-34.

Aebischer, P., A. F. Hottinger, et al. (1999). "Cellular xenotransplantation." Nat Med 5(8): 852.

Aebischer, P., N. A. Pochon, et al. (1996). "Gene therapy for amyotrophic lateral sclerosis (ALS) using a polymer encapsulated xenogenic cell line engineered to secrete hCNTF." Hum Gene Ther 7(7): 851-60.

Akins, P. T. and R. P. Atkinson (2002). "Glutamate AMPA receptor antagonist treatment for ischaemic stroke." Curr Med Res Opin 18 Suppl 2: s9-13.

Bath, P. M., R. Iddenden, et al. (2001). "Tirilazad for acute ischaemic stroke." Cochrane Database Syst Rev(4): CD002087.

Bonny, C., A. Oberson, et al. (2001). "Cell-permeable peptide inhibitors of JNK: novel blockers of beta-cell death." Diabetes 50(1): 77-82.

Borasio, G. D., W. Robberecht, et al. (1998). "A placebo-controlled trial of insulin-like growth factor-I in amyotrophic lateral sclerosis. European ALS/IGF-I Study Group." Neurology 51(2): 583-6.

Bradley, W. (2003). "Therapeutic trials in ALS." Amyotroph Lateral Scler Other Motor Neuron Disord 4(1): 6-7.

Carter, G. T., L. S. Krivickas, et al. (2003). "Drug therapy for amyotrophic lateral sclerosis: Where are we now?" Idrugs 6(2): 147-53.

Clarke, C. E. and K. D. Deane (2001). "Cabergoline versus bromocriptine for levodopa-induced complications in Parkinson's disease." Cochrane Database Syst Rev(1): CD001519.

Crosby, N., K. H. Deane, et al. (2003). "Amantadine in Parkinson's disease." Cochrane Database Syst Rev(1): CD003468.

Davalos, A., J. Castillo, et al. (2002). "Oral citicoline in acute ischemic stroke: an individual patient data pooling analysis of clinical trials." Stroke 33(12): 2850-7.

Davis, S. M., K. R. Lees, et al. (2000). "Selfotel in acute ischemic stroke : possible neurotoxic effects of an NMDA antagonist." Stroke 31(2): 347-54.

Desnuelle, C., M. Dib, et al. (2001). "A double-blind, placebo-controlled randomized clinical trial of alpha-tocopherol (vitamin E) in the treatment of amyotrophic lateral sclerosis. ALS riluzole-tocopherol Study Group." Amyotroph Lateral Scler Other Motor Neuron Disord 2(1): 9-18.

Dubois, B., I. McKeith, et al. (1999). "A multicentre, randomized, double-blind, placebo-controlled study to evaluate the efficacy, tolerability and safety of two doses of metrifonate in patients with mild-to-moderate Alzheimer's disease: the MALT study." Int J Geriatr Psychiatry 14(11): 973-82.

Ehrenreich, H., M. Hasselblatt, et al. (2002). "Erythropoietin therapy for acute stroke is both safe and beneficial." Mol Med 8(8): 495-505.

Eriksen, J. L., S. A. Sagi, et al. (2003). "NSAIDs and enantiomers of flurbiprofen target gamma-secretase and lower Abeta 42 in vivo." J Clin Invest 112(3): 440-9.

Etminan, M., S. Gill, et al. (2003). "Effect of non-steroidal anti-inflammatory drugs on risk of Alzheimer's disease: systematic review and meta-analysis of observational studies." Bmj 327(7407): 128.

Farlow, M. (2002). "A clinical overview of cholinesterase inhibitors in Alzheimer's disease." Int Psychogeriatr 14 Suppl 1: 93-126.

Gandolfo, C., P. Sandercock, et al. (2002). "Lubeluzole for acute ischaemic stroke." Cochrane Database Syst Rev(1): CD001924.

Horn, J. and M. Limburg (2001). "Calcium antagonists for ischemic stroke: a systematic review." Stroke 32(2): 570-6.

Imbimbo, B. P. (2002). "Toxicity of beta-amyloid vaccination in patients with Alzheimer's disease." Ann Neurol 51(6): 794.

Jankovic, J. and C. Hunter (2002). "A double-blind, placebo-controlled and longitudinal study of riluzole in early Parkinson's disease." Parkinsonism Relat Disord 8(4): 271-6.

Janus, C. (2003). "Vaccines for Alzheimer's disease: how close are we?" CNS Drugs 17(7): 457-74.

Kidwell, C. S., D. S. Liebeskind, et al. (2001). "Trends in acute ischemic stroke trials through the 20th century." Stroke 32(6): 1349-59.

Klivenyi, P., R. J. Ferrante, et al. (1999). "Neuroprotective effects of creatine in a transgenic animal model of amyotrophic lateral sclerosis." Nat Med 5(3): 347-50.

Lange, D. J., P. L. Murphy, et al. (1998). "Selegiline is ineffective in a collaborative double-blind, placebo-controlled trial for treatment of amyotrophic lateral sclerosis." Arch Neurol 55(1): 93-6.

Lees, K. R. (1997). "Cerestat and other NMDA antagonists in ischemic stroke." Neurology 49(5 Suppl 4): S66-9.

Lucetti, C., G. Gambaccini, et al. (2002). "Amantadine in Huntington's disease: open-label video-blinded study." Neurol Sci 23 Suppl 2: S83-4.

Miller, R. G., F. A. Anderson, Jr., et al. (2000). "The ALS patient care database: goals, design, and early results. ALS C.A.R.E. Study Group." Neurology 54(1): 53-7.
Miller, R. G., D. Moore, et al. (1996). "Placebo-controlled trial of gabapentin in patients with amyotrophic lateral sclerosis. WALS Study Group. Western Amyotrophic Lateral Sclerosis Study Group." Neurology 47(6): 1383-8.
Miller, R. G., J. H. Petajan, et al. (1996). "A placebo-controlled trial of recombinant human ciliary neurotrophic (rhCNTF) factor in amyotrophic lateral sclerosis. rhCNTF ALS Study Group." Ann Neurol 39(2): 256-60.
Mohandas, S., J. Mani, et al. (2002). "Neuroprotection for acute ischemic stroke : an overview." Neurol India 50 Suppl: S57-63.
Ricci, S., M. G. Celani, et al. (2002). "Piracetam for acute ischaemic stroke." Cochrane Database Syst Rev(4): CD000419.
Richard Green, A., T. Odergren, et al. (2003). "Animal models of stroke: do they have value for discovering neuroprotective agents?" Trends Pharmacol Sci 24(8): 402-8.
Saporito, M. S., E. M. Brown, et al. (1999). "CEP-1347/KT-7515, an inhibitor of c-jun N-terminal kinase activation, attenuates the 1-methyl-4-phenyl tetrahydropyridine-mediated loss of nigrostriatal dopaminergic neurons In vivo." J Pharmacol Exp Ther 288(2): 421-7.
Sareen, D. (2002). "Neuroprotective agents in acute ischemic stroke." J Assoc Physicians India 50: 250-8.
Schachter, S. C. and D. Tarsy (2000). "Remacemide: current status and clinical applications." Expert Opin Investig Drugs 9(4): 871-83.
Schilling, G., M. L. Coonfield, et al. (2001). "Coenzyme Q10 and remacemide hydrochloride ameliorate motor deficits in a Huntington's disease transgenic mouse model." Neurosci Lett 315(3): 149-53.
Schwartz, M., I. Shaked, et al. (2003). "Protective autoimmunity against the enemy within: fighting glutamate toxicity." Trends Neurosci 26(6): 297-302.
Scott, H. D. and K. Laake (2001). "Statins for the reduction of risk of Alzheimer's disease." Cochrane Database Syst Rev(3): CD003160.
Shults, C. W., D. Oakes, et al. (2002). "Effects of coenzyme Q10 in early Parkinson disease: evidence of slowing of the functional decline." Arch Neurol 59(10): 1541-50.
Stocchi, F. and C. W. Olanow (2003). "Neuroprotection in Parkinson's disease: clinical trials." Ann Neurol 53 Suppl 3: S87-97; discussion S97-9.
Trinh, N. H., J. Hoblyn, et al. (2003). "Efficacy of cholinesterase inhibitors in the treatment of neuropsychiatric symptoms and functional impairment in Alzheimer disease: a meta-analysis." Jama 289(2): 210-6.
Whone, A. L., R. L. Watts, et al. (2003). "Slower progression of Parkinson's disease with ropinirole versus levodopa: The REAL-PET study." Ann Neurol 54(1): 93-101.
Yamaguchi, T., K. Sano, et al. (1998). "Ebselen in acute ischemic stroke: a placebo-controlled, double-blind clinical trial. Ebselen Study Group." Stroke 29(1): 12-7.

6.1. ALZHEIMER DISEASE (AD)

6.1.1. Implemented clinical trials and cochrane review (AD)

Drug	Target / Mechanism	Trial / Phase	Population	Results	Cochrane review
Donazepil (Rogers et al. 1998; Rogers et al. 1998; Rogers and Friedhoff 1998; Burns et al. 1999; Birks et al. 2000)	Acetyl-cholinesterase inhibitor	Donezepil study group / Multinational trial (III)	473 / 818 AD	significant improvement of cognition and global function	modest improvement of cognitive function; Meta-analysis: modest beneficial impact on neuropsychiatric and functional outcome
Rivastigmine tartrate (Birks et al. 2000; Farlow et al. 2000)	Acetyl-cholinesterase inhibitor (preferentially subtype G1)	(III)	AD	significant better cognitive function	improvement of cognitive function, activities of daily living and severity of dementia; Meta-analysis: modest beneficial impact on neuropsychiatric and functional outcome
Galantamine (Raskind et al. 2000; Tariot et al. 2000; Wilcock et al. 2000; Olin and Schneider 2002)	Acetyl-cholinesterase inhibitor, nicotinergic AChRs-agonist	USA-1/10 (III)	978 / 636 AD	significant benefit in cognition, functional and behavioural symptoms	positive effects; Meta-analysis: modest beneficial impact on neuropsychiatric and functional outcome
		Gelantamine international 1 study group (III)	653 AD (525 completed)	slowed the decline of functions and cognition	
Tacrine (Knapp et al. 1994; Qizilbash et al. 2000)	Acetyl-cholinesterase inhibitor	The Tacrine study group (III)	663 AD	significant improvement in cognition and activity; unfavourable risk-benefit	no convincing evidence of benefit; high rate of gastrointestinal AEs

Drug	Target / Mechanism	Trial / Phase	Population	Results	Cochrane review
Metrifonate (Dubois et al. 1999; Raskind et al. 1999)	Acetyl-cholinesterase inhibitor	MALT / Metrifonate study group (III)	605 / 264 AD	significant improvement of symptoms	withdrawn due to side effects
Physiostigmine (Thal et al. 1996; Coelho and Birks 2001)		Physiostigmine study group (III)	366 AD	significant improvement of cognition and clinics global evaluation	no convincing benefit, high rate of withdrawals due to gastrointestinal adverse events (AEs)
Linopirdine (Rockwood et al. 1997)	enhances ACh release	(III)	382 AD	no clinically meaningful differences to placebo	
Estrogen (+Progesterone) (Hogervorst et al. 2002; Wooltorton 2003)	enhance ACh synthesis	WHIMS	4381 healthy woman	increased risk for dementia	5 trials (210 women with AD) were analysed, no evidence for benefit on AD
Propentophylline (Marcusson et al. 1997; Frampton et al. 2003)	Adenosine re-uptake inhibitor; phosphodiesterase inhibitor; calcium channel opener; ACh-agonist	The European Propentophyllin study group (III)	260 AD and nAD	AD significant better outcome	limited evidence of benefit in cognition, global function and activity
Risperidone (De Deyn et al. 1999)	5-HT2 and D2 receptor antagonist	(III)	344 AD and nAD	reduction in severity and frequency of behavioral symptoms like aggression	
Acetyl-L-Carnitine (ACL) (Thal et al. 1996; Hudson and Tabet 2003)	membrane stabilization; cholinergic activities	(III)	431 AD	subgroup may benefit	no clear evidence of benefit
Rofecoxib and Naproxen (Aisen et al. 2003)	COX-2 inhibitor	(III)	351 AD	no significant effect on cognition	

Drug	Target / Mechanism	Trial / Phase	Population	Results	Cochrane review
Selegiline (Sano et al. 1997; Birks and Flicker 2003)	selective MAO-B inhibitor; antioxidant	(III)	341 AD	slows the progression of disease	no evidence of clinically meaningful benefit in AD
α-Tocopherol (Sano et al. 1997; Tabet et al. 2000)	antioxidants / free radical scavengers	(III)	341 AD	slows the progression of disease	insufficient evidence; more trials are needed
Idebenone (Weyer et al. 1997)			300 AD	significant improvement of ADAS-total and Cog	
Gingko biloba extract (EGb 761) (Le Bars al. 1997; Le Bars and Kastelan 2000; Le Bars et al. 2000)	nootropic, exact mode of action is unclear	North American EGb study group (III)	309 AD / 244 AD / 214 AD	modest improvement of cognition and social function / no difference to placebo / not effective in dementia	uncertain clinical relevance
Memantine (Areosa and Sherriff 2003)	NMDA receptor antagonist, modulation of glutaminergic neurotransmission				possible benefit on cognition and global measures; more trials are needed
Piracetam (Flicker and Grimley Evans 2001)	acts at cell membranes and elevates cAMP levels				no evidence for benefit in dementia

6.1.2. Clinical trials in progress and early clinical stages (AD)

Drug	Target / Mechanism	Trial / Phase	Population	Results	
α-Tocopherol	antioxidant / free radical scavenger	(III)	10400 HV	in progress	
Naproxen	NSAID	ADAPT(III)	2625 HV	in progress	

39. PHARMACOLOGICAL STRATEGIES FOR NEURODEGENERATION

Drug	Target / Mechanism	Trial / Phase	Population	Results	Cochrane review
Flurbiprofen (Cirrito and Holtzman 2003)	NSAID, reduces amyloid β-42	II	200 AD	in progress	
Olanzapine	5-HT2 and D2 receptor antagonist	CATIE (III)	450 AD	in progress	
Simvastatin	increases α-secretase; HMG-CoA reductase inhibitors	CLASP(III)	400 AD	in progress	
Atovastatin		(II)	120 AD	in progress	
Folate; vitamin B6 and B12	reduces homocysteine	VITAL(III)	400 AD	in progress	
CX 516 (Ampalex)	AMPA receptor antagonist	(II)	40 AD / 160 nAD	in progress	
α-INF-2A	immunmodulator	(II)	AD	in progress	
NS 2330	release of acetylcholine, noradrenaline and dopamine		AD	in progress	
Neotrofin (AIT-082) (Grundman et al. 2003)	stimulates neuritogenesis and NGF production	(I)	36 AD	no significant effect	
AN-1792 (Imbimbo 2002)	A-β-42 amyloid vaccination			discontinued due to AE (meningo-encephalitis)	
Phenserin	acethycholinesterase inhibitor			in progress	
Nefiracetam; Nebracetam	m1- receptor agonist				

Drug	Target / Mechanism	Trial / Phase	Population	Results	Cochrane review
Melatonin	hormone for circadian sleep-wake cycle				
Trichlorfon	acetycholinesterase inhibitor				
Dapsone	anti-inflammatory				
Ibuprofen	non-selective COX inhibitor				
Cerebrolysin	neuronal growth factor				
Suritozole	inverse GABA agonist				
Nicotine	cholinergic agonist, presynaptic release of acetycholin				

6.2. PARKINSON'S DISEASE (PD)

6.2.1. Implemented clinical trials and cochrane review (PD)

Drug	Target / Mechanism	Trial / Phase	Population	Results	Cochrane review
Selegiline (1989; Olanow et al. 1998; Palhagen et al. 1998; Przuntek et al. 1999; Stocchi and Olanow 2003)	MAO-B antagonists	DATATOP (III)	800 PD	significant delay of requiring L-dopa or dopamine agonists	
		SELEDO (III)	116 PD		
		SINDEPAR (III)	100 PD		
		Swedish Parkinson Study Group (III)	157 PD		
Lazabemide (1996)		Parkinson study group	321 PD	delayed the need for L-dopa	
Amantadine (Crosby et al. 2003)	NMDA receptor antagonists	retrospective	PD	possibly lower progression	insufficient evidence
Remacemide (2000; Schachter and Tarsy 2000)		Parkinson study group (III)	200 PD	safe but ineffective	
Riluzole (Jankovic and Hunter 2002)	voltage dependent sodium channel	(III)	PD	safe but ineffective	
α-Tocopherol (1989)	antioxidants / free radical scavengers	DATATOP (III)	800 PD	safe but ineffective	
Coenzyme Q10 (Shults et al. 2002)		(III)	80 PD	slower progression	

Drug	Target / Mechanism	Trial / Phase	Population	Results	Cochrane review
Pramipexole (2000)	Dopamin (D2) agonists	CALM-PD SCECT (III)	301 PD	slower progression	
Ropinirole (Whone et al. 2003)		REAL-PET (III)	186 PD (162 analysed)	slower progression compared to L-dopa	
Cabergoline versus Bromocriptine (Clarke and Deane 2001)					similar benefits; cabergoline increased dyskinesia and confusion
Benzhexole, Orphenadrine, Benztropine, Bornaprine, Benapyrzine, Methixine (Katzenschlager et al. 2003)	Anti-cholinergic drugs				more effective in improving motor function; more neuropsychiatric and cognitive adverse events than Placebo

6.2.2. Clinical trials in progress and early clinical stages (PD)

Drug	Target / Mechanism	Trial / Phase	Population	Results
CEP-1347	MLK (mixed lineage kinase) inhibitor	III	800 PD	in progress
Donezepil	Acetyl-cholinesterase inhibitor	IV	28 dementia in PD	in progress
JP 1730	alpha-2 adrenergic receptor antagonist	II	30 PD	in progress
GM 1 Ganglioside	nerve cell membrane	II	150 PD	in progress
Quetiapine	D2 and 5-HT2 antagonist	IV	120 dementia in PD	in progress

Drug	Target / Mechanism	Trial / Phase	Population	Results	Cochrane review
EMD 128130	5-HT1A agonist	II	30 on/off PD	in progress	
Talampanel	glutamate receptor inhibitor	I and II	33 dyskinesia in PD	in progress	
Creatine	antioxidant / free radical scavenger				

6.3. AMYOTROPHIC LATERAL SCLEROSIS (ALS)

6.3.1 Implemented clinical trials and cochrane review (ALS)

Drug	Target / Mechanism	Trial / Phase	Population	Results	Cochrane review
Riluzole (Miller et al. 2002)	inhibition of glutamate release and post-synaptic glutamate effects; stabilization of voltage-dependent sodium channels	ALS/Riluzole study group (III)	>1100 ALS	modestly effective in prolonging survival of ALS	prolongs survival by about 2 month
α-Tocopherol (Desnuelle et al. 2001)	antioxidant / free radical scavenger	ALS riluzole-tocopherol study group (III)	289 ALS	no effect on survival and motor function	
Myothrophin rhIGF-1 (recomb. Human insulin growth factor 1) (Lange et al. 1996; Borasio et al. 1998)	nerve growth factor	European ALS/IGF-1 study group / North American ALS/IGF-1 study group (III)	183 / ?? ALS	may be modestly effective in ALS	may be modestly effective but evidence is insufficient
rhCNTF (recomb. ciliary neurotrophic factor) (Miller et al. 1996)	nerve growth factors	rhCNTF / ALS study group (III)	570 ALS	no effect in ALS	
BDNF (brain-derived neurotrophic factor) (Bradley 2003)					
GDNF (glial-derived neurotrophic factor)					

39. PHARMACOLOGICAL STRATEGIES FOR NEURODEGENERATION

Drug	Target / Mechanism	Trial / Phase	Population	Results	Cochrane review
Selegiline (Lange et al. 1998)	MAO-B antagonist and antioxidant		133 ALS	no significant effect	
Gabapentin (Miller et al. 1996)	reduce glutamate activity	WALS study group (II)	152 ALS	modestly effective in reducing loss of strength	
Branched-chain amino acids activate glutamate dehydrogenase	reduces glutamate plasma levels			no significant effect	
Dextrometorphan (Blin et al. 1996; Gredal et al. 1997)	NMDA receptor antagonist	(II)	45 / 49 ALS	no significant effect	
Verapamil (Miller et al. 1996)	voltage-dependent calcium channels	(II)	72 ALS	no significant effect	
Nimodipine (Miller et al. 1996)		(II)	87 ALS	no significant effect	

6.3.2. Clinical trials in progress and early clinical stages (ALS)

Drug	Target / Mechanism	Trial / Phase	Population	Results	Cochrane review
Vit C, Coenzym Q10; B-carotene, N-acetylcysteine	antioxidant / free radical scavenger			no clear beneficial effects	
Creatine (Klivenyi et al. 1999)	block mitochondrial passways		ALS	in progress	
Lamotrigene	voltage dependent sodium channel at glutamate synapse				

Drug	Target / Mechanism	Trial / Phase	Population	Results	Cochrane review
Neuroimmunophilin ligands	small molecular neurotrophic factors				
Minocycline	inflammatory effects; anti-apoptotic; caspase 1, 3 and iNOS (I and II) inhibitor				
Celecoxib	COX-2 inhibitor				
Topiramate	AMPA receptor				
Cannabinoids / Marijuana	antioxidant and neuroprotective				
Tamoxifen	inhibits glutamate reuptake				

39. PHARMACOLOGICAL STRATEGIES FOR NEURODEGENERATION 611

6.4. HUNTINGTON DISEASE (HD)

6.4.1 Implemented clinical trials and cochrane review (HD)

Drug	Target / Mechanism	Trial / Phase	Population	Results	Cochrane review
Remacemide (Schilling et al. 2001)	NMDA receptor antagonist	CARE-HD (III)	347 HD	no significant improvement	
Coenzym Q10 (Schilling et al. 2001)	antioxidant	CARE-HD (III)	348 HD	no significant improvement; possible slower progression	
Amantadine (Lucetti et al. 2002; Verhagen et al. 2002)	non-competitive NMDA receptor antagonist	(II)	8 HD	significant reduction of dyskinesia; no effect on neuropsychotic and psychiatric assessment	
			24 HD	significant reduction of dyskinesia	

6.4.2 Clinical trials in progress and early clinical stages (HD)

Drug	Target / Mechanism	Trial / Phase	Population	Results	Cochrane review
Minocycline (Smith et al. 2003)	inflammatory effects; anti-apoptotic; caspase 1, 3 and iNOS (I and II) inhibitor	(II)	63 HD	in progress	

6.5. ACUTE CEREBRAL ISCHEMIA (ACI)

6.5.1. Implemented clinical trials and cochrane review (ACI)

Drug	Target / Mechanism	Trial / Phase	Population	Results	Cochrane review
Nimodipine (Horn et al. 2001; Horn and Limburg 2001; Lopez-Arrieta and Birks 2002)	voltage-gated calcium channel antagonist	TRUST (III)	1215 ACI (683 discontinued)	no improvement of neurological outcome at 6 months	
		NEST (III)	880 AS (472 discontinued)	no improvement of neurological and functional outcome at 3 months	
		American nimodepine study group (III)	1064 ACI (434 discontinued)	no difference in mortality or neurological outcome at 21 days	
		INWEST (III)	295 ACI (108 discontinued)	unfavourable outcome in the nimodepine group	
		VENUS (premature termination) (III)	454 ACI (321 discontinued)	no improvement of functional outcome at 3 months	
		Nimodepine in acute ischemic strole (III)	ACI	no improvement of neurological and functional outcome at 6 months	
		German-Austrian stroke trial (III)	ACI	no improvement of neurological outcome at 21 days	

Drug	Target / Mechanism	Trial / Phase	Population	Results	Cochrane review
Flunarizine (Franke et al. 1996)	voltage-gated Ca^{2+} channel antagonist	FIST (III)	331 ACI (155 discontinued)	no improvement of neurological and functional outcome at 6 month	
Fosphenytoin (Sareen 2002)	voltage-gated sodium channel antagonist preventing pre-synaptic glutamate release; NO activity modulator	(III)	ACI	no improvement of functional outcome at 3 months	
Lubeluzole (Gandolfo et al. 2002)	voltage-gated sodium channel antagonist preventing pre-synaptic glutamate release	LUB-INT 5/9/13 (III)	Five trials involving a total of 3510 ACI patients were included	no reduction in mortality and no improvement of functional outcome at 3 months	no significant reduction of death in acute ischemic stroke; QTc prolongation
BMS-204352 (Jensen 2002)	calcium dependent and voltage-dependent potassium channel	(III)	1978 ACI	no improvement versus placebo	
Selfotel (CGS 19755) (Davis et al. 1997; Davis et al. 2000)	non-competitive NMDA-receptor antagonist	ASSIST	ACI	premature termination due to unfavourable risk-benefit ratio	
Cerestat (CNS 1102) (Lees 1997)		(III)	ACI	unfavourable risk-benefit ratio	
Eliprodil (Akins and Atkinson 2002)	antagonist at polyamide-site of NMDA-receptor	(III)	ACI	no improvement of functional outcome at 3 months	
Clomethiazole (Wahlgren et al. 1999)	GABA-A receptor agonist	CLASS (III)	1360 ACI	no improvement of functional outcome at 3 months	

Drug	Target / Mechanism	Trial / Phase	Population	Results	Cochrane review
Tirilazad mesylate (2000; Bath et al. 2001; van der Worp et al. 2002)	free radical scavenger	RANTTAS 1 / 2 (III) TESS 1 / 2 (III)	660 / 126 ACI 450 / 355 ACI	no improvement of functional outcome at 3 months	appears to worsen outcome in acute ischemic stroke
Ebselen (Yamaguchi et al. 1998)		Ebselen for acute ischemia stroke (III)	302 ACI	no improvement of functional outcome at 3 months	
Citicoline (CDP-choline) (Davalos et al. 2002)		Citicholine in acute stroke (III)	1372 ACI	no improvement of functional outcome at 3 months	
Ganglioside GM1 (Lenzi et al. 1994)	natural constituent of cell membranes	EST (III)	792 ACI	no improvement of neurological and functional outcome at 4 months	
		SASS (III)	287 ACI	no improvement of survival; neurological and functional outcome at 3 months	
Piracetam (De Deyn et al. 1997; Ricci et al. 2002)	acts at cell membranes and elevates cAMP levels	PASS (III)	927 ACI	no improvement of neurological outcome at 4 weeks	
Enlimomab (2001)	monoclonal antibody against ICAM	EAST (III)	625 ACI	unfavourable risk-benefit ratio	might have unfavourable effects on early death, not enough evidence so far

39. PHARMACOLOGICAL STRATEGIES FOR NEURODEGENERATION

6.5.2. Clinical trials in progress and early clinical stages (ACI)

Drug	Target / Mechanism	Trial / Phase	Population	Results	Cochrane review
Magnesium (Muir 2002)	blocks voltage gated ion channel of NMDA receptor non-competitive	Imgage S	ACI	in progress (data expected in 2003/4)	
Aminoguanidine	inhibits immunological nitric oxide synthase (iNOS)	Pilot studies			
N-acetaminide					
CPI-22 (PBN derivate	free radical scavenger	II			
Calpain inhibitors					
Kainate antagonist					
ZK200775	AMPA antagonists				
Antisense Oligonucleotides					
Opiod antagonists					
Serotonin antagonists					
GV 150526	Glycine site antagonist				

M. PTITO[1,2] AND R. KUPERS[2]

[1]*School of Optometry, Université de Montréal, Montreal, Canada*

[2]*The PET centre and CFIN, Aarhus University, Denmark*

40. MONITORING BRAIN DYSFUNCTION THROUGH IMAGING TECHNIQUES

Summary. We report on the contribution of positron emission tomography (PET) to the understanding of brain function through dysfunction. We present data of congenital blindness and show that in cases in which there was never any visual exposure, the visual cortex can be activated through another sensory system, suggesting cross-modal plasticity. We further present data on plastic changes in memory disorder and chronic pain processing. Although functional magnetic resonance imaging (fMRI) has taken the lead in neuroimaging, PET remains a valuable method for understanding brain functions. Its future largely lies in receptor and genomic mapping.

1. INTRODUCTION

With the introduction of modern brain imaging techniques such as Positron Emission Tomography (PET) and Functional Magnetic Resonance (fMRI), which form the scope of this chapter, a revolutionary change has occurred in our understanding of the functioning of the human brain in normal and pathological conditions. Before the introduction of these techniques, our knowledge of brain functioning was derived either from single unit recordings in patients during the course of a stereotactic intervention, or was based on the study of behavioral abnormalities in patients with specific brain lesions. For obvious reasons, these techniques have major limitations. First, single unit recordings are strongly limited in terms of the amount of brain tissue from which data can be collected. Micro-electrodes are usually only inserted in the area for neurosurgical intervention. In addition, the procedure is time-consuming and tedious, placing a burden on both patient and investigator. Although this technique may provide important information about neuronal responses in the area explored, the technical difficulty of performing single unit recordings in the operation theatre and the invasiveness of the procedure are serious drawbacks. Notwithstanding, this technique has made important contributions to our understanding of brain functioning. For instance, single unit recordings in patients with phantom pain have shown reorganization and altered neuronal responsiveness at the thalamic level. A major advantage of this technique compared to PET and fMRI is that it allows to draw conclusions on the type of neuronal responses in a particular area or condition (e.g. excitatory, inhibitory, abnormal after discharges etc.). The study of behavioral abnormalities in patients with brain lesions is the other classical technique, often used by neuropsychologists. By comparing behavioral or cognitive performance in patients with brain lesions with that in normal controls, information of the role of the specific brain area was inferred. A classical example in the neuropsychology literature is the case HM who has advanced our understanding of the role of the medial temporal cortex in episodic memory. The limitation of this technique is that it does not give us information about potential cortical or subcortical reorganization and how other brain areas may have taken over the function of the damaged part.

PET, and later fMRI, has offered a paradigm shift in the study of the brain in health and disease. First, these two techniques allow simultaneous data sampling from all parts of the brain. A difference in this respect with EEG is the much better spatial resolution and the fact that PET and fMRI also allow data sampling from subcortical structures. Second, because of their relatively low invasiveness, these methods can also be applied in normal healthy controls and in patients outside the context of a surgical intervention. Whereas PET studies dominated the field, fMRI has taken over the lead in neuroimaging. This for two reasons: First, in contrast with PET, fMRI does not involve the injection of radioactive substances in the body. As a consequence, the number of fMRI data sets that can be acquired in the same subject is principally unlimited. Second, the large availability of clinical MRI scanners made that the scientific community rapidly embraced this unique opportunity. Nowadays, only few activation studies are still performed with PET. This does not mean that PET has no longer a role in brain imaging. The future of PET clearly lies in its unique capacity to do receptor studies in vivo in the human brain. Whereas fMRI in the early days was merely used as a parallel technique to PET, measuring blood oxygenation instead of blood flow, the applications have diverged in recent years (see review by Turner and Jones, 2003). Besides activation studies in the classical sense, Magnetic Resonance Imaging (MRI) techniques are now also used for fiber tracking and voxel based morphometric studies. In this chapter, we will present data accumulated over the years on various sensory systems (vision, somesthesis and pain) in individuals suffering from various dysfunction's due to developmental brain anomalies or intractable epilepsy who underwent surgical therapy. We will moreover offer evidence from longitudinal studies of such patients for phenomena such as neural plasticity and sensory substitution.

2. BRAIN DYSFUNCTIONS

In the following sections we will report on the contribution of PET and fMRI imaging studies to 1) the understanding of brain function through dysfunction; 2) the visualization of functional recovery (plasticity) and 3) sensory substitution (cross-modal plasticity). We will present imaging data on reversible anterograde amnesia, recovery from developmental anomaly of both occipital lobes, effects of hemispherectomy on vision, sensory substitution in the congenitally blind and finally, pain.

2.1. Anterograde amnesia

The patient is a 53 year-old right-handed female with no previous neurological antecedents. She developed acute onset of memory problems that turned rapidly into a full anterograde amnesic syndrome. Upon arrival at the hospital, she was alert, oriented for place, person and time and quite cooperative. She did not present signs of dementia or speech impairment. She underwent a series of brain imaging examinations as well as a complete neuropsychological evaluation pre- and post-surgically. A T1-weighted MRI of the brain revealed a suprasellar craniopharyngioma with a diameter of 3.5 cm, which was elevating and compressing the mammillary bodies by 2.5 cm as well as the cerebral peduncles. The tumour did not reach the left hippocampus but was impinging upon the mesial surface of the right hippocampus by 4 mm. The temporal horns of the lateral ventricles were in their normal place on both sides and no structural abnormalities were seen in the temporal lobe, the amygdala, thalamus, fornix or frontal lobes (Figure 1A). This was confirmed by the results of a FDG-PET scan at rest which showed normal FDG uptake in this region (Figure 1B, upper portion).

The cystic tumour was surgically removed de visu without injuring the surrounding pituitary gland and hypothalamus. An MRI taken a month after the surgery confirmed the total removal of the tumour (figure 1 D). A post-operative FDG PET scan taken at the same time indicated a normal brain perfusion in the medial temporal region and upper brainstem (figure 1B, lower portion). The patient recovered well from the surgery. Her memory problems completely disappeared and she returned to her normal occupations. Formal neuropsychological testing was performed before and two months after surgery. The patient showed a selective overall memory deficit with a remarkable post-surgical recovery of neuropsychological functions in the normal range.

2.2. Brain imaging studies

In order to determine the brain areas involved in the memory deficit and recovery, the patient was scanned while she was performing a non-spatial associative learning test consisting of 10 faces each associated with a particular house (Kupers et al., in press). The task of the subject was to learn the specific pairings of the faces with the houses. One hour before the start of PET scans, the patient was shown the series of the paired associate stimuli and she was tested for retention. The PET study consisted of 12 sequential measurements (two conditions repeated six times each) of rCBF following the intravenous injection of 500 MBq $H_2^{15}O$ labelled water during the presentation of the visual stimuli (baseline and test conditions). Pre-operatively, the subtraction of the baseline condition from the test condition yielded significant activations in the anterior cingulate gyrus, the frontal lobe and the cerebellum. No significant rCBF increases were seen in any of the brain areas normally associated with memory functions such as the hippocampus, the parahippocampal areas or the thalamus (figure 1C). Post-operatively however, significant loci of activation were

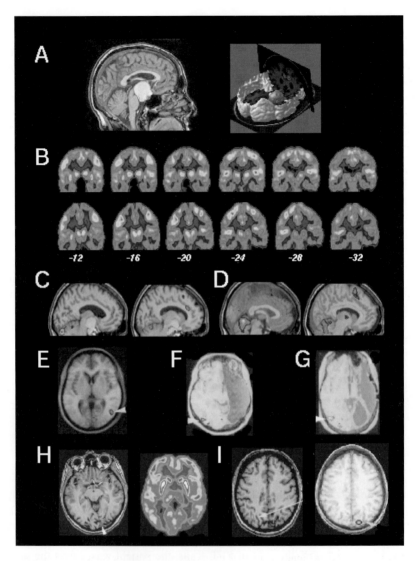

Figure 1. Anterograde amnesia patient. A- The tumor in situ (MRI and 3-D reconstruction). B- 18FDG perfusion scans before (top row) and after (lower row) surgery. C-D- PET activation scans showing significant rCBF increases pre- (C) and post- (D) surgery. Hemispherectomy: Activation sites following stimulation of the intact left visual field in normal subjects (E), the intact visual field in an hemispherectomized patient (F) and the blind visual field in the same patient (G). Note that activations in G are in the remaining hemisphere. Patient PK. H- MRI and 18FDG PET showing the bilateral anomaly of the occipital lobes and the preponderance of hypoperfusion in the right hemisphere. I- stimulation of the left visual hemifield (connected to the hemisphere with a larger lesion) activates the visual cortex (indicated by the yellow arrow) of the contralateral hemisphere (smaller lesion).

Observed in the cerebellum, frontal lobe and the anterior portion of the thalamus (figure 1D).

In light of previous anatomical data, it seems that the thalamus is actively involved in memory and that the mammillo-thalamic tract thereby plays an important role (see review by Van der Weirf et al., 2000). In our patient, PET activations confirm the importance of the mammillo-thalamic tract in the production of anterograde amnesia. First, MRI scans showed a compression of the right hippocampus and the mammillary bodies with no damage to the surrounding structures as evidenced by an FDG-PET scan at rest. This may have resulted in theperturbation of the mammillo-thalamic tract and the ensuing impaired memory functions. In the paired-associate learning task, the patient could not learn and retain information pertaining to the association between a house and a face. The pre-operative PET-activation results failed to show activation in memory-related structures. Post-surgically, our patient regained full memory functions, as evidenced by the neuropsychological and paired associate tests. At the same time, increased activation was observed in the anterior thalamic nuclei.

Overall, the combination between behavioural data and PET activation studies performed pre- and post-operatively on the same patient provide unique evidence that the mammillo-thalamic tract plays an important role in the ability to acquire new information and form new memories.

2.3 Hemispherectomy

Lesions of the primary visual cortex in humans induce a permanent loss of the contralateral visual field. Evidence has been accumulating that some residual visual functions remain in the hemianopic field (see review in Ptito et al., 2001). Patients with such lesions are usually unaware of visual stimuli presented in the blind field, although they respond correctly to them in a force-choice paradigm. This vision without awareness has been coined "blindsight" (see review by Ptito et al., 2001) and has received wide attention because it raised questions about the neural basis of visual awareness and the anatomo-functional organization of the visual system. It has been proposed that blindsight is mediated by the extrastriate areas of the lesioned hemisphere (mostly area V5 or MT) (see review by Stoerig and Cowey (1997) and that area 17 (V1) is primordial for conscious vision. This view has been challenged for reasons implying that residual vision within the scotoma might be due to methodological inadequacies (Faubert et al., 1999) or to residual tissue within the striate cortex (Fendrich et al, 2001). Hemispherectomy patients (Hd) present a unique opportunity to evaluate the neural substrates of residual vision in the blind hemifield since striate and extrastriate cortices have been ablated, leaving intact all projections from the retina to midbrain structures as demonstrated in post-mortem analysis in humans and in monkeys (Boire et al., 2001). Using PET and fMRI, we were able to demonstrate that visual information presented in the blind hemifield of these patients activated visual cortical areas of the remaining hemisphere. Hd patients viewed a black and white semi-circular ring with two embedded gratings moving vertically in opposite directions presented on a background of randomly moving dots (Bittar et al, 1999). This stimulus prevents Lambertian scatter (Faubert et al., 1999) and is optimal for activating structures forming the colliculo-pulvinar pathway. In normal subjects, the stimulation of one hemifield will activate visual cortical regions in the contralateral hemisphere (V1/V2, V3/V3A, lingual gyrus and V5) (Figure 1E). Stimulation of the intact hemifield of Hd patients produced several loci of activation exclusively in the occipital lobe of the remaining hemisphere, similar to those seen in normal control subjects (Figure 1F). Interestingly, stimulation of the blind hemifield yielded significant activations in visual structures (V1/V2; V3/V3A; V5) of the remaining hemisphere (Figure 1G). These PET data show for the first time that the remaining hemisphere in Hd patients does contribute to residual vision in the blind hemifield, either conscious or unconscious. These imaging results are supported by our own anatomical studies on hemispherectomized non-human primates, in which we demonstrated that retinal projections are still landing in the superior colliculus (SC) that appears remarkably intact both anatomically and metabolically (Boire et al., 2001). We have proposed a pathway that visual information takes from the blind hemifield towards the remaining hemisphere, known as the collicular hypothesis (Ptito et al, 2001). According to this hypothesis, the information entering the blind hemifield is routed to the ispilateral SC, then via the inter-collicular commissure to the contralateral SC and further to the contralateral Pulvinar of the thalamus (the one ipsilateral to the lesion being completely degenerated) to arrive finally in the cortex of the remaining hemisphere. The representation of both visual fields in the remaining hemisphere is a good example of brain plasticity. This rerouting could be best visualized by the newer technique of fiber tracking (tractography) which uses MRI to measure the magnitude and direction of the diffusion of water molecules in oriented fibre structures, therefore marking the fibres patterns. Preliminary results obtained in our group (and by others) are encouraging in this respect.

2.4 Neural plasticity

Brain imaging techniques are also valuable for evidencing plastic processes that result from developmental anomalies or from insults to brain tissue. The case of patient PK is unique and very eloquent in that respect. This patient showed a remarkable spontaneous reduction of her scotomas that were due to a bilateral anomaly of the occipital lobes. She was considered as clinically blind since the age of 5; her perimetry evaluation showing the loss of almost the entire visual field (except for a reduced quadrant in the right visual field including the midline). PK was ohptalmologically re-evaluated at the age of 26 and showed a remarkable recovery of vision (reduction of the hemianopic field) (Ptito et al, 1999a). To understand the process by which this patient had regained her vision, we submitted her to a series of brain imaging evaluations with ^{18}FDG-PET at rest, followed by $H_2{}^{15}O$ PET activation following stimulation of the left and right hemifields. The ^{18}FDG-PET investigation revealed an abnormal perfusion of the occipital lobes with a predominance in the right hemisphere (left visual field) (see figure 1H) which led us to believe that the recovery of the entire right visual field was under the control of the left hemisphere which had the smaller damage (Ptito et al., 1999b). We therefore proceeded to test this neural re-organization in a PET brain activation paradigm. PK was scanned while she was submitted to a series of visual stimulations in both hemifields. Irrespective of the locus of stimulation, activation sites were found only in the visual areas of the hemisphere showing the smallest damage during the ^{18}FDG-PET investigation (left hemisphere) (Figure 1 I and J). More importantly, the pulvinar on the side contrateral to the hemisphere with the largest damage showed a significant rCBF increase in rCBF. These PET results support the neural plasticity hypothesis put forth in animal studies on hemispherectomy concerning the contribution of the remaining hemisphere to residual vision and the pivotal role played by the colliculo-pulvinar pathway.

2.5 Sensory substitution in the blind

Using brain imaging techniques, it has been possible to demonstrate in man the existence of a phenomenon that is well studied in animals, namely sensory substitution (see Ptito et al., 2001; Bach-y-Rita, this volume). Sensory substitution refers to the replacement of one sensory input (vision, audition, somesthesis) by another, while preserving some key functions of the original sense. This cross-modal plasticity has been largely documented in animals as well as human models. For example, blind subjects show activation of primary and secondary visual cortical areas during Braille reading (reviewed in Pascual-Leone and Hamilton, 2001). Lesions in the visual cortex leads to Braille alexia emphasizing the role of the visual cortex in Braille reading in early blindness (see Pascual-Leone and Hamilton, 2001).

In recent years, many efforts have been devoted to develop devices that can convey visual information in blind subjects. Non-invasive techniques such as sensory substitution prosthesis via auditory or tactile (Sampaio et al, 2000) inputs have been successfully used. Invasive and more costly systems are also available and include implantable devices for electrical stimulation of the visual cortex, the optic tract or the retina. Although they offer a potential for sensory replacement, these methods require surgery and are very costly.

We have used a newly developed human-machine interface (Bach-y-Rita et al., this volume and Figure 2A), the Tongue Display Unit (TDU) that has proven its efficiency in the blind. The tongue has been chosen because of its conductance for electrical stimulation, its sensitivity, accessibility and large cortical representation within the somatosensory cortex. In our studies, we tested whether the tongue can convey somatosensory information to the visual cortex of congenitally blind subjects. We used PET during an orientation discrimination task of a letter applied to the tongue in normal subjects and congenitally blind subjects. Blind subjects demonstrated a significant rCBF elevation in visual cortical areas (Figure 2B, upper portion) whereas control subjects showed a significant rCBF decrease in the same areas (Figure 2B, lower portion). These results support the notion that cortical visual areas can be activated by another sensory modality and that the tongue is a viable organ to convey information to the visual cortex. The data are reminiscent of findings by Sadato et al (1996).

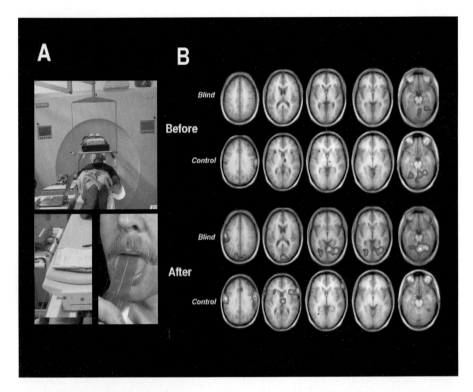

Figure 2. Cross-modal plasticity. A- Set-up and tongue stimulator. B- PET results showing pre- and post-training rCBF changes in normal controls and blind subjects. Note the activation of the visual cortex only in blind subjects after training (see text for details).

Investigated in blind and sighted subjects in a Braille tactile discrimination task. It thus seems that the prolonged use of the tongue, as is the case with the fingers in Braille readers, enhances the plastic processes that occur during sensory deprivation. Our PET results confirm psychophysical data obtained with tactile stimulation of the tongue showing this organ as a convenient medium to convey somatosensory information to the visual cortex, enabling blind people to « see ».

2.6. Imaging Pain

PET and fMRI studies have greatly furthered our understanding of the role of the cerebral cortex in normal and pathological pain processing. They have convincingly shown that upon application of a painful stimulus, activity increases in a number of cortical and subcortical structures, the so-called pain-matrix. Among the areas that show the most consistent response to a painful stimulus are the insula, the anterior cingulate cortex, thalamus, cerebellum and primary and secondary somatosensory cortices. The vast majority of the PET and fMRI studies have dealt with the processing of acute, experimental pain. Whereas the early studies largely focussed on mapping the areas that light up by a painful stimulus, more recent studies have used more sophisticated study-design that allow to evaluate the effect of cognitive and emotional factors on pain processing in the brain.

These studies have shown that the brain activation pattern produced by a painful stimulus depends as much on the psychological context (attention, anticipation, anxiety) under which the pain is presented as the stimulus parameters. For instance, it was shown that anticipation of a painful stimulation can produce activity in the same areas as produced by the painful stimulus itself. Further illustrations of the top-down modulation of pain come from studies in which attentive processes were manipulated. These studies showed that directing the attention away from a painful stimulus results in increased activity in the brainstem periaqueductal-grey (PAG) (Tracey et al., 2002). The increased activity in the PAG was shown to be inversely correlated with subject's pain ratings, indicative for a role of this structure in the endogenous modulation of pain.

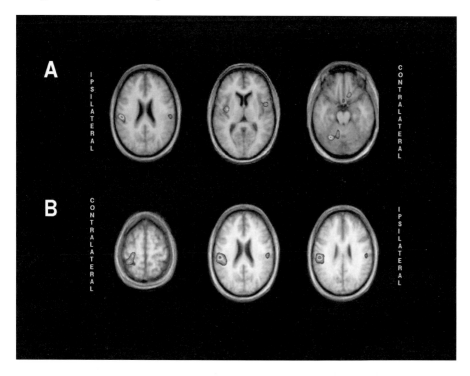

Figure 3: rCBF increases during brushing of the allodynic area (A) and the homotopic contralateral side (B) in patients with pain and allodynia after peripheral nerve lesion. The indications ipsilateral and contralateral are with reference to the site of brushing.

Whereas there is a relative consensus on the brain areas involved in the processing of acute pain, things are less clear for chronic forms of pain. The first studies on chronic pain gave evidence for altered processing in neuropathic pain conditions. More specifically, they showed a hypoactivity of the contralateral thalamus. However, some later studies suggested an amplification of the thalamic response to allodynic stimulation. To further address this question, we recently studied a series of nine patients with pain after peripheral nerve lesion. All patients suffered spontaneous pain and brush-evoked allodynia. Patients were scanned at rest, during allodynia, provoked by slight brushing of the affected nerve territory, and during brushing the homologous healthy contralateral area. All patients reported that brushing the affected side resulted in increased pain sensation. Allodynic brushing was associated with significant rCBF increases in the ipsilateral posterior insular and secondary somatosensory cortices, prefrontal cortex and cerebellum. No rCBF increases were observed in contralateral primary somatosensory cortex, thalamus or anterior cingulate cortex. This pattern of increased ipsilateral insular and secondary somatosensory cortical activation and the lack of anterior cingulate activation is at odds with the activation pattern observed in acute forms of pain and is indicative for cortical plasticity.

The results confirm our previous finding in a patient with neuropathic pain in whom we also observed insular and prefrontal rCBF increases and a lack of anterior cingulate, primary somatosensory cortical and thalamic activity (Kupers et al., 2000). Another chronic pain condition which has consistently been associated with an atypical brain activation pattern is cluster headache. Cluster headache is a debilitating form of chronic pain in which pain-free periods are alternated with cyclic bouts of extreme severe headache. PET water activation studies have shown a consistent rCBF increase in the hypothalamus during attacks of cluster headache (May et al., 1999). Other forms of headache such as migraine are not associated with hypothalamic activity. Studies using the technique of voxel-based morphometry (VBM) gave further support for a role of the hypothalamus in cluster headache. VBM is based on the analysis of structural T1-weighted MRI scans. The technique statistically compares anatomical differences in brain morphometry between groups of subjects and tries to correlate these with differences in behaviour or traits. Using this method, a significant structural difference in grey matter density in the hypothalamus was found in patients with cluster headache when compared to healthy volunteers. Together these findings argue for a pathogenic role of the hypothalamus in cluster headache. Phantom limb pain is another example of a chronic pain condition that has been associated with altered cortical processing. Studies have shown that patients with painful phantom limbs have an altered somatotopy in primary somatosensory cortex, indicative of plastic changes in parietal cortex. The altered maps in primary somatosensory cortex were abolished by treatments that successfully relieved the phantom pain. Moreover, no cortical reorganization was observed in amputees without phantom limb pain (Flor, 2000).

3. CONCLUSIONS

PET and fMRI have proved to be valuable tools for the study of the normal and diseased brain. fMRI has gained popularity in recent years in cognitive neuroscience because it is uniquely non-invasiveness, has good sensitivity and gives a good temporal and spatial resolution. It offers moreover a great potential for the study of brain development in children, the acquisition of skills and learning in subjects of any age. PET on the other hand will realize its full potential in the molecular imaging of the brain, either for research or as a diagnostic tool. Receptor and pathway imaging will allow for the understanding of neurochemical unbalances seen in many diseases such as dementia, schizophrenia, Parkinson and depression as well the development of new drugs.

REFERENCES

Bittar, R., Ptito, M., Faubert, J., Dumoulin, S.O. and Ptito, A. (1999). Activation of the remaining hemisphere following stimulation of the blind hemifield in hemispherectomized patients. NeuroImage, 10:339-346.

Boire, D., Theoret, H. and Ptito, M. (2001). The retinofugal projection system in the early hemicorticectomized green monkey. Prog. Brain Res., 134:379-398.

Faubert, J., Diaconu, V., Ptito, A. and Ptito, M. (1999). Residual vision in the blind field of hemidecorticated humans predicted by a diffusion scatter model and selective spectral absorption in the human eye. Vis. Res. 39: 149-157.

Fendrich, R., Wessinger, C.M. and gazzaniga, M. (2001). Speculations on the neural basis of islands of blindsight. Prog. Brain Res., 134:353-366.

Flor H, Phantom limb pain: characteristics, causes and treatments. Lancet Neurol 2000 (1): 182-189.

Kupers RC, Gybels JM, Gjedde A. Positron emission tomography study of a chronic pain patient successfully treated with somatosensory thalamic stimulation. Pain 2000 (87) :295-302.

Kupers, R; Fortin, A; Astrup, J; Gjedde, A and Ptito, M. Recovery of anterograde amnesia in a case of craniopharyngiome. Arch Neurol. (in press).

May A, Ashburner J, Buchel C, McGonigle DJ, Friston KJ, Frackowiak RS, Goadsby PJ. Correlation between structural and functional changes in brain in an idiopathic headache syndrome. Nat Med 1999 (5): 836-838.

Pascual-Leone, A. and Hamilton,R. (2001). The metamodal organization of the brain. Prog. Brain Res., 134: 427-446.

Ptito, M., Boire, D., Frost, D.O. and Casanova, C. (2001). When the auditory cortex turns visual. Prog. Brain Res., 134:447-458.

Ptito, M., Dalby, M. and Gjedde, A. (1999a). A case of abnormal development of both occipital lobes without cortical blindness. Acta Neurol. Scand., 99:252-254.

Ptito, M., Johanssen, P., Faubert, J. and Gjedde, A. (1999b). Activation of human extrageniculostriate pathway after damage to area V1. NeuroImage, 9:97-107.

Ptito A., Fortin, A. and Ptito, M (2001). "Seeing" in the blind hemifield of hemispherectomy patients. Prog. Brain Res., 134:367-378.

Sadato,, N., Pascual-Leone, A., Grafman, J, Ibanez, V., Deiber, M.P., Dold, J. and Hallett, M. (1996). Activation of the primary visual cortex by Braille reading in blind subjects. Nature, 380:526-528.

Sampaio, E., Maris, S. and Bach-y-Rita, P. (2001). Brain plasticity: 'visual' acuity of blind persons via the tongue. Brain Res. 908:204-207.

Stoerig, P. and Cowey, A. (1997). Blindsight in Man and Monkey. Brain, 120:535-559.

Tracey I, Ploghaus A, Gati JS, Clare S, Smith S, Menon RS, Matthews PM. Imaging attentional modulation of pain in the periaqueductal gray in humans. J Neurosci 2002 (22):2748-2752.

Turner,R. and Jones,T. (2003). Techniques for imaging neuroscience. Brit. Med. Bull. 65: 3-20.

Van der Weirf YD, Witter MP, Uylings HBM, Jolles J. (2000). Neuropsychology of infarction in the thalamus : a review. Neuropsychologia; 38 :613-627.

J.L. CANTERO AND M. ATIENZA

Laboratory Andaluz of Biology, Division of Neuroscience, University Pablo de Olavide, Seville, Spain

41. FUNCTIONAL ASSESSMENT OF HUMAN BRAIN WITH NON-INVASIVE ELECTROPHYSIOLOGICAL METHODS

Summary. Conventional electroencephalography (EEG) and event-related brain measurements (ERPs) provide unique non-invasive approaches to study *in-vivo* human cerebral functions with a temporal resolution of milliseconds. These techniques have proved to be helpful to assess cognitive abilities, differentiate between brain states, and support the diagnosis in neurological and psychiatric diseases. Human neurophysiological correlates of learning have also been revealed by task-dependent increases in gamma EEG coherence and amplitude modulations of specific brain evoked components at different levels of information processing. Changes in spontaneous and evoked brain responses can be also used to explore the role of sleep in both sensory function and memory consolidation. All together, these lines of evidence suggest that EEG and ERP techniques will remain omnipresent in those research and clinical environments devoted to assess certain human brain functions with non-invasive neurophysiological techniques.

1. INTRODUCTION

A broad range of methodological approaches in neuroscience has established that the brain is organized in multiple functional units at different spatio-temporal scales. It is commonly believed that cognition intimately depends on the functioning of the cerebral cortex. Thus, many important cortical functions result from the coordinated cooperation of different neuronal ensembles, and can be measured with recording techniques targeted at this level. The electroencephalogram (EEG) is the most inexpensive, non-invasive method for determining how multiple populations of neurons distributed over many cortical regions interact to control human behaviour and cognition in health and disease. Understanding the neural basis of cognition, therefore, requires knowledge of cortical operations at different organizational levels.

EEG visual inspection has been largely used in both research and clinical contexts to establish objective differentiation between different cerebral states ranging from active wakefulness to brain death. Traditionally, EEG has also been used in the diagnosis of some neurological, neurodegenerative, and psychiatric disorders (extensive reviews can be found in Niedermeyer and Lopes da Silva, 1993). Recent advances in quantitative analysis of EEG signals have provided important insights into neural mechanisms involved in pathological conditions such as Alzheimer's disease, schizophrenia, and epilepsy. The present chapter reviews these recent results derived from the application of these quantitative EEG analysis techniques to neurological and psychiatric disorders with high prevalence in the population.

EEG changes are not always spontaneous; sometimes they are related to particular sensorimotor and/or cognitive events. As these event-related changes are often much smaller than the ongoing EEG activity, it is necessary to average a high number of time-locked brain responses. The resulting waveforms are known as event-related potentials (ERPs).

Like spontaneous field potentials, some ERP — especially those occurring at subcortical level within the first 50 ms from stimulus onset — have been videly used in clinical practice (Chiappa, 1989). Recent evidence suggests that ERP components beyond 50 ms, which are mainly a reflection of cortical activation and cognitive function, may become valid tools for studying the cerebral substrate of human brain function and dysfunction. In the sections below, we review those late ERP components that have yielded the most promising results in the diagnosis and prognosis of some neurological and psychiatric disorders.

Determining neural mechanisms involved in integrating new inputs with information already present in our memory is one of the most intriguing issues in contemporary neuroscience. Learning or establishment of consolidated memories, requires a complex cascade of neural events occuring over different spatio-temporal scales. Recent evidence suggests that the slowest of these events are sleep-dependent. In other words, sleep seems to be necessary for some new memories to become stabilized (i.e., resistant to interference) and strengthened into longer-lasting forms of memory. This perspective of sleep as an experience-dependent phenomenon contrasts with those hypotheses that consider sleep a use-dependent mechanism directed at preserving neuronal synapses that have been underused during the prior wakefulness. Both EEG and ERP techniques have helped to advance knowledge of the neural mechanisms of memory consolidation and sleep functions, and will be reviewed in the two last sections of the present chapter.

2. ELECTROPHYSIOLOGICAL CORRELATES OF CEREBRAL ILLNESS: THE DISORDERED BRAIN

2.1. Detecting failures in spontaneous cortical dynamics

EEG generation stems from the short- and long-range interactions between neuronal populations located in different cortical regions. The formation of dynamic links between functionally specialized brain regions at different frequency codes has been proposed as the fundamental mechanism in the emergence of behavioural states and cognition (Varela et al., 2001). From this conceptual framework, it has been hypothesized that changes in EEG coherence patterns between cortical regions denote global changes in cortical function underlying structural abnormalities and behavioural deficits observed in different brain pathologies. This assumption is supported by previous studies reporting a drastic fall in the interhemispheric EEG coherence associated with genuine alterations in functional connectivity in patients with agenesis of the corpus callosum (Kuks et al., 1987; Nielsen et al., 1993), or after partial callosotomy (Montplaisir et al., 1990).

Alzheimer's disease (AD) involves senile plaques and neurofibrillary tangles distributed among the origins and terminations of long cortico-cortical association fibres, which has led the disease to be conceptualized as a neocortical disconnection syndrome (Morrison et al., 1986). This impairment in connectivity is associated with the death of pyramidal neurons that provide the cortico-cortical projections (Pearson et al., 1985). In support of this notion, different studies have shown aberrant patterns of long-range EEG coherence in AD during quiet wakefulness. For instance, Leuchter et al. (1992) reported a global decreased coherence between electrodes located in both the origins and terminations of long cortico-cortical fibre tracts in AD patients as compared with multi-infarct dementia patients and aged-paired control group. Subsequent studies have replicated these findings, but the reduced coherence in AD patients was restricted to the alpha band (e.g., Locatelli et al., 1998). Given that AD is a progressive neurodegeneration, long-range EEG coherence patterns may serve as a trait marker at the time of the first clinical assessment, to evaluate the effectiveness of some therapeutic strategies, and correlate changes over time with cognitive deterioration during the entire course of the disease.

Disruptions and/or abnormalities in cortico-cortical communication also correlate well with some symptoms typical of psychiatric disorders. Evidence suggests that EEG coherence might help in determining potential electrophysiological markers of schizophrenia and major depressive disorder. In general, intra- and interhemispheric coherences are abnormally greater in both schizophrenic patients (Merrin et al., 1989) and subjects at elevated risk (Mann et al., 1997) than in healthy controls. Although high levels of EEG coherence are hard to interpret in brain disorders, increased synchronization levels observed in schizophrenic patients might account for the failure to suppress or inhibit dominant associations (Anderson and Spellman, 1995).

While schizophrenic subjects show volume reduction in grey matter, depressed patients have extensive white-matter lesions as revealed by MRI scans. Some of these lesions have been associated with lower EEG coherence and with significantly poorer outcomes of treatment for depression at 2-year follow-up (Leuchter et al., 1997). Sleep-EEG coherence abnormalities were also recently reported in non-mentally-ill individuals at high risk for depression, pointing to sleep-EEG coherence markers as useful clinical indicators in family genetic studies of mood disorders (Fulton et al., 2000).

One of the major current fields in epilepsy research is focused on the prediction of seizures underlying intractable epilepsy. Spectral analysis methods are not sensitive enough to detect pre-ictal EEG changes. However, recent findings using measurements of non-linear changes in EEG signals allowed the anticipation of a seizure several minutes before it occurred (mean 7 min) in 25 out of 26 epileptic recordings (Le Van Quyen et al., 2001). Epileptic events are typically displayed as transient EEG patterns intermingled within the EEG background, which makes the use of non-stationary time-frequency approaches suitable for characterizing the spectro-temporal dynamics of the different stages of a seizure. Wavelet transforms are also appropriate to detect scalp EEG transients in the time-frequency domain, as revealed by recent results obtained with inter-ictal transient events (Senhadji and Wendling, 2002). In addition, time-frequency decomposition methods based on matching pursuit algorithms have proved to be a reliable tool for detecting seizure occurrence in long-term recordings, for differentiating seizures from artifacts on a multi-channel basis, and for examining patterns of seizure propagation as well (Jouny et al., 2003; Bergey et al., 2001).

2.2 Detecting failures in event-related cortical dynamics

ERPs provide real-time evidence of altered processing of external events. Among the extensive range of cognitive capabilities of the brain that can be impaired in different neurological and psychiatric disorders, ERPs have been demonstrated to be especially fruitful in evaluating sensory gating, sensory memory, perception, distractibility, and semantic processing.

The normal brain has the innate ability to filter out, or "gate", irrelevant sensory information. This automatic capacity of the brain's is generally measured by the attenuation of a positive vertex wave in response to the repetition of a tone within a pair. The positive waveform is known as P50, because it is recorded at about 50 ms from stimulus onset. Sensory gating abnormalities, as revealed by a reduced P50 suppression, have been tested mostly in schizophrenic patients, who suffer from — among other symptoms — dysfunction in the maintenance of attention and selective processing of sensory information. Typically, the activity elicited by the second stimulus in a pair is not attenuated in individuals with schizophrenia, unless they are taking atypical antipsychotic medication that improves sensory gating. These findings, together with the increasing understanding of the neural mechanisms involved in this phenomenon (Adler et al., 1998), make the P50 brain response a valid tool in the assessment of gating impairments in schizophrenic patients.

While the sensory gating paradigm evaluates the event-related brain response to repetition of the stimulus, other paradigms measure the brain-evoked response to discernible changes in a repetitive stimulus sequence. This procedure, named "oddball paradigm", has been used under passive (unattended) and active (attended) conditions. Each one of these versions provides relevant information on different stages of stimulus processing. For instance, when some repetitive aspect of auditory (or visual) stimulation is interrupted infrequently in any way, the change is automatically registered by the brain. This automatic change-detection is indexed by a negative ERP component called *mismatch negativity* (MMN). It is automatic because it can be elicited when attention is directed elsewhere. The MMN results from activation of modality-specific and frontal-unspecific generators to the incoming stimulus if memory traces for the repetitive aspects of stimulation are still active in sensory memory (for a recent review, see Näätänen et al., 2001). Aging appears to reduce the duration of auditory sensory memory, as revealed by the abnormal interstimulus-interval effect on MMN amplitude (Pekkonen, 2000). As expected from these results, the MMN amplitude is dramatically diminished in Alzheimer's disease, when the interstimulus interval is prolonged from 1 to 3 s (Pekkonen et al., 1994).

The MMN attenuation observed in schizophrenic patients seems to reflect their altered sound perception rather than a fast decay of the memory trace. In support of this notion, Javitt (2000) found that the reduced MMN amplitude and the poor performance of schizophrenic subjects correlated well with the increase in severity of their negative symptoms. Recent results further encourage using MMN as an index of the genetic predisposition to (Michie et al., 2002) and prognosis of schizophrenia (Javitt et al., in press).

The MMN is the ontologically earliest cognitive ERP component that can be recorded from the human brain. This makes it highly valuable as an early marker of developmental disorders in infants. Leppänen et al. (in press) have found that the discrimination accuracy of the duration of speech stimuli is significantly reduced in 6-month-old infants with dyslexic close relatives, as compared with control infants.

Results mentioned above are only some examples of the clinical applications of the MMN, but many other studies suggest that it could be a potential clinical tool in a broad range of populations (Näätänen, 2003). One of the most promising clinical applications is assessment of coma prognosis. Indeed, the MMN is considered the earliest indicator of awakening from coma (Kane et al., 1993).

As already mentioned, the MMN is typically elicited in oddball paradigms under unattended conditions. The same kind of paradigm may require attention; then, other late ERP components, like the P300 and the N400 become more interesting in evaluating further cognitive processes. Like the P50 and the MMN, the P300 and N400 have also been extensively used to evaluate anomalies in schizophrenia. The left temporal auditory P300 has shown abnormalities in both amplitude and topographical distribution not only in chronic patients but also at the first psychotic episode. The former deficit is positively correlated with the grey-matter volume of the underlying left superior temporal gyrus, and inversely correlated with the extent of delusions and thought disorders (for a review, see O'Donell et al., 1999). Likewise, studies with the N400, using sentence terminal words, which are either congruent or incongruent with the context of the preceding portion of the sentence, suggest that schizophrenic patients do not use context efficiently. In other words, the N400 amplitude to both congruent and incongruent sentence endings is often larger in schizophrenia than in healthy subjects (for a review, see McCarley et al., 1999).

All the ERP components reviewed above — the P50 attenuation response, the MMN, the P300, and the N400 — can even be measured during REM sleep (P50: Kisley et al., 2001; MMN: Atienza and Cantero, 2001; P300: Cote and Campbell, 1999, N400: Brualla et al., 1998). This finding is especially important because its allow the separation of state-dependent effects on these electrical brain responses and, hence, enhance the feasibility of these functional tests in clinical practice. Kisley et al. (2003) have already provided the first evidence demonstrating that cognitive dysfunction can be evaluated during REM sleep. They found that sensory gating impairment, indexed by P50 suppression, persists during REM sleep in schizophrenic patients. This finding encourages assessment of this and other cognitive functions during REM sleep in different clinical populations, especially to overcome the potential confusion in brain state.

3. ELECTROPHYSIOLOGICAL CORRELATES OF LEARNING: THE EMERGENT BRAIN PROPERTIES

Studying neural correlates of learning and memory is a prolific area of research in cognitive neuroscience. Non-invasive electrophysiological tools have provided an exceptional opportunity for determining formation of dynamic links mediated by the synchronization of neural activity in different frequency bands at short- and large-scale integration during acquisition and consolidation of new memories.

Both associative and procedural learning have been associated with changes in human brain dynamics of high-frequency activity. In the two cases, variations in the properties of neurons and in the functional organization of the cerebral cortex seem to account for the different stages of learning and memory consolidation (Gilbert et al., 2001). Evidence has been found for increases in the gamma (37-43 Hz) coherence pattern between different cortical regions involved in the processing of stimuli presented to different sensory modalities during an associative-learning procedure (Miltner et al., 1999). These findings strongly support the notion that associative learning stems from a progressive practice-dependent strengthening of cell assemblies, and point to gamma as the *electrophysiological dialect* involved in the establishment of new synaptic linkages. Similar enhancement in gamma coherence (around 40 Hz) was seen over posterior cortical regions in response to visual-stimulus feedback during an instrumental shaping procedure (Keil et al., 2001).

The increased gamma EEG response (around 36 Hz) has also been proposed as the main electrophysiological correlate of perceptual learning (Gruber et al., 2002). In that experiment, fragmented pictures of an object could easily be identified after the presentation of an unfragmented version of the same picture (see right panel of Figure 1). Results showed an increase in spectral gamma power at parietal electrode sites for identified pictures in a time window from 180 to 300 ms; this was not present when the picture was unidentified (upper traces in the left panel of Figure 1). Furthermore, neural activity recorded from these posterior electrodes was highly synchronized when perceptual learning took place (lower traces in the left panel of Figure 1). Functional coupling and decoupling between different cortical regions have also been reported to vary within different frequency ranges across the initial stages of motor learning. Thus, the improved behaviour observed during the acquisition of a motor co-ordination pattern was accompanied by a decrease in EEG coherence between the primary sensorimotor areas, and over the midline regions in the alpha and beta band, as well as by an enhanced interhemispheric connectivity between prefrontal areas within the gamma range (Serrien et al., 2003). Based on these results, across-time dynamic changes in cortico-cortical connections might reflect the experience-driven neural changes underlying the acquisition phase of procedural learning.

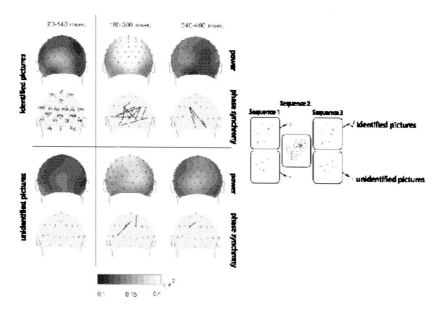

Figure 1. Right panel: Schematic representation of the three sequences of stimuli used in the experiment. Subjects identified pictures in the third sequence only if the stimulus matched the one presented in the immediately prior sequence. Left panel: Grand mean (N=9) topographical distribution and phase synchrony at 36.62 Hz wavelet for identified (top) and unidentified pictures (bottom). Modified with permission from Gruber et al. (2002).

Immediate gains in performance in procedural tasks are followed by additional gains, either with or without further training, over the following hours or days. These differences in the time course of perceptual learning can be explained by changes in brain functional organization within different time scales. The slower neural events responsible for the additional improvement are believed to include enlargement of the cortical areas representing the learned information, changes in the synchronization of pyramidal neurons, and a shift in the locus of representation from higher to lower areas of sensory processing, suggesting alterations in the brain mechanisms mediating attentional control (Gilbert, 1994).

ERP studies have shown not only that neurophysiological changes precede behavioural improvement (Tremblay et al., 1998), but also that different brain dynamics underlie the fast and slow components of perceptual learning (Atienza et al., 2002; Song et al., 2002). The fastest of these changes are revealed by a rapid increase in the amplitude of sensorimotor components of ERPs. However, the slow development of cortical dynamics is characterized by an enhancement of unspecific, frontal-evoked components, suggesting a progressive increase in both familiarity and automaticity. Indeed, subjects who were trained in a single session to discriminate two complex sound patterns showed further gains in performance several hours after training that matched specific changes in the cortical dynamics (Atienza et al., 2002). Some of these changes (the appearance of a slow positive wave at about 200 ms from stimulus onset 24 h after the training session) were seen for both the standard and the deviant pattern, and were interpreted as signalling a general increase in the familiarity with the sound patterns; while others (increase in the amplitudes of the MMN and P3a at 36 and 48 h, respectively) were specific for the deviant pattern (Figure 2). The latter changes are thought to be a reflection of processes mediating automatic shifts of attention towards a change in the sensory environment when attention is focused elsewhere. In other words, evidence from ERP studies suggests that the consolidation phase of learning not only improves performance when attention is focused on the task, but also increases the probability that a potentially meaningful stimulus outside the focus of attention can access consciousness for further processing. This is possible because the slow development of neural events evolving between sessions makes information processing more automatic, regardless of the focus of attention.

2. ELECTROPHYSIOLOGICAL CORRELATES OF SLEEP FUNCTIONS: THE SLEEPING — BUT NOT IDLING — BRAIN

The question of the function/s of sleep remains a biological enigma. Many studies have been designed to determine the role of sleep in brain restoration and recovery, energy conservation, memory consolidation, maintenance of synaptic balance, and so on. However, to date, there are no conclusive findings about the involvement of sleep in only one function. One plausible hypothesis postulates that sleep promotes changes from molecular to system levels serving to maintain the synaptic efficacy and brain integrity. Within this framework, sleep has been conceptualised as a use-dependent evolutionary mechanism designed to preserve brain function after the exposure to waking experiences (Krueger et al., 1995).

Only a few human studies have supported the use-dependent function of sleep. These studies have revealed modifications in human sleep architecture and cortical dynamics during the sleep period following a prolonged exposure to sensory stimulation (Kattler et al., 1994, Cantero et al., 2002a, 2002b). In the studies of Cantero et al.'s studies, passive exposure to six hours of unilateral auditory stimulation during wakefulness significantly increased the total duration of slow-wave sleep in the subsequent sleep period (Figure 3A). Only a few human studies have supported the use-dependent function of sleep. These studies have revealed modifications in human sleep architecture and cortical dynamics during the sleep period following a prolonged exposure to sensory stimulation (Kattler et al., 1994, Cantero et al., 2002a, 2002b). In the studies of Cantero et al.'s studies, passive exposure to six hours of unilateral auditory stimulation during wakefulness significantly increased the total duration of slow-wave sleep in the subsequent sleep period (Figure 3A). Modifications in the slow-wave sleep architecture were accompanied by an increase in the level of synchronization within the 8-14 Hz frequency range (Figure 3B) and EEG coherence between fronto-temporal regions to the detriment of the coherence between temporal and posterior cortical regions (Figure 3C).

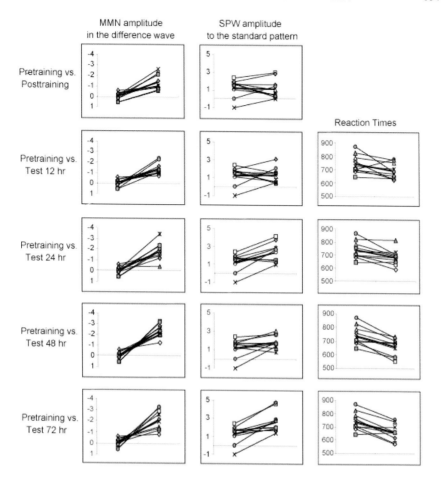

Figure 2. Superimposed individual (thin lines with different symbols) and group (thick line without symbols) results corresponding to the amplitude of the MMN elicited by the infrequent sound pattern (left column) and the slow positive wave (SPW) elicited by the frequent pattern (middle column) just before training in the discrimination task as compared with the different subsequent post-training tests for a period of 72 h. Mean reaction times in the different tests as compared with mean reaction times obtained in the last two blocks of the training session are shown in the right column. Modified from Atienza et al., (2002).

All electrophysiological changes were restricted to post-stimulation slow-wave sleep, and occurred independently of the side of auditory stimulation (Cantero et al., 2002b). These results strongly support the use-dependent hypothesis of sleep function, pointing to slow-wave sleep as a critical brain state for such synaptic reorganization after overloading of a specific sensory system.

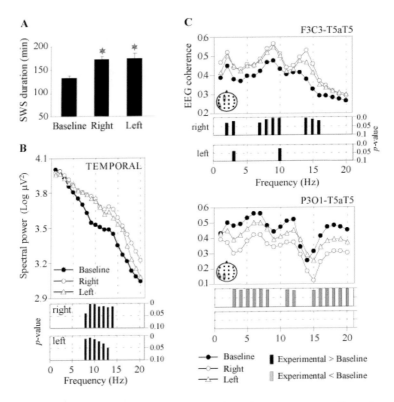

Figure 3. Use-dependent aspects of sleep. Prolonged exposure to unilateral auditory stimulation during wakefulness increased the duration of slow-wave sleep the subsequent night (A), produced a higher spectral power within the alpha and sleep spindle band in the auditory cortex (B), and caused complex changes in EEG coherence between auditory cortex and other sensory regions (C). A. Modified from Cantero et al. (2002a). B and C modified from Cantero et al. (2002b).

Another attractive hypothesis postulates that sleep serves to stabilize and strengthen new memories acquired during the prior wakefulness. Evidence supporting this notion comes mainly from neurophysiological studies in animals and behavioural studies in humans. To date, only one study in adult humans has provided insights into the effects of sleep on the cortical dynamics underlying the slow consolidation phase of perceptual learning (Atienza et al., in press). This study showed that posttraining sleep deprivation prevents part of the slow neuronal changes associated with learning consolidation, specifically those related with enhanced automaticity when attention is not focused on the task. Thus, the MMN - already evident immediately after training on an auditory discrimination task - evolved, together with the P3a, over the next two days in parallel with behavioural improvement (left panel of Figure 4). However, sleep deprivation prevented these changes in cortical dynamics from occurring (right panel of Figure 4). Whether the lack of these changes is reversible with the simple passage of time, without further training, has to be investigated in further studies. Likewise, it is necessary to distinguish between those neuronal changes occurring during sleep that account for the use- and experience-dependent aspects of sleep. Once these neural events have been disentangled, it will be feasible to study their relationship with the spatio-temporal dynamic changes of different memory systems.

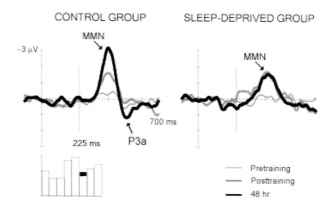

Figure 4. Grand average difference waves obtained after subtracting ERPs elicited by the frequent sound pattern from those elicited by the infrequent pattern before training in the discrimination task (pretraining), just after training (post-training), and 48 h later for both the control and the sleep-deprived group. The sound patterns consisted of different frequency tones that varied in the frequency of the sixth tone (at 225 ms from stimulus onset). Note the significant increase in the MMN/P3a at 48 h in the control group but not in those subjects who were deprived of sleep the night following training. Modified from Atienza et al. (in press).

5. CONCLUDING COMMENTS

Neuroelectric measurements play a major role in providing a realistic and functional view of brain electrical dynamics in human. Spontaneous EEG recordings and ERP studies have been progressively incorporated into the standard procedures of clinical diagnosis and prognosis in different brain disorders and cognitive dysfunctions. The effectiveness of these methods has been greatly improved with the inclusion of recent advances in quantitative EEG analysis and sophisticated cognitive paradigms to assess different brain functions and abilities in health and disease. Integration of data coming from other neuroimaging techniques with non-invasive electrophysiological recordings will in the near future provide unique insights into the description of the functional brain networks involved in human cognition.

REFERENCES

Adler, L.E., Olincy, A., Waldo, M., Harris, J.G., Griffith, J., Stevens, K., Flach, K., Nagamoto, H., Bickford, P., Leonard, S. & Freedman, R. (1998) Schizophrenia, sensory gating, and nicotinic receptors. Schizophr. Bull., 24, 189-202.

Anderson, M.C. & Spellman, B.A. (1995) On the status of inhibitory mechanisms in cognition: memory retrieval as a model case. Psychol. Rev., 102, 68-100.

Atienza, M. & Cantero, J.L. (2001) Complex sound processing during human REM sleep by recovering information from long-term memory as revealed by the mismatch negativity (MMN). Brain Res., 901, 151-160.

Atienza, M., Cantero, J. L. & Dominguez-Marin, E. (2002) The time course of neural changes underlying auditory perceptual learning. Learn. Mem., 9, 138-150.

Atienza, M., Cantero, J. L. & Stickgold R. (in press) Posttraining sleep enhances automaticity in perceptual discrimination. J. Cogn. Neurosci.

Bergey, G.K. & Franaszczuk, P.J. (2001) Epileptic seizures are characterized by changing signal complexity. Clin. Neurophysiol., 112, 241-249.

Brualla, J., Romero, M.F., Serrano, M. & Valdizan, J.R. (1998) Auditory event-related potentials to semantic priming during sleep. Electroencephalogr. Clin. Neurophysiol., 108, 283-290.

Cantero, J.L., Atienza, M. & Salas, R.M. (2002a) Effects of waking-auditory stimulation on human sleep architecture. Behav. Brain Res., 128, 53-59.

Cantero, J.L., Atienza, M., Salas, R.M. & Domínguez-Marin, E. (2002b) Effects of prolonged waking-auditory stimulation on electroencephalogram synchronization and cortical coherence during subsequent slow-wave sleep. J. Neurosci, 22, 4702-4708.

Chiappa, K.H. (1989). Evoked potentials in clinical medicine (2nd ed.). Raven Press, New York.

Cote, K.A. & Campbell, K.B. (1999) P300 to high intensity stimuli during REM sleep. Clin. Neurophysiol., 110, 1345-1350.

Fulton, M.K., Armitage, R. & Rush, A.J. (2000) Sleep electroencephalographic coherence abnormalities in individuals at high risk for depression: a pilot study. Biol. Psychiatry, 47, 618-625.

Gilbert, C. D. (1994) Learning. Neuronal dynamics and perceptual learning. Curr. Biol., 4, 627-629.

Gilbert, C. D., Sigman, M. & Crist, R. E. (2001) The neural basis of perceptual learning. Neuron, 31, 681-697.

Gruber, T., Muller, M.M. & Keil, A. (2002) Modulation of induced gamma band responses in a perceptual learning task in the human EEG. J. Cogn. Neurosci., 14, 732-744.

Javitt, D.C. (2000) Intracortical mechanisms of mismatch negativity dysfunction in schizophrenia. Audiol. Neurootol., 5, 207-215.

Javitt, D.C., Shelley, A.-M., Silipo, G., Lieberman, J.A. (in press) Deficits in AX-continuous performance test and mismatch negativity generation in schizophrenia: defining the pattern.

Jouny, C.C., Franaszczuk, P.J. & Bergey, G.K. (2003) Characterization of epileptic seizure dynamics using Gabor atom density. Clin. Neurophysiol., 114, 426-437.

Kattler, H., Dijk, D.J. & Borbely, A.A. (1994) Effect of unilateral somatosensory stimulation prior to sleep on the sleep EEG in humans. J. Sleep Res., 3, 159-164.

Kane, N.M., Curry, S.H., Butler, S.R. & Cummins, B.H. (1993) Electrophysiological indicator of awakening from coma. Lancet, 341, 688.

Keil, A., Muller, M.M., Gruber, T., Wienbruch, C. & Elbert, T. (2001) Human large-scale oscillatory brain activity during an operant shaping procedure. Brain Res. Cogn. Brain Res., 12, 397-407.

Kisley, M.A., Olincy, A. & Freedman, R. (2001) The effect of state on sensory gating: comparison of waking, REM and non-REM sleep. Clin. Neurophysiol., 112, 1154-1165.

Kisley, M.A., Olincy, A., Robbins, E., Polk, S.D., Adler, E.E., Waldo, M.C. & Freedman, R. (2003) Sensory gating impairment associated with schizophrenia persists into REM sleep. Psychophysiology, 40, 29-38.

Krueger, J.M., Obal, F., Kapas, L. & Fang, J. (1995) Brain organization and sleep function. Behav. Brain Res., 69, 177-185.

Kuks, J.B.M., Vos, J.E. & O'Brien, M.J. (1987) Coherence patterns of the infant sleep EEG in absence of the corpus callosum. Electroenceph. clin. Neurophysiol., 66, 8-14.

Leppänen, P.H.T., Pihko, E., Eklund, K.M., et al. (in press) Brain responses reveal speech processing differences in infants at a risk for dyslexia.

Leuchter, A.F., Newton, T.F., Cook, I.A., Walter, D.O., Rosenberg-Thompson, S. & Lachenbruch, P.A. (1992) Changes in brain functional connectivity in Alzheimer-type and multi-infarct dementia. Brain, 115, 1543-1561.

Leuchter, A.F., Cook, I.A., Uijtdehaage, S.H., Dunkin, J., Lufkin, R.B., Anderson-Hanley, C., Abrams, M., Rosenberg-Thompson, S., O'Hara, R., Simon, S.L., Osato, S. & Babaie, A. (1997) Brain structure and function and the outcomes of treatment for depression. J. Clin. Psychiatry, 58, 22-31.

Le Van Quyen, M., Martinerie, J., Navarro, V., Boon, P., D'Have, M., Adam, C., Renault, B., Varela, F. & Baulac, M. (2001) Anticipation of epileptic seizures from standard EEG recordings. Lancet, 357, 183-188.

Locatelli, T., Cursi, M, Liberati, D., Franceschi, M. & Comi, G. (1998) EEG coherence in Alzheimer's disease. Electroenceph. clin. Neurophysiol., 106, 229-237.

Mann, K., Maier, W., Franke, P., Roschke, J. & Gansicke, M. (1997) Intra- and interhemispheric electroencephalogram coherence in siblings discordant for schizophrenia and healthy volunteers. Biol. Psychiatry, 42, 655-663.

McCarley, R.W., Niznikiewicz, M.A., Salisbury, D.F., Nestor, P.G., O'Donnell, B.F., Hirayasu, Y., Grunze, H., Greene, R.W. & Shenton, M.E. (1999) Cognitive dysfunction in schizophrenia: unifying basic research and clinical aspects. Eur. Arch. Psychiatry Clin. Neurosci., 249, 69-82.

Merrin, E.L., Floyd, T.C. & Fein, G. (1989) EEG coherence in unmedicated schizophrenic patients. Biol. Psychiatry, 25, 60-66.

Michie, P.T., Kent, A., Stienstra, R., Castine, R., Johnston, J., Dedman, K., Wichmann, H., Box, J., Rock, D., Rutherford, E. & Jablensky, A. (2002) Phenotypic markers as risk factors in schizophrenia: neurocognitive functions. Aust. N. Z. J. Psychiatry, 34, S74-85.

Miltner, W.H., Braun, C., Arnold, M., Witte, H. & Taub, E. (1999) Coherence of gamma-band EEG activity as a basis for associative learning. Nature, 397, 434-436.

Montplaisir, J., Nielsen, T., Cote, J. Boivin, D., Rouleau, I. & Lapierre, G. (1990) Interhemispheric EEG coherence before and after partial callosotomy. Clin. Electroenceph., 21, 42-47.

Morrison, J.H., Scherr, S., Lewis, D.A., Campbell, M.J., Bloom, F.E., Rogers, J. & Benoit, R. (1986) The laminar and regional distribution of neocortical somatostatin and neuritic plaques: implications for Alzheimer's disease as a global neocortical disconnection syndrome. In A.B. Scheibel, A.F. Wechsler & M.A. Brazier (Eds.), The biological substrates of Alzheimer's disease. (pp. 115-131). Academic Press, Orlando.

Näätänen, R. (2003) Mismatch negativity: clinical research and possible applications. Int. J. Psychophysiology, 48, 179-188.

Näätänen, R., Tervaniemi, M., Sussman, E., Paavilainen, P. & Winkler, I. (2001) "Primitive intelligence" in the auditory cortex. Trends Neurosci., 24, 283-288.

Niedermeyer, E. & Lopes da Silva, F.H. (1993) Electroencephalography: Basic principles, clinical applications, and related fields (3rd ed.). Lippincott Williams & Wilkins, Baltimore.

Nielsen, T., Montplaisir, J. & Lassonde, M. (1993) Decreased interhemispheric EEG coherence during sleep in agenesis of the corpus callosum. Eur. Neurol., 33, 173-176.

O'Donnell, B.F., McCarley, R.W., Potts, G.F., Salisbury, D.F., Nestor, P.G., Niznukiewicz, M.A., Bamard, J., Shen, Z.J., Weinstern, D.M., Bookstein, F.L. & Shenton, M.E. (1999) Identification of neural circuits underlying P300 abnormalities in schizophrenia. Psychophysiology, 36, 388-398.

Pearson, R.C., Esiri, M.M., Hiorns, R.W., Wilcock, G.K. & Powell, T.P. (1985) Anatomical correlates of the distribution of the pathological changes in the neocortex in Alzheimer's disease. Proc. Natl. Acad. Sci. USA, 82, 4531-4534.

Pekkonen, E. (2000) Mismatch negativity in aging and in Alzheimer's and Parkinson's diseases. Audiol. Neurootol., 5, 216-224.

Pekkonen, E., Jousmaki, V., Kononen, M., Reinikainen, K. & Partanen, J. (1994) Auditory sensory memory impairment in Alzheimer's disease: an event-related potential study. Neuroreport, 5, 2537-2540.

Senhadji, L. & Wendling, F. (2002) Epileptic transient detection: wavelets and time-frequency approaches. Neurophysiol. Clin., 32, 175-192.

Serrien, D.J. & Brown, P. (2003) The integration of cortical and behavioural dynamics during initial learning of a motor task. Eur. J. Neurosci., 17, 1098-1104.

Song, Y., Ding, Y., Fan, S. & Chen, L. (2002) An event-related potential study on visual perceptual learning under short-term and long-term training conditions. Neuroreport, 13, 2053-2057.

Tremblay, K., Kraus, N., & McGee, T. (1998) The time course of auditory perceptual learning: neurophysiological changes during speech-sound training. Neuroreport, 9, 3557-3560.

Varela, F., Lachaux J.F., Rodriguez, E & Martinerie, J. (2001) The brainweb: phase synchronization and large-scale integration. Nature Rev. Neurosci., 2, 229-239.

M. HERDEGEN

Director of the Institute for Public Law and Director of the Institute for International Law, University of Bonn, Germany

42. LEGAL IMPLICATIONS OF THERAPEUTIC OPTIONS – STEM CELL PROCEDURES, CLONING AND PATENTS

1. INTRODUCTION

1.1. Regulatory challenges

Trying to keep pace with new perspectives for research, development of pharmaceuticals and medical treatment, the regulation of new therapeutic options in modern "bio-medicine" focuses on research involving embryos and embryonic stem cells[1] as well as cloning techniques. Many States (especially in Europe) have adopted or initiated legislation on the production of embryonic stem cells, their importation and their use. Somatic gene therapy is not a subject of major legal controversy[2]. The germ-line therapy is still addressed rather as an uncanny manipulation of the natural order than as a tangible option for treating serious hereditary defects[3]. An overarching issue is an adequate patent regime for inventive contributions in the area of bio-medicine.

In the field of legislative intervention two driving forces compete with each other:

---

[1] *The Royal Society* (ed.), Stem cell research and therapeutic cloning, Doc. 12/00, November 2000; *Nuffield Council on Bioethics*, Stem Cell Therapy: the ethical issues; *UK Department of Health* (ed.), Stem Cell Research: Medical Progress with Responsibility, June 2000, No. 2.8. According to figures released by the U.S. government, embryonic stem cell research may eventually lead to therapies that could be used to treat diseases that afflict approximately 128 millions Americans (Stem Cell Fact Sheet, http://www.whitehouse.gov/news/releases/2001/08/20010809-1.html [13.12.2002]).

[2] Somatic gene therapy raises some legal issues as to the release of genetically modified organisms (as contained in aerosols), see *M. Herdegen*, Rechtliche Regelung der somatischen Gentherapie am Menschen, in T. Herdegen/T. Tölle/M. Behr (eds.), Klinische Neurophysiologie, Heidelberg/Berlin/Oxford, 1997.

[3] An absolute prohibition governs the law in Austria (§ 74 of the *Genetic Engineering Act*), and Germany (§ 5 [1] of *Embryo Protection Act*). French law allows germ line therapy to cure serious hereditary defects, art. 16 (4) of the code civil.

(1) The medical and economic interests in allowing or stimulating advances in research tools and therapeutic treatment[4] and
(2) the claim to protect human dignity against the "instrumentalisation" of human life and the "commercialisation" of the human body.

In many States, especially in Germany and other countries of continental Europe, the constitutional underpinnings of these interests (fundamental guarantees covering human dignity as paramount value, the protection of human life and health as well as the freedom of research) govern legislative choices and inject a strong constitutional dimension into the political debate. However, these constitutional rules, especially the entrenchment of human dignity, were not tailored to address the subtle conflicts of interests raised by modern bio-medicine with all its techniques at the earliest stages of human life. Generally, these rules are far too indeterminate to prompt clear solutions and to pre-empt legislative options. Thus, traditional interpretation of human dignity offers little guidance on whether the embryo *in vitro* is a person which can claim dignity and if so, whether this claim to dignity bars the use of "spare" embryos for therapeutic purposes[5].

Moreover, confronted with uncertainty as to therapeutic perspectives, caught between strong interest groups and challenged by international competition, the legislator often resorts to nebulous terminology (such as "ethically sustainable" or "high-ranking research goals"). This strategy devolves the load of responsible decision to administrative bodies and advisory bodies.

The lack of pre-established social consensus and the imprecision in constitutional and other legal instruments enhances the influence of "scientific ethics" (especially "bio-ethics"), more in continental Europe than in Anglo-American systems. The impact of this "discipline" (even recognised as such in legal texts[6] stands in striking contrast to its weak methodological claims.

1.2. Limited international consensus

The quest for international standards has, so far, produced rather modest results. On global level, the emerging consensus is limited to
- a ban on *in vitro* fertilisation carried out specifically for research or other non-procreative purposes,
- a ban of reproductive cloning,
- the principle of informed consent as to the parents of an embryo used for the production of embryonic stem cell material.

Not even these parameters are universally entrenched in legal instruments.

---

[4] The *European Commission* estimates that the world-wide market potential of pharmaceutical biotechnology will amount to some 506 billion Euro in 2004 and assumes a constant increase to 818 billion Euro in 2010, *European Commission*, Communication from the Commission to the Council, the European Parliament, the Economic and Social Committee and the Committee of the Regions "Life sciences and biotechnology – A Strategy for Europe", COM (2002) 27 final, p. 7.

[5] See on the German Basic Law (art. 1 [1]) *Herdegen*, in: Maunz-Dürig, Grundgesetz, art. 1 Abs. 1, paras 62, 63.

[6] The *German Stem Cell Act* reserves a seat of the Central Ethics Commission to a representative of the discipline of ethics" (§ 8 [1]).

1.3 The European "Bioethics Convention" and its Additional Protocol

In Europe, some common ground has been gained with the Council of Europe's "Bioethics Convention". The Convention for the Protection of Human Rights and Dignity of the Human Being with regard to the Application of Biology and Medicine[7] adopted in 1997 has been ratified by a fair number of member States of the Council of Europe[8]. The Convention is designed to preserve human dignity, rights and freedoms, through a series of principles and prohibitions against the misuse of biological and medical advances. It provides for an "adequate protection" in the context with research on embryos in vitro (art. 18 [1]). Moreover the Convention bans the creation of embryos for research purposes (art. 18 [2]).

An additional protocol to the Convention, the so-called "Anti-Cloning Protocol" governs cloning techniques[9]. Art. 1 (1) of the Protocol prohibits any intervention seeking to create a human being genetically identical to another human being, whether alive or dead. Art. 1 (2) defines the term "human being genetically identical to another human being" as a human being sharing with another the same nuclear gene set. The term "human being" as the object of reproductive techniques banned by art. 1 refers to fully developed human life. Thus, the prohibition of art. 1 only covers human reproductive cloning. It does not restrict cloning techniques which do not aim at the implantation of the embryo and which are confined to the production of stem cell material[10], i.e. therapeutic cloning[11].

1.4. The Charter of the Fundamental Rights of the European Union

The Charter of Fundamental Rights of the European Union requires, that in the fields of medicine and biology in particular
- *the free and informed consent of the person concerned, according to the procedures laid down by law,*
- *the prohibition of eugenic practices, in particular those aiming at the selection of persons and*
- *the prohibition on making the human body and its parts as such a source of financial gain,*

are respected (art. 3 [2]).

---

[7] ETS No. 164.
[8] *Until today 31 member states of the* Council of Europe *have signed the convention, 15 of these states have also ratified the document. Among the States which have not ratified the Convention yet are France (signatory), Germany and the United Kingdom.*
[9] Additional Protocol to the Convention for the Protection of Human Rights and Dignity of the Human Being with regard to the Application of Biology and Medicine, on the Prohibition of Cloning Human Beings, *12 January 1998, ETS Nr. 168. The Protocol has been ratified by 13 States which have also ratified the Convention (not by Denmark and San Marino).*
[10] *Herdegen/Spranger*, in: Herdegen (ed.), Internationale Praxis Gentechnikrecht, Internationales Recht / Erläuterungen, Biomedizin-Übereinkommen und Klonprotokoll, para. 61 *et seq.*
[11] *Ackermann*, Therapeutisches Klonen und Stammzellgewinnung, in Bockenheimer-Lucius (ed.), Forschung an embryonalen Stammzellen, Köln, 2001, p. 27 (29).

2. STEM CELL PROCEDURES

The legal problems raised by stem cell procedures are closely related to the donor organism, i.e. an embryo created by in vitro fertilisation or an embryo generated by cloning (nuclear transfer technique) for the purpose of producing stem cell material ("therapeutic cloning"). As therapeutic cloning is defined by its finality, any regulation of this method will cover its stem cell production purposes[12]. By contrast, in vitro fertilisation on one hand and the production of stem cell material on the other hand are usually governed by separate legal regimes. A restrictive approach to the use of embryos for stem cell research governs the law in many States of Continental Europe. Elsewhere most technologically advanced States follow a rather liberal regime or entirely abstain from legislative intervention[13].

2.1. The "Bioethics Convention"

In addition to provisions on research with embryos (art. 18)[14], the Convention prohibits patents of the human body to be used for financial gains (art. 21).

2.2. Germany

In Germany, the 1990 Act on the Protection of Embryos prohibits the use of any embryo (defined as "fertilized human egg cells and totipotent cells taken from embryos") for purposes other than causing a pregnancy (§ 2 [2]). Over the last years, research with embryonic stem cells was the object of a particularly heated debate which focussed on imported stem cell material not covered by the 1990 Act. In June 2002 the German parliament (Bundestag), adopted the so-called "Stem Cell Act"[15]. According to § 1, the Act purports to ban as a matter of principle the importation and utilisation of embryonic stem cells, to prevent demand in Germany stimulating the production of embryonic stem cells or the production of embryos for that reason, and, finally, to determine the requirements for exceptionally permitting the importation and utilization of embryonic stem cells for research purposes. The Act explicitly pays tribute to human dignity, the protection of human life and health and the liberty of research (§ 1).

The Stem Cell Act provides for the import of embryonic stem cells and their subsequent utilisation as a narrowly confined exception and establishes an authorisation procedure (with the Robert Koch-Institute acting as the competent authority).

---

[12] See *infra*, 4.

[13] In Israel, the harvesting of human embryonic stem cells from "spare" embryos after *in vitro* fertilization is considered legal in the absence of legislative restraints, see *Ben-Or*, The Israeli Approach to Cloning and Embryonic Research, Heidelberg Journal of International Law (ZaöRV) 2000, 762 (768 *et seq.*).

[14] See *supra*, 1.3.

[15] Gesetz zur Sicherstellung des Embryonenschutzes im Zusammenhang mit Einfuhr und Verwendung menschlicher embryonaler Stammzellen, BGBl. 2002 Teil I, p. 2277 *et seq*.

The import of embryonic stem cells and their use can only be permitted, if
- *the cells were derived from cell lines established before 1 January 2002 in accordance with the relevant national legislation in the country of origin and have been kept in culture since then and stored using crykoconservation methods (section 2 no. 1 lit. a),*
- *the cells were derived from embryos created by medically-assisted in vitro fertilisation and definitely no longer destined for implantation without any genetic selection having taken place (section 2 no. 1 lit. b)*[16],
- *no compensation or other pecuniary benefit had been granted or promised in return for the donation of embryos for the purpose of stem cell derivation (section 2 no. 1 lit. c),*
- *other legal provisions, such as those of the German Embryo Protection Act, do not conflict with the importation and utilization of embryonic stem cells (section 2 no. 2) and*
- *the relevant stem cells have not obviously been derived in contradiction to major principles of the German legal system, with the important qualification that authorisation may not be refused on the grounds that the stem cells were derived from human embryos (section 3).*

§ 5 of the Act lays down the conditions for research using imported embryonic stem cells. Such research may only be authorised, if
- *scientific reasons have been given that the research serves "high-ranking research goals" and purports either to gain scientific knowledge in basic research or to increase medical knowledge for the development of diagnostic, preventive or therapeutic methods which are to be applied to humans (section 1 no. 1),*
- *the relevant issues have been clarified as far as possible through in vitro models using animal cells or through animal experiments (section 1 no. 2 lit. a) and*
- *the scientific knowledge to be obtained from the particular research project cannot be gained by using cells other than embryonic stem cells (section 1 no. 2 lit. b).*

According to § 6 (4) no. 3, authorisation of an embryonic stem cell research project is conditional upon an opinion issued by the Central Ethics Commission[17]. This opinion will evaluate the research goals and the origin of the embryonic stem cell line to be used in the project[18]. The Commission's view is not binding upon the competent authority[19]. The Federal Agency competent for authorisation and the Central Ethics Commission consider and assess whether the research projects are "ethically" sustainable in the light of the legal requirements (§ 6 [4] no. 2 and § 9).

---

[16] See BT-Drs. 14/8394, explanation p. 9.
[17] See also Section 8.
[18] BT-Drs. 14/8396, p. 10.
[19] *Gehrlein*, Das Stammzellgesetz im Überblick, NJW 2002, 3680 (3681).

As a multi-party compromise between a strict prohibition of embryonic stem cell techniques and a liberal approach, the German Stem Cell Act does not display a consistent regulatory philosophy. The rigid ban on domestic production of embryonic stem cells coupled with a cautious permission of research with imported stem cells invites the criticism of certain hypocrisy. The cut-off date of 1 January 2002 for the establishment of stem cell lines has some arbitrary element and bars German research from recurring on new cell lines with enhanced purity and stability. The legislator's fear of domestic demand stimulating the "consumption" of human embryos abroad seems to overrate the impact of German research interests.

The Act is a model of normative imprecision. The condition that permissible research serves "high-ranking research goals" turns a State authority into an arbiter on the value of research, an intervention colliding with the constitutional freedom of science and research (art. 5 [3] of the Basic Law[20]). The constitutionality of the act is, at least, doubtful. Under the new legislation, authorisation was granted for several research projects (on the migration and differentiation of human neural precursor cells in context with the developing brain [University of Bonn], the differentiation of embryonic stem cells into cardiomyocytes [University of Cologne] and on the differentiation of embryonic stem cells into cardiomyocytes to be transplanted into the myocardium after infarctions [University of Munich])[21].

2.3. Switzerland

The revised Swiss Constitution contains provisions concerning reproduction medicine and genetic engineering (art. 119). The draft Federal Act on Research on Spare Embryos[22] is strongly inspired by German legislation. However, unlike German law, the draft Act does not limit research to imported stem cell material. In March 2003 the Council of States (Ständerat) approved the bill (with some modifications); the bill now requires adoption by the National Council (Nationalrat). The draft provides for the generation of embryonic stem cells from embryos for research purposes, subject to authorisation by the Federal Agency for Health (art. 8) and to informed consent of the couple concerned (Art. 10). Research with embryonic stem cells requires an approving opinion by the competent ethics commission (art. 13) and it must fulfil rather strict scientific and ethical standards. In particular, the project must aim at an "essential" advance in knowledge about "grave illnesses" or about the human evolution biology and must be "ethically sustainable" (art. 6 [1] lit. a and lit. e). Import and export of embryonic stem cells are subject to authorisation (art. 17).

---

[20] *BVerfGE* 35, 79 (113).
[21] Register according to 11 of the Stem Cell Act, see
http://www.rki.de/GESUND/STEMCELL/STZG.HTM.
[22] Embryonenschutzgesetz No. 02.083, 20 November 2002.

2.4. France

France was among the first European countries to enact "bioethics legislation" in 1994. However, a specific legal regime governing stem cell research is still in the making. The conception of embryos in vitro is confined to procreative purposes[23]. Human embryos may not be used for industrial or commercial purposes.[24] French law does not ban the import of embryonic stem cells.

In early 2002, the French National Assembly approved the *Bill Concerning Bioethics*[25]. The draft Act provides that research on embryonic stem cells derived from spare embryos may be carried out for medical purposes, subject to authorisation of the Agency for Biomedicine[26], whilst cells derived from an aborted foetus may be used for diagnostic, therapeutic or scientific purposes[27]. The import of embryonic stem cells shall be subject to an authorisation granted by the Minister for Research[28]. In January 2003, the Senate presented some modifications[29]. In particular the Senate demands that research on embryonic stem cells derived from spare embryos must be susceptible of allowing major therapeutic progress.

2.5. United Kingdom

In contrast with Germany and France, the United Kingdom has opted for a rather proactive approach to biotechnology in general and embryonic stem cell procedures in particular. The legal framework governing the use of embryos in research is laid down in the Human Fertilisation and Embryology Act 1990. The 1990 Act allows research on human embryos, if the specific project fulfils the criteria established by the Act and is licensed by the Human Fertilisation and Embryology Authority.

According to the Act, research may be performed on spare embryos created in the course of *in vitro* fertilisation (section 11 [1] *lit.* c). As opposed to legislation in a number of European countries such as Germany and France prohibiting the creations of human embryos for research and therapeutic purposes, it is interesting to note that the Act also permits research on embryos which have been created for precisely this reason[30]. Initially, the 1990 Act permitted research on human embryos only if one of the following aims was being pursued: to promote advances in infertility treatment, to increase the knowledge about the cause of congenital disease, or of miscarriages, to develop more effective techniques of contraception, or to develop methods for detecting the presence of gene or chromosome abnormalities in embryos before implantation.

With the *Human Fertilisation and Embryology (Research Purposes) Regulations 2001*, the UK Government, following favourable resolutions of both Houses of

---

[23] Art. L. 2141-3 of the code de la santé publique.
[24] Art. L. 511-17, 2 of the code pénal.
[25] Texte Adopté No. 763, Projet de Loi relatif à la bioéthique, 22 January 2002.
[26] Draft art. L. 2153-3 of the code de la santé publique.
[27] Draft art. L. 1241-5 of the code de la santé publique.
[28] Draft art. L. 2151-3-1 of the code de la santé publique.
[29] Assemblée nationale (12ᵉ législ.): Projet de Loi no. 593.
[30] Between 1991 and 1998, 118 embryos have been created in the course of research, *Department of Health* (ed.), Stem Cell Research: Medical Progress with Responsibility, June 2000, p. 32.

Parliament, established three additional purposes in pursuance of which research on human embryos may be permitted:
- to increase knowledge about the development of embryos or about serious diseases or
- to enable any such knowledge to be applied in developing treatments for serious diseases[31].

With regard to therapeutic options, the last of these additional purposes is of particular interest.

2.6. USA

The regulatory approach pursued in the United States quite distinctly differs from the direction which the European legislation has taken. In particular, there is no legislative framework for embryonic stem cell research. The U.S. Government's intervention is confined to guidelines for government-funded research on embryonic stem cells. In his statement of 9 August 2001, President George W. Bush established the following criteria which scientists applying for government-funded research have to meet:
- Research must be confined to the use of cell lines established prior to the President's announcement.
- The donor embryo must have been created for reproductive purposes and was no longer destined for development as a human being.
- Informed consent with regard to the donation of the embryo must have been obtained and no financial inducements may have been offered in return for the donation of the embryo[32].

Following President Bush's decision, the National Institutes of Health have established a Registry listing the human embryonic stem cell lines which conform to the eligibility criteria[33].

2.7. Japan

In Japan, the Minister of Education, Culture, Sports, Science and Technology laid down guidelines for human stem cell research in August 2001[34]. The guidelines purport to ensure that the derivation or utilization of human stem cells always respect human dignity (preamble). These governmental guidelines allow research on

---

[31] 2001 No. 18, *The Human Fertilisation and Embryology (Research Purposes) Regulations 2002*, 24 January 2001.

[32] Remarks by the President of the United States of America on Stem Cell Research (http://www.whitehouse.gov/news/releases/2001/08/ 200110809-2.html#); *Office of the Press Secretary*, Official Fact Sheet Embryonic Stem Cell Research (http://www.whitehouse.gov/news/releases/2001/08/20010809-1.html).

[33] The registry can be accessed via the Internet at http://escr.nih.gov.

[34] *Guidelines for Derivation and Utilization of Human Embryonic Stem Cells* (http://www.mext.go.jp/a_menu/shinkou/seimei/2001/es/020101.pdf) based upon the *Report on the Human Embryo Research Focusing on the Human Embryonic Stem Cells* presented by the Subcommittee of Human Embryo Research, Bioethics Committee and the Council for Science and Technology (March 6, 2000).

stem cells harvested from "spare" embryos which have initially been created for the purposes of fertility treatment and which are surely intended to be discarded by its donors (art. 6 no. 1). The use of the embryos must be covered by the informed consent of the donors (art. 6 no. 2). Moreover, the guidelines require projects for establishing human stem cell lines and research on the cells to be approved and monitored by each institution's review board (art. 13) and reviewed by the Ministry of Education, Science, Technology, Sports and Culture (art. 16 section 1). The Ministry shall consult the "Bioethics and Biosafety Commission" (art. 16 section 3).

3. CLONING

In a much higher degree than embryonic stem cell procedures, cloning techniques raise serious normative and ethical concerns. They often collide with basic constitutional tenets. Reproductive cloning of humans is widely considered unethical. Many States ban this method with criminal sanctions. By contrast, quite a number of States allow therapeutic cloning.

3.1. Additional Protocol to the "Bioethics Convention"

The additional "Anti-Cloning Protocol" squarely prohibits human reproductive cloning[35] without banning the technique of cloning as such[36].

3.2. Germany

The German Embryo Protection Act of 1990 categorically forbids human reproductive cloning (§ 6). In February 2002, the German Parliament adopted a resolution against any form of cloning on a multi-party basis[37]. The resolution qualifies "any artificial creation of human embryos by cloning" irrespective of the applied technique and the underlying purpose as a violation of human dignity and calls for international conventions banning both reproductive and therapeutic cloning.

3.3. France

The French legislator has not directly addressed the question of reproductive or therapeutic cloning. According to an assessment published by the Comité Consultatif National d'Ethique, the 1994 "bioethics legislation", if interpreted widely, prohibits human reproductive cloning[38]. Apart from interdicting the production of embryos for research or industrial purposes, this legislation does not

---

[35] See *supra*, 1.3.
[36] *Herdegen/Spranger*, in: Herdegen (ed.), Internationale Praxis Gentechnikrecht, Internationales Recht/Erläuterungen, Biomedizin-Übereinkommen und Klonprotokoll, para. 58.
[37] BT-Drs. 15/463 (application); adopted by the Parliament, 20 February 2003 (28. Sitzung des Deutschen Bundestages), official Protocol: http://www.bundestag.de/aktuell/a_prot/2003/ap1528.html.
[38] *Comité Consultatif National d'Ethique*, Réponse au Président de la République au sujet du clonage reproductif, No. 54 – 22. April 1997, p. 20 *et seq*.

govern therapeutic cloning[39]. The proposed amendment to the bioethics legislation[40] would introduce an explicit prohibition of human reproductive cloning[41], but does not address therapeutic cloning. According to the French government, this legislative self-restraint shall defer to a broad political debate on the ethical acceptability of therapeutic cloning[42].

3.4. United Kingdom

The Human Reproductive Cloning Act 2001 explicitly bans human reproductive cloning: "A person who places in a woman an embryo which has been created otherwise than by fertilisation is guilty of an offence" (section 1 [1]). Research involving the creation or use of embryos by cell nuclear replacement is permissible under the Human Fertilisation and Embryology Act 1990, provided such research pursues one of the existing specified research purposes discussed above[43] and is undertaken during the first 14 days of the embryo's existence, the so-called pre-embryonic phase.[44]

3.5. USA

A bill aiming at the prohibition of human cloning (draft Human Cloning Prohibition Act of 2003) has been adopted by the House of Representatives in February 2003[45] and received in the Senate. The bill purports to ban all kinds of human cloning. It defines "human cloning" as "human asexual reproduction, accomplished by introducing nuclear material from one or more human somatic cells into a fertilized or unfertilized oocyte whose nuclear material has been removed or inactivated so as to produce a living organism (at any stage of development) that is genetically virtually identical to an existing or previously existing human organism" (sec. 301 of Title 18 United States Code). This bill – if enacted by Congress – would not only ban human reproductive cloning, but also proscribe therapeutic cloning.

---

[39] Article L. 2141-7; L. 2141-8 of the code de la santé publique. See also *Dictionnaire Permanent Bioéthique et Biotechnologie*, Embryon Humain *in vitro*, p. 822, No. 21.
[40] See *supra*, 2.4. at footnote 25.
[41] Draft art. 16-4-3 of the code civil.
[42] Communiqué de Presse, 26 Novembre 2001, Roger-Gérard Schwartzenberg: notre projet de loi su la bioéthique n'a pas retenu le clonage thérapeutique. See further, *Dictionnaire Permanent Bioéthique et Biotechnologie*, Bulletin 106 «Projet de loi relatif à la bioéthique», p. 7412.
[43] *Supra*, 2.5.
[44] *Spranger*, Legal Status and Patentability of Stem Cells in Europe, 21 Biotechnology Law Report, 105 at 108.
[45] Bill # H.R. 534.

3.6. Israel

The Human Cloning and Genetic Manipulation of Reproductive Cells Law of 1999 prohibits reproductive cloning (section 3) for a period of five years[46]. This legislative moratorium does not effect therapeutic cloning.

3.7. Japan

The Act Concerning Regulation Relating to Human Cloning Techniques and Other Similar Techniques categorically prohibits the transfer of a human somatic clone embryo, a human-animal amphimictic embryo, a human-animal hybrid embryo or a human-animal chimeric embryo into a uterus of a human or an animal (art. 3). The act does not specifically address therapeutic cloning. The provisions on the treatment of clone embryos not destined for implementation (art. 4 et seq.) seem to indicate that therapeutic cloning shall be governed by guidelines to be issued by the Minister of Education, Culture, Sports, Science and Technology.

4. PATENTS

4.1. The Patentability of biotechnological inventions in general

The protection of intellectual property is of vital importance for the further development of therapeutic options arising from biotechnological research: The perspective of an adequate legal protection of benefits provides an important incentive for the costly process of new techniques and products[47]. An excessive restriction of intellectual property protection would seriously stifle the necessary investments in research and development. On the other hand unduly brought patents risk to hinder research.

An invention may be patented if it meets the following three universally recognized conditions of patentability: The patent claim must refer to a process or a product which

- is novel (*i.e.* an invention is new, if it is not part of the 'state of the art')
- inventive (as opposed to a mere discovery, which was 'so easy that any fool could do it') and
- is capable of industrial application (*i.e.* making or using the invention in any kind of industry)[48]

Patents in the area of bio-medicine raise a number of specific legal issues:
- the patentability of substances identical in structure with elements of the human body,[49]

---

[46] *Ben-Or*, The Israeli Approach to Cloning and Embryonic Research, Heidelberg Journal of International Law (ZaöRV) 2000, 762 (766).

[47] See *Herdegen*, Die Erforschung des Humangenoms als Herausforderung für das Recht, JZ 2000, 633 (637).

[48] *Grubb*, Patents for Chemicals, Pharmaceuticals and Biotechnology, Oxford, 1999, p. 61.

[49] See *Grubb*, Patents for Chemicals, Pharmaceuticals and Biotechnology, Oxford, 1999, p. 246; *Herdegen*, Patents on Parts of the Human Body – Salient Issues under EC and WTO Law, 5 The Journal

- sufficient disclosure of substances to be patented,
- the scope of patents on substances (highly relevant for DNA sequences) and
- exclusions on grounds of morality and public order.

Within the framework of the WTO Treaties, art. 27 (1) of the *Agreement on Trade-Related Aspects of International Property Rights* (TRIPs) establishes the basic principle that inventions, whether products or processes, in all fields of technology shall be patentable[50].

Following the rapid technological progress in the field of biotechnology, the European Union has developed a comprehensive regular framework concerning the patenting of biotechnological inventions. The *EC Directive on the Legal Protection of Biotechnological Inventions* from 1998 (the so-called *Biotechnology Directive*)[51] constitutes a major breakthrough in favour of patents on human genes and other parts of the human body. In its article 3 (1) the Directive establishes the principle that biotechnological inventions are amenable to patent protection:

> "For the purposes of this Directive, inventions which are new, which involve an inventive step and which are susceptible to industrial application shall be patentable, even if they concern a product consisting of or containing biological material or a process by means of which biological material is produced, processed or used."

This clause does not only reflect well-established principles of patent law, it is also in line with WTO Law.

It follows from the general principles recited by art. 3 (1) of the Directive, that – as opposed to inventions – mere discoveries are not patentable. With regard to biological material which can be found in a natural environment, the need for an inventive step is of particular significance. In this context, particular attention has to be paid to Article 3 (2) of the Directive, which reads:

> "Biological material which is isolated from its natural environment or produced by means of a technical process may be the subject of an invention, even if it previously occurred in nature."

The inventive element in such cases lies either in isolating or synthesising the substance. Art. 3 (2) of the Directive clarifies, that – provided such an inventive element can be identified – biological matter found in nature is amenable to patent protection. Article 3 (2) thus clearly rejects the argument put forward by opponents to the legal protection of biotechnological inventions that biotechnological material occurring in nature could only be the object of a discovery and not of a (patentable) invention.

---

of World Intellectual Property 145, 149 *et seq.*; *id.*, Patent Protection for Genes and Other Parts of the Human Body under EC Law, European Biopharmaceutical Review Winter 2002; *Nuttfield Council*, The ethics of patenting DNA, 2002, p. 57; *Warren-Jones*, Patenting rDNA – Human and animal biotechnology in the United Kingdom and Europe, Oxon, 2001.

[50] *Herdegen*, Die Erforschung des Humangenoms als Herausforderung für das Recht, JZ 2000, 633 (637).

[51] Directive 98/44/EC of the European Parliament and of the Council of the European Union of 6 July 1998 on the legal protection of biotechnological inventions, O.J. L 213/13, 30 July 1998. See *Herdegen*, Patents on Parts of the Human Body – Salient Issues under EC and WTO Law, 5 Journal of World Intellectual Property 2 (145); see also *Herdegen*, Patenting Human Genes and other Parts of the Human Body under EC Biotechnology Directive, 3 Bio-Science Law Review 4 (133).

Article 5 of the Directive specifically deals with the patentability of substances of the human body. While Article 5 (1) reiterates the non-patentability of mere discoveries, Article 5 (2) clarifies that sequences of a human gene or other elements isolated from the human body through a technical process or synthesized can enjoy patent protection, regardless of a structural identity between the isolated element and its natural equivalent:

> "An element isolated from the human body or otherwise produced by means of a technical process, including the sequence or partial sequence of a gene, may constitute a patentable invention, even if the structure of that element is identical to that of a natural element."

Thus, the principle of non patentability applying to mere discoveries only excludes patents on the human body *in situ*.

In its *Utility Examination Guidelines* of December 2000[52] the *United States Patent and Trademark Office* (USPTO) rejected the suggestion generally to exempt DNA sequences from patentability.

4.2. Disclosure of functions

The requirement of utility implies a sufficiently precised disclosure of functions. This condition for patentability is of major relevance for DNA compositions. The necessary disclosure of functions bars random patent applications and hazy claims to speculative uses of DNA compositions. This criterion is justly emphasized both by the EC Biotechnology Directive and the USPTO *Utility Examination Guidelines*. The EC Biotechnology Directive requires full disclosure of the envisaged industrial application, if a sequence or a partial sequence of a gene is to be patented (art. 5 [3]).

In particular, patent applications must be sufficiently precise as to the functions of encoded proteins in terms of its biological activity. Thus, the EC Biotechnology Directive emphasizes in its recital 24:

> "... in order to comply with the industrial application criterion it is necessary in cases where a sequence or partial sequence of a gene is used to produce a protein or part of a protein, to specify which protein or part of a protein is produced or what function it performs".

4.3. The scope of patents on DNA sequences

Patents on genes and other DNA sequences (like the patent on the BRCA gene [3]) fuel legitimate concerns about exclusive rights extending to all possible functions and the use of all proteins encoded by the patented sequence.

Under the established principles of the patent law of most of the industrial countries patents on synthesized or isolated substances (compositions of matter) enjoy "absolute" protection: they cover all possible applications. This comprehensive protection was developed in context with physical or chemical substances whose inventive value lies in their structure as a "substance or

---

[52] Federal Register, vol. 66, no. 4, 1092 (January 5th 2001).

composition of matter". The extension of the traditional rules of "absolute protection" for patents on DNA sequences raises serious problems.

For a comprehensive protection for patents on DNA sequences potentially implies enormous dividends for the patent holder which may go well beyond his inventive contribution. The inventive value of DNA sequences does not flow from their biochemical structure as such, but from the encoded information. This distinguishes DNA patents from patents on other physical or chemical substances. The value of the encoded information is inseparably linked to the possibly multiple functions of the sequence. Once a DNA sequence as an information carrier enjoys comprehensive patent protection, the exclusive right conferred on the patent holder covers uses which rage to functions not described in the patent application and even totally unknown to the applicant. An exclusive right covering uses to initially unknown functions amounts to a kind of windfall profit which distorts the elementary symmetry between the inventive value and its reward by an exclusionary right.

This concern supports a "functional" limitation of patents on DNA sequences. Such patents should enjoy only "relative protection" within the context of the specific function disclosed in the patent application[53]. The EC Biotechnology Directive neither commands nor bars such an approach. It rather leaves the adequate protection of patents on DNA sequences to legislative implementation by each member state. The ongoing controversy in the Federal Republic of Germany and other EU member states illustrates the challenge for the national legislators. In France the Draft Act Concerning Bioethics adopted by the National Assembly[54] provided a categorical exemption of genes or gene sequences from patentability; this radical exclusion, clearly incompatible with the EC Biotechnology Directive, was modified by the Senate in terms of a protection confined to the specific function disclosed in the application. On the other hand the USPTO Utility Examination Guidelines are not receptive to a limitation of DNA patent claim scope to uses disclosed in the application.

4.4. Exclusion from patentability on grounds of morality: human cloning and stem cell techniques

Article 6 of the Directive imposes certain limits on the patentability of biotechnological inventions on ethical grounds. Article 6 (1) establishes that inventions shall be considered not patentable if their commercial exploitation would be contrary to *ordre public* or morality[55]. Article 6 (1) is a mere expression of conventional restrictions on patentability which can be found in a number of international treaties, such as in art. 27 (2) TRIPs and art. 53 *lit.* a EPC. Article 6 (2) contains a non-exhaustive list excluded from patentability enumerates, *inter alia*,

---

[53] *Herdegen*, European Biopharmaceutical Review, Winter 2002, p. 43 (44).
[54] *Infra*, Fn. 22.
[55] In this context, it is important to point out that the mere fact, that the exploitation is prohibited by law or regulation does not hinder patent protection being granted. See *Herdegen/Spranger*, in: Herdegen (ed.), Internationale Praxis Gentechnikrecht, Richtlinie (98/44/EG), para. 28.

processes for cloning human beings (*lit.* a), processes for modifying the germ line identity of human beings (*lit.* b) and uses of embryos for industrial or commercial purposes (*lit.* c).

The exclusionary clause on human cloning in Article 6 (2) *lit.* a must be construed narrowly. For the Directive's recital 41 defines a process for cloning human beings as "any process, including techniques of embryo splitting, designed to create a human being with the same nuclear genetic information as another living or deceased human being." Under this definition, human cloning is characterised by the aim of creating a human being. Therefore, the therapeutic cloning which only aims at the reproduction of single cells, tissues, or other parts of the human body remains patentable[56].

The exclusion of patents on uses of embryos for industrial and commercial purposes in art. 6 (2) *lit.* c is strictly confined to these purposes, as recital 42 of the Directive clarifies. Stem cell techniques serving procedures therapeutic purposes may be patented[57]. In 1999, the *German Patent Office* granted a patent on the production and therapeutic use of neural precursor cells[58]. For this invention a patent application is also pending before the *European Patent Office*.

---

[56] *Spranger*, Legal Status and Patentability of Stem Cells in Europe, 21 Biotechnology Law Report 105 at 111.

[57] *Herdegen*, Die Patentierbarkeit von Stammzellverfahren nach der Richtlinie 98/44/EG, GRUR International 2000, 859; *id.*, Patenting Human Genes and other Parts of the Human Body, 3 Bio-Science Law Review 102 at 105.

[58] Patent no. DE19756864C1. Greenpeace has announced a challenge to the patent which has been well-known since 1999, see Ärzte Zeitung of 27 February 2003. The legal grounds for the challenge seem somewhat hazy.

GLOSSARY

CHAPTER 1 (HOYER)

Adenosine Triphosphate: the main energy rich compound which ensures cellular and molecular work (gold standard). Physiologically, ATP derives from glucose metabolism only.

Advanced Glycation Endproducts: are formed from glucose or other sugars and the irreversible glycation of proteins in several reactions. AGE bind to receptors (RAGE) to stimulate intracellular expression of transcription and growth factors.

Amyloid Precursor Protein: is encoded on chromosome 21; glycoprotein with a single transmembraneous domain. Physiologically, APP is cleaved in two functionally important derivatives: secreted APP (APPs) and ßA4 (in low concentration). Abnormal cleavage results in increased concentration of ßA4 which is highly amyloidogenic and aggregates to neuritic plaques.

Apparent Diffusion Coefficient: represents a magnetic resonance parameter measured with the diffusion weighted imaging. ADC reflects water diffusion and cell swelling processes in tissue. Thus, ADC changes *in vivo* are caused by cell volume changes.

Glucose Transport Protein: a family of proteins ensuring glucose transport across membranes. The inactive form is located near the nucleus, the active form at the cell membrane.

Insulin / Insulin Receptor: the hormone formed in pancreatic beta cells and – with respect to the brain – in a subpopulation of pyramidal neurons. Binding of insulin to the Insulin Receptor: stimulates multiple intracellular and molecular reactions: glucose breakdown and energy production (ATP, see above), normal processing of APP (see above), activation state of glucose transport proteins (see above), and subsequently in signal transduction cascade.

Insulin Receptor Substrate: a family of proteins which conduct the insulin signal to numerous intracellular proteins to stimulate cell growth, cell repair etc.

Mixed Function Oxidation: biochemical process includes enzymatic and non-enzymatic reactions to inactivate aged (damaged) proteins. MFO increases in capacity during aging.

Hypothalamic-Pituitary-Adrenal Axis: functional morphobiologic entity regulating the release of cortisol from the adrenal gland, the HPA axis is controlled the hippocampus; as a whole, response to stress is mediated by both hippocampal and HPA axis function

Positron Emission Tomography: PET is a diagnostic technique for non-invasively imaging of radiopharmaceutical distribution inside the living body. Thus, PET allows a detection of metabolic and functional information about tissue and organs *in vivo*.

RAGE: receptor for Advanced Glycation Endproducts (see above). The derivative ßA4 of the amyloid precursor protein (see above) also binds to RAGE.

Transverse Relaxation Time: quantitative parameter in magnetic resonance imaging. Any excited magnetic moment relaxes back to equilibrium in a magnetic field. There are two components of this relaxation: longitudinal (T1) and transverse (T2). Severity of hippocampal atrophy correlates with prolongation of T2 relaxation time in several neurodegenerative diseases.

CHAPTER 2 (KACZMAREK)

Gene promoter: Regulatory region (encompassing approximately 100 base pairs, bp) involved in control of gene expression, and located immediately upstream the transcription inititiation site.

Transcription factors: Proteins capable to interact with specific DNA sequences as well as with other proteins to regulate the transcription.

Chromatin: Organization form of the genetic material - a complex of DNA, proteins and nascent RNA.

Histones: Positively charged (basic) proteins that form a complex with DNA to produce the essential level of chromatin organization. There are following major types of histones: H2A, H2B, H3, H4, and H1.

Nucleosome: Main unit of chromatin organization, composed of histone octamer suurounded by DANN.

Intron: An intervening DNA sequence present in the genomic DNA as well as in a form of RNA in the primary transcript, and separating exons.

Exon: A DNA sequence that codes for components of functional RNA. The same term defines also RNA regions coding for proteins.

Alternative/differential splicing: A process, by which introns are removed from the primary transcripts to produce the final RNA molecule. Alternative splicing allows to form RNAs of differential composition of exons, and thus different proteins encoded by the same gene.

mRNA editing: An enzymatic process allowing for a specific change in the mRNA sequence by replacing one nucleotide with another.

mRNA translocation: Specific delivery of defined mRNA molecules to selected cytoplasmic compartments, e.g., dendrites.

Polyadenylation: Addition of ca. 150-200 adenyl residues at the 3' mRNA end to produce the polyA tail.

CHAPTER 4 (MARTINEZ-MILLÁN)

PDZ: Derives from postsynaptic proteins PSD-95, Dlg and ZO1, which contain a consensus domain in which serine or threonine are crucial for associations with the COOH-terminus of GluR1.

CREB: *cAMP-response element binding protein*. A transcription factor that is phosphorylated by catalytic unit of protein kinase A together with recruited mitogen activated kinase. It activates genes that encode proteins important for growth and survival.

LTP: *Long-term potentiation*. Cellular model of memory that consists in an increase of the synaptic efficiency. It is experimentally achieved after a tetanizing discharge that is translated into enhanced synaptic potentials in response to stimuli of constant amplitude.

LTD: *Long-term depression*. Reduction in synaptic efficiency firstly observed in intracellular recordings of cerebellar Purkinje cells after concurrently stimulation of their parallel fibre and climbing fibre inputs. This effect of the climbing fibre on the transmission of signals from parallel fibres lasts from minutes to hours. This long-lasting change has also been observed in other central nervous structures such as hippocampus.

NMDAR: *NMDA-receptor*. Post-synaptic ion permeable (ionotropic) glutamate receptor activable by the glutamate analogue N-methyl-D-aspartate. It is formed by a combination of the constitutive NR1 subunit and the modulatory NR2A-D, which form an oligomeric voltage-dependent Ca^{2+}-channel. At basal membrane potentials this receptor is blocked by Mg^{2+} whose removal is a prerequisite for glutamate mediated Ca^{2+}-entry.

AMPAR: *AMPA-receptor*. Pre-synaptic ion permeable (ionotropic) glutamate receptor activable by the glutamate analogue α-amino-3-hydroxy-5-methyl-4-isoxazolepropionate. It is formed by a combination of GluR1-4 subunit, which form and oligomeric Na$^+$-channel. This receptor is no voltage-dependent and can also permeate Ca^{2+} when GluR2 subunit is absent.

MGluR: *Metabotropic glutamate receptor.* G-protein linked transmembrane receptor whose activation leads to G-protein activation and a subsequent elevation of second messengers such as cAMP. These receptors can be found either pre- or post-synaptically. At In comparison with ionotropic receptors, their action is slower and lasts for longer periods (seconds to minutes).

G-protein: Family of intracellular proteins, formed by α-, β- and γ-subunits, that links metabotropic receptors with the second messenger synthesizing machinery. G-protein term arises from the ability of these macromolecules to bind guanine nucleotides GTP and GDP.

Neurotrophins: According to the neurotrophic hypothesis, neurotrophins are supplied in limiting amounts by the target tissue and determine the number of afferents neurons that survive and differentiate; in such away that, only factors that fulfil these two conditions (e.g. NGF) can be considered neurotrophins. More recently, the neurotrophin concept has expanded to all of molecules with closely related structures that regulate neuron survival, development function and plasticity, despite of their concentration in nervous tissue (e.g. BDNF, NT-3, NT-4). Neurotrophins bind to tyrosine kinase receptors leading to gene expression changes by activating different signalling pathways.

Occlusion*:* Saturation of synaptic plasticity in a network should destroy the pattern of trace strengths corresponding to established memories and occlude new memory encoding.

Theta rhythm: Spikes of 4-7 Hz frequency that appears in hippocampus both spontaneously and when an animal moves in its environment.

CHAPTER 7 (ORTEGA)

Apoptosis: Morphological features (chromatin condensation, nuclear fragmentation, cell shrinkage and formation of apoptotic bodies) that characterized programmed cell death. Nowadays, the term apoptosis is also used to define the molecular events that participate in the process of programmed cell death.

Apoptosome: An oligomeric protein complex that consists of several molecules of Apaf-1 bound to cytochrome c and that is involved in the activation of caspase 9.

Bcl-2 proteins: A family of proteins that regulate apoptosis. Based on their functionality and sequence similarity these proteins are classified into three subfamilies. The Bcl-2 subfamily contains all four Bcl2-homology domains (BH1-4); the Bax/Bak subfamily lacks the BH4 domain and the BH3-only subfamily just share the BH3 domain. The Bcl-2 subfamily acts as negative regulators of apoptosis whereas the second and third subfamilies have pro-apoptotic functions.

Caspases: Family of aspartil-proteases that are the main executioners of programmed cell death upon their specific activation by various types of apoptotic stimuli. The precursor non-active forms of the caspases (procaspases) have an N-terminal domain that must be cleaved off for caspase activation.

ΔμH+: The proton electrochemical gradient or proton-motive force is generated across the inner mitochondrial membrane by the complexes of the respiratory chain that is further utilized for the synthesis of ATP by the H+-ATP synthase.

Encephalopathy: Pathology of grey and/or white matter of the brain.

Glutathione: Abundant cellular tripeptide (γ-Glu-Cys-Gly) that, by virtue of the free thiol group present in its Cys residue, protects cellular constituents from oxidative stress.

Heteroplasmy: The occurrence of wild-type and mutant populations of the mitochondrial genome (mtDNA) within the same cell.

Ischemic penumbra: The tissue region surrounding the necrotic core of an ischemic infarct in which neurons primarily die by apoptosis.

Metabolic stress: Condition in which the cellular availability of substrates, oxygen and other molecules required for energy production is below normal limits.

Permeabilty Transition Pore (PTP): Regulated mitochondrial megachannel. Formation of the PTP requires the association of inner and outer mitochondrial membrane proteins. Opening of the PTP promotes the collapse of the mitochondrial membrane potential hampering the production of biological energy and favours the generation of superoxide radicals and the release of mitochondrial proteins.

Reactive Oxygen Species (ROS): Oxygen compounds with a free radical that are produced as by-products of cellular oxidative metabolism. ROS are strong oxidants that damage proteins, lipids and DNA leading to cellular oxidative stress.

Substantia nigra: A part of the midbrain that contains dopamine-producing neurons. These neurons innervate the striatum and control body movements.

Respiratory Chain: The functional supramolecular structure composed of protein complexes that are located in the inner mitochondrial membrane and that transport electrons to reduce molecular oxygen. The transport of electrons is coupled to proton pumping out of the mitochondrial matrix to generate the proton electrochemical gradient that is used for the synthesis of ATP by mitochondria.

Zymogen: The inactive precursor of some proteins that is transformed into the biologically active protein by partial proteolysis.

CHAPTER 8 (PASCHEN)

Amyotrophic lateral sclerosis (ALS): Degenerative disease affecting motor neurons

Calcium release- activated calcium current (I_{CRAC}): Inward calcium current activated under conditions of ER calcium store depletion. Activation of I_{CRAC} is required to refill endoplasmic reticulum calcium stores.

Double-stranded RNA-dependent protein kinase (PKR): Protein kinase that phosphorylates the alpha subunit eukaryotic initiation factor 2, resulting in shut-down of translation.

Double-stranded RNA-dependent protein kinase-like endoplasmic reticulum kinase (PERK): Endoplasmic reticulum-resident protein kinase that phosphorylates the alpha subunit eukaryotic initiation factor 2, resulting in shut-down of translation. PERK is activated under conditions associated with endoplasmic reticulum dysfunction.

Endoplasmic reticulum (ER): Subcellular organelle playing a central role in calcium signaling and calcium storage. All memebrane and secretory proteins are folded and processed in this subcellular compartment.

Endoplasmic reticulum-associated degradation (ERAD): Stress response activated under conditions associated with endoplasmic reticulum dysfunction. Activates degradation of unfolded or misfolded proteins at the proteasome.

Growth arrest and DNA damage-induced gene 153 (GADD153): Expression of this gene is specifically activated under conditions associated with impairment of endoplasmic reticulum function.

Glucose-regulated protein 78 (GRP78): Endoplasmic reticulum-resident chaperon required for the folding of newly synthesized proteins. Under conditions associated with endoplasmic reticulum dysfunctioning, GRP78 is necessary to refold unfolded proteins accumulating in the lumen of the endoplasmic reticulum.

Sarcoplasmic/endoplasmic reticulum Ca^{2+}-ATPase (SERCA): Calcium pump that pumps calcium ions from the cytosol against a steppt concentration gradient back to the lumen of the endoplasmic reticulum. This is an energy requiring process.

Unfolded protein response (UPR): Highly conserved stress response of cells triggered under conditions associated with endoplasmic reticulum dysfunction. Activation of UPR results in shut-down of translation and activation of the expression of gene encoding endoplasmic reticulum stress proteins.

X-box binding protein-1 (XBP-1): Low-abundant protein in the brain. Under conditions of endoplasmic reticulum dysfunction, xbp-1 mRNA is processed. This results in the formation of a new protein that functions as transcription factor specific for endoplasmic reticulum stress genes including grp78.

CHAPTER 10 (GÖTZ)

Radicals: reactive compounds with single, unpaired electrons in a molecular orbital i.e. on a discrete energy level

ROS, reactive oxygen species: hydrogen peroxide, superoxide, hydroxyl radical

RNS, reactive nitrogen species: nitric oxide, peroxynitrite

Oxidative stress, nitrosative stress: Increased production of reactive oxygen or reactive nitrogen species. This may result in a dysbalance between oxidizing and reducing compounds in the cell leading to covalent modifications of biomolecules.

Lipid peroxidation: Process in which radicals abstract hydrogen atoms from unsaturated fatty acids in lipids followed by the addition of molecular oxygen. The lipid peroxides formed may further trigger radical chain reactions in membranes or may degrade to reactive aldehydes such as malondialdehyde and 4-hydroxy-2-nonenal.

Radical Scavengers: Compounds that react with radicals to stop radical chain reactions. Trapping of radicals can for example take place via oxidation of vitamin E or of vitamin C. The resulting radicals may be reduced back to the parent compounds by glutathione.

GSH, Glutathione. GSH is a low molecular weight water soluble tripeptide consisting of glutamate, cysteine and glycine. One major function of GSH is the conjugation of reactive electrophiles including aldehydes catalyzed by GSH-transferases, Another important function is the detoxification of organic hydroperoxides catalyzed by GSH-dependent peroxidases.

Alzheimer's disease: Chronic progressive neurodegenerative disease mainly affecting the limbic system, the allo- and neocortex leading to severe deterioration of cognitive functions, first described by Alois Alzheimer in 1906.

Parkinson's disease: Chronic progressive neurodegenerative disease affecting mainly the nigro-striatal system leading to severe akinesia, muscular rigidity and tremor, first described by James Parkinson in 1817.

CHAPTER 11 (DEL RÍO)

Theta rhythm: Spikes of 4-7 Hz frequency that appears in hippocampus both spontaneously and when an animal moves in its environment.

Chemoattraction/repulsion: Growth cone responses induced by permissive or non-permissive substrates or molecules acting at short or long-distance resulting in the advance or the backward movement of the axon.

Leading edge: Distal portion of the growth cone with a large number of filopodia and lamellipodia responsible for sensing the extracellular environment.

Perforant pathway: Also termed entorhino-hippocampal connection, the term refers to the axonal projections from the main (layer II-III) neurons of the entorhinal cortex which innervate the upper portion of the granule and pyramidal cell dendrites of the hippocampus.

Wallerian degeneration: Degeneration of the distal portion of the axon after axotomy of the nerve fiber.

Glial scar: Cellular and molecular response of the nervous tissue after injury which includes glial and meningeal cell proliferation and migration and the deposition of extracellular matrix molecules in the lesioned area.

CHAPTER 12 (ROSSUM)

Chemokines: cytokines with chemoattractive activities

Cytokines: small proteins and peptides with cell-regulatory functions and effects on the growth, survival, differentiation and functions of virtually any cell type

Interleukin: a family of more or less related cytokines which have been systematically named according to a nomenclature system after fulfilling certain criteria

Microglial activation: the process of transformation leading from a 'resting' cell status to increasingly active forms of microglia, accompanied by morphological, antigenic and especially functional adaptations

Resting microglia: the phenotypic form of microglia as seen in the healthy mature CNS tissue and as revealing a ramified cell shape with only low to moderate expression of cell surface antigens which will eventually increase upon stimulation

CHAPTER 13 (VICARIO-ABEJÓN)

BHLH: basic helix-loop-helix.

BMP: bone morphogenetic protein.

CBP: CREB binding protein

CNTF: ciliary neurotrophic factor.

CREB: cyclic-AMP-response-element-binding protein

EGF: epidermal growth factor.

FGF-2: fibroblast growth factor-2.

GFAP: glial fibrillary acidic protein; an intermediate filament protein present in many mature astrocytes.

GLAST: glutamate/aspartate transporter.

Hes: hairy/enhancer of split.

IGF-I: insulin-like growth factor-I

JAK: janus kinase.

LIF: leukaemia inhibitory factor.

MAPK: mitogen-activated protein kinase.

Neural stem cells: these cells have the capacity for self-renewal and the ability to generate neurons, astrocytes, and oligodendrocytes.

PCAP: pituitary adenylate-cyclase-activating polypeptide.

Progenitors: proliferating cells with a limited capacity for self-renewal, and are often unipotent or bipotent.

Precursor cells: precursor cells refers to a variety of immature cells, including NSCs and progenitor cells.

STAT: signal transducer and activator of transcription.

TGF: transforming growth factor

VIP: vasoactive intestinal peptide

CHAPTER 14 (DOMERQ)

Myelinating oligodendrocytes: postmitotic cells derived from oligodendrocyte precursors that migrate into developing white matter from germinal zones.

Myelin, a fatty insulating material composed of modified plasma membrane that ensheaths axons.

Satellite oligodendrocytes: perineuronally located cells of unknown function.

Glial restricted precursor: a bipotential cell originating astrocytes and oligodendrocytes.

Oligodendrocyte progenitors: cells at the earliest stages of the oligodendrocyte lineage.

Excitotoxicity: cell death by excessive activation of glutamate receptors.

Multiple sclerosis, the most common demyelinating disease.

Experimental autoimmune encephalitis: an animal model of multiple sclerosis.

AMPA and kainate receptors: ionotropic glutamate receptors expressed in oligodendrocytes.

Remyelination: myelin repair after demyelination.

CHAPTER 15 (ECHEVARRÍA)

Glial cell: Specialized cells of the nervous system that surround neurons, providing mechanical and physical support to and electrical insulation between neurons.

Myelin sheath: An insulating layer surrounding vertebrate peripheral neurons, which dramatically increases the speed of conduction and is made up of Schwann cells and oligodendrocytes. The cells wrap around the axons up to 50 times

Neural crest cells: Embryonic cells active during the early process of the neural tube development. They arise from the junction of the neural folds and the surface ectoderm. Initially, they are composed of a mass of cells in the presumptive dorsal midline of the neural tube. Subsequently, they split and migrate laterally and ventrally, dispersing widely into the lateral mesoderm. They give rise to almost all glial cells of the peripheral nervous system, branchial arch mesenchyme, sensory ganglia, adrenal medulla, and melanocytes.

Axon: A long process of a neuron carrying efferent (outgoing) action potentials (information from the cell body to target cells).

Cell lineage: Term used to describe cells with a common cellular ancestry that develop from the same type of identifiable immature cell.

Schwann-cell precursor: Neural-crest-derived cells specified to differentiate into Schwann cell and with proliferative activity.

Nodes of Ranvier: The exposed areas of a myelinated axon uncovered by the Schwann-cell sheath; they contain very high densities of sodium channels. Action potentials jump from one node to the next without involving the intermediate axon, a process known as saltatory conduction.

Trophic factor: Molecules with closely related structures that are known to support the survival of different classes of embryonic cells.

Axotomy: Transection or severing of an axon. This type of denervation is often used in experimental studies — on neuronal physiology and neuronal dependence of trophic factors from the target — aimed at understanding of nervous system diseases.

Connective tissue: A material made up of cells that form fibers in the framework providing a support structure for other body tissues.

Autocrine: Pharmacologically active molecules that are secreted by one cell for the purpose of altering its own functions (autocrine effect).

CHAPTER 16 (VECINO)

Determination: Commitment by an embryonic cell to a particular specialized path of development; it is a reflection of a change in the internal character of the cell.

GLAST: Predominant glutamate transporter in neurons.

Gliosis: Refers to changes observed in glial cells after damage and that may either support the survival of neurons or accelerate the progress of neuronal degeneration.

Glutamate: Small signaling molecule secreted by the presynaptic nerve cell at a chemical synapse to relay the signal to the postsynaptic cell.

Immune response: Response made by the immune system of a vertebrate when a foreign substance or microorganism enters its body.

MHC (Major histocompatibility complex): Complex of vertebrate genes coding for a large family of cell-surface proteins that bind peptide fragments of foreign proteins and present them to T lymphocytes to induce an immune response.

Microvilli: Thin cylindrical membrane-covered projections containing a core bundle of actin filaments.

Müller glia: The major glial cell type of the retina.

Neurotrophins: Polypeptide molecules that regulate the survival, development, and maintenance of specific functions in different populations of nerve cells.

NMDA (N-methyl-D-aspartate): Specific receptor of glutamate.

Phagocytosis: Special form of endocytosis in which large particles such as microorganisms and cell debris are ingested via large endocytic vesicles called phagosomes.

Stem cell: Relatively undifferentiated cell that can continue dividing indefinitely, throwing off daughter cells which can undergo terminal differentiation into particular cell types.

T-lymphocyte: Type of lymphocyte responsible for cell-mediated immunity; includes both cytotoxic T-cells and helper T-cells.

Transdifferentiation: Cellular process whereby a differentiated cell reverts to an undifferentiated state and proliferates to give rise to cells of a new phenotype.

Zonula adherens: Beltlike adherent junction that encircles the apical end of an epithelial cell and attaches it to the adjoining cell. A contractile bundle of actin filaments runs along the cytoplasmic surface of the adhesion belt.

CHAPTER 17 (DELGADO-GARCÍA)

Abducens internuclear interneurons: A population of neurons located in the abducens nucleus that share synaptic inputs with abducens motoneurons. Their role is to send a copy of horizontal eye motor commands to medial rectus motoneurons located in the contralateral oculomotor nucleus, via the medial longitudinal fascicle. The lesion of this fascicle produces in humans the syndrome of the internuclear ophthalmoplegia.

Anastomosis: The term refers to the surgical procedure to join two different nerve stumps, in order to facilitate a more reliable communication between them.

Classical (Pavlovian) conditioning: Pairing of a conditioned stimulus and an unconditioned stimulus, until the repeated association allows the conditioned stimulus to evoke a conditioned response that cannot be obtained by its sole presentation.

Compensatory mechanism: Role of a synergistic motor system, that compensate for the loss of the main neural motor system.

Neural elasticity: The term refers to the property of neural cells to return to their previous state; for instance, following their axotomy.

Neural plasticity: Changes in the effectiveness and micro-structure of synaptic connections, considered to be the basis of learning processes.

Neural regeneration: Processes involved in axonal guidance back to the targets following axotomy.

CHAPTER 19 (GONZÁLEZ-FORERO)

Axotomy: Cutting or severing an axon. It involves target disconnection that may lead to neuronal death or activation of a regeneration program in parent neurons.

Cross-modal plasticity: Plastic mechanism by which a brain area processes information of a different sensory modality after sensory deprivation.

Diaschisis: (from the Greek "shocked throughout"). Term introduced by von Monakow in 1914 to refer to the remote effects resulting from deafferentation after brain lesions.

Functional dedifferentiation: Loss of functional specialization acquired during development in mature neurons.

Functional reorganization: Implies that a system can change its functions qualitatively. Thus, a neural pathway might mediate a functional outcome that it does not ordinarily control.

Functional respecification: Induction of a different functional specialization by extrinsic factors after neuronal dedifferentiation.

Homeostatic plasticity: Mechanisms of plasticity that stabilize neuronal activity in order to maintain fundamental characteristics of neural circuits. Such mechanisms include changes in synaptic strength, neuronal excitability and synapse number.

Neurotrophic factors: Molecular family that increases cell survival, stimulates neurite outgrowth and branching during development or after injury, and maintains the differentiated phenotype of adult neurons.

Reactive synaptogenesis: Synapse formation induced following deafferentation, usually involving heterotypic collateral sprouting and establishment of new synaptic contacts on denervated postsynaptic membrane.

Regeneration: Regrowth of the axon to its original target sites.

Stem cells: Cells with the potential to differentiate into many different cell types.

CHAPTER 22 (ROSSI)

Axon regeneration: long-distance regrowth of a stem neurite from a severed neuritic stump.

Axon sprouting: budding of new processes from a neurite. Collateral sprouting occurs along the shaft, whereas terminal sprouting originates from the tip of an intact or severed neurite. The term sprouting is commonly referred to local outgrowth or structural remodelling of terminal arbours, as opposed to long-distance regeneration of the stem neurite.

Cell body response: cellular/molecular changes occurring in nerve cells following axotomy. This response includes morphological modifications (chromatolisis), changes in the metabolic state of the neuron, upregulation of growth genes, activation of pro or anti-apoptotic pathways, down-regulation of molecules involved in synaptic signalling.

Growth-associated genes: a specific set of genes whose coordinated activity is required to sustain long-distance elongation of stem neurites. The set includes axonal components (e.g. cytoskeletal proteins) and growth cone proteins involved in decoding navigation cues and processing the related signal transduction.

Intrinsic neuronal potential for neuritic growth: the actual capability of a neuron to initiate and sustain neuritic elongation. Most adult neurons are only competent for terminal arbour plasticity, whereas stem axon regeneration requires transcription of new genes.

Myelin-associated neurite growth inhibitory molecules: a set of proteins, including Nogo-A, Omgp and MAG, which are expressed by myelinating oligodendrocytes to inhibit neuritic growth and plasticity. The molecules are present on the surface of myelin sheaths and interact with a common receptor on the neuritic membrane.

Retrograde cues: molecules issued by target cells or glial elements along the axon that regulate the intrinsic growth potential of adult neurons. These factors or the relevant signals are retrogradely transported to the cell body where they modulate gene expression: positive cues promote growth processes, whereas negative ones hamper neuritic elongation and sprouting.

CHAPTER 23 (BERGMANN)

Aβ: Peptide produced by cleavage of β-APP; contributes to amyloid formation.

ALS: (Amyotrophic lateral sclerosis) Neurodegenerative disease with preferential loss of motoneurons in the cortex, brainstem and spinal cord.

Alzheimers disease: Age dependent progressive dementia characterised by degeneration of neurons in the neocortex and hippocampus.

AMPA: (α-Amino-3-hydroxy-5-methyl-4-isoxazol-proprionic acid) Excitatory amino acid, agonist for AMPA-type glutamate receptors.

Amyloid: Extracellular protein deposit, typically occurring in Alzheimer's disease.

Apoptosis: Programmed cell death associated with characteristic intracellular changes (vacuoles, chromatine condensation, DNA fragmentation).

β-APP: (β-amyloid precursor protein) Cleavage of β-APP produces Aβ. Mutations in the β-APP gene are known as cause of Alzheimers's disease.

Calpain: Ca^{2+}-dependent protease involved in apoptotic and necrotic cell death.

CaMKII (Calcium/calmodulin-dependent protein kinase II) Enzyme essential for induction of LTP and long-term memory storage by a mechanism involving autophosphorylation.

Caspases: Proteases with various substrates, activation of which represents a critical step in cell death.

CNS: Central nervous system.

EAAT2: Excitatory amino acid transporter 2. Glial glutamate transporter that removes glutamate from the synaptic cleft.

ER: (Endoplasmic reticulum) Ca^{2+} storing organelle.

Excitotoxicity: Glutamate mediated damage of postsynaptic neurons.

γ: Ca^{2+} transport rate, which reflects the combined action of slow uptake into organelles and Ca^{2+} transport across cellular membranes.

Glutamate: Most abundant excitatory neurotransmitter in the CNS; binds to ionotropic (NMDA, AMPA/kainate) and G-protein coupled (metabotropic) receptors.

Huntingtin: Product of the mutated human huntington gene in Huntingtons's disease.

IP$_3$: (Inositoltrisphosphate) Molecule that induces Ca^{2+} release from the ER by binding to IP$_3$ receptors.

Ischemia: Reduction or interruption of blood-flow.

K$_B$: Ca^{2+} buffering capacity of an added buffer, i.e. fura-2.

K$_d$: Dissociation constant of a Ca^{2+} buffer; indicates the concentration at which the buffer is half maximally saturated with Ca^{2+}.

k$_{on}$: Rate (concentration per time) at which a Ca^{2+} buffer binds Ca^{2+}.

K$_S$: Endogenous Ca^{2+} buffering capacity: relative fraction of bound vs. free Ca^{2+} ions in the cell.

LTP (Long term potentiation) Induction of a long lasting excitation of the postsynaptic membrane following short stimulation. Central mechanism for induction of synaptic plasticity.

Microdomain: Local accumulation of $[Ca^{2+}]i$ (up to 10μM), which can form around open Ca^{2+} channels or Ca^{2+} release sites.

MPP$^+$: N-methylpyridinium ion that selectively accumulates in dopaminergic neurons, where it inhibits the mitochondrial electron chain.

MPTP: (1-Methyl-4-phenyl-1,2,3,6-tetrahydropyridine) Neurotoxin that is oxidised to MPP$^+$ in the CNS. Used to induce Parkinson like symptoms in rodents.

Nanodomain: Zone of very high $[Ca^{2+}]$ (up to 100μM) in the direct vicinity (< 100nm) of an open Ca^{2+} channel.

Neurofibrillary tangles: Deposition of tau protein in soma and dendrites of neurons, typically occurring in Alzheimr's disease.

NMDA: (N-methyl-D-aspartate) Excitatory amino acid, agonist for NMDA-type glutamate receptors

Parkinson's disease: neurodegenerative disorder characterized by progressive degeneration of dopaminergic neurons in the substantia nigra.

PKA: (Protein kinase A) Protein kinase that is inducible by cAMP. Contributes to LTP and synaptic plasticity.

SOD1: (Co/Zn-super oxide dismutase) Enzyme that captures free radicals. Mutations in the SOD1 gene are cause of familial forms of ALS.

CHAPTER 24 (LÓPEZ-GARCÍA)

Neural stem cell: Proliferating cell located in the nervous tissue, which displays long-term self-renewal and multipotentiality.

Progenitor: Proliferating cell with limited self-renewal capacity and restricted potentially.

Ependyma – ependymal sulcus: Monostratified ciliated epithelia surrounding the adult cerebral ventricles. In lower vertebrates, some ependymal regions retaining neurogenic properties are known as sulci.

Radial glia – tanycyte: Embryonic astroglial cells with their somata located in the neuroepithelium. They display long straight processes that contact the pial surface and serve as guides for the migration of neurons to the nervous parenchyma. Radial glia cells may be the precursors of most cell types in the central nervous system.

Nestin: Intermediate filament protein, which is intensely expressed in, but not specific to, stem cells in diverse tissues, including the CNS.

Neuroblast – glioblast: Progenitors with restricted potentiality, which give rise only to neurons or glia respectively.

Subventricular zone: Active germinal zone of the central nervous system located in the subependyma of the lateral ventricles. This region persists in the postnatal forebrain and retains the capacity to generate glial cells, and interneurons that migrate into the olfactory bulb through the rostral migratory stream.

Subgranular zone: Germinal zone of the postnatal hippocampus, which contains mitotically active precursors capable of generating glial cells and granule neurons. This region is located in the border between the hilus and the granule-cell layer of the dentate gyrus.

Granule cells (neurons): Small neurons located in the olfactory bulb, dentate gyrus, and cerebellum.

Axonal regeneration: Phenomenon by which a neuron can re-grow its axon and re-establish synaptic contacts after the axonal projection has been transected or lesioned.

Neuronal regeneration: Mechanism by which neurons that have degenerated are replaced by newly generated ones. These new neurons integrate into the circuitry and re-establish functionality.

CHAPTER 26 (ARMENGOL)

Allograft: The donor is different from the host, but belongs to the same animal species.

Anisomorphic lesion: Central nervous system damage with considerable lost of neuronal elements and followed by a typical astroglial reaction.

Autograft: Donated tissue is from the host.

Autophagy: Mechanisms by which a cell eliminates its own constituents (e.g., aged mitochondria) by secondary lysosomes, when misrouting could evolve into complete cell destruction.

Blood-brain barrier: Specific and selective endothelial and astroglial interface that isolates the nervous system, avoiding noxa invasions.

Deep brain stimulation: Administration of electrical currents by electrodes stereotaxically placed into the brain to control or modulate the synaptic activity of functionally altered central nervous system circuits.

High frequency: Electrical stimulation of at least 100 or more Hertz by electrodes chronically implanted within the brain.

Neuronal cell death: Degeneration of neurons taking place normally during development, but which in the adult brain is the final effect caused by injury, neurodegenerative diseases or physiological aging processes. The term **necrosis** implies the degeneration of the cytoplasm followed by rupture of the cell nucleus.

The term **apoptosis** defines the onset of a proper genetic program that allows fragmentation of the cell nucleus.

Neurotransmitters: Molecules responsible for the chemical synaptic interneuronal communication. Disturbances of the synthesis or release of neurotransmitters, and/or the absence or alteration of their receptors, implies alterations in the normal brain functions.

Rejection: Host response, mediated by the immune system, which reacts to implanted material as to an injury, tending to its elimination.

Side-effect: Undesirable alteration or symptom elicited by medical or surgical treatment.

Stem cells: Undifferentiated multipotent cells capable of auto-generation and generation, by cell division, of a great variety of daughter-cell typologies. In healthy adult organisms, these cells give rise to progenitors of adult cells of those tissues undergoing continuous regenerative cycles (e.g., blood cells, skin cells, etc).

Stereotaxy: Surgical procedure using spatial co-ordinates, permitting the precise placement of electrodes within deep brain structures.

Transplant: (synonymous with **graft**). Surgical introduction of a healthy tissue within the host body to replace and restore the function of damaged organs.

Trophic factors: Proteins that stimulate the growth of neuronal branches during brain development and which in the adult brain play a key role in the survival and normal functioning of neurons. The most-relevant are NGF (nerve growth factor) – a member of the neurotrophins family [NT3, NT4, NT6, BDNF (brain-derived growth factor)], CNTF (ciliary neurotrophic factor), GDNF (glial-derived growth factor), IGF (insulin growth factor), etc. Their decrease or loss elicits apoptosis.

Xenograft: The donated tissue is from an animal species different to that of the host.

CHAPTER 28 (LISTE)

Primary cells (e.g. primary neurons): Cells isolated from fresh tissue and not sub-cultured (passaged).

Cell strains (secondary cultures, e.g. neurosphere cultures): Serially passaged, mortal cell cultures. Normally showing evolving properties with time in culture.

Cell lines (e.g. v-myc lines): Serially passaged, immortal cell cultures. Not transformed. Stable properties.

Totipotent cell: One able to generate a complete, living organism (e.g. the zygote).

Pluripotent cell: One able to generate all tissues in a living subject, but not to give rise to it (e.g., the ES cells).

Multipotent cell: A tissue-specific cell, able to generate cells from all types proper to a given organ, but not others. (A definition being questioned nowadays on the basis of trans-differentiation evidence).

Stem cell: Immortal immature cell with the ability to self-renew (through both symmetric and also asymmetric divisions after a given time point), and to give rise to differentiated cells, belonging to ecto-, meso- and endo-dermal cell lineages (pluri-potency).

Embryonic stem cell (ES cells): Stem cell derived from the inner cell mass of the pre-implantation blastocyst.

Adult stem cell. Stem cell present in postnatal tissues. As far as is known, it is not endowed with immortal properties. Multi-potent.

Neural stem cell (NSC): Stem cell in nervous system, able to give rise to glia and neurons — that is, multipotent. As far as is known, it is not endowed with immortal properties.

Neural precursor cell: Any mitotic cell that can give rise to other neural cells, not necessarily able to perpetuate itself through asymmetric divisions. Precursors may be comparable to transit-amplifying populations *in vivo*.

Neuronal / Glial progenitor cells: Last mitotic cells that are fully specified to produce glia or neurons, bot not both. Very limited mitotic potential. Generated from precursor cells.

Neuroblast or Glioblast: Neuronal or glial progenitor, respectively.

Neurosphere culture: Free-floating aggregates of a heterogeneous mix of neural stem/precursor/progenitor cells with ability to propagate to a limited extent in the presence of growth factors.

CHAPTER 29 (NAVARRO)

Axotomy: rupture or section of the axon, leading to loss of continuity and conduction of impulses through the distal axon.

Neurapraxia: focal block of impulse conduction, usually due to nerve compression.

Neurotmesis: complete transection of the peripheral nerve.

Wallerian degeneration: process of degradation of axonal and myelin debris after axotomy.

Axonal regeneration: regrowth and elongation of the severed axon tips.

Reinnervation specificity: adequate reinnervation of a target end organ by axons that originally served that organ.

Nerve graft repair: interposition of a nerve segment between the stumps of a transected nerve

Tubulization: implantation of a nerve guide to bridge a gap created in the peripheral nerve.

Artificial nerve graft: device designed to repair nerve gaps, as a substitute for natural nerve grafts.

CHAPTER 30 (BACH-Y-RITA)

Brain Plasticity: Ability of the brain to reorganize.

Volume transmission: Nonsynaptic diffusion neurotransmission.

Receptor plasticity: Up- or down regulation of synaptic and nonsynaptic neurotransmitter receptors.

Late rehabilitation: Rehabilitation provided 1 or more years post lesion.

Sensory substitution: Information from an artificial sensory receptor system (e.g., TV camera) provided to a person with a sensory loss (e.g., blindness) via an intact system (e.g., tactile).

Computer assisted motivating rehabilitation (CAMR): Using computer technology to provide rehabilitation programs that are of interest to the patient and that motivate compliance with the program.

CHAPTER 31 (ISENMANN)

AADC: aromatic-L-amino-acid decarboxylase

AAV: adeno-associated virus

AchE: acetyl choline esterase

Ad: adenovirus

ALS: amyotrophic lateral sclerosis, motor neuron disease

BBB: blood-brain barrier

BDNF: brain-derived neurotrophic factor

CMV: cytomegalovirus

CNTF: ciliary neurotrophic factor

FIV: feline immunodeficiency virus

GAD: glutamic acid decarboyxlase

GDNF: glial cell line-derived neurotrophic factor

GFAP: glial fibrillary acidic protein

GFP: green fluorescent protein

GTP CH 1: GTP-cyclohydrolase 1

HIV: human immunodeficiency virus

HSV: herpes simplex virus

IAP: inhibitor of apoptosis protein

Infection: transfer of genetic material by wild-type pathogens (e. g., viruses)

LacZ: gene for bacterial (E. coli) β-galactosidase

MBP: myelin basic protein

MoMLV: Moloney murine leukaemia virus

NSE: neuron specific enolase

NGF: nerve growth factor

SIV: simian immunodeficiency virus

SOD: superoxide dismutase

STN: subthalamic nucleus

TH: tyrosin hydroxylase

TK: thymidine kinase

Transduction: transfer of genetic material by viral vectors

Transfection: transfer of genetic material by non-viral vectors

VMAT-2: vesicular monoamine transporter-2

WPRE: woodchuck hepatitis virus posttranscriptional regulatory element

XIAP: X-chromosome linked inhibitor of apoptosis protein

CHAPTER 33 (KRARUP)

Wallerian degeneration: The specific set of processes that occur distal to the site of loss of axonal continuity, independent of its cause. These processes include the break-down of the axon and myelin and change of the microenvironment to enhance regeneration, including recruitment of macrophages, removal of debris, formation and secretion of cytokines and chemokines, and proliferation of Schwann cells.

Elongation: Outgrowth of the axon growth cone from the proximal nerve stump

Reinnervation: Reestablishment of functional contact between the regrown axon and the denervated target organ. If this connection occurs between alien axons and target organs, the function may not be clinically advantageous.

Maturation: Recovery of function and structure of the nerve fiber during and after elongation.

Regeneration: Collective term that covers elongation, reinnervation, and maturation.

Preferential motor reinnervation: This term, PMR, has been coined by T.M.E. Brushart to describe that motor axons preferably will reenter a Schwann cell tube that has previously housed a motor axon. This concept forms the first basis for the specificity of reinnervation and may depend on molecular interaction, humoral factors, and other mechanism.

Nerve conduction studies: The method of stimulating nerve to elicit action potentials that may be recorded from muscle (compound muscle action potential, CMAP) or from sensory nerve (compound sensory action potential, CSAP) and provide information about the integrity and number of functional motor and sensory nerve fibers during degeneration and regeneration.

CHAPTER 35 (WILMS)

MPTP: 1-methyl-4-phenyl-1,2,3,6-tetrahydropyridine, a neurotoxin highly selective for the substantia nigra of human and non-human primates. Primates treated with MPTP develop motor disturbances resembling those seen in idiopathic PD, including bradykinesia, rigidity and postural abnormalities.

Rotenone: pesticide and mitochondrial toxin that reproduces many features of Parkinson's disease in rats, including nigrostriatal dopaminergic degeneration and formation of alpha-synuclein-positive cytoplasmic inclusions in nigral neurons microglia: resident macrophages of the CNS

Neuromelanin: dark-coloured pigment produced in the dopaminergic neurons of the human substantia nigra

Lewy bodies: proteinaceous intraneuronal inclusions, a common pathological hallmark of sporadic PD. They are composed of α-synuclein, ubiquitin and proteasomal subunits.

α-synuclein: Small, soluble protein expressed primarily in neural tissue, where they are seen mainly in presynaptic terminals. Normal cellular functions has not been determined yet. Mutations in alpha-synuclein are associated with rare familial cases of early-onset Parkinson's disease, and the protein accumulates abnormally in Parkinson's disease

Ubiquitin/proteosomal pathway: proteolytic pathway of misfolded proteins and their detoxification in eukaryotic cells.

Parkin: The gene causing autosomal recessive juvenile Parkinson's disease was designated *parkin* and encodes a protein of 465 amino acids, which acts as a E3 ubiquitin ligase.

CHAPTER 37 (ALBERCA)

Aβ amyloid: A substance generated from the proteolytic cleavage of a transmembrane protein called β-amyloid precursor protein (APP) that accumulates in the brain and is believed to play a key role in AD pathogenesis

Alzheimer's disease: A clinicopathological syndrome described by Alois Alzheimer in 1906, characterized clinically by dementia and pathologically by the existence of both neurofibrillary tangles and senile plaques.

Apolipoprotein E: A component of several classes of plasma, CSF, and cell proteins, which are related to remodelling and sprouting of neurons, synaptic plasticity, and maintenance of neuronal integrity and cholinergic activity. Three alleles on the long arm of chromosome 19 code for the isoforms Apo E2, Apo E3, and Apo E4 (which is a risk factor for AD).

Cholinergic system: A neurotransmission system that arises from the nucleus basalis of Meynert, diffusely projects to the cerebral cortex and employs acetylcholine as neurotransmitter. Cholinergic deficit correlates with dementia in AD.

Dementia: A chronic condition with multiple cognitive deficits (including memory defect and at least one of the following: aphasia, apraxia, agnosia, and dysexecutive functioning). Dementia represents a decline from previous state and is severe enough to impair occupational or social functions. Dementia is the clinical landmark of Alzheimer's disease

Delusions: Mistaken beliefs that are held with unshakeable conviction.

Mild Cognitive Impairment: A syndrome of isolated pathological memory loss, which in most instances represents "preclinical" Alzheimer's disease.

Minimental State Examination: A screening cognitive tool universally employed to detect dementia

Hallucinations: Perceptions which are not based on external stimuli. They can occur in any modality of perception.

Neurofibrillary tangle: Cytoskeletal abnormality in which neurons contain paired helical filaments composed mainly of abnormal aggregations of phosphorylated tau.

Protein "tau": Group of translation products of a single gene with six major isoforms that belong to the group of microtubule associated proteins. They are abnormally phosphorylated in AD and aggregate into paired helical filaments.

Senile plaque: Elementary extracellular lesion with a core formed by pleated Aβ peptides surrounded by degenerating axons, dystrophic dendrites and microglia.

CHAPTER 40 (PTITO)

Positron Emission Tomography (PET): None-invasive imaging technique that measures the distribution of positron emitting radioisotopes in the body. PET measures the three-dimensional distribution of tracers radiolabelled with positron emitting radioisotopes such as oxygen-15, carbon-11 and fluorine-18. It is assumed that an increased flow to a brain area is an indication of increased function. This is because the active areas use more oxygen and metabolites and produce more waste products. So an increased blood flow is necessary to supply the former and remove the latter.

Functional Magnetic Resonance (fMRI): None-invasive technique, based on MRI, to accurately localize brain activity by measuring the Blood oxygen level dependent or BOLD response. fMRI BOLD signal is based upon the diffusion of hydrogen nuclei (protons) in tissue water in a local inhomogeneous magnetic fields caused by intrinsic contrast agent (i.e., deoxyhemoglobin). Deoxyhemoglobin is paramagnetic and so a blood vessel containing this product placed in a magnetic field alters that field in its locality. It is possible to map this blood flow based on changes in local magnetic fields in different brain regions. Consequently, a decrease in the relative amount of deoxygenated haemoglobin will actually cause an increase in the fMRI signal.

BOLD : Blood oxygen level dependent response.

Magnetic Resonance Imaging (MRI): A strong magnetic field is passed through the patient's head, causing some molecules to emit radiowaves. Different molecules emit energy at different frequencies. The MRI scanner is tuned to detect the concentration of radiation from hydrogen molecules and use this information to prepare pictures of slices of the brain.

Tractography (fibre tracking): Usually diffusion is related to the random thermal motion of molecules. With MRI, we can measure the magnitude and direction of the diffusion of water molecules. When there is oriented fibre structure, the diffusing water molecules will move preferentially parallel to the fibres direction therefore marking the fibres patterns.

Sensory substitution: The replacement of one sensory input by another, while preserving some key functions of the original sense.

Cross-modal plasticity: The adaptation of certain aspects of sensory cortical identity instructed by another type of sensory stimulation that differs from the original one.

Behavioural recovery: Restoration of adequate physiological capabilities to survive adequately in the environment.

Blindsight: Ability of a person, apparently totally blind, to discern subconsciously the location, form, and size of objects, thereby strongly ignoring having "seen" something.

Phantom limb: Sometimes referred to as "stump hallucination": it is the subjective sensation, not arising from an external stimulus, that an amputated limb is still present.

Hemispherectomy: Removal of an entire cerebral hemisphere. Most frequently done for control of seizures for patients with severe damage to the hemisphere that has resulted in loss of function and seizures.

CHAPTER 41 (CANTERO)

Cortical dynamics: Electrophysiological activity generated by interconnected neural networks located in different layers of the cortex. These cortical mechanisms have been largely proposed as neurophysiological foundations of some unique cognitive capabilities present in higher mammals and humans.

EEG coherence: EEG measure of the linear association between two derivations in the frequency domain. It varies between 0 and 1, and is mathematically analogous to a cross-correlation coefficient. This quantitative EEG technique has considerable clinical utility and can directly reflect neural network connectivity and neural network dynamics.

Human brain rhythms: Levels of neuronal synchronization within different frequency ranges associated with different behavioral correlates. Most important EEG frequency bands in humans are: delta (0.5-3 Hz), theta (3.5-7.5 Hz), alpha (8-12), sleep spindles (12-14 Hz), beta (15-30 Hz), and gamma (31-70 Hz). It is important to mention here that the limits between beta and gamma bands are not quite clear to date, as revealed by the different frequency ranges employed in similar studies.

Memory consolidation: The neural representation underlying the memory trace of an event changes with the simple passage of time, becoming less fragile with respect to behavioural interference. As a result of this strengthening of memory, performance becomes less dependent on voluntary attention and the brain engages new regions during the recall process.

Procedural learning: Gains in performance on perceptual and/or motor tasks as a result of experience-dependent changes in the properties of neurons and in the functional organization of the cerebral cortex.

Sensory gating: The brain's normal ability to automatically inhibit the response to unimportant stimuli.

Spectral analysis: Quantitative EEG technique based on a linear decomposition into harmonic waves with constant amplitudes. This analysis technique provides information about the levels of neural synchronization underlying electrodes placed on cortical regions of interest in specific frequency ranges. Measurement units are $\mu V^2/Hz$ (spectral power).

Wavelets: Mathematical function used to describe the temporal course of a neurophysiological process in the frequency domain. It can be also useful to recover weak EEG signals from noise. The most important difference with fast Fourier techniques is the absence of limitations with the window size, making it more flexible to study cerebral processes with different temporal resolutions.

LIST OF CONTRIBUTORS

Roman Alberca
Department of Neurology
Alzheimer's Disease Unit
Virgen del Rocío
41012 Sevilla
Spain

Håkan Aldskogius
Uppsala University Biomedical Center
Department of Neuroscience
PO Box 587
751 23 Uppsala
Sweden

José A. Armengol
Department of Anatomy
University of Sevilla
Carretera de Utrera km 1
41013 Sevilla
Spain

Mercedes Atienza
Laboratorio Andaluz de Biología
University Pablo de Olavide
Carretera de Utrera km 1
41013 Sevilla
Spain

Paul Bach-Y-Rita
Departments of Orthopedics and Rehabilitation Medicine
University of Wisconsin
Madison, WI 53706
USA

Mathias Bähr
Department of Neurology
University Hospital Göttingen
Robert-Koch-Strasse 40
37075 Göttingen
Germany

LIST OF CONTRIBUTORS

Beatriz Benitéz-Temi
Department of Physiology and Zoology
University of Sevilla
Avda. Reina Mercedes, 6
41012 Sevilla
Spain

Friederike Bergmann
Center of Physiology
University of Göttingen
Humboldtallee 23
37073 Göttingen
Germany

Johannes Bufler
Department of Pharmacology and Toxicology
University of Hannover
Carl-Neuberg-Strasse 1
30623 Hannover
Germany

Jose L. Cantero
Division of Neuroscience
University Pablo de Olavide
Carretera de Utrera Km 1
41013 Sevilla
Spain

Valentin Ceña
Centro Regional de Investigaciones Biomédicas
University Castilla-La Mancha
02071 Albacete
Spain

Jose Maria Cuezva
Department of Molecular Biology
University Autónoma de Madrid
28049 Madrid
Spain

Fernando De Castro
Institute of Neuroscience
University of Salamanca
37007 Salamanca
Spain

Rosa Rodriguez de la Cruz
Departments of Physiology and Zoology
University of Sevilla
Avda. Reina Mercedes, 6
41012 Sevilla
Spain

José M. Delgado-García
Division of Neuroscience
University Pablo de Olavide
Carretera de Utrera Km 1
41013 Sevilla
Spain

Reinhard Dengler
Department of Pharmacology and Toxicology
University of Hannover
Carl-Neuberg-Strasse 1
30623 Hannover
Germany

Günther Deuschl
Department of Neurology
University of Kiel
Schittenhelmstrasse 10
24105 Kiel
Germany

Maria Domercq
Department of Neurosciences
University del País Vasco
48940 Leioa, Bizkaya
Spain

Diego Echevarría
Institute of Neuroscience
University Miguel Hernandez
Carretera de Valencia, km 87
03550 Alicante
Spain

M. Fernández
Centro Regional de Investigaciones Biomédicas
University Castilla La Mancha
Avda. De Almansa, s/n
02071 Albacete
Spain

Michael Frotscher
Institute for Anatomy and Cell Biology
University of Freiburg
Albertstrasse 17
79104 Freiburg
Germany

Minoca García
Department of Cellular Biology
University del País Vasco
48940 Leioa, Bizkaia
Spain

Gonzal García del Caño
Department of Neuroscience
University del País Vasco
48940 Leioa, Bizkaia
Spain

Manfred Gerlach
Department of Child, Youth Psychiatry and Psychotherapy
University of Würzburg
Füchsleinstrasse 15
97080 Würzburg
Germany

Iumaculada Gerrikagoitia
Department of Neurosciences
University of the Basque Country
48940 Leioa, Bizkaia
Spain

David Gonzáles-Forero
Department of Physiology and Zoology
University of Sevilla
Avda. Reina Mercedes, 6
41012 Sevilla
Spain

C. González-García
Centro Regional de Investigaciones Biomédicas
University Castilla La Mancha
Avda. De Almansa, s/n
02071 Albacete
Spain

Mario E. Götz
Institute of Pharmacology
University of Kiel
Hospitalstrasse 4
24105 Kiel
Germany

Aqués Gruart
Division of Neuroscience
University Pablo de Olavide
Carretera de Utera km 1
41013 Sevilla
Spain

Carola A. Haas
Institute of Anatomy and Cell Biology
University of Freiburg
Albertstrasse 17
79104 Freiburg
Germany

Uwe-Karsten Hanisch
University of Göttingen
Institute of Neuropathology
Robert-Koch-Strasse 40
37075 Göttingen
Germany

Matthias Herdegen
Faculty of law
University of Bonn
Adenauerallee 24-42
53113 Bonn
Germany

Thomas Herdegen
Institute of Pharmacology
University of Kiel
Hospitalstrasse 4
24105 Kiel
Germany

Siegfried Hoyer
Institute of Pathology
University of Heidelberg
Im Neuenheimer Feld 220
69120 Heidelberg
Germany

Stefan Isenmann
Department of Neurology
University of Jena
Philosophenweg 3
07740 Jena
Germany

Jan Johnson
Department of Anatomy and Developmental Biology
University College London
Gower Street
London WC1E 6BT
UK

Joaquin Jordán
Centro Regional de Investigaciones Biomédicas
University Castilla La Mancha
Avda. De Almansa, s/n
02071 Albacete
Spain

Ole Steen Jørgensen
Department of Pharmacology
University of Copenhagen
Rigshospitalet 6102
2100 Copenhagen
Denmark

Leszek Kaczmarek
Nencki Institute
Pasteur 3
02-093 Warsaw
Poland

Bozena Kaminska
Department of Cellular Biochemistry
Nencki Institute
Pasteur 3
02-093 Warsaw
Poland

Bernhard U. Keller
Center of Neurophysiology
University of Göttingen
Humboldtallee 23
37073 Göttingen
Germany

Pawel Kermer
Department of Neurology
University of Göttingen
Robert-Koch-Strasse 40
37075 Göttingen
Germany

Deniz Kirik
Department of Physiological Sciences
University of Lund
22184 Lund
Sweden

Christian Krarup
Department of Clinical Neurophysiology
Rigshospitalet
Blegdamsvej 9
2100 Copenhagen
Denmark

Ron Kupers
Aarhus University Hospital
PET Center
Norrebrogade 44
8000 Aarhus
Denmark

Ludwin Ley
Institute of Pharmacology
University of Kiel
Hospitalstrasse 4
24105 Kiel
Germany

Isabel Liste
Center of Molecular Biology Severo Ochoa
University Autónoma de Madrid
Campus Cantoblanco
28049-Madrid
Spain

Carlos López-García
Department of Neurobiology
University of Valencia
Dr. Moliner, 50
46100 Burjassot, Valencia
Spain

Salvador Martínez
Institute of Neurosciences
University Miguel Hernandez
Carretera de Valencia, km 87
03550 Alicante
Spain

Luis Martínez-Millán
Department of Neuroscience
University of the Basque Country
48940 Leioa, Bizkaia
Spain

Alberto Martínez-Serrano
Center of Molecular Biology Severo Ochoa
University Autónoma de Madrid
28049 Madrid
Spain

Carlos Matute
Department of Neurosciences
University del País Vasco
48940 Leioa Vizcaya
Spain

E. Montes
Department of Neurology
Alzheimer's Disease Unit
Virgen del Rocío
41012 Seville
Spain

Juan Nacher
Neurobiology, Cell Biology Department
University of Valencia
Dr. Moliner, 50
46100 Burjassot
Spain

Xavier Navarro
Institute of Neurosciences
University Autónoma de Barcelona
08193 Bellaterra
Spain

Annette Skræp Nielsen
Department of Neurorehabilitation,
Copenhagen University Hospital
Hvidovre Hospital 231
2650 Hvidovre
Denmark

Jochen H.M. Prehn
Department of Physiology
Royal College of Surgeons in Ireland
123 St Stephen's Green
Dublin 2
Ireland

Manuel Nieto-Sampedro
Institute Cajal
Avda. Doctor Arce, 37
28002 Madrid
Spain

A. D. Ortega
Department of Molecular Biology
Centro de Biología Molecular Severo Ochoa
University Autónoma de Madrid
28049 Madrid
Spain

Wulf Paschen
Laboratory of Molecular Neurobiology
Max-Planck-Institute for Neurological Research
Gleuelerstrasse 50
50931 Köln
Germany

Konstanze Plaschke
Department of Anesthesiology
Im Neuenheimer Feld 110
69120 Heidelberg
Germany

Angel M. Pastor
Departament of Physiology and Zoology
University of Sevilla
Avda. Reina Mercedes, 6
41012 Sevilla
Spain

Maurice Ptito
School of Optometry
University of Montréal
C.P 6128, Succ. Centre Ville
Montreal, Quebec H3C 3J7
Canada

J. Alexandre Ribeiro
Laboratory of Neurosciences
Av. Prof. Egas Moniz
1649-028 Lisbon
Portugal

Jose Antonio del Río
Department of Cellular Biology
University of Barcelona
Josep Samitier, 1-5
08028 Barcelona
Spain

Ferdinando Rossi
Department of Neuroscience
Rita Levi Montalcini Centre
University of Turin
Corso Raffaello 30
10125 Turin
Italy

Eduardo Soriano
Department of Cellular Biology
University de Barcelona
Parque Científico
08028 Barcelona
Spain

Denise van Rossum
Institute of Neuropathology
University of Göttingen
Robert-Koch-Strasse 40
37075 Göttingen
Germany

Elena Vecino
Department of Cellular Biology
University del País Vasco
48940 Leioa, Vizcaya
Spain

Enrique Verdú
Institute of Neurosciences
University Autónoma de Barcelona
08193 Bellaterra
Spain

Carlos Vicario Abejón
Centro de Investigaciones Biologicas
Calle Ramiro de Maeztu, 9
28040 Madrid
Spain

Hartmut Wilms
Department of Neurology
University of Kiel
Schittenhelmstrasse 10
24105 Kiel
Germany

Christian Winkler
Department of Neurology
Medical School of Hannover
Carl-Neuberg-Strasse 1
30625 Hannover
Germany

Maria J. Yusta-Boyo
Centro de Investigaciones Biologicas
Calle Ramiro de Maeztu, 9
28040 Madrid
Spain

INDEX

α-tocopherol 145, 155, 275, 280, 282, 593, 594, 596, 602, 605, 608
α-synuclein 90, 108, 136, 144, 543, 544, 546, 581
12-O-tetradecanoyl phorbol-13-acetate (TPA) 152
1-Methyl-4-phenyl-1,2,3,6-tetrahydropyridine (MPTP) 144, 161, 374, 494, 530-534, 539-543, 594
Aβ amyloid 160, 188
abducens internuclear interneurons 270, 271
abducens internuclear neurons 268-270, 342, 343
abducens nucleus 268, 271, 286, 289, 341
aberrant splicing of glutamate receptors 123
acetyl choline esterase (AchE) 13, 501, 566, 570, 571, 593
acetylcholine 3, 7, 11, 13, 15, 18, 19, 128, 485, 570, 603
-- action 20
-- receptors 292
acetyl-CoA 2, 3, 7
activation of RNA polymerase 23
activator protein 1 (AP-1) 151, 163
acute neurodegenerative disorder 109
acute oxidative stress insults 99
AD, free radical research 135
ADC changes 12
ADC increase 12
adenine nucleotide translocator (ANT) 99, 102, 105, 532, 533
adeno-associated virus (AAV) 490, 491, 493-495, 502
adenosine triphosphate (ATP) 2, 3, 5-7, 10, 11, 13-16, 18-20, 26, 39, 86, 87, 97, 99-101, 103, 106, 109, 117, 137, 142, 153, 154, 185, 186, 189, 370, 385, 527, 531-534
adenovirus 416, 490, 491, 502, 558
adrenalectomy 7, 387

adrenocorticotropic hormone (ACTH) 17
adult mammal brain 260, 262
adult stem cell 205, 413
advanced glycation end products (AGEs) 14
afferent parallel fibre afferents 54
AGE receptor (RAGE) 14
age-associated brain diseases/disorders 1, 2, 8, 403
age-related behavioral changes 11
age-related disorders 1, 2, 12, 198, 483
aging 8-20, 68, 136, 141, 144, 156, 198, 211, 389, 413, 483, 569, 632
aging brain 1, 2, 8, 13, 17, 20, 228
-- as risk factor for neurodegeneration 8
aging cells 543
aging hippocampus 390
agrin 315, 325
AIDS dementia 197
aldynoglia 323, 332, 334, 335
allograft 451, 459, 461
allosteric enzymes hexokinases 2
alpha-amino-3-hydroxy-5-methyl-4-isoxazolepropionate receptor (AMPAR) 49, 52, 54, 55
alternative grafts 459
alternative splicing 27, 123, 128, 129
Alzheimer disease (AD) 1, 2, 11, 16 69, 89, 90, 91, 108, 109, 118, 135, 136, 140, 144, 148, 150, 152, 153, 155, 158-160, 185, 188, 192, 194, 196, 197, 200, 201, 211, 275, 282, 300, 304, 329, 365, 371, 372, 374, 376, 377, 391, 394, 413, 419, 484, 485, 490-493, 496, 501, 504, 505, 537, 563-572, 580, 581, 589, 592-594, 600, 603, 630, 631, 633
Alzheimer, A. 10
amacrine cells 246, 255
amyloid 158, 183, 201, 282, 564, 565, 594
amyloid deposits 201, 563, 565

amyloid fibrils 544
amyloid precursor protein (APP) 4, 5, 11 89, 90, 135, 158, 196, 563, 564
amyotrophic lateral sclerosis (ALS) 108, 118, 129-131, 194, 196-198, 369, 371-373, 376, 377, 492-494, 501, 504, 505, 575-585, 591, 595, 596, 608, 609
anastomosis 263-265, 267, 271, 476
androgen receptor (AR) gene 126
anisomorphic lesions 324, 419
anoxia 14, 247, 249, 250, 276, 324
antigen-presenting cell (APC) 188, 190-192
anti-inflammatory cytokines 425
antioxidant agents 245, 532
antioxidant mechanisms 99
antioxidant protectors 275, 280
antioxidants 86, 108, 109, 136, 141, 142, 144, 146, 150, 152, 154, 159-162, 275, 276, 278-282, 328, 329, 425, 428, 505, 555, 565, 566, 570, 593, 596, 597, 602, 605, 607-611
antioxidative therapeutic strategies 159
apical microvilli 246
apolipoprotein (APO) 20, 135, 158
apolipoprotein E (APOE) 163, 185, 282, 396
apoptosis 18, 24, 79-92, 95, 99, 101-109, 126, 143, 150, 151, 153, 169, 189, 206, 219, 223, 238, 276, 277, 303, 309, 333, 371, 374-376, 425, 429, 432, 433, 496, 501, 504, 505, 527, 528, 533, 534, 544, 563, 590, 592, 594-596
apoptosis activating factor 1 (APAF 1) 82, 163
apoptosis inducing factor (AIF) 81, 87, 101, 103, 105, 151, 534
apoptosis pathways 81, 82, 101, 103, 120, 235, 533, 534, 595, 596
apoptosis-like processes 71
apoptosome 81, 82, 101, 105, 151, 533, 534

apoptotic cell death 71, 79, 80, 83, 87, 89, 101, 105, 108, 109, 150, 371, 387, 429, 530, 534, 542, 596
apoptotic cells 99, 101, 103
apoptotic machinery of mitochondria 101
apparent diffusion coefficient (ADC) 12, 20
arachidonic acid (AA) 50, 113, 145, 247, 317, 318, 528
arachidonic acid metabolites 328
aromatic-L-amino-acid decarboxylase (AADC) 494, 501
arterial hypoglycemia 14
arterial hypotension 14
arterial hypoxemia 14
artificial nerve graft 451, 461, 467, 469, 470
asbestos 152
ascorbic acid 141, 160, 161, 275, 280
astroblasts 328, 330, 331
astrocyte generation 203-206, 209
astrocyte stimulation 158
astrocyte surface 331
astrocyte-restricted progenitor cells 204
astrocytes 10, 52, 58, 62, 129, 135, 140, 153, 158, 174, 183, 191, 192, 194, 203-207, 208-212, 215, 217, 218, 220, 222, 226, 227, 239, 245, 247, 251, 252, 256, 279, 282, 30, 308-3310, 313, 324, 325, 327, 329-334, 336, 338, 344, 345, 367, 386, 414, 433, 486, 541, 590
-- as neural stem cells 210
-- differentiation of 203-207, 208-210, 330
-- fibrous 203, 204, 216, 329, 331
-- perivascular 161
-- protoplasmic 203, 204
-- retinal 248
-- role in brain damage and repair 210
-- swelling 324
astroglyosis 10, 226, 330, 542

atomic force microscopy 544
autocrine 172, 187, 192, 231, 237, 239, 240
autograft 407, 451, 459, 461, 466, 515
-- repair 458, 459, 461, 467
automatic movements 334, 335
autophagy 81, 82, 88, 91
axis cylinder 33
axon disease 61
axon dysfunction 61, 132
axon guidance 165-167, 170
-- cues 168, 175
-- molecules 165, 166, 168, 173, 174, 176
axon sprouting 59, 310, 315, 318, 523
axon sprouts 310, 313, 315, 318, 325, 330, 462, 467, 520
axon(al) growth 165, 237, 240, 301, 302, 312, 323, 326, 327, 329, 331, 338, 343, 349, 350, 356-358, 413, 451, 455, 457, 461, 464-466, 514, 517, 523
axon(al) injury 34, 36, 63, 64, 66, 73, 286
axon(al) regeneration 165, 211, 231, 240, 287, 289, 291, 302, 304, 325, 327, 333, 334, 336, 338, 349, 350, 355, 385, 412, 451, 453-455, 457-459, 461, 467, 468, 470, 485, 495, 511, 513, 514, 557, 570, 591
axonal arbours 57
axonal damage 73, 224, 228, 286, 331, 452, 490
axonal degeneration 61, 63-74, 236
axonal density 56
axonal disintegration 70, 71
axonal fragmentation 64
axonal integrity 61, 63, 71, 74
axonal membrane 454
axonal metabolism 62, 64, 113
axonal pathology 73
axonal projection 34
axonal re-growth 301
axonal terminals 113, 264, 268, 310, 317, 494

axonal transport 42, 61-63, 68, 69, 71, 97, 287, 350, 352, 455, 487, 582, 590
axons 33-36, 38, 39, 41-43, 47, 62-64, 66, 68, 69, 71, 72, 92, 93, 135, 156, 158, 165, 168-175, 189, 215-224, 226-228, 231-233, 237, 240, 246, 260, 262, 263, 265, 271, 286, 288, 289, 301, 302, 310, 311, 3131, 314, 324, 326, 327, 331, 332, 340, 342, 343, 351-358, 383, 385-387, 389, 397, 406, 407, 410-412, 451-462, 464, 466, 469, 511, 513-520, 523, 550, 554, 557, 578, 580, 582
-- invertibrate 42
axotomy 57, 58, 236-239, 261, 262, 268, 270, 271, 286-290, 301, 350, 351, 356, 358, 432, 455, 457, 458, 492

basal ganglia 142, 334, 417, 418, 537-540
basic helix-loop-helix (bHLH) 204, 221
-- transcription factors 203, 204, 208, 209
Bcl-2 family members 83, 84, 86, 91, 420, 454
Bcl-2 proteins 87, 89, 91, 101-103, 105, 118, 434, 485, 486, 495, 501, 532, 595
behavioral abnormalities 8, 617, 618
behavioral dysfunction 285
behavioral significance 334
behaviorally significant locomotion 334
behavioural recovery 289, 407, 449
β carotene 145, 281, 596
β-scission 145, 146
β-amyloid aggregation 135, 158, 564
β-amyloid precursor protein (β-APP) 108, 109, 163, 374
βA4 in neurons 10
β-amyloid 530
β-amyloid peptide 69, 89, 108, 282

β-NG-1 234, 236
bFGF 219, 223, 249, 251, 254, 287, 400
binding of the transcription factors 23
bipolar neurones 34
bipolar neurons 255
bis-(2-aminoethyl0-amine-N,N,N',N'-pentaacetic acid (DTPA) 141, 163
blindsight 621
blood-brain barrier (BBB) 1, 10, 141, 161, 162, 183, 186, 187, 196, 203, 209, 224, 254, 278, 414, 419, 487, 492, 506, 508, 531, 566, 589-591
blood-cerebrospinal fluid-barrier 183
bone morphogenetic proteins (BMP) 203-206, 209, 210
Braille reading 239, 298, 473, 623, 624
brain aging 10, 12
brain amyloid deposits 565
brain blood flow regulation 11
brain damage 135, 158, 200, 210, 211, 262, 276, 277, 285, 286, 298, 299, 303, 304, 391, 403, 408, 423-425, 428, 473-476, 503-509
brain diffusion 12
-- compartation 11
brain disorders 79, 80, 92, 111, 396, 426, 483, 631, 639
brain functions 10, 15, 27, 137, 262, 391, 392, 404, 617, 618, 629, 630, 636, 639
brain glucose metabolism 2, 19
brain insults 79, 103, 185, 423
brain ischaemia 99, 483, 527-529, 533, 534
brain plasticity 298, 425, 473, 474, 477, 478, 622
brain, social life in the 165
brain's putative malleability 260
brain-derived neurotrophic factor (BDNF) 51
-- antibodies 50, 51, 54, 73, 126, 185, 191, 194, 237-240, 251, 287, 294, 301, 350, 354, 355, 388, 389, 485, 486, 495, 586, 595, 608

brainstem nuclei 334, 538
breathing 345
burden of life 1, 2, 8, 9, 20

CA^{2+} signalling 48
Cajal-Retzius cells 395, 398
calcium 61, 70-73, 102, 108, 111-113, 116, 118, 130, 131, 142, 143, 151, 152, 161, 223, 240, 247, 313, 365, 372, 418, 486, 501, 527, 528, 531, 532, 534, 565, 579, 580, 597
calcium activity 112, 113
calcium channel 51, 54, 70, 113, 223, 366, 367, 371, 376, 377, 506, 528, 530, 597, 601, 613
calcium ion (Ca2+) 29, 48-51, 53, 55, 70, 88-90, 109, 136, 142, 143, 157, 163, 223, 228, 309, 313, 314, 317, 319, 365, 367-378, 527, 530-533
calcium kinases 316
calcium release- activated calcium current (I_{CRAC}) 112
calcium/calmodulin-dependent protein kinase II (CaMKII) 112, 370
Calpain 73, 81, 85, 87, 88, 90-93, 137, 142, 143, 150, 371, 528, 533, 597
-- inhibitor 73, 88, 615
cAMP-response element-binding protein (CREB) 49, 54, 123, 125-127, 206, 317, 333, 388
CAN 43
cancer 141, 166, 323, 488, 496, 504
capping 23, 25
CARD 82, 101
caspase activated DNAses (CAD) 151
caspase-dependent pathway 87, 103
caspase-independent pathway 103
caspases 79
CAT 140, 146, 153-155, 157, 163
cataractogenesis 141
catecholamines 4, 44, 136, 153, 156, 157, 544

cell biology of learning 307
cell body 33-38, 40, 41, 44, 51, 61-63, 68, 72, 80, 101, 113, 153, 156, 175, 219, 231, 249, 253, 262, 264, 291, 310, 312, 350-354, 357, 385, 394, 406, 451, 455, 457, 511, 523, 542, 544, 557, 578
-- response 68, 350-352, 354
-- shrinkage 101
-- stress 34
cell cycle 16-18, 27, 97, 103, 149, 219, 223, 240, 254, 256, 413, 445, 590, 592
-- activity 6, 17
-- G0/G1 phases 18
-- regulation 6, 19, 151, 254
-- S/G2/M phases 18, 19
cell death 71, 79-92, 95, 100-103, 107, 109-113, 117-120, 126, 129, 136, 142, 144, 150, 151, 155, 174, 184, 189, 191-193, 196, 197, 199, 220, 223, 224, 234, 301, 367, 370, 371, 373, 381, 384, 387, 391, 413, 428, 429, 431-434, 485, 493, 504, 507, 509, 527, 528, 530-534, 542, 544-546, 557, 570, 579, 596
-- in chronic neurodegenerative diseases 136
-- signals 98
-- machinery 80
-- mechanisms 86, 485
-- pathway 79-81, 86, 87, 102, 103, 109
cell lineage 232-234, 413
cell lines 119, 151, 413, 439-442, 445-447, 488, 556, 591, 595, 647, 648, 650, 651
cell morphology 11
cell nuclei 24, 26, 253, 337, 652
cell proteases 101, 103
cell protein mutations 107
cell strains 440, 442, 445
cell-cell communication 151
cell-cell interactions 203, 204
cell-cycle arrest 103, 219, 223
cellular aging, mechanisms of 144

cellular calcium homeostasis 102
cellular mechanisms in mammalian brain 1
cellular redox equilibrium 150
central dorsal rhizotomy repair with EC transplants 338
cerebellar granule cells 34, 41, 151, 167
cerebral atrophy 11, 563, 567
cerebral cortex 2, 3, 5, 8, 11-16, 19, 48, 51, 108, 204, 209, 256, 260, 272, 384-386, 391, 414, 417, 496, 538, 624, 429, 634
cerebral defence system 181
cerebral glucose metabolism 14
cerebral glucose utilization 7
cerebral insulin receptor 5
cerebral oligemia 15
cerebral oxidative/energy metabolism 14
cerobrospinal fluid (CSF) 4, 16, 158, 159, 185, 186, 279, 280, 372, 487, 495, 567-569
ceruloplasmin 279
cervical rhizotomy 340
chemical carcinogens 148
chemoattraction/repulsion 165, 169
chemokines 135, 158, 186, 188, 190, 191, 193, 194, 196
chemorepulsive proteins 301
chemotactic activities 313
chemotactic cytokines 455
chemotactic effects 542
chemotactic factors 314
chemotactic gradients 187, 190
chemotropic factors 314
chemotropic hypothesis 165
chemotropic molecules 170
cholinergic system 10, 572
Chromatin 23-26, 36, 41, 80, 82, 83, 87, 88, 101, 132, 150, 215
chronic exposure to lead 16
chronic neuromuscular disorder 129
chronic progressive cell degeneration 144
Cicero, Marcus Tullius 1, 2, 20

ciliary neurotrophic factor (CNTF) 203-207, 210, 212, 220, 239, 240, 251, 254, 287, 416, 467, 495, 501, 586, 595
citrate synthase 13
Classical (Pavlovian) conditioning 48, 266
climbing 54, 345
-- fibers 408, 410
clinical MRI scanners 618
CMAp 337
CNAp 337
CNS 3, 10, 12, 14, 36, 43, 65, 88, 166, 181-184, 186, 191-193, 195-201, 215, 217-219, 224-227, 231, 260, 261, 264, 268, 285, 287, 289, 291, 295, 296, 298, 301-303, 307, 308, 309, 311, 312, 314-316, 318, 323-330, 332-338, 341, 345, 349-354, 357, 358, 372, 389, 385, 388, 391, 404, 413, 414, 419, 425, 484, 485, 487, 388, 490, 492, 495, 496, 517, 520, 613
-- astrocytes 191
-- axis 495
-- axons 43, 174, 301
-- cells 184, 192, 197, 198, 268, 329, 496, 541
-- complications 194
-- concentration of noradrenaline in 10
-- conditions 483, 498
-- defence potential 188
-- development 203
-- disorders/diseases 190, 404, 439, 580
-- gene therapy 487, 494
-- gene transfer 486, 488
-- homeostasis 181, 184
-- infection 193, 200
-- inflammation 83
-- injury 174, 186, 299, 335, 338, 341, 345
-- lesions 174, 289, 395, 323, 328, 332, 497
-- lesions repair 308, 325, 326, 345
-- macroglia 333, 334

-- myelin 217, 327, 352, 354
-- neurons 262, 269, 270, 301-303, 314, 352-355, 497
-- pathways 352
-- pathologies 485, 504
-- phenotypes 493
-- plasticity 325
-- regeneration 175, 336
-- sprouts 310
-- tissue 442, 595
-- trauma 326, 404
coenzyme Q10 161, 605
collapsin response mediator proteins (CRMPs) 170
collateral sprouting 57, 58, 581
collateral sprouts 310, 327
compensatory mechanism 259, 265-267, 270-272, 285, 291, 292, 452, 520
compensatory motor systems 267
computer assisted motivating rehabilitation (CAMR) 478
cones 167, 169, 172, 237, 240, 246, 309, 311, 318, 330, 355, 455, 514
connective tissue 125, 231, 232, 240, 324, 329, 452, 455, 464
contemporary neuroscience 259, 630
copper zinc dismutase reaction 278
copper-containing cytochromes 139
copper-zinc superoxide dismutase (CuZn SOD) 277, 577
cortical disconnections 11, 631
cortical neurons 9, 292, 294, 374, 391, 393, 394, 396
corticospinal axons 302, 327, 412
corticospinal fibers 327, 343
corticospinal systems 358, 580
corticospinal tract 173, 295, 327, 343, 344, 355, 580
corticospinal transection 341
cortisol 8, 10, 16, 17, 249
-- action 20
-- after stress 10
-- production 17
-- treatment 8

CREB binding protein (CBP) 27, 123, 125, 126, 206, 207, 209, 210, 333
Creutzfeldt-Jakob disease 89, 91, 194
cross-modal plasticity 298, 617, 623, 624
cyclic ADP-ribose (cADPR) 113
cyclic-AMP-response-element-binding protein (CREB) 26, 49, 54, 123, 125, 126, 206, 317, 388
cyst formation 341, 344, 345
cysteine 143, 146, 148, 152, 158, 223, 250
-- proteases 79, 82, 88, 91, 433, 597
-- thiols 151, 152
cysteine aspartyl-specific proteases 504
cystometrography 338, 339
cytochrome a 13, 139
cytochrome a3 13, 139
cytochrome b 39, 140
cytochrome b_5 139
cytochrome b_{562} 139
cytochrome b_{566} 139, 140
cytochrome c 81-83, 86-89, 91, 97, 100-102, 120, 139, 151, 371, 534, 579
cytochrome c_1 139
cytochrome P-450 139
cytochrome oxidation 24, 39, 276, 531, 553, 579
cytokines 47, 52, 85, 135, 143, 151, 158, 183, 185, 186, 191, -194, 196, 199, 200, 225, 228, 239, 251, 254, 328, 388, 425, 455, 514, 541, 542, 565
-- pathway 432
cytomegalovirus (CMV) 486, 489
cytoplasm 39-40
cytoskeleton 34, 37, 41, 54, 61, 84, 159, 167, 187, 221, 222, 240, 332, 366, 393, 400, 455, 590

DCC/UNC receptors 166
death-inducing signalling complex (DISC) 90, 81, 102
deep brain stimulation 494

deficits of the gaze-stabilizing and postural reflexes 292
Del Rio Hortega, P. 215, 216
delta-notch 203, 204
delusions 572, 633
dementia 1, 2, 9, 107, 126, 128, 162, 197, 228, 483, 504, 505, 544, 545, 563, 564, 566, 569, 571, 572, 576, 577, 600-602, 606, 619, 627, 631
dendrites 9, 23, 29, 33-36, 38, 41, 44, 54, 57, 58, 92, 131, 175, 268, 310, 311, 313, 314, 331, 399, 455 566, 582
-- apical 394
-- cortical 168
-- distal 41
-- granule 398
-- primary 34
-- proximal 41
dendritic mRNA translation 131
dentatorubralpallidoluysian atrophy (DRPLA) 124, 126
-- protein 126
deoxyribonucleic acid (DA) 140, 156, 157, 163
diabetes 141
dialysate of the hypothalamus of aged rats 10
diaschisis 295
Dieters, K. 33
differentiation 103, 150, 152, 190, 192, 203-207, 208-210, 212, 215, 217-224, 226, 227, 299, 231, 233-236, 239-241, 250, 255, 285, 286, 289, 304, 313-315, 330, 404, 407, 413, 440, 441, 443, 444, 446, 448, 499, 569, 629, 648
diffusion-weighted MR imaging 11
dioxins 152
dioxygen 276, 277
directional factors 314
disuse hypersensitivity 291
DNA base pairs (bp) 24
DNA damage 85, 101, 103, 148, 375, 444

DNA fragmentation 83, 87, 101, 151, 534
DNA repair 82, 103, 148, 149, 151, 384
dorsal rhizotomy 291, 336, 338, 340
double-stranded RNA-dependent protein kinase (PKR) 114, 117
double-stranded RNA-dependent protein kinase-like endoplasmic reticulum kinase (PERK) 114

early to long-term potentiation transition 49
EC migration 335
EC tranplants 327, 335-344, 346
electron microscopy 15, 33, 42, 66, 86, 91
elongation 23, 25, 97, 123, 124, 132, 165, 166, 17, 291, 301, 302, 328, 349, 351, 353-355, 457, 476, 505, 514-517, 519
embryogenesis 234-236
embryonic stem cell (ES cells) 447, 448, 591, 643, 644, 646-651
encephalopathy 106, 433, 542
end feet 203
endonuclease G 81, 87, 101, 103
endonucleases 114, 136, 142, 157, 371, 528, 533, 534
endoplasmic reticulum (ER) 3, 5,7, 10, 18, 19, 36-41, 89, 111-119, 139, 253, 279, 367, 370, 374, 528-530
-- dysfunction 111, 112
-- stress 89
endoplasmic reticulum-associated degradation (ERAD) 89, 91, 111, 113-116
energy failure 15, 63, 71, 111, 116
energy production 20, 150, 160, 527
energy transduction apparatus of mitochondria 98
energy turnover 20
engramma 316
eNOS 143
ensheating cells (Ecs) 302, 323, 327, 332, 335, 336, 338-346

entorhinal cortex lesion (ECL) 312
environmental modifications 260
enzymes 2, 3, 5, 7, 13, 16, 25, 27, 30, 38, 63, 83, 88, 98-100, 137, 139-144, 146, 149, 151-154, 158, 162, 201, 221, 280-282, 367, 370, 371, 407, 416, 475, 483-485, 494, 501, 528, 534, 541, 545
ependymal canal 217
ependymal stem cells 194, 245, 255, 256, 266, 383, 386
ependymal sulcus 385
Eph family 53, 175
Eph receptor guidance system 172
Eph receptors 172-175
ephrins 53, 165, 166, 172, 173, 175
epidemiology 8, 504, 575
epidermal growth factor (EGF) 191, 207, 330, 440
epidermal growth factor receptor (EGFR) 207, 330
epileptic seizures 88, 92, 111, 116, 119, 567, 572
ER lumen 112-116, 119
ER protein processing 113
ERAD, induction of 115
estrogen receptors (ER) 332-334
-- homeostasis 111, 112, 116-118
ethylenediamine tetraacetic acid (EDTA) 141, 163, 367
etiology 70, 149, 195, 199, 215, 224, 225, 227, 300, 398, 404, 577
excitatory amino acid transporter (EAAT2) 2 372
excitatory amino acids (EAA) 163, 200, 249, 250, 505, 527, 529, 579
-- mission 431
-- neurotransmitters 247, 431
-- receptor ligand binding 475
-- receptors antagonists 431
-- release 527, 534
excitatory postsynaptic potentials (EPSPs) 50, 57, 408
excitotoxicity 87, 107, 109, 129, 193, 199, 223, 225, 228, 247, 372,

374, 425, 427, 428, 430, 432, 485,
501, 505, 532, 579, 595, 598
Exon 126-130, 486
expansion of trinucleotide 123
experimental autoimmune encephalitis
(EAE) 224, 225
external stimuli-driven gene expression
in the brain 23
extracellular matrix (ECM) 15, 54, 58,
59, 62, 165, 167, 174, 186, 211, 218-
220, 222, 227, 237, 240, 302, 328,
329, 345, 395, 452, 457, 462, 464,
466-468, 590
extracellular space volume and
geometry 11

facial paralysis 265, 266, 476
fatty acid peroxides 135, 136, 158
fatty acids 98, 139, 144-146, 153, 158-
160, 186, 237, 280, 310, 428, 529
feline immunodeficiency virus (FIV)
489, 493
Fenton chemistry 140
Fenton reaction 145, 275
fibroblast cells 174
fibroblast growth factor (FGF)145
-- (FGF-1) 327
-- (FGF-2) 207, 208, 212, 220, 239,
251, 350, 467
fibroblasts 191, 202, 237, 301, 324,
328, 329, 336, 452, 469, 462, 466,
488, 496, 497, 566
fibrous astrocytes 203, 204, 216, 324,
329, 330, 331
flavin adenine dinucleotide (FADH2)
99, 100, 139, 163
flavin adenine mononucleotide (FMN)
139, 140, 163
flavoproteins 139, 140, 170
forelimb function 340
formation of radicals 135
Fos proteins 27, 340
fragile x mental retardation protein
(FMRP) 123, 130-132
free metal ions 275
Freidrich's ataxia 106

fructose-1,6-diphosphate 13
functional dedifferentiation 286
functional magnetic resonance
imaging (fMRI) 11, 617, 618, 621,
624, 626
functional recovery 58, 74, 92, 259,
261, 262, 267, 270-272, 285-287,
289-291, 293-295, 297-299, 301,
303, 304, 327, 328, 335, 339, 355,
408, 412, 431, 448, 452, 457-459,
464, 473, 485, 511, 523, 551, 553,
554, 559, 560, 618
functional reorganization 267, 285,
294-298, 300, 302
functional respecification 262
functional significance of neuronal
gene expression 29

GABA-mediated inhibitory
postsynaptic potentials (IPSP) 55
γ-aminobutyric acid (GABA) 3, 12,
18, 129, 190, 246, 294, 373, 391,
410, 416, 418, 448, 494, 501, 538,
539, 550
ganglion cells 34, 68, 167, 172-174,
218, 240, 246, 248-251, 253, 291,
301, 352, 354, 382, 489, 492
gap-junctions 44, 222, 308, 590
gene expression 3, 19, 23-31, 47, 49,
54, 62, 89, 113, 124-126, 150, 152,
165, 209, 234, 236, 240, 254, 255,
334, 357, 416, 441, 444, 484, 486-
490, 497, 502, 553, 560, 597, 620
gene expression profile 19, 20
gene promoter 26, 84, 123, 124, 440
general neuronal structure 34
glia 3, 10, 20, 109, 174, 182, 183,
187-189, 192-194, 199, 203, 204,
208-211, 215, 218, 233, 234, 245-
247, 249-251, 253, 255-257, 307-
310, 315, 323, 325-327, 332-334,
338, 383, 385, 386, 389, 390, 411,
413, 434
glia cells 2-4, 7
glia limitans 308, 324, 330

glial cell line-derived neurotrophic factor (GDNF) 73, 185, 239, 287, 407, 416, 467, 485, 486, 492, 499, 501, 549, 555-560, 595, 608
glial fibrillary acidic protein (GFAP) 10, 20, 203-207, 208, 210-212, 254, 256, 329, 331, 332, 334, 344, 386, 414, 443, 486
-- induction 254
glial restricted precursor (GRP) 218, 220
glial scar 174, 211, 212, 287, 323, 324, 329-331, 344, 345, 423, 424, 568, 589
glial-restricted progenitors 204
gliobastoma 498
-- cells 497
-- patents 496
gliosis 10, 12, 211, 212, 225, 253-255, 315, 331, 344, 345, 423, 424, 568, 589
globus pallidus (Gpi) 142, 538
glutathione peroxidase 90, 100, 136, 163, 275, 278, 281, 529, 543, 595
glucocorticoids (GC) 4, 7, 16, 200, 388, 389
-- and HPA-axis 16
-- in the brain 7
glucose 2, 13
glucose transport protein (GLUT) 20
-- GLUT 1 248
-- GLUT 3 2
-- GLUT 4 2
-- GLUT 5 2
glucose uptake 2, 7
glucose/energy metabolism 5
glucose-regulated protein 78 (GRP78) 113, 115, 117-119
glutamate 3, 10, 27-29, 47-50, 52, 53, 55, 92, 107, 118, 128, 131, 185, 193, 198, 199, 223, 225, 226, 228, 229, 245, 247-250, 309, 317, 329, 365, 367, 372, 373, 375, 423, 424, 427, 428, 430-432, 494, 505, 527-529, 538, 579, 580, 594, 595, 597, 608-610, 613

glutamate antagonists 485, 506
glutamate homeostasis 225
glutamate measurements 49
glutamate metabolism 129
glutamate/aspartate transporter (GLAST) 204, 248, 254
glutamate receptors (GR) 49, 52, 91, 108, 123, 131, 142, 223-225, 246, 292, 317, 367, 372-375, 387, 388, 430, 486, 505, 528, 533, 579, 607
glutamate toxicity 129, 228, 249, 371-373, 430, 433, 579
glutamate transport 129, 159, 373
glutamate transporters 129, 228, 247-249, 367, 372, 495, 579
glutamic acid decarboyxlase (GAD) 415, 418, 494, 501
glutamine synthetase 249, 254
gluthathione (GSH) 99, 135, 136, 141, 145, 148, 152-155, 157, 163, 223, 248-250, 278, 529, 530, 532, 543
gluthathione disulfide (GSSG) 99, 152-155, 162, 163, 278
gluthathione disulfide reductase (GSSG-Rd) 153, 163
gluthathione peroxidase (GSH-Px) 99, 100, 136, 163, 275, 278, 281, 529, 543, 595
glycerophosphocholine 158
glycolytic key enzymes hexokinase 5
glycoproteins 113, 128, 174, 188, 222, 239, 302, 352, 489, 597, 659
Golgi apparatus (GA) 3, 5, 7, 10, 18, 19, 38, 39, 41, 42
-- and the smooth Endoplasmic Reticulum and Lysosomes (GERL) 38
Golgi silver staining method 215
Golgi, C. 33
G-protein 50, 54,
GPX 99, 100
granule cells 9, 35, 40, 48, 51, 58, 151, 167, 175, 303, 311, 312, 383, 387, 389, 397-399, 409

green fluorescent protein (GFP) 52, 415, 489
growth arrest and DNA damage-induced gene 153 (GADD153) 117, 118
growth factors 14, 19, 51, 58, 61, 62, 64, 73, 79, 83, 107, 113, 168, 185, 191, 207, 212, 220, 223, 237, 239, 245, 250, 251, 254, 287, 313, 317, 319, 330, 354, 382, 384, 387, 388, 411, 432, 467, 468, 470, 484, 501, 514, 517, 549, 555, 566, 586, 593, 595, 604, 608
-- FGF-2 239
-- insulin 191, 208, 220, 239, 287, 382, 388, 389, 467, 495,
-- PDGF-AA 239
growth-associated genes 349, 352-356
GTP-cyclohydrolase 1 (GTP CH 1) 494, 501, 560
guanosine monophosphate (GMP) 143, 163, 317

Haber-Weiss reaction 277, 279
hairy/enhancer of split (Hes) 208, 440, 442, 447
Hallervorden-Spatz disease 142
hallucinations 558, 572
Hamburger, V, 79, 313
health problems 1, 2, 483
Hebb, Donald 316
Hebb's principle 48
Hebbian plastic mechanism 285
Hebbian rules 291
Hebbian synapse 316
hebbosomes 318
helix-loop-helix 26, 204, 217, 255
hereditary vestige 289
herpes simplex virus (HSV) 489-491, 497, 501, 502
heteroplasmy 97
hindlimb function 339
hippocampal connections 170, 175, 176
hippocampus 4, 5, 7, 9, 12, 14-16, 19, 48, 50-53, 55, 58, 135, 155, 158, 171, 173, 175, 205, 209, 211, 261, 295, 311, 313, 315-317, 331, 370, 374, 377, 381-390, 393, 395-398, 425, 429-432, 491, 496, 504, 523, 619, 620
histidine 146
histone acetylation 25, 26, 84
histone deacetylases (HDAC) 25, 83
histone methyltransferases (HMT) 25
histone modifications 26
-- acetylation 25
-- methylation 25
-- phosphorylation 25
-- ubiquitination 25
histone octamer 24
histone species 24
histone tails 25, 84
histone transacetylases (HAT) 25, 83
histones 24-27, 83, 149
homeodomain 26, 221
homeostatic plasticity 290, 293
horizontal cells 246
hormononal changes 260
host rejection 468
human immunodeficiency virus (HIV) 64, 194, 197, 489, 491, 502, 591
human neural stem cells (hNSC) 440, 444, 446
human stem cells (hSC) 440, 650
Huntingtin 89, 107, 124-126, 365, 373, 505
Huntington's disease (HD) 89, 91, 107, 118, 123-125, 127, 142, 304, 329, 365, 371-374, 376, 377, 404, 416, 419, 420, 492, 502, 504, 505, 568, 589, 593, 596, 611
hydrogen peroxide (H_2O_2) 99, 100, 137, 138, 140, 141, 148-156, 159, 163, 275, 277-280, 282, 529, 543, 564
hydroxyl radical 106, 135, 137, 140, 147, 163, 275, 279-282, 529
hyperexcitability 47, 434

hypertrophic fibrous astrocytes 329, 331
hypoglycemia 88, 111, 247, 324
hypotension 506
hypothalamic hormone secretion 313
hypothalamic neurons 292, 332
hypothalamic-neurohypophyseal system 332
hypothalamic-pituitary-adrenal axis 20
hypothalamus 58, 128, 153, 310, 332, 384, 554, 620, 626
hypoxia 85, 70, 103, 109, 111

immune response 184, 225, 227, 228, 252, 303, 415, 489, 497, 565
immunity 182, 188, 191, 251, 459, 502, 592
inappropriate target reinnervation 289, 301
increasing life expactancy 1, 2, 483
inducible transcription factors (ITFs) 26
infection 71, 183, 186, 190, 192, 194, 197, 199, 200, 224, 225, 227, 391, 480, 490, 568, 579
inflammation cytokines 455
inhibitor of apoptosis protein (IAP) 83, 102
injury-induced retrogade degeneration 301
inOS 143, 162, 597, 610, 611, 615
insulin (I) 3
insulin action 20
insulin production 3
insulin receptor (IR) 4, 20
insulin receptor distribution 3
insulin receptor regulation 4
insulin receptor signaling in the brain 3
insulin receptor substrate (IRS)-1 4
insulin/insulin receptor (I/IR) 6, 7
insulin/insulin receptor (I/IR) function 10
insulin/insulin receptor signal transduction 6, 7, 10, 18

insulin-like growth factor-I (IGF-I) 73, 191, 208-209, 287, 388, 389, 467, 495, 595
interfascicular cells 215
interferon-γ (IFN-γ) 186, 192, 193
interleukin (IL) 58, 91, 186, 212, 239, 388, 423, 432, 455, 542, 586
internuclear interneurons 268, 270, 271
interpositus neurons 261
intracellular signalling 151, 173, 174, 199
Intron 27, 106, 128-130, 132
ionizing radiation 148, 152
iron metabolism, deficiency in 141
iron-containing cytochromes 139
ischaemia/ischemia 36, 39, 61, 71, 85, 91, 92, 111, 116, 117, 119, 142-144, 159, 161, 162, 192, 199, 215, 228, 229, 247-251, 254, 324, 365, 371, 372, 375, 377, 387, 389, 429, 432, 433, 452-454, 464, 475, 506, 527-530, 532-534, 592, 597, 614
ischemia-reperfusion injury 141
ischemic penumbra 89
ischemic period 454
isomorphic lesions 324

Jagged-Notch-Hes pathway 221
James, William 307
janus kinase (JAK) 205
janus kinase/signal transducer and activator of transcription (JAK-STAT) pathway 205, 210
Jun proteins 28, 152

kainate receptors 131, 223, 225, 228, 229, 388
K_B 369
K_d 366, 367
k_{on} 366
K_S 368, 369, 373

lactate formation 13

LacZ 492
laminin 166, 314, 336, 459, 466, 469
large nucleolated cell body (soma) 5, 33, 36, 58, 92, 262, 311, 352, 455, 543, 582
late rehabilitation 476, 479
L-dopa 300, 539, 560, 594, 605, 606
lead (Pb) 16, 21, 63
leading edge 165
Leigh syndrome 106
leucine zipper 26, 152, 210
leukaemia inhibitory factor (LIF) 204-207, 210, 239, 351, 440, 442, 455
leuko-araiosis 11
leukocyte infiltration 328, 329
levator palpebrae muscle 265
Levi-Montalcini, R. 79, 262, 263, 313
Lewy body (LB) 136, 155, 157, 163, 507, 542, 544-546, 568, 581
liberation of pro-inflammatory cytokines 328
light microscopy 15, 33, 39, 44, 64, 65, 68
limbar rhizotomy 339
lipid metabolism 135, 158
lipid peroxidation (LPO) 99, 135, 136, 141, 144, 146, 147, 149, 158, 163, 280, 281, 486, 529
-- in AD 158
lipofuschin 39, 41
-- autofluorescence 39
-- granules 39, 42
-- -- accumulation of 39
lipopolysaccharide (LPS) 185-187
-- receptor 192, 193
lipoprotein receptor 282, 396
lipoproteins 160, 161, 237, 280, 282
locus coeruleus (LC) 139, 156, 163
long-term depression (LTD) 47, 53, 54, 55, 317-319
long-term potentiation (LTP) 47-53, 57, 163, 314, 316-318, 345, 370, 389
lumbosacral rhizotomy 338
lysosomes 39-41, 139, 151

macroglia 308, 332-334

magnetic resonance imaging (MRI) 11, 21, 568, 569, 583, 617-620, 622
-- scans 496, 626, 631
major histocompatibility complex (MHC) 188, 190, 192, 193, 252
mammalian central nervous system 204
manganese superoxide dismutase (MSOD) 140, 154, 278
Marinesco 307
maturation 54, 96, 98, 132, 184, 189, 191, 215, 217, 219, 221, 227, 234, 235, 239, 302, 310, 314, 315, 352, 448, 456, 459, 465, 523, 551
maturation of nerve fibers 523
medial longitudinal fascicle transection 341
medial longitudinal fascicles (MLFs) 269-271, 341, 342
membrane bound ribosomes 38
membrane proteins 3, 5, 105, 106, 113, 151, 159, 233, 280, 314
-- oxidation 154
Mendelian inheritance 104
metabolic processes 14, 34, 141
metabolic stress 108
metabolism 1, 2, 5-7, 13-16, 18-20, 39, 62-64, 73, 89, 107, 108, 111-113, 116, 123, 128, 129, 135, 139, 141, 143, 154, 155, 158, 161, 248, 278, 373, 374, 431, 483, 492, 493, 501, 528, 529, 542, 565, 581, 582
metabotropic glutamate receptors 50, 55, 113, 319, 373, 430
metabotropic receptors 317, 430, 528
metal-catalyzed oxidation (MCO) 14, 20
metallothioneins 275, 279
metaplasticity 55
methionine 146, 277, 282, 545
MgluR 50, 51, 54, 55, 132, 373, 430
microdomain 222, 366, 368, 370, 371

microglia 80, 148, 181, 183, 184, 186-201, 215, 225, 226, 28, 245, 251-253, 266, 308, 309, 324, 328, 329, 386, 433, 551, 565, 583, 592, 595
-- activated 200
-- aging of 190, 198
-- a sorce of soluble mediators 191
-- as antigen-presenting cell (APC) 190
-- as macrophage-like cell 189
-- as part of CNS defence potential 188
-- as sensor for CNS homeostasis 184
-- as target of pharmacological interference 198
-- chronic activation of 196
-- excessive acute activation of 195
microglia response factor (MRF) 193
microglia--astroglial interactions 58
microglial activation 158, 160, 181-183, 185-188, 190, 195-201, 225, 228, 433, 542, 543, 595
microglial activity 186, 196, 199, 200
microglial cells 58, 135, 143, 158, 181, 182, 184, 190, 191, 194, 196, 211, 226, 542
microglial chemotaxis 542
microglial dysregulation 194
microglial environment 182
microglial mechanisms 196
microglial misbehavior 195
microglial phagocytosis 150
microglial production 199, 200
microglial products 197, 199
microglial stimulus 196
microglyosis 542
microtubule-associated proteins 62, 394
-- tau protein 6, 68, 108, 128
mild cognitive disorders 160
mild cognitive impairment (MCI) 567
Minea 307
minimental state examination (MMSE) 568, 571
misregulated alternative splicing 128

mitochondria 24, 39, 41, 43, 44, 66, 71, 82, 83, 86, 87, 89, 92, 95-99, 101-109, 112, 139, 140, 143, 144, 149-151, 154, 161, 275-277, 367, 370, 372, 373, 376, 504, 528-534, 541, 565, 596
mitochondrial apoptosis pathway 81, 82
mitochondrial cell activity 107
mitochondrial channels 102
mitochondrial megapore 102
mitochondrial proteins 39, 96, 97, 101, 103, 104, 106, 150, 277, 527, 530
mitogen-activated protein kinase (MAPK) 167, 200, 205, 543
-- activation 167
-- pathway 205
mixed function oxidation (MFO) 14
MOG 174
Mohr-Tranebjerg syndrome 106
molecular layer 311, 312, 398, 399, 409-411
molecular mechanisms in mammalian brain 1
Moloney murine leukaemia virus (MoMLV) 488
monoamine oxidase (MAO) 140, 153, 163, 594-596, 602, 605, 609
morphobiology 9, 10
morphologic abnormalities in mitochondria 108
morphologic abnormalities with aging 11
morphological changes 9, 187, 316, 318, 455
morphological criteria 79-81
morphological peculiarities of the neuron 33
motor learning 54, 259, 260, 267, 634
motor neurons 107, 108, 118, 129, 172, 197, 209, 217, 218, 220, 239, 334, 371, 411, 495, 519, 575-585
-- disease 494, 495, 575, 585, 587
MPP$^+$ 541

mRNA 2-4, 6, 23-25, 27-29, 42, 51, 63, 96-98, 106, 114, 117, 119, 123, 124, 126, 128-132, 154, 234, 236-239, 248, 395, 398, 445, 484, 487
-- editing 29
-- export from the nucleus to the cytoplasm 23
-- localization 23
-- survival in the cytoplasm 23
mtDNA 24, 95-98, 104, 106, 107, 149, 150, 163
-- mutations 103, 104, 106, 149
Müller glia 208, 209, 245, 250, 255, 256
-- cells 245, 249, 253, 254, 256
-- role in neuroprotection 245
multi system atrophy 142
multiple sclerosis (MS) 64, 190, 192, 194, 215, 224-228, 239, 413, 428, 504
multipolar neurones 34, 36
multipotent cell 203, 204, 413
multivesicular bodies 39, 41
muscle wasting 129, 583
myelin 59, 64, 65, 128, 141, 174, 192, 215, 216, 218, 221-227, 231-237, 239, 240, 302, 324, 325, 327, 332, 352-354, 357, 452, 455, 512, 513, 523, 580, 581
myelin basic protein (MBP) 220, 237, 486
myelin glial cell 231
myelin ovoids 64, 236
myelin sheaths 36, 43, 215, 222, 227, 231, 232, 236, 237, 240, 241, 404, 420, 453, 454, 456, 459
myelin-associated axonal growth inhibitors 59
myelin-associated proteins 202, 357
myelinating cells 215, 221, 227, 504
myelinating diseases 194, 215, 219, 224, 229
myelinating oligodendrocytes 215, 217, 221, 227

myelination 11, 43, 62, 128, 216, 217, 221, 222, 227, 229, 231, 235, 239, 240, 241, 459, 514
myelin-specific genes 234, 240
myoblasts 6
myoclonus epilepsy with ragged-red fibres (MERRF) 104

NAD^+-isocitrate dehydrogenase 13
NADH 97, 99, 139, 140, 153, 154, 161, 163
NADPH 39, 139, 141, 143, 144, 146, 153, 154, 163, 278, 279
nanodomain 366, 368
nausea 292, 558, 560, 571
necrosis 52, 74, 79, 80, 82, 86, 109, 150, 186, 276, 324, 433, 475, 485, 504, 542, 590, 595, 596
neocortex 303, 334, 372, 384, 387, 392, 393, 395, 396, 404, 475, 491, 528
neocortical neurons 9, 384
nestin 211, 212, 218, 386, 387, 389, 396, 409, 411
netrins 166-168, 170, 174, 175, 219
neural connections 47, 165, 175, 240
neural connectivity 263
neural damage 260, 334, 477
neural parenchyma 165
neural pathways 286, 294, 300
neural plasticity 259, 260, 307-310, 324, 325, 345, 618, 622
neural precursor cell 211, 648, 657
neural precursor marker 386
neural precursors 488
neural progenitor cells 209, 210
neural protection 427
neural record 316
neural regeneration 174, 175, 256, 261
neural remodeling in CNS of adult mammals 260
neural repair 387, 423, 424
neural response to brain 326

neural response to spinal chord trauma 326
neural response to the lesion 335
neural retina 255
neural stem cell (NSC) 92, 204, 206, 207, 208, 209, 211, 255, 256, 304, 328, 330, 413, 439, 443
neural-crest stem cells 232, 233, 236
neurapraxia 452
neuritogenic activity 314, 315, 318
neuritogenic factors 314, 328, 331
neuritogenic proteins 325
neuroblast 211, 325, 382, 393, 419
neurod 255
neurodegeneration, oxidative stress hypothesis of 136
neurodegenerative disorders/diseases (ND) 11, 12, 16, 68, 69, 79, 89-92, 95, 98, 103, 104, 107, 108, 109, 118, 120, 123, 124, 126, 136, 140-144, 154, 159, 162, 196-198, 211, 224, 225, 247, 275, 279, 282, 286, 302, 202, 329, 365, 371, 373-375, 377, 378, 404, 405, 408, 413, 415, 416, 419, 420, 432, 439, 447, 449, 485, 487, 488, 494, 496-498, 503-506, 537, 546, 579-581, 589, 590, 592, 593, 595, 596, 629
neurofibrillary tangles 69, 108, 157, 374, 545, 573, 566, 567, 581, 631
neurohypophysis 310, 332
neurologic(al) disorders/diseases 104, 128, 131, 497
neuromelanin (NM) 136, 155-157, 542
neuron as cell for reception and transmission of electrochemical information 33
neuron specific enolase (NSE) 486
neuronal / glial progenitor cells 210, 383, 389
neuronal cell adhesion molecules (NCAMs) 54, 188, 210, 222, 237, 240, 332
neuronal cell adhesion molecules (NCAMs) in synaptic plasticity 54

neuronal cell death 85, 86, 89, 91, 107, 111, 112, 119, 120, 373, 387, 391, 413, 546
neuronal deafferentation 291
neuronal growth factor (NGF) 51, 63, 73, 79, 87, 185-191, 192, 194, 237-239, 287, 301, 313, 314, 332, 333, 336, 343, 350, 351, 353, 432, 467, 496, 501, 593, 603
neuronal insulin/insulin (I/IR) signal transduction cascades 10, 18, 19, 76
neuronal insulin, alpha/beta-subunits of 4
neuronal protection against oxidative damage 275, 277, 279, 281
neuronal regeneration 166, 381, 383, 385, 387, 389,
neuronal stress 36
neurones, specific morphological features 34
neurons, morphological peculiarities 33
neuroprotection 5, 160-162, 245, 247, 249-251, 255, 257, 372, 376, 377, 418, 430, 433, 491, 492, 497, 506, 533, 592, 594
neurotmesis 453
neurotransmitters 15, 44, 131, 132, 136, 193, 203, 222, 246, 247, 262, 292, 299, 308, 387, 388, 404, 405, 415, 416, 428, 431474, 527
neurotrophic 5, 57, 107, 182, 183, 186, 204, 220, 231, 287, 291, 301, 313, 314, 328, 331, 344, 416, 460-462, 466, 485, 486, 501, 566, 591
neurotrophic factors 44, 51, 73, 108, 126, 187, 189, 191, 194, 204, 211, 237, 239, 250, 251, 254, 263, 268, 287, 289, 301, 309, 313, 314, 318, 326-329, 335, 345, 346, 357, 388, 416, 419, 423, 432, 455, 456, 464, 467, 468, 470, 485, 495, 496, 505, 549, 555, 556, 558, 595, 608, 610
neurotrophins 47, 51, 52, 57, 58, 176, 250, 287, 294, 313, 333, 356

-- BDNF 237-239, 354, 395
-- neurturin 239
-- NGF 192, 238, 239
-- NT-3 237-239
-- NT-4/5 237-239
Nissl bodies 34, 38, 39, 41, 455
Nissl, F. 38
nitric oxide (NO) 50, 71, 99, 116, 135, 137, 143, 186
nitric oxide synthase (NOS) 71, 74, 116, 142, 163, 529, 597
nitrogen free radicals 328
nitrosative stress 135, 144, 165
N-methyl-D-aspartate (NMDA) 163, 247, 387, 430
N-methyl-D-aspartate type glutamate receptor (NMDAR) 48
nNOS 142-144, 162
nodes of Ranvier 62, 66, 310, 609
non-heme iron-sulfur proteins 139
non-neuronal cells 61, 62, 64, 69, 92, 113, 174, 459, 460, 462
noradrenaline action 20
normal adult brain 2, 512
notch/delta signalling pathway 208, 208
NSAID 200, 565, 602, 603
nuclear factor κB (NFκB) 151, 163, 542
nuclear membrane breakdown 101
nuclear-encoded genes 106
nucleic acid damage, by ROS 148
nucleolus 34, 36, 41
nucleoside diphosphate (NDP) 38
nucleosome 24, 25, 83, 84
nucleus 23, 25, 26, 34, 36, 37, 38, 49, 58, 68, 87, 96, 98, 124, 126, 130, 131, 142, 149, 152, 161, 173, 205-207, 208, 253, 264, 268, 270, 271, 286, 288, 289, 296, 297, 310, 332, 339, 341, 383, 417, 418, 477, 488, 491, 494, 496, 538, 540, 550, 566, 580, 591

occlusion 15, 20, 21, 57, 109, 228, 248, 325, 375, 454, 528, 591, 592, 596
oculomotor reinnervation 341
olfactory bulb 3, 4, 168, 172, 173, 205, 211, 327, 332, 336, 341, 381, 382, 384, 388
oligodendrocyte differentiation into myelinating oligodendrocytes 221
oligodendrocyte disease 224
oligodendrocyte excitotoxicity 225
oligodendrocyte migration 218
oligodendrocyte number, control 219
oligodendrocyte origin 217
oligodendrocyte precursor cells (OPCs) 215, 217, 218, 220, 226
oligodendrocyte progenitors 169, 171, 217
oligodendrocyte vulnerability 222
oligodendrocytes 43, 62, 128, 209, 215, 217-229, 231, 239, 245, 246, 251, 302, 332, 334, 349, 352, 413, 414, 433, 442, 486
oligodendrocytes express neurotransmitter receptors 222
Omi/HtrA2 83, 101, 102
OML 312
omnipotent cells 304
orbicularis oculi (OO) muscle 264, 265, 512
oxidants 86, 99, 137, 138, 150-153, 159, 163, 276, 282, 529, 547
oxidation of proteins 146
oxidative catalysts 279
oxidative damage 100, 106, 108, 148, 149, 160, 275-277, 281, 282, 374
-- to lipids 276, 281
-- to proteins 281
oxidative energy 1, 2, 483
-- metabolism 7, 13, 18
-- related metabolism 2
oxidative injury 129, 250, 565, 570
oxidative mechanisms in neurodegenerative diseases 282

oxidative stress 142
oxygen as oxidant 276
oxygen free radicals 39

pancreatic acinar cells 38
paraneoplastic opsoclonus myoclonus ataxia (POMA) 128
Parkin 90, 119, 543-546
Parkison disease (PD) 1, 2, 8, 16, 89, 91, 92, 108, 119, 135, 148-150, 153-157, 160, 161, 163, 194, 196-198, 278, 304, 329, 365, 371, 372, 374-377, 391, 404, 406, 413, 416, 417, 419, 439, 440, 447, 483-485, 490, 493, 494, 498, 501, 504-508, 537-543, 546, 549-560, 568, 580, 581, 589, 591-595, 605, 607, 627
pathogenic mechanism of neurodegenerative diseases 123
pathological condition of SAD 11
pathology 11, 64, 66, 68, 71, 73, 108, 123, 144, 157, 224, 225, 308, 326, 372, 374, 397, 430, 462, 485, 489, 490, 492, 542, 54, 545, 565-567, 572, 578, 580, 581, 592-594, 596
PD, free radical research 136
PDZ 52, 173
perforant pathway 48, 92, 386
perinatal stress to the fetus/offspring 19
peripheral nervous system (PNS) 43, 61, 62, 165, 168, 231, 233, 237, 240, 259, 262, 271, 285, 287, 291, 295, 302, 314, 323, 324, 336, 350, 352-355, 358, 451, 468, 469, 511
permeability transition pore (PTP) 102
peroxynitrite 135, 137, 144, 163, 529, 542, 578
phagocytosis 141, 148, 183, 189, 190, 194, 195, 197, 201, 245, 253, 264, 565, 592
phantom limb 296, 297, 626
phantom pain 297, 617, 626
pharmacotherapy 299, 429, 483, 484, 549
phosphocholine 158

phosphoethanolamine 158
phosphofructokinase 2, 5, 13
phospholipid metabolites 158
photochemical spinal cord damage 344
pituicytes 310, 332, 333
pituitary adenylate-cycle-activating polypeptide (PACAP) 208
plasticity in the brain of aged rats 9
plastic—regenerative mechanisms, relationship 261
polio-araiosis 11
polyadenylation 23, 25, 29
polyglutamine 90, 118, 123, 124, 126, 127
polyribosomes 38, 39, 41, 117, 131
polysialylated-neural-cell-adhesion molecule (PSA-NCAM) 221, 332
POMA disease antigen, Nova-1 128
positron emission tomography (PET) 11, 21, 295, 298, 303, 568, 569, 583, 606, 617-624, 626, 627
postmitotic cells 215, 253, 302
post-synaptic neural elements 38, 260
post-transcriptional gene regulation 28
POU domain 26, 235
precursor cells 204, 205, 207, 208-212, 215, 218, 235, 303, 388, 439, 443, 446, 447, 449, 648, 657
preferential motor reinnervation 457, 519
preganglionic lumbosacral rhizotomy 339
pre-mRNA processing 123, 124, 130
presynamptic axon terminals 260
primary neurons 119, 150
private microenvironment 260
progenitors 169, 171, 204, 217, 218, 223, 233, 255-257, 382, 386, 388, 440, 447, 488
progressive supranuclear palsy 142, 568

pro-inflammatory cytokines 192, 228, 328, 541, 565
proteases 73, 79-82, 84, 85, 87, 88, 90, 91, 101, 135, 136, 142, 148, 150, 157, 183, 186, 194, 199-201, 239, 328, 371, 433, 486, 504, 527, 528, 534, 564, 590, 597
proteasome 102, 111, 116, 136, 148, 157, 192, 485, 545, 546
protein kinase (PK) 6, 21, 26, 49, 51, 78, 152, 167, 200, 234, 317
protein kinase A (PKA) 317, 367, 370
protein kinase C (PKC) 5, 52, 317
protein synthesis 27, 38, 41-43, 49, 51, 52, 54, 62, 89, 114-117, 123, 130-132, 135, 141, 158, 187, 311, 317, 318, 351
protein transcription 28
protooncogenes 150
PSDs 315, 318
Purkinje cell dendrites 54
Px 140, 153-155, 157, 163
pyramidal neurons 9, 10, 155, 294, 303, 311, 312, 375, 631, 635, 659
pyruvate dehydrogenase (PDH) 2, 5, 13, 21, 162
pyruvate kinase 2, 146
pyruvate production 13

quinolinic acid 328
quinone toxicity 141, 155

radial glia 204, 208, 209, 211, 218, 245, 256, 334, 383, 385, 386, 389-392, 396, 398, 411
radiation injury 141
radical scavengers 61, 74, 159, 254, 485, 594
radicals 14, 39, 71, 99, 100, 107, 116, 135, 137-141, 144-146, 150, 157, 159, 161, 194, 199, 223, 254, 277, 279-282, 328, 329, 335, 346, 367, 425, 486, 529, 530, 542
Ramon y Cajal, S. 33, 215, 237, 238, 259, 307, 316, 331, 336, 338, 404, 513, 519

reaction atrophy 457
reactive astrocytes 10, 174, 211, 212, 227, 324, 327, 329-331, 336, 338, 345
reactive astrocytic gliosis 580
reactive gliosis 10, 210-212, 253-255
reactive nitrogen species (RNS) 99, 135, 138, 150, 152
reactive oxygen species (ROS) 81, 99, 100, 106, 109, 136, 137, 141-144, 146, 148-154, 156-159, 163, 275-277, 279-282, 376, 528, 529, 531, 533, 544, 546, 590
-- consequences of 144
reactive synaptogenesis 57-59, 291, 307, 311-314, 325
reactive thiol species 279
receptor plasticity 473, 475
receptor trafficking 52
redox cycling 151
redox equilibrium 144, 150, 152
reeling signal transduction pathway 398
reflex movements 334
regeneration 61, 154, 165, 166, 174, 175, 182, 195, 211, 231, 232, 236, 237, 240, 245, 256, 259-261, 268, 270, 271, 280, 287, 289, 291, 301, 302, 304, 325-328, 331-334, 336, 342, 343, 349-358, 381-383, 385-387, 411, 412, 425, 451-454, 457-462, 464-470, 485, 495, 511-514, 516, 517, 519, 522, 523, 530, 549, 555, 557
regenerative sprouting 57, 458
regenerative sprouts 310
regional cerebral blood flow (rCBF) 3, 11, 468, 619, 620, 622-626
reinnervation 262-265, 287, 289, 291, 301, 312, 340, 341, 355, 357, 358, 451-453, 457, 458, 464, 511, 513, 514, 517-523, 551-555, 558, 581, 583
rejection of the graft 419, 459

relaxation of the chromatin structure 23
remyelination 219, 226-228, 239, 523
replacing cells 302
repulsion 165, 167-171, 173
respiratory chain 3, 13, 96, 99, 100, 103, 107, 149, 541, 542
resting microglia 184
retina 59, 92, 173, 208, 209, 219, 245-257, 381, 382, 384, 477, 489, 621, 623
retrogade degeneration 63, 65, 301, 457
retrogade signalling 286
retrograde cues 351
retrograde messenger 49, 50, 55, 318
retrogade rhizotomy 291, 336-341
retrogade transport 455, 489, 591
rhizotomy repair results 327
ribonucleic acids 148, 149
Ricinus communis agglutinin II 268
RNA binding proteins 128, 130, 131
RNA polymerase II (RNA pol II) 23, 25, 27, 97, 123, 127
RNA processing 25, 123, 129-131
RNA transcription 123
Rotenone 541
rough endoplasmic reticulum (RER) 38, 39

sarcoplasmic/endoplasmic reticulum Ca^{2+}-ATPase (SERCA) 112, 116, 367, 370
satellite oligodendrocytes 215
scar tissue 301, 453
Schwann cells 36, 62, 174, 215, 231-235, 237-241, 302, 325-327, 332-334, 336, 338, 342, 351, 452-460, 462, 464, 467-470, 514, 515, 517, 520
Schwann-cell development 234, 235
Schwann-cell lineage 233
Schwann-cell precursor 234-236
sciatic nerve 326, 336, 337, 339, 463
-- resection repair 336
-- section 289

secondary neuronal death 323, 324, 326, 328, 329, 331
selenium deficiency in animals 278
semaphorins 165, 166, 168, 170, 172, 174, 175
senescence 8, 9, 12, 13, 103, 198, 442, 445
senile plaque 140, 564, 566, 567, 631
sensory substitution 477, 618, 623
-- in the blind 623
serotonin 131, 136, 140, 300, 388, 448, 572, 586, 597, 615
serum responsive element (SRE) 26
shadow plaque 227
Sherrington, C. 33, 417
side-effect 200, 268, 300, 329, 403, 419, 430, 483, 484, 487, 494-496, 506, 549, 554, 558-560, 565, 572, 590, 592-596
signal transducer and activator of transcription. (STAT) 205-206, 209, 210
signal transduction 150-152
-- pathways 23, 82, 114, 115, 117, 120, 152, 508
simian immunodeficiency virus (SIV) 489, 502
singlet oxygen 277, 280, 282
Slit/roundabout guidance molecules 167, 170
Smac/DIABLO 83, 101, 102, 534
small nuclear ribonucleoproteins (snRNPs) 129
social life in the brain 165
sonic hedgehog (Shh) 208-210, 217, 218
Sp-1 transcription factor 123-125, 127
spasticity 129, 345, 474, 480, 582, 586
Speidel, C. C. 309, 311
spinal and bulbar muscular atrophy (SBMA) 126
spinocerebellar ataxias (SCAs) 123, 124, 126

spiral muscular atrophy 123, 129, 505, 577, 584
splicing 23-25, 98, 123, 128, 129, 131
-- defects 129
-- differential 27, 28
spontaneous nystagmus 292
sporadic Alzheimer disease (SAD) 8, 11, 14, 483
sprouting 57-59, 174, 289, 291, 295, 297, 301, 307, 310, 312-315, 318, 338, 342, 353, 355-358, 407, 523, 556
-- fibers 397, 398, 458, 556, 558-560
STAT phosphorylation 209
stem cell therapy 590
stem cells 203, 256, 276, 304, 307, 403-405, 413-415, 419, 439, 443, 444, 448, 449, 498, 502, 570, 589-591, 595, 643-646, 651-655, 657
stepwise 4-vo 15
stress conditions during aging 14
striatonigral degenerations 142
stroke 9, 88, 92, 104, 109, 111, 161, 162, 194, 199, 215, 228, 229, 302, 304, 377, 387, 391, 408, 433, 473, 475-480, 484, 490, 491, 497, 501, 503-506, 527-529, 587, 591, 592, 597, 612-614
stroke (brain ischemia) 1, 2, 483
structural plasticity 47, 57-59, 353
subgranular zone 303, 383, 386-388
Substantia nigra 107, 136, 139, 163, 278, 303, 374, 418, 485, 494, 507, 537-545, 549, 550, 556, 559, 594
subthalamic nucleus (STN) 417, 418, 494, 538-540, 550
subventricular zone (SVZ) 217, 256, 303, 333, 382, 387, 413, 414
superoxide dismutase (SOD) 86, 138, 140, 146, 154, 155, 157, 163, 486, 495, 501, 575-579, 582, 586
superoxide ion radical 277
superoxide production 139
superoxide radical 99, 100
survival of motor neurons (SMN) protein 123, 129, 131

sympathetic tone 11, 20
synapse 29, 33, 34, 41, 44, 47-49, 52, 53, 57, 58, 130-132, 203, 223 246, 291, 292, 294, 295, 308, 309 312, 313, 315-318, 339, 358, 398 408-410, 434, 451, 475, 564, 570 609, 630
-- disconnection 313
-- formation 28, 310, 312, 315, 326
-- maturation 54, 314
-- renewal 307, 309, 310-316, 318, 325
synaptic apoptosis 92
synaptic cleft 41, 44, 49, 50, 247
synaptic complex 36, 41
synaptic connection 9, 30, 260, 262, 289, 405, 408, 550, 553
synaptic contacts 36, 44, 54, 175, 261, 264, 287, 291, 301, 303, 308, 310, 315, 326, 386, 387, 405
synaptic function 10, 29, 131, 286, 290
synaptic membranes 53, 54
synaptic memory 92
synaptic neuropile 4, 51
synaptic plasticity 5, 10, 47, 49, 51-55, 57, 123, 308, 309, 313, 314, 316, 326, 327, 370, 420, 425
synaptic potentiation 318
synaptic release machinery 30, 47
synaptic silencing 47
synaptic strenghtening 52
synaptic stripping 186
synaptic structure 52, 54, 123, 131
synaptic terminal 33, 36, 38, 41-44, 58, 67
synaptic transmission 19, 50-53, 61, 62, 247, 290, 317, 373, 475
synaptic vesicles 41, 44, 310, 315, 543
synaptic weakening 47
synaptotagmin 318
synuclein 365

T2 relaxation 12
tanycytes 332-334, 383

tau pathology 158, 567
tau phosphorylation 563, 569
tau-protein 6, 7, 11, 18, 19, 68, 69, 108, 128, 135, 158, 222, 396, 566, 581, 594
Tello, J.F. 326
Tello's strategy 327
teloglial cells 36
terminal or ultraterminal sprouts 310
tetanic stimulation 48, 49, 51, 53, 55
TF deficiencies in polyglutamine-mediated diseases 127
TgFα 226
TGF-β 185, 191, 193, 197, 201, 212, 219, 239, 287, 303, 304, 335, 435, 558
thalamocortical neurons 538, 539
therapeutic intervention 74, 111, 198, 212
Theta rhythm 55
thiamine pyrophosphatase (TPP) 38
thiobarbituric acid reactive substances (TBARS) 163
thymidine kinase (TK) 497, 501
T-lymphocyte 224, 252
tocophenols 135, 159-161
totipotent cells 646
TPA response element (TRE) 152, 164
tractography (fibre tracking) 622
transcription factors 14, 23-28, 83, 90, 91, 96, 97, 103, 115, 118, 123-127, 132, 150-152, 199-203, 205-207, 208-210, 217, 218, 221, 231, 233, 235, 241, 317, 333, 486, 542
transcriptional activators 123-126, 209
transcriptional apparatus 25
transcriptional control 25
transcriptional dysfunctions 123
transcriptional induction 93, 94
transcriptional machinery 25, 83, 127
transcriptional regulation 151, 166
transcriptional upregulation 85, 119
transcriptosomes 123
transdifferentiation 255
transfection 355, 502

transforming growth factor (TGF) 407
translation 23, 29, 50, 89, 96, 97, 104, 114, 115, 123, 130-132, 148, 491, 503
transmission failure 15
-- theory 15
transplant 297, 301, 303, 326, 328, 337, 340, 342, 357, 405, 412, 415, 419, 441, 446, 551
transverse relaxation time (T2) 21
trauma 58, 61, 63,66, 71-73, 92, 109, 119, 181, 183, 186, 194, 247, 302, 326, 328, 329, 331, 391, 404, 408, 423-425, 428, 429-432, 434, 453, 484, 485, 487, 490, 497, 501, 503, 504, 564
traumatic brain injury 88, 107, 117, 386, 423-435, 476
traumatic death 91
traumatic insults 70, 72
traumatic lesions 225, 228, 260, 261, 286, 294, 323
tricarboxylic acid cycle (TCAC) 3, 21
trigeminal nucleus 264, 297
trophic factors 9, 47, 79, 101, 102, 237, 249, 263, 276, 286, 287, 314, 315, 353, 354, 405, 413, 415, 447, 457, 466, 467, 570, 596
trophins 51
tryptophan 146, 277
tubulization 336, 461, 462, 465-468
tumor necrosis factor alpha (TNF) 52, 74, 82, 102, 186, 189, 192, 542, 586, 595
tumor promotors 148
two-photon microscopy 52
tyrosin hydroxylase (TH) 494

ubiquitin/proteosomal pathway 112, 118, 119
ultra terminal sprouting 318
unfolded protein response (UPR) 89, 90, 111, 113
-- induction of 114

unfolded proteins 111, 113-116, 118, 119, 544
unipolar neurones 34
unmyelinated axons 36, 462
urinary bladder function 338

vasoactive intestinal peptide (VIP) 185, 208
vasoconstriction 11, 428
vasodilation 11
ventricular enlargement 11
vertigo 292
vessel occlusion (Vo) 15, 21, 454
Vimentin 203, 212, 332, 334
Virchow, 215, 308
VMAT-2: vesicular monoamine transporter-2 494
volume transmission 473-475
voluntary movements 334, 538, 549
vomiting 292, 571

Waldeyer, H. 33
Wallerian degeneration 63, 66, 69, 72, 73, 236, 239, 324, 330, 451, 454, 455, 458, 470, 511, 513, 514, 517, 523, 580
weakness 129, 494, 582, 583, 585
white-matter damage 215, 228, 229
Wilson's disease 278
woodchuck hepatitis virus posttranscriptional regulatory element (WPRE) 486

X-box binding protein-1 (XBP-1) 115, 117, 119
X-chromosome linked inhibitor of apoptosis protein (XIAP) 83, 485, 492, 494, 501

zinc finger 26, 221, 235
zone of penumbra 324, 328
Zonulae adherens 246
Zymogen 101